Planets and Life: The Emerging Science of Astrobiology

Astrobiology involves the study of the origin and history of life on Earth, planets and moons where life may have arisen, and the search for extraterrestrial life. It combines biology, biochemistry, paleontology, geology, planetary sciences, and astronomy. This textbook brings together world experts in each of these disciplines to provide the most comprehensive coverage of the field currently available. Topics cover the origin and evolution of life on Earth, the geological, physical, and chemical conditions in which life might arise, and the detection of extraterrestrial life on other planets and moons. The book also covers the history of our ideas on extraterrestrial life and the origin of life, as well as the ethical, philosophical, and educational issues raised by astrobiology. Written to be accessible to science students and scientists from diverse backgrounds, this text will be welcomed by advanced undergraduates and graduates who are taking astrobiology courses, as well as by practicing scientists who desire a comprehensive introduction to this emerging and exciting field.

WOODRUFF SULLIVAN is Professor of Astronomy and Adjunct Professor of History at the University of Washington (UW). His interests are in astrobiology, the search for extraterrestrial intelligence (SETI), the history of astronomy, and gnomonics. He is Chair of the Steering Group of the UW's interdisciplinary graduate Astrobiology Program.

JOHN BAROSS is a Professor in the School of Oceanography, UW. His research focuses on thermophilic microorganisms from volcanic environments, the origin and evolution of life, life on other planets and moons, and microbial ecology. He is a founding member of UW's Astrobiology Program.

Planets and Life

The Emerging Science of Astrobiology

Edited by

Woodruff T. Sullivan, III

and

John A. Baross
University of Washington

CAMBRIDGE
UNIVERSITY PRESS

CAMBRIDGE UNIVERSITY PRESS

Cambridge, New York, Melbourne, Madrid, Cape Town, Singapore, São Paulo

Cambridge University Press
The Edinburgh Building, Cambridge CB2 8RU, UK

Published in the United States of America by Cambridge University Press, New York

www.cambridge.org
Information on this title: www.cambridge.org/9780521531023

First published 2007

Printed in the United Kingdom at the University Press, Cambridge

A catalog record for this publication is available from the British Library

ISBN 978-0-521-82421-7 hardback
ISBN 978-0-521-53102-3 paperback

Où finit le teléscope, le microscope commence.
Lequel des deux a la vue la plus grande?
[Where the telescope ends, the microscope begins.
Which of these has the grandest view?]

Victor Hugo (*Les Misérables*, 1862)

If the planets be inhabited, what a scope for folly;
if they not be inhabited, what a waste of space.

Thomas Carlyle (1795–1881)

Contents

Contributors

John Armstrong, Department of Physics, Weber State University, Ogden, Utah, USA

Stanley M. Awramik, Department of Geological Sciences, University of California, Santa Barbara, California, USA

John A. Baross, School of Oceanography, University of Washington, Seattle, Washington, USA

Steven A. Benner, Department of Chemistry, University of Florida, Gainesville, Florida, USA

André Brack, Centre de Biophysique Moléculaire, CNRS, Orleans, France

Donald E. Brownlee, Department of Astronomy, University of Washington, Seattle, Washington, USA

Roger Buick, Department of Earth and Space Sciences, University of Washington, Seattle, Washington, USA

Paul Butler, Department of Terrestrial Magnetism, Carnegie Institute of Washington, Washington, DC, USA

Diane Carney, Department of History, University of Washington, Seattle, Washington, USA

David Catling, Astrobiology Program, University of Washington, Seattle, Washington, USA, and Department of Earth Sciences, University of Bristol, Bristol, UK

Christopher F. Chyba, Department of Astrophysics and Woodrow Wilson School of Public and International Affairs, Princeton University, Princeton, New Jersey, USA

Mark Claire, Department of Astronomy, University of Washington, Seattle, Washington, USA

Carol E. Cleland, Philosophy Department and the Center for Astrobiology, University of Colorado, Boulder, Colorado, USA

George D. Cody, Geophysical Laboratory, Carnegie Institute of Washington, Washington, DC, USA

Pamela G. Conrad, Planetary Science and Life Detection Section, Jet Propulsion Laboratory, Pasadena, California, USA

David W. Deamer, Department of Chemistry and Biochemistry, University of California, Santa Cruz, California, USA

Jody W. Deming, School of Oceanography, University of Washington, Seattle, Washington, USA

Steven J. Dick, History Division, Office of External Relations, NASA Headquarters, Washington, DC, USA

John Edwards, Department of Biology, University of Washington, Seattle, Washington, USA

Hajo Eicken, Geophysical Institute, University of Alaska, Fairbanks, Alaska, USA

Jelte Harnmeijer, Department of Earth and Space Sciences, University of Washington, Seattle, Washington, USA

Julie A. Huber, The Marine Biological Laboratory, Woods Hole, Massachusetts, USA

Bruce M. Jakosky, Laboratory for Atmospheric and Space Physics and Department of Geological Sciences, University of Colorado, Boulder, Colorado, USA

James F. Kasting, Department of Geosciences, Pennsylvania State University, University Park, Pennsylvania, USA

Monika Kress, Department of Physics and Astronomy, San Jose State University, San Jose, California, USA

John A. Leigh, Department of Microbiology, University of Washington, Seattle, Washington, USA

Jonathan Lunine, Lunar and Planetary Laboratory, University of Arizona, Tucson, Arizona, USA

Andrew McArthur, The Marine Biological Laboratory, Woods Hole, Massachusetts, USA

Christopher P. McKay, Space Science Division, NASA Ames Research Center, Moffet Field, California, USA

Kenneth J. McNamara, Department of Earth and Planetary Sciences, Western Australian Museum, Perth, Australia

David J. Patterson, The Marine Biological Laboratory, Woods Hole, Massachusetts, USA

Randall Perry, Planetary Science Institute, Tucson, Arizona, USA

Cynthia B. Phillips, Carl Sagan Center for the Study of Life in the Universe, SETI Institute, Mountain View, California, USA

Nicolas Pinel, Department of Microbiology, University of Washington, Seattle, Washington, USA

Margaret S. Race, Carl Sagan Center for the Study of Life in the Universe, SETI Institute, Mountain View, California, USA

Alonso Ricardo, Simches Research Center, Massachusetts General Hospital, Boston, Massachusetts, USA

Bashar Rizk, Lunar and Planetary Laboratory, University of Arizona, Tucson, Arizona, USA

John D. Rummel, Office of Space Science, NASA Headquarters, Washington, DC, USA

Matthew O. Schrenk, Geophysical Laboratory, Carnegie Institute of Washington, Washington, DC, USA

James H. Scott, Geophysical Laboratory, Carnegie Institute of Washington, Washington, DC, USA

Robert Shapiro, Department of Chemistry, New York University, New York, USA

Mitchell L. Sogin, The Marine Biological Laboratory, Woods Hole, Massachusetts, USA

David A. Stahl, Department of Civil and Environmental Engineering, University of Washington, Seattle, Washington, USA

James T. Staley, Department of Microbiology, University of Washington, Seattle, Washington, USA

Woodruff T. Sullivan, III, Department of Astronomy, University of Washington, Seattle, Washington, USA

Jill Tarter, Carl Sagan Center for the Study of Life in the Universe, SETI Institute, Mountain View, California, USA

Steve Vance, Department of Earth and Space Sciences, University of Washington, Seattle, Washington, USA

Peter D. Ward, Department of Earth and Space Sciences, University of Washington, Seattle, Washington, USA

Llyd Wells, The Center for Northern Studies at Sterling College, Craftsbury Common, Vermont, USA

Frances Westall, Centre de Biophysique Moléculaire, CNRS, Orleans, France

Preface

The emerging field of astrobiology encompasses a daunting variety of specialties, from astronomy to microbiology, from biochemistry to geology, from planetary sciences to phylogenetics. This is both exciting and frustrating – exciting because the potential astrobiologist is continually exposed to entirely new ways to look at the world, and frustrating because it is difficult to understand new results when venturing outside the confines of one's own specialty. There are now many excellent popular books on astrobiology, but a scientist wants more details and more sophistication than these afford. Where can an astronomer without any formal biology since high school learn the basics of cellular metabolism? Or the principles of evolution? Or notions about alternative forms of life? And where can a microbiologist with little physics and no astronomy learn the basics of how a planetary atmosphere works? Or how the Earth formed? Or how planets are detected around other stars? This book is designed to fill these needs.

We have endeavored to cover all the important aspects of astrobiology at an advanced level, yet such that *most* of the contents in *every* chapter should be understandable to anyone versed in any relevant science discipline. We envision our youngest readers to be science majors near the end of undergraduate study or the beginning of graduate study. And at the other extreme, we aim to serve scientists who haven't taken an academic course for forty years, but are intrigued by the nascent field of astrobiology. Readers should find this volume a challenging, yet accessible, way to get "up to speed" and "up to date" on the many facets of astrobiology.

While we anticipate that this book will be used for courses in astrobiology, and while it has a definite pedagogical aim, it is *not* in fact a textbook in the usual sense. A textbook, unlike this volume, would have chapter-end problems, uniformity of style, methodical development of concepts, and less idiosyncratic coverage of the various aspects of astrobiology. On the other hand, a single-author textbook of astrobiology, which must range over a multitude of disciplines, cannot carry the authority of the chapters herein.

Astrobiology at the University of Washington (UW) began with a graduate seminar offered in 1996, organized by us and also called "Planets and Life." This led three years later to establishing a graduate Astrobiology Program (depts.washington.edu/astrobio), largely funded by the IGERT (Integrative Graduate Education and Research Traineeship) program of the US National Science Foundation (NSF). As we struggled with how best to train astrobiologists in a limited time, the need for a book such as the present volume became evident. Our students obtain "normal" Ph.D.s in their home departments, but on top of that they complete various requirements for a Certificate in Astrobiology (see Chapter 28). The goal of the UW Astrobiology Program is to produce scientists who are firmly rooted in one of the disciplines that contributes to astrobiology, but also to expose each student to the other relevant disciplines: their fundamental questions and how they approach answers, their culture, and their terminology. We create intellectual and social connections between the disciplines to foster the asking of entirely new sorts of questions and their eventual answer. This all flies in the face of modern science, which usually drives its participants to ultra-specialization. If astrobiology is to succeed, however, the proverbial pendulum needs a push back towards synthesis, not unlike the synthesis that spawned molecular biology a half century ago (Section 28.1).

The individual chapters have been written by experts in the various specialties and leaders in shaping astrobiology today. We have asked authors to focus not on a standard review of their field, but on an introduction to the basics of the field and how problems are approached. In this way the scientific and educational value of the chapters should endure longer than usual. We have not tried to enforce too strict a uniformity in the styles of writing except for our Prime Directive: all jargon must be defined and most of each chapter must be understandable by nonexpert scientist readers. Whether we have succeeded will be judged by biologists learning about the subsurface ocean world on Europa, and by astronomers grappling with the RNA world on early Earth.

Sociologists chart the rise of new scientific disciplines through the appearance of textbooks, journals, and degree programs. So it is with emerging astrobiology. Below we give an annotated list of the astrobiology books and other resources that might be used as advanced texts or references at this time (although some are unfortunately far too expensive for even a professorial budget, let alone a student's!). The books are in order of usefulness and financial accessibility.

An endeavor such as this book does not happen without the aid of many persons and institutions. We and the UW Astrobiology Program have profited from the support of the University of Washington, NSF, and NASA. We also especially thank Arthur Whiteley and his eponymous Center for providing a marvellous scholarly retreat in the San Juan Islands where major portions of this book were edited and written. The NASA Astrobiology Institute and the UW directly supported the production of this book. Our astrobiology graduate students commented on chapter drafts (and are co-authors of three chapters), and have in essence taught us how we need to teach this new discipline. Linda Khandro and Nancy Quensé have also provided essential help through the years.

We thank especially those authors who have done what we asked and when we asked, and yet then been patient with wayward fellow authors and the editors in the face of delays. Cambridge University Press has also exhibited the patience of Job. WTS is extremely grateful for the supportive environment created for "The Book" by Barbara Sullivan. JAB would like to acknowledge the support of and helpful discussions with Jody Deming.

Resources for advanced astrobiology

Books

Existing astrobiology textbooks for *non*-science majors are listed in footnote 2 near the end of Chapter 28.

Lunine, J. (2005). *Astrobiology: A Multi-Disciplinary Approach*. San Francisco: Addison Wesley. Comprehensive and challenging textbook by a planetary scientist for advanced science undergraduates and graduate students.

Schulze-Makuch, D. and Irwin, L. N. (2004). *Life in the Universe*. New York: Springer. Excellent treatment of the nature of life, its properties, alternative types of life, and biosignatures; suitable for nonexperts.

Chela-Flores, J. (2001). *The New Science of Astrobiology: From Genesis of the Living Cell to Evolution of Intelligent Life*. Dordrecht: Kluwer. Idiosyncratic look by a physicist at life in a cosmic context; suitable for all scientists.

Horneck, G. and Baumstark-Khan, C. (eds.) (2002). *Astrobiology: The Quest for the Conditions of Life*. Berlin: Springer. Most chapters (by individual authors) are an outgrowth of a 2001 workshop on astrobiology held in Germany; uneven coverage and levels.

Gargaud, M., Barbier, B., Martin, H. and Reisse, J. (eds.) (2005, 2006). *Lectures in Astrobiology* (Vols. I and II). Berlin: Springer. Mammoth volumes of technical chapters (by individual authors) mostly on the early Earth, the origin of life, and other possible habitats for life; based on lectures in 1999 and 2001 at French summer schools.

Ehrenfreund, P., Irvine, W. M., Owen, T., *et al.* (eds.) (2004). *Astrobiology: Future Perspectives*. Dordrecht: Kluwer. Technical chapters, mostly covering the context of life chemically (on Earth and in space) and geologically; based on a 2003 workshop in Switzerland.

Seckbach, J. (ed.) (2004). *Origins: Genesis, Evolution and Diversity of Life*. Dordrecht: Kluwer. Technical chapters covering a broad range of astrobiology, especially the origin of life.

Journals

Astrobiology (2001–). Mary Ann Liebert, Inc.

International Journal of Astrobiology (2002–). Cambridge University Press.

Other

Mix, L. (chief ed.) (2006). The Astrobiology Primer: an Outline of General Knowledge. *Astrobiology*, **6**, 735–813. Graduate students and postdocs in astrobiology have written short sections on the basics of each subfield of astrobiology (e.g., four pages on "life's basic components"); very nice resource to accompany the present volume.

Chyba, C. F. and Hand, K. (2005). Astrobiology: The study of the living Universe. *Ann. Rev. Astron. Astrophys.* **43**, 31–74. Best current review article.

Des Marais, D. J., Allamandola, L. J., Benner, S. A., *et al.* (2003). The NASA Astrobiology Roadmap. *Astrobiology*, **3**, 219–35. NASA's bible for what astrobiology is and what research should be done.

Sullivan, W. T., III (2006). depts.washington.edu/ astrobio/research/references.html. Annotated bibliography of approximately 100 books on all aspects of astrobiology; categorized as textbooks, popular, scholarly, and historical.

Prologue

A new synthesis: the Biological Universe

A remarkable shift in our scientific world picture is taking place, potentially as fundamental in its consequences as the new views put forth by Copernicus in the sixteenth century, or by Darwin in the nineteenth. Although astronomers have long been involved with the prospects for extraterrestrial life, their fundamental task since Newton has been to apply physics to a lifeless Universe. On the other hand, biologists have pursued their studies for centuries in cosmic isolation, meaning that biology considered life on Earth, with no attention paid to its cosmic context. Today, however, both camps are recognizing fruitful and exciting avenues of research created by a new synthesis. Biology is vastly enriched when attention is paid to a broader context for life as we know it, as well as the possibilities for other origins of life. And astronomy is coming to realize that the themes of cosmic, galactic, stellar, and planetary evolution, which have become central over the past century, must now also incorporate biological origin(s) and evolution(s).[1] Historian of science Steven Dick (1996) has hailed this new synthesis as the *Biological Universe*.[2]2 Although astronomy and biology are its two primary poles, many other disciplines are also vital components, in particular Earth and planetary sciences.

Astrobiology has become the rubric for this new synthesis, but it has also been called *bioastronomy*, *cosmobiology*, *biocosmology*, etc.[2] It has received a tremendous boost over the past decade because empirical science can for the first time powerfully address three questions that have always been fundamental to humans.

Fundamental questions

What is life?

What types of life differing from our own are possible? Our understanding of molecular genetics has allowed new insights into the mechanics of how life works, and results from studies on the origin of life have begun to elucidate the question of what separates a system of chemical reactions from a living entity.

What is the course of life?

How did life come to be? When and where did life first arise on Earth? How does life evolve? What are the limits of Earth life? What are the possibilities for future life? How might the answers to these questions change for other, extraterrestrial, locations? Is biology an inevitable stage of cosmology? Are the origins and evolutions of the physical Universe and of life part of a larger whole?

Are we alone?

This experimental aspect of astrobiology has become possible only with the technology of the past fifty years. New types of telescopes, as well as robotic visits to other planets, have meant that, guided by studies of Earth life, we can now actively search other worlds for evidences of extant life, of fossils, or of technology indicative of intelligence.

---

[1] Note that the astronomer's usage of the term *evolution* refers to the change with time of an entity (such as a star), analogous to the ageing of an individual organism. On the other hand, the biologist's meaning for *evolution* is in the context of a specific theory of how life as a whole changes over time. Yet it is interesting how the two concepts are today melding, both within biology ("Evo Devo"; see Chapter 10) and in terms of the Biological Universe.

[2] In this Prologue superscript numerals in square brackets refer to chapter numbers that deal with the indicated topics.

Planets and Life: The Emerging Science of Astrobiology, eds. Woodruff T. Sullivan, III and John A. Baross. Published by Cambridge University Press. © Cambridge University Press 2007.

These three questions in turn inform the root question:

Who are we?

This question of course has many aspects beyond science, but our yearning for an answer drives much of today's astrobiology. Where did we come from? Are we here by chance or necessity? Where are we heading?

The public is also deeply interested in these questions, which can be a two-edged sword. Modern science is happy to have political and financial support (sometimes for billions) and educators find astrobiology a compelling lure to attract students to science. On the other hand, the beliefs of many of the public regarding extraterrestrial life, especially intelligent life, have less to do with science and more to do with wishful fantasies and Hollywood tales. This is often embarrassing to the astrobiologist. Historian Karl Guthke (1990) has rightly called extraterrestrial intelligent life society's modern myth, in the sense of a widespread quasi-religious belief satisfying deep needs.

Recent discoveries

Specific scientific developments over the past decade have helped create a sense of legitimacy for astrobiology, as has strong institutional support from the NASA Astrobiology Institute, founded in 1997.[2] The discovery of planets circling other stars, now numbering over 200 and since 1995 climbing ever higher, has made our own solar planetary system just one of a myriad.[21] The announcement in 1996 of evidence for fossil life from Mars in the meteorite ALH 84001 ironically is now thought to have been premature, but on the other hand the next claim could be right: subsequent studies have shown that microbial life could indeed travel and survive between planets on a rock rocket.[18] At about the same time, the NASA Galileo mission to Jupiter discovered strong evidence for a huge ocean of liquid water beneath the icy crust of the moon Europa – a bizarre potential habitat for life that has revolutionized our thinking about niches for life.[19] On Mars, the case for liquid water in the past has grown ever stronger over the past decade, culminating with the Mars Rovers (2004–) returning stunning geological evidence of extensive flowing water on the surface at Meridiani Planum.[18] The ESA/NASA Cassini/Huygens mission to Saturn (2005 and ongoing) has revealed unprecedented details of the amazing world of Titan, where today methane cycles through phases not unlike water's hydrological cycle on Earth, and the rich organic chemistry may teach us much about the prebiotic Earth.[20] And most recently, on the very day the manuscript for this book was sent off to the publishers, evidence was announced of organic compounds and geyser action on Saturn's small moon Enceladus, with possible liquid water just below the ice-covered surface (*Science*, 10 March 2006).

On the biological side, it has become increasingly appreciated how robust microorganisms can be in their adaptations right here and now on Earth, engendering further optimism that the extreme conditions of other planets and moons may be less of a deterrent than we had thought.[14,15] Moreover, new molecular and culturing techniques are being applied to decipher the physiologies of the "unknown majority" of microorganisms sampled in most Earth environments, about which we had been ignorant since they had no analogues among the characterized organisms in culture. For example, recent sleuthing of this kind has revealed novel photosynthesizing microorganisms, unique metabolic pathways for using carbon dioxide and methane as carbon sources, the importance of hydrogen as an energy source in many extreme environments, and unusual strategies for growth and survival in extremely nutrient-deprived environments such as the deep sea and deep marine sediments. It is likely that new organisms will be discovered that will expand our current view about the physiological versatility of organisms to grow and survive under environmental conditions not found on Earth. In addition, ever more powerful genetic and mathematical techniques now allow insights into the evolutionary history of life that, when combined with the geological record, reveal the complex manner in which new species and new adaptations arose.[10–13]

A science of optimism

Astrobiology is the scientific discipline of optimism – optimism that the grand questions are not only fundamental, but tractable; optimism that life has a good chance to exist elsewhere; and optimism that such life can be detected. Astrobiology tends to attract scientists who enjoy discussing radical ideas about science and are excited about the search for life elsewhere despite great difficulties and probably many decades at best before success.[28] Carl Sagan deserves special mention in this regard (Fig. 2.7). From 1960 onwards he was the consummate astrobiologist and an eloquent popularizer of science, often receiving great criticism for taking on each of these vital roles. We profit today from his lifelong belief in the field and his battle for its respectability. It is a shame that he did not live long enough to witness today's efflorescence of astrobiology. We commend his many excellent and inspiring books to younger scientists who no longer have the opportunity to experience the man directly.

There is a considerable scientific, philosophical and theological literature proclaiming the uniqueness of life (particularly human-like life) and its likely restriction to Earth among the billions of planets. The arguments expressed follow three primary lines: (1) the origin of life is so complex that it is not likely to occur more than once in the history of the Universe; (2) we and our Earth are too "special" to have evolved more than once (the Rare Earth hypothesis);[1,3] and (3) if intelligent life exists elsewhere, why have we not been notified (the Fermi Paradox)?[26] We cannot accept these arguments and the resulting a posteriori conclusion that we are lonely hunters trying to find meaning in a Universe of chance. The question of whether or not other intelligent life exists or has existed in the Universe may never be answered, but knowing the answer is not necessary in our search for life's meaning. How could we be lonely when we are intimately enmeshed in a Cosmos that inspires us in our search to understand how intelligent life arose on Earth from the same elements and physical principles that spawned the entire Universe? Should there be any doubt that at least in this sense we are not alone in the Universe?

Contingency or natural outcome?

Contingency has been billed as one of the most important arguments for our uniqueness as an intelligent species and our planet's uniqueness in supporting intelligence. The late paleontologist Stephan Jay Gould expressed this idea in *Wonderful Life* (1989) by positing that a "rewinding and replaying of the tape of life" to life's origin would result in a completely different outcome after 4 Gyr of evolution, and very likely not including the likes of us. An example of his views:

> Since dinosaurs were not moving toward markedly larger brains, and since such a prospect may lie outside the capabilities of reptilian design, we must assume that consciousness would not have evolved on our planet if a cosmic catastrophe had not claimed the dinosaurs as victims. In an entirely literal sense, we owe our existence, as large and reasoning mammals, to our lucky stars.   Gould (1989)

Gould misses the point that catastrophic events, including killer impacts and other events that cause radical changes in atmospheric chemistry and temperature,[3,4,17] are features of the environment, natural outcomes of physical laws. These events occur and cause long-term effects frequently enough that evolution has selected traits that allow organisms to survive these events and in most instances take advantage of new habitats thus created. Moreover, we do not know if intelligent mammals or some other animal lineage would have evolved alongside the dinosaurs – why does intelligent life have to look like us? Catastrophic extinction events are as natural as tides changing or temperatures shifting. There would be no life-supporting planets without impact events.[3]

While Gould was only concerned with our own emergence on Earth, similar arguments have been made about the improbability of a second instance of life. But how can we confidently make this assertion when we do not even know how life originated on Earth or if there are many different ways to make life?[6–9] Indeed, we cannot even agree on a definition of life.[5] What we do not *yet* understand, however, should not be considered as *never* scientifically understandable. These conundrums and others illustrate how biology is in many ways hamstrung by having only one example of its primary phenomenon, a problem that astrobiology modestly aims to solve. The history of science has shown that the acquisition of a second data point, $n = 2$, or at least a great extension of $n = 1$ to previously untapped realms, often leads to a more general theory and profound breakthroughs. Witness the extension of Newtonian gravity to general relativity, the discovery of viruses in the early twentieth century ($n = 1.5$?), the realization over the past decades that dark matter and dark energy dominate over "normal" matter in the Universe, and, as mentioned, the new extrasolar planetary systems, most of which are nothing like our familiar Solar System.[21]

The origin of life is very likely a solvable problem and one that could possibly become much easier if we found life elsewhere, whether that life form were identical to Earth life or quite different. Conversely, the more we understand about the origin of life on *this* planet, the better our chances to recognize and detect elsewhere biochemical precursors to life or early stages in the evolution of the cell. Understanding the conditions here that favored the origin of life and its evolution will also greatly aid in identifying candidate planets and moons that now have or have had the potential to generate life.[18–22]

An astrobiological optimist believes that the origin and evolution of life has less to do with chance and more to do with principles of nature, including many we do not yet fully understand. In response to Gould, fellow paleontologist Conway Morris (1998) expresses the view that there may be limits to the number of evolutionary possibilities. For example, the emerging science of Evolution/Development ("Evo Devo") has altered our perception of convergent evolution (separate evolution of similar traits) by showing that groups of regulating genes, ubiquitous in all animal lineages, control the evolution of animal form and organ development (Carroll, 2005).[10] Evo Devo also offers plausible mechanisms for explaining the onset of the Cambrian explosion. The incredibly diverse body forms that appeared very quickly among the newly emerged animals [16] are believed to be a consequence of the evolutionary appearance of these key regulating genes.

Moreover, even at the level of organic chemistry, there appear to be rules that allow prediction of the kinds of complex organic compounds that can be formed and the conditions that favor their formation. These rules should inform the definitions of biosignatures that we will employ in our search for evidence of life elsewhere.[22,23] It is interesting that life also restricts the number of possible biochemical structures.[6,7] For example, cholesterol theoretically has 256 possible stereoisomer structures, yet life only synthesizes one. There are also rules in biochemistry restricting the number of possibilities for how macromolecules become synthesized and then fold and twist into complex three-dimensional structures.[3] It is likely that some of the early chemical steps that can lead to the development of life are very common in the Universe and, depending on the environmental conditions of the planet and its evolution, sometimes (commonly?) lead to complex biochemical structures and living cells. The key is the degree of habitability of the planet – whether or not conditions evolve that favor the development of ever-greater complexity, from organic compounds to organisms.

The bio-friendly Universe

Further extending the notion that life is a natural, perhaps inevitable, product of cosmic evolution, we come to the *Anthropic Principle*, the "weak" form of which, a self-selection principle, says that the Universe must be bio-friendly (to use cosmologist Paul Davies's felicitous term), since we sentient beings have

succeeded in arising within its confines and insist on writing books about it (Barrow and Tipler, 1986). Further investigation reveals a remarkable panoply of physical constants (such as the gravitational constant *G*) and cosmological properties (such as the size of fluctuations in the cosmic background) whose values seem *very* "finely tuned," i.e., slightly different values would yield a Universe where life (especially intelligent life) as we know it *could not* exist. For example, the very existence of long-lasting stars rests on a delicate balance that can be thrown off by adjusting any one of many physical values. As with the biochemistry discussed above, it appears that the Universe *must* be like it is in its various physical details, or we would not be viewing it! This brand of thinking, called the Strong Anthropic Principle, is pre-Copernican, in that we revert to a Cosmos all set up for our benefit – a scientist of the early twenty-first century finds it unsettling (Falk, 2004). As the physicist Freeman Dyson (1979) says:

> The more I examine the Universe and study the details of its architecture, the more evidence I find that the Universe in some sense must have known that we were coming.

A recent theory that explains the bio-friendliness of the Universe in a post-Copernican manner has been developed by James Gardner (2003). Inspired by work of several leading cosmologists over the past decade, Gardner has developed a scheme whereby an entire universe evolves along with its intelligent life. This intelligence eventually develops (over huge times compared to the present age of the Universe) the ability to fabricate new universes that are designed to be even better at producing intelligent life. The physical laws and constants of any universe are its blueprint or "DNA," as established by its "parent," so they are necessarily designed to optimize life and intelligence. This is heady stuff, to be sure, but astrobiology fosters debates over concepts like this – for example, is there a way to test this theory? Is such thinking the best of scientific creativity, or does it dangerously border on religion and just-so stories? In either case, we see the intimate links, today as throughout the past, between cosmology and astrobiology.

Analogies to Earth and its life

The one model of a living planet that we have, Earth, provides us with a set of key characteristics that have affected and continue to shape the diversity and

---

[3] A recent study indicates that even the origin of the genetic code might have little to do with contingency and more to do with rules of chemistry (Copley *et al.*, 2005).

complexity of life. Earth as a tectonically active, aquatic planet produces volatile gases such as carbon dioxide, methane, hydrogen, and hydrogen sulfide, which in the early stages of its history resulted in a greenhouse atmosphere.[4] This atmosphere helped maintain a liquid water ocean and otherwise contributed to the chemical events that led to the origin of life and later provided the chemical energy sources for Earth's fledgling life forms.[9,11] The continuing interactions between organisms and their environments resulted in complex biogeochemical cycles and periods of relatively stable atmospheric and ocean geochemical conditions. The evolution of photosynthesis and the accumulation of oxygen in the atmosphere are thought to have been the key to the evolution of complex animals.[4,12] Imposed on this were catastrophic events including impacts by giant bolides, volcanoes, magma plumes and possibly "snowball Earth" periods which resulted in major extinction events and changes in habitat conditions that led to selection of different communities of organisms.[16,17]

Earth may, however, be just one of many possible models for planets that can evolve complex life.[27] We know only one kind of life:[27] (a) it is carbon-based, (b) it requires liquid water, and (c) it has evolved in a Darwinian sense under particular environmental conditions into complex organisms that ask questions, experience love and hate, and commune with gods. But we do not know how wise it is to assume that planets outside of our perception of a habitable zone[3] could not harbor life, particularly life as we *don't* know it. Our practical search for extraterrestrial life is focused on aquatic planets and moons because of the possibility that they can support Earth-like life. Of course, this does not prevent the astrobiology community from thinking about ways in which carbon-based life could thrive in non-aqueous solvents such as exist on Titan[20] or in the hot, sulfuric acid atmospheric of Venus. Even more radical is the possibility of "weird life" on extremely hot, cold, or gas giant planets (Feinberg and Shapiro, 1980; Schulze-Makuch and Irwin, 2004).

Astrobiology rests on a Great Analogy, that between the one known case of a life-bearing planet and other potential extraterrestrial loci for life.[1,2] To a philosopher an argument by analogy is inductive and can never logically prove anything. It can only be relatively strong or weak depending on the number and quality of properties listed to be analogous/similar versus those that are contra-analogous/dissimilar (and of course all properties must be deemed relevant). But in most realistic cases there are numerous properties to consider, and the sticky issue is how to *weigh* the competing arguments. The philosopher John Stuart Mill in his *System of Logic* (1843) discussed arguments by analogy in general and then gave as an example none other than the question of whether the Moon or planets were inhabited. Mill concluded that, despite many positive attributes of the Moon, its apparent lack of water trumped all other properties and it therefore was not likely inhabited. For the planets, however, he came to a different conclusion that is well for us to note. He opined that there was so little information on, say, Mars that we should not come to *any* conclusion based on analogy.

So is the Great Analogy an unsound foundation for astrobiology? Not at all, says Mill, for analogical reasoning suggests scientific experiments that can provide further insight (Crowe, 1986: 231–2). This in fact is what we do in astrobiology as we make our analogies and argue about the relative weights to apply, then recast the analogies based on new knowledge of Earth life, then generate new working hypotheses and consider new extraterrestrial sites as analogues, and so on. Astrobiology rekindles the relationship between philosophy and science in this and many other ways.[1,2,5]

The promising future

Ultimately, analogies do not satisfy. Astrobiologists, arch empiricists, therefore go out and explore. This is what makes the field so exciting, for new technologies mean that, if public support continues, by 2025 we will have returned samples of rocks from Mars, carefully chosen to optimize the chances for finding evidence of past or extant life.[18,22] We will have landed on Europa, perhaps even have drilled through its ice.[19] We will have searched several million Sun-like stars for extraterrestrial radio signals of intelligent origin (SETI).[26] We will have detected Earth-sized planets of moderate temperatures around many stars, and in some cases determined the composition of their atmospheres.[21] We will have made significant progress in understanding the history of life on Earth, and perhaps as well the origin of life.

We will have made great strides in discovering the physiological versatility of the microbial world and how evolution has experimented with more possibilities then can now be imagined. The microbial communities in sub-seafloor crust will have been found to have an ancient origin and the capacity to thrive in the absence of chemical nutrients derived from sunlight-driven photosynthesis. Through studies of viruses associated with organisms that live in the most extreme environments, we will have made significant advances in understanding the origin of viruses and their pivotal roles in the origin of life, as well as in biocomplexity and diversity. We will likely have been able to create carbon-based life forms that grow in environments beyond the bounds of Earth environments, allowing us to better understand the ultimate limits for carbon-based life. Finally, it will have been recognized that evolution is an essential feature of *all* life, on Earth or elsewhere.

While this volume only touches on the broader philosophical questions that astrobiology engenders,[5,24,25] inherent in many chapters is a sense of the profound implications if life, particularly intelligent life, were discovered elsewhere in the Universe.[26] Will we have found strong evidence for extraterrestrial life within the lifetime of our younger readers? This of course is not knowable, but a positive answer will certainly thrust us into a new philosophical framework. And a negative answer, if the past be any guide, will still have yielded a great deal of ancillary information about life on Earth and how to better define the search.

Successful or not in finding extraterrestrial life, astrobiology will make important contributions to helping us answer that root question: *who are we?*

We have found a strange footprint on the shores of the unknown. We have devised profound theories, one after another, to account for its origins. At last, we have succeeded in reconstructing the creature that made the footprint. And lo! It is our own.

Arthur Eddington (*Space, Time and Gravitation*, 1920)

We shall not cease from exploration
And the end of all our exploring
Will be to arrive where we started
And know the place for the first time.

T. S. Eliot (*Four Quartets*, No. 4, 1942)

REFERENCES

Barrow, J., and Tipler, F. (1986). *The Anthropic Cosmological Principle*. Oxford: Oxford University Press.

Carroll, S. B. (2005). *Endless Forms Most Beautiful: the New Science of Evo Devo and the Making of the Animal Kingdom*. New York: W. W. Norton.

Conway Morris, S. (1998). *The Crucible of Creation*. Oxford: Oxford University Press.

Copley, S. D., Smith, E. and Morowitz, H. J. (2005). A mechanism for the association of amino acids with their codons and the origin of the genetic code. *Proc. Natl. Acad. Sci.*, **102**, 4442–4447.

Crowe, M. J. (1986). *The Extraterrestrial Life Debate, 1750–1900*. Cambridge: Cambridge University Press.

Dick, S. J. (1996). *The Biological Universe: the Twentieth Century Extraterrestrial Life Debate and the Limits of Science*. Cambridge: Cambridge University Press.

Dyson, F. (1979). *Disturbing the Universe*. New York: Harper & Row.

Falk, D. (2004). The Anthropic Principle's surprising resurgence. *Sky & Telescope*, **107**, No. 3, 43–48.

Feinberg, G. and Shapiro, R. (1980) *Life Beyond Earth: the Intelligent Earthling's Guide to Life in the Universe*. New York: William Morrow.

Gardner, J. (2003). *Biocosm – the New Scientific Theory of Evolution: Intelligent Life is the Architect of the Universe*. Portland, Oregon: Inner Ocean Publishing.

Gould, S. J. (1989). *Wonderful Life*. New York: Norton.

Guthke, K. S. (1990). *The Last Frontier: Imagining Other Worlds from the Copernican Revolution to Modern Science Fiction*. Ithaca, NY: Cornell University Press.

Schulze-Makuch, D. and Irwin, L. N. (2004). *Life in the Universe: Expectations and Constraints*. Berlin: Springer-Verlag.

Part I
History

1 History of astrobiological ideas

Woodruff T. Sullivan, III and Diane Carney

University of Washington

1.1 Overview

1.1.1 Why history?

The core questions of astrobiology are not new. They have always been asked and are central to Western intellectual history. How did life begin? How has it changed? What is the relation of humans to other species? Does life exist elsewhere? If so, where might it be and what is it like? Although these questions are ancient, what *is* new are the tools at hand to search for answers, ranging from robotic spacecraft to genome sequencing, from electron microscopes to radio telescopes. These tools and other factors (see the Prologue and Chapter 2) appear to have brought astrobiology to a point where it is gelling into something qualitatively different – our first sound attack on these questions. But is this so? Or is today no different from any other time in the past few centuries?

In every era, including our own, scientists can do no more than tackle questions with the best tools available, apply the best insight they can muster, and struggle to fashion a consensus as to the nature of the world. In this manner our understanding has progressed, for example, from the "animalcules" that van Leeuwenhoek described three hundred years ago to the richness of contemporary microbiology. To understand such a thread as it meanders through history, we need to document more than the accumulation of facts. When evaluating a given episode, historians of science look carefully at evidence of not only the science itself, but also of the larger enveloping context. At each step along the way, the scientific enterprise has always been shaped by metaphysics, doctrines, and predilections as received from philosophy, religion, and society. But, you say, this is all irrelevant to *today's* science – have we not rid ourselves of such prejudices and biases? Think again! Try to get funding for a project working on the hypothesis that Earth was visited by intelligent

beings 100,000 years ago, or that DNA does not carry the esssence of the genetic code, or that microbes populate the Venusian surface, or that *Homo sapiens* are the predestined outcome of the evolutionary process. Without arguing for the validity of any of these notions, we only wish to demonstrate that prejudices and biases are still very much with us. History has shown that the greatest breakthroughs often are made by those who somehow recognize, resist, and surmount the prejudices of their own time. History of science tells great stories and is intellectually fascinating, but it can also be "useful" to practicing scientists in providing some insight as to how the chalklines of today's playing field have come to be where they are.

Studying the development of astrobiological ideas has a particular allure because the associated science has often been "on the edge" of uncertain *epistemological*[1] status. Witness evolutionary biologist George Gaylord Simpson's famous remark in the 1960s that exobiology (as the field was then known) was the only science that had yet to prove that its subject matter existed! Furthermore, scientific views on these fundamental questions have often touched on religious doctrine, creating political and social pressures. Many scientists are attracted to astrobiology precisely *because* it is on the edge of our current knowledge in exciting ways; historians of science also find this fascinating as they try to understand how the practice of science has worked.

1.1.2 Synopsis

Through 2500 years of Western history astrobiological ideas have moved from the realm of natural philosophy to (Christian) theology to science; from pure metaphysics to empiricism. Ideas on extraterrestrial life have

---

[1] *Epistemology* is the branch of philosophy dealing with knowledge, in particular how we justify that we *know* something.

Planets and Life: The Emerging Science of Astrobiology, eds. Woodruff T. Sullivan, III and John A. Baross. Published by Cambridge University Press. © Cambridge University Press 2007.

ranged from us as the special product of all creation to the plurality of worlds, a Universe in which every star is a Sun with peopled planets. Ideas on the origin of life have shifted from pre-existence theory and spontaneous generation everywhere all the time to a series of chemical and physical events in the distant past. Our cosmological worldviews have also shifted, from geocentric to heliocentric to no special location in the Universe. Historian Steven Dick (1996; Sections 2.1 and 2.4) has stressed the intimate connection between cosmological ideas and attitudes toward extraterrestrial life. We will see that as our ideas have evolved regarding the cosmos and where we fit in it, our rating of the prospects for extraterrestrial life and its nature have also changed.

This chapter has two parts. The first (Sections 1.2–1.9) sweeps over the history of ideas on extraterrestrial life, and the second (Sections 1.10–1.15) covers the history of ideas on spontaneous generation and the origin of life. In each part we endeavor to give the overall development of the ideas leading to today's astrobiology, but details can be afforded for only a few illustrative "episodes." For further details and entry into the literature the best accounts are the books by Dick (1982), Crowe (1986), Dick (1996), Dick and Strick (2004), Guthke (1990), Farley (1974), Fry (2000), and Strick (2000), which have been invaluable resources to us. This chapter takes the story through the first half of the twentieth century, while Chapter 2 focuses on the second half of the twentieth century. In addition, many individual chapters also cover developments of the past few decades as the basis for current views.

1.2 Peopled worlds in antiquity and the Middle Ages

1.2.1 The atomists versus Aristotle

There is a tendency to think of the ancient Greek philosophers as a mostly unified group with views veering little from Plato and Aristotle, whose ideas came to dominate Western philosophy. But in fact, over the centuries in which Greek philosophy flourished (sixth to third centuries BC) there were many schools of thought.[2]

---

[2] One of the more famous cases of a minority cosmological view was that of Aristarchus of Samos, who in the third century BC developed a heliocentric system with a rotating Earth and the planets moving around a central Sun. The theory was rejected on very rational grounds, namely the lack of any wind that would be caused by a moving Earth, and also the lack of any shifting in the stars' positions as a consequence of the Earth moving about.

One minority school was *atomism*, active in the fifth through third centuries BC, and represented most prominently by Leucippus, Democritus, and Epicurus. For the atomists the cosmos was infinite in extent, and completely filled with an infinite number of microscopic *atoms* (the Greek word means "indivisible"), all continually in motion and suffering collisions that produced all observed physical and chemical effects, whether the taste of sweetness or the formation of the Moon. Since collisions of these atoms had caused the formation of our Earth, atomists saw no reason why such processes should not be active elsewhere. This then implied that there were an infinite number of worlds being created (and destroyed) all the time. These worlds were of a great variety – some had moons, others not; some were forming, others were dying; some were peopled, others not. Thus a theory of the small-scale structure of matter led logically to the existence of extraterrestrial life of many kinds.

The atomist and Epicurean philosophy was passed to later Europe primarily through the long first century BC poem *De Rerum Natura* (*On the Nature of Things*) by the Roman philosopher Lucretius. He argued for the uniformity of nature and its tendency to complete all possible processes to the fullest. The historian Arthur Lovejoy (1936) has called this the *principle of plenitude* and it was to be central in many later arguments, especially when applied by Christian theologians with regard to God's will. Lucretius argued that nothing was unique:

> It is in the highest degree unlikely that this earth and sky is the only one to have been created... Nothing in the Universe is the only one of its kind, unique and solitary in its birth and growth... You are bound therefore to acknowledge that in other regions there are other earths and various tribes of men and breeds of beasts. *De Rerum Natura*, Book II, lines 1055–7, 1074–8 (Latham, 1951).

But despite the atomists, the philosophies that were to overshadow all others were those of Plato and his student Aristotle (fourth century BC). Of most relevance for the present discussion is Aristotle's *De Caelo* (*On the Heavens*), in which he laid out his familiar scheme of a finite cosmos with a spherical Earth at the center. Mundane materials were composed of the corruptible four elements – earth, air, fire, and water – while celestial realms were made of a perfect fifth element, the *quintessence*, or *æther*. Each of the four elements had its own inherent "natural motion" striving to take it to its "natural place": earth and water downwards, air and fire upwards, and æther in eternal circles. The Earth was surrounded by heavenly spheres, each of which carried one of the seven planets (the five naked-eye planets of today plus the Sun and Moon). The outermost (but still finite) sphere carried the stars. Although Aristotle did not directly comment on the possibility of extraterrestrial life on any of the known planets, his metaphysical system allowed for the existence of only these seven planets. Nor could one have a second, similar cosmos outside of ours because of a logical contradiction: if one had *two* sets of nested spheres, a given mass in either one would be conflicted as to which of the two centers controlled its natural motion.

1.2.2 The scholastics

Beginning in the eleventh century AD, in the great universities and monasteries of the Middle Ages, there thrived *scholasticism*, a catchall term for the philosophical systems and arguments of Christian intellectuals who sought to reconcile the Bible and Aristotle, theology and philosophy, faith and reason. The possible existence of a *plurality of worlds*, as the topic came to be called, was batted back and forth as one of the central areas of doctrinal debate. In the thirteenth century, Thomas Aquinas craftily argued that yes, God's omnipotence meant that he certainly *could* have created many worlds, but that in fact he had not because there was more goodness in a unitary, perfect world (ours) than in many imperfect worlds. These sorts of things were not idle chatter – one's opinions on such matters did matter. For instance, as a result of battles between the Faculties of Arts and of Theology at the University of Paris, in 1277 Bishop Etienne Tempier issued a *Condemnation of 219 Propositions*, adherence to any of which was grounds for excommunication. Two hundred and eighteen of these heresies concerned the intelligence of angels, the mobility of God, the motions of the heavens, the nature of the soul, the relative degree of happiness in this life compared to another, whether pleasure in sexual acts impeded use of the intellect, whether Christian law impeded learning, etc. Amidst all this, Heresy Number 27 read: "That the first cause [God] cannot make more than one world," a teaching that many were espousing based on strict Aristotelian principles of a single, Earth-centered cosmos. But establishment Christianity, modifying many of Aristotle's principles, reasoned at this time that outside of *our* set of spheres, there could well be other worlds, perhaps extending indefinitely

(given the immensity of God), perhaps even obeying different sets of laws.

Almost all of these commentators on the plurality of worlds did not, however, address the question of *life* on those worlds. One exception was the theologian Cardinal Nicholas of Cusa (on the Moselle River), who espoused the cause of extraterrestrial life in his marvellously titled *On Learned Ignorance* (1440), in which the word *learned* should be construed as pronounced both possible ways – the student both learns the boundaries of his ignorance and becomes wiser for having done so. This book is most famous for its prescience (to modern eyes) in stating that "the world-machine has its center everywhere and its circumference nowhere," characteristics of Cusa's boundless Universe and God. Cusa also argued that other planets would be innumerable, and would have inhabitants very different from Earth's. The nature of these inhabitants would be determined by the influence ("in-flowing") of the stars, just as for life on Earth.

1.3 Copernicanism: Earth is a planet

The Church (and its favorite natural philosophers such as Aristotle) continued to control the establishment for centuries more, but the Renaissance and the Protestant Reformation sparked ever more unorthodox thinking. In 1543, while on his deathbed, the Polish canon Nicolaus Copernicus oversaw the final proofs of his magnum opus *De Revolutionibus Orbium Cælestium* (*On the Revolutions of the Heavenly Spheres*), a complex treatise arguing that the Sun, not the Earth, is the center of the Universe. His motivation was not that the dominant theory of Ptolemy[3] was inadequate to explain the data, but rather a philosophical predilection that the Sun, that Great Luminary and source of all heat and light, should be identified with God. His new geometric scheme could adequately reproduce the planets' sky positions as well as Ptolemy's, but its details were not, as is often erroneously stated, any simpler (Kuhn, 1957: Chapter 5). This shift from a geocentric to a heliocentric system was far more than a mathematical transformation – its philosophical and cosmological, and indeed astrobiological, implications were, and continue to be, profound. The Earth now

---

[3] Claudius Ptolemy, second century AD Alexandrian scholar, brought the Aristotelian scheme to its apotheosis in his *Almagest*, which became the authority for all of astronomy and cosmology for 1400 years.

was displaced from the center of the cosmos and it *moved* at a stupendous velocity (a very problematic notion at the time, but one that eventually led to physics as we know it). Furthermore, the Earth was now relegated to being only one among six planets, all orbiting the Sun. Gone was the Aristotelian dichotomy between the mundane and the celestial. Finally, Copernicus removed the stars to much greater distances,[4] and thus allowed the possibility that the Universe was vast, perhaps infinite. The stars could then be imagined as other suns, which then, by the principle of plenitude, were expected to have their own systems of circling planets (a proposition that took 452 years to verify with the discovery of the first extrasolar planet – see Chapters 2 and 21).

The implications for extraterrestrial life were enormous. Since the other planets in our system were now analogous to Earth in all respects, why should they not be inhabited, too? And likewise for the putative planets attendant to distant stars. The Earth became typical, not special, and this notion has since become enshrined as the *Copernican Principle*. It has become dogmatic for astronomy and cosmology ever since; still today, scientific models or theories that place the Earth (or the Sun, or (later) our Galaxy) in any kind of unusual situation are at best strongly suspect, often not even debated. The Copernican Principle has also been influential to this day in the realm of extraterrestrial life and indeed in all of biology, but with a more checkered history.

European culture, which had done very well over its entire history with a geocentric cosmos, was not easily swayed by one book, and it took most of two centuries before all educated persons accepted that Earth had indeed moved off-center. Giordano Bruno (1548–1600) was a flamboyant Italian Dominican monk who left his order, traveled widely, and made enemies wherever he went. His *On the Infinite Universe and Worlds* (1584) was the first major study to grab Copernicus's (and Lucretius's) ideas and run with them full tilt. Arguing that there was no absolute truth and that all things (and locales) were relative, Bruno described a Universe boundlessly filled with stars and their populated planets, suffused by an infinite God. These non-Aristotelian ideas and heretical teachings, including that Jesus Christ was not divine, caught the attention of the Inquisition in Rome and eventually led to Bruno's death by burning. We should not then be surprised that his contemporary Galileo did not seriously touch the question of extraterrestrial life. Another contemporary, Johannes Kepler, safely ensconced in Germany away from the Inquisition, did, however, argue strongly for an inhabited Moon and other bodies. In his *Somnium* (*Dream*) (1634), which details a trip to the Moon, Kepler deduces from observations the nature of the Moon's environment and inhabitants. In a similar vein the English clergyman John Wilkins in 1638 published the widely read *Discovery of a World in the Moone*. He argued for the Copernican system and for an inhabited Moon, for "as their world is our Moone so our world is their moon" (cited by Dick 1982: p. 100). The notion of extraterrestrial beings was catching on among academics, but not until much later did a popular book give much broader currency to the idea.

1.4 Plurality of worlds: Fontenelle and his *Conversations* (1686)

In the late seventeenth century Bernard le Bovier de Fontenelle wrote a slight book that had enormous influence on the reading public in Europe. *Conversations on the Plurality of Worlds* (*Entretiens sur la pluralité des mondes*) was published in Paris in 1686 and over a century ran through almost a hundred editions in many languages. By coincidence it appeared within a year of Newton's masterpiece *Principia*, but there the similarities end. In the traditional manner Newton wrote a complex treatise in Latin, aimed at his fellow natural philosophers (the term then used for scientists). Fontenelle, on the other hand, invented a new genre, writing his 100-page volume in the vernacular and with an engaging, witty style aimed at a broad audience. As he says in his Preface, "I've tried to treat philosophy in a very unphilosophical manner."[5] Under the guise of a series of moonlit conversations with a charming but unschooled marquise, Fontenelle lays out the latest in astronomical knowledge and argues strongly for the existence of inhabitants not only on the planets we know, but also on presumed planets circling every star in the sky. Because of this book's long influence, as well as its delightful arguments and style, we will discuss it here in some detail.

---

[4] Copernicus was forced to place the stars at a very large distance because otherwise the Earth's annual orbital motion would have caused perceptible annual shifts in the apparent positions of all stars, which in fact were not observed.

[5] All quotations are from the 1990 translation by H. A. Hargreaves (Berkeley: University of California Press).

Fontenelle (1657–1757) was a playwright and writer of some success in Parisian circles, but by far his most famous work, written when he was only 29, was *Conversations*. This served as his entrée to popularity as well as to the Academy of Sciences, where he soon became Perpetual Secretary, a post he held for over four decades. His eulogies (*éloges*) for deceased members were widely read and admired for their ability to capture the essentials of both personality and scientific contributions. Fontenelle was also a central figure in that mainstay of the French Enlightenment, the fashionable *salon* circuit, consisting of regular intellectual gatherings run by aristocratic women. The device in *Conversations* of dialogue centered on a woman is thus no surprise.

The book has chapters devoted to each of six evening lessons between a philosopher and his aristocratic hostess at a country chateau. Fontenelle immediately captures our interest with his description of a beautiful young woman with a vivacious intelligence, albeit little knowledge of the natural world. She and the philosopher engage in a lively repartee, often flirtatious, that deals with many basic philosophical and cosmological questions of the day. At the start we find a brilliant description of the nature of scientific investigation (then called philosophy):

"All philosophy," I told her, "is based on two things only: curiosity and poor eyesight;[6] if you had better eyesight you could see perfectly well whether or not these stars are solar systems, and if you were less curious you wouldn't care about knowing ... The trouble is, we want to know more than we can see ... So true philosophers spend a lifetime not believing what they do see, and theorizing on what they don't see, and it's not, to my way of thinking, a very enviable situation." (p. 11)

Fontenelle is much taken by the revelations garnered through the seventeenth century's premier "mathematical instruments" (as they were known), the telescope and the microscope. The telescope had made (habitable) worlds out of planets and the microscope, especially in the recent works of the Dutchman Antony van Leeuwenhoek, had uncovered microcosms in a drop of water (Section 1.12).

There are as many species of invisible animals as visible. We see from the elephant down to the mite;

there our sight ends. But beyond the mite an infinite multitude of animals begins for which the mite is an elephant, and which can't be perceived with ordinary eyesight. We've seen with lenses many liquids filled with little animals that one would never have suspected living there ... Even in very hard kinds of rock we've found innumerable small worms, living in imperceptible gaps and feeding themselves by gnawing on the substance of the stone ... Even if the Moon were only a mass of rocks, I'd sooner have her gnawed by her inhabitants than not put any there at all. (pp. 44–5)

The crux of Fontenelle's reasoning rests in the Copernican picture of the Earth itself as only one among the planets circling the Sun, implying that the stars are themselves other suns. Although fully a century and a half had passed since Copernicus, his ideas were still known to only a portion of European readers. Fontenelle then employs the principles of plenitude and of the uniformity of nature to assert that the existence of all these other suns surely implies that, just as for our Sun, they have their own planetary retinues (Fig. 1.1). Likewise, these planets surely have their own inhabitants, as does our Earth. Our world exhibits a profound fecundity and diversity and it would certainly be wasteful of Nature to accomodate all these other locales without populating them. Yet in the end Fontenelle realizes that he may have extrapolated too far:

"Listen, Madame," I answered, "since we're inclined to keep mixing foolish lovetalk with our serious conversation, the logic of mathematics is like that of love. You can't grant a lover the least favor without soon having to grant more, and still more, and in the end it's gone awfully far. Well, if you grant a mathematician the least principle, he'll draw a conclusion from it that you must grant him too, and from that conclusion another, and in spite of yourself he'll lead you so far you'll have trouble believing it." (p. 64)

These ideas were dangerous in a nation ruled by a strong king (Louis XIV) and a Church that still officially banned Copernicus's teachings – Fontenelle only escaped censorship because of good connections and the lighthearted, sometimes veiled manner in which he treated unorthodox ideas. Although he never once mentioned the role of God, he also made no explicitly antireligious arguments. To further cover his bets, in the preface he points out that it should not be concluded that the probable inhabitants of planets are

[6] The original French felicitously reads: *l'esprit curieux et les yeux mauvais* ("a curious spirit and bad eyes").

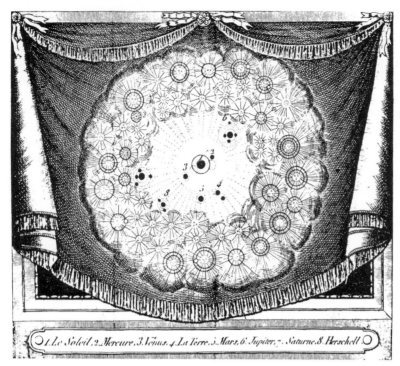

FIGURE 1.1 The frontispiece for Fontenelle's influential book *Conversations on the Plurality of Worlds* (1686), which went through over a hundred editions. The Sun is circled by numbered planets (some of which have moons circling them); but note that all of the distant stars *also* have planetary orbits girding them. This version is from an 1821 French edition, and includes the new eighth planet, called for a while (in France) Herschel, after William Herschel, who discovered what came to be called Uranus in 1781.

1. Le Soleil, 2. Mercure, 3. Vénus, 4. La Terre, 5. Mars, 6. Jupiter, 7. Saturne, 8. Herschell.

"sons of Adam," that is, men. Such men would then need Salvation, which would raise vexing theological questions. Rather, Fontenelle says that Nature's intrinsic diversity will guarantee that they are wholly *unlike* men. Fontenelle had no desire to risk martyrdom.

Fontenelle was an important transition figure between the so-called Scientific Revolution and the Age of Enlightenment, when rationality reigned and intellectuals were the heroes. *Conversations* influenced a whole genre of utopian novels and imaginary voyages – witness Christiaan Huygens's posthumous *Cosmotheoros* of 1698,[7] Jonathan Swift's *Gulliver's Travels* of 1726, and Voltaire's *Micromégas* of 1752.[8] Fontenelle's ideas on extraterrestrial life were the ineluctable result of this new exciting process, called *science*, that he saw as key to society's progress. By applying reason, always tempered

with a healthy skepticism, to the latest observations, natural philosophers were producing a Universe in which the Earth and its inhabitants were typical, not uniquely special. Geocentrism was *passé*. This stance led to a relativism radical for its time, but one that rings modern to the ears of astrobiologists in the early twenty-first century. As he wrote:

> The same desire that makes a courtier want to have the most honorable place in a ceremony makes a philosopher want to place himself in the center of a world system, if he can. He's sure that everything was made for him, and unconsciously accepts that principle which flatters him. (p. 17)

And when the marquise asks whether, despite Earth being so small compared to Jupiter, we can be seen from the Jovian realm:

> There'll be astronomers on Jupiter who, after taking great pains to construct excellent telescopes,... will finally discover in the heavens a tiny planet that they've never seen before... [but] they wouldn't have the faintest suspicion that it could be inhabited. If anyone were to think of it, heaven knows how all Jupiter would laugh at him. It's possible we're the cause of philosophers being prosecuted there who have tried to insist that we exist. (p. 57)

[7] *Cosmotheoros, or New Conjectures Concerning the Planetary Worlds, Their Inhabitants and Productions*, was very much modeled on *Conversations*, although with much more technical analysis. For example, Huygens argued that any planets circling other stars would be unobservably faint, and showed in detail how to estimate the distance to the stars (he calculated 0.4 light-years for Sirius, ~20 times too close, but his method was sound).

[8] *Micromégas* is a satire aimed at the pretensions of humans. The title character is a 120,000-foot tall inhabitant of a planet of the star Sirius who tours the planets of our solar system, finding varied inhabitants wherever he goes.

The marquise muses:

> I could imagine with pleasure these telescopes aimed at us, as ours are toward them, and the mutual curiosity with which the planets consider one another and ask among themselves, "What world is that? What people live on it?" (p. 57)

On another occasion the philosopher answers a query as to what sort of beings might inhabit the Moon:[9]

> Honestly, Madame, I've no idea. If it could be that we were rational, yet weren't men, and if besides we happened to live on the Moon, could we possibly imagine that down here in this place there were bizarre creatures who called themselves the human race? Would we be able to fantasize something that has such mad passions and such wise reflections; a life so short and views so long;... such a strong desire for happiness and such a great inability to achieve it?... We look at ourselves incessantly, and we're still guessing at how we're made. (p. 32)

Fontenelle also consistently applied his skepticism to more than claims about the natural world. In a warning to his readers (also applicable to the reader of this present chapter!), he compared historic "facts" with scientific ones, arguing that neither should be considered true or false, but rather colored in many shades of epistemic gray. He reckoned that the existence of Alexander the Great had sufficient evidence that it should be considered more probable than the existence of planetary inhabitants, but that the evidence for many other accepted points of history was in fact *less* than that for extraterrestrials.

1.5 Natural theology

As modern science took shape in the seventeenth century and extended its influence, a majority of its practitioners were either Christians or *deists*, who believed in a God that created the Universe, set it running according to natural laws, and thereafter did not interfere. Many reconciled their religious beliefs and their findings from natural philosophy by learning of God (and even proving his existence) through study of his handiwork manifest in the "Book of Nature." As the

mechanical Universe of Isaac Newton took hold in the eighteenth century, this approach became known as *physico-theology* or later *natural theology*, taking its place alongside *scriptural* or *revealed theology* based on the Bible. Although the great philosophers David Hume and Immanuel Kant presented cogent logical analyses against these "arguments from Design," natural theology had an important influence on mainstream science throughout Europe and America as late as the mid-nineteenth century,[10] with particular persistence and strength in Britain. The English poet Alexander Pope expressed the spirit of the age in his *Essay on Man* (1734).

> He, who thro' vast immensity can pierce,
> See worlds on worlds compose one universe,
> Observe how system into system runs,
> What other planets circle other suns,
> What vary'd Being peoples every star,
> May tell why Heav'n has made us as we are.

Numerous books tied together natural theology and the plurality of worlds. An early one was William Derham's *Astro-Theology: or a Demonstration of the Being and Attributes of God, from a Survey of the Heavens* (1715). Derham, a countryman and follower of Newton's, argued that the more magnificent and fruitful the Universe (as evidenced by extraterrestrial inhabitants), the greater was God's demonstrated glory and providence (Dick 1982: 151–4). A German study called *Hydrotheologie* (*Water Theology*) in 1734 by J. A. Fabricius pointed out that ice's lower density than water, allowing aquatic creatures to survive in cold weather, was a clear example of divine prescience (Brooke 1991: 197). A century later Thomas Chalmers, a Scottish minister, wrote an influential treatise entitled *A Series of Discourses on the Christian Revelation, Viewed in Connection with the Modern Astronomy* (1817). One argument of particular interest to today's astrobiology concerned microscopic realms, which, Chalmers said, revealed worlds and "tribes of animals" every bit as unknown and vast and fascinating as those seen in telescopes. Infinity in one direction was balanced by infinity in the other. Since God's beneficence had applied to these realms even before we were aware of them, so God cared for humans even though we might be insignificant on a cosmic scale.

---

[9] In the end Fontenelle concluded that the Moon, unlike the planets, was *not* inhabited, because he took it to have no atmosphere. Conversely, based on various observational evidence, Kepler and Wilkins had earlier concluded the opposite, that the Moon indeed *did* have an atmosphere and was therefore likely inhabited.

[10] Even today, those who promote the necessity, based on scientific findings, for so-called Intelligent Design are very much working in this tradition. For a remarkable example, see *The Privileged Planet* by Gonzalez and Richards (2004).

One minority view in the mid-nineteenth century came from the Cambridge don and polymath William Whewell, whose *Of the Plurality of Worlds* (1853) argued that Earth and its intelligent life were probably unique. Whewell used the latest astronomical data to point out that most stars seemed to be in binary systems, making orbits unstable and conditions on any planets highly variable. Many stars also were variable in intensity and seemed to be of lower mass than the Sun. Furthermore, physical conditions on the other planets in our own system were extreme, not at all suitable for life. Whewell's book sparked a storm of negative reviews and rejoinder books whose titles tell all: *The Universe No Desert; The Earth No Monopoly* (1855) by William Williams of Boston, for example, and *More Worlds than One: the Creed of the Philosopher and the Hope of the Christian* (1854) by David Brewster, a leading English physicist.

1.6 Two nineteenth-century revolutions

Attitudes toward extraterrestrial life were profoundly affected during the nineteenth century by developments in astronomy, geology, and biology. By century's end the standard picture was of a Solar System that formed long ago, of an Earth that had a long history that could be read through the study of rocks and fossils, of each planet having a (finite) history and future, and of many stars confirmed (through their chemical make-up) to be other suns, likely therefore to have their own life-bearing planets. On the biological side, Charles Darwin's theory of evolution by natural selection (1859) revolutionized biology and provided an entirely new context in which to think about the origin of life and about extraterrestrial life.

1.6.1 The nebular hypothesis and the start of astrophysics

The notions of a *changing world*, as well as of a *very old world*, entered the geological and astronomical worlds around the turn of the eighteenth into the nineteenth century. Before this time the world was considered largely static since the Creation about 6,000 years ago (or at least since Noah's Flood not too long thereafter). Geologists like the Scot James Hutton and later Charles Lyell (a close friend of Charles Darwin) came to startling new conclusions based on detailed fieldwork and a new principle of *Uniformitarianism* – in order to explain the present state of Earth, one should appeal to no more than the processes we *now* observe

on Earth (e.g., erosion and sedimentation from rivers), not to past catastrophes such as the Flood (*Catastrophism*). But to build mountains and continents at today's estimated rates necessarily implied previously unimaginable lengths of time, counted in the millions of years and probably much longer – as Hutton famously put it: "we find no vestige of a beginning, no prospect of an end." Furthermore, fossil animals and plants were found to correlate well with sedimentary strata, implying that the past had also witnessed an ever-changing suite of species – a profound change in the *living* world, too. For the first time, the Earth and its life were perceived as not in stasis since Creation, but they had a scientific *history*. In 1837 Lyell even found these past paleontological worlds in some sense better than considering life on other planets.

> Geology ... has demonstrated the truth of conclusions scarcely less wonderful [than the astronomer's], the existence on our own planet of many habitable surfaces, or worlds as they have been called, each distinct in time, and peopled with its peculiar races. (Crowe, 1986: 223)

On the astronomical side, a changing Cosmos was likewise coming into its own. Pierre-Simon Laplace (1749–1827), sometimes called the "Newton of France," was a mathematician and natural philosopher who analyzed the details of how planets gravitationally influence each other's orbits, in the process largely inventing the field of celestial mechanics. He had been able to mathematically demonstrate that the Solar System was extremely stable, i.e., that the perturbations on each planet's orbit from its fellows did not lead to disastrous changes with time, only oscillatory changes. But how did the planets come to be? In 1796 Laplace proposed in his masterpiece *Exposition du Système du Monde* (*Introduction to the System of the World*) (which remained authoritative for the next half-century) what became known as the *nebular hypothesis*. He was impressed with the cataloguing of thousands of nebulae[11] by German/English astronomer William Herschel (1738–1822), the greatest observer of his age and perhaps of all time, as well as with Herschel's

---

[11] *Nebula* in Latin means mist or cloud, and was applied to anything that looked diffuse in a telescope, unlike the sharpness of a star. Although obsolete, the term survives today in the names of a huge variety of objects, e.g., Orion nebula (now known to be hot gas and young stars), Crab nebula (a supernova remnant), planetary nebula (hot gas ejected by an old low-mass star (nothing to do with a planet!)), and Andromeda nebula (a galaxy).

speculation that these fuzzy patches consisted of a shining fluid out of which stars formed. Laplace sketched the idea that the Sun and planets formed from a cloud of hot gas collapsing under its own gravity. The Sun formed in the center and had a vast residual atmosphere surrounding it. The individual planets formed because the collapsing cloud eventually broke up into a series of gaseous rings, each of which gradually cooled off and coagulated into a planet. Presuming that the initial cloud slightly rotated, conservation of angular momentum would mean that all the planets (and their moons) would rotate and revolve in the same sense and in a flattened, aligned manner, i.e., all planetary orbits would lie closely in one plane (as observed). If the stars were other suns, Laplace's scheme boded well for a plurality of (inhabited) worlds.

Schaffer (1989) has shown how the nebular hypothesis lay dormant until the 1830s when it was resurrected,[12] mainly for political purposes more than astronomical ones: the notion of the formation of the Sun and planets by natural law, a type of astronomical *progress*, was used to legitimize the goals of the British reform movements, which sought to show that *social* progress was also natural and inevitable. The key book in this regard was *Views of the Architecture of the Heavens* (1837) by John Nichol, a Scottish political economist and astronomer. This book influenced the philosopher Herbert Spencer (see Section 1.7.1), as well as Charles Darwin.[13] By 1850, however, leading British scientists, for fear of their credibility, had disavowed the link between the nebular hypothesis and political matters.

A key aspect of the nebular hypothesis as it was developed during the nineteenth century (especially once thermodynamics became established in the 1840–70 period) was that the outer gaseous rings would cool off fastest (being farther from the Sun) and therefore form the first (molten) planets, which would continue to cool off and eventually solidify. Based on this model and the current measured rate of heat loss, Joseph Fourier (in 1819) and, starting in the 1850s, William Thomson (later Lord Kelvin) calculated that the Earth simply could not be as old as the geologists had deduced. The physicists' derived values

ranged up to 300 Myr, but no higher. Such ages had all the authority of physics, but were nevertheless unacceptable to the geologists and of great concern to Darwin (Section 1.6.2), who required a much longer time to effect evolutionary change (Brush, 1996).

Using ideas that eventually became enshrined in the Second Law of Thermodynamics, one could also see that the Earth would continue to cool down until the inevitable high-entropy "Heat Death." As Thomson put it in 1852:

> Within a finite period of time past, the earth must have been, and within a finite period of time to come the earth must again be, unfit for the habitation of man as at present constituted.   (Brush, 1996: 10)

Like geology before, physics and astronomy were furnishing the Earth with a past history (as well as a finite future). Moreover, calculations for the Sun indicated that it could be no older than 20–30 Myr – assuming it was powered by the gravitational energy that its material lost as it collapsed.[14] On these ideas one concluded that outer and/or smaller planets were today both older and cooler than inner planets – Mars was older, cooler, and nearer death than Earth, and the Moon was completely dead. In a popular book (Langley, 1884: 167–72) one finds the striking juxtaposition of three illustrations meant as analogues: the extremely wrinkled hand of an old woman, a withered apple, and mountainous terrain on the Moon! This thinking would prove important in late-nineteenth-century ideas about life on Mars (Section 1.7).

Another major trend in astronomy affecting attitudes toward the prospects for extraterrestrial life was the rise of what came to be called astrophysics. Until the mid-nineteenth century astronomy was concerned almost exclusively with measuring and trying to understand the positions and changes in positions of planets and stars – there was little else one could do with a telescope. In 1815, however, German optician Joseph Fraunhofer had first analyzed in detail hundreds of dark absorption lines in the solar spectrum, produced by passing sunlight through a slit and a prism. These lines were shown to indicate, by comparison with laboratory flame sources, the presence in the Sun of familiar elements such as sodium and magnesium.

---

[12] The term *nebular hypothesis* was not actually coined until 1833, by William Whewell in England.

[13] Historians debate the degree of influence that early-nineteenth-century ideas of astronomical evolution had on Darwin's thinking, but most conclude there is definitely a sibling relationship, if not a maternal one. (Brush, 1996: 62–75.)

[14] It would not be until the twentieth century that these problematic lifetimes for the Earth and Sun would be made obsolete by new discoveries: (1) an additional source of heat in the Earth's interior (from radioactive elements), and (2) nuclear reactions powering the Sun via conversion of mass into energy.

Spectroscopy of *stars*, however, did not really take off until the 1860s, once better spectroscopes could be attached to telescopes and one better understood the physics of how lines were produced (Kirchhoff's Laws of 1859). Moreover, by the end of the nineteenth century, photography had matured to a sufficient state that a star's spectrum could be recorded on a photographic plate, allowing much more accurate measurements. For instance, one could now measure a star's radial velocity (through Doppler shifts in line wavelengths, first done in 1868). Also, it was found that many of the nebulae exhibited *bright* lines in their spectra (not the dark lines seen in stellar spectra), strong evidence that they consisted of a hot gas. This was consistent with the nebular hypothesis, which required the presence of such gas, and many argued that such nebulae were planetary systems in formation.

This "new astronomy," also called "celestial physics" or "astro-physics," was vital for the question of extraterrestrial life because it now allowed one in principle to determine the composition of planetary atmospheres, of stars, and of nebulae that were producing stars. Analogies between the life-supporting Sun and other stars were also greatly strengthened when it became apparent that many other stars were indeed Sun-like in their spectra – why should they not have planets and life, too? The thread of life seemed to be woven into the fabric of the Cosmos. As the American astronomer Samuel Langley wrote in his book *The New Astronomy* (1884: 222):

> This wonderful instrument [the spectroscope] of the New Astronomy can find the traces of poison in a stomach or analyze a star ... The ancients were nearly right when they called man a microcosm, or little universe ... You and I are ... but children of the sun and stars ..., having bodies actually made in large part of the same things that make Sirius and Aldebaran. They and we are near relatives.

1.6.2 Darwin's evolution

Charles Darwin's (1809–1882) theory of evolution by natural selection revolutionized biology and provided an entirely new context in which to think about the origin of life on Earth and about its chances elsewhere. Within two decades after the publication of *On the Origin of Species by Natural Selection, or The Preservation of Favoured Races in the Struggle for Life* (1859), evolution by natural selection was the dominant paradigm in all biological fields, despite no

consensus on the mechanisms of heredity or population variation (this would not come until the early twentieth century with Mendelian genetics). Natural theology, the reigning tradition of more than two centuries, was turned on its head: instead of a cat being a good mouser because God in his wisdom designed the cat to have agility and sharp claws *in order to* catch mice, Darwin now said that cats exist only *because* they catch mice well.

Darwin's theory raised a ruckus in Victorian England and continues to do so today in various segments of society. The theory contradicted many readings of the Bible by requiring (a) no Designer, and (b) eons of past time as supplied by the new geology. Furthermore, organisms were continually being modified, not created in the past in a perfect manner (whether all at once, or "instantaneously, as needed, along the way" as many paleontologists held before Darwin). Finally, humans were not *apart* from the natural world, but were *of* it, descended from the apes. With regard to extraterrestrial life, the question shifted from what might be the Divine Purpose in having other stars and planets, to whether or not life could originate in some other locale, as it apparently had long ago on Earth. Furthermore, with an evolutionary scheme, once life did begin somewhere, by analogy one might expect that, given enough time, complex life and intelligence would develop.

Darwin himself did not directly comment on extraterrestrial life,[15] and felt that the question of the origin of life was too difficult to tackle,[16] but his effect on both fields has been profound. By the end of the nineteenth century, when Darwin's evolution was combined with the nebular hypothesis and Earth's long past and future, as well as with schemes of social evolution that grew out of Darwinism (Section 1.7.1), one obtained a new past and a new future that looked very different from earlier views, whether scientific or religious. The following chronology, first established at this time, is still our view in the early twenty-first century:

---

[15] However, the co-inventor of evolution through natural selection, Alfred Russel Wallace, weighed in with two books in his 80s, strongly arguing for the uniqueness of humans (whom he held to be *not* part of standard evolution on Earth). These were *Man's Place in the Universe* (1903) and *Is Mars Habitable?* (1907), both available in full at www.wku.edu/~smithch (also see Dick (1996: 38–49).

[16] Although see his famous "warm little pond" quotation in Section 6.4.5.3 or in Strick (2000: 92–3).

FIGURE 1.2 The frontispiece (painted by Lancelot Speed) to *Nebula to Man* by Henry Knipe (London, 1905). This elaborately illustrated book tells the history of the Universe and, through the paleontological record, of life on Earth, all in (bad) verse. Note the Darwinian apes lurking in the background.

Gaseous nebula → molten planet → cooled, solid planet → simple life → complex life → intelligent life → advanced civilization → death of civilization and species.

Figure 1.2, from a 1905 volume, wonderfully encapsulates all but the last stage.

1.7 Enter Mars

The episode of the martian canals is at once one of the strangest and most fascinating in the history of ideas on extraterrestrial life. For a thirty year period (∼1880–1910) many of the world's best observational astronomers were embroiled in the question of whether or not Mars had long linear, canal-like features and, if so, whether or not they were dug by intelligent beings desperate to save their civilization (see above). Debates involved the merits of the steadiness of air at various observatories, the efficacy of different telescopes, and the skill and style applied to drawings of Mars. The public, meanwhile, was not in doubt, urged on by a

slew of popular science books and fictional accounts involving Martians. Although by 1910 almost all astronomers felt that the case against the Martians (certainly) and the linear features (almost certainly) was closed, they were embarrassed that a minority of astronomers and the public hung on long after. And the popular legacy never went away and still lingers, for even today it is "the Martians" that are linked to extraterrestrial intelligent life, invasion by aliens, and "Little Green Men" (Markley, 2005). The putative canals and intelligent beings on Mars also fostered science fiction such as H. G. Wells's *War of the Worlds* (1898) and Kurd Lasswitz's *Auf Zwei Planeten* (1897), which began the long tradition of extraterrestrials in fiction that continues to this day.

The influence of these late-nineteenth-century episodes extended right up until the space age began in 1960, when possible vegetation on Mars remained central to scientific arguments regarding extraterrestrial life (Section 1.8.3). One of NASA's key goals was to search for life, and the place to search was Mars, which was eventually realized with the Viking mission of 1976 (Sections 2.2.1 and 18.5.2). And now in the twenty-first century it is ironic that Mars remains of great astrobiological interest, although connections with the reasonings of a century ago or of a half-century ago are tenuous (Chapters 2 and 18).

1.7.1 Areography[17] and the Martians' *canali*

Giovanni Schiaparelli (1835–1910) was director of the small Brera Observatory in Milano, Italy, but nevertheless established himself as a leading international researcher on comets, meteors, and the planets. In 1877, when Mars was particularly close to the Earth and therefore favorable for observation,[18] Schiaparelli used his 22-cm diameter refractor to map out the features of Mars with unprecedented detail. Such mapping involved great patience at the telescope eyepiece, monitoring the shimmering image of the planet until

---

[17] *Areography* (from Ares, the Greek name for Mars) is the Martian equivalent of geography. Lane (2005) discusses the close connections between late-nineteenth-century geography and the martian maps and mapmakers discussed in this section.

[18] Mars is at its closest, or *opposition* (relative to the Sun), every 26 months and therefore the history of Mars observations (and NASA missions) has a biennial character. Furthermore, its relatively eccentric orbit means that especially favorable oppositions occur approximately every 15–16 years, when its angular size can be as large as 25″. Favorable oppositions occurred in 1877, 1892, and 1909. (The last one was August 2003.)

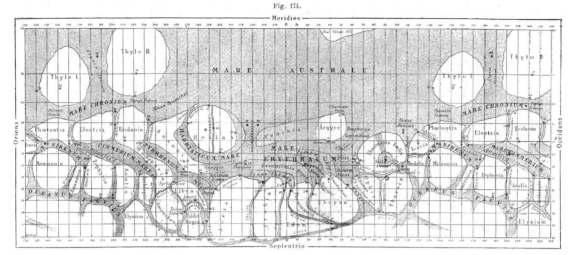

Fig. 174.

TRIANGULATION DE L'ARÉOGRAPHIE, PAR M. SCHIAPARELLI, EN 1877, POSITIONS DE 62 POINTS MESURÉS, ET CARTE NOUVELLE.

FIGURE 1.3 Schiaparelli's first published map of Mars, based on observations at the opposition of 1877 from Milano's Brera Observatory (as published in Flammarion (1892: 293)).

atmospheric conditions relented and a sharp image popped into view for a second or two,[19] whereupon he quickly noted the shape, size, and (usually low) contrast of features (often tiny and/or narrow), estimated their apparent positions on the martian disk with respect to a network of 62 control points, noted the time (for it was known that Mars rotated once every 24.6 hours), and then quickly sketched what he had seen. After six months he produced what in effect was a new Mars (Fig. 1.3). Schiaparelli (1878) established categories for the dark features with Latin names such as *mare* (sea, as on the Moon), *lacus* (lake), and *sinus* (bay). Specific names came from Mediterranean geography, history, and mythology. Many of these names are still with us, e.g., Sinus Sabaeus, Syrtis Major (after the Gulf of Sidra in modern Libya), Elysium, Mare Acidalia, Tharsis, Chryse (where Viking 1 landed in 1976), and Hellas ("Greece," now recognized as Mars's largest known impact feature). Although he cautioned that these hydrological names were meant only as descriptions and not interpretations, over ensuing decades neither he nor his peers heeded this warning.

Among his categories were *canali*, an Italian word that indicates either (natural) *channels* or (constructed) *canals*.[20] But in English within a few years they uniformly became *canals* – canal-building was one of the high-tech activities of the day, with the Suez Canal having been finished just eight years before. Schiaparelli catalogued several dozen of these thin, dark, linear features arrayed in a complex network over his bright continents, connecting various other "acquatic" features:

> The complex pattern of dark lines that join the blobs that we consider as seas is another argument in favor of the hypothesis [that the martian maria are like Earth's seas]. The color of these lines must be from the same cause as that of the seas, and the lines can only be connecting *canali* or straits.[21]

Over the decade of Mars's next five oppositions, Schiaparelli (1888) modified, refined, and elaborated his map until he had produced Fig. 1.4. The shapes of the dark and light regions (and the polar caps) were finely detailed and of great interest, but it was the unprecedented network of *canali* that caught the attention of both astronomers and the public.[22] By the 1882

[19] The shimmering effect of the atmosphere, which astronomers call *seeing* (as opposed to the *transparency* (clearness)), arises because light rays pass through a column of air in which temperature and density, and thus the refractive index, are ever varying. A changeable refractive index in turn means that a telescope's image of any small detail on a planet dances around and becomes blurred with the images of any adjacent details. Good seeing is when the magnitude of the dancing is minimal, allowing the finest detail to be discerned. Consistently superior seeing is one of the prime desiderata when selecting an observatory site.

[20] The term *canale* was first used to describe a few martian features in 1859 by the Italian Jesuit astronomer Angelo Secchi (1859: 59–60).

[21] Schiaparelli (1878), as quoted (in French) in Flammarion (1892: 298).

[22] Earlier observers had seen linear features, several of which corresponded to Schiaparelli's, but none had recorded anything like his complex network. (Flammarion, 1892; Sheehan, 1996: 75–7)

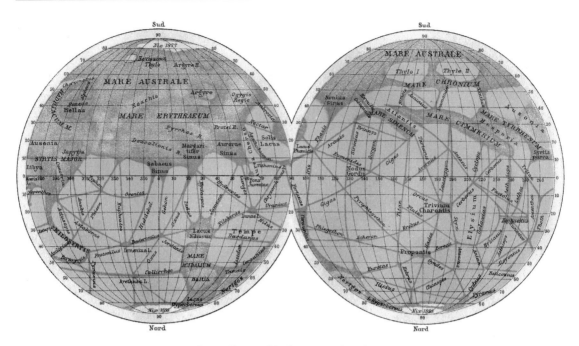

Carte d'ensemble de la planète Mars
avec ses lignes·sombres non doublées
observées pendant les six oppositions de 1877-1888
par J.V. Schiaparelli

FIGURE 1.4 Schiaparelli's second map of Mars, based on observations over 1877–88. This is the map where martian *canali* first appeared in strength (as published in Flammarion (1892: 296)).

opposition Schiaparelli had identified 60 canals, and observed another phenomenon: many canals now appeared as *two* parallel lines, a process he called "gemination" ("twinning"). Over the years he also recorded changing "coastlines," which he interpreted as seasonal or topographic changes affecting the extent of the water.[23]

The canals of Schiaparelli, who became known as the "prince of Mars observers," became well known primarily as a result of the French astronomer and author Camille Flammarion (1842–1925). Over a sixty-year period he was in many ways the "Carl Sagan" of his era as a prolific author, founder of an astronomical society, celebrity, and strong advocate for extraterrestrial intelligent life. At the age of twenty Flammarion (1862) wrote *Plurality of Inhabited Worlds* (*La pluralité des mondes habités*), which ended up going through more than forty editions throughout his life. Written in an exuberant, confident style, the book made many

standard arguments for extraterrestrial life, but also included a healthy dose of teachings on the transmigration of souls from planet to planet. This was the first of a long string of Flammarion books that dealt with astronomy, science fiction, and spiritualism; for instance, in 1883 he wrote *Lumen*, a tale in which a soul travels away from Earth faster than light and is thus able to observe history running backwards. Eventually Flammarion established his own observatory outside of Paris, dedicated to planetary observations. When Schiaparelli's results came along, Flammarion enthusiastically interpreted them as strong evidence for a race of Martians who had built the canals. He did this even though he himself, a skillful observer, had only been able to see a few of the broadest canals. In 1892 he published a 600-page compilation of all extant observations of Mars[24] and, after considering various possible natural explanations (such as ice crevasses and geological faults), concluded:

---

[23] Schiaparelli believed in water on Mars, but over his career was primarily agnostic about whether there was life, although he did write one paper in 1895 that speculated on the nature of possible societies on the planet.

[24] *La Planète Mars* (1892), was joined by Volume II in 1909; a planned third volume never did appear. These volumes are filled with drawings of Mars, and are fascinating to peruse even if one does not read French.

Truly, the more one looks at these maps, these drawings, the less one feels them to be the blind work of nature ... It is natural to conclude that Mars is inhabited just as is the Earth, and that its humanity, whatever it may be, must be more advanced than us ... Is the layout of these rectilinear canals that double in certain seasons the result of the ingenuity of these unknown beings?... Whereas the piercing of the Alps, the Isthmus of Suez, and the Isthmus of Panama ... seem today to be colossal enterprises for our science and technology, in the future they will be no more than child's play for us...

The globe of Mars must be almost levelled by the centuries, and there is very little water. That which will happen on Earth in several million years must have already happened on Mars.[25]   (Flammarion 1892: 581, 586, 589, 587)

The canal story crossed the Atlantic when Percival Lowell, a rich American with many interests including astronomy, headed out to the Arizona Territory and set up a new observatory to check out the canals for himself during the 1894 opposition. He was to come to the same conclusions as Flammarion and similarly write a popular book that sensationally spread the word.

Percival Lowell (1855–1916; Fig. 1.5) was a member of the so-called Boston Brahmins that included lots of money (the cotton mill town of Lowell, Massachusetts), social and philanthropic aristocracy, and academic prestige (his brother Lawrence was President of Harvard University, while sister Amy and cousin James Russell Lowell were well-known poets). His Harvard education was equally on the mathematical/physical and humanities sides. For a while he was a successful businessman, but rebelled against this expected role as the elder son and for a long period made many trips to the Far East, eventually publishing four successful travelogues of Japan and Korea, including analysis of Oriental culture. He marvelled at and praised aesthetic aspects of that culture, but felt that it was intellectually an example of "survival of the unfittest," meaning that cultural evolution had died and reached a stasis so that Japan was as dead as the Moon (Sheehan, 1988: 161).

In early 1894 Lowell's third and last career began when astronomy came to the fore. He had always been

FIGURE 1.5 Percival Lowell (1855–1916), an aristocratic observer in the frontier desert. (Lowell Observatory archives.)

extremely interested in the topic (even taking a largish telescope along with him on one trip to Japan), and was a member of the board of visitors of Harvard Observatory. But now, having pursued one exotic culture, he set off to find another. He soon arranged a joint expedition with two Harvard astronomers to set up an observatory in a remote site in the US West, chosen to have optimum observing conditions. Lowell offered to supply his fortune and himself to establish an observatory to study the physical conditions of the planets, especially Mars. Just before heading out West, Lowell told the Boston Scientific Society:

If the nebular hypothesis is correct,... then to develop life more or less distantly resembling our own must be the destiny of every member of the solar family which is not prevented by purely physical considerations, size and so forth, from doing so ...

The most self-evident explanation [of the canals] ... is probably the true one; namely, that in

---

[25] Flammarion's picture is that erosion eventually wears down all topography on a planet, and that the quantity of surface water steadily decreases as it drains into the interior. Furthermore, a vast network of canals could only exist on a planet of relatively low relief.

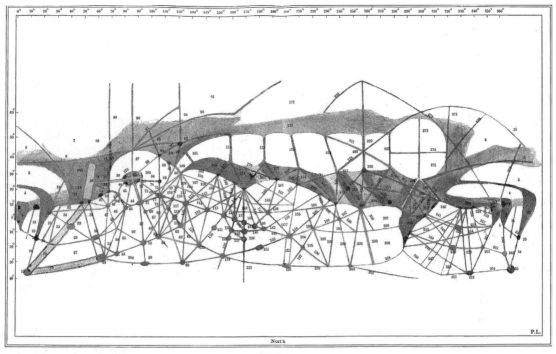

MAP OF MARS
on Mercator's Projection

LOWELL OBSERVATORY
Flagstaff, A. T. 1895

FIGURE 1.6 Lowell's map of Martian canals published in his book *Mars* (1895), based on observations from Flagstaff, Arizona, during the opposition of 1894.

them we are looking upon the result of the work of some sort of intelligent beings. (Strauss, 2001: 154; Sheehan, 2005: 109)

Within a short time Lowell and his Harvard colleagues settled on the 2100-m altitude frontier town of Flagstaff, Arizona, and only five months before Mars's opposition in October 1894, they commenced observations with a 46-cm diameter refractor. One year later Lowell had data in hand to be convinced not only that Schiaparelli's canals were real, but, taking matters one step further, that they were conclusive evidence of an advanced intelligence on a dying planet. By late 1895 he published *Mars*, the first of several popular books setting forth his arguments, and over the next several years turned the site into the permanent Lowell Observatory (still existing as an excellent institution).

There had been some controversy about how much to believe of Schiaparelli's maps, but his reputation as a skilled observer was such that those who could not confirm for themselves various details, including the canals, took that to indicate deficiencies in either their telescope, their site, or their observing skill. But now Lowell was considerably raising the ante. Lowell's map

(Fig. 1.6) showed a dense network of canals that he argued could not be naturally produced in any way and must be "supernatural," the product of engineering. He came to this conclusion because of their straightness, their uniform width, the fact that they did not run "anywhither," and the frequent occurrence of *three* lines neatly intersecting (he called these nodes "oases," where agriculture was possible) (Lowell, 1895: 148–52).

Lowell's final step was to cast these observations within the sweeping philosophy of Herbert Spencer, an influential nineteenth-century British philosopher whose ten-volume *Synthetic Philosophy* provided a framework to combine the physics of the nebular hypothesis (Section 1.6.1) with evolutionary principles not just of biology, but also societies (e.g., Spencer coined the term "survival of the fittest" and saw capitalism as inevitable in its efficiency) (Strauss, 2001: Chapter 5). Spencer and allies, sometimes called *cosmic philosophers* and sometimes *social Darwinists*, proposed that the fundamental trend of nature was an innate evolution towards more complexity from lifeless matter (Laplace's nebular hypothesis was central) to organic entities to intelligent beings to ever more advanced societies. Progress was inevitable. Darwin's

brand of evolution was only one part of this framework, and some aspects of Spencerianism could in fact
be more Lamarckian than Darwinian. All conventional
disciplines were seen as a hierarchy of subdisciplines of
cosmology, with astronomy at the top (being the most
inclusive), geology a subfield of astronomy, biology
likewise of geology, followed in turn by psychology,
sociology, and ethics. (Where would Spencer have
placed astrobiology?)

As a cosmic philosopher and social Darwinist,
Lowell found the purpose for the canals: because Mars
was at an evolutionary state more physically advanced
than Earth (in accord with the nebular hypothesis), it
was drying up. The Martians, who were much more
socially and intellectually advanced, had dealt with this
calamity by building a network of canals in order to
bring annual spring snowmelt from the poles to the
lower desert latitudes (Lowell, 1895: 206–9). There was
no doubt about how to assign the evolutionary hierarchies of (a) (the dead) Moon versus Mars versus Earth,
or (b) Japanese versus Westerners versus Martians.
Observing Martian society provided a "cosmoplanetary
breadth of view" not unlike how travels to the Orient
allowed him to better understand Western ways (Strauss,
2001: 61).

As he said in a poem written about Mars during his
first few months in Flagstaff (Sheehan, 1988: 181–2):

One voyage there is I fain would take
While yet a man in mortal make;
Voyage beyond the compassed bound
Of our own earth's returning round ...
Surface through which half the oceans have sunk,
Their once broad bosoms already shrunk ...[26]
And Mars like our moon through space shall roll,
One waterless waste from pole to pole,
A planet corpse, whence has sped the soul.

The public liked what they read, but professional
astronomers on the whole felt that Lowell's conclusions were wholly unwarranted. Even ignoring his sensational interpretations, many astronomers distrusted
the skills and objectivity of this dilettante who was only
starting serious astronomy in his late 30s; when he
retorted that they, the established professionals, could
not see his canals because their instruments or sites
or observing skills were deficient (lack of "attentive

perception"), they became only more incredulous. But
Lowell was a skilled, persuasive writer and speaker,
and for the rest of his life was able to keep the idea
alive in the minds of the public and of a minority of
astronomers despite many criticisms. After all, he
argued, the establishment did not at first believe
Galileo's moons of Jupiter, nor Huygens's rings
around Saturn (Lowell, 1895: 140).

At subsequent Mars oppositions for the next three
decades many astronomers strained to see the canals,
with varying degrees of success; most, however, just
ignored the whole intractable topic. Eugène-Michel
Antoniadi (1870–1944), who was an excellent drafts-
man, started his career with Flammarion as a believer,
but became convinced otherwise by decades of nights
spent drawing Mars as viewed with the Meudon
Observatory 83-cm refractor (located outside of
Paris). His massive compilation of results (Antoniadi,
1930: 28) squarely stated that the canals were not real:

No one has ever seen a true canal on Mars, and thus
the more or less linear "canals" of Schiaparelli do
not exist as canals nor as geometric lines. But they
do have a basis in reality ...

Antoniadi went on to say that there was a tremendous amount of small-scale, irregular mottling of the
martian surface, and he could see how each of
Schiaparelli's canals plausibly had been based either
on "connecting the dots" (Fig. 1.7) of this texture, or
on a border between lighter and darker regions. When
the seeing was *poor* (but not when it was good), he
himself could even perceive the canal lines with his
own eyes. Although deferential to Schiaparelli, he had
no mercy on Lowell and his maps, saying they were
illusory and resembled nothing so much as a fly's worst
nightmare.

Antoniadi's conclusions settled the canals question
for most researchers, but many still wondered what was
real and what was not on the martian surface. At the
start of the Space Age in the late 1950s, the US Air
Force published a canals map (by Lowell's successor
Earl C. Slipher) as the best official reference for planned
exploration of Mars (Sheehan, 1996: 147). Even later,
Carl Sagan felt it necessary to apply the *coup de grâce*
once NASA's Mars-orbiter mission Mariner 9 made
the first all-planet photographic mapping. After making detailed comparisons (Fig. 1.7) of the positions
of canals with photographs having 1 km resolution
(Earth-based observations may achieve ~80 km resolution when seeing is at its very best and Mars at its
closest), Sagan and Fox (1975: 609) declared:

[26] A common mechanism to explain Mars's desiccation was that great
surface cracks had developed as the planet shrivelled like an old
apple (such cracks could be seen on the Moon), allowing the oceans
to drain into the interior.

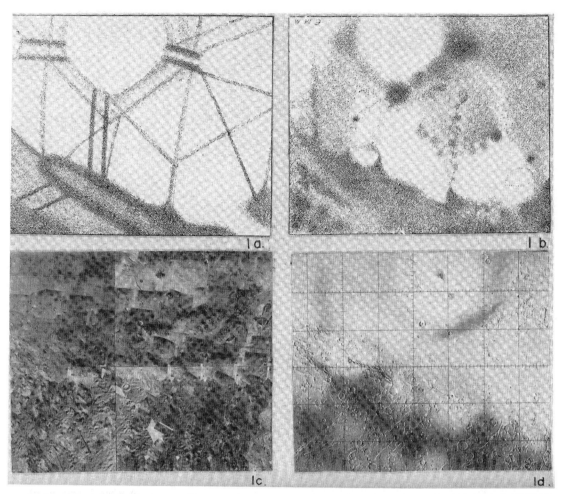

FIGURE 1.7 A comparison of the same ∼4,000 km-sized equatorial region (Elysium) on Mars as observed by (a) Schiaparelli (1877–90), (b) Antoniadi (1909–26), (c) images from the Mariner 9 spacecraft (1972), and (d) cartography based on Earth-based observations and Mariner 9. Sagan and Fox (1975) made this and other detailed comparisons in order to lay the century-old martian canals to rest once and for all. (From Sagan and Fox, 1975.)

The vast majority of the canals appear to be largely self-generated by the visual observers of the canal school, and stand as monuments to the imprecision of the human eye–brain–hand system under difficult observing conditions.[27,28]

1.7.2 Martians in fiction

Following Lowell's first book in 1895, the British novelist H. G. Wells (who was trained as a biologist under

[27] William Sheehan's book *Planets and Perception* (1988) discusses many of the psychological, physiological, and physical issues involved in preceiving fine details in a shimmering image (in the cold at 5 a.m.).

[28] Besides the citations explicity made, this section has drawn from Crowe (1986: Chapter 10) and Dick (1996: Chapter 5).

T. H. Huxley) was inspired to write *The War of the Worlds* (1898), which took Lowell's dying civilization, made it malevolent, and had it attack Earth for *Lebensraum*:

At most, terrestrial men fancied there might be other men upon Mars, perhaps inferior to themselves and ready to welcome a missionary enterprise. Yet across the gulf of space, minds that are to our minds as ours are to those of the beasts that perish, intellects vast and cool and unsympathetic, regarded this earth with envious eyes, and slowly and surely drew their plans against us . . .

A big grayish rounded hulk, the size, perhaps of a bear, was rising slowly and painfully out of the cylinder . . . It glistened like wet leather. Two large dark-coloured eyes were regarding me steadfastly . . . There was a mouth under the eyes, the lipless brim of which quivered and panted, and dropped saliva.[29]

The story closes with a strong astrobiology lesson regarding transfer of lifeforms between planets (Chapter 25): the Martians, impervious to Earth's most powerful physical weapons, nevertheless in the end succumb to inadvertent biowarfare waged by our microbial diseases, to which they are not immune.

A related and contemporaneous novel was *Auf Zwei Planeten* (*On Two Planets*) by Kurd Lasswitz (1897), which had a similar influence on the Continent as *The War of the Worlds* had in the English-speaking world. There is also interplanetary warfare in this German novel, but the Martians, although very advanced scientifically and socially, are physically similar to humans. Guthke (1990: Chapter 5) and Dick (1996: Chapter 5) discuss these seminal works in science fiction, as well as the many fictional treatments of extraterrestrial life (often on Mars) that followed. The trail that starts with Wells and Lasswitz[30] prominently includes Olaf Stapledon (*Last and First Men*, 1930), C. S. Lewis (*Out of the Silent Planet*, 1938), Ray Bradbury (*The Martian Chronicles*, 1950), Arthur C. Clarke (*Childhood's End*, 1953) and Stanislaw Lem (*Solaris*, 1961). Over the last fifty years films have had far more influence on popular culture than books and the extraterrestrial life theme continues: witness *2001: a Space Odyssey* (1968), *Star Wars* (1977), *Close Encounters of the Third Kind* (1977), *Alien* (1979), *ET: the Extraterrestrial* (1982), and *Contact* (1997).

It is important to pay attention to this litany of twentieth-century science fiction because it illustrates a vital synergy between the imaginations of the writer and of the scientist (Dick, 1996: 266). The writer often takes contemporary science and extrapolates to fictional realms, but in turn the scientist (or budding scientist) often reads the science fiction and is thereby inspired to ask questions that she or he would never have otherwise raised. Astrobiology and astrobiologists do not operate in a vacuum, and have been mightily influenced by science fiction.

---

[29] Wells's story has had many adaptations, but the original novella remains the best version and is highly recommended.

[30] Hillegas (1975) discusses both Wells and Lasswitz as well as their nineteenth-century precursors (for example, Jules Verne) and contemporaries. There were perhaps a dozen other Mars romances written in the late nineteenth century.

1.8 Ideas on extraterrestrial life, 1900–60

1.8.1 Extraterrestrial life becomes rare, then rebounds

The expected number of planets associated with other stars plummeted in the first decades of the twentieth century when the nebular hypothesis for the formation of our Solar System (and therefore presumably other planetary systems) went out of favor. It was replaced by a hypothesis in which planets would be rarely formed around stars, rather than as a natural byproduct of the very formation of the star. This new scheme was developed around 1905 by geologist Thomas C. Chamberlin and astronomer Forest R. Moulton of the University of Chicago and involved the close passage of one star past another, causing material to be projected out from one star into a spiral form. This hot spiral gas then developed dense knots in it, *planetesimals*, that collapsed into cold solid form and steadily grew into a planet by accreting other planetesimals. Chamberlin and Moulton preferred this to the nebula hypothesis because it seemed a more plausible way for a planet to hold on to an atmosphere, since it would not form in a molten state, but at a much colder temperature; it also better explained why most of the angular momentum in the Solar System is in the giant planets, not the Sun.

Chamberlin and Moulton also could point to the many catalogued nebulae of spiral form as examples of forming planetary systems. This connection, however, was short-lived, for in the years after 1909 spectral measurements of the Doppler shifts of about a dozen bright examples of these spiral nebulae, expressly made to investigate the supposed forming planets, instead revealed that they were all receding at very large speeds (hundreds of km s^{-1}) away from the Sun! These data were acquired by Vesto M. Slipher through painstaking multi-night exposures at none other than the Lowell Observatory, and quickly ruled out the spirals as (a) a class having anything to do with planets, and (b) taking part in the motions of the system of stars (which moved typically at tens of km s^{-1}). Thus began a remarkable 20-year period that culminated with Edwin Hubble establishing that the spiral nebulae were all well outside our Milky Way, i.e., separate, so-called "island universes," and then discovering the relationship between the distance to a spiral nebula (or galaxy) and its speed of recession (Hubble's Law). By 1930 the expanding Universe and modern cosmology had been born.

Once the spirals were removed from the Chamberlin–Moulton planetesimal hypothesis, one had no actual objects in the sky indicative of the process, but had to calculate how often two stars would either collide or have a close passage. This calculation quickly showed that such collisions would be extremely rare, simply because the ratio of a star's diameter to the typical spacing between stars is so small ($\sim 5 \times 10^{-8}$). The inference then was that planets were rare, no matter how pre-Copernican it might seem. A related theory was developed in England starting in 1915 by astrophysicist James Jeans and geophysicist Harold Jeffreys, except the physics was far more detailed in what became known as the *tidal theory*, because it was now gravitational tidal forces that pulled hot material out of the nearly colliding stars. The probabilities of a tidal event, however, did not improve – at one point, Jeans worked out that within our Milky Way system there would be on average only one encounter in 30 Gyr, and yet a guess for the age of the Universe, although unknown, was no more than 10 Gyr. The resulting extreme rarity of extrasolar planets and therefore of life was the accepted situation in astronomy and with the public for quite a while, in particular because Jeans wrote several best-seller popular books around 1930 in which he made statements like:

> Astronomy does not know whether or not life is important in the scheme of nature, but she begins to whisper that life must necessarily be somewhat rare. (Dick, 1996: 176)

The Chamberlin–Moulton planetesimal theory and the Jeans–Jeffreys tidal theory, however, did not survive the 1940s. First, in the late 1930s it was shown that the hot gas drawn out in either picture would dissipate before it could ever form planets. Second, in wartime Germany (while working on the German atomic bomb project) physicist Carl Friedrich von Weizsäcker (1944) published what in essence was a new nebular hypothesis. The Solar System again formed from a collapsing gas cloud, but the chief difference was that von Weizsäcker started off with much more mass than in the final system. He supposed that most of the nebula's initial mass was dissipated once solid bodies had formed. This then explained why the planetary system had lost almost all its hydrogen and helium, whereas the Sun was 99% hydrogen and helium. Furthermore, he employed magnetic forces to transfer angular momentum from the Sun to the planets. After World War II von Weizsäcker's picture was developed further and soon became the basis of the *solar nebula theory*

that still obtains today (Section 3.9). And with regard to the total number of planets and therefore the likelihood of extraterrestrial life, one now was back to the nineteenth-century situation, where planets might well accompany almost all stars. Table 2.2 in the following chapter nicely summarizes the huge change in informal estimates of the number of planetary systems in the Milky Way – before the war, it was fewer than 100; afterwards, 10^6 to 10^{11}!

1.8.2 Claims for finding extrasolar planets

The first claim for a planet orbiting a normal star was not 51 Pegasi in 1995 – there were many earlier, but none stood up to detailed scrutiny (Section 2.2.2; Fig. 2.6; Chapter 21). In this section we briefly describe these earlier claims and their demise; Dick (1996) tells the story in detail.

In the 1930s several astronomers developed a sensitive photographic technique to seek slight, long-term wobbles over decades in the sky position of stars that had no visible companions. Binary star systems, in which one *sees* two stars mutually orbiting about their center of mass, are very common, but the game here was to catch a star apparently orbiting about "nothing." Application of Newton's orbital laws then allowed one to deduce the mass of the dark companion – ideally it would be much less than the smallest visible stars (then $\sim 0.15\ M_\odot = 150\ M_j$, where $M_\odot$ is the Sun's mass and M_j is Jupiter's mass), perhaps even approaching Jupiter's mass.

Danish-American astronomer Kaj Strand was carrying out such a program, except that he was looking for a *third*, dark companion accompanying two visible orbiting stars. In 1943 he published evidence that the binary system 61 Cygni (which is only 11 lt-yr distant) had a dark companion of only 16 M_j that revolved around one of its stars – as he said, "Planetary motion has been found outside the solar system." The planet's orbit was 2.4 AU in size and had a period of 4.9 yr. This was quickly followed by a similar claim from another group of an 8–12 M_j planet around another star. These results were largely accepted by the astronomical community – planets had at last been found. It took decades of follow-up work eventually to show, however, that the claimed wobbles could not be reproduced by other observers.

Two more claims were made in the early 1960s, both from astronomers at Swarthmore College in Pennsylvania. In 1960 Sarah Lippincott found a 10 M_j companion to a star, but she never mentioned the word "planet" in her low-key publication. Her Dutch-American colleague Peter van de Kamp (1901–1995), however, had no such reservations when he trumpeted his own planet of 1.6 M_j, by far the smallest planet ever claimed (van de Kamp, 1963). The planet orbited the second closest star to the Sun, called Barnard's star, with a period of 24 years and an orbit size similar to Jupiter's around the Sun. Furthermore, the orbital wobbles were based on a remarkable one-man series (although aided by dozens of student helpers) of 2400 photographic plates taken since 1936 (see Fig. 2.4). For the next decade this result seemed unimpeachable, but then so did Richard Nixon at first. Because van de Kamp had only observed slightly more than one orbital period, it was reassuring when in the late 1960s he not only confirmed his earlier result, but in fact "improved" on it by stating that his data could now be teased to reveal not one, but *two* orbiting planets, with masses of 1.1 and 0.8 M_j and periods of 26 and 12 yr. It was clearly difficult at first for other astronomers to independently check on this result, but during the 1970s strong doubts were raised by researchers in van de Kamp's own group who re-analyzed his data, as well as by outsiders. The final consensus was that van de Kamp, despite his expertise and painstaking care to calibrate his plates uniformly and accurately over the decades, had not succeeded in doing so. In particular, changes in his telescope's properties when the lens was removed in 1949 for cleaning, as well as minute variations in plate sensitivity, were enough to cause spurious wobbles. By 1980 the astronomical community was both bereft of its extrasolar planets and very suspicious of any new claims. One can then appreciate the gauntlet of scrutiny that had to be run by the claim for 51 Pegasi in 1995 (Fig. 2.6; Chapter 21).

1.8.3 Vegetation on Mars

Although the canals of Mars were discounted by a very high fraction of professional astronomers after 1910, and even more so after 1930, their presence was still felt. Studying Mars, and by extension other planets, had become not an altogether reputable line of work. Besides, planets seemed mundane when compared with the possibilities in applying the new nuclear physics and quantum mechanics to stars (e.g., white dwarf stars, nuclear fusion as the source of a star's luminosity), or in using galaxies to open up new vistas. Thus the fraction of telescope time devoted to planets was small, as was overall research in understanding the planets before the

1960s (Dick, 1996: 115). Planet work that was done, however, still focused on Mars because of its potential for (non-intelligent) life and, unlike any other planet, its rich and variable surface markings that could be mapped.

Dick (1996: 105–26) provides a blow-by-blow account of the ever-changing assessments over the 1920–60 period of the changeable surface markings, temperatures, pressure, and atmospheric composition of Mars. By 1930 new vacuum thermocouple technology attached to large reflector telescopes at Lowell and Mt. Wilson Observatories had allowed reasonably accurate measurements of Mars's surface temperatures – it was a cold place, but, especially away from the poles, not a whole lot worse than Siberia, occasionally even rising above 0 °C. There was much less consensus in trying to suss out the total pressure of the martian atmosphere – measurements were in the 50–100 mbar range.[31] Spectroscopy of the atmosphere was extremely difficult, primarily because one had to look through the Earth's own atmosphere, which has thousands of its own spectral lines, including the very constituents (H_2O, O_2, CO_2, etc.) that were sought on Mars. But despite turn-of-the-century claims for H_2O,[32] nothing could be detected except CO_2, which Gerard Kuiper (1905–1973), a Dutch-American astronomer who was the leader of American solar system astronomy in the post-World War II era (Doel, 1996), detected in 1947 with a new PbS infrared spectrometer (Fig. 1.8).[33] Mars was not only cold, but very dry, and it certainly did not have a thriving, widespead, photosynthesizing, Earth-like plant (or animal) community.

Kuiper was one of the main proponents in the pre-Space Age era of the presence of vegetation on Mars. Amongst professionals this was a mainstream idea (unlike Lowell's canals, say, in their time) and it was based on various lines of evidence. First, the physical conditions on Mars, although deathly for most Earth life, did not rule out the hardiest organisms. Secondly, ever since Schiaparelli, astronomers had observed changes in martian coloration that correlated with the seasons in intriguing ways. In particular, as spring came on, observers often saw a "wave of darkening" (some even saw a pale greenish tint in it) that started at

---

[31] The correct value is 6 mbar (1,000 mbar = Earth atmospheric pressure).

[32] Water was not reliably detected in the martian atmosphere until 1963, and then at a very low level.

[33] Kuiper had also recently detected CH_4 on Saturn's moon Titan; see Section 20.3.

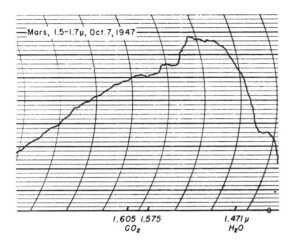

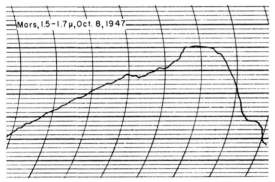

FIGURE 1.8 The near-infrared (1.6 μ) spectrum of Mars in which Gerard Kuiper detected carbon dioxide in the atmosphere of Mars using the 2.1-m McDonald Observatory telescope in 1947. The carbon dioxide bands are the two slight dips to the left of the peak of the continuum spectrum, which represents the blackbody emission of Mars. (From Kuiper, G. P. (1949). *The Atmospheres of the Earth and Planets*, p. 360, University of Chicago Press.)

Earthbound plants, and even published spectra of an alga and a lichen to bring home his point. Even before these data appeared, Kuiper had argued in the early 1950s for a lichen-like plant in terms of not having chlorophyll (people had unsuccessfully searched for its spectrum) and able to survive in cold deserts.[34]

Throughout this time, parallel work was going on in the Soviet Union in the Astrobotany (!) Section of the Kazakh Academy of Sciences, led by Gavriil A. Tikhov (1875–1960) of Alma-Ata Observatory. For five decades Tikhov had been observing Mars; after World War II, he began measuring its colors with low-resolution spectra, and then comparing these to spectra of plants such as tea shrub, cabbage, filbert, and especially polar and high-altitude plants. Like Lowell's work in the West, Tikhov was strongly criticized in the Soviet Union as he marched to the beat of a different drummer (Dick, 1996: 122). Yet once NASA was created and began exobiology research, it soon translated into English all of Tikhov's publications on Mars and plants, previously little known in the West.

By the early 1960s the probable presence of simple plant life on Mars was the consensus amongst astronomers and those biologists involved in NASA's new effort of exobiology (Section 2.2). Frank Salisbury, an American plant physiologist wrote a review in *Science* in 1962 entitled "Martian biology." In this he laid out all the evidence for plant life on Mars, argued that higher plants fit the data better than lichen, and then produced the remarkable Fig. 1.9 to illustrate possible martian biogeochemical cycles (e.g., these had to keep free oxygen out of the atmosphere and confine water to the atmosphere). Molecular biologist and Nobelist Joshua Lederberg (1960), who coined the term *exobiology*, stated that "the most plausible explanation of the astronomical data is that Mars is a life-bearing planet." Astronomer Carl Sagan (1961), after carefully weighing all the evidence, opined that "the evidence taken as a whole is suggestive of life on Mars." But he also made the excellent point that it was questionable, if the tables were reversed, whether an observer on Mars would *unambiguously* be able to detect life on Earth!

Looking more broadly, we note that the 1900–60 era is the first one we have encountered where the main concern with possible extraterrestrial life is with something other than *intelligent* life. The martian canals had

a pole (often described as a "melting water-ice cap") and moved toward the equator. Third, would not the whole martian surface be a reddish color as a result of the frequent large-scale dust storms unless there were some kind of regenerative mechanism that could modify dusted regions, i.e., plants pushing up through the dust? Fourth, infrared spectra taken by American astronomer William Sinton (of Lowell Observatory), first at the opposition of 1956 and then of much better quality with the famous Palomar 200-inch telescope in 1958, indicated the presence of features at 3.4–3.7 μm that he ascribed to vibration bands of C–H bonds in organic molecules (Fig. 2.1). Moreover, these absorption features were much stronger in the darker areas such as Syrtis Major than in the bright "desert" regions (e.g., Amazonis). In his article in *Science* Sinton (1959) interpreted his spectra as similar to those in

---

[34] In 1965 Sinton essentially withdrew his claim, saying that two of his claimed three features actually originated in HDO in the Earth's atmosphere.

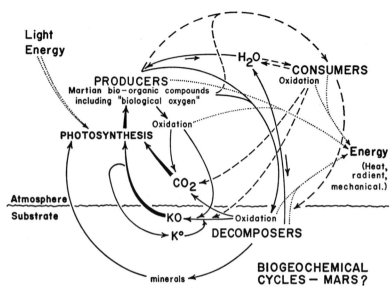

FIGURE 1.9 Martian biogeochemical cycles as speculated by botanist Frank Salisbury in 1962. Arguing that Mars had higher plant life, he envisioned global cycles that would keep free oxygen out of the atmosphere and confine water to the atmosphere. This kind of enthusiasm for plant life on Mars led to the development of the Viking Mission of the 1970s. (Courtesy AAAS and *Science* magazine.)

"gone," but there was still great interest in "lower" forms of life, unlike the plurality of [intelligent] worlds in previous centuries, and unlike SETI (next section) that was just about to start up. William Whewell (1853) had expressed most people's opinion when he noted that "a belief in unintelligent living things will have little interest for most persons" and that, when considering extraterrestrial life, morals were what mattered, not whether there were other tenants such as corals, fishes and creeping things.

1.9 The beginning of SETI

SETI (Search for ExtraTerrestrial Intelligence[35]) came about as a marriage of radio astronomy and interest in extraterrestrial life. Radio astronomy, the study of radio waves naturally emitted by astronomial sources, grew out of the intensive development of radar during World War II. By the mid-1950s it was well established around the world as an important contributor to astronomy – witness the discoveries of hundreds of distant radio sources for which no optical counterparts could be found, giant bursts of radiation from the Sun and from Jupiter, and a spectral line of hydrogen that for the first time allowed mapping of the entire Milky Way (Edge and Mulkay, 1976; Sullivan, 1984). In the late 1950s radio astronomy research in the

United States significantly trailed that in Britain and Australia, but there were nevertheless several American research groups, among which was the fledgling National Radio Astronomy Observatory (NRAO), nestled in the radio quietude of the Appalacians at Green Bank, West Virginia. NRAO had only been in existence two years when in 1958 Frank Drake (Fig. 1.10), who had just obtained his Ph.D. in Astronomy from Harvard University, signed on as its third staff scientist.

Drake had been fascinated since childhood with the possibilities of extraterrestrial life. As a graduate student he learned radio astronomy and worked out that a powerful (1 MW) military radar of that time could handily communicate with a similar system at a distance of 10 light years (lt-yr).[36] And now he had access to an 85-ft (26-m) diameter dish that could do the job. In early 1959 he broached the idea of doing a search for extraterrestrial radio signals of intelligent origin with NRAO's interim director, Lloyd Berkner, a leading ionospheric physicist who had been instrumental in founding the Observatory. Berkner gave the go-ahead to the audacious idea so long as (1) it was kept

[35] The acronym SETI became standard in the 1970s; before then the field was either called "interstellar communications" or (especially in the Soviet Union) CETI (Communication with ExtraTerrestrial Intelligence).

[36] Discussion of communicating with Mars via radio, and even claims of received signals and attempts to transmit at opposition (especially 1924), were common in the early days of "wireless" (1900–30). These attempts were all at frequencies below 30 MHz and no calculations of expected signal strengths were ever made. Perhaps the ensuing 30-yr hiatus before Drake was necessary for the field to mature such that one could indeed calculate that, with current technology, there was some hope of success. See Dick (1996: 401–14) for a full accounting.

FIGURE 1.10 Frank Drake, the founder of the modern Search for Extraterrestrial Intelligence. Drake poses here a few years after Project Ozma in front of a 91-meter diameter radio telescope at the National Radio Astronomy Observatory. (Courtesy NRAO.)

quiet, to avoid hordes of press and criticism from other scientists,[37] and (2) it could be done with equipment useful for more conventional studies.[38] Support from the top continued after the appointment shortly later of Otto Struve (1897–1963), scion of a Russian family of astronomers extending back for over a century, as permanent director. Struve brought prestige with his record of 35 years as a leader of American astronomy, although he had never done any radio research himself. He had long believed that intelligent life was abundant in the Universe, based partly on arguments that most

low mass stars such as the Sun must be accompanied by planets.[39]

While the receiver and radio telescope were being built, the 19 September issue of *Nature* arrived in Green Bank with an article entitled "Searching for interstellar communications" by Giuseppe Cocconi and Philip Morrison (1959), two high-energy physicists at Cornell University. One can imagine the consternation with which Drake and Struve read this short paper arguing that searches should be undertaken for possible extra-terrestrial civilizations around nearby stars using radio telescopes tuned to a wavelength of 21 cm – precisely what Drake was planning! Cocconi and Morrison argued that of all the various electromagnetic bands, radio waves between 1 MHz and 10 GHz were most technically feasible for interstellar communications when operating from "under" a planetary atmosphere. Within this range they focused on using microwaves because of minimal interference from the natural galactic background and from the host star of any putative transmitter. Finally, within the microwave band they argued there lay a "unique, objective standard of frequency, which must be known to every observer in the Universe: the outstanding radio emission line at 1420 Mc/sec ($\lambda = 21$ cm) of neutral hydrogen."[40] Any transmitting civilization seeking contact would choose this frequency for their beacon on the assumption that any emerging civilization with the requisite radio technology would also know about its existence. Drake too was planning to search at this frequency, but solely for the pragmatic reason that NRAO needed such a receiver anyway for studying naturally emitting interstellar hydrogen gas. Cocconi and Morrison also argued for searching the nearest solar-like stars for narrowband signals with pulse modulation (as Drake was planning); their short list of candidate stars even included the two that Drake was planning to observe. Although conceding that their notions were speculative, Cocconi and Morrison emphasized the importance of a positive detection and concluded with a stirring appeal:

[37] Although his bosses Berkner and Struve were supportive, Drake's colleagues John Findlay and David Heeschen thought that looking for extraterrestrials was a waste of time (Drake and Sobel, 1992: 27–35). Struve wrote a few years later: "It was greeted suspiciously at first by most scientists, who made it the butt of jokes and an object of scathing criticism" (Struve and Zebergs, 1962: 110). A tape of a talk that Drake gave on the then-ongoing Project Ozma to his fellow radio astronomers in April 1960 (published as Drake, 1961) tellingly records his opening ice-breaker: "I think I'm here to provide the comic relief for this shindig."

[38] In an article at that time Struve justified why the project was allowed at NRAO. His first four reasons dealt with technology spinoff and other uses of the equipment; only his fifth stated that it was worthwhile doing such a search on its own merits (Struve, 1960: 23).

[39] Struve's argument was based on the observed rotation velocities of stars and the conservation of angular momentum. Stars with masses similar to or lower than that of the Sun were known to have much lower rotational velocities than those of higher mass; the inference was then that the "missing" angular momentum had been taken up by (unseen) accompanying planets, as in our own solar system (Dick, 1996: 195–7).

[40] The 21-cm line of neutral atomic hydrogen is a hyperfine transition resulting from a flip in the relative orientation of the magnetic moments of the atom's proton and electron. It was first detected in interstellar regions in 1951 and ever since has been a vital tool for radio astronomers.

"The probability of success is difficult to estimate; but if we never search, the chance of success is zero." In subsequent decades this oft-quoted sentence became central dogma to those who contributed to SETI.

Drake now redoubled his efforts and Struve decided to remove the mantle of secrecy so that NRAO would receive due credit. In November 1959 Struve therefore announced the planned NRAO observations during his Compton Lectures at the Massachusetts Institute of Technology,[41] which of course led to the press pestering everyone at Green Bank. But the publicity did have a positive side in that the head of a Boston electronics firm donated one of his latest radio receivers, a so-called parametric amplifier, which improved detection of signals by a factor of five. Drake also now hastened to write an article for the January 1960 issue of the popular magazine *Sky & Telescope*: "How can we detect radio transmissions from distant planetary systems?" He called his endeavor Project Ozma, after a princess of Oz (in the children's books by Frank Baum) who ruled a land "far away, difficult to reach, and populated by exotic beings" (Drake, 1960: 142). Drake described his motivations in a short popular book *Intelligent Life in Space*:

> Man is only now emerging from his childhood[42] and preparing to take a place among the community of galactic civilizations that may exist ... [By contacting other civilizations we will gain] knowledge of the many ways in which other intelligent beings utilize their intellects for practical and philosophical pursuits and the cultivation of beauty ... Someday, from somewhere out among the stars, will come the answers to many of the oldest, most important, and most exciting questions mankind has asked.   (Drake, 1962: 106–11)

Observations began in April 1960. On the expectation that any signal would be confined to a small range of frequencies, Drake's receiver responded at any time to an extremely small band (100 Hz), requiring exceptional stability in the electronics (Fig. 1.11). This band was tuned over successive frequencies, on the supposition that any beacon's signal would be

FIGURE 1.11   The control room of the 85-ft (26-m) dish at Green Bank, West Virginia, during Project Ozma in 1960. (Courtesy NRAO.)

located on an orbiting planet and thus undergo Doppler shifts (from the 1420.4057 MHz rest frequency of the line). At any given frequency he spent 100 s of observation time, before moving on to an adjacent frequency. Targets were the two closest Sun-like, single stars in the northern sky, τ Ceti and ε Eridani (both at ~10 lt-yr). The receiver's output was directed to a strip chart recorder, loudspeaker, tape recorder, and digital voltmeter (a novelty at that time) whose values were written down every minute by the telescope operator.

On the very first day, observations of ε Eridani caused the most excitement of the entire run when the dish suddenly picked up very intense pulses (eight per second), which however lasted for only a few minutes. Once his heart stopped pounding, Drake was left with combined amazement and frustration, for all he could do to test its authenticity was to observe ε Eridani every subsequent day and hope the signal would repeat. He decided to augment the two "control" channels he already had by adding a third, a small horn and receiver stuck out the lab's window. If a signal were manmade interference, it would likely flood into the window system as well as into the dish proper; if it was truly from ε Eridani, then it would appear only in the main receiver and dish. After about ten days, the pulses indeed reappeared, but alas also in the window system; the best guess was that it was airborne military radar, but in any case it was manmade. With that candidate signal dismissed, daily observations of background noise soon grew, well, boring. Then the finicky receiver broke and, after a six-week gap, Drake resumed for the month of June and accumulated 200 hours of data before closing up shop. Although his negative results were never "properly" written up in a

---

[41] The published lectures end by discussing the possibilities of extraterrestrial life and the principles of radio searches, but although Cocconi and Morrison are mentioned, Drake and Ozma are curiously omitted (Struve, 1962: 157–9).

[42] This theme of the end of humankind's childhood was prominently espoused in the 1950s by the science fiction writer Arthur C. Clarke, most notably in *Childhood's End* (1953).

refereed journal, he informally published the result that he could have detected, if it had existed near either star, a 1 MW transmitter coupled to a 600-ft (180-m) diameter dish.

As he began observations Drake (1960) wrote that his chances of success were somewhere between 25% and one in a million, and that "it appears probable that this project or a similar one will someday succeed." His lack of a signal, however, only whetted his interest, and to this day he remains optimistic. Cocconi and Morrison's paper and Drake's Project Ozma were remarkable in their simultaneity, outlandishness, and courage. They set long-lasting precedents as SETI slowly grew through subsequent decades (Section 2.2.4). Narrowband and pulsed signals in the vicinity of 21 cm wavelength are still the parameters of choice for SETI. Nearby solar-like stars are still being searched. Manmade interference is still the main operational bugaboo to any SETI program (Chapter 26).

Most importantly, these developments in 1959–60 continued an old quest of humans, but now using a tool of seemingly much greater power. The identities of three visitors to the Project Ozma control room are telling as to this broader significance. Each man learned of the project through press accounts and felt compelled to journey to rural West Virginia. One was Barney Oliver (who later would become a leader in the field), a noted physicist in the electronics industry who was fascinated with the technical aspects of SETI. The second visitor was John Lear, science reporter for the prestigious magazine *Saturday Review*, who passed this story on to the public as a great scientific advance. And the last was Father Theodore Hesburgh, S. J., theologian and president of Notre Dame University, who dared not limit God's power and looked upon the question of extraterrestrial life as another means to know and understand God through his works, arguments we have earlier discussed. Finally, Drake's boss Struve (1960: 23) argued that science was at a new epistemological horizon, now that manmade emanations from Earth (and supposedly other planets) competed with natural ones:

> The conclusion is of great philosophical interest [that it is probable that a good many of the billions of planets in the Milky Way support intelligent forms of life]. I believe that science has reached the point where it is necessary to take into account the action of intelligent beings, in addition to the action of the classical laws of physics . . . This constitutes a drastic and a challenging departure from the past.[43]

1.10 Introduction to the history of ideas on the origin of life

One of the central challenges in modern astrobiology is understanding the origin of life on Earth and the possible origins of life elsewhere in the Universe (Chapters 6–9). While today we treat the origin of life on Earth as having occurred only in the remote past, and as distinct from the regeneration of new life forms from living beings, this has not always been the case. Since the times of ancient Greece, the dominant idea for the origin of living beings has been spontaneous generation, whether for complex organisms like mice and even humans, simpler organisms like worms and insects, or microorganisms imperceptible to the naked eye. Even decades after Louis Pasteur's experiments, often cited as the definitive refutation of the theory of spontaneous generation, debate over the issue persisted. We will examine this celebrated "crucial experiment" in its contemporary framework and find that what appears as decisive empirical evidence from today's perspective was not unanimously viewed as such in the past. Furthermore, by tracing the history of ideas surrounding spontaneous generation, we gain a valuable perspective on our current grappling with the persistent origin of life problem. The current endeavors of astrobiologists to solve the origin of life problem are only the latest episode in a long research tradition aimed at answering an ancient question.

The spontaneous generation controversy, essentially a debate over the origins of life, has a long and colorful history. At times during the debate, religious, political, and social implications of spontaneous generation had significant influence right along with experimental evidence. At other times, no amount of experimental evidence could settle an issue at hand; only an advance in instrumental capabilities or a new discovery settled the disagreement between camps. This situation of *under-determination*, where empirical evidence cannot settle a debate due to the different intellectual frameworks of the protagonists, is an important issue for the historian of science. In the following discussion we will consider aspects of the spontaneous generation and origin of life debates that highlight how multifaceted the path of science can be.

---

[43] Amongst other sources, this section has profited from accounts in Drake (1985) and in Drake and Sobel (1992), interviews by WTS of Frank Drake and other principals, as well as research by Dick (1993, 1996: 414–31).

1.11 Early ideas on spontaneous generation of life

The question of the origin of the world, or *cosmogony* (from the Greek κοσμο + γονι, 'world' + 'begetting') is an ancient human inquiry. Imaginative creation stories across cultures are among the very oldest known oral and written traditions. These mythological tales account for the origin of inanimate matter and living beings by invoking the action of deities upon some form of primal matter. Such stories satisfy the questions of *who* or *what* was behind the origin of life, and perhaps *why*, but it is generally agreed that it was not until the sixth century BC that *rational* cosmogonies attempted to explain *how* the present conditions came into existence (Greene, 1992).

Written records of these early philosophers are incomplete, and most of what is known about their ideas is derived only from later writers. For example, in the sixth century BC, Xenophanes wrote about his contemporary, Anaximander of Miletus, who had authored a very imaginative cosmogony that was the first to include ideas about the origin of man:

> Anaximander said that the first living creatures were born in moisture, enclosed in thorny barks; and that as their age increased they came forth on to the drier part and, when the bark had broken off, they lived a different kind of life for a short time.
>
> Further he says that in the beginning man was born from creatures of a different kind; because other creatures are soon self-supporting, but man alone needs prolonged nursing. For this reason he would not have survived if this had been his original form.
>
> Anaximander of Miletus conceived that there arose from heated water and earth either fish or creatures very like fish; in these man grew, in the form of embryos retained within until puberty; then at last the fish-like creatures burst and men and women who were already able to nourish themselves stepped forth.   (Kirk *et al.*, 1983: 141)

Anaximander is one example of the Greek *monists* who sought to answer the question of the basic *physis* ("material") of the Universe by invoking only one primary substance to describe the original material as well as the variety of observable forms. For Anaximander, this primary substance was a distinct, primal material, which he called the *apeiron* ("indefinite" or "boundless"). For other Greek monists the primary substance was common, for example, earth, water, air, or fire, while dualist philosophers chose pairs of substances such as earth and moisture, or air and fire. While not always free of deities, these rational cosmogonies did not invoke the direct action of gods in the creation of the world. In addition, these cosmogonical systems were distinct from early cosmologies; they described *how* the world came into being, rather than explaining the facets of the present world.

In contrast to the monists, the fifth century BC pluralist philosophy of Empedocles held that the four basic substances (earth, air, fire, and water) together account for the origin and composition of all forms by mixing in various proportions. His system was then incorporated into Aristotle's philosophical framework that dominated thought for two thousand years (Section 1.2.1). From antiquity to the early seventeenth century, the origin of life through spontaneous generation was generally accepted and was supported by everyday observations. It was commonly accepted that heat and moisture, together acting on materials such as soil and bark, could spontaneously form plants, worms, and insects. Mud and decaying meat were regarded as particularly effective sources of spontaneous generation (Fry, 2000: 10). Aristotle postulated the spontaneous generation of both plants and animals from such starting materials; he observed that insects and worms sprang forth from morning dew and rotting material and that a variety of creatures including mice, frogs, and fish were generated by moist soil or muddy river beds (Oparin, 1938: 6). Aristotle's teachings on the subject were widely accepted and were long supported by the Christian Church, as evidenced by Saint Augustine's attribution of spontaneous generation to the will of God (Fry, 2000: 18).

In medieval times, Aristotle's doctrine was supported by more evidence for spontaneous generation. Tales and recipes describing the spontaneous generation of birds and mammals from a variety of sources abounded. The influential teachings of sixteenth century Swiss physician and alchemist Paracelsus (1493–1541) included that the generation of life forms was only an extension of a universal chemical transmutation. He held that:

> The generation of all natural things is twofold: one which takes place by Nature without Art, the other which is brought about by Art, that is to say, by Alchemy, though, generally, it might be said that all things are generated from the earth by the help of putrefaction.[44]

---

[44] This and the following quotations by Paracelsus (ca. 1570) are from Waite (1894: 120).

Paracelsus documented a series of observations of the spontaneous generation of mice, frogs, and eels from a variety of starting materials, and even provided the exact recipe for producing a *homunculus*, a little human:

> Let the semen of a man putrefy by itself in a sealed cucurbite [glass vessel] with the highest putrefaction of the *venter equinus* [horse's womb] for forty days, or until it begins at last to live, move, and be agitated, which can easily be seen. After this time it will be in some degree like a human being, but, nevertheless, transparent and without body. If now, after this, it be every day nourished and fed cautiously and prudently with the arcanum of human blood, and kept for forty weeks in the perpetual and equal heat of a *venter equinus*, it becomes, thenceforth a true and living infant, having all the members of a child that is born from a woman, but much smaller. This we call a homunculus; and it should be afterwards educated with the greatest care and zeal, until it grows up and begins to display intelligence.

Paracelsus regarded this process as a "miracle and marvel of God, an arcanum [secret] above all arcana, and deserves to be kept secret until the last times, when there shall be nothing hidden, but all things shall be made manifest."

In the seventeenth century natural philosophers added more empirical evidence in favor of spontaneous generation, but now with different recipes. The Flemish physician Jean-Baptiste van Helmont (1577–1644) was known for his exact and critical experimental techniques; in addition to discovering several gases, he made important contributions to understanding plant growth. The results of his experiments supported spontaneous generation:

> For if you press a piece of underwear soiled with sweat together with some wheat in an open mouth jar, after about 21 days the odor changes and the ferment coming out of the underwear and penetrating through the husks of the wheat, changes the wheat into mice. But what is more remarkable is that mice of both sexes emerge (from the wheat) and these mice successfully reproduce with mice born naturally from parents … But what is even more remarkable is that the mice which came out were not small mice … but fully grown (van Helmont, 1682).

Other empirical evidence from a variety of mice-producing experiments within jars of soiled rags and wheat further improved the epistemic status of the theory of spontaneous generation.

1.12 The seventeenth- and eighteenth-century spontaneous generation debate

The seventeenth-century French philosopher and mathematician René Descartes (1596–1650) explained all phenomena using material and mechanical means. For the generation of a living being, Descartes only required that putrefying matter be agitated by the action of heat. Early in the century, when Descartes's mechanical philosophy was flourishing, the theory of spontaneous generation enjoyed a wide following and was still in accord with empirical evidence. By the middle of the seventeenth century, however, theories of spontaneous generation were brought into question as purely mechanical or materialistic explanations seemed increasingly inadequate to explain the origin of living beings. New anatomical discoveries made it ever more difficult to support the generation of complex organisms through mere actions of heat and motion. In addition, "spontaneous" generation was ruled out for those who accepted English physician William Harvey's 1651 theory, *omne vivum ex ovo*, "all life comes from an egg." Meanwhile, another theory became popular, approved by the Church and avoiding the unpalatable notion of life proceeding from inanimate materials. *Pre-existence theory* held that a new plant or animal developed from a pre-exisiting germ that had been present in the parent *and* all preceding ancestors, right back to God's creation of the world.

In the 1670s, however, Dutchman Antony van Leeuwenhoek (1632–1723) began writing to the newly founded Royal Society in London about his advances in the magnification and resolving power of microscopes (Fig. 1.12) and his discovery of an active, living world teeming with tiny and seemingly simple forms that he called *animalcules*. With the discovery of these microscopic forms, debates over spontaneous generation largely shifted to the microcosm – what was the origin of these forms? Did they have a role in the spontaneous generation of larger organisms?

The Italian physician Francesco Redi (1626–1698) was the first to call the theory of spontaneous generation into question using experimental evidence. In his 1668 *Generation of Insects*, Redi held that:

> The earth, after having brought forth the first plants and animals at the beginning by order of the Supreme and Omnipotent Creator, has never since produced any kind of plants or animals, … and everything … that she has produced, came solely from the true seeds of the plants and animals

FIGURE 1.12 Although Leeuwenhoek used only single-lens microscopes, in the late seventeenth and eighteenth centuries compound microscopes such as this Marshall-type (ca. 1730) were widely used.

themselves ... And although it be a matter of daily observation that infinite numbers of worms are produced in dead bodies and decayed plants, I feel, I say, inclined to believe that these worms are all generated by insemination.   (Farley, 1974: 14)

Redi demonstrated that boxes of many different types of dead animal material all produced the *same* kind of flies hatching from maggots, thus challenging the notion that decaying meat spontaneously produced worms. He also noted the presence of eggs on the surface of decaying material: "These eggs make me think of those deposits dropped by flies on meats, that eventually became worms, a fact ... well known to hunters and to butchers, who protect their meats in summer

from filth by covering them with white cloths" (Farley, 1974: 14). From a modern perspective, Redi's experiments are often hailed (e.g., in textbooks) as the beginning of the refutation of spontaneous generation theory. However, Redi's demonstration was confined to only one case of commonly observed spontaneous generation, that of worms arising from decaying meat, and his experiments may not have had the controls usually attributed to them.[45] Redi himself still maintained that certain other insects or worms could arise spontaneously, given special conditions within living matter. For example, he claimed that spontaneous generation was the best explanation for parasitic worms in the intestines of animals and for insects within fruits. For all other cases, however, he held to the orthodox idea of the time: pre-existence theory (Fry, 2000: 24–7).

If Redi's demonstrations raised doubt about spontaneous generation of one type, Irish priest John Turberville Needham's (1713–1781) experiments in the middle of the eighteenth century, along with his later collaboration with French aristocrat Comte de Buffon (1707–1788), provided *support* for the spontaneous generation of microorganisms. Needham demonstrated in 1748 that if mutton gravy was sealed in a glass vessel with tree resin and heated to boiling, the gravy contained many organisms visible by microscope:

> [I] took a quantity of Mutton-Gravy hot from the fire, and shut it up in a phial, closed with a cork so well masticated, that my precautions amounted to as much as if I had sealed my phial hermetically. I thus effectually excluded the exterior air, that it might not be said my moving bodies drew their origin from insects, or eggs floating in the atmosphere. I would not instil any water, lest, without giving it as intense a degree of heat, it might be thought these productions were conveyed through that element ... My phial swarmed with life ...
>
> Let it suffice for the present to take notice, that the phials, closed or not closed, the water previously boiled, or not boiled, the infusions[46] permitted to teem, and then placed upon hot ashes to destroy their productions, or proceeding in the vegetation without intermission, appeared to be so nearly the same, that, after a little time, I neglected every precaution of this kind, as plainly unnecessary.   (Farley, 1974: 24)

[45] Findlen (1993) has analyzed Redi's experiments in the context of his demonstrations before his patrons (the Medici court), showing how different in design they were from modern experiments.

[46] An *infusion* is the extract obtained by boiling a substance in water.

Needham and Buffon argued that sealing and boiling their vessels demonstrated that the microorganisms did not originate from eggs or the air, but from the gravy itself, which they believed had been sterilized by heating, but still maintained active "organic molecules" that could self-assemble into simple organisms (Strick, 2000: 6).

In a similar set of experiments in 1765, the Italian cleric Abbé Lazzaro Spallanzani (1729–1799) sealed the glass tubes shut instead of using a tree resin, and boiled the fluid in the vessel for a prolonged period of time. Upon later analysis with his microscope, Spallanzani reported that he found no microorganisms present:

> In hermetically sealed vessels, provided that the enclosed air has not been exposed to heat, one is not always sure of preventing the birth of animals in boiled infusion. If, on the contrary, this air has been strongly heated, no animals will be born, at least when no new air is permitted to enter. That is to say air which has not been exposed to heat, is indispensible for the production of animals. And as it will not be easy to prove that there are no small eggs floating in the volume of air contained in the flasks, it seems to me that the existence of these eggs is always suspected and that heat has not destroyed entirely the fear of their existence in infusions.   (Farley, 1974: 25)

Spallanzani performed hundreds of similar experiments with different starting materials and with similar attention given to sealing the infusions from the air and heating them for a prolonged period of time. In none of these cases did he report seeing microorganisms after boiling. He harshly criticized Needham and Buffon as well as their experimental methods. Never observing anything like "organic molecules," Spallanzani claimed that Needham and Buffon were not careful with their preparations in the laboratory. Believing that the germs of microorganisms were carried in air, he concluded that his method of sealing glass vessels from air was superior to those of his rivals.

No experiment could have settled this issue – the scientists operated in different intellectual frameworks, and designed their experiments and procedures with different emphases. For Spallanzani, whose germs were in the air, the central issue was pre-sterilization by heating of his infusions and glassware; for Buffon and Needham, whose germs were in the infusions, the central issue was preserving their "active molecules." To revive their theory despite Spallanzani's potentially

damning evidence, Needham and Buffon needed only to claim (and did!) that Spallanzani had over-heated his infusions, thus destroying what Needham called the "vegetative force" in the infusion (Farley, 1974: 25). Spallanzani never was able to convince Needham, Buffon, or their followers of any flaws in their assumptions and experimental methods. Bolstered by Needham and Buffon's work, spontaneous generation enjoyed a wide following well into the nineteenth century.[47] The existence of parasitic worms, as well as microscopic "active molecules" widely observed in blood, tissues, and various infusions, were also obstacles to complete refutation of spontaneous generation (Strick, 2000: 10).

The epistemological problem with refuting the theory of spontaneous generation is clear. To support the theory, one needs only to show that spontaneous generation *sometimes* occurs, perhaps even only in special cases. On the other hand, to refute the theory, one must prove that spontaneous generation *never* occurs, a much more difficult task. Thus, the theory of spontaneous generation has been applied to different classes of organisms at different times. When its application to one type of organism was refuted, in each case the theory survived through the re-application of the idea to different organisms. Later, we will see that the theory, in the form of twentieth century ideas about the origin of life, has survived to this day through a re-application of a different kind – not in terms of the applicable organisms, but in terms of *when* spontaneous generation is believed to have occurred.

1.13 Pasteur vs. Pouchet: relegation of life's origin to the past

Spallanzani's attempt to dissuade supporters of spontaneous generation through his numerous and carefully executed experiments failed to convince ardent supporters of the theory. Proponents were able to find special cases (or just *one* special case) for which experimental results supported the theory of spontaneous generation. For over a century, the debate stood at this standstill until interest picked up again in the middle of the nineteenth century.

Félix Pouchet (1800–1872), a French biologist and director of the Natural History Museum in Rouen, in 1859 published *Hétérogénie, ou, Traité de la Génération*

---

[47] It was in criticizing this work that T. H. Huxley, a strong opponent of spontaneous generation, stated his famous line in 1870: "The great tragedy of science [is] the slaying of a beautiful hypothesis by an ugly fact." (Strick, 2000: 7)

FIGURE 1.13 An artist's rendition of Pasteur cultivating microbes in his underground "hothouse" laboratory at the Ecole Normale, Paris (1883). (Edgar Fahs Smith Collection, University of Pennsylvania Library.)

Spontanée (*Heterogeneity:*[48] *a Treatise on Spontaneous Generation*), a 700-page tome that presented detailed evidence supporting spontaneous generation of micro-organisms. Unlike the earlier views of Buffon and Needham who believed that it was the *whole organism* that sprang forth spontaneously, Pouchet held that it was the *eggs* or *seeds* of the organism that were generated. He wrote:

> Spontaneous generation does not produce an adult being; it proceeds in the same manner as sexual generation which, as we will show, is initially a completely spontaneous act by which the plastic force brings together in a special organ the primitive elements of the organism.   (Farley, 1974: 97)

Another central issue for Pouchet was that life only generated from previously living material. He assumed the first generation of a living thing on Earth operated under "divine inspiration," but since then, spontaneous generation had always occurred, as evidenced by his laboratory experiments involving various extracts such as from hay. Pouchet's experimental methods reflected his assumption that the vital life force was in the materials themselves rather than the air; he worked with organic infusions and showed that the generation of life forms was often preceded by the fermentation or decomposition of organic material. In practice, his experiments were similar to Needham's and Buffon's from a century earlier.

Pouchet's book caught the attention of the French Academy of Sciences, which soon after offered a prize for experiments that would shed light on the question of spontaneous generation. The award was granted in 1862 to Louis Pasteur (1822–1895; Fig. 1.13) for his experiments and arguments weighing against spontaneous generation. Pasteur's carefully designed and logically presented experiments clarified many questions pertaining to the spontaneous generation debate. Taken as a whole, his experiments dealt a serious blow to the theory of spontaneous generation and mark the final turn in the debate's long history. Although a few die-hard supporters clung to the theory for a few decades more, by the turn of the twentieth century the theory of spontaneous generation had gasped its last breath.

Pasteur (1861) reported the results of numerous experiments that left no room for skepticism regarding one significant claim: air is a source of contamination. He performed multiple experiments demonstrating that microorganisms are found aplenty in the air, as well as on the surfaces of glassware and other laboratory equipment and reagents. Pasteur set out to demonstrate that air was contaminated with microorganisms by filtering air through a plug of gun cotton (nitrated cellulose). After a time he removed the plug, dissolved it in a mixture of ether and alcohol, and under a microscope viewed the extract, which contained scores of microorganisms. In another experiment he sterilized a solution in a flask by heating, then introduced air which had also been heated and re-cooled. No microorganisms formed in this case, either. He then introduced into the solution a cotton plug that had been used to filter air as described above, and observed that

[48] *Heterogenesis* was the type of spontaneous generation in which life could arise only from organic matter that was alive or derived from an organism. This concept was opposed to *abiogenesis*, in which life could arise from inorganic matter.

the solution soon started to cloud with microorganisms. He thus showed that heating had not destroyed any "vital force" of the fluid as supporters of Needham and Buffon claimed. Pasteur also described a "swan neck" flask experiment in which a sterilized solution was open to air at ambient temperature through a side arm of the flask that had kinks to trap microorganisms. Again, the solution remained sterile until the neck of the flask was broken off, allowing normal, unaltered air to contaminate the flask and causing microorganisms to grow (Farley, 1974: 100–105; Conant, 1953).

These experiments served two purposes: they cast great doubt on the theory of spontaneous generation, while simultaneously explaining the source of earlier controversies in the spontaneous generation debate. Pasteur himself admitted the difficulty of making further claims:

> More fortunate than inventors of perpetual motion, the champions of spontaneous generation will, for a long time yet, be privileged to arouse the attention of the scientific world. In the mathematical sciences it can be demonstrated that a given proposition is not and could not be so, but the sciences of nature are not so well devised. Mathematicians can reject unread all memoirs concerned with the squaring of the circle or perpetual motion. The question of spontaneous generation, on the contrary, is always capable of inflaming public opinion. For, in the present state of science, it is impossible to prove a priori that there can be no self-creation of life apart from the pre-existence of similar living forms.   (Conant, 1953: 60)

Pasteur is often celebrated as the scientist who *disproved* spontaneous generation, but in the above words he himself admits the impossibility of disproving the theory. Pasteur's hypothesis that spontaneous generation does not occur is the basis of modern microbiology, but it has not become so through decisive evidence from any one of his experiments.[49] Only through the gradual elimination of each claim of spontaneous generation (some of which occurred *after* Pasteur's landmark work) has there been an acceptance of our modern conception that, on Earth today, life comes only from life.

Once one adopts an intellectual framework assuming that spontaneous generation never occurs, another immediate and important question appears: how did the *first* living thing arise? There are only a few logical possibilities besides divine intervention. Life may have been delivered to Earth from an outside source (*panspermia*), but although this was proposed by such luminaries as physicists William Thomson (in 1871) and Hermann von Helmholtz (in 1875), and Nobelist chemist Svante Arrhenius (in 1908), it of course begs the question of origins (Crowe, 1986: 400–06). Another possible solution is that life may have been generated on the Earth in the distant past not as a relatively rapid event, but as the culmination of a lengthy process. Under this scenario, one might also hold that there was a type of spontaneous generation, but of something simpler than a whole, living microorganism – perhaps of a system that might be described as on the way to living, half-living, or nearly living. This option only became available once the concept reigned of a very old Earth, where physical and chemical conditions were very different in its youth. As discussed in Section 1.6, this was made possible by developments in geology (Uniformitarianism), astronomy (the nebular hypothesis), and biology (Darwinian evolution).

Thus, we see that the theory of spontaneous generation has not only been applied to different organisms (until refuted in each case), it has also been applied to different periods of time on the Earth. With the relegation of spontaneous generation to a single occurrence in the distant past, the problem was recast into its modern form and given a new name: the *origin of life* problem. At the same time, the problem's center of gravity shifted from the domain of biologists and microbiologists to the purview of chemists, biochemists, and physicists who study the conditions and mechanisms on the early Earth.

1.14 Oparin and Haldane: origin of life as the domain of chemistry and physics

In the 1920s A. I. Oparin and J. B. S. Haldane independently proposed theories that life originated only once in Earth's distant past and that this origin could be understood using only the laws of chemistry and physics. Their treatment of the origin of life on Earth as a legitimate, scientific question revitalized interest, spurred intellectual attention, and later (starting in the 1950s) inspired considerable scientific inquiry. Together, their theories became known as the Oparin–Haldane hypothesis, which is seen by today's researchers on origin of life as the beginning of their tradition. In the introduction to their book *Origins of*

---

[49] Conant (1953) gives a close analysis of the ambiguities in both Pouchet's and Pasteur's experiments vis-à-vis their claims.

ПРОЛЕТАРИИ ВСЕХ СТРАН, СОЕДИНЯЙТЕСЬ!

А.И.ОПАРИН

ПРОИСХОЖДЕНИЕ ЖИЗНИ

„МОСКОВСКИЙ РАБОЧИЙ"
1   9   2   4

FIGURE 1.14 A.I. Oparin's 1924 tract, *Proiskhozhdenie zhizny* (*Origin of Life*), proposed that the origin of life on the early Earth could be explained using only the laws of chemistry and physics.

Life: The Central Concepts, Deamer and Fleishaker (1994) claim that "As a scientific research field, origins of life is quite young. It can be said to date from the 1924 publication of Oparin's 36-page tract *Proiskhozhdenie zhizny* (*The Origin of Life*; Fig. 1.14)."[50]

We have seen that scientific inquiry into the origin of life problem dates back centuries before Oparin's publication, but it was always termed the *spontaneous generation* debate. Deamer and Fleishaker's claim that the *origin of life* field is distinct from earlier work does

not mean that previous traditions were less scientific, but rather that they involved erroneous hypotheses (e.g., that mice or microorganisms can generate spontaneously today in a finite length of time). So important is this distinction that the subject has been given a new name and we no longer speak of (or attempt to gain funding for) investigating spontaneous generation. "Origin of life" today carries with it the assumptions that we are speaking of events in the *distant* past and that *today* life comes only from existing life, not through spontaneous generation. The origin of life approach also allows for a longer process for the emergence of life than spontaneous generation does.

This seminal shift is correctly attributed to Oparin and Haldane, but note that they were nevertheless connected to their predecessors. For example, Oparin reviewed Aristotelian and Medieval Christian ideas on the origin of life problem (as well as more recent history), while Haldane mentioned Redi and Pasteur. Both recognized that they were contributing new ideas to a long intellectual tradition. With regard to an origin for life in the distant past, they also had more immediate predecessors in the German physiologist Eduard Pflüger, who argued in 1875 for organic synthesis of life with cyanogen being a key radical; the German naturalist and philosopher Ernst Haeckel, who in 1904 pressed strongly for carbon as the key to life and its origin; and the French naturalist Edmond Perrier, who in 1920 also attributed life to a chemical synthesis from ordinary molecules (Raulin-Cerceau, 2004).

Aleksandr Ivanovich Oparin (1894–1980) (Fig. 1.15) was a prominent Soviet biochemist at Moscow State University who was interested in a wide range of biochemical problems. J. D. Bernal (1967: 240), a British physicist important in early molecular biology and origin of life research, commented that Oparin's 1924 tract

contains in itself the germs of a new programme in chemical and biological research ... Oparin's programme does not answer all the questions, in fact, he hardly answers any, but the questions he asks are very effective and pregnant ones and have given rise to an enormous amount of research.

Still today, eighty years later, Oparin's and Haldane's approach forms the basis of our own attack on the origin of life problem. Oparin (1924: 214) wrote:

The more closely and accurately we get to know the essential features of the processes which are carried out in the living cell, the more strongly we become

[50] Deamer and Fleishaker (1994:31–71) provide an English translation of Oparin's tract, as well as Haldane's 1929 article (pp. 73–81). Oparin's tract, which is very readable (at the level of a *Scientific American* article today), is also available in English in Bernal (1967: 199–234), accompanied by Bernal's commentary and updating.

FIGURE 1.15 A. I. Oparin in 1946.

convinced that there is nothing peculiar or mysterious about them, nothing that cannot be explained in terms of the general laws of physics and chemistry.

Oparin's approach was reductionist and materialistic, the latter not surprising given that it arose in the early years of the Soviet Union.[51] He did not think that living material contains any special characteristics distinguishing it from non-living material. Rather, it is characterized by its special combination of chemical and physical properties.

Referencing Pasteur's work, Oparin started with the assumption that all life comes from life and posed the question of how the *very first* living thing might have arisen. He considered the theory of panspermia, that the Earth was seeded with life from outer space, but concluded that the theory was problematic for many reasons, including that it begs the basic question of origins: "If life could originate at some point in the Universe, there is no reason to suppose that it could not originate on the Earth as well" (Oparin, 1924: 206). He proposed that there is no difference between the world of the living and the dead, noting the many important biomolecules that had been synthesized,

beginning with Friedrich Wöhler's 1828 artificial synthesis of urea. For Oparin, the central problem in understanding the origin of life was to understand when the first living thing was formed, as defined by his so-called *special properties*: definite structure and organization, metabolism, and response to stimulation. Oparin supported the possibility that on the early Earth a long chemical evolution preceded the formation of any system he would label as "living." In his greatly expanded work *Origin of Life* (1938: 250–1), he described:

> ... the gradual evolution of organic substances and the manner by which ever newer properties, subject to laws of a higher order, were superimposed, step by step upon the erstwhile simple and elementary properties of matter ... The tremendously long intervals of time separating the single steps in this process make it impossible to reproduce the process as it occurred in nature under available laboratory conditions. There still remains, however, the problem of the artificial synthesis of organisms but for its solution a very detailed knowledge of the most intimate, internal structure of living things is essential.

Oparin then encouraged biologists, chemists, and physicists to contribute to solving the increasingly interdisciplinary problem of the origin of life on Earth, a tradition that has continued to the present day.

John Burdon Sanderson Haldane (1892–1964), a prominent British biochemist, geneticist and physiologist, independently proposed a similar theory of the origin of life in 1929, before Oparin's work became known in the West.[52] Haldane was a prolific scientist and writer on topics as diverse as animal biology and the role of science in western civilization, but it was the problem of the origin of life that captivated his interest for almost 50 years (Clark, 1968: 86). In Haldane's 1929 article he proposed that life might have been synthesized from the simple minerals and organic compounds on the prebiotic Earth:

> Now, when ultra-violet light acts on a mixture of water, carbon dioxide, and ammonia, a vast variety of organic substances are made, including sugars and apparently some of the materials from which proteins are built up ... In this present world, such substances, if left about, decay – that is to say, they are destroyed by microorganisms. But before the

---

[51] In fact several important figures in origin of life research in the 1920–70 period were Marxists: besides Oparin, there were Haldane, Bernal, and the virologist N. W. Pirie. Marxist materialistic principles were in accord with their scientific ideas, although the nature and existence of direct links is debated (Graham, 1987; Fry, 2000: 78).

[52] Oparin's ideas were little known in the West until the appearance in 1938 of an English translation of his 1936 book *Origin of Life*.

origin of life they must have accumulated till the primitive oceans reached the consistency of hot dilute soup.   (Haldane, 1929: 7–8)

Like Oparin, Haldane believed that chemical and physical processes, alone, could account for the origin of life on the early Earth. Haldane speculated about the conditions on prebiotic Earth and suggested possible mechanisms for the formation of simple organic substances in his famous "hot dilute soup."

While Oparin emphasized long chemical evolution as a key to understanding the origin of life, Haldane focused on the problem of drawing the line between the living and the dead in a chemical or biochemical sense. He highlighted the discovery a decade earlier of bacteriophage (the first known viruses; this type destroys bacteria), and the problems encountered in determining whether or not it was living. He commented that "the bacteriophage is a step beyond the enzyme on the road to life, but it is perhaps an exaggeration to call it fully alive. At about the same stage on the road are the viruses which cause such diseases as smallpox, herpes, and hydrophobia" (Haldane, 1929: 7). Thus, Haldane suggested some kind of continuum between the living and the dead, and moved forward in his discussion using terms like "half-living" and "nearly alive" to describe his hypothetical, life-like systems.

Haldane (1929: 9–10) ended with the same admission as Oparin: his conclusions

"will remain [speculative] until living creatures have been synthesized in the biochemical laboratory ... Until that is done the origin of life will remain a subject for speculation. But such speculation is not idle, because it is susceptible of experimental proof or disproof ...

   "How did the first such system on this planet originate?" This is a historical problem to which I have given a very tentative answer on the not unreasonable hypothesis that a thousand million years ago matter obeyed the same laws that it does today ... The biochemist knows no more, and no less, about this question than anyone else. His ignorance disqualifies him no more than the historian or the geologist from attempting to solve a historical problem.

Indeed, Haldane was no more disqualified from suggesting a new approach to the origin of life problem than historians are disqualified from weaving a story of the problem's past. The independently formulated ideas of Haldane and Oparin became known as the Oparin–Haldane hypothesis in the 1930s, even though

the two men did not meet until the 1960s (Clark, 1968: 250). By then their hypothesis had been extremely influential and a new research tradition had been spawned. When the Space Age and NASA emerged in the late 1950s, the Oparin–Haldane outlook was the standard in the new field of exobiology (Section 2.2.3; Dick and Strick, 2004).

1.15 An early twenty-first-century approach to the origin of life

Today's scientists utilize two conceptually different strategies that were described by Oparin and Haldane to attack the problem of the origin of life on Earth; the reader will encounter both of these strategies in subsequent chapters. The first is a *reductionist* approach that takes life as we know it and asks, "what is the simplest system that might be called living?" The second is a *synthetic* approach that takes a prebiotic environment as a starting place and asks the same question. Both approaches run into cases which challenge assumptions about the boundaries between the living and the dead. A working definition for life is requisite to clearly chalk the lines between the living and the non-living, but in fact definitions of life are notoriously problematic (Chapter 5).

The synthetic approach received great attention starting in 1953 when Harold Urey and his graduate student, Stanley Miller, showed that amino acids, one of the building blocks of life, could be synthesized from a reducing atmosphere (Miller, 1953; Section 6.2.3). Since that time, examples of each of the four main classes of biomolecules – proteins, nucleic acids, carbohydrates, and lipids – have been synthesized from materials and conditions plausibly present on the early Earth (Deamer and Fleishaker, 1994: 133–226). Information about habitable zones in the Universe, formation of planets and atmospheres, delivery of organics to habitable bodies, sources of energy, and (geo)chemical cycles also contribute to the synthetic approach. For the Earth the synthetic approach depicts a long chemical evolution that led to the first living system. It is still contested whether this first system was comprised of a metabolic system, an information system, both of them, or neither (something else or something simpler?).

The reductionist approach was suggested by Haldane when he discussed the discovery of bacteriophage. He recognized that phages are much smaller than cells, yet still much larger than molecules, which must limit the smallest possible size of a living entity. In general one can begin with a living cell and contemplate what simplest system the living cell could be reduced to. In the minds of some, a metabolic system would remain. In the minds of others, it would be a system of replication. Discussions of a possible "RNA World" and its relation to this aspect of the problem are found in Sections 6.5 and 8.3.1. The reductionist approach is also reflected in attempts to find a record of the earliest life on Earth. Such a record can be sought within geology (in the form of fossils) or in biology (in the form of phylogeny). For example, information about life's deepest roots may suggest the study of certain extant extremophiles or of possible places of refuge on the Earth during past episodes of extinction.

The displacement of the origin of the first living thing to different times and to different environments with *special* conditions on an early Earth bears resemblance to the displacement of the theory of spontaneous generation to different organisms and to different kinds of *special* conditions today. The theory of spontaneous generation had a long tenure owing to the difficulty of complete refutation. Indeed, one could argue that the theory *still* exists within the modern approach to the origin of life, and that it ultimately cannot be refuted. But in fact the hypothesis that life originated on the early Earth *can* be tested, for example, by attempting to generate a "living" system in the laboratory under presumed early Earth conditions. However, success in this endeavor of course requires that we know what it means to be "living" and that we know what the conditions on the early Earth were like. We may also look elsewhere in the Universe for clues about the origin of life on Earth or on other planets, satellites and planetary systems defined as "habitable." We can also study the delivery of raw materials to habitable places. Closer to home we may investigate the fossil record on the Earth or perhaps we may look to Mars for clues.

Although we have a long way to go toward understanding life's origin, one of the central questions of astrobiology, we are optimistic, as was Oparin (1924: 234):

> What we do not know today we shall know tomorrow. A whole army of biologists is studying the structure and organization of living matter, while a no less number of physicists and chemists are daily revealing to us new properties of dead things. Like two parties of workers boring from the two opposite ends of a tunnel, they are working towards the same goal. The work has already gone a long way and very, very soon the last barriers between the living and the dead will crumble under the attack of patient work and powerful scientific thought.

1.16 Conclusion

We have followed the two threads of extraterrestrial life and origin of life from ancient times to the mid-twentieth century, and now pass the historical baton to Steven Dick in Chapter 2. There we will see that it was the dawning of the Space Age in the late 1950s that brought these two communities together and led to NASA inventing a discipline called *exobiology* (the direct ancestor of astrobiology), one of whose prime concerns was investigating the Oparin–Haldane hypothesis not just for the early Earth, but also for other feasible planetary environments. And as we today do astrobiology, we still find ourselves, like Fontenelle's philosopher (Section 1.4), at that tantalizing margin between curiosity and poor eyesight. As Kevin Zahnle (2001: 213) stated in a marvellous and entertaining review of the history of life on Mars:

> In some ways the debate has really moved little since the days of Flammarion and Lowell. The most interesting information remains right at the limit of resolution, be it meters in satellite images of gullies, or nanometers in microscopic images of magnetite crystals. Always life on Mars seems just beyond the fields that we know.

It is indeed a bold and provocative enterprise that scientists tackle when they test the world and develop views on topics as fundamental as the role of life in the Universe. As we try to understand how the Copernican principle applies to the *biological* world, the issues become even more profound than for Copernicus, for the uniqueness of us humans, as well as our form of life, is more deeply vested in our psyche than is the uniqueness of any physical aspect of our home planet.

Our best efforts in science are surely steadily improving in their usefulness and their verisimilitude to the natural world, but we should not forget the many historical examples of strong influence from the prevailing cultural milieu. To emphasize this is not to denigrate either yesterday's or today's science, but only to acknowledge the essential humanity woven through a powerful way to understand the world.

REFERENCES

Antoniadi, E.-M. (1930). *La Planète Mars, 1659–1929*. Paris: Librairie Scientifique Hermann (trans. by P. Moore (1975) as *The Planet Mars*. Shaldon, Devon, UK: Keith Reid).

Bernal, J. D. (1967). *The Origin of Life*. Cleveland: World Publishing Company.

Brooke, J. H. (1991). *Science and Religion: Some Historical Perspectives*. New York: Cambridge University Press.

Brush, S. G. (1996). *Nebulous Earth: the Origin of the Solar System and the Core of the Earth from Laplace to Jeffreys*. Cambridge: Cambridge University Press.

Clark, R. (1968). *The Life and Work of J. B. S. Haldane*. Oxford: Oxford University Press.

Cocconi, G. and Morrison, P. (1959). Searching for interstellar communications. *Nature*, **184**, 844.

Conant, J. B. (ed.) (1953). *Pasteur's and Tyndall's Study of Spontaneous Generation*. Cambridge, MA: Harvard University Press.

Crowe, M. J. (1986). *The Extraterrestrial Life Debate, 1750–1900*. Cambridge: Cambridge University Press.

Deamer, D. W. and Fleishaker, G. R. (eds.) (1994). *Origins of Life: the Central Concepts*. Boston: Jones and Bartlett.

Dick, S. J. (1982). *Plurality of Worlds: the Origins of the Extraterrestrial Life Debate from Democritus to Kant*. Cambridge: Cambridge University Press.

Dick, S. (1993). The search for extraterrestrial intelligence and the NASA High Resolution Microwave Survey (HRMS): historical perspectives. *Space Sci. Rev.*, **64**, 93–139.

Dick, S. J. (1996). *The Biological Universe*. Cambridge: Cambridge University Press.

Dick, S. J. and Strick, J. E. (2004). *The Living Universe: NASA and the Development of Astrobiology*. New Brunswick, NJ: Rutgers University Press.

Doel, R. E. (1996). *Solar System Astronomy in America: Communities, Patronage, and Interdisciplinary Science, 1920–1960*. Cambridge: Cambridge University Press.

Drake, F. (1960). How can we detect radio transmissions from distant planetary systems? *Sky & Telescope*, **19**, 140–143.

Drake, F. (1961). Project Ozma. *Physics Today*, **14**, 40–46 (Apr.).

Drake, F. (1962). *Intelligent Life in Space*. New York: Macmillan.

Drake, F. (1985). Project Ozma. In *The Search for Extraterrestrial Intelligence*, eds. K. I. Kellermann and G. A. Seielstad, pp. 17–26. Charlottesville, VA: NRAO.

Drake, F. and Sobel, D. (1992). *Is Anyone Out There?: the Scientific Search for Extraterrestrial Intelligence*. New York: Delacorte.

Edge, D. O. and Mulkay, M. J. (1976). *Astronomy Transformed: the Emergence of Radio Astronomy in Britain*. New York: Wiley.

Farley, J. (1974). *The Spontaneous Generation Controversy from Descartes to Oparin*. Baltimore, MD: Johns Hopkins University Press.

Findlen, P. (1993). Controlling the experiment: rhetoric, court patronage and the experimental method of Francesco Redi. *History of Science*, **31**, 35–64.

Flammarion, N. C. (1862). *La Pluralité des Mondes Habités*. Paris: Gauthier-Villars.

Flammarion, N. C. (1892, 1909). *La Planète Mars et ses Conditions d'Habitabilité* (2 Vols.). Paris: Gauthier-Villars.

Fry, I. (2000). *The Emergence of Life on Earth*. New Brunswick, NJ: Rutgers University Press.

Gonzalez, G. and Richards, J. W. (2004). *The Privileged Planet: How Our Place in the Cosmos Is Designed for Discovery*. Washington, DC: Regnery Publishing.

Graham, L. R. (1987). *Science, Philosophy, and Human Behavior in the Soviet Union*. New York, NY: Columbia Universitiy Press.

Greene, M. T. (1992). *Natural Knowledge in Preclassical Antiquity*. Baltimore, MD: Johns Hopkins University Press.

Guthke, K. S. (1990). *The Last Frontier*. Ithaca, NY: Cornell University Press.

Haldane, J. B. S. (1929). The origin of life. *The Rationalist Annal*. pp. 3–10 (reprinted in Bernal (1967: 242–249) and in Deamer and Fleishaker (1994: 73–81)).

Hillegas, M. R. (1975). Victorian "extraterrestrials." In *The Worlds of Victorian Fiction*, ed. J. H. Buckley, pp. 391–414. Cambridge, MA: Harvard University Press.

Kirk, G. S., Raven, J. E., and Schofield, M. (1983). *The Presocratic Philosophers: a Critical History with a Selection of Texts*. Cambridge: Cambridge University Press.

Kuhn, T. S. (1957). *The Copernican Revolution: Planetary Astronomy in the Development of Western Thought*. Cambridge, MA: Harvard University Press.

Lane, K. M. D. (2005). Geography of Mars: cartographic inscription and exploration narrative in late Victorian representations of the red planet. *Isis*, **96**, 477–506.

Lederberg, J. (1960). Exobiology: approaches to life beyond the Earth. *Science*, **132**, 393–400.

Lowell, P. (1895). *Mars*. Boston: Houghton Mifflin.

Lucretius (trans. 1951). *On the Nature of the Universe* (trans. by R. E. Latham of *De Rerum Natura*. New York: Penguin.

Markley, R. (2005). *Dying Planet: Mars in Science and the Imagination*. Durham, NC: Duke University Press.

Miller, S. L. (1953). A production of amino acids under possible primitive Earth conditions. *Science*, **117**, 528–529.

Oparin, A. I. (1924). *Proiskhozhdenie zhinzy*, trans. by A. Synge as "The origin of life" in Bernal (1967: 199–234), and reprinted in Deamer and Fleishaker (1994: 31–71).

Oparin, A. I. (1936). *Vozhiknovenie zhizny na aemle*, trans. as *Origin of Life* (1938), New York: Macmillan; reprinted (1952) New York: Dover. Final edition (third) in 1957 as *The Origin of Life on Earth*. New York: Academic Press.

Pasteur, L. (1861). Mémoire sur les corpuscules organises qui existent dans l'atmosphère. *Annales de Chimie et de Physique (ser. 3)*, **64**, 5–110.

Raulin-Cerceau, F. (2004). Historical review of the origin of life and astrobiology. In *Origins: Genesis, Evolution and Diversity of Life*, ed. J. Seckbach, pp. 17–33. Dordrecht: Kluwer.

Sagan, C. (1961). On the origin and planetary distribution of life. *Radiation Research*, **15**, 174–92.

Sagan, C. and Fox, P. (1975). The canals of Mars: an assessment after Mariner 9. *Icarus*, **25**, 602–12.

Salisbury, F. B. (1962). Martian biology. *Science*, **136**, 17–26.

Schaffer, S. (1989). The nebular hypothesis and the science of progress. In *History, Humanity and Evolution*, ed. J. R. Moore, pp. 131–64. Cambridge: Cambridge University Press.

Schiaparelli, G. V. (1878). Osservazioni astronomiche e fisiche sull'asse di rotazione e sulla topographia del pianeta Marte. *Atti della Roy. Accademia dei Lincei*. Memoria 1, Ser. 3, Vol. 2. Also available in *Le Opere di G. V. Schiaparelli* (1930) (10 vols.). (1968 reprint), New York: Johnson Reprint Corporation.

Schiaparelli, G. V. (1888). In *l'Astronomie* (Jan., Feb., Mar., and Apr.), as cited and reproduced in Flammarion (1892: 296, 436). Also available in *Le Opere di G. V. Schiaparelli* (1930) (10 vols.). (1968 reprint), New York: Johnson Reprint Corporation.

Secchi, P. A. (1859). *Sistema Solare*. Roma: Tipografia delle Belle Arti.

Sheehan, W. (1988). *Planets and Perception: Telescopic Views and Interpretations, 1609–1909*. Tucson: University of Arizona Press.

Sheehan, W. (1996). *The Planet Mars: a History of Observations and Discovery*. Tucson: University of Arizona Press.

Sheehan, W. (2005). To Mars by way of Noto. *Sky & Telescope*, **110**, 108–111.

Sinton, W. M. (1959). Further evidence of vegetation on Mars. *Science*, **130**, 1234–1237.

Strauss, P. (2001). *Percival Lowell: the Culture and Science of a Boston Brahmin*. Cambridge: Harvard University Press.

Strick, J. E. (2000). *Sparks of Life: Darwinism and the Victorian Debates over Spontaneous Generation*. Cambridge, MA: Harvard University Press.

Struve, O. (1960). Astronomers in turmoil. *Physics Today*, **13**, 18–23.

Struve, O. (1962). *The Universe*. Cambridge, MA: MIT Press.

Struve, O. and Zebergs, V. (1962). *Astronomy of the 20th Century*. New York: Macmillan.

Sullivan, W. T., III, (ed.) (1984). *The Early Years of Radio Astronomy: Reflections Fifty Years after Jansky's Discovery*. Cambridge: Cambridge University Press.

Van de Kamp, P. (1963). Astrometric study of Barnard's star from plates taken with the 24-Inch Sproul refractor. *Astron. J.*, **68**, 515–21.

Van Helmont, J. B. (1682). *Opera Omnia*. New York: Readex Microprint Landmarks of Science Microcards (1969).

Von Weizsäcker, C. F. (1944). Über die Enstehung des Planetensystems. *Zeitschrift f. Astrophysik*, **22**, 319–55.

Waite, A. E. (ed.) (1894). *The Hermetic and Alchemical Writings of Aureolus Philippus Theophrastus Bombast of Hohenheim, called Paracelsus the Great*. London: Elliott.

Whewell, W. (1853). *The Plurality of Worlds*. In 2001 reprint, ed. M. Ruse. Chicago: University Chicago Press.

Zahnle, K. (2001). Decline and fall of the martian empire. *Nature*, **412**, 209–13.

FURTHER READING AND SURFING

Dick (1982), Crowe (1986), Dick (1996), Dick and Strick (2004), and Guthke (1990) are the key historical and literary studies on our ideas of extraterrestrial life from ancient times to the present.

Fry (2000), Farley (1974), and Strick (2000) are the best overviews of the history of the development of concepts of the origin of life.

Wells, H. G. (1898). *The War of the Worlds*. London: William Heinemann. A classic and a page-turner (forget about Tom Cruise).

www.jstor.org (available through most universities). Scanned pages of the complete run of the world's oldest scientific journal, *Philosophical Transactions of the Royal Society*, from its first issue in 1665. For example, check out Antony van Leeuwenhoek's papers in the 1690–1720 era, such as: "Observations on the seeds of cotton, palm, or date-stones, cloves, nutmegs, goose-berries, currans, tulips, cassia, lime-tree: on the skin of the hand, and pores, of sweat, the crystalline humour, optic nerves, gall, and scales of fish: and the figures of several salt particles, etc." Vol. 17, pp. 949–60 (1693).

Shklovskii, I. S. and Sagan, C. 1966. *Intelligent Life in the Universe*. San Francisco: Holden-Day. Status of the field at the end of this chapter's time period. Still in print and well worth reading not only for history, but also for the light it sheds on today's scientific thinking.

Geison, G. L. (1995). *The Private Science of Louis Pasteur*. Princeton: Princeton University Press. After 10,000 pages of Pasteur's lab notebooks, correspondence, etc. first became available in the 1980s, Geison produced this excellent and iconoclastic scientific biography.

Browne, E. J. (1996). *Charles Darwin: Voyaging*. Princeton: Princeton University Press.

Browne, E. J. (2002). *Charles Darwin: the Power of Place*. New York: Alfred A. Knopf. This two-volume biography is the one to read.

2 From exobiology to astrobiology

Steven J. Dick
NASA Headquarters

The rise of exobiology, the study of the origin of life and of possible life outside the Earth, was intimately related to the birth of the Space Age, and particularly to the birth of the US National Aeronautics and Space Administration (NASA) in 1958. By providing the means to enter space, NASA placed exobiology into the arena where an age-old problem could be empirically tested with *in situ* observations and experiments. Moreover, in pursuit of exobiology, on the ground NASA funded experiments on the origin of life, revived planetary science, sponsored theoretical and observational studies on planetary systems, and assembled the flagship program in the Search for Extraterrestrial Intelligence (SETI) – all conceptual elements of the budding discipline. Over four decades, at a relatively small but steady level of funding punctuated by the landmark Viking mission to Mars, the American space agency put into place the conceptual, institutional, and community structures necessary for a new discipline, leaving no doubt of NASA's status as the primary patron of exobiology. The interest in the search for life, however, knew no national boundaries. Especially in the post-Viking era, international involvement grew in the field that also became known as "bioastronomy," and that was transformed at the end of the century into a broadened "astrobiology" effort.

The ability to search for life beyond Earth did not guarantee its adoption as a program within NASA or any other space agency. The will to pursue exobiology was an issue in science policy, and the fact that NASA quickly incorporated the problem into its activities was motivated not just by a compelling scientific puzzle but also by public interest in the problem. Indeed, the search for life beyond Earth became more than just another NASA program; at times it served as a driving force for the space program. It offered the chance to answer the question of humanity's place in the Universe, to place *Homo sapiens* in a cosmic context just as Darwin had placed us in a terrestrial context. In this chapter, we analyze the conceptual formation of exobiology against the background of cosmic evolution, detail NASA's leading role as a major patron, discuss when practitioners began to view the study of life beyond Earth as a discipline, and describe the discipline's transition at the end of the century from exobiology to astrobiology. The earlier history (before 1960) of ideas on extraterrestrial life and on the origin of life is covered in Chapter 1. We conclude with comments on the significance of astrobiology for science and culture. Details of this chapter are available in *The Living Universe: NASA and the Development of Astrobiology* (Dick and Strick, 2004) and, more broadly for the entire twentieth century, in *The Biological Universe* (Dick, 1996).

2.1 Cosmic evolution as a context for exobiology

As the Space Age began, the concept of *cosmic evolution* – the connected evolution of planets, stars, galaxies *and* life – provided the grand context within which the enterprise of exobiology was undertaken. The intellectual basis for this guiding principle of cosmic evolution had its roots in the nineteenth century (Chapter 1). Crowe (1986) and Schaffer (1989) have shown how a combination of Laplace's nebular hypothesis and Darwinian evolution gave rise to the first tentative expressions of parts of this worldview. The philosophy of Herbert Spencer extended it to the evolution of society, and some Spencerians, including Percival Lowell, extended evolutionary principles to life on other planets (Strauss, 2001). In England and America Richard A. Proctor, and in France Camille Flammarion, also spread Darwinian ideas to illuminate the question of life on other worlds.

Planets and Life: The Emerging Science of Astrobiology, eds. Woodruff T. Sullivan, III and John A. Baross. Published by Cambridge University Press. © Cambridge University Press 2007.

Such a set of general ideas was a long way from a research program. In the first half-century of the post-Darwinian world, cosmic evolution did not find fertile ground among astronomers, who were hard-pressed to find evidence, other than spectroscopic confirmation of the widely assumed "uniformity of nature." Astronomers recognized and advocated parts of cosmic evolution, as in the study of stellar evolution (Hale, 1908). But even Lowell (1909) limited his evolution of worlds to physical, not biological evolution, martian canals notwithstanding. For the most part, biologists were also reluctant cosmic evolutionists. Alfred Russel Wallace (1903), co-founder with Darwin of the theory of natural selection, held that the Universe was static, humanity's position unique, and Earth likely the only inhabited planet. Ten years later Harvard biochemist Lawrence J. Henderson (1913) investigated how the environment on Earth became fit for life, and concluded "the properties of matter and the course of cosmic evolution are now seen to be intimately related to the structure of the living being and to its activities; they become, therefore, far more important in biology than has been previously suspected. For the whole evolutionary process, both cosmic and organic, is one, and the biologist may now rightly regard the universe in its very essence as biocentric." Clearly, Henderson grasped the essentials of cosmic evolution, used its terminology, and believed his research into the fitness of the environment supported it. But the idea was largely stillborn, and would lie dormant for almost a half century.

Cosmic biological evolution first had the potential to become a research program in the 1950s and 1960s when its cognitive elements had developed enough to become experimental and observational sciences, and when the researchers in these disciplines first realized they held the key to a larger problem that could not be resolved by any one part, but only by all of them working together. Harvard College Observatory Director Harlow Shapley was an early proponent of this concept, and already in 1958 spoke of it in now familiar terms (Shapley, 1958). The Earth and its life, he asserted, are "on the outer fringe of one galaxy in a universe of millions of galaxies. Man becomes peripheral among the billions of stars in his own Milky Way; and according to the revelations of paleontology and geochemistry he is also exposed as a recent, and perhaps an ephemeral manifestation in the unrolling of cosmic time." Shapley went on to elaborate his belief in billions of planetary systems, where "life will emerge, persist and evolve." Shapley's belief was unproven then, and

TABLE 2.1 Cosmic evolution as conceived by NASA's SETI program, 1979 (ordering: left column first, downwards)

Big Bang	Precambrian biology
Galaxies	Complex life
Stars	Intelligent life
Biogenic elements	Cultural evolution
Planets	Civilizations
Chemical evolution	Science and technology
Origin of life	Study of life in the Universe

remains to be proven today. The transition from belief to proof is tantamount to discovering whether cosmic evolution commonly ends with planets, stars, and galaxies, or with life, mind, and intelligence. Put another way, does cosmic evolution produce not only a physical Universe, but also a *biological Universe* (Dick, 1996)?

The idea of cosmic evolution spread rapidly over the next 40 years, both as a guiding principle within the scientific community and as an image familiar to the general public (Sagan, 1980; Reeves, 1984). NASA enthusiastically embraced, elaborated, and spread the concept of cosmic evolution from the Big Bang to intelligence as part of its SETI and exobiology programs in the 1970s and 1980s (Table 2.1). And when in 1997 NASA published a "Roadmap" for its new "Origins" program, it described the goal of the program as "following the 15 billion year long chain of events from the birth of the universe at the Big Bang, through the formation of chemical elements, galaxies, stars, and planets, through the mixing of chemicals and energy that cradles life on Earth, to the earliest self-replicating organisms – and the profusion of life" (NASA, 1997). With this proclamation, cosmic evolution became the organizing principle for most of NASA's space science effort.

Today, cosmic evolution is elaborated in ever greater theoretical and descriptive detail (Delsemme, 1998; Chaisson, 2001), and the biological possibilities are viewed as playing out on an incomparably larger stage than the static Universe conceived by Wallace and most astronomers a century ago. The scientific problem now is viewed as determining whether those possibilities are real, or only wishful thinking.

2.2 The conceptual formation of exobiology

In the 1950s and 1960s four scientific fields – planetary science, the search for planetary systems, origin of life studies, and SETI – converged to give birth to the field of exobiology. At first quite separate in terms of researchers, techniques, and goals, these fields over four decades gradually became integrated, in large measure because of the scientific and public desire to search for life beyond Earth. During the early Space Age planetary science produced ground-based claims of vegetation on Mars and spacecraft exploration of the planets began. The search for planetary systems remained an embryonic activity, and produced its first modern claim of extrasolar planets in 1963. Origin of life studies began their modern era with the Urey–Miller experiments in 1953. SETI received its underpinnings with a seminal article in *Nature* in 1959, and the first radio telescope search for artificial extraterrestrial signals in 1960. These studies followed their own trajectories, but gradually became inextricably intertwined as parts of the larger scientific problem of life in the Universe. And at the present time it is inconceivable that they should be treated separately, and the synergy among them is fueling a vigorous research program.

2.2.1 Planetary science

By the mid-1950s the field of planetary science, until then a small-scale interdisciplinary field, was beginning to reawaken (Doel, 1996). By that time, too, the idea of life in the Solar System had a long history stretching back to its underpinnings as an implication of the Copernican revolution, which made the Earth a planet and the planets potential Earths (Dick, 1982, 1996; Crowe, 1986: Chapter 1). Especially since Percival Lowell's claims for canals constructed by intelligent beings on Mars, that planet had become a focus for the idea of life beyond Earth (Sections 1.7 and 1.8). Although that aspect of Lowell's work was largely discredited, and although techniques in observational planetary science had made only slow improvement (Schorn, 1998), the belief in a harsher, but still Earth-like, Mars with vegetation was still very much alive at mid-century. At that time astronomers believed Mars had an atmospheric pressure of about 85 millibars at its surface, ten times thinner than Earth's. Atmospheric oxygen and water vapor had not been detected with certainty, but Dutch-American astronomer Gerard Kuiper (1949) had used early near-infrared techniques to discover carbon dioxide, one of the principal gases in the process of photosynthesis. Surface temperatures were known to be harsh, but not impossibly so in terms of primitive life.

Seasonal vegetation across parts of Mars was commonly accepted, based on visual and photographic observations showing unmistakable seasonal changes on the surface as the polar caps melted, spreading a wave of darkening (Slipher, 1927; Barabashev, 1952; Kuiper, 1955). In the Soviet Union an Astrobotanical (!) Section of the Academy of Sciences of the Kazakh Republic, dedicated to studies of martian (compared to terrestrial) vegetation, was led in the postwar decade by astronomer Gavriil A. Tikhov. The second edition of the standard astronomy textbook of the time (Russell, Dugan, and Stewart, 1945), was pessimistic about the existence of even primitive animal life, but asserted that the existence of vegetation was "more likely than not." Astronomer Gerard DeVaucouleurs (1954) injected a note of caution when he wrote in his definitive summary of martian physical properties that the problem of life on Mars "is still, to a large extent, beyond the limits of our positive knowledge and can only be the subject – either way – of vague speculation in which general 'principles' of a metaphysical nature have always to be taken as a guide." Nevertheless, the scientific approach to the problem of life on Mars was soon bolstered by spectroscopic evidence (Fig. 2.1) for vegetation by American astronomer William Sinton (1957, 1959), one of many claims that demonstrate the problematic nature of observation when science functions at the limits of its technical capabilities – as it almost always has in exobiology.[1]

Such was the situation at the beginning of the Space Age. Even before planetary spacecraft became an imminent possibility with the launch of Sputnik in 1957, the field of space medicine had precipitated what Lowell Observatory Director Albert Wilson characterized as "the first American symposium in Astrobiology" (Wilson, 1958). Here, at a meeting chaired by German-American space medicine pioneer Hubertus Strughold, the French planetary astronomer Audouin Dollfus spoke on the nature of the martian surface, Sinton highlighted his claims for spectroscopic evidence of vegetation, and Strughold and his colleagues reported on the behavior of microorganisms under simulated martian conditions. By 1959 the integration of this work in space medicine with the more general field of space science was being discussed in the Lunar and Planetary Exploration Colloquia, held in various locations in California. A seminal series of discussions in which Kuiper and numerous influential scientists participated, these meetings included papers

---

[1] See Section 1.8.3 for more details.

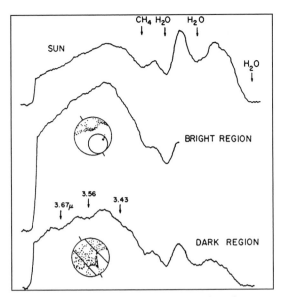

FIGURE 2.1 Sinton's infrared spectroscopic evidence for vegetation on Mars, obtained on the Palomar 200-inch telescope. The top curve shows a solar spectrum, with superimposed absorptions by methane and water in the Earth's atmosphere. The middle curve shows a spectrum of a bright desert area of Mars, where no vegetation was expected. The bottom spectrum, obtained when the spectrograph slit was placed over one of the dark areas of Mars, shows three apparent absorption features (indicated by arrows) that were interpreted as due to vegetation. The evidence turned out to be spurious; the absorptions were actually due to deuterated water (HDO) in the Earth's atmosphere, as Sinton himself published six years later. With permission, from Sinton (1959: 1234). Copyright 1959 AAAS.

with titles like "Advances in Astrobiology" (Strughold, 1959). In the wake of Sputnik, the immediate concern that biologists raised was the contamination of the planets with terrestrial microorganisms, or back-contamination of Earth by extraterrestrial organisms – an indication of how seriously they took the possibility of life beyond Earth. Already in late 1957 geneticist Joshua Lederberg, soon to receive the Nobel Prize, brought the issue of contamination before the scientific community (Dick, 1996). This led to standards for spacecraft sterilization that were taken very seriously; the three-week quarantine of the early Apollo astronauts returning from the Moon reflects how seriously the problem of back contamination of Earth was taken (Fig. 24.2).

Besides the "practical" concerns of space medicine and spacecraft contamination, the space program reinvigorated planetary science in general (Tatarewicz, 1990), and the search for life in particular. Lederberg himself (Fig. 2.2) argued that spaceflight furnished a unique method for studying the origin and evolution of

FIGURE 2.2 Joshua Lederberg (1925–), pioneer in exobiology (he coined the term), shown about 1962 in his laboratory at Stanford University.

life beyond Earth, and he coined the term *exobiology* for this study (Lederberg, 1960).[2] Curiously, the Soviet Union did not emphasize the search for life on Mars as part of its space program, despite Oparin's pioneering experimental work (Section 1.14) and a general interest in exobiology (Oparin, 1975). Although a Soviet discovery of life on Mars would have been an even greater coup than a successful outcome for the much-vaunted "Moon race," Soviet interest focused more on SETI, perhaps because ground-based radio astronomy was cheaper and the payoff of a discovery even greater.

Within two years of its founding, however, NASA embraced exobiology. In 1960 it set up an Office of Life Sciences, and authorized Caltech's Jet Propulsion Laboratory to study the type of spacecraft needed to land on Mars and search for life. In the same year, NASA set up its first life sciences laboratory at its Ames Research Center in Mountain View, California, destined to become the premier laboratory for exobiological experiments. Support for such work received a strong endorsement from the National Academy of Sciences in 1962 when its Space Science Board set the search for extraterrestrial life as "the prime goal of space biology" (Space Science Board, 1962). The Board had no illusions as to the importance of this work when it remarked that "it is not since Darwin – and

[2] Lederberg also coined the word *esobiology* for Earth's own brand of biology, but this term never caught on.

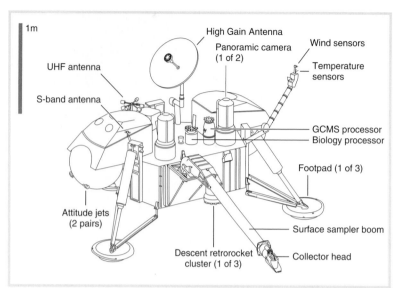

FIGURE 2.3 The Viking lander, a complex machine incorporating eight experiments, landed on the surface of Mars in July, 1976, followed by another lander in September. The biology processor (labeled to the lower right) was within a small canister of volume 0.03 m^3. (For scale: diameter of the lander body is ~3 m.)

before him Copernicus – that science has had the opportunity for so great an impact on man's understanding of man." By 1964 the Board reinforced its earlier report, saying "the exploration of Mars – motivated by biological questions – does indeed merit the highest scientific priority in the nation's space program over the next decade" (Pittendrigh *et al.*, 1966).

With such an endorsement it is not surprising that the search for life on Mars became one of NASA's most important and visible goals, even if ground-based evidence related to such life was tentative at best. Although Sinton's observations were called into question and eventually proven spurious (Rea, O'Leary, and Sinton, 1965), further observations such as the detection of 25 mbar of water vapor in the martian atmosphere in 1963 continued to fuel interest in the spacecraft exploration of Mars. Following a series of reconnaissance missions in the 1960s and early 1970s, the culmination of the search for life in the Solar System was the landing of two Viking spacecraft (Fig. 2.3) on the surface of Mars in 1976 (Ezell and Ezell, 1984). The Viking mission (two Mars orbiters and two landers costing ~$4 billion in present dollars) undertook 13 separate investigations, including meteorology, seismology, chemistry, imaging, and physical properties of the planet Mars. Although more was learned about Mars from these investigations than all previous efforts combined, the public and scientific focus was on the biology experiments and the conclusions of the biology science team, headed by biologist Harold Klein of NASA's Ames Research Center. With Viking the theory and practice of exobiology met for

the first time, with high stakes for science, for humanity, and for the fledgling science.

The Viking biology package, limited to 15 kg and about 0.03 m^3, included three distinct experiments using different philosophies, environmental conditions, and detectors. All three experiments sought to detect metabolic activities, and are discussed in detail in Sections 18.5.2 and 23.2. Once on the surface of Mars, these experiments yielded immediate surprises when two of them gave positive results. Yet for most investigators these results were trumped by the gas-chromatograph mass spectrometer, which detected no organic molecules down to parts per billion – without organic molecules there could be no life. The consensus among scientists still today is that the reactions apparently indicative of life were caused by a highly oxidized surface material.

Members of the Viking biology team drew varying conclusions about life in the Universe. Gilbert Levin of Hazleton Labs, Inc. believed he had found life, based on one of the ostensibly positive results. Biologist Norman Horowitz (1986) of Caltech concluded that "it is now virtually certain that the Earth is the only life-bearing planet in our region of the Galaxy." Klein (1977) believed that the experiments may have been flawed and the results therefore indecisive. In general, although scientists were much more pessimistic about life on Mars in the post-Viking era, many felt that the question had not been definitively answered. It was not the first or last episode in the extraterrestrial life debate illustrating the limits of science. The inevitable result was a decline in interest in exobiology.

Twenty years later, the difficulties of the search for martian life were once again displayed when McKay *et al.* (1996) announced the discovery of possible fossils and other evidence for life in the martian meteorite ALH84001 found in Antarctica. The "Mars rock fossils" revived the debate about life on Mars, and although the original claims have not been validated by the scientific community (Section 18.5.3), other research *has* indicated that the transport of such fossils or even spores between planets is quite possible (Section 18.4.4). Even with a specimen that could be subjected to sophisticated analytical techniques, the ensuing controversy did not resolve the question of martian life. But it did superbly illustrate that interdisciplinary work was necessary to answer exobiology's questions. In the wake of the Mars rock, tantalizing evidence continued to amass from a variety of directions. The recent Mars Global Surveyor and Mars Odyssey missions have both indicated that water ice still exists in plentiful amounts just below the surface, and the Mars Exploration Rovers have found excellent evidence for plentiful liquid water below and on the surface in the past (Chapter 18). Thus, during the past forty years, the debate over life on Mars has shifted from vegetation to organic molecules, and then to fossils, even as the discovery of water has fueled the belief that life might still exist.

If Mars has not yet proven dead, even more surprising are the possibilities beyond Mars. The confirmation in the mid-1990s that the jovian moon Europa was covered by a thick sheet of ice, and that it was likely that a water ocean existed beneath the ice, has brought attention to the possibilities of life beyond what had been considered the habitable zone of our Solar System (Chapters 3 and 19). Moreover, many believe the existence of organics in the atmosphere of the saturnian moon Titan holds clues to the origin of life (Chapter 20). At the start of the millennium we thus find several bodies in the Solar System of intense exobiological interest. Planetary science has not only contributed to this interest, but has itself been largely revived and is being increasingly driven by the search for life.

2.2.2 Planets beyond the Solar System

The search for planets outside our Solar System was less centered on NASA than was the search for life within the Solar System. Such distant planets were not known to exist, and even if they did, would not come under NASA's space flight charter. This left several possibilities: ground-based studies related to other planetary systems, theoretical studies related to the

formation of planetary systems, or studies of space missions that could detect planets or protoplanets, if not visit them. In fact, NASA adopted all these strategies, but, ambiguous about its ground-based role, did it with none of the urgency that characterized the space flight possibilities of searching for life on Mars.

A prerequisite to interest in searching for planetary systems was theoretical support that they might exist. Such support long predated the Space Age, but had ebbed and flowed during the twentieth century (see Section 1.8.1 for more details). That numerous planets did exist beyond our Solar System was an implication of Laplace's nebular hypothesis (Section 1.6.1) – if planets really formed as the byproduct of stellar evolution, then they should be extremely common. The nebular hypothesis was eclipsed for the first four decades of the century by the Chamberlin–Moulton and then the Jeans–Jeffreys theories that planets formed by close encounters between stars, the so-called tidal theory in which material was pulled out of the stars to form planets. Because such close encounters would be extremely rare events, planetary systems would be extremely rare. Only in the 1940s, when the tidal theory was shown to be flawed and the nebular hypothesis came back into vogue, could an abundance of planetary systems once again be postulated. During the 15-year period from 1943 to 1958, the commonly accepted number of planetary systems in the Galaxy went from 100 to one billion (Table 2.2)! The turnaround involved many arguments, from the observations of a few possible planetary companions (see, e.g., Strand, 1943), to binary star statistics, the nebular hypothesis, and stellar rotation rates. Helping matters along was the dean of American astronomers, Henry Norris Russell, whose *Scientific American* article "Anthropocentrism's demise" enthusiastically embraced numerous planetary systems based on just a few observations (Russell, 1943). Definitive evidence, however, would be much more elusive, for it turned out that Russell's declaration was fifty years premature.[3]

Meanwhile, as NASA was preparing spacecraft for Mars, the Dutch-born American astronomer Peter van de Kamp announced his discovery of a planet of mass 1.6 times that of Jupiter orbiting around nearby Barnard's star. This discovery (van de Kamp, 1963) also proved controversial; van de Kamp did not detect the planet itself, but the gravitational effects of the planet on the motion of its parent star (Fig. 2.4). Van

---

[3] Further details of these developments, as well as those in the following paragraph, are in Sections 1.8.1 and 1.8.2.

TABLE 2.2 Estimates of frequency of planetary systems, 1920–1961

Author	Argument	No. of planetary systems in galaxy	No. of habitable planets in galaxy
Jeans (1919, 1923)	Tidal theory	Unique	1
Shapley (1923)	Tidal theory	"Unlikely"	"Uncommon"
Russell (1926)	Tidal theory	"Infrequent"	"Speculation"
Jeans (1941)	No. of stars	10^2	–
Jeans (1942)	Improved tidal	One in six stars	Abundant
Russell (1943)	Companions	Very large	$>10^3$
Page (1948)	Weizsäcker	$>10^9$	$>10^6$
Hoyle (1950)	Supernovae	10^7	10^6
Kuiper (1951)	Binary star statistics	10^9	–
Hoyle (1955)	Stellar rotation	10^{11}	–
Shapley (1958)	Nebular hypothesis	10^6–10^9	–
Huang (1960)	Stellar rotation	10^9	10^9
Hoyle (1960)	Stellar rotation	10^{11}	10^9
Struve (1961)	Stellar rotation	$>10^9$	–

Source: adapted from Dick (1996: 199)

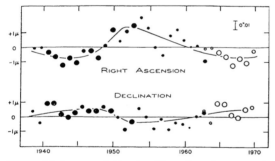

FIGURE 2.4 Peter van de Kamp's evidence for a planet around Barnard's star made use of the classical astrometric method for planet detection. Van de Kamp reported that the star underwent minute, periodic gravitational perturbations of a few hundredths of an arcsecond over three decades. The two plots show the star's measured east–west and north–south relative positions (in microns) on photographic plates taken with a 24-inch refractor. With permission from Elsevier Science Ltd.

de Kamp had been observing Barnard's star and others since the 1930s to search for minute motions that might indicate a planet; even after decades of observation, these motions were at the limits of detectability. Because of his technique, van de Kamp's claims could not easily be proven or disproven; although eventually shown to be spurious, they were accepted throughout the 1960s and 1970s and carried exobiological hopes beyond the Solar System.

Van de Kamp's solitary work can hardly be considered an indication that "planetary systems science"

had been born, although theoretical studies flourished, some of them funded by NASA. Even in 1978, after a conference on planet formation dominated by theorists, the editor of *Protostars and Planets* wrote that "with this book we hope to stimulate a new discipline" involving star and planet formation (Gehrels, 1978). By the time of the second conference in 1984 the editor wrote that the elements of the new discipline (which he dubbed "planetary systems science") "are now beginning to emerge." The detection of planets remained elusive, however, and as long as theory dominated, planetary systems science (like exobiology) risked being labeled a science without a subject.

Van de Kamp's observational results not only revived belief that planets might be common, but stimulated much thought about planetary detection techniques. In the late 1970s NASA discussed planetary systems first as part of its SETI program, then as part of the expansion of planetary science to other solar systems, and finally in connection with its Origins program. Even as these studies were proceeding, a surprising discovery was made in 1983 bearing on the existence of other planetary systems. NASA's Infrared Astronomical Satellite (IRAS) found during its calibration tests that the bright star Vega had an "infrared excess." Additional observations showed that the source of this excess infrared emission was a ring of dust surrounding Vega, and by mid-1984 some 40 such "circumstellar disks," or "protoplanetary systems,"

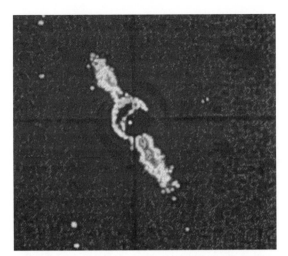

FIGURE 2.5 Image of a disk, at the near-infrared wavelength of one micron, around the star β Pictoris (1984), early evidence for circumstellar material perhaps related to planet formation. The star itself is occulted with a circular mask in order better to see the faint disk. (Courtesy B. Smith, R. Terrile, and JPL.)

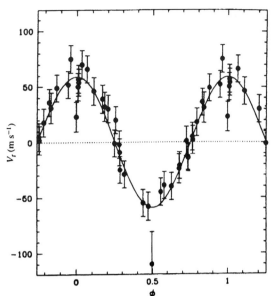

FIGURE 2.6 The first evidence for an extrasolar planet orbiting a normal star. The sinusoidal curve represents a line-of-sight variation in the radial motion of the star 51 Pegasi of $\pm$ 59 m/s as the star is tugged one way and then another over 4.2 days by the inferred planet of mass of at least 0.5 times the mass of Jupiter. The phase ϕ of the orbit varies from 0 to 1 over one period. Reprinted with permission from Mayor and Queloz (1995).

had been found, depending on the interpretation given to the infrared excess. Although not planets themselves, the discoverers saw the Vega phenomenon as "the first semidirect evidence that planets are indeed common in the universe" (Aumann *et al.*, 1984). By late 1984 one of the IRAS objects, β Pictoris, had been imaged by a ground-based optical telescope, producing one of the most famous images in 1980s astronomy (Fig. 2.5). In the following fifteen years, the dream of a planetary systems science rapidly moved towards reality with ground-based discoveries of further circumstellar disks, Hubble Space Telescope's discovery of disks around dozens of young stars in the Orion nebula (Fig. 3.3), the detection of long-sought brown dwarf stars, and – most of all – the detection of extrasolar planets.

The discovery of actual planets had little to do with NASA. In 1992 planets were found around pulsars (Wolszczan and Frail, 1992), using a radio astronomy timing technique. Although pulsars are exotic neutron stars not likely to harbor habitable planets, optimists declared that if planets could form around pulsars, they could form around normal stars also. Only in 1995 did the Swiss team of Michel Mayor and Didier Queloz (1995) announce the first planet around a Sun-like star, 51 Pegasi, using a radial velocity technique that detected the gravitational effect of the planet on the star (Fig. 2.6). By now over 200 planets have been found by the radial velocity technique, many of them by the American team of Geoff Marcy and Paul Butler (Chapter 21). All are giant planets, some in highly

eccentric orbits, others very close to their parent star. Although the American team had received minimal support from NASA before 1995, for the most part they had struggled on in the face of skepticism. Their success, however, emboldened other teams, led to significant funding by NASA and others, and spawned new discoveries.

With extrasolar planets in hand, there is no doubt that planetary systems science now exists. Although Earth-sized planets remain beyond the limits of detection (Earth's mass is only 1/300 of Jupiter's), a variety of spacecraft are on the drawing boards designed not only to make such a discovery, but also to detect evidence of life by observing the spectroscopic signatures of any extrasolar planetary atmospheres. As planetary systems science has developed, the ultimate motivation has long been clear: "The search for planets is the first step in the search for extraterrestrial intelligence, as the surface of a planet is probably the only viable location for the origin and evolution of life" (Scargle, 1988).

2.2.3 Origin of life

The existence of planets, even planets with Earth-like conditions, was one thing, but to posit planets with life was quite another. The chances for life-bearing planets depended on a better understanding of the origin of life on Earth, and that in turn depended on developments in biochemistry. By 1958 a beginning had been made, based on the Oparin–Haldane theory of chemical evolution. The Russian biochemist Aleksandr Ivanovich Oparin and the British biologist J. B. S. Haldane had independently suggested that life originated on Earth by chemical evolution in a hot dilute soup under conditions of a primitive reducing atmosphere (Section 1.14). Their early ideas provided a basis for experimentation, beginning with the famous Urey–Miller experiment in 1953 (Section 6.2.2), in which amino acids – the building blocks of life – were synthesized under possible primitive Earth conditions, specifically a reducing atmosphere rich in methane and ammonia. Their success set off numerous experiments around the world to verify another step in cosmic evolution.

It was in this milieu that the Space Age began. The small but steady stream of money that NASA put into exobiology and the life sciences went largely for research on the origin of life, or related experiments in life detection. In the early 1960s protein chemist Sidney Fox, biochemist Melvin Calvin, Lederberg, and biologist Wolfgang Vishniac were only the most prominent among a rapidly expanding number of biological researchers. By 1963 NASA's life sciences expenditures (including exobiology) had reached $17 million. The $100 million spent on the Viking biology experiments was closely related to origin of life issues, since an informed search for life required a definition of life and a knowledge of its origins. Even though exobiology saw a slump in the 1980s in the aftermath of the Viking failure to detect life, NASA kept it alive with a grant program of $10 million per year, and with the largest exobiology laboratory in the world at its Ames Research Center.

Origin of life studies became integrated with the search for extraterrestrial life in several ways. First, the ability to go to Mars prompted biological scientists, primarily biochemists and geneticists, to consider more seriously how origin of life studies related to outer space. Miller and Urey (1959) saw the relevance of space to their work, arguing that the discovery of life beyond Earth was a testbed for theories of the origin of life. In the Soviet Union Oparin (1975) wrote extensively on the origin of life in space, and called exobiology and the origin of life "the inseparable connection." Theories and experiments on the origin of life were increasingly seen in an extraterrestrial context; in this sense origin of life was one of the few areas in which

biological science could aspire to universality. Second, the discovery of amino acids in carbonaceous chondrite meteorites, and of complex organic molecules in interstellar molecular clouds, comets, and interplanetary dust, forced biological interest into the extraterrestrial realm (Section 3.8). Finally, difficulties with the theories of origin of life on Earth inspired re-introduction of the old notion of *panspermia*, in which it was postulated that life might have come from outer space (Crick and Orgel, 1973).

The origin of life on Earth and in space also shared philosophical issues. Old problems such as chance, necessity, and the nature of life, already recognized in the terrestrial realm, were magnified in the extraterrestrial realm. The crucial question for exobiology was whether life would arise wherever it could, or whether the Earth was a fluke. The contingency or necessity of life would be one of the great scientific and philosophical questions of cosmic evolution (Chapters 1, 5, and 6). Already in 1940, when the British Astronomer Royal Harold Spencer Jones (1940) wrote *Life on Other Worlds*, he remarked that "it seems reasonable to suppose that whenever in the Universe the proper conditions arise, life must inevitably come in to existence." Harvard biochemist George Wald (1954) proclaimed the Oparin–Haldane process a natural and inevitable event, not just on our planet, but on any planet similar to ours in size and temperature. In the Soviet Union Oparin teamed with V. Fesenkov to write *Life in the Universe* (1956, translated into English in 1961), which expressed this same view of the inevitability of life.

Over the past quarter century theories of the origin of life proliferated, with various implications for exobiology (Chapters 6, 8, and 9). Furthermore, the discovery of life in extreme environments – around deep sea hydrothermal vents, in deep underground rock, in conditions of great salinity and acidity, has fostered a new appreciation for the tenacity of life, and broadened our idea of the conditions under which life might originate on another planet, or on Earth (Chapters 14 and 15). As the possibilities of panspermia have become more widely accepted, spurred on by the Mars rock controversy and the realization that material does transfer between planets (Section 18.4.4), some researchers believe that exobiology may be the key to the origin of life on Earth. Skeptics commented that panspermia schemes unsatisfactorily beg the question by pushing life's ultimate origin further back in cosmic evolution. Nonetheless, surely knowledge that life originated on Earth by panspermia would be an important breakthrough! As for life originating here on Earth, experiments on prebiotic synthesis continue, enlivened by the new hypothesis of an "RNA world" (Sections 6.5 and 8.3.1). During the past two decades both theory and experiments on the origin of life have become an integral part of the field of astrobiology.

2.2.4 Search for Extraterrestrial Intelligence (SETI)

Beyond the origin of life loomed the even more difficult question of the possible evolution of intelligence elsewhere in the Universe. Although likewise hampered by a lack of understanding of how this had happened on Earth, discussion of contact with such intelligence was spurred on by the landmark paper by the physicists Giuseppe Cocconi and Philip Morrison (1959), who pointed out that radio wavelengths were the optimal part of the spectrum, in terms of atmospheric transparency and lack of interference, for the detection of transmissions from extraterrestrial intelligence. Furthermore, they emphasized that detection of such transmissions was feasible with radio telescope technology already in hand.[4] Frank Drake's Project Ozma at the National Radio Astronomy Observatory (NRAO) in Green Bank, West Virginia, the following year ushered in a series of projects around the world to detect such transmissions. Moreover, in 1961 Drake, supported by NRAO director Otto Struve, convened the first conference on interstellar communication at Green Bank (Cameron, 1963). Although a small conference attended by only eleven people including Struve, there were representatives from the astronomy community (Carl Sagan, Su Shu Huang, and Drake) and the biological community (Melvin Calvin), along with Cocconi and Morrison, physicist and engineer Bernard M. Oliver, and dolphin communications specialist John C. Lilly. Early interest in SETI therefore originated entirely outside of NASA.

It was at the Green Bank meeting that the now-famous Drake Equation was formulated (Section 26.4). This equation – aspiring to estimate the number of technological civilizations in the galaxy – soon became the icon of cosmic evolution, embodying in one compact expression not only the astronomical and biological aspects of cosmic evolution, but also its cultural aspects. Drake and others in the field recognized then (and still do now) that this equation is best viewed

FIGURE 2.7 Carl Sagan (1934–1996), as an Assistant Professor of Astronomy at Harvard, circa 1962.

as a way of organizing our ignorance. Best known are the astronomical parameters; for example, based on the current sample, the fraction of Sun-like stars with (large) planets is estimated to be between 10 and 20% (Chapter 21). Unfortunately the other parameters are very poorly known, and become increasingly uncertain in the biological and cultural arena. The value of the lifetime of a technological civilization is often seen as dominating the uncertainties in the overall calculation. Little can be said about its value, which incorporates not only the success or failure of cultural evolution, but also the outcome of that evolution.

The interest in intelligent life beyond Earth was by no means solely an American phenomenon. With the approval of the USSR Academy of Sciences astrophysicist Iosef Shklovskii wrote, for the fifth anniversary of Sputnik, his book *Universe, Life, Mind* (1962). Carl Sagan (Fig. 2.7) first became prominent with his elaboration of this book, appearing in 1966 as *Intelligent Life in the Universe*, a book that became the bible for those drawn to the life in the Universe theme (Shklovskii and Sagan, 1966). Nor was Shklovskii's book an isolated instance of Soviet interest. As early as 1964 Soviet scientists convened meetings on extraterrestrial civilizations, funded their own observing programs, and published extensively on the subject (Tovmasyan, 1967). After the demise of the Soviet Union and the accompanying economic problems in the early 1990s, however, observational programs dwindled.

Despite a slow start, it was once again NASA that eventually spearheaded the flagship SETI program. In particular, in the late 1960s interest developed not from the exobiology program, but independently at Ames Research Center. Under the guidance of John Billingham, an MD working in space medicine, a SETI

[4] Cocconi and Morrison's paper and Drake's Project Ozma are discussed in detail in Section 1.9.

program slowly grew over the 1970s and 1980s, having been given a jump-start by Project Cyclops, a summer design study that was overly ambitious but very influential (Oliver and Billingham, 1972). Finally, in 1990 a major ten-year observational project was approved under the leadership of astronomer Jill Tarter. First observations, using an analyzer having an unprecedented 14 million frequency channels, began in October 1992, the quincentennial of Columbus's landfall in the New World (Dick, 1993). Congressional politics, however, axed funding for this project in 1993. Altogether SETI had expended some $55 million, largely in the design and construction of specialized instrumentation and software. Major remnants of the NASA SETI program nonetheless survived, under the aegis of the non-profit SETI Institute in California, founded in 1984 with Frank Drake as its President. With private funding, the SETI Institute has not only carried out detailed searches since 1994, but also secured funding for the Allen Telescope Array, an array of some 350 small radio dishes that will operate as a dedicated SETI facility (Section 26.8.3). Smaller SETI endeavors are carried out around the world by a variety of institutions (Tarter, 2001; Table 26.2). Thus this last, most far-reaching aspect of exobiology remains alive. Though no longer government-funded and not an official part of NASA's astrobiology program, there is no doubt that SETI is an intellectual component of the new discipline.

2.2.5 From exobiology to astrobiology

In the post-Viking era, exobiology languished. The two Viking spacecraft had been expensive, and although they produced a massive amount of scientific data, many viewed their somewhat-disputed conclusion that no life existed on Mars as an exobiological failure, or at the least a major disappointment. NASA's exobiology program did continue in the post-Viking era at a subsistence level, but many felt the program lacked energy, direction, and new ideas.

Such was the situation when a deep organizational restructuring at NASA in 1995 precipitated a rebirth of this field under a new name. Faced with budget cuts, Administrator Daniel Goldin ordered a review that would eliminate redundancy by realigning the roles of NASA's centers. Some viewed the expertise in life, Earth, and space sciences at the Ames Research Center as a fragmented mission, whose parts might best be parceled out to others. Ames management and scientists responded by creating a redefined exobiology

program that would provide integration and a new identity. Since the early 1960s Ames had always been NASA's center for exobiology, and in 1995 Ames argued that NASA should assign to it a newly strengthened endeavor centered on a "life in the Universe" theme. NASA agreed, termed the revamped endeavor *astrobiology*, and officially designated Ames as the lead NASA Center for astrobiology. Astrobiology involved much more than renaming a discipline; it was much more broadly defined than exobiology, and was to include research in cosmochemistry, chemical evolution, the origin and evolution of life, planetary biology and chemistry, formation of stars and planets, and expansion of terrestrial life into space. The birth of astrobiology was spurred on by the landmark events of 1995–1996, including the martian meteorite ALH 84001, the announcement of the extrasolar planet around 51 Pegasi, and the Galileo spacecraft's new evidence for an ocean beneath the ice on Europa.

NASA's strategic plan for 1996 used the term *astrobiology* for the first time anywhere in a NASA document, and focused on three key questions. Astrobiology was "the study of the living Universe" to be sure, but in particular it was seen as providing the scientific foundation for studying the origin and distribution of life in the Universe, the role of gravity in living systems, and the study of the Earth's atmosphere and ecosystems. These three programs were already in existence, but astrobiology was to go beyond those, asking questions that required the sharing of their resources and striking out in new directions as well.

Over the next few years Ames sponsored a variety of workshops to define the new area, and by 1998 produced a so-called "Astrobiology Roadmap." This document identified four major principles.

- Astrobiology is multidisciplinary, and achieving its goals will require the cooperation of different scientific disciplines and programs.
- Astrobiology encourages planetary stewardship, through an emphasis on protection against biological contamination and recognition of the ethical issues surrounding the export of terrestrial life beyond Earth.
- Astrobiology recognizes a broad societal interest in our subject, especially in areas such as the search for extraterrestrial life and the potential to engineer new life forms adapted to live on other worlds.
- In view of the intrinsic excitement and wide public interest in the subject, astrobiology includes a strong element of education and public outreach.

TABLE 2.3 NASA Astrobiology Institute members (through 2002)

Institution	Research focus
11 institutions announced May 1998	
Arizona State University	Organic synthesis
Carnegie Institution of Washington	Life in hydrothermal systems
Harvard University	Geochemistry and paleontology
Pennsylvania State University	Coevolution of Earth's biota
Scripps Research Institute	Self-replicating systems
University of California, Los Angeles	Paleomicrobiology; early ecosystems
University of Colorado	Origin/habitability of planets; RNA catalysis; philosophical aspects
Marine Biological Laboratory, Woods Hole	Microbial diversity; origins of proteins
Ames Research Center	Planet formation; Earth–biosphere interaction
Jet Propulsion Laboratory	Biosignatures of life
Johnson Space Center	Biomarkers in rocks
Four additional institutions announced March 2001	
Michigan State University	Earth analogues to life on Mars and Europa
University of Rhode Island	Extremophiles in deep biosphere
University of Washington	Earliest life on Earth; extrasolar planetary life
Jet Propulsion Laboratory	Recognizing biospheres of extrasolar planets
Associate Members (formal agreement between NASA and the host government agency)	
Centro de Astrobiologia, Torrejon de Ardoz, Spain	
Australian Centre for Astrobiology, Sydney, Australia	
Affiliate Members (no formal agreement with host government agency)	
United Kingdom Astrobiology Forum and Network, Cambridge, UK	
Grupement des Recherches en Exobiology, Paris, France	

The roadmap also laid out much more specific goals and objectives for astrobiology, grouped according to the three fundamental questions that had been enunciated early in the development of the concept.

How does life begin and evolve?
Does life exist elsewhere in the Universe?
What is life's future on Earth and beyond?

It was an ambitious research program for the next century.

An additional feature of astrobiology was the formation of an Astrobiology Institute. This idea also grew from constrained budgets, as well as Goldin's desire that NASA should leverage its contacts with the academic community for scientific research and do less in-house research. It was to be a "virtual institute," with its members geographically dispersed. Its members were not individuals, but organizations, ranging from industry, universities, and non-profit groups to NASA centers and other government agencies. The virtual institute members would be tied together by wideband

Internet connections; by personnel exchanges; by workshops, seminars and courses; and by sharing common research interests. The resulting research would complement work carried out by individuals funded by the Exobiology and Evolutionary Biology programs.

In 1998 NASA announced the selection of eleven academic and research institutions as the first members of the Astrobiology Institute, and billed it as "launching a major component of NASA's Origins Program." These winners included five universities, three research institutions, and three NASA centers (Table 2.3). These institutions – engaged in an extraordinary variety of research – shared only $4 million for the first year, but by 2002 annual funding had grown to $40 million for fifteen members. In 1999 Nobelist and virologist Baruch S. Blumberg was named head of the Institute, headquartered at Ames. Astrobiology, Goldin remarked, was "the cornerstone to NASA's mission in the new millennium. ... Quite possibly the rewards from this pursuit of Astrobiology may eclipse the societal and economic benefits of all prior NASA activity." He characterized the understanding of the origin and evolution of life to the generational effort of cathedral building a thousand years before, and hoped to bring a new level of knowledge to biology as had been done for physics during the previous fifty years.

Thus, forty years after the formation of NASA, and three years after the crisis at Ames, exobiology was revitalized under the rubric of astrobiology. The contrast between exobiology as conceived forty years ago and the new astrobiology was striking, although they both shared the core concerns of origin of life research and the search for life beyond Earth. But astrobiology placed life in the context of its planetary history, encompassing the search for planetary systems, the study of biosignatures, and the past, present and future of life. Astrobiology added new techniques and concepts to exobiology's repertoire, raised multidisciplinary work to a new level, and included study of the history of Earth's life and of present organisms. Blumberg was fond of comparing the new field to the Lewis and Clark exploration, and pointed out that it was different from most science in that instead of becoming ever more specialized, it should become increasingly generalized, making use of many specialties to tackle a very broad set of questions.

2.3 A new discipline

In the late 1950s, the study of cosmic evolution was not at all a connected research program. Planetary science, the search for planetary systems, origin of life studies, and SETI remained largely separate, undertaken by different researchers. In terms of technique, research programs, and even goals, the planetary spectroscopy of Kuiper and Sinton had little in common with van de Kamp's astrometric studies of stellar motions or Drake's radio astronomy, aside from the shared general culture of astronomy. And all three of these were far removed from the biochemists in their laboratories studying the origin of life. It took forty years for these disparate fields (and many others) to come together in the new astrobiology.

2.3.1 Declarations of discipline status

In 1955, when Struve pondered the use of the word *astrobiology* to describe the broad study of life beyond the Earth, he explicitly decided against a new discipline: "The time is probably not yet ripe to recognize such a completely new discipline within the framework of astronomy. The basic facts of the origin of life on

Earth are still vague and uncertain; and our knowledge of the physical conditions on Venus and Mars is insufficient to give us a reliable background for answering the question" of life on other worlds (Struve, 1955). But five years later the imminent birth of *exobiology was* palpable when Lederberg (1960) coined the term and set forth an ambitious agenda based on space exploration. Speaking from a different point of view, Shklovskii shortly thereafter proclaimed "we are witnessing the inception of a new science, which occupies a boundary position between astrophysics, biology, engineering and even sociology" (Shklovskii, 1965). Over the next twenty years numerous such proclamations of a new discipline were made. By 1979 Billingham wrote that "over the past twenty years, there has emerged a new direction in science, that of the study of life outside the Earth, or exobiology. Stimulated by the advent of space programs, this fledgling science has now evolved to a stage of reasonable maturity and respectability."

Although one still hears echoes of evolutionist George Gaylord Simpson's statement that exobiology "is a 'science' that has yet to demonstrate that its subject matter exists," a statement he accompanied with the complaint that "ex-biologists" were siphoning off funding for more realistic research (Simpson, 1964), it seems clear that over the last forty years astrobiology has emerged as a discipline by almost any definition of the term (also see Chapter 27).

2.3.2 Building the new discipline

While cognitive content is the most important element of any new discipline, funding and the creation of a coherent research community are also essential. We have already seen how NASA has been the major patron of exobiology. But other agencies, including the European Space Agency, are also now funding research similar to NASA's Origins and Astrobiology programs, including the Darwin spacecraft to search for planetary systems and the unsuccessful Beagle 2 lander portion of the Mars Express mission. Moreover, groups in Spain, France, the UK, and Australia are officially affiliated with NASA's Astrobiology Institute (Table 2.3), and the first "European Workshop on Exo/Astrobiology" was held in Italy in 2001, organized by the European Exobiology Network. The NASA Astrobiology Institute has also sponsored large annual conferences since 2000.

A coherent community of researchers has been built over four decades through major conferences on extraterrestrial life (Table 2.4), supported by NASA, the

TABLE 2.4 Selected conferences on extraterrestrial life and origin of life (through 2001)

Date	Sponsor location	Remarks
1961	US National Academy of Sciences	Green Bank Conference where Drake Equation originated
1963	NASA/Jet Propulsion Lab	Exobiology
1964	Armenian Academy of Sciences	First Soviet meeting on extraterrestrial civilizations
1964–65	US National Academy Of Sciences (Space Science Board)	Produced *Biology and the Exploration of Mars* (Pittendrigh *et al.*, 1966)
1970	NASA Ames	First NASA interest in SETI, resulting in *Interstellar Communication* (Ponnamperuma and Cameron, 1974)
1971	US and USSR Academy of Sciences	First of a series of US–USSR Decennial SETI Conferences (Byurakan, Armenia)
1975–76	NASA Ames	Workshops leading to SETI (Morrison *et al.*, 1977)
1979	NASA Ames	*Life in the Universe* (Billingham, 1981)
1979	International Astronomical Union	*Strategies for the Search for Life in the Universe* (Papagiannis, 1980) Gave birth to IAU Bioastronomy Commission and triennial conferences (not listed) beginning in 1984
1970s–present[a]	International Society for the Study of the Origin of Life (ISSOL)	International conference on the origin of life Triennial meetings (not listed) increasingly integrating exobiology
1991–92	NASA Ames/SETI Institute	*Social Implications of the Detection of an Extraterrestrial Civilization* (Billingham *et al.*, 1999)
1997	—	Fifth Trieste Conference on Chemical Evolution
1997	NASA	Sixth Symposium on Chemical Evolution and the Origin of Life
2000	NASA Astrobiology Institute	First astrobiology science conference
2001	European Exobiology Network/ European Space Agency	First European workshop on exo/astrobiology, indicating increasing European interest in astrobiology

Note:
[a] The ISSOL conferences and their predecessors have been the principal forum for the exchange of scientific information on problems related to the origin and evolution of life in the Universe. The series started with the Symposium on the Origin of Life on the Earth in 1957 (Moscow) and has continued up to the 1999 (San Diego, USA) meeting that was attended by more than 300 persons from 25 countries.
Source: Adapted from Dick (1996).

Academies of Science of the USA and of the USSR, by professional societies, and by the SETI Institute. Increasingly, the subject and its community of researchers were recognized through representation in established scientific societies. By the early 1980s a new Commission on Bioastronomy was formed in the prestigious International Astronomical Union; the International Society for the Study of the Origin of Life increasingly incorporated exobiology into its meetings; and a variety of other societies in a broad range of disciplines also embraced exobiology. Already in 1968 the journal *Origins of Life* (now *Origins of Life and Evolution of the Biosphere*) provided a common professional outlet for origin of life research. In 2001–02 two new journals devoted to the more general field of astrobiology began publication: *Astrobiology* (published in the US) and *International Journal of Astrobiology* (published in the UK).

Astrobiology was institutionalized in other ways as well, not least by becoming more respectable in the sense that students could enter the field without fear of being marginalized. Courses sprang up at numerous universities, and at least one (the University of Washington in Seattle) began a graduate program in astrobiology (Chapter 28). NASA offered postdoctoral positions, began an Astrobiology Academy for summer students, and instituted a far-reaching public outreach program in astrobiology. Public and Congressional support for the Origins program, which incorporates "life in the Universe" as its central theme, has remained high. Astrobiology is entering the twenty-first century as a thriving enterprise, providing the framework for an expansive research program and drawing in young talent to ensure its future. Full-fledged university departments of Astrobiology are a strong possibility for the twenty-first century.

2.4 The broader significance of astrobiology

In the public mind, more than in the scientific discipline, astrobiology is dominated by extraterrestrial intelligence. Humans seem to have been "captured by aliens" in the sense that belief in extraterrestrials is common, despite lack of proof (Achenbach, 1999). At the turn of the millennium polls showed that 54% of Americans believe extraterrestrial intelligence exists, 30% believe the aliens have visited Earth, and 43% believe UFOs are real. One percent (2.8 million people!) claim to have encountered an alien, but only one in five would board an alien spacecraft if invited (Peebles, 1994; Fox, 2000). Such interest is expressed in popular culture in a variety of ways, notably in an increased interest in alien science fiction. For more than a century science fiction has explored a full range of alien scenarios with greater or lesser degrees of intelligence (Fig. 2.8). H. G. Wells's *The War of the Worlds* expressed one stunning possible outcome of alien encounters (Section 1.7.2). The works of Olaf Stapledon and Arthur C. Clarke, including *Star Maker*, *Last and First Men*, *Childhood's End*, and *2001: a Space Odyssey* played out the opposite, positive outcome. The Polish science fiction author Stanislaw Lem represents yet a third choice: in *Solaris* and *His Master's Voice* he argued that we may be unable to comprehend, much less communicate with, extraterrestrials. By the late twentieth century these themes had been elaborated in ever more subtle (and sometimes not so subtle) form, and were brought to a much broader public via film. Aliens figure in some

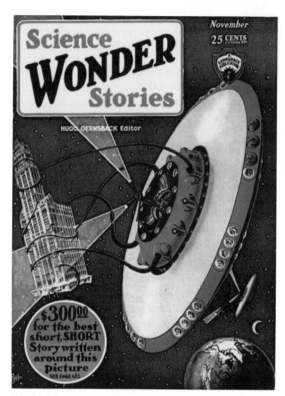

FIGURE 2.8 Extraterrestrial intelligent creatures have always dominated popular culture, but not astrobiology's scientific research program. This cover from Hugo Gernsback's *Science Wonder Stories* for November 1929 depicts powerful aliens in a saucerlike spacecraft. Copyright 1929 by Gernsback Publications.

of the most popular films of all time: *Star Wars*, *Close Encounters of the Third Kind*, *Contact*, *Independence Day*, *Men in Black*. It is no exaggeration to say that in the modern world extraterrestrials pervade popular culture; some analysts have even called extraterrestrials our modern mythology or religion (Guthke, 1990).

Beyond popular culture, astrobiology – considered in its most robust sense as the part of cosmic evolution encompassing everything from the formation of planets to the origin, evolution and future of life – raises deep questions of societal impact at many levels. NASA itself has sponsored a number of studies identifying the impact of astrobiological ideas on society. One of the earliest, a free-ranging discussion involving astronomers Sagan and Richard Berenzden, physicist Philip Morrison, Harvard biochemist George Wald, and theologian Krister Stendahl, was more notable for its diversity of opinion than for any consensus (Berenzden, 1973). In conjunction with the launching of the NASA SETI program, in 1991–1992 NASA sponsored a series of workshops on the cultural aspects of SETI (Billingham *et al.*, 1999). And shortly after the

Astrobiology Institute was launched, the study was broadened to include those concerns (Harrison *et al.*, 2002).

Dick (2000b) and Harrison *et al.* (2002) have surveyed some of the societal impacts of astrobiology in the context of the three main areas of NASA's astrobiology roadmap. The first area, the study of the origin and evolution of life on Earth, illuminates our place in nature and raises fundamental questions: "What is life?"; "Is there a cosmic imperative for life imbedded in the laws of Nature?"; and "What is the role of chance and necessity in the origin and evolution of life?" Scientists involved in this research have already generated considerable discussion on these questions. In his classic book *Chance and Necessity*, Jacques Monod (1971), who received the Nobel Prize for his work on the genetic regulation of protein synthesis, concluded that life at the molecular level was "the product of an enormous lottery presided over by natural selection, blindly picking the rare winners from among numbers drawn at utter random." Life was not inevitable and had no destiny, he argued. "The universe was not pregnant with life, nor the biosphere with man. Our number came up in the Monte Carlo game." Others disagreed. Christian de Duve, equally familiar with the intricate machinery of life from his Nobel Prize-winning work on the structure and functions of the cell, viewed life as a cosmic imperative, with trillions of biospheres throughout the Universe. These biospheres, including the Earth, are part of a cosmic cloud of "vital dust" that exists because "the universe is what it is. Avoiding any mention of design, we may, in a purely factual sense, state that the Universe is constructed in such a way that this multitude of life-bearing planets was bound to arise" (De Duve, 1995: 292). Answers to this and other philosophical questions will be strongly influenced by which theory for the origin of life holds sway at any time (Davies, 2000). Life arising from panspermia carries notably different implications than if it arose on Earth, whether in Darwin's warm pond, in hydrothermal vents, or in a hot deep biosphere.

The second area in the astrobiology roadmap raises questions regarding the implications of the actual discovery of extraterrestrial life. Although the cultural impact of discovering primitive life has received little scholarly attention, the expected short-term reaction to contact with intelligence beyond Earth has been discussed in detail (Tarter and Michaud, 1990), and policy issues regarding a response to an extraterrestrial communication remain under consideration (Michaud, 1998). Different approaches to the long-term problems of contact have been explored in a variety of studies. Billingham *et al.* (1999) reported the results of a NASA-sponsored workshop in which experts discussed implications of contact from the point of view of science, history, human behavior, political policy, education, and the media. Dick (1995) argued that historical analogues form a useful basis for discussion, not in the form of the usually disastrous physical and cultural contacts on Earth, but by studying the transmission of knowledge across cultures and the reception of scientific world views. In his book *After Contact* psychologist Albert Harrison (1997) has pioneered another approach to extraterrestrial contact by applying living systems theory. This approach relies on a systems theory in which Earth's organisms, societies, and supranational systems are discussed as analogues to aliens, alien civilizations and the "Galactic Club." It offers promise of bringing the social sciences into SETI in a substantive way. Finally, the Foundation for the Future, which focuses on the question of the state of humanity a thousand years hence, sponsored a meeting on the implications of deciphering a message with high information content (Tough, 2000).

Philosophical, theological and moral questions of discovering extraterrestrial intelligence also will be of *practical* importance in the event of contact. The problem of objective knowledge, or "extraterrestrial epistemology," has been broached by philosopher of science Nicholas Rescher and artificial intelligence pioneer Marvin Minsky. Rescher (1985) argues that extraterrestrial science and ways of knowing about the Universe will be distinctly different from ours. Minsky (1985), on the other hand, argues that any extraterrestrials will think like us, despite their different origins, because intelligent problem solvers must evolve information processing brains to respond to the same environmental constraints. Ruse (1985) has tackled some of the ethical considerations, including whether there is a "deep structure" to morality in the way that Chomsky found a deep structure to linguistics. Lacking such a deep structure, "Is rape wrong on Andromeda?", as Ruse provocatively puts it. Is there an absolute "right" and "wrong"? How should we behave toward extraterrestrials, beyond the glib answers given in Hollywood movies? Theological issues have a long history in the extraterrestrial life debate (Crowe, 1986; Dick, 1996, 1998; Vakoch, 2000; Chapter 1). They run the gamut from doctrinal issues in particular religions to scriptural and more general concerns. In Christianity, for example, the relevance of the Incarnation and Redemption of Christ to extraterrestrials has been an issue long discussed. Dick (2000a) presents a wide variety of opinions from scientists, philosophers and theologians.

The future of life on Earth and beyond, the third area of the NASA astrobiology roadmap, has implications best known today in terms of planetary protection, and the problems of forward and back contamination (Chapters 24 and 25). These have been given prominent attention because the problems are immediate and the potential implications catastrophic. The ethical questions of planetary protection, however, have only begun to be explored. Other cultural issues are raised when we consider moving beyond Earth and sustaining either artificial life or "normal" life in bioengineered ecosystems, in its grandest vision involving *terraforming* an entire planet. In the twenty-first century the issue of terraforming Mars will likely become real. As McKay (2000) points out, we may soon be faced with extending the principles of environmental ethics to Mars. Movement off planet Earth (Finney and Jones, 1984) also raises an entire spectrum of social issues associated with space exploration, in terms of manned or unmanned exploration, space colonization, and societal spending priorities. Considering the future of life on Earth and beyond raises the daunting issue of where our species wants to go in its cultural evolution.

Finally, both scientists and historians of science have treated the idea of a Universe full of life as a kind of world view, similar in status to the Copernican and Darwinian world views. Oliver and Billingham (1972) termed it a *biocosmology*[5] and Dick (1989, 1996, 1998) has argued that this *biological Universe* could be viewed as an emerging cosmology. In elaborating his views of cosmic evolution Shapley (1958) saw the idea of extraterrestrial life as the "Fourth Adjustment" in humanity's view of itself since the time of ancient Greece. Struve (1962) agreed, seeing extraterrestrial life as one of astronomy's three great revolutions.

Astrobiology can be seen as a world view, as one of the landmark questions of human thought, or as an essential element in the question of humanity's place in nature. But no matter how it is viewed, astrobiology undoubtedly has the potential to impact society no less than the other great revolutions of science.

---

[5] The term *biocosmology* has not caught on, nor the related *cosmobiology*, first used in 1952 in a talk to the British Interplanetary Society by the evolutionary biologist J. D. Bernal.

REFERENCES

Achenbach, Joel. (1999). *Captured by Aliens: the Search for Life and Truth in a Very Large Universe*. New York: Simon and Schuster.

Aumann H. H., Beichman, C. A., Gillet, F. C. *et al.* (1984). Discovery of a shell around alpha lyrae, *Astrophys. J.*, **278**, L23–L27.

Barabashev, N. P. (1952). *A Study of the Physical Conditions of the Moon and Planets*. Kharkov: (publisher unknown).

Berenzden, R. (1973). *Life Beyond Earth and the Mind of Man*. Washington: NASA.

Billingham, J. (ed.) (1981). *Life in the Universe*. Cambridge, MA: MIT Press.

Billingham, J., Heyns, R., Milne, D. *et al.* (1999). *Social Implications of the Detection of an Extraterrestrial Civilization*. Mountain View, CA: SETI Press.

Cameron, A. G. W. (ed.) (1963). *Interstellar Communication: a Collection of Reprints and Original Contributions*. New York: W. A. Benjamin, Inc.

Chaisson, E. (2001). *Cosmic Evolution: the Rise of Complexity in Nature*. Cambridge, MA: Harvard University Press.

Cocconi, G. and Morrison, P. (1959). Searching for interstellar communications. *Nature*, **184**, 844.

Crick, F. H. and Orgel, L. E. (1973). Directed panspermia. *Icarus*, **19**, 341–346.

Crowe, M. J. (1986). *The Extraterrestrial Life Debate, 1750–1900*. Cambridge: Cambridge University Press.

Davies, P. (2000). Biological determinism, information theory and the origin of life, in Dick (2000a).

De Duve, Christian (1995). *Vital Dust: Life as a Cosmic Imperative*. New York: Basic Books.

Delsemme, A. (1998). *Our Cosmic Origins: from the Big Bang to the Emergence of Life and Intelligence*. Cambridge: Cambridge University Press.

DeVaucouleurs, G. H. (1954). *Physics of the Planet Mars: an Introduction to Areophysics*. London: Faber and Faber.

Dick, S. (1982). *Plurality of Worlds: the Origins of the Extraterrestrial Life Debate from Democritus to Kant*. Cambridge: Cambridge University Press.

Dick, S. (1989). The concept of extraterrestrial intelligence – an emerging cosmology. *Planetary Report*, **9**, 13–17.

Dick, S. (1993). The search for extraterrestrial intelligence and the NASA High Resolution Microwave Survey (HRMS): historical perspectives. *Space Sci. Rev.*, **64**, 93–139.

Dick, S. (1995). Consequences of success in SETI: lessons from the history of science, in *Progress in the Search for Extraterrestrial Life*, ed. G. Seth Shostak. San Francisco: Astronomical Society of the Pacific, 521–532.

Dick, S. (1996). *The Biological Universe: the Twentieth Century Extraterrestrial Life Debate and the Limits of Science*. Cambridge: Cambridge University Press.

Dick, S. (1998). *Life on Other Worlds*. Cambridge: Cambridge University Press.

Dick, S. (ed.) (2000a). *Many Worlds: the New Universe, Extraterrestrial Life and the Theological Implications*. Philadelphia: Templeton Press.

Dick, S. (2000b). Cultural aspects of astrobiology: a preliminary reconnaissance at the turn of the millennium. In *Bioastronomy '99: a New Era*, eds. G. A. Lemarchand and K. J. Meech, San Francisco: Astronomical Society of the Pacific.

Dick, S. J. and Strick, J. R. (2004). *The Living Universe: NASA and the Development of Astrobiology*. Piscataway, NJ: Rutgers University Press.

Doel, R. E. (1996). *Solar System Astronomy in America: Communities, Patronage, and Interdisciplinary Science, 1920–1960*. Cambridge: Cambridge University Press.

Ezell, E. C. and Ezell, L. N. (1984). *On Mars: Exploration of the Red Planet, 1958–1978*. Washington, DC: NASA/Government Printing Office.

Finney, B. and Jones, E. (1984). *Interstellar Migration and the Human Experience*. Berkeley: University of California Press.

Fox, Cynthia (2000). The search for extraterrestrial life. *Life Magazine* (March, 2000), 46–56.

Gehrels, T. (1978). *Protostars and Planets: Studies of Star Formation and of the Origin of the Solar System*. Tucson: University of Arizona Press, preface, xviii.

Guthke, K. S. (1990). *The Last Frontier*. Ithaca, NY: Cornell University Press.

Hale, G. E. (1908). *The Study of Stellar Evolution*. Chicago: The University of Chicago Press.

Harrison, A. A. (1997). *After Contact: the Human Response to Extraterrestrial Life*. New York: Plenum.

Harrison, A. A, Connell, K., and Schmidt, G. (2002). *Workshop on the Societal Implications of Astrobiology* (held at NASA Ames Research Center, November 16–17, 1999). Moffet Field: NASA Ames Research Center. Executive Summary at astrobiology.arc.nasa.gov/workshops/societal/.

Henderson, L. J. (1913). *The Fitness of the Environment*. Cambridge, MA: Harvard University Press, reprinted with an Introduction by Harvard biologist George Wald, Gloucester, MA, 1970, 312.

Horowitz, N. H. (1986). *To Utopia and Back: the Search for Life in the Solar System*. New York: W. H. Freeman.

Klein, H. P. (1977). The Viking biological investigation: general aspects. *J. Geophys. Res.*, **82**, 4677–4680.

Kuiper, G. P. (ed.) (1949). The atmospheres of the Earth and planets: papers presented at the fiftieth anniversary symposium of the Yerkes Observatory, September, 1947. Chicago: University of Chicago Press.

Kuiper, G. P. (1955). On the martian surface features. *Publ. Astron. Soc. Pacific*, **67**, 271–282.

Lederberg, J. (1960). Exobiology: approaches to life beyond the Earth. *Science*, **132**, 393–400.

Lowell, P. (1909). *The Evolution of Worlds*. New York: The Macmillan Company.

Mayor, M. and Queloz, D. (1995). A Jupiter-mass companion to a solar-type star. *Nature*, **378**, 355.

McKay, C. (2000). Astrobiology: the search for life beyond the Earth. In S. Dick, *Many Worlds: The New Universe, Extraterrestrial Life and the Theological Implications*. Philadelphia: Templeton Press, 45–58.

McKay, D., Gibson, E. K., Jr., Thomas-Keprta, K. L., *et al.* (1996). Search for past life on Mars: possible relic biogenic activity in martian meteorite ALH84001. *Science*, **273**, 924–930.

Michaud, M. A. G. (1998). Policy issues in communicating with ETI. *Space Policy*, **14**, 173–178.

Miller, S. and Urey, H. C. (1959). Organic compound synthesis on the primitive Earth. *Science*, **130**, 245–251.

Minsky, M. (1985). Why intelligent aliens will be intelligible. In Regis (1985), pp. 117–128.

Monod, J. (1971). *Chance and Necessity*. New York: Knopf.

Morrison, P. M., Billingham, J. and Wolfe, J. (1977). *The Search for Extraterrestrial Intelligence*. Washington, DC: NASA (SP-419).

NASA, origins subcommittee of the space science Advisory Committee, (1997). *Origins: Roadmap for the office of Space Science Origins Theme.*

Oliver, B. M. and Billingham, J. (1972). *Project Cyclops: a Design Study of a System for Detecting Extraterrestrial Intelligence*. Washington, DC: NASA.

Oparin, A. I. (1975). Theoretical and experimental prerequisites of exobiology. In M. Calvin and O. Gazenko (eds.), *Foundations of Space Biology and Medicine*, vol. **1**, 321–367.

Oparin, A. I. and Fesenkov, V. G. (1961). *Life in the Universe*. (English trans. of 1956 Russian original.) New York: Twayne Publishers.

Papagiannis, M. (ed.) (1980). *Strategies for the Search for Life in the Universe*. Dordrecht: Reidel.

Peebles, C. (1994). *Watch the Skies!* Washington and London: Smithsonian Institution Press.

Pittendrigh, C. S., Vishniac, W., and Pearman, J. P. T. (eds.) (1966). *Biology and the Exploration of Mars*. Washington: National Research Council.

Ponnamperuma, C. and Cameron, A. G. W. (eds.) (1974). *Interstellar Communication: Scientific Perspectives*. Boston: Houghton Mifflin.

Rea, D. G., O'Leary, B. T., and Sinton, W. M. (1965). The origin of the 3.58 and 3.69-micron minima in the infrared spectrum. *Science*, **147**, 1286–1288.

Reeves, H. (1984). *Atoms of Silence: an Exploration of Cosmic Evolution*. Cambridge, MA: MIT Press. Translated from *Patience dans l'azur: l'évolution cosmique*. Paris: Editions du Seuil.

Regis, E. Jr. (ed.) (1985). *Extraterrestrials: Science and Alien Intelligence*. Cambridge: Cambridge University Press, 117–128.

Rescher, N. (1985). Extraterrestrial science. In Regis (1985), 83–116.

Ruse, M. (1985). Is rape wrong on Andromeda? An introduction to extraterrestrial evolution, science and morality. In Regis (1985), 43–78.

Russell, H. N. (1943). Anthropocentrism's demise. *Sci. Am.* (July, 1943), 18–19.

Russell, H.N., Dugan, R.S., and Stewart, J.Q. (1945). *Astronomy: a Revision of Young's Manual of Astronomy*. Boston and New York: Ginn and Company.

Sagan, C. (1980). *Cosmos*. New York: Random House.

Scargle, J. (1988). Planetary detection techniques: an overview. In G. Marx (ed.) *Bioastronomy – the Next Steps*. Dordrecht: Kluwer.

Schaffer, S. (1989). The nebular hypothesis and the science of progress. In *History, Humanity and Evolution: Essays for John C. Greene*, J.R. Moore (ed.). Cambridge: Cambridge University Press, 131–164.

Schorn, R.A. (1998). *Planetary Astronomy: from Ancient Times to the Third Millennium*. College Station: Texas A&M University Press.

Shapley, H. (1958). *Of Stars and Men: Human Response to an Expanding Universe*. Boston: Beacon Press.

Shklovskii, I.S. (1962). *Vselennaia, Zhizn, Razum* (*Universe, Life, Mind*). Moscow (Publisher not known).

Shklovskii, I.S. (1965). Multiplicity of inhabited worlds and the problem of interstellar communications. In *Extraterrestrial Civilizations*, ed. G.V. Tovmasyan. Erevan: Akademii Nauk Armyanskoi SSR. English translation, Jerusalem: Israel Program for Scientific Translations, 1967, 5.

Shklovskii, I.S. and Sagan, C. (1966). *Intelligent Life in the Universe*. San Francisco: Holden-Day.

Simpson, George Gaylord (1964). The non-prevalence of humanoids. *Science*, **143**, 769–775.

Sinton, W.M. (1957). Spectroscopic evidence for vegetation on Mars. *Astrophys. J.*, **126**, 231–239.

Sinton, W.M. (1959). Further evidence of vegetation on Mars. *Science*, **130**, 1234–1237.

Slipher, E.C. (1927). Atmospheric and surface phenomena on Mars. *Publ. Astron. Soc. Pacific*, **39**, 209–216.

Space Science Board, National Academy of Sciences (1962). *A Review of Space Research: the Report of the Summer Study Conducted under the Auspices of the Space Science Board of the National Academy of Sciences at the State University of Iowa, June 17–August 10, 1962*, 9–2, 9–3.

Spencer Jones, Sir H. (1940). *Life on Other Worlds*. New York: MacMillion, 57.

Strand, K. (1943). 61 Cygni as a triple system. *Publ. Astron. Soc. Pacific*, **55**, 29–32.

Strauss, D. (2001). *Percival Lowell: the Culture and Science of a Boston Brahmin*. Cambridge, MA: Harvard University Press.

Strughold, H. (1959). Advances in astrobiology. *Proceedings of the Lunar and Planetary Exploration Colloquium*, **1**, no. 6, 1–7.

Struve, O. (1955). Life on other worlds, *Sky and Telescope*, **14**, 137–146.

Struve, O. (1962). *The Universe*. Cambridge, MA: MIT Press.

Tarter, J. (2001). The Search for Extraterrestrial Intelligence (SETI). *Ann. Rev. Astron. Astrophy.*, **39**, 511–548.

Tarter, J.C. and Michaud, M.A.G. (eds.) (1990). SETI post-detection protocol. *Acta Astronautica*, **21**, 69–154.

Tatarewicz, J.N. (1990). *Space Technology and Planetary Astronomy*. Bloomington: Indiana University Press.

Tough, A. (ed.) (2000). *When SETI Succeeds: the Impact of High-Information Contact*. Bellevue, Washington: Foundation for the Future.

Tovmasyan, G.M. (ed.) (1967). *Extraterrestrial Civilizations*. Jerusalem: Israel Program for Scientific Translations.

Vakoch, D. (2000). Roman Catholic views of extraterrestrial intelligence: anticipating the future by examining the past. In Tough (2000).

Van de Kamp, P. (1963). Astrometric study of Barnard's star from plates taken with the 24-inch Sproul refractor. *Astron. J.*, **68**, 515–521.

Wald, G. (1954). The origin of life. *Scientific American* (August, 1954).

Wallace, A.R. (1903). *Man's Place in the Universe*. New York: McClure, Phillips and Company.

Wilson, A.G. (ed.) (1958). Problems common to the fields of astronomy and biology. *Publ. Astron. Soc. Pacific*, **70**, 41–78.

Wolszczan, A. and Frail, D.A. (1992). A planetary system around the millisecond pulsar PSR 1257+ 12. *Nature*, **355**, 145–147.

FURTHER READING AND SURFING

Achenbach, Joel (1999). *Captured by Aliens: the Search for Life and Truth in a Very Large Universe*. New York: Simon and Schuster. A readable account of extraterrestrials in popular culture today.

Chaisson, E. (2001). *Cosmic Evolution: the Rise of Complexity in Nature*. Cambridge, MA: Harvard University Press. A coherent and up-to-date theory of cosmic evolution, from the Big Bang to cultural evolution.

Darling, David (2001). *Life Everywhere: the Maverick Science of Astrobiology*. New York: Basic Books. A popular treatment of developments in astrobiology.

Deamer, D. W. and Fleishaker, G. (1994). *Origins of Life: the Central Concepts*. Boston: Jones and Bartlett. The standard reader for pioneering origins of life papers.

Dick, S. (1996). *The Biological Universe: the Twentieth Century Extraterrestrial Life Debate and the Limits of Science*. Cambridge: Cambridge University Press. The standard history, complete with numerous references.

Dick, S. (1998). *Life on Other Worlds*. Cambridge: Cambridge University Press. An abridgement and update of *The Biological Universe*.

Dick, S. J. and Strick, J. R. (2004). *The Living Universe: NASA and the Development of Astrobiology*, Piscataway, NJ: Rutgers University Press. Much more detail about this chapter's topic.

Goldsmith, D. (1980). *The Quest for Extraterrestrial Life: a Book of Readings*. Mill Valley, CA: University Science Books. An excellent collection of readings from ancient times until the 1970s.

Sleep, N. (2005). pangea.stanford.edu/courses/gp205/webbook.html. An engaging and insightful short book about astrobiology today and in the past; used for a course at Stanford University.

Part II
The origin of Earth-like planets and atmospheres

3 Formation of Earth-like habitable planets

Donald E. Brownlee, *University of Washington*
Monika E. Kress, *San Jose State University*

3.1 Introduction

Habitable planets are those bodies that provide environments, materials and processes that are advantageous for the formation and long-term evolution of life. Understanding the processes that lead to the formation of such planets is a central issue in astrobiology. We are of course handicapped in this quest since Earth is the only example of a planet with proven habitability – the only one known to have provided thermal, chemical, and other physical conditions that allowed life to form and survive for ~3.5 Gyr.

This chapter emphasizes the formation of *Earth-like* planets, those with environments capable of supporting complex life comparable to Earth's plants and animals. The focus on life comparable to Earth's multicellular organisms is partly due to the practical consideration that we better understand the environmental constraints of such life. Despite this restricted focus, note that most astrobiologists consider that the dominant form of life in the Universe, as it has been on Earth over most of its history, is probably far simpler and more rugged, analogous to Earth's bacteria and archaea[1] (Section 3.2).

This interpretation of habitability is highly Earth-centric and assumes that life elsewhere is similar to terrestrial life and requires environments similar to those of terrestrial organisms. The actual cosmic limits of life are of course unknown, but the Earth-centric view is a reasonable, albeit conservative, place to start. Until there are detailed data on other inhabited planets, discussions of extraterrestrial life would be prudently biased by what is known from our Earth experience. The Earth-centric view of life provides experience-based ground rules, but we know that at the same time it suffers from myopia because life in the Universe, with its varied environmental conditions and evolutionary histories, is surely more diverse than our earthly example.

Note that when we bring in the factor of time, even Earth does not meet the minimal needs of animal habitability over much of its lifetime. Life began 3.5–4.0 Ga but animals appeared only ~0.6 Ga. And also considering the future, because of the Sun's increasing intensity (see below), animals will likely exist for only ~10% of the full 12 Gyr expected lifetime of the Earth and Sun (Ward and Brownlee, 2003).

In order for Earth-like life to exist and evolve to complexity analogous to animals and plants, we assume that three minimum conditions must be met: (1) a solid planet must be present in a "habitable zone" (Section 3.3) around a central star, (2) the planet must provide the essential materials for life (liquid water and carbon compounds), and (3) the planet must provide suitable environments for the origin and long-term support of life. In this chapter we examine the properties of Earth's formation that played important roles in satisfying these conditions and generalize these properties to the formation and maintenance of other Earth-like habitable worlds. First, however, we briefly examine the possibility of microbial life on non-Earth-like bodies.

3.2 Habitable bodies for microbial life

Small bodies and non-Earth-like planets seem to be unlikely hosts for evolved life analogous to Earth's multicellular organisms, but they conceivably could support subsurface, Earth-like microbial life. Although this chapter focuses on planets that could host complex surface or near-surface life, it is important to remember that short-lived environments suitable for rugged microbial communities might exist

---

[1] Bacteria and Archaea are single-celled microorganisms comprising two of the three domains of all Earth life (along with the Eukarya).

Planets and Life: The Emerging Science of Astrobiology, eds. Woodruff T. Sullivan, III and John A. Baross. Published by Cambridge University Press. © Cambridge University Press 2007.

almost anywhere in planetary systems. Various Earth microbes can live in a wide range of environments with "extreme" ranges of temperature, pH, salinity, pressure, and other factors that far exceed the limits of multicellular life (Chapter 14). They can also remain dormant for extended periods of time, disperse over long distance, and resist global extinction. While it is relatively safe to predict that the Earth is presently the only solar system body that supports plants and animals, a similar statement cannot be confidently made for microbial life.

Surface water is likely to be a critical requirement for complex life but it is possibly not needed for certain microbial communities such as the archaeans discovered kilometers deep in the Columbia River basalt of the Northwest USA (Stevens and McKinley, 1995). These are purported (but not yet conclusively proven) to live in isolation from the surface with metabolic energy derived from hydrogen produced by the oxidation of ferrous iron in basalt. This interpretation is controversial but, if correct, analogous life could now exist or have existed in the past in a wide range of bodies that do not normally have surface water, but do have damp subsurface regions. Over the full history of the Solar System, warm wet environments potentially suitable for microbial organisms have existed, at least for limited periods of time, inside a wide range of Solar System bodies. These include planets like Mars (Chapter 18) and Pluto, many of the Solar System's moons (for example, Europa (Chapter 19) and Titan (Chapter 20)), and even asteroids and some comets.

So-called *primitive meteorites* (unchanged since the earliest stages of the Solar System) contain *hydrated* minerals (those containing bound water) and other evidence that they were at one time warm and wet. These rocks originally resided inside asteroid *parent bodies*[2] (Section 3.9.2) and it is likely that these asteroidal interiors were once suitable for the long-term survival of microorganisms, at least in the sense that they provided warm damp environments. The asteroidal wet environments lasted for millions of years inside bodies of diameter less than 1,000 km, but only in the early Solar System. They were probably warmed by an internal heat source such as the radioactive decay of

short-lived isotopes such as ^{60}Fe and ^{26}Al that existed in the early Solar System. To date, however, there is no compelling evidence that life ever existed in any asteroid.

3.3 Habitable zones for Earth-like planets

A planet that can have liquid water on its surface, and therefore might harbor either simple or complex life, is considered to reside within a star's *habitable zone* (HZ), the range of distances from a star that allows temperatures and other conditions suitable for life. The intensity of a star's radiation that heats planetary surfaces declines as the inverse square of the distance between the star and a planet. Too far from the star and surface water will freeze down to great depths, as appears to be the case today for much of Mars, whose orbital distance of 1.5 AU means that it receives only 43% as much solar intensity as does Earth.[3] Too close and water-driven "greenhouse effects" (see below) will cause surface water to be lost to space and/or the surface to be heated far above the survival limits of known organisms. The intensity of sunlight at Venus's orbit of 0.7 AU is twice that at Earth, and its high surface temperature was therefore too high to retain water. In the present Solar System the inner edge of the HZ is at ~0.9 AU, while the outer edge is less well determined – estimates range from inside the orbit of Mars (1.5 AU) to beyond (further discussion is below and in Section 4.2). Besides radiation intensity, many other factors also influence habitability in ways that are not well understood, including planet size and the abundances of water and elements that are essential for life (particularly carbon, hydrogen, and nitrogen).

The HZ concept is fundamentally important to astrobiology although its application is open to interpretation because a large number of atmospheric and planetary processes complicate the issue. For example, what if Mars were moved to Earth's orbit and clearly into the HZ? Would Mars – with only 11% of the Earth's mass, a hundred times less volcanic activity, and no plate tectonics – become sufficiently Earth-like to support complex life? Would it develop a thicker atmosphere and enough greenhouse warming to

---

[2] Most meteorites are believed to be fragments of asteroids, *parent bodies* that formed in the region between Mars and Jupiter and grew to only a few hundred km in size. Many of these originally contained small amounts of ice. During the early history of the Solar System, all of the meteorite parent bodies were heated to varying degrees, some to the melting point of ice and others to that of rock.

[3] The *astronomical unit* (AU) is the mean distance between the Earth and Sun and is a handy and widely used "meterstick" for planetary distance scales; note, however, that the Earth's orbit is slightly non-circular, and the Earth–Sun distance therefore varies by ±1.7%. Appendix B contains a table of key properties of all the planets.

provide above-freezing temperatures? Would it have oceans and land interacting in ways that regulate surface temperature and preserve adequate levels of CO_2? Might entirely other processes play important roles in influencing its surface environments? Issues like these mean that location in the HZ is not at all a guarantee that a planet will be adequately habitable for the long term survival and evolution of life; nor does location *outside* the HZ always mean that life is impossible.

In an influential early study, Hart (1979) calculated that the Sun's present HZ was extraordinarily narrow: planets interior to 0.958 AU would experience runaway greenhouse warming and those beyond 1.004 AU would experience runaway glaciation. Hart's estimate of an HZ width of <5% of Earth's orbital radius was remarkable at the time and also highlighted the "faint young Sun paradox": how did Earth manage not to freeze over 3–4 Ga? This might be expected because over their main sequence lifetimes Sun-like stars brighten at a rate of ~10% per Gyr (Section 4.2 and Fig. 4.4), effectively causing a star's HZ to migrate outwards with time (see below). A changing, thin HZ thus "hosts" a planet (which usually remains in a relatively fixed orbit) for only a short time.

Hart's estimates, however, did not take into account Earth processes such as thermostatic regulation by the carbonate-silicate weathering cycle (Walker *et al.*, 1981; Kasting, 1988) and other complicating factors such as the likely presence of CH_4 in the early atmosphere (Section 4.2.3). With further study, estimates of the present outer bound of the HZ moved outwards because it was realized that a lower-temperature planet also has less removal of atmospheric CO_2 by silicate weathering, thus allowing volcanically liberated greenhouse gases to accumulate and produce increased warming (Section 4.2.1). Kasting (1988) showed that the present inner edge of the HZ is at 0.95 AU, a value similar to Hart's, but a limit caused not by runaway greenhouse warming, but rather by the onset of a moist greenhouse effect, a less severe means for a planet to lose its oceans to space.

The *moist greenhouse effect* works as follows. At 0.95 AU the Sun's intensity is only 10% higher than presently seen by Earth, but this small change leads to substantially more water vapor in the stratosphere of a planet. From the stratosphere, water molecules eventually reach high enough altitudes that they are dissociated by solar ultraviolet photons into hydrogen and oxygen atoms. The hydrogen, with its lower mass and higher thermal speed, ultimately

escapes into space, while the oxygen, sixteen times heavier, remains in the atmosphere.[4] Kasting's modeling indicates that this slightly higher solar intensity leads to orders of magnitude higher concentration of water vapor at high altitudes, where it is lost to space. Once started, the moist greenhouse process can completely deplete a planet's surface water, as well as inject huge amounts of oxygen into the atmosphere. This process may have happened on Venus, and, because of the Sun's steadily increasing output of radiation, will also occur in Earth's future. We expect that such ocean loss is a major and inevitable milestone in the evolutionary history of any HZ planet, and leads to major changes in climate, plate tectonics, and habitability (Ward and Brownlee, 2003).

The present *outer* limit of the HZ is much more difficult to estimate. Kasting *et al.* (1993) estimated 1.37 AU based on the point where CO_2 begins to condense in the atmosphere; or 1.67 AU based on the maximum greenhouse effect. Complicating effects include the influence of clouds, particularly the effects of condensed CO_2 ice (Forget and Pierrehumbert, 1997; Mischna *et al.*, 2000). Clouds can both cool a planet by reflecting sunlight back into space, and, particularly in the case of CO_2 clouds, warm a planet by retarding the loss of outgoing infrared radiation.

The width of an HZ influences both the probability of finding a planet within the zone and the time span that the planet resides in the zone as the star's radiation output changes. Planets in our Solar System are logarithmically spaced, with each planet very roughly 70% farther from the Sun than its inner neighbor. This regular spacing is likely a common consequence of planet formation since today's planets are the ultimate survivors of their formation processes, i.e., they represent the bodies that were *not* ejected out of the planetary system or assimilated into other planets or the Sun (Section 3.9). Wide spacing of planets is also necessary for long-term orbital stability. Planets cannot orbit too close to each other for long before mutual gravitational effects perturb their orbits into unstable configurations. We would thus expect that any planets in or near the HZ would commonly have well-spaced orbits. If the HZ of a star is broad enough, a planetary system

---

[4] Although this effect occurs on Earth today, the loss rate is trivial, only a few meters of surface water per Gyr. The present loss rate is low because cold temperatures at the top of the troposphere cause water vapor to condense out before reaching the stratosphere, thus leaving higher altitudes dry (a so-called "cold trap").

might have several planets within it, but if it is too narrow, then it is possible that no planets would fall within the zone.

Planets are complex entities with many interacting and not well understood processes. The ranges (even among Earth-like planets) of variations in properties such as planet mass, carbon, and water content, and plate tectonics are unknown. Franck *et al.* (1999, 2001a) have attempted to extend our understanding of the HZ by what they call an Earth systems approach. Using a model that includes life, carbonate-silicate cycling, and plate tectonics, they estimate the limits of the HZ over time for solar-like stars. In their models, the outer edge of the HZ is determined when CO_2 on an Earth-like planet declines to 10 ppm, the minimum level for photosynthesis by terrestrial plants (the present CO_2 level on Earth is 375 ppm). Their dynamic model implies that the outer edge of the HZ varies with time in a manner that is remarkably different from earlier estimates that showed both the inner and outer edges of the HZ simply moving outward as the Sun brightens with age. Their model instead shows that, as the inner edge moves outwards, the HZ outer edge moves *inward* with time. This results from increased removal of atmospheric CO_2 by increased silicate weathering rates as the star brightens. As shown

in Fig. 3.1, for a one solar mass star[5] the outer HZ edge starts at >1.8 AU and moves to less than 1.0 AU after 6 Gyr. Considering that the Sun and Solar System are now 4.6 Gyr old (Section 3.9.2), Earth on this model has 1.4 more Gyr before habitability ends (*not* the ~5 Gyr that is usually claimed, based on the remaining main sequence lifetime of the Sun (see below)). For the case of an Earth orbiting a star like the Sun, note that the *width* of the HZ remains at ~0.8 AU over a period of 3.5 Gyr, and then dramatically narrows and ultimately vanishes.

3.4 Other factors in habitability

A variety of other factors besides location in the habitable zone can influence the possibility of advanced life on a planet. A few are listed below and discussed in detail by Ward and Brownlee (2000):

- planet mass,
- abundance of potassium for geothermal heat,
- presence of plate tectonics,
- amount of water and carbon,
- presence of a large moon,
- strength of magnetic field,
- presence of giant planets,
- star with long main-sequence lifetime,
- star with appropriate chemical composition.

The quantitative role of many of these factors is poorly understood. Earth-like planets are complex bodies with many interacting systems and possible evolutionary pathways. As an example, planetary mass, potassium abundance, and plate tectonics are properties that surely influence habitability but there is no existing theory for predicting their significance. Mass and potassium abundance[6] influence the generation, transport and loss of a planet's internal heat, and therefore the existence and character of plate tectonics. We do not know the range of planetary mass and potassium that still permit Earth-like plate tectonics to occur. The Earth is highly depleted in the volatile element potassium relative to its abundance in primitive meteorites. How would Earth differ if it had eight times as much

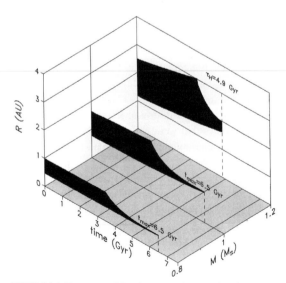

FIGURE 3.1 Estimates of the changes with time of the habitable zone (HZ) for stars with 0.8, 1.0, and 1.2 solar masses (M). *R* is the distance from the star in astronomical units (AU). For the Sun the HZ is a band from 0.9 to 1.7 AU for the first 3.5 Gyr, but then shrinks to zero over the next 3 Gyr. The results are from the geodynamic modeling of Franck *et al.* (2001a).

[5] One solar mass is 2.0×10^{33} g and is a standard unit of mass in astronomy. This is ~1,000 times the mass of Jupiter, which in turn is ~300 times the mass of the Earth.

[6] The radioactive isotope ^{40}K (which decays mainly to ^{40}Ca with a half-life of 1.4 Gyr) has been the main source of heating of the Earth's interior over its lifetime.

potassium? Would habitability be enhanced or hindered by the much larger internal heat source?

Another important factor is the amount of water and carbon. If the Earth had just a little more surface water, no land would be exposed and Earth's atmospheric cycles involving land and water would not exist. How much water does a habitable planet need, and what determines whether it resides on the surface or deep in the mantle? The optimum amount of carbon is also an open question. Some is needed but what happens if a planet has too much? Would a more carbon rich planet be able to maintain habitable atmospheric conditions (Gaidos, 2000)?

A fundamental aspect of habitability is the stability of a planet's star. For long-term habitability, a parent star must have a relatively stable output intensity. The longest duration of such constancy in a star's lifetime occurs when the star is "on" the so-called *main sequence*, a stage in a star's evolution[7] when its source of energy is the conversion of mass to energy through hydrogen fusing into helium in its core. This main-sequence lifetime turns out to be ~90% of the entire lifetime of a star, since other stages go relatively rapidly: (1) the earlier stages of star formation (Section 3.6), and (2) the later stages of becoming a red giant and eventually a white dwarf star.

The main-sequence lifetime depends primarily on a star's mass, and can be estimated as follows. Theory and observations agree that the luminosity L (total power output, which is equal to the star's energy generation rate) varies as about the fourth power of the star's mass M. The lifetime T is proportional to the available nuclear fuel that can be "burned" (which is about 0.1 solar masses for the Sun) and inversely proportional to the energy generation rate (L). We therefore have

$$T \propto M/L \propto M^{-3}.$$

For example, a star just twice as massive as the Sun is 16 times more luminous and only able to generate energy at this prodigious rate for 1/8 of the Sun's 11 Gyr lifetime as a main sequence star. A planet orbiting a star only 40% more massive than the Sun experiences a main sequence lifetime of ~4 Gyr and thus its habitability would end before the 4 Gyr that passed from Earth's formation to the appearance of animals.

Also, as we have seen for the Sun, stars increase in luminosity by a factor of ~2.5 even during their sojourn on the main sequence. More massive stars experience this increase more rapidly and pass more quickly through the stage where their luminosity is optimal for surface water to exist on any Earth-like planet.

Other astronomical factors that influence habitability are the parent star's chemical composition and its location in a galaxy. All elements heavier than helium, called *metals* in astronomical jargon, were made via nuclear reactions inside stars. The abundance of these has steadily increased with time due to a cycle of element production in stellar cores followed by ejection of freshly synthesized material back into interstellar space, from which increasingly metal-rich stars are eventually formed. The range of heavy element abundances in observed stars varies by a factor of ~10,000, with the Sun near the upper part of the range and most main sequence stars in our Galaxy within a factor of ten of the Sun. The oldest stars (nearly as old as the 13.7 Gyr age of the Universe) are metal deficient because their building materials were less processed and enriched. Within our Milky Way galaxy[8] we observe a distinct trend of composition with distance from the Galaxy's center (which is ~30,000 light-years away from the Sun). Stars closer to the center have systematically higher metal contents while those beyond the Sun's orbit have lower. Because planets like Earth are primarily made of these "metals," planets with solid surfaces around metal-poor stars are likely to be smaller on average. Near the Galaxy's center, however, a planetary system's disruptive encounters with other stars are more frequent and the probability of serious effects from astronomical events such as supernovae is higher (Section 17.5.6). Life on planets around such stars would be subjected to increased harassment by impacts and radiation effects from other stars. This has led to the proposal that there is a "galactic habitable zone" (Gonzalez *et al.*, 2001), a broad range of distances from the center of our Galaxy where the long-term survival of life on a planet is more favorable.

---

[7] In astronomy the term *stellar evolution* refers to what a biologist would call *ageing*, namely the changes in a given star over its lifetime, not the star spawning new modified generations.

[8] The terms *Galaxy* (capitalized) and *Milky Way* are used interchangeably in astronomy, although *Milky Way* can also refer more specifically to the band across the sky that is the most obvious manifestation of the Galaxy. Uncapitalized *galaxy* is a generic term referring both to the Milky Way and the estimated >10^{11} other such assemblages of stars, each with 10^7 to 10^{13} stars. Our Galaxy consists of ~4×10^{11} stars in a disk of diameter ~1×10^5 light-years.

3.5 Other planetary systems

The formation of our Solar System's planets, satellites, comets, and asteroids was a natural, albeit not completely understood, consequence of the formation of the Sun. Although many aspects of planet formation are generally agreed upon, there is no standard theory that starts with a set of input parameters and predicts the key properties of a resulting planetary system, even in a probabilistic sense. Our Solar System has small *rocky planets* (also called *terrestrial planets*) close to the Sun and large gaseous planets farther away, all having fairly circular orbits confined rather closely to a single plane. Is this a typical planetary system? The discovery of planets around other stars (*extrasolar planets* – fully discussed in Chapter 21) has shown how diverse planets and planetary systems can be. To date, extrasolar planet detection has largely been limited by technology to planets more massive than Neptune (17 Earth masses). There is also an observational bias toward detecting large planets that orbit close to their parent stars. But even with these biases in mind, the detected planets and planetary systems are remarkably different from those in our own Solar System. Notably, nearly all known extrasolar planets are giant planets orbiting either very close to their stars or in moderately elliptical orbits. A sobering aspect of this work is that few of the known extrasolar planetary systems are likely to harbor Earth-like planets with the orbital stability we think necessary for long-term habitability. In most cases, gravitational interactions from the observed giant planets would have long ago perturbed the orbits of any Earth-like planets out of the habitable zones of their stars. Although all planetary systems discovered to date are different from our own, it is important to emphasize that it is nevertheless still not known how typical or atypical our Solar System is.

The discovery of over 200 extrasolar planets implies that planet formation is relatively common. Further evidence comes from disks of *dust*[9] often found around

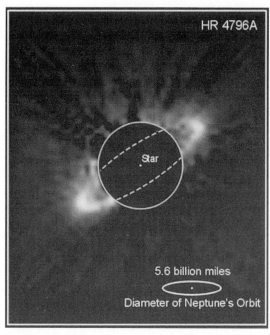

FIGURE 3.2 Starlight reflected from the debris disk of dust orbiting the star HR 4796 A, whose light has been blocked by an occulting disk (the black circle). It is likely that this dust is generated by the gradual disintegration of comets orbiting the star. Neptune's orbit size (diameter of 60 AU) is shown for scale. (NASA Hubble Space Telescope image by B. Smith and G. Schneider.)

Sun-like stars, particularly those less than 500 Myr old. Regions near these stars are observed to emit infrared radiation from circumstellar dust that is slightly warmed by starlight. In some cases the disks can also be observed directly in reflected starlight (Fig. 3.2). Most of the dust in these so-called "debris disks" cannot have been inherited directly from the interstellar medium since dust grains are destroyed rather quickly by processes such as collisions and light pressure effects. There must be a continuous source of new dust, and the standard explanation is that, like in our Solar System, the dust stems from the disintegration of larger bodies. Most of the dust is believed to be generated by objects similar to those in our Kuiper belt of comets, which are leftover planetesimals of dust and ice that orbit the Sun in a flattened annular region just beyond Neptune (at 30 AU). The inferred presence of comets around these other stars then implies that the growth processes that lead to planets (Section 3.9) also occur. Thus there is good reason to believe that the majority of stars are orbited by planetary bodies.

---

[9] *Dust* particles are ubiquitous in interstellar and interplanetary space (for example, look at the dark areas in the Milky Way visible with your naked eye). They have sizes of 0.1–1.0 μm and are presumably composed of condensable elements with relative abundances similar to those found in the Sun. The elemental composition of the particles is dominated by Mg, Si, and Fe, elements that that are totally condensed into interstellar solids, and O and C, elements that are not fully condensed and divided between gas and solid phases.

3.6 Star formation

Before turning to the formation of habitable planets per se, we consider the concurrent formation of the parent star and other members of its planetary system. Some of our best clues to this era 4.6 Ga are ancient Solar System materials such as meteorites, asteroids, comets, and cosmic dust. In the following sections we discuss the interstellar medium (the feedstock of new stars and planets) and the environments and processes that transformed interstellar matter into a planetary system around a young star. We describe how collapsing gas and dust became concentrated in a short-lived disk in which planets and other bodies formed. Finally, we describe the results of numerical models simulating the formation of Earth-like planets, as well as more specifically Earth's formation and the sources of its organic materials, its ocean, and its atmosphere.

Earth and the other bodies of the Solar System formed from the *solar nebula*, a short-lived, flattened disk of gas and dust orbiting the newly formed Sun. The gas and fine dust particles in the solar nebula dissipated within a few million years, while newly formed rocks and larger planetesimals persisted longer. The concept of planets forming from a solar nebula goes back centuries to the time of Kant and Laplace (Chapter 1), but ideas on the nature of the nebula and its processes have of course considerably changed. Disks surrounding forming or new stars are now known to be a common and important aspect of star formation and in fact much of a star's mass passes through the inner regions of a disk during its formation.

The material that formed the Sun and planets was originally gas and dust contained in an interstellar region where hundreds to thousands of stars formed over a period of only a few tens of millions of years. We observe analogous regions where stars are currently forming in *giant molecular clouds*, regions of interstellar gas and dust containing thousands to millions of times as much mass as the Sun, spread over 10 to 100 light-years. These molecular clouds have average densities of 10^2 to 10^3 molecules cm^{-3} and temperatures as low as 10 K. They are called "molecular" because their hydrogen is mainly in the form of molecular hydrogen, although they contain many other compounds as well (Section 3.8.1). These environments contain a mix of young stars, protostars, neutral gas and dust, and regions of hydrogen ionized by ultraviolet light from newly formed hot stars.

Models indicate that a star forms when a high-density region in a molecular cloud reaches a state where the forces due to thermal gas pressure, magnetic pressure, and turbulence are not sufficient to withstand gravitational collapse of the interstellar material. The collapse proceeds quickly, and by 0.1–1.0 Myr a dense, hot core (called a *protostar*) forms and is surrounded by a disk of still-accreting material. When the central temperature of the protostar reaches $\sim 1 \times 10^6$ K, deuterium-to-helium fusion is possible, but the deuterium is soon used up; further collapse of the nascent star's core eventually leads to a central temperature of $\sim 10 \times 10^6$ K, which allows hydrogen-to-helium fusion, as in today's solar core. At this point the nuclear energy output creates a pressure that withstands any further gravitational collapse and one has a "proper" star at the beginning of the relatively stable, main sequence stage of its life.

Meanwhile, also forming from the start has been a large, thin disk of material, 100 to 1000 AU in diameter, orbiting the central star – the solar nebula that leads to planets. Disks take shape around forming stars because much of the infalling material has too much angular momentum to fall directly to the center of mass of the system (the forming star) and therefore takes up roughly circular orbits around the center.[10] So much dust initially surrounds the young star that most or all of the star's visible light is adsorbed and reradiated at longer infrared wavelengths. The nebular disk has strong radial density and temperature gradients as well as complex processes involved in the transport of mass, momentum and energy. The mass of the disk is uncertain: at least 2% of the forming star's mass, but perhaps much more. Finally, after only ~ 1 to 10 Myr the gas and dust of the nebula are swept away by strong winds from the new star (we observe such winds in young stars (T Tauri-type) today) and/or ablated by ultraviolet radiation from nearby young stars, stunting most further formation of gas-rich planets.

Star formation is fast on astrophysical timescales, yet too slow to observe in real time for any given object. The study of star formation thus entails (1) observing dense cores within interstellar clouds as sites of potential star formation, (2) computational modeling of the

---

[10] Today's planets contain $> 98\%$ of the Solar System's angular momentum (the remainder being in the Sun), although they have only 0.2% of the total mass.

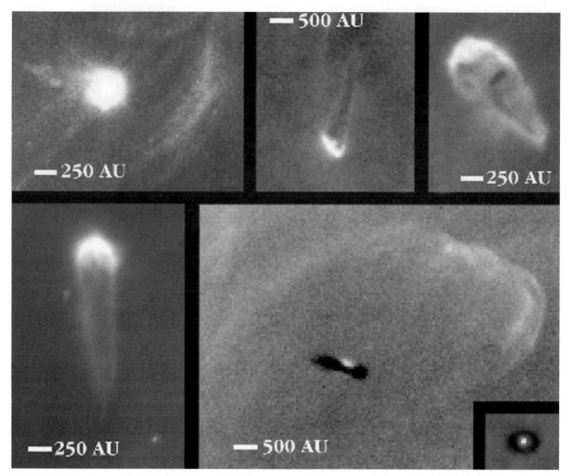

FIGURE 3.3 Young stars with disks in the Orion nebula. The top row and the image on the lower left show disks of gas and dust being ablated by ultraviolet light from nearby stars. The two images at the lower right are silhouettes where disks associated with young stars block light from the background hot gas. Note that the ~500 AU size of these disks is much larger than the region of ~50 AU over which we find planets today in our Solar System. (NASA Hubble Space Telescope images by J. Balley, D. Devine, and R. Sutherland.)

physics of gas and dust undergoing gravitational collapse, and (3) observing newly formed stars and the surrounding gas and dust of the parent cloud in which they are embedded.

Most stars form together in large groups called *clusters* where influences from neighboring stars may affect planet formation. The closest star-forming locations to us today are the Taurus region, a small cluster of low mass stars, and the Orion region, which contains thousands of young stars and includes the famous Orion nebula. Since large clusters such as Orion have more massive and brighter stars than Taurus, planet formation may vary in significant ways. Most stars, including the Sun, appear to form in Orion-like clusters. The presence of supernova debris such as ^{60}Fe in the early Solar System suggests that the Sun formed in a high-mass cluster where supernovae are much more common

(Ovellette *et al.*, 2005; also see Section 17.5.6). We observe the disks of presently forming stars in Orion as silhouettes against the glow of hot background gas (Fig. 3.3). One of the effects of nearly simultaneous formation of a large group of stars is that gas in the outer regions of a planetary disk can be ablated away within a million years by the enormous amounts of ultraviolet radiation generated by neighboring hot, massive stars. Similarly, intense *particle* winds from nearby massive stars may inhibit planet formation by stripping gas from the outer parts of disks where gas giants like Jupiter form.

3.7 Building materials: element abundances

Interstellar gas is largely hydrogen and helium while most of the other elements reside in submicron grains

TABLE 3.1 Element abundances (% by number)

Atomic no.	Element	Solar abundance	Entire Earth	Earth's atm.	Bacterium (dry)	Bacterium	Human
1	H	93.4		2–3 (var)	55	63	62.2
2	He	6.5		0.0005			
6	C	0.03		0.03	28	9	14.2
7	N	0.011		78	7	2	1.1
8	O	0.06	50	21	9	26	21.9
12	Mg	0.003	16				
13	Al	0.0002	1.1				
14	Si	0.003	14				
15	P	0.00003	0.08		0.7	0.2	0.17
16	S	0.001	1.6		0.2	0.1	0.03
26	Fe	0.002	15				
28	Ni	0.0001	1.1				

Notes:
1. All abundances are % *by number*. Elements are listed by atomic number.
2. The list here of solar abundances, often called "cosmic abundances," includes 12 of the top 17 elements; omitted are Ne (sixth in abundance), Ar (11th), Ca (13th), Na (14th), and Cr (16th).
3. The list here of Earth abundances includes the top seven plus P (11th).
4. The list here of Earth's atmosphere abundances includes the top 7 except Ar (0.91%, fourth) and Ne (0.0018%, sixth).
5. The list here of Bacterium abundances is for *E. coli*, with and without H_2O included (Neidhardt *et al.*, 1990:3). All of the top seven are included except K (sixth).
6. The list here of *Homo sapiens* abundances includes the top seven except Ca (0.25%, fifth) (Sterner and Elser, 2002: 3). By weight, humans are 60% water, 36% organic molecules, and 4% "ash."

of dust. The production of solid Earth-like planets requires heavy elements ("metals") like O, Mg, Si, and Fe that reside in dust. A necessary step leading to the formation of planets with solid (or liquid) surfaces is therefore the process of making these heavy elements. By numbers of atoms Earth is 50% O, 16% Mg, 15% Fe, and 14% Si, with all other elements at or below the 1% level (Table 3.1). The formation of the chemical elements began 13.7 Ga during the birth of the Universe, the Big Bang. This event produced most of the H and He in the Universe[11] but essentially none of the metals. As previously mentioned, heavier elements were later produced via nuclear reactions in the hot dense cores of stars. These elements are produced inside stars and are ultimately released into the interstellar medium in the last stages of stellar evolution, when some stars eject a significant fraction of their mass into space. The net effect of the cycle of star formation, evolution and death is to continuously cycle matter in and out of stars, enriching the interstellar medium in the heavy elements. Our Galaxy had already been enriched in the heavy elements for over 8 Gyr by the time the Sun formed 4.6 Ga.

The ingredients of the Solar System are a fossil record of the element abundances in our region of the Milky Way 4.6 Ga. These so-called *solar abundances* are important because they tell us about the original material from which our Solar System formed. Many of Earth's properties, including its mass, its internal heat component generated by radioactivity, and its core are influenced by the abundances of its constituent elements. The abundances of the elements in the Sun (Table 3.1) are often called *cosmic abundances*, but this is inappropriate terminology. Although many stars do have metal abundances roughly similar to those of the Sun, there is a broad range depending on a star's age, its location in a galaxy, and the type of its parent galaxy. The most metal-poor stars have heavy element abundances relative to hydrogen that are a hundred thousand times less than those of the Sun. The most metal-rich stars are not greatly richer than the Sun but the Sun itself is actually metal-rich by ~25% compared to its neighboring stars. It has been found that stars with detected

[11] The amount of He that has been produced by all the main-sequence stars over the entire past of the Universe amounts to only a small fraction of the initial amount produced in the first three minutes after the Big Bang.

planets are systematically metal-rich (Chapter 21; Laws *et al.*, 2003). It may well be that this enhanced composition aided the planet formation process.

The relative abundances of the elements in the interstellar medium are similar to those of the Sun: for every 10^6 H atoms, there are $\sim 10^5$ He atoms, 700 O atoms, 300 C atoms, ~ 100 N and Ne atoms, and 30–40 atoms each of Si, Mg, and Fe. Note that the major elements required by all known life forms (Table 3.1) are among the most abundant elements throughout the cosmos. On the other hand, it is remarkable that many cosmically trace elements play major roles in terrestrial biology and even in Earth's physical evolution.

3.8 Organic compounds before Earth formed

The organic compounds used to form Earth's first life were either produced on Earth or created elsewhere and carried to Earth. Some of these organic materials formed in the solar nebula, but others undoubtedly formed in the interstellar medium and later survived the formation of the Solar System. They were incorporated into comets, asteroids and other planetesimals and delivered to the young Earth as components of meteorites and dust, a process that continues today. Meteorites and dust serve as carriers of this "cosmic manna," providing thermal protection for high-speed passage into planetary atmospheres and a means of "soft landing" on the surface. Because of their potential importance to the formation of life on habitable planets, we will discuss interstellar molecules in some detail.

3.8.1 Organic chemistry in the interstellar medium

Interstellar clouds come primarily in two flavors, diffuse and molecular. *Diffuse interstellar clouds* have densities of $100 \, \mathrm{cm}^{-3}$ and temperatures of 70–200 K.[12] Ultraviolet radiation from distant luminous stars penetrates through these clouds, dissociating most chemical bonds in gas phase molecules. As a result, diffuse clouds contain mostly simple gas-phase radicals and ionized species (e.g., H, CN, OH, CH^+, and Si^+) and

a few types of strongly bound compounds, some quite complex, that are resistant to ultraviolet (e.g., CO and *polycyclic aromatic hydrocarbons* (*PAHs*). About 10–30% of the carbon in the interstellar medium is thought to be in the form of PAHs (Fig. 3.4; soot is a familiar example on Earth).

Molecular clouds, on the other hand, have densities of 10^4 to $10^6 \, \mathrm{cm}^{-3}$, temperatures of only 10 K, and contain a rich suite of small molecules, mostly detected with radio dishes via microwave and millimeter spectroscopy of their rotational lines. There are now over 120 molecular species known, the largest being $HC_{11}N$ and the most interesting to an astrobiologist being glycine ($C_2H_5NO_2$), the simplest of life's amino acids (see discussion in the next section). Figure 7.13 illustrates the structures of many interstellar molecules. The molecular clouds present conditions radically different from Earth's atmosphere. In our atmosphere, solar ultraviolet radiation drives photochemical reactions (such as those that produce ozone and smog), temperatures allow thermally driven chemical reactions, and high densities allow efficient three-body reactions. In molecular clouds, however, none of these conditions are met. Ultraviolet radiation cannot drive an interesting chemistry because it is absorbed in the cloud cores by the dust, and temperatures and densities are extremely low.

The chemistry in molecular clouds is instead driven by ubiquitous interstellar cosmic rays, which are energetic particles (atomic nuclei, not electromagnetic radiation) that can penetrate deep within the clouds. A major driving force behind the gas phase chemistry is the ionization of H_2 molecules by cosmic rays, opening up otherwise inaccessible pathways to chemical complexity. Densities are so low that reactions involving collisions of three gas-phase molecules essentially never occur. Interesting chemical reactions do take place, however, on the surfaces of dust grains, which in effect serve as a "third body." Furthermore, volatile compounds (e.g., H_2O, CO_2, CH_4, NH_3, CO, N_2) form icy mantles on dust grains. These ices are observed via their absorption lines along lines of sight toward forming stars still deeply embedded in their parent molecular cloud (Fig. 3.5).

As reactions between ions and molecules proceed, minute differences in rates between isotopes such as ^{15}N and ^{14}N become apparent. The difference in rates depends on $\mathrm{e}^{-\Delta T/T}$, where $k\Delta T$ (k is Boltzmann's constant) is the difference between the activation energies of a reaction involving ^{15}N versus one with ^{14}N. A typical value for ΔT is 9 K. At room

---

[12] Compare this density with that of the Earth's atmosphere, which is $10^{19} \, \mathrm{cm}^{-3}$ at sea level. Spectroscopy of the interstellar medium's rarefied gas is only possible because of the correspondingly long column lengths. (Quoted densities are in terms of the number of atoms and/or molecules.)

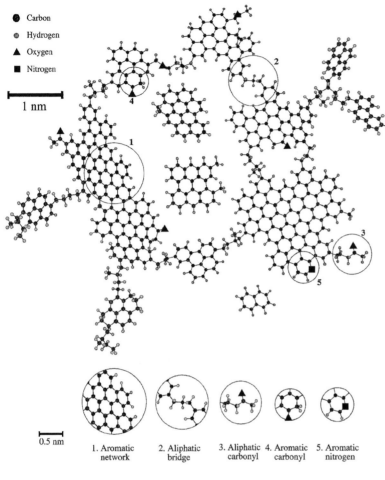

FIGURE 3.4 The structure of a polycyclic aromatic hydrocarbon (PAH) that matches a 3.4 μm infrared spectral feature, attributed to CH bonds, observed in molecular clouds in the interstellar medium. (Pendleton and Allamandola, 2002)

- ● Carbon
- ○ Hydrogen
- ▲ Oxygen
- ■ Nitrogen

1 nm

0.5 nm

1. Aromatic network 2. Aliphatic bridge 3. Aliphatic carbonyl 4. Aromatic carbonyl 5. Aromatic nitrogen

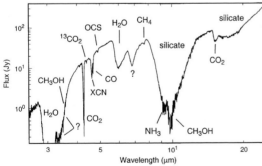

FIGURE 3.5 Infrared spectrum of a young stellar object embedded within a cold dense molecular cloud, showing absorption features caused by the indicated solid-state materials in the cloud. The dominant materials are silicates and ices with some similarity to the solar system's comets. (Gibb et al., 2000.)

temperature of 300 K, $e^{-\Delta T/T}$ is close to unity, and the reactions proceed at almost equal rates. By contrast, at 10 K as in a molecular cloud, this factor is significantly less than unity, leading to reactions that distinguish between isotopes, a process called *fractionation*. Such

isotopic signatures are indeed found in primitive Solar System materials such as cosmic dust and meteorites. For example, organic compounds in some primitive materials have a $^{14}N/^{15}N$ ratio up to a factor of two smaller than the atmospheric value of 273 that is used as a comparison.

3.8.2 Amino acids in the interstellar medium and meteorites

Although over 70 amino acids have been identified in the Murchison meteorite (see Figs. 7.10, 7.11, and 7.12),[13] not all of the amino acids used by life on Earth are present. Early studies showed that the

[13] Section 7.4 gives more details about the organic compounds found in Murchison and other meteorites, as well as a separate discussion of their significance. The Murchison meteorite fell in Australia in 1969.

meteorite (unlike life) contained equal amounts of left- and right-handed varieties of amino acids, although Cronin and Pizarello (1997) have reported a very slight preference for left-handed isovaline in the Murchison and Murray meteorites. Terrestrial contamination can be ruled out by the large fraction of these compounds that are nonbiological (for example alpha-amino iso-butyric acid).

How did extraterrestrial amino acids form? The existence of these compounds in meteorites suggests that they formed in the meteorite's parent body, an asteroid (Section 3.2). The Murchison meteorite shows signs of once having contained liquid water, now present in hydrated minerals, and other evidence of liquid water alteration. Thus, the amino acids have been presumed to be the product of aqueous chemistry long ago in the parent body. But other factors are not consistent with this idea. Liquid water in the meteorite was somewhat depleted in deuterium relative to hydrogen, whereas the amino acids are enriched in deuterium, which is more suggestive of an interstellar heritage (Sandford, 2002).

Bernstein *et al.* (2002) have demonstrated in the lab that glycine and other amino acids can form under the conditions that characterize dense, dark interstellar clouds. They irradiated ices, analogous to those observed in interstellar clouds, with ultraviolet light, and produced the amino acids glycine, alanine, and serine. Thus, one possible pathway to form amino acids is via photochemistry of interstellar ices.

Glycine and other complex organics may form via gas-phase reactions as well as in ices. Kuan *et al.* (2003, 2004) detected two dozen radio-wavelength spectral lines of glycine in *hot cores* (temperatures of $\sim$100 K) – such dense, hot regions of molecular clouds, warmed by a nearby massive young star, are excellent hunting grounds for organic molecules. In the hot cores ice-mantled dust grains are evaporating after absorbing photons from newly formed massive stars. Evaporation of these ice mantles introduces fresh chemical constituents into the gas phase, where ion–molecule chemistry is already in progress. Theoretical models show that complex organics may be either a product of ion–molecule chemistry or of ice-mantle evaporation followed by gas-phase reactions.

Purines and pyrimidines have also been identified in the Murchison meteorite. The nitrogen heterocycle pyrimidine, which is related to the DNA bases cytosine, thymine, and uracil, has also been searched for in the same hot cores in which glycine was detected.

These observations were unsuccessful, likely due to the extremely low gas-phase abundance of pyrimidine, which, like amino acids, is vulnerable to being destroyed by ultraviolet radiation (Kuan *et al.*, 2003).

While compounds like pyrimidine and glycine do not survive interstellar conditions unless they remain protected in interstellar dust grains, PAHs are extremely resistant to destruction by ultraviolet radiation. The strong aromatic bonds between carbon atoms in these compounds allow them to absorb ultraviolet radiation and reradiate the energy by fluorescing in the visible and infrared. PAHs are also chemically very stable, and together these attributes make them the most likely compounds to survive the transition from interstellar matter to meteorites. Indeed, meteoritic PAHs tend to have similar deuterium enrichments to those found in interstellar material, implying that they survived all solar nebula conditions. PAHs are found in a wide variety of interstellar environments, from the extended atmospheres of carbon-rich evolved stars, to the diffuse interstellar clouds and the general interstellar medium.

Interstellar chemistry is highly nonequilibrium: abundances cannot be determined using thermodynamic equilibrium calculations because the physical conditions change over reaction timescales. From astronomical observations, which give only a snapshot of the chemistry at a particular time, we must infer what the initial conditions were and how they are presently changing. For example, because amino acids are particularly vulnerable to destruction by ultraviolet radiation, they must remain in the icy mantles of interstellar dust grains in order to be incorporated into larger bodies (e.g., comets and asteroids; Ehrenfreund *et al.*, 2001).

Another mystery regarding the origin of amino acids in meteorites is the great disparity in their abundances in different meteorites. The Tagish Lake meteorite (which fell on a frozen lake in Canada in 2000) is a carbonaceous chondrite with evidence of aqueous alteration, as is Murchison; however Tagish Lake contains only trace amino acids (800 ppb, and even these appear to be terrestrial contaminants), whereas Murchison contains 16,900 ppb amino acids (which are definitely extraterrestrial). Murchison and Tagish Lake must be samples of different parent bodies. If all carbonaceous chondrite parent bodies inherit the same material from the interstellar medium, this disparity would suggest that processes within the solar nebula and/or within the parent bodies greatly altered the original inventory of organics inherited from interstellar space.

3.8.3 Processing within the solar nebula

With these interstellar chemical processes in mind, we now consider processing within the solar nebula. Some materials survived nebular processing while others were totally reworked with all original chemical bonds broken. The nebula had a radial temperature gradient, and in its inner regions temperatures were hot enough (>1700 K) to vaporize all original dust. The nebula is often envisioned as static such that each radial distance can be assigned a single temperature. This is a useful model but a simplification: for example, temperatures surely decreased with height above or below the nebula mid-plane, they surely varied at any given place as the nebula evolved with time, and it is also likely that local temperatures fluctuated on short timescales. As an additional complication, the compositions and condensation temperature of solids are influenced by their O/H and C/H ratios, which can be greatly enhanced in dense parts of a disk due to the concentration and subsequent vaporization of ice and organic-bearing particles. For example, ice particles that settle to the warm mid-plane and vaporize can in some cases cause the H_2O/H_2 ratio of mid-plane gas to locally increase by orders of magnitude, greatly affecting the oxidation state of expected condensates.

The elemental composition of meteorites and the inner planets appears to have been influenced by the effects of condensation or vaporization and it is apparent that much of the inner Solar System was initially very hot or otherwise led to heavy processing of the interstellar material. Early on, parts of the inner nebula would have been so hot that only the most refractory compounds could exist as solids: the first condensed solids (when the temperature fell below ~1700 K) were minerals rich in Al, Ca, and Ti. As the nebula cooled, silicates and metals composed of Fe, Ni, and other iron-loving trace elements could condense into solids. The temperature dependence of compounds that could form solids is called the *condensation sequence* and is usually estimated by the assumption of chemical equilibrium between vapor and condensed solids. Figure 3.6 presents the results of standard calculations for the condensation sequence. The sequence involves not only actual condensation but also chemical back-reactions between condensed solids and nebular gas. For example, iron condenses in the cooling gas first as a metal (right-hand side of Fig. 3.6), but then later reacts with H_2S vapor to form FeS when the temperature drops further by several hundred degrees. At even lower temperatures, exposed iron surfaces should react

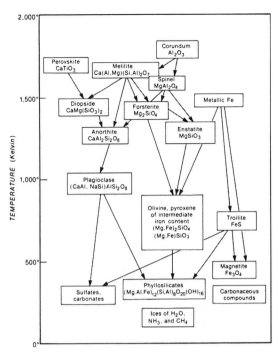

FIGURE 3.6 The estimated condensation sequence of solid grains in the solar nebula based on the assumption of chemical equilibrium between vapor and condensed phases. (Wood, 1990.)

with H_2O vapor to form Fe_3O_4. Likewise, the first abundant silicates contain only Mg and no Fe because all of the Fe condenses first as metal. But at lower temperatures, silicates containing oxidized iron become stable under equilibrium conditions. The composition of the condensed solids and the nebular gas changes with temperature, and as the nebula cools or grains are transported to different temperature regions, many grains are transformed to new phases.

What actually occurred in the solar nebula probably followed this condensation scheme at the highest temperatures, but at lower temperatures, where reactions are slower, departures from equilibrium become significant. For a solid to be in equilibrium with nebular gas, interior atoms must be close to a surface and there must be time to allow adequate diffusion from the surface inwards. This does not occur when solids become larger than dust or accrete into larger bodies, so models must include estimates for the growth rate of grain and rock sizes in order to understand the nebular chemistry. Even nebular gas has difficulty maintaining equilibrium compositions during the brief lifetime of the solar nebula (1–10 Myr). For example, we observe that comets formed near the edge of the nebula disk have moderate amounts of frozen CO but not much frozen CH_4. Yet under equilibrium in the cool temperatures of the outer disk, most gaseous carbon should have transformed to CH_4. Possible explanations are: (1) the nebula cooled too rapidly for the chemical reactions to occur, (2) cometary volatiles derive from the interstellar ices (see Figure 3.5), or (3) chemistry in the outer Solar System was similar to that in the cold dense molecular cloud cores.

Even with its limitations, the condensation sequence is a very useful concept and its predictions are generally consistent with the large-scale compositional differences between outer and inner Solar System bodies, as well as the fractionation patterns exhibited in meteorites by groups of volatile or refractory elements. However, the condensation predictions unfortunately do not provide clear insight into the formation of carbon- or water-bearing solids, materials of key importance to astrobiology. If equilibrium condensation was the only way for nebular carbon to form solids, it would have occurred only at very low temperatures, say where methane could condense at the extreme outer edge of the planetary system. The high carbon content of certain meteorites, and the presence of organic compounds in comets and on the surfaces of satellites, indicates that some interstellar organics survived, or were minimally altered, as they were incorporated into Solar System bodies.

Incorporation of water into solids is of major interest to astrobiology. Under nebular conditions, water condenses directly to ice at temperatures below 170 K. The solar distance where this happened is called the *snow line*, the nebular boundary beyond which water vapor condenses to ice. Most models indicate that the snow line occurred at ~2.5–3.0 AU, well outside the orbit of Mars (at 1.5 AU) and inside that of Jupiter (5.2 AU), and near the outer parts of the main asteroid belt. But note that if chemical equilibrium occurred, water would have been incorporated into minerals at much higher temperatures and much closer to the Sun. For example, nebular water vapor could have reacted with condensed silicates at ~400 K to form hydrated silicates such as the mineral serpentine ($Mg_6Si_4O_{10}(OH)_8$). This type of reaction would have made water bound in rocks available in the warmer regions of the nebula, leading to much higher water abundances than we see today for Earth and its fellow terrestrial planets. It seems that water remained predominantly in the gas phase and did not react with silicates in the <2 AU region. Again it appears that the truncated lifetime of the nebula inhibited this reaction – there was simply not enough time for it to

occur.[14] How then did Earth get its water, the most essential ingredient to life? It was probably delivered from bodies that formed near or beyond the snow line, as discussed in the following section.

3.9   Formation of planets

Planetary bodies in the solar nebula formed by at least two different methods. The rocky planets, Pluto, asteroids, comets, and the various moons all seem to have formed by accretion of solid materials. The *accretion* process is fundamentally the collision and partial sticking of materials ranging in size from a few nanometers to thousands of kilometers. Although Uranus and Neptune do not have solid surfaces, their element composition is also consistent with formation largely by accretion of solid materials. These two planets, at the edge of the Solar System, are often called *ice giants* because they are believed to have formed largely by accretion of icy solids as well as rocky materials. The *gas giants* Jupiter and Saturn, however, clearly did not form by accretion of solid materials because they are largely composed of H and He, elements that under solar nebula conditions cannot be efficiently incorporated into solids. Hydrogen can clearly form condensable compounds like H_2O but almost all of the nebula's hydrogen was in the form of H_2. The solar abundance of hydrogen is so high that only a tiny fraction of the hydrogen could be bound to oxygen, even if all of the nebular oxygen was in water molecules. As discussed in the previous section, the rocky planets, the ice giants, and the gas giants represent differing degrees of fractionation from solar composition (and from the original collapsing interstellar cloud). The gas giants are the closest approximation to solar composition, the ice giants are highly depleted in hydrogen and helium, and the rocky planets have the most fractionated compositions as they are depleted in many of the elements that largely resided in icy or gaseous materials in the solar nebula.

3.9.1   Gas giant and ice giant formation

Ice and gas giants are not themselves habitable because they do not have surfaces. Even in their warm and wet interior layers it is hard to imagine a mechanism that would allow organisms to maintain a given altitude so as to linger in an adequately stable temperature regime.

---

[14] Hydrated silicates *are* abundant in some meteorites, but they are generally believed to have formed inside asteroids by reaction of silicates with liquid water originally delivered as ice.

The giant planets do, however, have large moons that might harbor life – although quite unlike Earth, moons like these could in fact be the most common life-supporting habitats in the Universe. For example, the wet warm subsurface regions of Jupiter's moons may be habitable for microorganisms (Section 3.2, Chapter 19), but we do not know if life in such environments could ever evolve past the microbial stage. On the other hand, *surface* life definitely seems unlikely on moons, particularly those outside the habitable zone (HZ), because of extreme cold and the severe radiation environment caused by high-energy particles within the giant planets' magnetospheres.

Giant planets influenced not only the outer Solar System, but conditions in the inner Solar System, too. They scattered volatile-rich solid materials (water ice, organic compounds, methanol, carbon monoxide, etc.) from the outer planetary regions into the HZ region. This is extremely important because, as discussed in the previous section, HZ materials that formed *in situ* (e.g., silicate minerals and iron) were highly depleted in volatiles because higher nebular temperatures in the HZ prevented volatile compounds from condensing or becoming incorporated into solid material. Giant planets are also critically important because their proximity to an HZ can preclude the long-term presence of Earth-like planets in the HZ. In our own Solar System, the failure of a planet to have formed at the asteroid belt's orbital radius, as well as the small size of Mars, are commonly attributed to the deleterious effects of Jupiter forming close by. If a giant planet's orbit is noncircular, it can even more effectively perturb nearby planets. The gravitational effects of giant planets can eject Earth-like planets into interstellar space, or cause them to impact other planets or even the Sun.

There are two major concepts regarding the formation of gas giants – either a two-step process involving solids *accretion* followed by gas accretion, or formation by *disk instabilities* that led to rapid gravitational collapse. The first concept is that gas giants (note that Jupiter's mass is ~300 Earth masses) first accreted a roughly 10–15 Earth-mass core of icy and rocky solid materials through collisions (of the type discussed in the next section), at which point the gravity of the core would be sufficient to retain nebular gas (which was nearly 99% hydrogen and helium by mass). In this concept Uranus (15 Earth masses) and Neptune (17 Earth masses) just reached the needed core mass, but due to slower growth in the outer regions of the solar nebula, a vigorous gas accretion phase did not occur because nebular gas was cleared away (as mentioned in

FIGURE 3.7 Rapid formation of Jupiter-mass protoplanets by gravitational instability in a numerical simulation of the solar nebula disk. Such protoplanets orbit the central star until gravitational interactions among them lead to either coalescence or ejection, resulting in only a few survivors. (Mayer *et al.*, 2002.)

Section 3.6) before enough core mass could be accreted from solid materials. This resulted in gaseous planets that were highly depleted in hydrogen and helium compared to Jupiter and Saturn.

This two-stage model explains the apparent heavy element cores in Jupiter and Saturn, but it has problems forming the planets as quickly as required. Any model must form gas and ice giants before the disk is cleared, and observations of young stars indicate that disks persist no longer than 1–10 Myr. A successful model must also form Jupiter quickly enough so that its gravitational "greediness" can retard the formation of Mars (which would presumably have otherwise been much larger than its 0.1 Earth mass), as well as totally abort the otherwise likely formation of an Earth-sized planet at the distance of the asteroid belt.

To produce giant planets more quickly, an alternative disk instability model makes gas giant planets via rapid gravitational collapse (Boss, 2003; Mayer *et al.*, 2002). If the disk density is high enough, numerical simulations indicate that local instabilities can lead to gravitational collapse of Jupiter-mass clumps of gas (Fig. 3.7). The process is *very* rapid, producing protoplanets within only a few hundred years! These proto-Jupiters are at first extended low-density objects and then take millions of years to radiate away their thermal energy and shrink to near-final size. Note that even today such a cooling process continues at a very low rate – Jupiter is observed to emit twice as much energy as it receives from the Sun. A weakness of the instability model is that it produces Jupiters with what appear to be too small an abundance of heavy elements, although the composition of gas giants is still an open question. It also cannot directly produce planets like ice giants because gravitational collapse cannot yield compositions highly depleted in the hydrogen and helium that make up almost all of the solar nebula. Boss (2003) has suggested that the ice giants might have originally formed as gas giants via the core-accretion model, but then suffered stripping of their outer layers, rich in hydrogen and helium, by ultraviolet radiation from nearby stars.

Gas giant planets are now known to be present around >5% of nearby solar-like stars; estimates that take detection limitations into account suggest that the percentage of stars that actually have giant planets is ~25–50% (Chapter 21). It is likely that formation of gas and ice giants is commonly accompanied by formation of small rocky planets. This is significant in that these large planets are thought to play major roles in perturbing volatile-rich materials from the outer disk to planets forming in the HZ. Over its entire history our planet has been bombarded by outer Solar System debris that was transferred to Earth-crossing orbits through stochastic gravitational perturbations by the giant planets. An important aspect of the known extrasolar planets is that many of them are "hot Jupiters" that appear to have migrated inwards when planetary formation was still occurring (Lin *et al.*, 1996). Such a migration would gravitationally eject any previously existing HZ terrestrial planets either into the star or back out into interstellar space. This cleaning-out process does not, however, necessarily exclude the presence of HZ planets with Earth-like habitability. Some recent studies suggest that after a migration episode there may still remain time to form a slightly later generation of terrestrial planets exterior to the orbit of a close-in hot Jupiter (Mandell *et al.*, 2007).

3.9.2 Terrestrial planet formation

As discussed at the beginning of this chapter, long-term habitability for complex life implies an Earth-like planet in a stable orbit in a stellar HZ. The Solar System's terrestrial planets formed by accretion of solid materials in the inner regions of the solar nebula. The accretion process involved an untidy aggregation process of the accretion of ever larger bodies, starting with dust and ending with planets. However simple this might seem, the processes involved were actually complex, involving growth, destruction, scattering, nebular evolution, and the gravitational interplay of many bodies. The accretion process that formed terrestrial planets can be considered to have four stages (Table 3.2).

TABLE 3.2 Formation of terrestrial planets by accretion

Stage A	Gas to dust – dust to rocks
	Settling to mid-plane of disk
	Increase in rock/gas ratio
	Chondrule formation
Stage B	Collisions and growth
	Rocks to km-sized planetesimals
	Heating of inner planetesimals
Stage C	Orbits circular, nearly in one plane
$\sim 10^5$ yr	Accretion within feeding zones
	Planetesimals grow to lunar-sized bodies
	Gravitational focusing dominates
	Runaway growth of largest planetesimals
Stage D	Growth to Mars-sized and larger bodies
$\sim 10^8$ yr	(embryos)
	Scattering between feeding zones
	Disk clears at $\sim 10^6 - 10^7$ yr (giant planets fully formed)
	Severe impacts on planets-to-be, e.g., Earth's Moon formed by impact
	Terrestrial planets fully formed

Stage A

The first generation of solids included nebular condensates, interstellar grains that had survived vaporization, and solids that had been highly processed in the nebula (Section 3.8.3). The initial, small grains were swept along with the dominating gas motions. Over time, collisions, sticking and other growth processes led to formation of larger and larger particles. When particles reached centimeter size, their ratio of particle surface area to mass became low enough that they began to decouple from the gas motion and settle towards the mid-plane of the nebula (Stage A in Table 3.2). In addition to concentration by just settling, particles may also have been concentrated by turbulence (Cuzzi *et al.*, 2001). The density of solids at the mid-plane became concentrated by factors of up to thousands, creating the environment for rapid accretion to meter and then kilometer and larger bodies.

Stage B

Our window into the earliest stages of planet formation is through *meteorites*, the objects that survive passage through the Earth's atmosphere and are thus available for detailed laboratory study. Many meteorites are well-preserved fragments of ancient solar nebula objects that escaped accretion into planets. Meteorites, dated by a variety of methods based on the decay of long-lived radioactive isotopes such as ^{87}Rb, ^{40}K, ^{147}Sm, ^{235}U, and ^{238}U, are the oldest known rocks, and most date very closely to 4.56 Ga, when they apparently accreted to form asteroid-sized bodies. This then is the basis for the remarkably precise date of Solar System formation: 4.56 billion years ago.

There are over a dozen classes of meteorites, but the major distinction is between those that have condensable-element compositions similar to the Sun and those that do not. In the latter class, called *differentiated meteorites*, groups of chemically related elements have been systematically depleted or enhanced by processes that occurred inside their original parent bodies – these meteorites came from asteroids that underwent at least partial melting that allowed large-scale separation of elements according to their geo-chemical properties. Classic types of differentiated meteorites are *irons* and *basaltic achondrites*. The irons are composed of metallic iron and are enriched in siderophile (iron loving) elements such as Pt, Ir, and Os that readily enter metallic iron in planetary interiors. In several ways the iron meteorites are compositionally analogous to Earth's core. Many basaltic achondrites are similar to terrestrial basalt and are depleted in siderophile elements but enriched in lithophile elements like Al, Ti, Ba, and U (which concentrate in the crusts of planets).

Differentiated meteorites are rare, however, and the most common meteorites impacting Earth are *chondrites*, so named because they contain mysterious spherical particles called *chondrules* (Fig. 3.8). Chondrites are primitive meteorites that most closely approximate solar composition and do not have large systematic depletions as described above. This implies that they never underwent melting inside parent bodies and that their element compositions have therefore been determined by nebular processes rather than parent-body processes. Different classes of chondrites have systematic depletions or enrichments of volatile or refractory elements that are believed to be related to their formation histories. *Carbonaceous chondrites* have the highest carbon abundances and the highest abundances of volatile elements such as Zn, Cd, and Bi. Examples such as the Murchison meteorite are famous for their organic materials, in particular amino acids (Section 3.8.2).

The properties of nebular solids must have varied significantly with distance from the center of the nebula. In the cold outer regions of the nebular disk, where comets formed, the solids may have been dominated by ice and relatively unchanged interstellar grains. In the warm inner regions, the solids were likely either

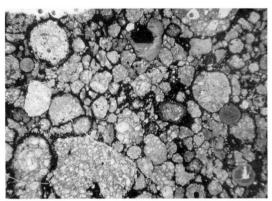

FIGURE 3.8. The interior of the primitive chondrite meteorite Semarkona shows components that before accretion were individual particles orbiting the Sun 4.6 Ga in the solar nebula. This polarized transmitted light image of a thin section shows that the meteorite is composed almost entirely of millimeter-sized chondrules and chondrule fragments, with only small amounts of filling material. Minerals present in chondrules imply that they were heated to their melting points by a transient heat source, possibly shock waves in the solar nebula gas. Image width is 5 mm.

secondary condensates or other materials that underwent significant chemical processing in the nebula. Today's meteorites are samples of bodies that formed in the transition zone between the outer and inner planets, and present to us a remarkable melange of materials with different chemical, physical, and mineralogical proprieties. The chondrules found in meteorites are perhaps the most perplexing contributors to this melange (Fig. 3.8). They are millimeter-sized glass and crystalline silicate materials whose contents indicate that they must have been heated above their melting points (>1300 °C) when they were individual particles in the solar nebula (beginning of Stage B in Table 3.2). To produce these chondrules, at least some regions of the nebula must have had short-lived energetic events or environments capable of efficiently converting most of the local matter into molten silicate droplets of millimeter size. In these regions, as much as 75% of the mass of the solid material must have been converted to millimeter chondrules, which then became the primary building blocks for assembling larger bodies. The origin of chondrules is not well understood, although transient melting by nebular shocks is an increasingly popular explanation (Desch *et al.*, 2002).

Chondrules and components called "calcium aluminum inclusions," the latter forming in even more severe environments, appear to have formed in extreme conditions in the inner solar nebula. It was therefore a surprise to find fragments of these materials when the

Stardust spacecraft returned the first samples from a comet (Brownlee *et al.*, 2006). Their presence inside a body that formed near Pluto shows that there was a moderate level of mixing of dust and rocks between the inner and outer solar nebula during Stages A and B.

Stage C

As bodies became larger (now called *planetesimals*), gravity ultimately began to dominate and strongly influence growth and scattering. Accretion is often envisioned as initially occurring by accumulation of solid materials in annular *feeding zones*, each zone ultimately forming a single planet (Stage C in Table 3.2). Objects that were not destroyed by high-speed collisions grew at a uniform rate: analogous to being steadily painted by a spray gun, the growth rate ds/dt of an object of radius s is initially constant. Both ds/dt and the mass accretion rate per unit area of the object depend on the product of the volume density ρ of matter[15] approaching the body and its relative approach speed V. These two parameters in turn can be related to the local surface density Σ of the nebular disk and its local angular rotation rate ω (radians per second).

Surface density, the integrated column density perpendicular to the plane of the disk, is a fundamental property of any disk. It decreases with radial distance from the center of the solar nebula, and its value at a distance r can be estimated from considering the mass of the planet that formed at that distance: a minimum value for Σ is the total mass in the annular ring needed to form a particular planet divided by the surface area of the ring. In this way the minimum surface density of solid materials at Earth's orbital distance is estimated to have been ~ 20 g cm^{-2}. For an estimated thickness of the disk $h \sim 0.1$ AU, the volume density at that time was thus $\rho = \Sigma / h \sim 10^{-11}$ g cm^{-3}.

From basic physics, the orbital angular velocity ω at a distance r depends only on the mass of material M inside the body's orbit: $\omega \propto M / r^{3/2}$. The Earth, at 1 AU from the Sun of mass 2×10^{33} g, travels at $\omega \sim 2 \times 10^{-7}$ radians/s (an orbital velocity of 30 km/s) to maintain its circular orbit in the Sun's gravitational field. The mean *relative* velocity V for accretion, however, is the *dispersion* of orbital speeds about their mean, i.e., the typical relative velocity of bodies at a given distance from the Sun. If all bodies had perfectly circular orbits in exactly the same plane, the relative velocity of nearby particles would be zero. With increasing inclination angles (with

respect to the plane of the disk) and eccentricity of orbits, however, the relative velocities increase. An approximation to this relative velocity at a particular distance r from the center is to multiply the orbital velocity (ωr) by h/r, the thickness-to-radius ratio of the nebula. This ratio also represents the fractional component of a body's total orbital speed in the vertical direction, as well as a typical angle (in radians) at which an inclined orbit crosses the disk mid-plane. For example, a body orbiting at 1 AU with a circular orbit inclined by 5° encounters a body with 0° orbital inclination with a collision speed of $\sim 10\%$ of the orbital speed, or ~ 3 km/s. The half thickness of a disk composed of a large number of such bodies would thus be about 10% of its radius, or 0.1 AU.

Combining all these relations, one finds that the growth rate of the accreting body's radius is

$$\frac{ds}{dt} \propto \rho V \propto \left(\frac{\Sigma}{h} \right) \cdot \left(\frac{\omega r h}{r} \right) \propto \Sigma \omega.$$

Since Σ and ω are properties of the disk at a given distance from the center, all bodies within a given feeding zone initially grow at the same rate. Note that the growth process is much slower at greater distances, where the disk is less dense and rotation rates slower. If we did not know that the outer solar system existed, we might well have predicted that it could not exist!

When a body becomes large enough, *gravitational focusing* becomes important. Particles that would miss a gravity-free body are deflected onto hyperbolic paths that intersect the growing body, resulting in collisions and enhanced accretion. The gravitational enhancement factor f is the effective increase in the cross-sectional "gathering" area of the accreting body and is equal to L^2/S^2, as defined in Fig. 3.9. It can also be derived from the conservation of energy and angular momentum as

$$f = 1 + \frac{V_{esc}^2}{V^2}.$$

The *escape velocity* V_{esc} refers to the speed that any particle would need to acquire to escape the gravitational influence of the accreting body from its surface.[16] For rocky bodies of a given density, it is directly proportional to the size of the body. For instance, when Earth had grown to half its final size,

---

[15] This volume density is the total mass of material divided by the total volume of space; it is *not* the density of the solid material itself.

[16] The relevance of escape velocity to a problem of accretion can be understood by noting that the escape velocity is also the velocity (just before impact) of an *infalling* body that starts from rest at a great distance.

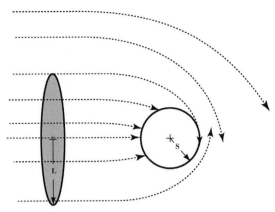

FIGURE 3.9 The gravitationally enhanced growth rate of a body accreting solid nebular debris is proportional to L^2/S^2, the ratio of areas of the gravitational capture cross section and the geometric cross section. Particles passing through the gray disk impact the growing body while those outside the disk miss and are scattered into other orbits.

its escape velocity was roughly half its present value of 11.2 km/s. If the approach velocity is appreciably smaller than the escape velocity, it can be seen that the enhancement factor is large. Numerical simulations show that gravitational focusing effects cause the largest bodies within a feeding zone eventually to dominate and grow much faster than all the others, consuming or scattering the bulk of the smaller debris.

Stage D

The strong gravitational interactions at the end of Stage C and throughout Stage D have two important effects: (1) the final stages of accretion involve the formation of many ~0.01 to 0.1 Earth-mass bodies (bodies of Moon to Mars size called *planetary embryos*) in each feeding zone, and (2) especially important for astrobiology, collisions with these large bodies transport materials from one feeding zone to the next.

The production of large embryos means that the assembly of the "last survivor" (the ultimate planet) in a feeding zone involves collisions with several bodies of planetary or near-planetary size; for example, formation of the Earth probably involved impact with several bodies in the 1,000 to 5,000 km diameter range. Collisions of such huge bodies would inject heat and compounds such as water directly into the interiors of growing planets. More generally, the enormous violence of such collisions probably plays fundamental and somewhat stochastic roles in determining the initial heat content, interior structure and even composition of young planets. Such impacts could lead to loss of volatiles and oceans, but could also inject water deep

into the interior where it could be absorbed by hot, high-pressure rock. In contrast, if a planet like Earth grew mostly through low-energy impacts of, say, 10 km-sized bodies, most of the energy of each impact would be radiated back into space and the impactor's volatile content (e.g., water and carbon) would be lost by ejection. Also, note that the favored model for the formation of the Moon invokes the early impact with Earth of an embryo greater than the mass of Mars.

The scattering of materials between feeding zones weakens the concept of feeding zones as separate entities. It appears that much of Earth's mass indeed came from our feeding zone between Venus and Mars, but important accretion also occurred from objects outside our zone. Even today, we suffer the long-lived "tail" of this accretion process as over 30,000 tons/yr of cometary and asteroidal material pummel the Earth.

An important aspect of this process for astrobiology is the stochastic nature of delivery of water, carbon, and other outer solar system materials to habitable zone planets. In our Solar System the source of the Earth's water is a major conundrum. As discussed in the previous section, it is likely that feeding zones near the Earth's distance from the Sun were dominated by solids that were intrinsically water-free and depleted in volatile materials. This is dramatically illustrated by comparing the water and carbon contents of the Earth with those of carbonaceous chondrite meteorites, which formed beyond 2.5 AU. Some of these meteorites have bulk carbon contents of >4% and water contents (in hydrated silicates) of >10%, fully two orders of magnitude higher than the bulk Earth. If critical building blocks of life such as carbon and water are carried by only a few large bodies from distant, colder feeding zones, then the amount of material transferred to a particular HZ planet can greatly vary from one case to the next. If water were delivered by a flux of many small particles, then it would be more predictable, but if it arrives in a few Moon- or Mercury-sized packages then the magnitude of the total delivery will be highly variable in different cases. The efficiency of the delivery will also depend on the size of the object, its place of origin, impact speed, location of impact on the planet and other factors. Numerical simulations by Chambers (2001) and Raymond *et al.* (2003) indicate that HZ planets as a group acquire a very large range in water content, driven purely by chance (Fig. 3.10). In their simulations of the formation of Earth-like planets, most ended up with much more water than Earth has today in its oceans. But note that Earth has an undetermined amount of water in its

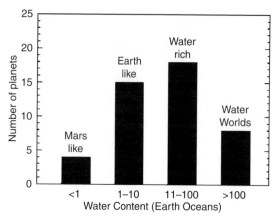

FIGURE 3.10 Water contents (measured in units of Earth's oceans) of 42 terrestrial planets formed by accretion in numerical simulations. The delivery of water by Moon- to Mars-sized bodies formed in the outer cold nebula is highly stochastic and results in habitable zone planets whose water contents vary by orders of magnitude. (Raymond *et al.*, 2003.)

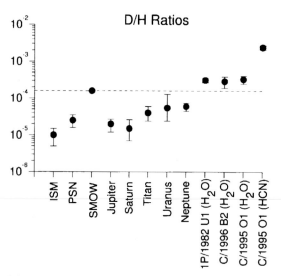

FIGURE 3.11 Deuterium-to-hydrogen (D/H) ratio of comets (the four rightmost points) compared to Earth's oceans. SMOW is Standard Ocean Water, ISM is interstellar medium, and PSN is the solar nebula. The higher D/H values of the comets indicate that comets with similar compositions could not have been the dominant contributors to Earth's oceans. (Huebner, 2002.)

mantle, and it may have lost oceans of water early in its history, say during formation of the Moon.

An important clue to the origin of Earth's water and perhaps other volatile compounds is the deuterium-to-hydrogen (D/H) ratio of the oceans. As seen in Fig. 3.11, the oceans are isotopically lighter than the few comets where D/H has been measured. Since any

loss of water over Earth's history preferentially loses H, the discrepancy at the time of Earth's formation was probably even worse. Comets that formed near the edge of the Solar System (such as the four for which we have measured D/H ratios) are therefore not likely to be the major source of Earth's water, but ice-bearing comets that formed in the warmer regions nearer to the solar nebula's snow line could match the oceans' D/H ratio.

3.10 Comments and conclusions

The formation and evolution of habitable planets is complex and poorly understood. For some habitability issues just a broad understanding may be acceptable, but for others, fine detail is needed. A goal is to understand planets well enough that they can be modeled and predictions can be made about their formation and evolution. But the reality is that, like people, planets are very complex and a reasonable understanding of their functions is truly a formidable task.

A prime example is the composition of terrestrial planets. They often are considered to have similar compositions except for differing amounts of oxygen due to variations in the mean oxidation state of iron. It is also often stated that planet compositions can be simply made by mixing x percent of nebular component A, y percent of component B, etc. As discussed above, a popular idea for Earth's composition is that it is a mix of volatile-poor materials from its feeding zone and a salting of volatile-rich material from the outer Solar System. Drake and Righter (2002) conclude, however, that the compositions of Earth and other terrestrial planets cannot be modeled as a simple mix of local and distant components. For example, Earth and its rocky neighbors are distinguished from each other by distinctly different Al/Si and Mg/Si ratios, oxygen and osmium isotopic compositions, and other element ratios. They suggest that the Earth's bulk composition does not admit of any delivered mixture of known Solar System materials. It appears that many unknown effects are involved in determining a planet's final composition, which of course is one of its most basic properties and one that surely affects habitability.

The formation and evolution of habitable planets is a central issue in astrobiology because life, at least as we know it, needs a home to provide its environmental needs. Planets like Earth are oases in a Universe that is otherwise quite hostile to life. Stars, the interstellar medium, comets, asteroids, most moons, giant gaseous planets, and even most terrestrial planets do not even come close to providing stable Earth-like environments that could harbor life analogous to terrestrial organisms. It is likely that terrestrial planets are abundant in the cosmos, but their habitability by complex or even the simplest organisms remains an open question and an important challenge to astrobiology.

Great strides have been made in the understanding of planets and some aspects of their habitability, but much remains to be done. The dramatic discovery of large numbers of extrasolar planets has not only opened up new opportunities to understand the astronomical role of planets, but has also forged bonds of common interest between astronomical, planetary, and biological sciences. Planet discovery has so far been largely limited to giant planets, but we will soon know how common terrestrial planets are, where they are located, and perhaps something about their atmospheres. The future of this field is bright but also limited by the inherit difficulty of telescopic study of faint and distant Earth-like planets orbiting other stars. Even for the closest stars, a view comparable to Galileo's first observation of our own Solar System's planets is far beyond the grasp of current capabilities. The most detailed information about habitable bodies will probably continue for a long time to be dominated by the study of Solar System bodies.

We continue to increase our knowledge of the Earth and other Solar System bodies, but one great challenge for the science of astrobiology is how to apply what we can study in detail in our own backyard to the cosmos as a whole. On the other hand, it is remarkable how *little* is known about Earth's physical and biological past. Examples include details of the long-term chemical evolution of the atmosphere, the mix of oceans and land, the roles of volcanic activity, and the workings over time of the Earth's many cycles involving the interplay of life, land, ocean, atmosphere, and even interior processes. There also has been only limited investigation into the long-term futures of Earth and the other terrestrial planets. Basic processes such as life's adaptation to increasing solar brightness, the loss of oceans, the end of plate tectonics, the biology of an ocean-free Earth, the cooling of the core, the future evolution of the atmosphere, and the response of life to very severe asteroid impacts all raise fundamental questions for astrobiology that have barely been studied.

In addition to more work on Earth, one of the most significant tasks of astrobiology is to continue the exploration of Mars. Detailed comparison of Mars and Earth should provide improved understanding of what is involved in the formation and maintenance of habitable planets. The ongoing plans of international investigations of Mars will, in the coming decades, provide considerable insight into the history and workings of this small and different, but nonetheless most Earthlike, fellow planet. This new information should also provide strong evidence for or against the past or present existence of life on Mars. Newly discovered sediments on Mars that appear to have been saturated with water for prolonged periods (Section 18.8) provide remarkable sites for future searches for chemical, isotopic, or even direct fossil evidence for life. Proven habitability for any type of organism on Mars or elsewhere would greatly improve our understanding of the nature of planets that could reasonably be an "abode of life," as Percival Lowell called Mars a century ago. It is difficult to generalize with only one example of life at hand, but astrobiology doggedly seeks to find that second case.

References

Bernstein, M. P., Dworkin, J. P., Sandford, S. A., Cooper, G. W., and Allamandola, L. J. (2002). Racemic amino acids from the ultraviolet photolysis of interstellar ice analogues. *Nature*, **416**, 401–403.

Boss, A. P. (2003). Rapid formation of outer giant planets by disk instability. *Astrophys. J.*, **599**, 577–581.

Brownlee, D. E., Tsou, P., Aléon, J. *et al.* (2006). Comet 81P/Wild 2 under a microscope. *Science*, **314**, 1711–1716.

Crowin, J. R. and Pizzarello, S. (1997). Enantiomeric excesses in meteoritic amino acids. *Science*, **275**, 951–955.

Cuzzi, J. N., Hogan, R. C., Paque, J. M., and Dobrovolskis, A. R. (2001). Size-selective concentration of chondrules and other small particles in protoplanetary nebula turbulence. *Astrophys. J.*, **546**, 496–508.

Chambers, J. E. (2001). Making more terrestrial planets. *Icarus*, **152**, 205–224.

Desch, S. J. and Connolly, H. C. (2002). A model of the thermal processing of particles in solar nebulashocks: application to the cooling rates of chondrules. *Meteoritics and Planetary Science*, **37**, 183–207.

Drake, M. J. and Righter, K. (2002). Determining the composition of the Earth. *Nature*, **416**, 39–44.

Ehrenfreund, P., Bernstein, M. P., Dworkin, J. P., Sandford, S. A., and Allamandola, L. J. (2001). The photostability of amino acids in space. *Astrophys. J.*, **550**, L95–L99.

Forget, F. and Pierrehumbert, R. T. (1997). Warming early Mars with carbon dioxide clouds that scatter infrared radiation. *Science*, **278**, 1273.

Franck, S., Kossacki, K., and Bounama C. (1999). Modeling the global carbon cycle for the past and future evolution of the Earth system. *Chemical Geology*, **159**, 305–317.

Franck, S., Block, A., von Bloh, W., Bounama, C., and Schellnhuber, H. J. (2001). Planetary habitability: is Earth commonplace in the Milky Way? *Naturwissenschaften*, **88**, 416–426.

Gaidos, E. J. (2000). A cosmochemical determinism in the formation of Earth-like planets. *Icarus*, **145**, 637–640.

Gibb, E. L., Whittet, D. C. B., Schutte, W. A., *et al.* (2000). An inventory of interstellar ices toward the embedded protostar W33A. *Astrophys. J.*, **536**, 347–356.

Gonzalez, G., Brownlee, D., and Ward, P. (2001). The Galactic Habitable Zone: Galactic chemical evolution. *Icarus*, **152**, 185–200.

Hart, M. (1979). Habitable zones about main sequence stars. *Icarus*, **37**, 351–357.

Huebner, W. F. (2002). Composition of comets: observations and models. *Earth, Moon and Planets*, **89**, 179–195.

Kasting, J. F. (1988). Runaway and moist green house atmospheres and the evolution of Earth and Venus. *Icarus*, **74**, 472–494.

Kasting, J. F., Whitmine, D. P., and Reynolds, R. T. (1993). Habitable zones around main sequence stars. *Icarus*, **101**, 108–128.

Kuan, Y., Charnley, S. B., Huang, H., Tseng, W., and Kisiel, Z. (2003). Interstellar glycine. *Astrophys. J.*, **593**, 848–867.

Kuan, Y., Charnley, S. B., Huang, H., Kisiel, Z., Ehrenfreund, P., Tseng, W., and Yan, C. (2004). Searches for interstellar molecules of potential prebiotic importance. *Adv. Space Res.*, **33**, 31–39.

Laws, C., Gonzalez, G., Walker, K. M., Tyagi, S., Dodsworth, J., Snider, K., and Suntzeff, N. B. (2003). Parent stars of extrasolar planets. VII. New abundance analyses of 30 systems. *Astron. J.*, **125**, 2664–2677.

Lin, D. N. C., Bodenheimer, P., and Richardson, D. C. (1996). Orbital migration of the planetary companion of 51 Pegasi to its present location. *Nature*, **380**, 606–607.

Mandell, A. M., Raymond, S. N., and Sigurdsson, S. (2007). Formation of Earth-like planets during and after giant planet migration. *Astrophys. J.*, **660**, 823–844.

Meyer, L., Quinn, T., Wadsley, J., and Stadel, J. (2002). Formation of giant planets by fragmentation of protoplanetary disks. *Science*, **298**, 1756–1759.

Mischna, M. A., Kasting, J. F., Pavlov, A., and Freedman, R. (2000). Influence of carbon dioxide clouds on early martian climate. *Icarus*, **145**, 546–54.

Neidhardt, F. C., Ingraham, J. L., and Schaechter, M. (1990). *Physiology of the Bacterial Cell: A Molecular Approach*. Sunderland, MA: Sinauer Association.

Ovellette, N., Desch, S. J., Hester, J. J., and Leshin, L. A. (2005). A nearby supernova injected short-lived radionucleides into our protoplanetary disk. In *Chondrites and the Protoplanetary Disk*, eds. A. N. Krot, E. Scott, and B. Reipurth, pp. 527–538. Tucson: University of Arizona Press.

Pendleton, Y. J. and Allamandola, L. J. (2002). The organic refractory material in the diffuse interstellar medium: mid-infrared spectroscopic constraints. *Astrophys. J. Supp. Series*, **138**, 75–98.

Raymond, S. N., Quinn, T. R., and Lunine, J. L. (2003). Making other Earths: dynamical simulations of terrestrial planet formation and water delivery. *Icarus*, **168**, 1–17.

Sandford, S. A. (2002). Interstellar processes leading to molecular deuterium enrichment and their detection. *Planet. Space Sci.*, **50**, 1145–1154.

Sterner, R. W. and Elser, J. J. (2002). *Ecological Stoichiometry: the Biology of Elements from Molecules to the Biosphere*. Princeton: Princeton University Press.

Stevens, T. O. and McKinley, J. P. (1995). Lithoautotrophic microbial ecosystems in deep basalt aquifers. *Science*, **270**, 450–454.

Walker, J. C. G., Hays, P. B., and Kasting, J. F. (1981). A negative feedback mechanism for the long-term stabilization of the Earth's surface temperature. *J. Geophys. Res.*, **86**, 9776–9782.

Ward, P. W. and Brownlee, D. E. (2000). *Rare Earth; Why Complex Life is Uncommon in the Universe*. New York: Copernicus Books.

Ward, P. W. and Brownlee, D. E. (2003). *Life and Death of Planet Earth: How the New Science of Astrobiology Charts the Future of Our World*. New York: Times Books.

Wood, J. A. (1990). Meteorites. In *The New Solar System*, eds. K. Beatty and A. Chaikin. Cambridge, MA: Sky Publishing.

FURTHER READING AND SURFING

Beatty, J. K., Peterson, C. C., and Chaikin, A. (eds.) (1999). *The New Solar System* (fourth edn.). Cambridge, MA: Sky Publishing. Detailed, well illustrated, authoritative, and very readable.

Brandt, J. C. and Chapman, R. D. (2004). *Introduction to Comets* (second edn.). Cambridge: Cambridge University Press. Excellent overview.

Canup, R. M. and Righter, K. (eds.) (2000). *Origin of the Earth and Moon*. Tucson: University of Arizona Press.

De Pater, I. and Lissauer, J. J. (2001). *Planetary Sciences*. Cambridge: Cambridge University Press. Excellent, comprehensive graduate-level text.

Doyle, L. R. (1996). *Circumstellar Habitable Zones*. Menlo Park, CA: Travis. Proceedings of an important conference; very well edited.

Ehrenfreund, P., Irvine, W., Becker, L., *et al.* (2002). Astrophysical and astrochemical insights into the origin of life. *Reports of Progress in Physics*, **65**, 1427–1487. Good technical review.

Lauretta, D. S. and McSween, H. Y. (eds.) (2006). *Meteorites and the Early Solar System*. Tucson: University of Arizona Press.

Lodders, K. and Fegley, B. (1998). *Planetary Scientist's Companion*. New York: Oxford University Press. Modest-sized, but comprehensive compilation of solar system data.

4 Planetary atmospheres and life

David Catling, *University of Washington**
James F. Kasting, *Pennsylvania State University*

Earth is *not* the only body in the Solar System that is habitable. Life as we know it requires liquid water and free energy gradients, both of which probably also exist on Mars and Europa, although liquid water on those bodies is restricted to the subsurface. Earth is, however, the only planet in the Solar System that has liquid water at its surface. Similar planets may exist around other stars (Chapter 21) and would be of profound interest for two reasons. First, biology on such planets might resemble life on Earth. Second, the biosphere on such planets would interact with the planet's atmosphere and could modify it in a way that may be detectable remotely. Today, life may be thriving on Mars or Europa but its discovery will require subsurface exploration. In contrast, we might be able to tell whether a distant Earth-like planet is inhabited by measuring the spectrum of its atmosphere.

Thus, from an astrobiological standpoint, one of the most fundamental characteristics of a planet is its surface temperature T_s. If T_s is not within the range in which liquid water can exist, remotely detectable life will probably not exist there. Consequently, the first part of this chapter is concerned with planetary surface temperatures. The constraint on temperature is not as obvious as $0 < T_s < 100\,°C$. Water boils at $100\,°C$ at Earth's surface because the overlying atmospheric pressure is 1 bar ($= 10^5\,N/m^2 = 10^5$ Pascal) and because the atmosphere is not in equilibrium with water at this temperature. If Earth's entire surface were at $100\,°C$, the oceans would still not boil because the overlying atmospheric pressure would then be 2 bars (1 bar N_2-O_2 and 1 bar H_2O). Thus, planetary atmospheres act like pressure cookers, with gravity taking the place of the pressure cooker's lid. In general, liquid water is stable all the way up to a temperature of $374\,°C$ ($647\,K$) for pure water, or even higher for salt water.

More practical upper limits on T_s for a habitable planet can be considered. The currently known lower limit for microbial activity is $\sim -20\,°C$ (in very salty solutions; Chapter 15). The currently known upper limit for prokaryotic microbial life is $121\,°C$ (Kashefi and Lovley, 2003) while that for more complex eukaryotic life is $60\,°C$ (Rothschild and Mancinelli, 2001) (Chapter 14). The latter temperature is also about the point where a 1-bar atmosphere like Earth's rapidly begins to lose its water because the molecules can move efficiently to the upper atmosphere. This process involves energetic solar ultraviolet photons dissociating (breaking up) water molecules at the top of the atmosphere, after which the light H atoms escape to space (Section 4.2.2).

In summary, from about $-20\,°C$ to $60\,°C$ would seem to be a probable temperature range of interest for habitable planets populated by organisms similar to those found on Earth.

Planetary habitability is also affected by atmospheric composition. There are several factors to consider here, which are best discussed within the context of our own planet Earth, for which we have considerable data. Was a certain atmospheric composition necessary for the origin of life? How did atmospheric composition change once life had originated? When and why did O_2 become an abundant atmospheric constituent? Could we infer the presence of life by remotely analyzing Earth's atmosphere, now or in the past? And did the availability of O_2 set the tempo for the evolution of complex life? These questions will form the basis of the second half of the chapter.

4.1 Fundamentals of global climate

We begin by outlining a few fundamental concepts governing planetary climates. These will prove useful in understanding how Earth's climate has evolved and why the climates of Mars and Venus evolved so differently.

* Currently at the University of Bristol.

Planets and Life: The Emerging Science of Astrobiology, eds. Woodruff T. Sullivan, III and John A. Baross. Published by Cambridge University Press. © Cambridge University Press 2007.

4.1.1 Planetary energy balance and the greenhouse effect

Earth is warmed by absorption of visible and near-infrared radiation from the Sun and is cooled by emission of thermal infrared radiation. If we treat the Earth as a *blackbody radiator*[1] with *effective temperature* T_e, we can equate the emission of thermal infrared radiation F_{IR} (given by the Stefan–Boltzmann law) with the average solar radiation absorbed by the Earth. Let us denote the annual mean solar flux at Earth's orbit by S ($=1366 \pm 3 \, W/m^2$). The Earth, with radius R, presents an area πR^2 normal to the solar beam and has a total surface area $4\pi R^2$. Thus, the globally averaged flux of solar radiation received per unit area is $S/4$, but a fraction of that flux, called the *albedo* A ($\cong 0.3$ for Earth), is reflected back to space and does not warm the Earth. Thus, the planetary energy balance (energy flux out = energy flux in) is

$$F_{IR} = \sigma T_e^4 = \frac{S}{4}(1 - A), \qquad (4.1)$$

where σ is the Stefan–Boltzmann constant. Solving for Earth's effective temperature yields $T_e = 255 \, K$. Earth, however, is *not* a blackbody, but instead has an atmosphere that warms the surface by the *greenhouse effect*. Thus, the globally averaged surface temperature T_s is in fact $\sim 288 \, K$. The difference between T_e and T_s gives the magnitude of the greenhouse effect:

$$\Delta T_g \equiv T_s - T_e = 33 \, K. \qquad (4.2)$$

In the greenhouse effect, the atmosphere warms as it absorbs upwelling infrared radiation from the Earth below. Because the atmosphere is warm it radiates. Some of this radiation is emitted downwards towards the Earth. Consequently, the Earth's surface is warmer than it would be in the absence of an atmosphere because it receives energy from two sources: the Sun and the heated atmosphere. The additional heating from the atmosphere is called the greenhouse effect.

In Earth's atmosphere today, the two most important contributors to greenhouse heating are CO_2 and H_2O. H_2O is responsible for about two-thirds of the

warming, although it acts in a different manner than CO_2 because it is near its condensation temperature, at least in the lowest layer of the atmosphere (the *troposphere*, up to 8 km at the poles and 17 km at the equator). CO_2 accounts for most of the remaining third of the greenhouse effect. The remaining ~ 2–$3 \, K$ comes from CH_4, N_2O, O_3, and various human-produced chlorofluorocarbons (CFCs). Although the total of these trace constituents is small, they are very efficient greenhouse gases: for example, each CH_4 molecule is ~ 20 times more effective than a CO_2 molecule in the modern atmosphere (Schimel *et al.*, 1996). This is because they absorb within the 7.5–12 μm wavelength region, an otherwise transparent "window" through which most of the Earth's thermal infrared radiation energy escapes to space.

Equations (4.1) and (4.2) show that the mean planetary temperature T_s depends on only three factors: (a) the solar flux (set by astronomical geometry and solar physics), (b) the albedo, and (c) the greenhouse effect. Any changes in T_s during Earth's history (e.g., as evidenced by low-latitude glaciation) can only be understood by appealing to changes in one or more of these factors. The most difficult factor to estimate well is the planetary albedo because $\sim 80\%$ of it is caused by clouds. Clouds can be observed and parameterized in Earth's present atmosphere, but their properties for other atmospheres are uncertain. Climate calculations for early Earth or for other Earth-like planets are thus subject to considerable uncertainty.

4.1.2 Climate feedbacks and feedback loops

Water vapor acts as a feedback on the climate system because it is near its condensation temperature. As the atmosphere cools, the saturation vapor pressure (the maximum vapor pressure possible before water condenses) drops. If the relative humidity[2] remains constant, then the water vapor concentration in the atmosphere will decrease proportionately. Less water vapor results in a smaller greenhouse effect, which in turn results in further cooling. Just the opposite happens if the climate warms: atmospheric H_2O increases, thereby increasing the greenhouse effect and amplifying the initial warming.

The interaction between water vapor and surface temperature can be expressed by means of a feedback diagram such as Fig. 4.1a. Here, the boxes represent

[1] A *blackbody radiator* is an ideal body that reradiates all of the electromagnetic energy that it absorbs. The spectral distribution of this radiated energy depends only on the temperature of the body and is given by the Planck distribution formula. If a body's spectrum is only approximately that of a blackbody (as is typical for planets and stars), one can still define the *effective temperature* as the

[2] The *relative humidity* is the ratio of vapor pressure to saturation vapor pressure, usually expressed as a percentage.

(a)

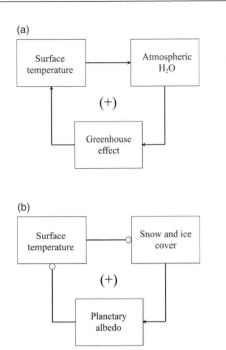

(b)

FIGURE 4.1 Feedback loops for the Earth's climate. (a) The water vapor positive feedback loop. (b) The H_2O and snow-and-ice albedo positive feedback loop.

three components of a simplified climate model: surface temperature, atmospheric H_2O, and the greenhouse effect. The arrows connecting them represent *positive couplings*, meaning that an increase (decrease) in one component causes a corresponding increase (decrease) in the next. All the couplings are positive in this diagram, so it constitutes a *positive feedback loop*. This particular feedback loop is very important for Earth's climate, essentially doubling the effect of any climatic perturbations such as changes in solar flux or in atmospheric CO_2.

A second important feedback loop is shown in Fig. 4.1b. This loop describes the interaction between surface temperature and the fraction of Earth's surface covered by snow and ice. The lines with circular endings represent *negative couplings*, e.g., an *increase* in surface temperature causes a *decrease* in snow and ice cover. The second coupling is positive, however, because more snow and ice causes more reflection of sunlight (compared to rock or vegetation) and thus increases the albedo. But an increase in albedo causes a decrease in surface temperature, so the third coupling is again negative. Two negatives make a positive in this type of diagram, so the overall feedback loop is again *positive*. The snow-and-ice albedo feedback loop has played a major role in the advances and retreats of the ice sheets over the past 2 Myr.

Positive feedbacks are unstable because they tend to make a parameter such as T_s greatly change, rather than stay near an equilibrium value. Since Earth's climate system is stable, there must also be negative feedbacks for T_s. The most basic negative feedback (not shown) is the interaction between surface temperature and the outgoing infrared flux F_{IR}. As T_s increases, F_{IR} increases, which is a loss of energy leading to a cooler surface (lower T_s). This creates a negative feedback loop that is so fundamental that it is often overlooked, although it is the reason that Earth's climate is stable on short timescales. On long timescales, however, factors that influence climate can change, and we need to look for something else to ensure stability. We will argue below that the most important long-term climate feedback involves the interaction between atmospheric CO_2 and surface temperature. Let us consider what factors affect climate over the long term and see why negative feedbacks are required to stabilize it.

4.2 The faint young Sun problem

Detailed models of the Sun's evolution indicate that it, like other stars, gets brighter as it ages in its *main sequence* phase. During this long-lasting, relatively stable phase, the Sun produces energy by fusing four hydrogen nuclei ^1H (4 protons) into one ^4He nucleus deep within its core. Ionized matter in the interior of the Sun behaves like an ideal gas. Thus, the relationship between its pressure P and temperature T is $P = nkT$, where n is the number density of all particles (electrons and nuclei) and k is Boltzmann's constant. But when hydrogen nuclei fuse to form helium, n decreases slightly. This causes contraction as the weight of overlying material presses inward. As the core contracts, gravitational acceleration g increases because $g \propto (\text{radius})^{-2}$. Core pressure P increases with g because $P = $ force per unit area $= (\text{mass per unit area in a column}) \times g$, where the amount of overlying material remains constant. Thus, T must also increase to maintain a pressure balance and the core is heated. Higher core temperatures cause fusion reactions to proceed faster, so the Sun produces more energy. The solar luminosity (total power output) must therefore increase with time. Standard models predict that the Sun's luminosity was about 30% less when it formed 4.6 Ga and that it has increased roughly linearly with time ever since (Fig. 4.2). This theoretical prediction is considered robust because it arises from basic physics, namely nuclear fusion and hydrostatics, and is

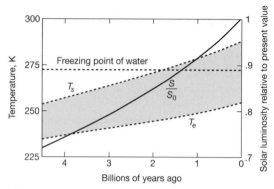

FIGURE 4.2 Calculations of the surface temperature of the Earth over its lifetime, illustrating the "faint young Sun problem." T_s is the surface temperature assuming a present-day atmospheric composition, T_e is the planetary equilibrium temperature, S is the solar luminosity at times in the past and S_0 is the present solar luminosity. (From Kasting et al., 1988.)

consistent with observations of many other solar-like stars of various ages.

Although an understanding of increasing solar luminosity began in the 1950s, it was not until much later that Sagan and Mullen (1972) noted the implications for planetary climates. If one lowers the value of S by 30% in Eq. (4.1), holding the albedo and greenhouse effect constant for simplicity, one finds that T_e drops to 233 K and T_s to 266 K = −7 °C, below the freezing point of water. If one then repeats this calculation with a climate model that includes the positive feedback loop involving water vapor, the problem becomes even more severe. The dashed curves in Fig. 4.2 show T_e and T_s calculated using a one-dimensional, radiative-convective climate model, assuming constant CO_2 concentrations and fixed relative humidity (Kasting et al., 1988). The results are similar to those predicted earlier by Sagan and Mullen: T_s is below the freezing point of water prior to ∼2 Ga. Especially once the snow/ice-albedo feedback loop is taken into account, this would seem to imply that the Earth was then globally glaciated. However, geologic evidence tells us that liquid water and life were both present back to certainly 3.5 Ga and maybe even earlier. The presence of an ocean at 4.3 Ga has been deduced from ancient zirconium silicate minerals (zircons) (Wilde et al., 2001; Mojzsis et al., 2001). These zircons are enriched in heavy oxygen (^{18}O), which can result from interaction with liquid water at low temperature. The inference from the zircons is that the crust from which they formed interacted with an early liquid ocean, not a frozen one.

How can the faint young Sun problem be solved? The most likely solution involves a greater greenhouse effect in the past. A drastic decrease in cloudiness, leading to much less reflected light (a lower albedo), would also solve the problem (Rossow et al., 1982), but this seems unlikely because the ancient climate appears to have been, if anything, even warmer than today, promoting evaporation and cloud formation. Instead, there are good reasons to believe that the faint young Sun problem is best solved by the presence of abundant greenhouse gases in the early atmosphere.

4.2.1 The carbonate-silicate cycle and CO_2-climate feedback

Carbon dioxide is the second most important greenhouse gas today. Over long timescales, CO_2 is controlled by the *carbon cycle*, which has several different parts. The more familiar part is the *organic carbon cycle* in which plants (and many microbes) convert CO_2 and H_2O into organic matter and O_2 by photosynthesis. Organic matter on average can be represented as "CH_2O," so the overall reaction for (oxygenic) photosynthesis can be written as

$$CO_2 + H_2O \rightarrow CH_2O + O_2. \qquad (4.3)$$

Photosynthesis is almost entirely balanced by the twin processes of respiration and decay, both of which are the reverse of the above reaction.

There are two reasons, however, why the organic carbon cycle cannot be the primary control on CO_2 levels over long timescales. First, the living biosphere is not a large carbon reservoir – it contains only about as much carbon as does the atmosphere. Second, any small imbalance that occurs when organic carbon is buried in sediments cannot be maintained for very long because it is controlled by a negative feedback loop involving atmospheric O_2 (e.g., Catling and Claire, 2005). Basically, an increase in organic carbon burial causes an increase in atmospheric O_2, which in turn causes a decrease in organic carbon burial. Moreover, this cycle is controlled more by oxygen than by climate, so it cannot contribute appreciably to climate stability. Indeed, this cycle may have destabilized climate on more than one occasion, leading to possible global glaciation episodes (Section 4.2.4).

The part of the carbon cycle most important to long-term climate is the *inorganic carbon cycle*, also called the *carbonate–silicate cycle*. Beginning at the left-hand side of Fig. 4.3, CO_2 dissolves in rainwater

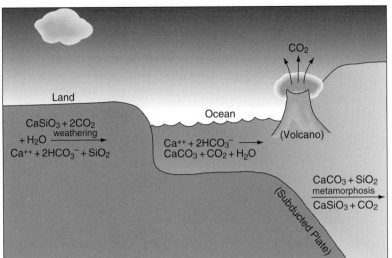

FIGURE 4.3 Earth's carbonate-silicate cycle.

to form carbonic acid, H_2CO_3. Carbonic acid is a weak acid – it is the acid in soda pop – but over long times it is strong enough to dissolve silicate rocks. We will focus here on calcium silicates, which can be represented by the simplest silicate mineral, wollastonite ($CaSiO_3$). The products of this silicate *weathering*, including calcium (Ca^{++}) and bicarbonate (HCO_3^-) ions and dissolved silica (SiO_2), are transported by rivers to the ocean. There, organisms such as foraminifera use the products to make shells of calcium carbonate ($CaCO_3$), preserved, for example, as limestone. Other organisms such as diatoms and radiolarians make shells out of silica.[3] When these organisms die, they fall into the deep ocean. Most of the shells redissolve, but a fraction survive to be buried in sediments on the seafloor. The combination of silicate weathering plus carbonate precipitation can be represented chemically by the reaction

$$CO_2 + CaSiO_3 \rightarrow CaCO_3 + SiO_2. \qquad (4.4)$$

If this was all that was happening, we would not have a complete cycle, and all of Earth's CO_2 would end up in carbonate rocks. But the seafloor is not static and is continuously being created at the mid-ocean ridges and subducted at plate boundaries where oceanic plates slide beneath less dense continental plates. When this happens, the carbonate sediments are

carried down to depths with high temperatures and pressures. Under these conditions, reaction (4.4) reverses itself: carbonate minerals recombine with silica (which by this time is in the form of quartz) to reform silicate minerals, releasing CO_2 in the process. This reaction is termed *carbonate metamorphism*.[4] The CO_2 released from carbonate metamorphism re-enters the atmosphere, thereby completing the carbonate–silicate cycle. This cycle has a timescale of $\sim$200 Myr and replenishes all of the CO_2 in the combined atmosphere–ocean system every $\sim$0.5 Myr; at any time almost all of the carbon ($>$99.99%) is in the Earth's crust.

Let us now consider the implications of this cycle for the faint young Sun problem. Suppose, for the sake of argument, that the oceans were completely frozen over. In this case, the *hydrological cycle*[5] would have almost completely shut down. Some H_2O would still cycle by sublimation and snowfall, but surficial liquid water would be completely absent. Weathering of rocks requires liquid water to proceed at an appreciable rate, so the rate of silicate weathering would drop drastically. Volcanic activity, however, would continue unabated. Figure 4.3 makes it appear as if the volcanic CO_2 source might disappear, too, but in fact CO_2 would continue to be pumped into the atmosphere because (1) there are other types of volcanism that

---

[3] Although the presence of life enhances the carbonate–silicate cycle on Earth, it would operate even on a sterile planet. Calcium carbonate would still become incorporated into ocean sediments through nonbiological chemistry when the mineral simply gets saturated in seawater.

[4] *Metamorphism* refers to any geological process at depth leading to structural, mineralogical or chemical changes in rocks. In this chapter metamorphism (along with volcanism) is important as a source of *outgassing* to the atmosphere.

[5] The *hydrological cycle* refers to the cycling of H_2O between liquid, solid, and vapor phases as a result of evaporation, sublimation, rain, snow, etc.

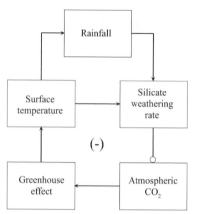

FIGURE 4.4 The negative feedback loop involving the silicate weathering rate, CO_2, and climate.

release CO_2 (e.g., mid-ocean ridge and hotspot volcanism); and (2) the mean lifetime of seafloor material is ~60 Myr between creation and subduction, and so there would be a long delay before all subducted material was carbon free. Thus, volcanic CO_2 would accumulate in the atmosphere until the greenhouse effect eventually became strong enough to melt the ice, allowing the hydrological cycle to resume. For a completely frozen Earth, about 0.3 bars of CO_2 (~1,000 times today's atmospheric amount) would be required to do this. At present rates of volcanism, this much CO_2 would take only ~10 Myr to accumulate.

Although there may indeed have been a few short times in Earth's history when such an extreme scenario actually occurred (the evidence for this is discussed at the end of Section 4.2.4), over the longest timescales this CO_2–climate negative feedback loop has also probably played a major role in *stabilizing* Earth's climate (Fig. 4.4). The key is that any change in temperature affects the silicate weathering rate, which counteracts that change. For example, a lowering temperature leads to less rain and a lower silicate weathering rate, which produces less atmospheric CO_2 and less greenhouse effect, thereby causing the temperature to rise.

4.2.2 Failure of climate stabilization on Mars and Venus

One way of evaluating the importance of the CO_2–climate feedback loop is by examining what happened to Venus and Mars. Neither planet has a habitable surface according to our previous definition. Venus has a mean surface temperature of 460 °C, well above the critical point for water, while Mars's mean temperature is −55 °C, well below freezing (Table 4.1).

What went wrong with the carbonate–silicate cycle on these planets?

Venus was close enough to the Sun that its higher temperature led to loss of water by photodissociation followed by escape of hydrogen to space (see Section 4.4.2 for a description of hydrogen escape for Earth). Once the water was lost, silicate weathering could not occur, so volcanic CO_2 simply accumulated in Venus's atmosphere and produced the hot, dry planet that we see today.

In the case of Mars, the situation was strikingly different. Mars formed farther from the Sun, so that for liquid water to exist a stronger atmospheric greenhouse effect would have been needed to warm its surface. It can be shown that gaseous CO_2 and H_2O by themselves could not have kept Mars warm enough early in its history when the Sun was less bright (Kasting, 1991; Colaprete and Toon, 2003). However, Mars had another, more serious problem: Mars is only 10% of Earth's mass and 15% of Earth's volume. Smaller objects cool down more rapidly, and thus widespread volcanism on Mars ceased long ago. Without volcanism, there was no mechanism for recycling CO_2, and what CO_2 there was should have accumulated in the crust. However, in spite of several spectroscopic searches, carbonate minerals have not been found in abundance on Mars. One explanation for this lack is that the early martian atmosphere (greatly aided by low martian gravity) was cumulatively blasted away to space by comet or asteroid impacts (Melosh and Vickery, 1989). For our purposes, the lesson to be drawn from Mars is that size matters: a small, geologically inactive planet is not likely to develop a stable climate or to hold on to its atmosphere.[6]

4.2.3 Complications to the faint young Sun story: methane

The explanation provided above for how early Earth avoided freezing is probably true to some extent. But silicate weathering feedback is not the only factor that helped keep the early Earth warm. As we discuss later, atmospheric O_2 concentrations are thought to have been low up until ~2.4–2.3 Ga. In a low-O_2 atmosphere, greenhouse gases such as methane (CH_4) could conceivably have been more abundant. In their 1972 paper, Sagan and Mullen proposed that high

---

[6] See Chapter 18 for more discussion of Mars and Chapter 3 for discussion of the notion of a *habitable zone* for planets over a certain range of distance from a star.

TABLE 4.1 Properties of Venus, Earth and Mars and their atmospheric compositions. Note the large disparity in the atmospheric pressures of the planets when interpreting their atmospheric composition. For example, 3.5% N_2 in the atmosphere of Venus represents 3.3 bars of N_2 (four times that of the Earth), indicating a planet that is either more volatile-rich than the Earth or much more efficiently outgassed. ppm = parts per million. (Sources: Lodders and Fegley (1998); Pollack (1991).)

Parameter	Venus		Earth		Mars	
Mean surface pressure (bar)	95.6		1.0		0.006	
Mean surface temperature (K)	735		288		218	
Mass relative to Earth (5.97×10^{24} kg)	0.815		1.0		$0.107 \approx 1/9$	
Mean radius relative to Earth (6371 km)	0.950		1.0		$0.532 \approx 1/2$	
Key gases in atmosphere (by volume)	CO_2	96.5%	N_2[*]	78.084%	CO_2	95.32%
	N_2	3.5%	O_2[*]	20.946%	N_2	2.7%
	SO_2	150 ± 30 ppm (22–42 km)	H_2O	0.1 ppm–4%	Ar	1.6%
		25 to 150 ppm (12–22 km)		(varies)	O_2	0.13%
	Ar	70 ± 25 ppm	Ar	9340 ppm	CO	0.08%
	H_2O[¶]	30–70 ppm (0–5 km)	CO_2[*§]	~280 ppm	H_2O	0.03%
	CO[¶]	45 ± 10 ppm (cloud top)		(pre-industrial)		(varies)
		17 ± 1 ppm (12 km)		~380 ppm	NO	~100 ppm
	He	12 (+24/−8) ppm		(year 2006)	Ne	2.5 ppm
	Ne	7 ± 3 ppm	Ne	18.18 ppm	Kr	0.3 ppm
			^{4}He	5.24 ppm		
			CH_4[*]	1.7 ppm		
			Kr	1.14 ppm		
			H_2[*]	0.55 ppm		
			N_2O[*]	~230 ppb		
			CO[*]	125 ppb		

Notes:
[¶] Altitude-dependent.
[*] Under varying degrees of biological influence.
[§] CO_2 levels increased at a rate of ~1.9 ppm/yr over 1995–2005 due to fossil fuel burning by humans.

concentrations of ammonia (NH_3) were the solution to the faint young Sun problem. This idea now seems unlikely because NH_3 is easily dissociated by ultraviolet photons unless shielded by atmospheric O_2 and O_3. CH_4, though, is not subject to this problem, as it is broken up only by photons with wavelengths below ~145 nm, and this has led to the suggestion that CH_4 may have been an important greenhouse gas on the early Earth (Kiehl and Dickinson, 1987; Lovelock, 1988). Indeed, photochemical models (e.g., Pavlov et al., 2001) predict a ~10,000–20,000-year lifetime for CH_4 in a low-O_2 atmosphere, as opposed to ~10 years today. The present biological methane flux could have supported an atmospheric CH_4 *mixing ratio*[7] in the

[7] The *mixing ratio* is the fraction of particles of a given type in a gas.

past of several hundred ppm (Pavlov et al., 2001), instead of the scant 1.7 ppm today (*ppm is parts per million*).

By itself this of course does not prove that CH_4 was abundant during the Archean eon (>2.5 Ga), because we have assumed that methane was produced at the same rate then as now. This is plausible, however, since microbes that make methane (*methanogens*) are thought to be evolutionarily ancient (Chapter 10). Virtually all methanogens can subsist by way of the reaction $CO_2 + 4\ H_2 \rightarrow CH_4 + 2\ H_2O$. Furthermore, the energy sources used by methanogens – H_2, CO_2, acetate, and formate – are all thought to have been abundant during the Archean. Thermodynamic considerations suggest that methanogens should have converted most of the available atmospheric H_2 into CH_4

by this pathway (Kasting *et al.*, 2001). H_2 mixing ratios without any biological processes present are thought to have been of the order of 10^{-3} ($=1,000$ ppm), so CH_4 mixing ratios after methanogens evolved should have been about half that, given that the total number of hydrogen atoms would be preserved. This estimate is consistent with the one obtained in the previous paragraph, although the argument used is entirely different.

If CH_4 was an abundant constituent of the Archean atmosphere, then the greenhouse effect could have been large even in the absence of high CO_2 concentrations. Calculations of global mean surface temperatures (Pavlov *et al.*, 2000) for the Late Archean (2.8 Ga), when the solar flux was ~80% of its present value, indicate that if the CH_4 mixing ratio was 10^{-4}–10^{-3}, as suggested above, then the surface could have been warm even if the mixing ratio of CO_2 was no higher than today. Furthermore, most methanogens are thermophilic (heat-loving), and those with higher optimum growth temperatures reproduce faster than those that grow best at lower temperatures. This creates a positive feedback loop that may have tended to keep the Archean climate warm. This positive feedback was limited, though, by an additional complication: when CH_4 becomes as or more abundant than CO_2 in a planet's atmosphere, photochemical models (e.g., Pavlov *et al.*, 2001) indicate that a hydrocarbon haze is formed similar to that observed today on Saturn's moon Titan (Chapter 20). This haze, in turn, can cool the planet's surface via the *anti-greenhouse effect*. In the anti-greenhouse effect, solar radiation is absorbed and reradiated high in the atmosphere without ever reaching the surface. The net effect is the same as having a higher albedo. Thus, if the haze became too thick on early Earth, the surface would have cooled, and conditions would have become less favorable for thermophiles such as methanogens.

All of this suggests that the Archean climate may have been regulated by a negative feedback loop similar to that depicted in Figure 4.5. Starting from the box labeled surface temperature (T_s) at the left, an increase in T_s increases atmospheric CH_4 from methanogens. At the same time T_s tends to decrease atmospheric CO_2 through silicate-weathering feedback. Both factors increase the atmospheric CH_4/CO_2 ratio, thereby promoting the formation of hydrocarbon haze. This haze, though, then cools the surface as soon as its transmission of visible light drops below 70–80%. Thus, the overall feedback of both sides of the loop is negative. So, one hypothesis (albeit speculative) is that Archean

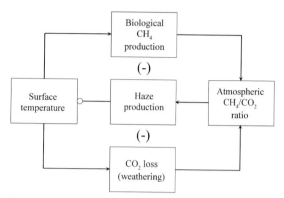

FIGURE 4.5 A negative feedback loop involving CO_2, biologically produced CH_4, and organic haze formation that may have operated in the Archean age on Earth (>2.5 Ga).

climate was stabilized by the presence of a thin organic haze layer.

An alternative hypothesis is that after oxygenic photosynthetic microbes had evolved (see Section 4.5.4), a CH_4-mediated climate could have been stabilized in negative feedback with O_2 (Catling *et al.*, 2001). In this feedback, an increase of CH_4 causes greenhouse warming, which increases rain and continental weathering rates and, hence, sedimentation rates in the ocean. Many field measurements show that organic burial rates strongly correlate with sedimentation rates (Betts and Holland, 1991). These increased organic burial rates would then elevate O_2 fluxes (via $CO_2 + H_2O = CH_2O$ (buried) $+ O_2$), and lower CH_4 levels by enhanced microbial oxidation in the water column or by atmospheric oxidation.

4.2.4 End of the methane greenhouse: the first Snowball Earth

We end our discussion of planetary climates at the point on which the remainder of the chapter is focused: the rise of O_2 in the Earth's early atmosphere. As discussed below, a variety of different geological indicators suggest that atmospheric O_2 increased abruptly from near zero prior to ~2.4–2.3 Ga to much higher values afterwards. If methane was an important contributor to the greenhouse effect during the Late Archean, then the rise of O_2 must have been significant for the climate. Indeed, the Canadian geologist Stuart Roscoe first suggested that drastic climate change occurred around the time of the rise of O_2. He observed rocks in the Huronian sequence just north of Lake Huron in southern Canada (Roscoe, 1973; Prasad and Roscoe, 1996). The sequence spans 2.45–2.2 Ga and contains evidence for glaciation, in fact three periods of glaciation, in the form of three layers of clusters of unconsolidated rock fragments bound up in a matrix (diamictites). Striations and dropstones provide further evidence of glaciers. The diamictite layers also appear to be separated by periods of intense weathering.

Roscoe also observed that the glacial layers were sandwiched by deposits indicating low O_2 conditions below and high O_2 levels above (more details are in Section 4.6.1.1). Thus, the glaciations and the rise of atmospheric O_2 appear to be related. This timing, of course, makes sense if CH_4 was an important part of the atmospheric greenhouse at this time. The rise of O_2 would have eliminated most of the methane by enabling more efficient oxidation of methane in the atmosphere,[8] as well as by constraining the environments in which methanogens could survive. The loss of methane then led to much lower temperatures and glaciation. The fact that there were two initial glacial episodes, followed by a third, even larger glaciation, suggests that both the atmospheric redox state and the climate may have been oscillating during a 0.2–0.25 Gyr period.

Evidence for glaciation at $\sim$2.4–2.2 Ga is also found in Australia, India, Russia, Scandinavia, South Africa, and elsewhere in North America (Eyles and Young, 1994). However, apart from the Transvaal in South Africa, three discrete glacial layers are not generally evident, perhaps because the rock record from this early time is poorly preserved. Furthermore, some glacial deposits in South Africa are interbedded with rocks that apparently formed at low latitudes, according to paleomagnetic analysis.[9]

It is now also known that there were three glacial episodes between 720 and 580 Ma, of which the first two were low-latitude (Hoffman and Schrag, 2002; Halverson, 2005; Xiao and Kaufman, 2006). These two glaciations, as well as the earlier ones, have been dubbed *Snowball Earth* episodes by Joe Kirschvink (1992, 2000). In these episodes, Earth's surface may have frozen over entirely to a depth of a kilometer or more. Snowball Earth glacial deposits are also overlain in some places by thick layers of carbonate rocks ("cap carbonates") with unusual textures indicating rapid deposition and unusual ratios of carbon isotopes. These cap carbonates are thought to have formed when CO_2 that built up during the ice-covered-surface period was later removed in an episode of rapid carbonate and silicate weathering (as observed). In essence, the "thought experiment" described earlier in the chapter with respect to the faint young Sun problem appears to have actually happened on several separate occasions, although the question of whether the Earth froze over entirely remains controversial.

4.3 The coevolution of atmospheric oxygen and life

We now turn for the remainder of the chapter to discussing the partial pressure of atmospheric oxygen (pO_2) through time, which has been a major factor in the evolution of life on Earth. This section gives an overview of the history of atmospheric oxygen (summarized in Fig. 4.6), while later sections supply the details.

Animals and multicellular plants require O_2 to live. The atmosphere started out with virtually no oxygen ($pO_2 \sim 10^{-13}$ bar) before life existed but now contains about 21% O_2 by volume ($pO_2 \sim 0.21$ bar). Nursall (1959) first suggested that the appearance of macroscopic animals in the fossil record reflected the emergence of an oxygen-rich atmosphere able to support the higher energy requirements of animal metabolism. Preston Cloud (1988) championed a similar view, pointing out aspects of the geologic record that indicated secular increases in the level of O_2. Today, the evidence is strong that not until about 2.4–2.3 Ga was pO_2 high enough to oxidize exposed continental rocks ubiquitously. Some also argue that there was a second increase in pO_2 around 0.8–0.6 Ga (Canfield and

---

[8] The first step in the atmospheric destruction of methane is an attack by hydroxyl OH. After the rate-limiting first step ($OH + CH_4 = CH_3 + H_2O$), a series of subsequent reactions produces net oxidation of methane: $CH_4 + 2O_2 = CO_2 + 2H_2O$. In today's troposphere, production of OH is linked to tropospheric ozone (O_3): ozone photolyzes and produces an excited oxygen atom that produces OH from water vapor (i.e., $O + H_2O = 2OH$). In turn, tropospheric ozone production is tied to the presence of O_2 in the atmosphere because the oxygen atoms within O_3 originate from O_2. In the Archean, in the absence of a stratospheric ozone layer, short wavelength, energetic ultraviolet light would have penetrated into the troposphere, allowing OH to be produced directly from H_2O photolysis ($H_2O + h\nu = H + OH$). But without O_2 to remove the H, the OH would recombine with the H and not be such an effective oxidant for CH_4 as it is today. Thus, levels of O_2 control the CH_4 abundance, albeit indirectly, through oxidation mediated by OH.

[9] The field lines of Earth's predominately dipole magnetic field are approximately perpendicular to the surface near the poles and parallel near the equator. Igneous rocks containing iron minerals such as magnetite (Fe_3O_4) become magnetized in the direction of the prevailing magnetic field when they cool. Thus, rocks in which the

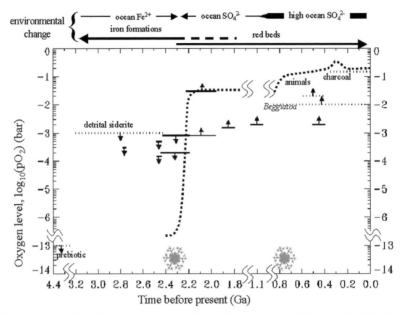

FIGURE 4.6 The history of atmospheric O_2 – note the indicated breaks in the timescale of the plot. The thick dashed line shows a possible evolutionary path for atmospheric O_2 that satisfies geochemical and biological constraints, as well as modeling results. Dotted upper and lower limits show the duration of geochemical and biological constraints, such as the occurrence of detrital siderite ($FeCO_3$) in ancient riverbeds. Arrows indicate upper and lower bounds on the level of oxygen. Unlabeled solid-line upper and lower bounds are from the evidence of specific paleosols (ancient soils), with the length of each line showing the uncertainty in the age of each. Bounds on pO_2 from paleosols are taken from Rye and Holland (1998). Biological lower limits on pO_2 are based on estimates for the requirements of the marine sulfur-oxidizing bacteria *Beggiatoa* (Canfield and Teske, 1996), and also the requirements of macroscopic animals that appear after 0.59 Ga (Runnegar, 1991). The upper bound on the level of pO_2 in the prebiotic atmosphere at $\sim$4.4 Ga (shortly after the Earth had differentiated into a core, mantle, and crust) is based on photochemical calculations. A "bump" in the oxygen curve at 0.3 Ga (in the Carboniferous) is based on the model of Berner *et al.* (2000). Snowflake symbols indicate the occurrence of episodes of low-latitude glaciation (Snowball Earth events), which appear to have some broad correlation with the oxygen history of the atmosphere.

Teske, 1996; Canfield, 1998). A very small number of researchers have continually questioned the idea that pO_2 has changed at all (e.g., Ohmoto, 1997), but the great preponderance of evidence indicates a significant increase in O_2 at $\sim$2.4–2.3 Ga (Section 4.6).

Paleontological evidence is also consistent with the hypothesis that a rise in O_2 at 2.3 Ga enhanced biological evolution. For example, the oldest known fossils of possible eukaryotic origin, found in shales from Michigan dated at 1.87 Ga, are remains of spirally coiled organisms resembling the extinct photosynthetic alga *Grypania spiralis* (see Section 12.2.1). Furthermore, molecular *phylogeny* (the reconstruction of evolutionary lineages using genes) suggests that several anaerobic lineages were lost in the range 2.5 to 1.6 Ga (the Paleoproterozoic) (Hedges *et al.*, 2001). This is consistent with the toxicity of O_2 to obligate anaerobes.[10]

The rise of O_2 may also be linked to the emergence of multicellular life. Fossils that are 4–5 mm long are found in northern China at 1.7 Ga, and may be multicellular (Shixing and Huineng, 1995). Then by 1.4 Ga, similar, often larger, carbonaceous fossils become abundant worldwide in marine sediments.

Another pivotal evolutionary event was the appearance of macroscopic animal fossils (Ediacara – Section 16.2.4.4) at $\sim$575 Ma. Similar fossils appear on six continents in strata that lie above rocks from a glacial episode at $\sim$580 Ma. Their appearance may be linked to a second rise in O_2 marked by increased sulfate levels in the ocean (Canfield and Teske, 1996; Knoll and Carroll, 1999; Fike *et al.*, 2006; Canfield *et al.*, 2007). Certainly pO_2 levels must have been at least 10% of present pO_2 to support large animal metabolism. The Neoproterozoic era (1,000–542 Ma) also bears many curious similarities to the Paleoproterozoic discussed above. Both eras are characterized by Snowball Earth episodes (Section 4.2.4), significant changes in the biota, and large oscillations of sedimentary carbon isotopes.

---

[10] *Anaerobes* are microorganisms that do not require O_2, as opposed to *aerobes* that require O_2; *obligate* (an)aerobes *must* have (a lack of) O_2. *Oxic* and *anoxic* environments refer to a presence or lack of O_2.

Finally, ever since since the rapid rise of complex animals starting in 542 Ma, pO_2 has probably always been 0.2 ± 0.1 bar. Charcoal is found in continental rocks from 350 Ma onwards, which indicates an O_2 mixing ratio of at least 15% for all recent epochs because wood cannot burn below this O_2 level (Lenton and Watson, 2000). The colonization of the land by plants began around ~450 Ma. Subsequently large amounts of organic carbon were buried on the continents in the Carboniferous (360–300 Ma), possibly because lignin, a structural compound in woody plants, was difficult to decompose until organisms like fungi evolved the means to do so. Because organic carbon burial is accompanied by a release of O_2 (Eq. (4.3)), organic burial may have led to peak pO_2 ~0.3 bar at ~300 Ma (Berner *et al.*, 2001). This may explain the presence of giant Carboniferous insects, such as 70 cm wingspan dragonflies, which must have relied on the diffusion of O_2 for respiration (Dudley, 1998).

After this quick overview of the history of atmospheric O_2, we now proceed though it again more thoroughly.

4.4 The prebiotic atmosphere

What was the atmosphere like before life appeared? The *prebiotic* atmosphere was in all likelihood "weakly reducing,"[11] composed primarily of N_2, CO_2, and water, with relatively small quantities of H_2, CO, and CH_4, and negligible O_2 (Fig. 4.7) (Walker, 1977; Kasting and Brown, 1998). Pioneering laboratory experiments that produced complex organic molecules (such as amino acids) in CH_4-NH_3-H_2 atmospheres led some researchers to suggest that the early atmosphere was highly reducing to explain how life originated, but this view has now been largely abandoned (Chapters 6–9). The early atmosphere accumulated

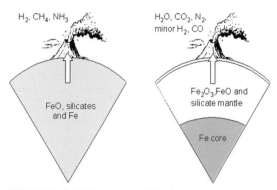

FIGURE 4.7 (Left) During 4.4–4.5 Ga when the mantle was Fe-rich during core formation, reduced volcanic gases were introduced into the atmosphere. (Right) From 4.4 Ga onwards, after the Earth had differentiated into a core, mantle, and crust, a weakly reducing mixture of volcanic gases fed the atmosphere.

from volcanic *degassing*, which arose from volcanic gases associated with melts and metamorphic gases from hot rocks that did not melt. But the major volatile gases introduced into today's atmosphere are oxidized species such as H_2O, CO_2, and N_2 rather than reduced forms such as H_2, CO or CH_4, and NH_3. The reduced/oxidized gas ratio (H_2/H_2O, CO_2/CO, etc.) in volcanic gases, in particular, depends on the degree of oxidation in the upper mantle, the source region for such gases (Holland, 1984).

The basic composition of the upper mantle was set 4.4 Ga. The heat produced by impacts during Earth's accretion melted the planet, and molten metals (mostly iron) separated from the surrounding mixture of silicates and oxides and sank to form the Earth's core. Siderophile ("iron-loving") elements segregated into the core and some were either radioactive or the decay products of radionuclides, leaving clues about the time of core formation that suggest it occurred within the first 100 Myr of Earth history (Halliday *et al.*, 2001). This by itself does not tell us the oxidation state of the mantle at that time. However, the molten mantle almost certainly contained water, some of which would have dissociated to produce hydrogen and oxygen. If the hydrogen became lost by volcanic outgassing followed by escape to space or by being incorporated into the metallic core, the overall mantle oxidation state would increase (Fig. 4.7). Exactly how fast this oxidation occurred is currently a matter of debate. We shall return to this topic later in the chapter because even small changes in volcanic gas composition could have been important for the origin of life and for later atmospheric evolution.

[11] Chemical reactions that involve an exchange of electrons, or more accurately, that change the degree of electronegativity (e.g, electrons becoming bound strongly to the oxygen nucleus, which is much more electronegative than, say, either C or H), are called reduction–oxidation or *redox* reactions. In such a reaction a *reducing agent* (often called an *electron donor* in biology) is *oxidized* while simultaneously an *oxidizing agent* (*electron acceptor*) is *reduced*. A "reducing" gas is composed of reducing agents, which for the Earth's atmosphere are typified by H_2, CH_4, CO, and NH_3, whereas an oxidizing mixture consists of N_2, CO_2, and/or H_2O. The redox state of an atmosphere is then the result of these opposing reactions; oxidizing atmospheres are generally "hydrogen poor" and reducing atmospheres "hydrogen rich." Note that oxygen does *not* have to be involved to oxidize something, although it often is.

4.4.1 O_2 in the prebiotic atmosphere

The weakly reducing nature of volcanic emissions must have greatly limited pO_2 in the prebiotic atmosphere because reduced gases would have dominated over the only abiotic source of free O_2, namely *photolysis* of water (breaking up the molecule from irradiation by light) *and* the subsequent escape of hydrogen to space. (By itself, photolysis of H_2O or CO_2 does not provide a net source of oxygen because the oxygen soon recombines.) Hydrogen escape and O_2 production associated with water vapor destruction is severely limited because little water vapor is able to rise to the upper atmosphere because water condenses at the top of the troposphere. Such condensation is called *cold-trapping*, borrowing a laboratory term for cooling a gas to condense out water vapor. Consequently, the abiotic production rate of O_2 is very small.

It is useful to think of an oxidizing atmosphere as "hydrogen poor" and a reducing atmosphere as "hydrogen rich." Even a small excess of H_2 tips the balance. Hydrogen exerts a control on the level of O_2 through a series of photochemical reactions that add up to a net reaction

$$O_2 + 2H_2 \rightarrow 2H_2O. \tag{4.5}$$

To estimate pO_2 for the prebiotic atmosphere, we take outgassing rates as greater on early Earth than today because of increased heat flow from a hotter, more radioactive interior. Assuming that H_2 outgassing rates were in fact 3–5 times higher, detailed photochemical models indicate that the prebiotic atmosphere's pO_2 was only $\sim 10^{-13}$ bar (Kasting, 1993).

4.4.2 The escape of hydrogen to space and H_2 in the prebiotic atmosphere

The severe depletion of noble gases in the composition of the Earth provides evidence that Earth did not retain gaseous volatiles from the original solar nebula (Chapter 3). Thus, Earth's hydrogen does not derive from accreted H_2, and instead was accreted in an oxidized form in solids such as water ice (H_2O), water of hydration in silicates (-OH), or hydrocarbons (-CH).

The escape of hydrogen from the Earth (albeit via atmospheric gases other than H_2O) is important in the history of atmospheric O_2 (Section 4.7). When hydrogen escapes to space, the Earth as a whole is irreversibly oxidized. It is immaterial whether the hydrogen is transported through the atmosphere as H_2, H_2O, HCN, NH_3, or any other H-bearing compound. For example, when hydrogen emanates from volcanoes and subsequently escapes to space, the upper mantle is oxidized through reactions such as $3FeO + H_2O = Fe_3O_4 + H_2$. Similarly, when hydrogen originates from metamorphic gases in the crust, the crust is oxidized.

The abundance of hydrogen in the atmosphere is set by a balance between H_2 outgassing (from volcanism and metamorphism) and escape of hydrogen to space. Because hydrogen atoms are so light, there are always some at the top of the atmosphere traveling fast enough (>11 km s^{-1}) to escape Earth's gravity, even today (Fig. 4.8).[12] Extreme ultraviolet radiation from the Sun is absorbed and converted into heat in the *thermosphere*, which on the modern Earth stretches from 85 km to $\sim$500 km altitude and has temperatures of $\sim$1,000 to 2,500 K. Although the average hydrogen atom thermal velocity is insufficient to achieve escape, the high temperature at the top of the thermosphere induces a significant high-velocity "tail" in the Maxwellian distribution of velocities of hydrogen atoms, from which many atoms escape. This process is termed *thermal escape* because it represents escaping atoms from the thermal velocity distribution. However, *non-thermal escape* mechanisms are even more important, accounting for 60–90% of the H atom escape. Non-thermal escape results from hydrogen atoms being boosted above escape velocity by photolytic, electronic, impact, or ionic reactions.

There are two principal bottlenecks as hydrogen atoms (either free or as part of molecules later to be photolyzed) work their way upwards and eventually escape. The first is the cold trap mentioned previously and the second is a slow diffusion process through the thermosphere. The thermosphere is analogous to a membrane with a certain partial pressure of hydrogen below and a vacuum above – the flux through the membrane is determined by the partial pressure below.

Calculations show that the amount of escaping H today is only 9×10^{10} mol H yr^{-1} (for details see Kasting and Brown, 1998); about half of these atoms originate from water vapor that makes its way above

---

[12] (1) In a mixed gas at a certain temperature, each species of atom or molecule can be characterized by an average kinetic energy of $0.5\,m\,v^2$, where m is the mass of the species and v is its velocity. Those species of lower mass on average have higher velocities. (2) Basic dynamics shows that escape of an object from the gravity of another does not depend on the mass of the escaping object, only on its velocity. Thus one must achieve a velocity of 11 km/s to escape Earth's gravity whether launching a rocketship or a hydrogen atom.

FIGURE 4.8 An ultraviolet image of Earth by NASA's Dynamics Explorer I spacecraft at 16 500 km altitude above 67°N latitude. The faint glow beyond the edge of the planet is entirely due to Earth's extended hydrogen atmosphere (*geocorona*) as seen in resonantly scattered Lyman-α (121 nm) solar radiation. Energetic hydrogen atoms in the geocorona are escaping to space. Features on the Earth's disk (dayglow from the sunlit atmosphere, the oval ring of the aurora borealis, and equatorial airglow) are due to emission from atomic oxygen and molecular nitrogen. Isolated points of light are background stars that are bright in the ultraviolet.

the cold trap and the other half from biogenic methane, which is not cold-trapped. For early Earth, an outgassing rate 5 times higher than today can be balanced with a hydrogen escape flux to space that depends on the hydrogen mixing ratio. This calculation yields a hydrogen mixing ratio of $\sim10^{-3}$ for the prebiotic atmosphere, compared to today's value of 5.5×10^{-7}.

4.4.3 Atmospheric synthesis of formaldehyde and hydrogen cyanide

Formaldehyde

Reactions in the weakly reducing prebiotic atmosphere may have produced molecular precursors to life. One of the early self-replicating chemical systems may have been based on ribonucleic acid (RNA), a hypothesized biochemistry known as "RNA World" (Chapters 6–8). RNA consists of chains of monomers called nucleotides, each comprising a phosphate molecule, a ribose (sugar)

molecule, and a nitrogen-containing base. The phosphate molecule would likely be derived from phosphate released in weathering of rocks, whereas the other components might have had an atmospheric origin. Ribose is $C_5H_{10}O_5$, which can be formed from five molecules of formaldehyde, H_2CO. Photochemical reactions in weakly reducing CO_2-rich atmospheres are predicted to produce large quantities of formaldehyde (Pinto *et al.*, 1980), which is soluble and would rain out. Subsequent spontaneous reactions could allow formaldehyde to form sugars, including ribose.

Hydrogen Cyanide

The simplest nitrogen-containing base is adenine, $C_5H_5N_5$, which is a chain of five hydrogen cyanide molecules (HCN). Substantial HCN can form in a primitive atmosphere with a few tens of parts per million of methane (Zahnle, 1986). N atoms can be produced by ionization of N_2 in the ionosphere ($N_2 + h\nu = N_2^+ + e^-$) followed by dissociative recombination ($N_2^+ + e^- = N + N$). The N atoms then flow down into the stratosphere where they combine with fragments produced in CH_4 photolysis to make HCN:

$$CH_2 + N \rightarrow HCN + H; \qquad (4.6)$$

$$CH_3 + N \rightarrow HCN + H_2. \qquad (4.7)$$

However, the rate constants for these reactions and their products have not yet been studied experimentally. Also, there may not have been enough prebiotic methane. Much of the methane entering today's atmosphere is biogenic. Today's abiotic source from mid-ocean ridges is $\sim1.5 \times 10^{10}$ mol CH_4 yr^{-1} and would only produce 0.5 ppm CH_4 in a weakly reducing atmosphere. This would result in only 5.3×10^8 mol HCN yr^{-1} (Kasting and Brown, 1998).[13] For comparison, consider extraterrestrial delivery. *Today's* flux of micrometeorites would produce only 1% as much HCN if all the N were converted to HCN by heating during atmospheric entry, but the flux may have been $\sim10^2$–10^3 times higher before 4 Ga.

---

[13] Fluids emanating from a new, off-axis hydrothermal vent field on the mid-Atlantic ridge (Lost City) are different in composition and in particular have high CH_4 concentrations (Kelley *et al.*, 2005). If other off-axis vent fluids have similar compositions, the resulting global CH_4 source today could be as high as 1×10^{11} mol/yr, but it is not clear how this process would scale for early Earth. A second possible source of abiotic CH_4 is impacts. Condensing particles in impact plumes catalyze the conversion of CO into CH_4 and CO_2 (Kress and McKay, 2004).

4.5 Effects of primitive life on the atmosphere

Today's biosphere affects the cycling of every major element of importance to biology, such as C, N, O, P, and S, and it is likely that early life behaved similarly (Chapter 12). Life modulates the cycle of carbon, the second most abundant volatile after water, by removing carbon from the atmosphere to synthesize organic matter. Life also modulates the cycle of Earth's third most abundant volatile, nitrogen, by extracting it from the air to make ammonium ions, and then recycling it back to N_2 in air. In fact, when different metabolisms evolved they must have affected several important volatiles, and we now consider H_2, CO_2, N_2, and O_2.

4.5.1 Methanogens and methane production

The simultaneous presence of reduced and oxidized gases in the early atmosphere would have provided a source of energy for primitive life because such gases can be catalyzed to react together. As discussed in Section 4.2, methanogens derive energy from hydrogen and carbon dioxide:

$$4H_2 + CO_2 \rightarrow CH_4 + 2H_2O. \qquad (4.8)$$

They use the energy to synthesize cell material from inorganic nutrients. Thus the early biosphere would have increased atmospheric CH_4 at the expense of more metabolically desirable H_2 (Section 4.2.3). Assuming that methanogens were thermodynamically energy-limited, $\geq 90\%$ of the H_2 would have been converted to CH_4 (Kasting *et al.*, 2001), and H_2 levels would have dropped from a prebiotic mixing ratio of 10^{-3} to $\sim 10^{-4}$. CO_2 levels would have also decreased because CH_4 is a powerful greenhouse gas, which would have led to increased warming and loss of CO_2 via increased continental weathering in the carbonate–silicate cycle (Section 4.2.1).

4.5.2 Biological nitrogen fixation and atmospheric N_2

Despite abundant atmospheric N_2, nitrogen is often a limiting nutrient for life. Life would fall apart without proteins, DNA and RNA, all of which contain N, an essential structural element. Nitrogen is contained in the peptide bond-CO-NH- that links amino acids along a protein. Peptide bonds are so mechanically effective that chemists utilize them to make plastics like nylon. Yet few microorganisms can metabolize nitrogen

unless it is *fixed* from the atmosphere, i.e., acquired and turned into a more soluble form. So how did the very first organisms obtain nitrogen? In the anoxic prebiotic atmosphere (Section 4.4), N_2 would have been oxidized with CO_2 by lightning:

$$N_2 + 2CO_2 \rightarrow 2NO + 2CO. \qquad (4.9)$$

NO then gets converted to soluble nitrosyl hydride (HNO). Thus, early lightning would have provided a modest flux of fixed nitrogen to the ocean that may have been important for the origin of life. The above reaction depends on the abundance of CO_2. For CO_2 mixing ratios ranging from 0.04 to 0.5, the rate of fixation would be 2.6×10^9 to 3×10^{11} g N yr^{-1} at Earth's present lightning rate (Navarro-Gonzale *et al.*, 2001). In the ocean, the end products of dissolution would have been NO_3^- and NO_2^- (Mancinelli and McKay, 1988). The latter can be reduced by ferrous iron (Fe^{2+}) dissolved in the early ocean to produce ammonia, NH_3 (via $6Fe^{2+} + 7H^+ + NO_2^- = 6Fe^{3+} + 2H_2O + NH_3$), some of which would flux to the atmosphere. NH_3 would then be quickly photolyzed to N_2 and H_2, completing the cycle (Fig. 4.9a). Another way that nitrogen could have been fixed abiotically was through HCN synthesis in atmospheres containing trace levels of CH_4 (Section 4.4.3). HCN is hydrolyzed in solution to form NH_4^+ (ammonium).

Nitrogen fixation is strictly anaerobic and its origin perhaps even predates the divergence of the three domains of life. It may have been derived from an earlier biochemical function such as an enzyme for detoxifying cyanides like HCN present in the early environment (Section 4.4). In any case, once anaerobes developed nitrogen fixation, NH_4^+ would likely have become the dominant combined form of nitrogen in the ocean. In the absence of oxygen, early nitrogen fixation may been more effective and nitrogen recycling diminished relative to the present day. This would have resulted in a lower atmospheric pressure than today, but currently there are no constraints from geologic data on total atmospheric pressure.

After oxygenic photosynthesis arose (next section), nitrification would have become important. *Nitrification* happens when ammonium ions (NH_4^+) are oxidized to nitrite (NO_2^-) and then nitrate (NO_3^-) by aerobic bacteria. *Denitrification* is the microbial reduction of NO_3^- to N_2O and N_2. Denitrifiers use organic matter as a reductant, and inhabit anoxic zones such as continental margin sediments. Thus, after the ~ 2.4–2.3 Ga rise of O_2, the nitrogen cycle was probably similar to today's (Fig. 4.9b). However, the evolution of the nitrogen cycle on early Earth is

(a)

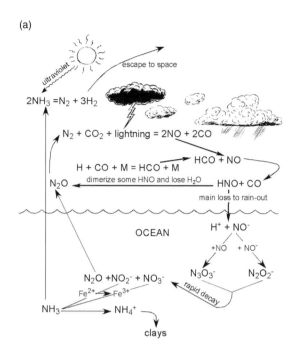

FIGURE 4.9(a) The major pathways of the prebiotic nitrogen cycle. NO formed by lightning from N_2 and CO_2 gets converted to more soluble forms such as HNO. Dissolved ferrous iron in the ocean allows reduction of soluble nitrates or nitrites to ammonia, although some ammonium may leak into clays. The ammonia is photolyzed in the atmosphere back to N_2. (After Mancinelli and McKay, 1988.) (b) The modern nitrogen cycle. In the atmosphere 99.99% of the nitrogen is in the form of N_2 and 99.6% of the rest is N_2O. Other gases such as ammonia (NH_3) are rapidly removed. In the ocean, 95% of the nitrogen is in the form of dissolved N_2, while the remainder is roughly split between NO_3^- and dead organic matter. Biological processes dominate the fluxes of nitrogen in the modern cycle.

(b)

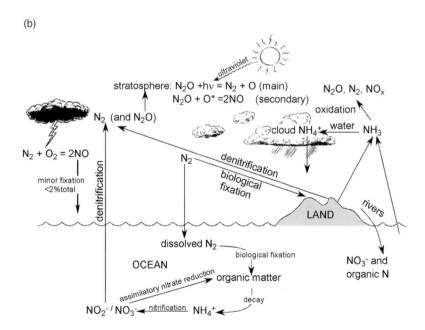

uncertain because few data exist for nitrogen isotopes from ancient organic matter (Section 12.2.5). The early nitrogen cycle is thus a fruitful area for future research.

4.5.3 Origin of oxygenic photosynthesis

The most significant biological event in the history of the Earth's atmosphere was the evolution of oxygenic photosynthesis. Photosynthesis is the metabolic process of making organic carbon from CO_2 using the energy of sunlight. *Oxygenic photosynthesis* is the process whereby water is split using the energy of sunlight, and the extracted hydrogen is used to reduce CO_2 to organic carbon (Eq. (4.3)). Knowing when oxygenic photosynthesis first appeared is important for understanding the evolution of O_2. As discussed in

Section 12.3.3, organic molecules in the geological record and certain fossils (stromatolites) provide good evidence that oxygenic photosynthesis had evolved by 2.7 Ga.

In *anoxygenic photosynthesis* no O_2 is released and microorganisms use another electron donor (e.g., Fe^{2+}, H_2S, H_2, S) instead of H_2O to reduce CO_2 to organic carbon. For example, anoxygenic photosynthesis using H_2S is

$$2H_2S + CO_2 + h\nu \rightarrow CH_2O + H_2O + 2S. \qquad (4.10)$$

Anoxygenic photosynthesis predates the evolution of oxygenic photosynthesis (Blankenship and Hartman, 1998; Xiong *et al.*, 2000). *Cyanobacteria* are eubacteria that evolved the capability of oxygenic photosynthesis (although they can also accomplish anoxygenic photosynthesis if reductants are readily available). Plants and other modern photosynthetic eukaryotes all have ancestors that acquired photosynthesis when a cyanobacterium became resident inside a larger cell and the whole cell evolved into a single organism with an intracellular, photosynthesizing structure (now called a chloroplast) descended from the cyanobacterium.

The first successful oxygenic photosynthesizers must have overcome the toxic effects of O_2. But what drove microbes to develop defenses against O_2 in an anoxic world? Also, if oxygenic photosynthesis, which extracts hydrogen from water, perhaps developed in response to a lack of electron donors such as H_2 and H_2S, what oxidants were available before oxygenic photosynthesis to remove such reductants? An answer may reside in the early atmosphere (McKay and Hartman, 1991), where a series of photochemical reactions produce H_2O_2, a powerful oxidant. The net effect is:

$$2H_2O \rightarrow H_2O_2 + H_2. \qquad (4.11)$$

H_2O_2 is soluble and, after rainout, could have accumulated locally, for example, in lakes. Alternatively, non-atmospheric H_2O_2 may have been produced by surface-mediated reactions between H_2O and pyrite (FeS_2) in anoxic environments. The presence of H_2O_2 would have exerted a selective pressure for anoxygenic photosynthetic organisms to evolve enzymes, such as peroxidase, which reduces H_2O_2 to H_2O, and catalase, which converts $2H_2O_2$ to $2H_2O$ and O_2. These enzymes are analogues to structures that facilitate electron transfer in today's oxygenic photosynthesis. Hence, H_2O_2, perhaps from the atmosphere, may have promoted the evolution of oxygenic photosynthesis.

4.5.4 Oxygen oases in an anoxic environment

Globally low levels of *atmospheric* O_2 in the Archean (Section 4.6) do not imply that O_2 levels were everywhere and always low. Underground or underwater oases or strata where pO_2 was locally high were no doubt common, in the same way that today one easily finds places (e.g., muds) where H_2S or CH_4 is locally high despite the O_2-rich atmosphere.

O_2 in microbial communities probably triggered evolution in symbiotic heterotrophs. Methanotrophy (methane oxidation) is one such innovation that was probably global by ~2.8 Ga (Hayes, 1994), although methanotrophs may have utilized sulfate rather than O_2. Free-living aerobic bacteria that were the ancestors of mitochondria (structures in eukaryotes that perform aerobic respiration) probably evolved in response to locally available O_2. Much of the early photosynthetic O_2 would then have been efficiently respired by symbiotic aerobes. O_2 oases may have also facilitated the evolution of eukaryotes with mitochondria. Phylogeny suggests that eukaryotes acquired mitochondria before chloroplasts (Knoll, 1992), perhaps before the rise of O_2.

4.6 Geological evidence for the rise of oxygen

Geological evidence suggests that O_2 rose at ~2.4–2.3 Ga, and again at ~0.8–0.6 Ga (Fig. 4.6). There are no pristine samples of air trapped in rocks from these times, so the history of atmospheric O_2 has to be reconstructed from indirect tracers left behind by the chemical action of O_2. On the continents, weathering processes release chemicals from rocks in a manner that depends upon atmospheric CO_2 and O_2 concentrations. Also atmospheric gases are mixed into the oceans, so that the chemistry of marine sediments depends on atmospheric composition.

4.6.1 Non-marine environments

Paleosols[14] indicate that at ~2.4–2.3 Ga atmospheric pO_2 rose from <0.0008 bar to >0.002 atm (Rye and Holland, 1998). Paleosols before ~2.4–2.3 Ga show that iron was leached during weathering by the reaction

---

[14] *Paleosols* are the lithified remains of ancient soils. They are useful for studies of the atmosphere because we know they were in contact with it at one time.

"FeO" $+ H_2O + 2CO_2 \rightarrow Fe^{2+} + 2HCO_3^{-}$,   (4.12)

where "FeO" represents ferrous iron bound in unweathered material, but that after $\sim$2.2 Ga, iron was rendered immobile. Ferrous iron (Fe^{2+}) is soluble whereas oxidized (ferric) iron (Fe^{3+}) is not. Consequently, iron is flushed through a soil or porous rock if rainwater has little dissolved oxygen but is immobile with oxygenated rainwater. The degree of weathering also depends on the rainwater's pH, which is related to pCO_2 because CO_2 dissolves in water to make carbonic acid. Consequently, paleosols constrain the O_2/CO_2 ratio in the atmosphere rather than O_2 independently, requiring a separate estimate of pCO_2. The pCO_2 can also be deduced from paleosol chemistry (Section 4.2.3). Cerium also indicates low pO_2 because it changes from Ce^{3+} to Ce^{4+} to form cerianite (CeO_2) in oxic weathering. Thus the presence of Ce^{3+}-rich phosphates in a 2.6–2.45 Ga granite paleosol implies a largely anoxic atmosphere (Murakami *et al.*, 2001). Figure 4.6 includes a compilation of pO_2 values deduced from paleosol studies.

Detrital grains are sedimentary minerals that never completely dissolve in weathering. Such grains in pre-2.4 Ga riverbeds commonly contain reduced minerals that would only survive at low pO_2 (Rasmussen and Buick, 1999). Grains of pyrite (FeS_2), uraninite (UO_2), and siderite ($FeCO_3$) place upper bounds on Archean pO_2 of roughly 0.1 bar, 0.01 bar, and 0.001 bar, respectively. Given their roundness, such grains were transported long distances in aerated waters. In oxic waters, uraninite dissolves to form soluble U^{6+} ions, pyrite oxidizes to sulfate (SO_4^{2-}) and ferric iron (Fe^{3+}), and siderite oxidizes to produce ferric iron (Fe^{3+}). After about 2.2 Ga the occurrence of reduced detrital grains becomes rare and restricted to locally anoxic environments.

Red beds provide further evidence for atmospheric redox change. They derive from windblown dust or river-transported particles coated with red-colored hematite (Fe_2O_3). After $\sim$2.3 Ga, red beds are ubiquitous. Occasionally, red beds that appear to be very old actually result from much more recent oxidation. For example, the deep burrowing of subterranean termites, which have existed for at least 100 Myr, has sometimes allowed the penetration of O_2 into groundwater.

4.6.2 Marine environments

A transition in O_2 levels at $\sim$2.4–2.3 Ga is consistent with the distribution of *banded iron formations* (BIFs)

through time. BIFs are laminated, marine sedimentary rocks containing $\geq$15 wt% iron, usually with alternating iron-rich and silica-rich layers. They occur prominently in the Archean ($>$2.5 Ga), and then decline in abundance and disappear after $\sim$1.8 Ga. The iron in large part originated from hydrothermal input into the anoxic, deep ocean, probably mostly at mid-ocean ridges (Holland, 1984). Ocean upwelling then carried Fe^{2+} to the continental shelves where it was oxidized to insoluble Fe^{3+}, typically forming magnetite (Fe_3O_4). However, even in today's oxic world, there is a significant flux of reduced iron to the ocean in river particulates and windblown dust, so there was probably also such a contribution to Archean BIFs (Canfield, 1998). Many researchers believe that it was microbes that oxidized Fe^{2+} to Fe^{3+}. For the late Archean BIF in Hamersley, Australia, the estimated iron deposition rate requires an ancient oxygen supply of $\sim$1% of the global, modern O_2 flux from the burial of organic carbon. The eventual disappearance of the BIFs may have happened when the deep ocean became oxygenated as pO_2 rose to $\sim$1/3 to 1/2 of present levels.

Alternatively, Canfield (1998) has proposed that iron disappeared in the oceans when sulfate concentrations became sufficient to increase the rate of microbial sulfate reduction. Because sulfate reduction produces sulfide (S^{2-}), the deep ocean would then have been swept free of iron by the precipitation of insoluble pyrite (FeS_2), perhaps continuing until the 0.8–0.6 Ga rise in O_2.

4.6.2.1 *Marine carbon isotopes*

Marine carbon isotopes indicate drastic environmental change in the Paleoproterozoic era (2.5–1.6 Ga). The average $\delta^{13}C_{in}$ value of carbon entering *into* the atmosphere–ocean system from volcanism, metamorphism, and weathering is about −5.5‰[15] (Holser *et al.*, 1998). On timescales that are long compared to the carbon residence time in the ocean, $\sim$10^5 yr, the number of ^{12}C atoms and ^{13}C atoms entering the system must equal the number exiting the system either as carbonate or organic carbon sediments, implying

$$\delta^{13}C_{in} = f_{carb}\delta^{13}C_{carb} + f_{org}\delta^{13}C_{org},   (4.13)$$

---

[15] The notation $\delta^{13}C = -5.5$‰ means that the isotope ratio $^{13}C/^{12}C$ is depleted by 5.5 parts per thousand relative to a standard calibration value. Metabolic reactions in organisms preferentially use lighter isotope molecules because reactions using them are very slightly energetically favored. Therefore these and other *fractionation* ratios in the geologic record are used as indicators of biotic processes. Also see Section 12.2.3.

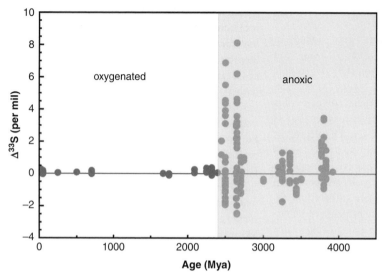

FIGURE 4.10 $\Delta^{33}S$ vs. time. $\Delta^{33}S$ expresses the deviation, in parts per thousand, from a standard sulfur isotope mass fractionation line: $\Delta^{33}S \equiv \delta^{33}S - 0.515 \, \delta^{34}S$. Samples younger than 2.1 Ga are considered consistent with fractionation by purely mass-dependent processes. In contrast, the samples older than 2.0–2.5 Ga are mass-independent. (Farquhar *et al.*, 2000.)

where f_{carb} is the fraction of carbon buried in carbonate minerals with a global average isotopic composition $\delta^{13}C_{carb}$, and f_{org} is defined similarly. One mole of buried organic carbon generates one mole of O_2 (Eq. (4.3)), so f_{org} provides the O_2 global production rate relative to the amount of carbon entering the atmosphere and ocean. Throughout geologic time, $\delta^{13}C_{org}$ is about −30‰ and $\delta^{13}C_{carb}$ is about 0‰. Solving Eq. (4.13) with $\delta^{13}C_{in} = -6$‰ gives f_{org} about 0.2. In other words, ∼20% of the CO_2 coming into the ocean–atmosphere system is fixed biologically and exits as buried organic carbon, whereas the remaining 80% exits as carbonate carbon.

The largest excursions in $\delta^{13}C_{carb}$ in Earth history occurred between 2.4 and 2.1 Ga, with positive and negative oscillations between +10‰ and −5‰ (see Fig. 12.8). But the Snowball Earth episodes that also occurred during 2.4–2.2 Ga (Section 4.2.4), combined with sparse $\delta^{13}C$ data, complicate the interpretation of $\delta^{13}C_{carb}$ fluctuations. Positive excursions of $\delta^{13}C_{carb}$ can be interpreted as resulting from increased organic burial on a global scale, leading to pulses of O_2. However, given the geologically short residence time of atmospheric O_2 (today, a few million years), pulses of organic burial would merely cause atmospheric O_2 to rise and decay, and cannot be responsible for the permanent rise of O_2 (Section 4.7.1). The $\delta^{13}C$ record is further discussed in Section 12.2.3.

4.6.2.2 Marine sulfur isotopes

Marine sulfur isotopes indicate an increase in sulfate concentrations at ∼2.3 Ga consistent with a rise of O_2

(Fig. 12.9). Archean sulfides display $^{34}S/^{32}S$ that cluster around the unfractionated mantle value ($\delta^{34}S = 0$‰), implying Archean oceans with <0.2 mM sulfate, compared to 28.9 mM in today's surface seawater. Lack of sulfate is consistent with low pO_2, which would induce little oxidative weathering of sulfides, limiting the river supply of sulfate to the oceans. Further discussion is in Section 12.2.4.

Measurements of other sulfur isotope ratios, $\delta^{32}S$ and $\delta^{33}S$, in addition to $\delta^{34}S$, show a major change in the sulfur cycle occurring between 2.45 and 2.09 Ga, most probably related to the rise of O_2 (Fig. 4.10) (Farquhar *et al.*, 2000). Sulfur isotopes in rocks older than ∼2.4 Ga are peculiarly fractionated. But in younger rocks, ^{32}S, ^{33}S, and ^{34}S obey "mass-dependent" fractionation, in which the difference in abundance between ^{33}S and ^{32}S is approximately half that between ^{34}S and ^{32}S. Such fractionation is produced by many aqueous chemical and biochemical reactions, such as sulfate reduction. In contrast, sulfur isotopes in pre-2.4 Ga rocks show large "mass-independent" fractionation, which is thought to result solely from gas-phase photochemical reactions such as photolysis of SO_2. The isotopes in this case are still fractionated by mass, but their relative abundances deviate from what is expected in mass-dependent fractionation. In a high-O_2 atmosphere, sulfur gases are rapidly oxidized to sulfuric acid, H_2SO_4, which rains out as dissolved sulfate, SO_4^{2-}. In the absence of O_2 and an ozone layer in the early atmosphere, shortwave ultraviolet light penetrated into the lower atmosphere and the photochemistry of S-bearing atmospheric species would produce mass-independent signatures (Pavlov and Kasting, 2002).

From $\sim$0.8–0.6 Ga, sulfides are increasingly found with ^{34}S-depletions exceeding the fractionation threshold of sulfate reducing bacteria. This has been interpreted as indicating a second rise of pO_2 from $\sim$1–3% to greater than 5–18% of present levels (Canfield and Teske, 1996). This can be explained if sulfide was re-oxidized at the sediment–water interface to SO_4^{2-} and reduced again by bacteria, cyclically increasing the isotope fractionation. Possibly, O_2 increased to the point where it penetrated marine sediments, making the deep ocean aerobic for the first time (Canfield, 1998). This could supply the increase of O_2 thought necessary to explain the appearance of macroscopic animals in the fossil record after 590 Ma (Knoll and Carroll, 1999).

4.7 Models for the Earth's atmospheric O₂ history

There is still no consensus about why atmospheric O_2 levels increased in the manner indicated by the geologic record. Before discussing various hypotheses, we introduce some general principles in understanding how the number of moles of O_2 (R_{O2}) changes with time. The rate of change of R_{O2} in the atmospheric reservoir is

$$\frac{d(R_{O2})}{dt} = F_{source} - F_{sink}, \qquad (4.14)$$

where F_{sink} is the removal flux of O_2 from the atmosphere (in moles yr^{-1}) due to numerous oxidation reactions, and F_{source} is the source flux of oxygen. After the advent of oxygenic photosynthesis (and of corresponding respiration processes), F_{source} was dominated by the burial flux of organic carbon. Most organic carbon (today, $\sim$99.9%) is rapidly oxidized via respiration, producing no net O_2. But the small leak of organic carbon to sediments, where it is segregated from O_2, contributes in effect to F_{source}. F_{sink} is due to several oxidation processes:

$$F_{sink} = F_{volcanic} + F_{metamorphic} + F_{weathering}. \qquad (4.15)$$

These loss terms are the reaction of O_2 with various reductants: reduced volcanic gases ($F_{volcanic}$), reduced metamorphic gases ($F_{metamorphic}$), and reduced material on the continents ($F_{weathering}$).

The F_{source} and F_{sink} fluxes depend on R_{O2} in complicated, nonlinear ways. If we knew these dependencies explicitly (and we do not) we could substitute Eq. (4.15) into Eq. (4.14) and integrate to derive R_{O2} as a function of time. A simplifying assumption is that

at any instant, R_{O2} will roughly be in "steady state," which means that O_2 will have accumulated in the atmosphere to some value of R_{O2} where the O_2 sink (F_{sink}) will be about equal to the O_2 source (F_{source}). In this case, $d(R_{O2})/dt = 0$ in Eq. (4.14). Such a balance of source and sinks is how O_2 remains constant today.

The secret to how O_2 levels have evolved entails understanding how F_{source} and F_{sink} have altered over Earth's history. We argue below that the evidence is most consistent with the idea that the rise of O_2 was the consequence of the sink from reduced gases diminishing relative to the source of O_2. In the Archean, F_{sink} was dominated by rapid losses of O_2 to reduced volatiles such as H_2. Under such conditions, Eq. (4.14) was balanced in steady state at very low values of R_{O2}. After the rise of O_2, oxidative weathering ($F_{weathering}$) became a more important sink, and the steady-state balance was at significantly higher values of R_{O2}. To use an analogy, the amount of O_2 in the atmosphere can be likened to the water level in a bathtub. Even with water gushing out of the tap at a steady rate, the equilibrium level of the water in the bathtub depends on the size of the plughole. Similarly, the standing amount of O_2 in the air does not just depend on the source (the organic burial rate) but on the subsequent fate of the O_2 (the loss of O_2 to reductants).

4.7.1 The delay between the origin of cyanobacteria and the rise of O₂

Oxygenic photosynthesis surprisingly originated by 2.7 Ga (Section 4.5.3), 0.3–0.4 Gyr *before* the rise of O_2 (Section 4.6). One explanation is that a massive pulse of organic burial caused the rise of O_2, as evidenced by the large carbonate isotope excursions during 2.4–2.1 Ga (Section 4.6.2.1). However, given the geologically short residence time of O_2 ($\sim$2–3 Myr, even today) a pulse of organic burial would mean that O_2 would return to its previous low levels once burial and oxidation of previously buried carbon had re-equilibrated. For high O_2 to persist, a secular shift in source and sink fluxes of O_2 must occur. A second hypothesis is that as geothermal heat declined due to the decay of radioactive materials inside the Earth, the flux of volcanic gases dwindled, lessening the sink on O_2. However, increased past volcanic outgassing would have also injected proportionately more CO_2. Carbon isotopes from 3.5 Ga onwards show that $\sim$20% of the CO_2 flux into the biosphere was fixed biologically and buried as organic carbon with the remainder buried as carbonate (Section 4.6.2.1). Consequently, increased outgassing in the past, on its own, cannot explain the oxic transition because O_2 production due to organic burial would have paralleled O_2 losses. A third explanation of the rise of O_2 takes account of the problem with the previous idea by invoking a gradual, irreversible shift of outgassed volatiles from reduced to oxidized. This explanation is probably the most viable. Before discussing how and why the redox state of outgassed volatiles might change, we first consider the sinks for O_2 in the modern atmosphere.

4.7.2 Modern sinks for O_2: reduced gases and oxidative weathering

The ultimate source of O_2 comes from burying organic carbon. However, burial of other (non-detrital) redox species can also effectively generate or consume oxygen. For example, marine sulfate is microbially reduced to sulfide, which is buried with overall reaction $2Fe(OH)_3 + 4H_2SO_4 = 2FeS_2 + 15O_2 + 7H_2O$. Today, the burial of organic matter and pyrite (FeS_2) each contribute about 50% to F_{source} in Eq. (4.14). The burial of ferrous iron also effectively adds oxygen ($2Fe_2O_3 = 4FeO + O_2$), whereas the burial of sulfate, which requires that SO_2 is oxidized, effectively removes O_2 (Holland, 2002).

About 80–90% of F_{source} is removed in oxidative weathering, while the remainder reacts with reduced outgassed volatiles. But in the Archean, the balance must have been different. First, the lack of red beds and presence of reduced detrital minerals means that oxidative weathering was small. Second, the scarcity of marine sulfate means that the sulfide burial rate was lower (consistent with low C/S in Archean shales). Third, the similar rate of organic carbon burial relative to carbonates means that the O_2 sink from reduced gases must have been larger than today. But how could reduced gases, which apparently account for only 10–20% of the O_2 sink today, have consumed nearly all the O_2 produced in the Archean?

4.7.3 Methane, hydrogen escape, and metamorphic gas fluxes

Catling *et al.* (2001) argue that the sink on O_2 from reductants emanating from the Earth was greater in the Archean because oxidation of the crust due to hydrogen escape may have been important. Low O_2 would stabilize biogenic methane to an abundance $\sim 10^2$–10^3 ppm. This methane level would promote rapid escape of hydrogen to space, oxidizing the

Earth, and it would also counteract the fainter Sun by greenhouse warming (Section 4.2.3). Thus, elevated Archean methane couples the "faint young Sun" and "rise of O_2" problems.

In the crust, the moles of excess oxygen locked up in minerals, such as sulfate or iron oxides, greatly exceeds reduced carbon (tabulated in Catling *et al.*, 2001). This can only be explained by a net hydrogen loss because otherwise photosynthesis would have produced organic carbon and oxygen in equimolar quantities (Eq. (4.3)). Thus, if the Archean crust oxidized due to hydrogen escape, the H_2/CO_2 ratio in metamorphic gases would drop, and O_2 would no longer be overwhelmed by reaction with CH_4 or H_2. At this point, atmospheric O_2 would accumulate until balanced by oxidative weathering, and CH_4 levels would fall, inducing global cooling (Section 4.2.4).

In this model, microbial communities mediate the production of CH_4 (and ultimately hydrogen escape) via

$$2CH_2O \rightarrow CH_4 + CO_2, \tag{4.16}$$

where the organic matter derives from photosynthesis. Consequently, addition of Eq. (4.16) to twice Eq. (4.3) gives the overall reaction

$$CO_2 + 2H_2O \rightarrow CH_4 + 2O_2. \tag{4.17}$$

Although Eq. (4.17) mostly goes in the reverse direction, CH_4 can accumulate because O_2 reacts faster with outgassed H_2 and CO than it does with biogenic methane. A geological flux of H_2 is still needed to support high levels of CH_4. If we add twice Eq. (4.5) to Eq. (4.17), we get $4H_2 + CO_2 \rightarrow CH_4 + 2H_2O$. Thus, dominance of CH_4 over O_2 requires excess H_2 supplied by the Earth regardless of the pathway of Eq. (4.16). This means that when hydrogen escapes to space via biogenic methane, the place where the hydrogen originated (the crust or the mantle (Section 4.4.2)) must ultimately be oxidized.

4.7.4 Excess hydrogen in the Archean

Holland (2002) has proposed a hypothesis for why the Archean atmosphere had excess hydrogen. He notes that the net reduction path of the biosphere is

$$CO_2 + 2H_2 \rightarrow CH_2O + H_2O. \tag{4.18}$$

Even if O_2 is generated by oxygenic photosynthesis (Eq. (4.3)), it is destroyed by Eq. (4.11) with hydrogen; Eqs. (4.3) + (4.11) produce the net reaction of Eq. (4.18). H_2 that remains in volcanic gas after the reduction of 20% of its carbon to CH_2O (as inferred from sedimentary C isotopes) is available for the biosphere to reduce sulfur gases (principally SO_2) into sulfide. Sulfide then accumulates principally as sedimentary pyrite (FeS_2). Throughout the past 540 Myr, there has been enough H_2 left over from Eq. (4.18) to convert about 2/3 of the sulfur into pyrite (FeS_2), while the rest exits into sediments as sulfates. However, if Archean outgassing was H_2-rich, all of the sulfur could have been reduced to sulfide. In this case, there would be excess H_2 and the atmosphere would necessarily become "hydrogen-rich" and anoxic. Compared to modern volcanic gases, the increase in the proportion of hydrogen needed to flip to such an anoxic state would be a factor of ~2.4. Holland (2002) discusses possible ways in which this increased H_2 outgassing may have happened.

In summary, the most plausible explanation for why the late Archean atmosphere was anoxic is that excess reductants scavenged O_2. Hydrogen escape would have then oxidized the Earth, lowering the sink on O_2 until an oxic transition occurred. Explaining the second rise of O_2 at 0.8–0.6 Ga (Section 4.6.2.2) remains an area of future research. Perhaps methane persisted at a level of order about tens to a hundred ppm, and cumulative hydrogen escape over a further billion years helped shift another redox buffer beyond a critical threshold.

4.8 Formation of an ozone ultraviolet shield

An important consequence of the rise in O_2 at 2.4–2.3 Ga was the creation of an ozone (O_3) layer, which shielded the surface from biologically harmful solar ultraviolet (UV) radiation (but note that subterranean and marine organisms at depth do not require such protection). Radiation below a wavelength of ~200 nm is strongly absorbed by CO_2. For example, on Mars the CO_2-rich atmosphere prevents UV below 200 nm from reaching the surface, whereas biologically harmful radiation (in the 200–300 nm range) gets through (Cockell *et al.*, 2000). Today on Earth we are largely protected from radiation in the 200–300 nm range because of our atmosphere's ozone layer.

Ozone today is formed in the *stratosphere* (ranging from 20 to 50 km altitude) from photochemical reactions involving O_2. Absorption of radiation below 240 nm dissociates O_2 into O atoms:

$$O_2 + h\nu \rightarrow O + O, \qquad (4.19)$$

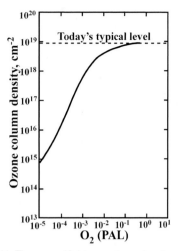

FIGURE 4.11 The ozone (O_3) layer column abundance as a function of pO_2 in the atmosphere. O_2 is expressed as a ratio to the Present Atmospheric Level (PAL). A typical ozone column abundance today is shown by the dotted line. The ozone layer's protective absorption of ultraviolet light became significant at pO_2 of ~0.01 PAL, which occurred ~2.4–2.3 Ga. (After Kasting and Donahue, 1980.)

and the O atoms react with other oxygen molecules to form ozone:

$$O + O_2 + M \rightarrow O_3 + M. \qquad (4.20)$$

Here M denotes any air molecule, usually N_2, which acquires the excess energy liberated by Eq. (4.20) and dissipates it through collisions with other air molecules. Ozone absorbs UV at longer wavelengths (200–310 nm) than O_2, and in the process the ozone is dissociated into O_2 and O and the absorbed energy heats the atmosphere. This is why the stratosphere increases in temperature with height, unlike the troposphere, where temperature decreases with height.

Photochemical models (Fig. 4.11) show that an atmosphere with $pO_2 \geq 0.002$ bar (1% of present) creates an ozone layer that absorbs most harmful UV (Kasting and Donahue, 1980). Such a level occurred on Earth at ~2.4–2.3 Ga.

4.9 Advanced life and O₂ on other Earth-like planets

The high O_2 content of Earth's atmosphere sets our planet apart from all others in the Solar System. But does Earth's O_2-rich atmosphere differ from all other planets in the Galaxy or Universe? Is O_2 required for higher life? And would a high O_2 atmosphere, which results from biology on Earth, be diagnostic of life on another planet?

4.9.1 Is O_2 required for animal-like life?

There are several good reasons for believing that extraterrestrial multicellular life (comparable to animals on Earth) would metabolize molecular oxygen. Life anywhere in the Universe must be constrained by the chemistry possible within the periodic table. Advanced extraterrestrial life, like its terrestrial counterpart, would require substantial energy for chemical, electrical, osmotic, and mechanical work. Fluorine is the most energetic oxidant (per electron transfer) available in the periodic table, and aqueous chlorine is similar to oxygen. However, F is useless because it spontaneously explodes on contact with organic material! Similarly, Cl forms bleach in water. O_2 differs because its bond provides greater stability than the weak single bonds of the halogens. Given that F_2 and Cl_2 are not realistic oxidants, free O_2 in a planetary atmosphere allows life to utilize the greatest energy source per electron transfer. Oxygen is also plentiful in the cosmos, third in abundance behind H and He. Furthermore, oxygen occurs as a gas in the temperature–pressure range of liquid water. A solid (e.g., sulfur) or liquid terminal oxidant would have a confined distribution and thereby limit possible habitats. No such restriction applies with gaseous O_2, which can be ubiquitous throughout an atmosphere and within oceans. Consequently, given the constraints of the periodic table, aerobic respiration is the most energetic chemistry for advanced life. Thus high atmospheric O_2 is probably a necessary precursor for animal-like life anywhere (Catling et al., 2005).

Consequently, the controls on the timescale for the rise of O_2 may be critical for the viability of advanced life on an extrasolar planet. For example, if a large, more chemically reducing planet around a Sun-like star took $\sim$10 Gyr to undergo a significant rise in O_2 (much more than Earth's $\sim$3.9 Gyr to reach the Cambrian), advanced life could be precluded because the star would have already evolved to a much more luminous red giant (Chapter 3).

4.9.2 Spectroscopic detection of life on extrasolar planets

In the future, life could be detected on extrasolar planets by remotely sensing their atmospheres (see Chapter 21 for plans to do this in the next decade). Compared to Venus and Mars, Earth has an anomalous atmosphere: water-rich, with strong ozone absorption at a wavelength of 9.6 μm (Fig. 4.12). Such an

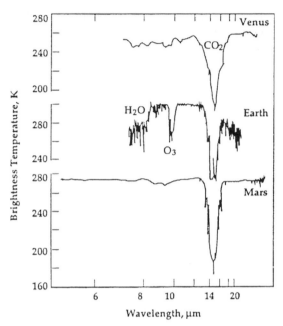

FIGURE 4.12 Remote sensing spectra obtained by orbiting spacecraft for Venus, Earth, and Mars. Brightness temperature is a measure of the temperature in the atmosphere or on the surface from where the emission originates. All atmospheric spectra show significant absorption due to CO_2 at 15 μm (Table 4.1). The spike in the center of the Earth's 15 μm band is characteristic of a warm layer of CO_2 in the stratosphere. The Earth's stratospheric ozone (O_3) layer is revealed by its absorption feature at 9.6 μm. Earth's spectrum also shows the presence of considerably more H_2O vapor than either Mars or Venus, indicative of Earth's oceans. (After Hanel et al., 1975.)

atmosphere is diagnostic of life because the ozone implies a continuous biological source of O_2 (Sagan et al., 1993). If photosynthetic life went extinct, O_2 would decrease to less than 1% of present levels in only $\sim$10–15 Myr by reacting with reduced surface materials and reduced gases. Joshua Lederberg and James Lovelock first suggested that the presence of gases that are far removed from equilibrium is diagnostic of life (Lederberg, 1965; Lovelock, 1965; 1978). The problem, however, is that hot planetary interiors and stellar radiation, independent of life, provide other sources of free energy that produce disequilibrium, so discriminating life becomes a question of degree.

Ozone absorption at 9.6 μm is a good surrogate for the presence of O_2. Even at pO_2 only 1% of the present, the ozone column depth would be about a third of the present global average value (Fig. 4.11). Thus, astronomers on a distant planet with a powerful spectrometer-telescope could have deduced the presence on Earth of O_2 over the past 2.4–2.3 Gyr. But could substantial

O_2 arise in the absence of life (Leger *et al.*, 1999)? Today, Venus has <1 ppm of O_2 in its atmosphere, but it may have had much more in the past. Early Venus is thought to have lost its oceans due to a severe greenhouse effect (Section 4.2.2). With a wet upper atmosphere, the rate of water-induced hydrogen loss became correspondingly very high. Thus, Venus could have had a large abiotic production of O_2 from dissociation of water (noting that O_2 build-up may have been self-limiting by thwarting hydrogen escape). Earth could experience a similar fate when the oceans boil 1–2 Gyr from now as the Sun heats up. In observing extrasolar planets, we must therefore estimate the likelihood of such scenarios from orbital elements and the age and luminosity of the parent star (Chapter 21).

To detect life on planets resembling early Earth *before* O_2 became abundant, one could look for the strong 7.6 μm absorption band of biogenic methane. However, abiological processes can also produce abundant methane. The atmosphere of Titan, Saturn's largest moon, has 4.9% CH_4 below 8 km altitude. The surface temperature of Titan is ~95 K – too cold for liquid water and conventional life – so it is believed that CH_4 emanates from geothermal processes in the subsurface (Chapter 20). Overall, the presence of methane as a spectroscopic biomarker is less certain than the presence of ozone.

4.10 Summary and conclusions

An Earth-like planet with liquid water on its surface is the most likely type of planet to be inhabited and to possess a biosphere that is detectable through remote atmospheric spectroscopy. Thus, a key aspect of a planet's biological potential is whether the planet's surface temperature is conducive to liquid water. The average surface temperature of a planet is determined by the planet's albedo, the atmospheric greenhouse effect, and the energy flux from its parent star. Negative feedbacks in a climate system tend to stabilize a planet's surface temperature within a particular range. Over geological time, the most important negative feedback for regulating the Earth's climate is the CO_2-climate feedback arising from the carbonate–silicate cycle. The rate of silicate weathering reactions, and therefore the consumption of CO_2, increases with the amount of CO_2 in the atmosphere. Given a source of CO_2 from volcanism and metamorphism that is independent of the amount of atmospheric CO_2, the climate tends to reach equilibrium at some equable level of CO_2 where the CO_2 sink balances the CO_2

source. However, such climate regulation failed on Venus, which was too close to the Sun and lost its water from dissociation and escape processes. Without water to remove CO_2 during weathering, CO_2 accumulated on Venus. Mars, being a small planet, shut down tectonically and failed to provide the necessary replenishment of CO_2 to the atmosphere. Mars probably also lost much of its atmosphere during early impact bombardment. For both reasons, Mars slipped into a permanent ice age.

Over geological history, the Earth's climate has been greatly affected by changes in the chemistry of the atmosphere coupled to the evolution of life. Today, apart from argon, all of the quantitatively important gases are at least in part biologically controlled (Table 4.1). Oxygen, in particular, has no significant abiological source. Consequently, before life evolved, Earth's atmosphere had negligible O_2. The prebiotic atmosphere probably consisted mainly of N_2, CO_2, and water, with relatively small amounts of H_2, CO, and CH_4. Reactions in such an atmosphere may have synthesized formaldehyde (H_2CO) and hydrogen cyanide (HCN), the basic building blocks for sugars and the nitrogen-containing bases required for RNA. Once life arose, microbes would have consumed H_2 and transformed it into CH_4. As a powerful greenhouse gas, CH_4 would have lowered the amount of CO_2 through temperature-dependent consumption of CO_2 in weathering. Thus, before O_2 became abundant in Earth's atmosphere, atmospheric models suggest that biogenic methane would have been stable at much higher levels than today and therefore was a more important greenhouse gas than CO_2.

The most significant biological event for the history of Earth's atmosphere was the evolution of oxygenic photosynthesis, a metabolism that dates back to at least 2.7 Ga. However, atmospheric O_2 remained below part per million levels until ~2.4–2.3 Ga. The history of O_2 can be characterized in terms of its chemical adversary: the collection of reducing, hydrogen-bearing atmospheric gases. Before the rise of O_2, hydrogen-bearing gases like CH_4 and H_2 must have tipped the redox balance in the atmosphere against O_2. A plausible mechanism for redox change is that the escape of excess hydrogen to space inexorably oxidized the Earth over geological time, shifting the redox balance in favor of an oxidizing atmosphere. Once O_2 finally rose, it created an ozone layer, shielding the surface from harmful ultraviolet radiation.

Given that atmospheric O_2 has no significant source other than from life, life on extrasolar Earth-like planets can be detected using ozone as a proxy for abundant oxygen. On Earth, the amount of oxygen in the atmosphere must have set the tempo for the evolution of complex biology to some degree. Chemical arguments suggest that complex life forms elsewhere will probably also utilize oxygen in their metabolism. Thus, understanding the redox history of habitable planets and their atmospheres is critical for the question of whether humans are an improbable accident and the sole sentient beings in the universe or whether we are more typical. Planetary atmospheres thus run like a thread through some of the most important questions in astrobiology: life's origins, planetary habitability, and the distribution of complex life.

References

Berner, R., Petsch, S. T., Lake, J. A., *et al.* (2000). Isotope fractionation and atmospheric oxygen: implications for phanerozoic O2 evolution. *Science*, **287**, 1630–1633.

Betts, J. N. and Holland, H. D. (1991). The oxygen content of ocean bottom waters, the burial efficiency of organic carbon, and the regulation of atmospheric oxygen. *Palaeogeog. Palaeoclim. Palaeoecol.*, **97**, 5–18.

Blankenship, R. E. and Hartman, H. (1998). The origin and evolution of oxygenic photosynthesis. *Trends Biochem. Sci.*, **23**, 94–97.

Canfield, D. E. (1998). A new model for Proterozoic ocean chemistry. *Nature*, **396**, 450–453.

Canfield, D. E., Poulton, S. W., and Narbonne, G. M. (2007). Late-Neoproterozoic deep-ocean oxygenation and the rise of animal life. *Science*, **315**, 92–95.

Canfield, D. E. and Teske, A. (1996). Late Proterozoic rise in atmospheric oxygen concentration inferred from phylogenetic and sulphur isotope studies. *Nature*, **382**, 127–132.

Catling, D. C. and Claire, M. W. (2005). How Earth's atmosphere evolved to an oxic state: a status report. *Earth Planet. Sci. Lett.*, **237**, 1–20.

Catling, D. C., Zahnle, K. J., and McKay, C. P. (2001). Biogenic methane, hydrogen escape, and the irreversible oxidation of early Earth. *Science*, **293**, 839–843.

Catling, D. C., Glein, C. R., Zahnle, K. J., and McKay, C. P. (2005). Why O2 is required by complex life on habitable planets and the concept of planetary "oxygenation time". *Astrobiology*, **5**, 415–38.

Cloud, P. (1988). *Oasis in Space: Earth History from the Beginning*. New York: Norton.

Cockell, C. S., Catling, D. C., Davis, W. L., Snook, K., Kepner, R. L., Lee, P., and McKay, C. P. (2000). The ultraviolet environment of Mars: biological implications past, present, and future. *Icarus*, **146**, 343–359.

Colaprete, A. C. and Toon, O. B. (2003). Carbon dioxide clouds at high altitude in the tropics and in an early dense Martian atmosphere. *J. Geophys. Res.*, **108**, 6-1, doi:10.1029/2002JE001967.

Dudley, R. (1998). Atmospheric oxygen, giant paleozoic insects, and the evolution of aerial locomotor performance. *J. Exp. Biol.*, **201**, 1043–1059.

Eyles, N. and Young, G. M. (1994). Geodynamic controls on glaciation in Earth's history. In *Earth's Glacial Record*, eds. M. Deynous *et al.*, pp. 1–28. New York: Cambridge University Press.

Farquhar, J., Bao, H. M., and Thiemens M. (2000). Atmospheric influence of Earth's earliest sulfur cycle. *Science*, **289**, 756–758.

Fike, D. A., Grotzinger, J. P., Pratt, L. M., and Summons, R. E. (2006). Oxidation of the Ediacaran Ocean. *Nature*, **444**, 744–747.

Halliday, A. N., Lee, D. C., Porcelli, D., Wiechert, U., Schonbachler, M., and Rehkamper, M. (2001). The rates of accretion, core formation and volatile loss in the early Solar System. *Philos. T. Roy. Soc.*, **A359**, 2111–2134.

Halverson, G. P., Hoffman, P. F., Schrag, D. P., Maloof, A. C., and Rice, A. H. N. (2005). Toward a Neoproterozoic composite carbon-isotope record. *GSA Bull.*, **117**, 1181–1207.

Hanel, R. A., Conrath, B. J., Jennings, D. E., and Samuelson, R. E. (1975). *Exploration of the Solar System by Infrared Remote Sensing*. New York: Cambridge University Press.

Hayes, J. M. (1994). Global methanotrophy at the Archean-Proterozoic transition. In *Early Life on Earth*, ed. S. Bengtson, pp. 220–236. New York: Columbia University Press.

Hedges, S. B., Hsiong, C., Kumar, S., Wang, D. Y.-C., Thompson, A. S., and Watanabe, H. (2001). A genomic timescale for the origin of eukaryotes. *BMC Evolutionary Biol.*, **1**, 4, doi:10.1186/1471-2148-1-4.

Hoffman, P. F. and Schrag, D. P. (2002). The Snowball Earth hypothesis: testing the limits of global change. *Terra Nova*, **14**, 129–155.

Holland, H. D. (1984). *The Chemical Evolution of the Atmosphere and Oceans*. Princeton, NJ: Princeton University Press.

Holland, H. D. (2002). Volcanic gases, black smokers, and the great oxidation event. *Geochim. Cosmochim. Acta*, **66**, 3811–3826.

Holser, W. T., Schidlowski, M., MacKenzie, F. T., and Maynard, J. B. (1988). Geochemical cycles of carbon and sulfur. In *Chemical Cycles in the Evolution of the Earth*, ed. C. B. Gregor, R. M. Garrels, F. T. MacKenzie, and J. B. Maynard, pp. 105–173. New York: John Wiley.

Kashefi, K. and Lovley, D. R. (2003). Extending the upper temperature limit for life. *Science*, **301**, 934.

Kasting, J. F. (1991). CO2 condensation and the climate of early Mars. *Icarus*, **94**, 1–13.

Kasting, J. F. (1993). Earth's early atmosphere. *Science*, **259**, 920–926.

Kasting, J. F. and Brown, L. L. (1998). The early atmosphere as a source of biogenic compounds. In *The Molecular Origins of Life*, ed. A. Brack. New York: Cambridge University Press.

Kasting, J. F. and Donahue, T. M. (1980). The evolution of atmospheric ozone. *J. Geophys. Res.*, **85**, 3255–3263.

Kasting, J. F., Pavlov, A. A., and Siefert, J. L. (2001). A coupled ecosystem-climate model for predicting the methane concentration in the Archean atmosphere. *Origins of Life Evol. Biosph.*, **31**, 271–285.

Kasting, J. F., Toon, O. B., and Pollack, J. B. (1988). How climate evolved on the terrestrial planets. *Sci. Am.*, **256**, 90–97.

Kelley, D. S., Karson, J. A., Früh-Green, G. L., *et al.* (2005). A serpentinite-hosted ecosystem: the lost city hydrothermal field. *Science*, **307**, 1428–1434.

Kiehl, J. T. and Dickinson, R. E. (1987). A study of the radiative effects of enhanced atmospheric CO2 and CH4 on early Earth surface temperatures. *J. Geophys. Res.*, **92**, 2991–2998.

Kirschvink, J. L. (1992). Late Proterozoic low-latitude glaciation: the Snowball Earth. In *The Proterozoic Biosphere*, ed. J. W. Schopf and C. Klein, pp. 51–52. New York: Cambridge University Press.

Kirschvink, J. L., Gaidos, E. J., Bertani, L. E., Beukes, N. J., Gutzmer, J., Maepa, L. N., and Steinberger, R. E. (2000). Paleoproterozoic snowball Earth: extreme climatic and geochemical global change and its biological consequences. *Proc. Nat. Acad. Sci.*, **97**, 1400–1405.

Knoll, A. H. (1992). The early evolution of eukaryotes: a geological perspective. *Science*, **256**, 622–627.

Knoll, A. H. and Carroll, S. B. (1999). Early animal evolution: emerging views from comparative biology and geology. *Science*, **284**, 2129–2137.

Kress, M. E. and McKay, C. P. (2004). Formation of methane in comet impacts: implications for Earth, Mars, and Titan. *Icarus*, **168**, 475–483.

Kump, L. R., Kasting, J. F., and Barley, M. E. (2001). Rise of atmospheric oxygen and the "upside-down" Archean mantle. *Geochem. Geophys. Geosyst.*, **2**, doi:10.1029/2000GC000114.

Lederberg, J. (1965). Signs of life: criterion system of exobiology. *Nature*, **207**, 9–13.

Leger, A., Ollivier, M., Altwegg, K., and Woolf, N. J. (1999). Is the presence of H₂O and O₃ in an exoplanet a reliable signature of a biological activity? *Astron. Astrophys.*, **341**, 304–311.

Lenton T. M. and Watson A. J. (2000). Redfield revisited: 2. What regulates the oxygen content of the atmosphere? *Global Biogeochem. Cycles*, **14**, 249–268.

Lodders, K. and Fegley, B. (1998). *The Planetary Scientist's Companion*. New York: Oxford University Press.

Lovelock, J. E. (1965). A physical basis for life detection experiments. *Nature*, **207**, 568–570.

Lovelock, J. E. (1978). Thermodynamics and the recognition of alien biospheres. *Proc. Roy. Soc. Lon.*, **B189**, 167–181.

Lovelock, J. E. (1988). *The Ages of Gaia*. New York: Oxford University Press.

Mancinelli, R. L. and McKay, C. P. (1988). The evolution of nitrogen cycling. *Orig. Life Evol. Biosphere*, **18**, 311–325.

McKay, C. P. and Hartman, H. (1991). Hydrogen peroxide and the origin of oxygenic photosynthesis. *Origins of Life*, **21**, 157–163.

Melosh, H. J. and Vickery, A. M. (1989). Impact erosion of the primordial atmosphere of Mars. *Nature*, **338**, 487–489.

Mojzsis, S. J., Harrison, T. M., and Pidgeon, R. T. (2001). Oxygen-isotope evidence from ancient zircons for liquid water at the Earth's surface 4,300 Myr ago. *Nature*, **409**, 178–181.

Murakami, T., Utsunomiya, S., Imazu, Y., and Prasad, N. (2001). Direct evidence of late Archean to early Proterozoic anoxic atmosphere from a product of 2.5 Ga old weathering. *Earth Planet. Sci. Lett.*, **184**, 523–528.

Navarro-Gonzalez, R., McKay, C. P., and Mvondo, D. N. (2001). A possible nitrogen crisis for Archaean life due to reduced nitrogen fixation by lightning. *Nature*, **412**, 61–64.

Nursall, J. R. (1959). Oxygen as a prerequisite to the origin of the metazoa. *Nature*, **183**, 1170–1172.

Ohmoto, H. (1997). When did the Earth's atmosphere become oxic? *The Geochem. News*, **93**, 12–13, 26–27.

Pavlov, A. A. and Kasting, J. F. (2002). Mass-independent fractionation of sulfur isotopes in Archean sediments: strong evidence for an anoxic Archean atmosphere. *Astrobiology*, **2**, 27–41.

Pavlov, A. A., Kasting, J. F., and Brown, L. L. (2001). UV-shielding of NH3 and O2 by organic hazes in the Archean atmosphere. *J. Geophys. Res.*, **106**, 23267–87.

Pavlov, A. A., Kasting, J. F., Brown, L. L., Rages, K. A., and Freedman, R. (2000). Greenhouse warming by CH4 in the atmosphere of early Earth. *J. Geophys. Res.*, **105**, 11981–90.

Pinto, J. P., Gladstone, C. R., and Yung, Y. L. (1980). Photochemical production of formaldehyde in the earth's primitive atmosphere. *Science*, **210**, 183–185.

Pollack, J. B. (1991). Kuiper Prize Lecture: Present and past climates of the terrestrial planets. *Icarus*, **91**, 173–198.

Prasad, N. and Roscoe, S. M. (1996). Evidence of anoxic to oxic atmospheric change during 2.45-2.22 Ga from lower and upper sub-Huronian paleosols. Canada. *Catena*, **27**, 105–121.

Rasmussen, R. and Buick, R. (1999). Redox state of the Archean atmosphere: evidence from detrital heavy minerals in ca. 3250–2750 Ma sandstones from the Pilbar Craton, Australia. *Geology*, **27**, 115–118.

Roscoe, S. M. (1973). The Huronian Supergroup: a Paleophebian succession showing evidence of atmospheric evolution. *Geol. Soc. Can. Spec. Pap.*, **12**, 31–48.

Rossow, W. B., Henderson-Sellers, A., and Weinrich, S. K. (1982). Cloud feedback: a stabilizing effect for the early Earth? *Science*, **217**, 1245–1247.

Rothschild, L. J. and Mancinelli, R. L. (2001). Life in extreme environments. *Nature*, **409**, 1092–1101.

Runnegar, B. (1991). Oxygen and the early evolution of the Metazoa. In *Metazoan Life Without Oxygen*, ed. C. Bryant. New York: Chapman and Hall.

Rye, R. and Holland, H. D. (1998). Paleosols and the evolution of atmospheric oxygen: a critical review. *Amer. J. Sci.*, **298**, 621–672.

Sagan, C., Thompson, C. R., Carlson, R., Gurnett, D., and Hord, C. (1993). A search for life on Earth from the Galileo spacecraft. *Nature*, **365**, 715–721.

Sagan, C. and Mullen G. (1972). Earth and Mars: evolution of atmospheres and surface temperatures. *Science*, **177**, 52–56.

Schimel, D., Alvez, D., Enting, I., *et al.* (1996). Radiative forcing of climate change. In *Climate Change, 1995*, ed. J. T. Houghton *et al.*, pp. 65–131. New York: Cambridge University Press.

Shixing, Z. and Huineng, C. (1995). Megascopic multicellular organisms from the 1700-million-year-old Tuanshanzi Formation in the Jixian area, north China. *Science*, **270**, 620–622.

Walker, J. C. G. (1977). *Evolution of the Atmosphere*. New York: Macmillan.

Wilde, S. A., Valley, J. W., Peck, W. H., and Graham, C. M. (2001). Evidence from detrital zircons for the existence of continental crust and oceans on the Earth 4.4 Gyr ago. *Nature*, **409**, 175–178.

Xiao, S. and Kaufman, A. J. (2006). *Neoproterozoic Geobiology and Paleobiology*. Dordrecht: Springer.

Xiong, J., Fischer, W. M., Inoue, K., Nakahara, M., and Bauer, C. E. (2000). Molecular evidence for the early evolution of photosynthesis. *Science*, **289**, 1724–1730.

Zahnle, K. J. (1986). Photochemistry of methane and the formation of hydrocyanic acid (HCN) in the Earth's early atmosphere. *J. Geophys. Res.*, **91**, 2819–2834.

FURTHER READING

De Pater, I. and Lissauer, J. (2001). *Planetary Sciences*, Cambridge: Cambridge University Press. Background to planetary atmospheres aimed at graduate students and researchers.

Kump, L. R., Kasting, J. F., and Crane, R. G. (2004). *The Earth System*, second edn. Upper Saddle River, NJ: Pearson Education. A readable background to Earth system science aimed at a general undergraduate audience.

Walker, J. C. G. (1977). *Evolution of the Atmosphere*. New York: Macmillan. Somewhat outdated, but still a very readable scientific introduction to atmospheric evolution.

Walker, J. C. G. (1986). *Earth History: the Several Ages of the Earth*. Boston: Jones & Bartlell Publishers. Written for the layman, this book describes how Earth's atmosphere and surface evolved to produce a habitable planet.

Yung, Y. L. and DeMore, W. B. (1999). *Photochemistry of Planetary Atmospheres*. New York: Oxford University Press. Planetary atmospheres with a focus on the photochemistry, aimed at researchers.

Part III
The origin of life on Earth

5 Does 'life' have a definition?

Carol E. Cleland, *University of Colorado*
Christopher F. Chyba, *Princeton University*

5.1 Introduction

The question "What is life?" is foundational to biology and especially important to astrobiologists who may one day encounter utterly alien life. But how should one approach this question? One widely adopted strategy among scientists is to try to define 'life.'[1] This chapter critically evaluates this strategy. Drawing from insights gained by philosophical investigations into the nature of logic and language, we argue that it is unlikely to succeed. We propose a different strategy, which may prove more fruitful in searches for extraterrestrial life.

We begin in Section 5.2 by reviewing the history of attempts to define 'life,' and their utility in searches for extraterrestrial life. As will become apparent, these definitions typically face serious counterexamples, and may generate as many problems as they solve.

To explain why attempts to define 'life' are fraught with so many difficulties, we must first develop the necessary philosophical background. Therefore, in Sections 5.3 and 5.4 we discuss the general nature of definition and of so-called theoretical identity statements. Section 5.5 then applies the material developed in these sections to the project of defining 'life.' We argue that the idea that one can answer the question "What is life?" by defining 'life' is mistaken, resting upon confusions about the nature of definition and its capacity to answer fundamental questions about natural categories (Cleland and Chyba, 2002).

To answer the question "What is life?" we require not a definition but a general theory of the nature of living systems. In the absence of such a theory, we are in a position analogous to that of a sixteenth-century investigator trying to define 'water' before the advent of molecular theory. The best she or he could do would be to define it in terms of sensible properties, such as its being wet, transparent, odorless, tasteless, thirst quenching, and a good solvent. But no amount of observational or conceptual analysis of these features will reveal that water is H_2O. Yet, as we now know, "H_2O" is the scientifically most informative answer to the question "What is water?" Analogously, in the absence of a general theory of the nature of living systems, analysis of the features that we currently associate with life is unlikely to provide a particularly informative answer to the question "What is life?"

5.2 Attempts to define 'life'

The history of attempts to define 'life' is very long, going back at least to Aristotle, who defined 'life' in terms of the capacity to reproduce (Aristotle, *De Anima* 415^{a22}–415^{b2}; Matthews, 1977; but see also Shields, 1999). To this day, there remains no broadly accepted definition of 'life' (Chyba and McDonald, 1995). The scientific literature is filled with suggestions; decades ago Sagan (1970) catalogued physiological, metabolic, biochemical, Darwinian (which he called "genetic"), and thermodynamic definitions, along with their counterexamples. There have been many other attempts[2] (see, e.g., Schrödinger, 1945; Monod, 1971; Feinberg and Shapiro, 1980; Dyson, 1985; Kamminga, 1988; Fleischaker, 1990; Joyce, 1994, 1995; McKay, 1994; Shapiro and Feinberg, 1995; Bedau, 1996; Rizzotti *et al.*, 1996; Adami, 1998; Kauffman, 2000; Conrad and Nealson, 2001; Harold, 2001; Schulze-Makuch *et al.*, 2002). All typically face important problems, in

---

[1] Single quotation marks around a word indicate that it is being mentioned as opposed to being used. Definitions provide one example. Another example is the claim that 'life' has four letters; contrast this with the very different claim that life originated on Earth around four billion years ago.

[2] Lahav (1999: 117–21) compiles 48 definitions of life (with citations) offered from 1855 to 1997.

Planets and Life: The Emerging Science of Astrobiology, eds. Woodruff T. Sullivan, III and John A. Baross. Published by Cambridge University Press. © Cambridge University Press 2007.

that they include phenomena that most are reluctant to consider alive, or exclude entities that clearly are alive (Chyba and McDonald, 1995).

Consider a few attempted definitions by way of illustration (Sagan, 1970). A *metabolic* definition, for example, might be based on the ability to consume and convert energy in order to move, grow, or reproduce. But fire, and perhaps even automobiles, might be said to satisfy some or all of these criteria. A *thermodynamic* definition might describe a living system as one that takes in energy in order to create order locally, but this would seem to include crystals, which like fire would not generally be considered alive. A *biochemical* definition would be based on the presence of certain types of biomolecules, yet one must worry that any such choice could in the future face exceptions in the form of systems that otherwise appear alive but are not made of our particular favored molecules. *Genetic* or *Darwinian* definitions are now more generally favored than any of these other definitions, but these too face drawbacks and will be discussed in detail in Section 5.2.2 below.

Another approach has been not so much to define life as simply to list its purported characteristics (e.g., Mayr, 1982; Koshland, 2002). But essentially the same difficulties arise in this approach; for example, Schulze-Makuch *et al.* (2002) present a list of nonbiological parallels to various supposedly distinguishing criteria of life such as metabolism, growth, reproduction, and adaptation to the environment.

Nevertheless, the philosophical question of the definition of 'life' has increasing practical importance, as laboratory experiments approach the synthesis of life (as measured by the criteria of some definitions), and as greater attention is focused on the search for life on Mars (Chapter 18) and Jupiter's moon Europa (Chapter 19). In particular, definitions of 'life' are often explicit or implicit in planning remote *in situ* searches for extraterrestrial life. The design of life-detection experiments to be performed on Europa (e.g., Chyba and Phillips, 2001; 2002) or Mars (e.g., Nealson and Conrad, 1999; Banfield *et al.*, 2001; Conrad and Nealson, 2001) by spacecraft landers depends on decisions about what life is, and what observations will count as evidence for its detection (Chapters 22 and 23). This is clearly illustrated by the story of the Viking mission's search for life on Mars.

5.2.1 Viking's search for life on Mars

The Viking mission's search for life on Mars in the mid-1970s remains the only dedicated *in situ* search for extraterrestrial life to date. Details of the experiments performed and their results are given in Sections 18.5.2 and 23.2. The basic approach was to conduct experiments with the martian soil to test for the presence of metabolizing organisms, and indeed the results of the labeled release experiment in particular were not unlike what had been expected for the presence of life (Levin and Straat, 1979; Levin and Levin, 1998). But in the end, The Viking biology team's consensus was for a nonbiological interpretation (Klein, 1978; 1979; 1999), strongly influenced by the failure of the Viking gas chromatograph mass spectrometer (GCMS) to find any organic molecules to its limits of detection in the soil with sample heating up to 500 °C (Biemann *et al.*, 1977). This instrument had not been intended to conduct a "life-detection" experiment, but de facto did so, implicitly employing a biochemical definition. Moreover, the GCMS would not have detected as many as $\sim 10^6$ bacterial cells per gram of soil (Klein, 1978; Glavin *et al.* 2001; Bada 2001), and it now appears that oxidation of meteoritic organics on the martian surface may have produced nonvolatile organic compounds that would not have been easily detectable (Benner *et al.*, 2000). Correctly interpreted or not, the result was psychologically powerful: no (detected!) organics, no life. Chyba and Phillips (2001; 2002) have presented a list of lessons to be learned from this experience – one lesson is that any *in situ* search for extraterrestrial life should employ more than one definition of life so that results can be intercompared.

Of course, if there were really *one* correct, known definition of 'life,' this would be an unnecessary strategy. Currently, it is the Darwinian definition that seems most accepted. We examine this definition below, but shall see that rather than providing us with an unassailable definition, it instead presents fresh dilemmas.

5.2.2 The Darwinian definition

Darwinian (sometimes called *genetic*) definitions of 'life' hold that life is "a system capable of evolution by natural selection" (Sagan, 1970). One working version that is popular within the origins-of-life community is the "chemical Darwinian definition" (Chyba and McDonald, 1995), according to which "life is a self-sustained chemical system capable of undergoing Darwinian evolution" (Joyce, 1994, 1995). Joyce (1994) explains that "the notion of Darwinian evolution subsumes the processes of self-reproduction, material continuity over a historical lineage, genetic variation, and natural selection. The requirement that the system be self-sustained refers to the fact that living systems contain all the genetic information necessary for their own constant production (i.e., metabolism)." The chemical Darwinian definition excludes computer or artificial "life" through its demand that the system under consideration be "chemical"; it also excludes biological viruses, by virtue of the "self-sustained" requirement.

Some researchers (e.g., Dawkins, 1983; Dennett, 1995), on the other hand, do not restrict Darwinian evolution to chemical systems, explicitly leaving open the possibility of computer life. This reflects the functionalist view (e.g., Sober, 1992) that Darwinian evolution is a more general process that can be abstracted from any particular physical realization. In this view, it is not the computer that is alive but rather the processes themselves. The artificial vehicle of the computer, produced by human beings, has a status no different from that of the artificial glassware that might be used in a laboratory synthesis of organic life. It is thus not surprising that, according to this view, "living" systems or ecosystems can in fact be created in a computer (e.g., Rasmussen, 1992; Ray, 1992).

Yet this too may seem unsatisfactory: a computer simulation of cellular biochemistry is a simulation of biochemistry, and not biochemistry itself. No computer simulation of photosynthesis, for example, is actually photosynthesis since it does not yield authentic carbohydrates; at best, it yields simulated carbohydrates. So why should a computer simulation of "life" be called life itself, rather than a *simulation* of life? On the functionalist view, the simulation *is* life, because life is an abstract process independent of any particular physical realization.

There are further problems with Darwinian definitions, in addition to the quandary regarding computer "life." It is possible (though not generally favored among current theories of the origin of life on Earth) that early cellular life on Earth or some other world passed through a period of reproduction without DNA-type replication, during which Darwinian evolution did not yet operate (e.g., Dyson, 1985; Rode, 1999; New and Pohorille, 2000; Pohorille and New, 2000). In this hypothesis, protein-based creatures capable of metabolism predated the development of exact replication based on nucleic acids. If such entities were to be discovered on another world, Darwinian definitons would preclude them from being considered alive.

There is an additional simple objection to the Darwinian definition, namely that individual sexually reproducing organisms in our DNA-protein world do not themselves evolve, so that many living entities in our world are not, by the Darwinian definition, examples of "life." The Darwinian definition refers to a *system* that at least in some cases must contain more than one entity; with this reasoning Victor Frankenstein's unique creation (Shelley, 1818), for example, is not "life" even though it is a living entity. But this resolution needs to be explained as more than an ad hoc move to shave from the definition bedeviling entities that we would otherwise call examples of "life," but which cause trouble for a particular definition.

Finally, there is a practical drawback to Darwinian definitions. In an *in situ* search for life on other planets, how long would we wait for a system to demonstrate that it is "capable" of Darwinian evolution, and under what conditions (Fleischaker, 1990)? This objection, however, is not decisive in itself, since an operational objection is not an objection in principle, and ways (see Chao, 2000) might be found to operationalize the definition.

We have focused on Darwinian definitions because they are currently in vogue, especially in light of the great successes of the RNA world model for the origin of life (Gilbert, 1986; Chapters 6 and 8). Nevertheless, as we have discussed, all of the popular versions of the Darwinian definition face similar severe challenges.

5.3 Definitions

To understand why attempts to define 'life' prove so difficult, we now develop the philosophical background for the nature of definition. Definitions are concerned with language and concepts. For example, the definition " 'bachelor' means unmarried human male" does not talk about bachelors. Instead, it explains the meaning of a *word*, in this case 'bachelor,' by dissecting the *concept* that we associate with it. As this example illustrates, every definition has two parts. The *definiendum* is the expression being defined ('bachelor') and the *definiens* is the expression doing the defining.

5.3.1 Varieties of definition

Many different sorts of things are commonly called "definitions." In this section we will discuss only those that are relevant to understanding the problem of providing a scientifically useful definition of 'life'; for more on definitions, see, for example, Audi (1995).

Lexical definitions report on the standard meanings of terms in a natural language. Dictionary definitions provide a familiar example. Lexical definitions contrast with *stipulative definitions*, which explicitly introduce new, often technical, meanings for terms. The following stipulative definition introduces a new meaning for an old term: 'work' means the product of the magnitude of an acting force and the displacement due to its action. Stipulative definitions are also used to introduce invented terms, e.g., 'electron' (means basic unit of electricity), or 'gene' (means basic unit of heredity). Unlike lexical definitions, stipulative definitions are arbitrary in the sense that rather than reporting on existing meanings of terms, they explicitly introduce new meanings.

Another familiar type of definition is the *ostensive definition*. Ostensive definitions specify the meaning of a term merely by indicating a few (ideally) prototypical examples within its *extension*; the extension of a term is the class of all the things to which it applies. An adult who explains the meaning of the word 'dog' to a child by pointing to a dog and saying "that is a dog" is providing an ostensive definition. Someone who defines 'university' as "an institution such as the University of Colorado, Stanford University, Universidad de Guadalajara, and Cambridge University" is also providing an ostensive definition.

Operational definitions provide an important related form of definition. Like ostensive definitions, operational definitions explain meanings via representative examples. They do not, however, directly indicate examples, but instead specify *procedures* that can be performed on something to determine whether or not it falls into the extension of the definiendum. An example of an operational definition is defining 'acid' as 'something that turns litmus paper red.' The definiens specifies a procedure that can be used to determine whether an unknown substance is an acid. Operational definitions are particularly important for our discussion since many astrobiologists, e.g., one of these authors (Chyba and McDonald (1995), McKay (1994), Nealson and Conrad (1999), and Conrad and Nealson (2001)) have called for the use of operational definitions in searches for extraterrestrial life. The problem with operational definitions is that they do not tell one very much about what the items falling under the definiendum have in common. The fact that litmus paper turns red when placed in a liquid doesn't tell us much about the nature of acidity; it only tells us that a particular liquid is something called 'acid.' In other words, operational definitions differ from ostensive definitions primarily in the manner in which they pick out the representative examples of items falling under the definiendum, namely, indirectly by means of "tests," as

opposed to lists or gestures (the 'dog' example). We will return to this important point later.

The most informative definitions specify the meanings of terms by analyzing concepts and supplying a noncircular synonym for the term being defined. In philosophy, such definitions are known as *full* or *complete definitions*. But because philosophers sometimes use these expressions to designate more fine-grained distinctions, we shall use the term *ideal definition*.

5.3.2 Ideal definitions

Ideal definitions explain the meanings of terms by relating them to expressions that we already understand. It is thus important that the definiens make use of neither the term being defined nor one of its close cognates; otherwise the definition will be *circular*. Defining 'line' as "a linear path" is an example of an explicitly circular definition, while an implicitly circular definition is defining 'cause' as "something that produces an effect." Someone who does not understand the meaning of 'cause' will also not understand the meaning of 'effect' since 'effect' means something that is caused. Many lexical definitions suffer from the defect of circularity, which is why philosophers dislike dictionary definitions.

The definition of 'bachelor' (as "unmarried human male") with which we began this discussion provides a salient illustration of an ideal definition. It is not circular since the concept of being unmarried, human, and male does not presuppose an understanding of the concept of bachelor. The definiens thus provides an informative analysis of the meaning of 'bachelor.' An ideal definition may thus be viewed as specifying the meaning of a term by reference to a logical conjunction of properties (being unmarried, human, and male), as opposed to representative examples (ostensive definition), or a procedure for recognizing examples (operational definition). The conjunction of descriptions determines the extension of the definiendum by specifying necessary and sufficient conditions for its application. A *necessary condition* for falling into the extension of a term is a condition in whose absence the term does not apply and a *sufficient condition* is a condition in whose presence the term cannot fail to apply.

Most purported ideal definitions face borderline cases in which it is uncertain as to whether something satisfies the conjunction of predicates supplied by the definiens. A good example is the question of whether a ten-year-old boy is a bachelor. Moreover, even if one resolves such cases by adding additional conditions (e.g., adult) to the definiens, there will always be other borderline cases (e.g., the status of eighteen-year-old males). Language is vague. This is brought forcefully home by the classic example of trying to distinguish a bald man from a man who is not bald in terms of the number of hairs on his head. The fact that we cannot specify a crisp boundary does not show that there is no difference between being bald and not being bald. Ideal definitions that specify both necessary and sufficient conditions are rare. Nevertheless, ignoring the problem of borderline cases, we can often construct fairly satisfactory approximations. If definitions of 'life' faced nothing more serious by way of counterexamples than borderline cases (e.g., viruses), there might not be insurmountable problems. But they have more serious problems. Just as good definitions of 'bachelor' or 'bald man' must deal with, respectively, forty-year-old unmarried men and men sporting thick heads of hair, so good definitions of 'life' must deal with quartz crystals and candle flames, which are (presumably) clearly not alive.

5.4 Natural kinds and theoretical identity statements

Ideal definitions specify meanings by providing a "complete" (within the constraints of vagueness) analysis of the concepts associated with terms. They work well for terms such as 'bachelor' or 'fortnight' or 'chair,' which designate categories whose existence depends solely on human interests and concerns. Indeed, it is hard to imagine a better answer to the question "What is a bachelor?" than "an unmarried, adult human male."

Ideal definitions do not, however, supply good answers to questions about the identity of *natural kinds* – categories carved out by nature, as opposed to human interests, concerns, and conventions.[3] This issue is particularly important for our purposes since it seems likely (but not certain) that 'life' is a natural kind term – that whether something is living or nonliving represents an objective fact about the natural world. Consider, for example, trying to answer the

FIGURE 5.1. The distillation of *aqua vitae*, a form of 'water.' From *Das Buch zu Distillieren* by Hieronymus Braunschweig (Strassburg, 1519). (From Roberts (1994: 100); courtesy British Library.)

question "What is water?" by defining the natural kind term 'water.' One could try to define 'water' by reference to its sensible properties, features such as being wet, transparent, odorless, tasteless, thirst quenching, and a good solvent. (This is analogous to some suggested definitions of 'life,' e.g., that of Koshland (2002).) Unlike the definition of 'bachelor,' however, this definition of 'water' is not simply a matter of linguistic convention. Nevertheless, reference to a list of sensible properties cannot exclude things that superficially resemble water but are not in fact water. As an example, the alchemists, impressed by water's powers as a solvent, identified nitric acid and mixtures of hydrochloric acid as water, the former being known as *aqua fortis* ("strong water") and the latter as *aqua regia* ("royal water"); *aqua vitae* ("water of life") was a mixture of alcohols (Roberts, 1994; Fig. 5.1). Even today we commonly classify as 'water' various liquids that greatly differ in their sensible properties, e.g., salt

---

[3] Some philosophers of science (known as "anti-realists") reject, to greater or lesser degrees, claims that there are knowable, mind-independent facts, entities, or laws. We cannot engage with this literature here; for an introduction see Audi (1995) and references therein.

water, muddy water, and distilled water. Which of the sensible properties (e.g., transparency or tastelessness) of the various things called 'water' are the important ones? Five hundred years ago Leonardo da Vinci (1513) expressed this dilemma well:

> And so it [water] is sometimes sharp and sometimes strong, sometimes acid and sometimes bitter, sometimes sweet and sometimes thick or thin, sometimes it is seen bringing hurt or pestilence, sometimes health-giving, sometimes poisonous. So one would say that it suffers change into as many natures as are the different places through which it passes. And as the mirror changes with the colour of its object so it changes with the nature of the place through which it passes: health-giving, noisome, laxative, astringent, sulphurous, salt, incarnadined, mournful, raging, angry, red, yellow, green, black, blue, greasy, fat, thin.

Without an understanding of the intrinsic nature of water, there is no definitive answer to the question "What is water?" Given an understanding of the molecular structure of matter, however, such quandaries disappear. Water *is* H_2O – a molecule made of two atoms of hydrogen and one atom of oxygen. H_2O is what salt water, muddy water, distilled water, and even acidic solutions have in common, despite their obvious sensible differences. The identification of water with H_2O explains why liquids (e.g., nitric acid) that (in some ostensibly important ways) resemble water are not water; their molecular composition is more than H_2O alone. Furthermore, the identification explains the behavior of what we call 'water' under a wide variety of chemical and physical circumstances. The identification holds regardless of whether the water is in any of its familiar solid, liquid or vapor phases, and it will hold equally well in less familiar high-pressure solid phases. Indeed, before the advent of modern chemistry, it was not widely recognized that ice, steam, and liquid water are phases of the same kind of stuff. Some ancient Greeks (for example, Anaximenes) believed that steam was a form of "air" (Lloyd, 1982: 22). As late as the late seventeenth century, ice and water were thought to be different "species." The Aristotelian view of water as one of the four basic elements out of which all matter is constructed only began to fall into disfavor in the late eighteenth century with work such as Antoine Lavoisier's paper (1783) entitled "On the nature of water and on experiments that appear to prove that this substance is not

properly speaking an element, but can be decomposed and recombined." It took more than analysis of sensible properties to definitively settle questions about the proper classification of such ostensibly different substances as ice and steam.

Notice that the identification of water with H_2O does not have the character of an ideal definition. It cannot be viewed as explicating the concept that has historically been associated with the term 'water' since that concept encompasses stuff varying widely in chemical and physical composition. Moreover, in daily discourse we still use the word 'water' for things that are not pure H_2O. The claim that water is H_2O began as a testable empirical conjecture (situated within Lavoisier's new theoretical framework for chemistry), and it is now considered so well confirmed that most scientists characterize it as a fact. Nevertheless, it remains a scientific hypothesis. It is conceivable (even if extraordinarily unlikely) that we may someday discover that current molecular theory is wrong in some important respect and that water is not H_2O, just as Planck and Einstein showed a century ago that the wave theory of light was incomplete and that light also behaves like a particle. If the claim that water is H_2O represented an ideal definition, we could not admit the possibility that water might not be H_2O any more than we can conceive of a married bachelor or a month-long fortnight.

It is sometimes claimed that *theoretical identity statements* such as "water is H_2O," "temperature is mean kinetic energy," and "sound is a compression wave" represent stipulative definitions. On this view they amount to nothing more than linguistic decisions to take familiar terms from common language and give them wholly new technical meanings within the context of a currently accepted theory (Nagel, 1961). The prima facie problem with this account is that it prevents us from making sense of the idea that these statements tell us something new about the stuff designated by the old familiar terms ('water,' 'temperature,' 'sound'). Rather than learning something new, in this view we are merely attaching new concepts (identifying descriptions) to old terms, and hence only changing the way we talk about the world. One might be tempted to say that this is the way language works: if one changes the concept associated with a word radically enough, then one is no longer talking about the same thing. However, such an approach, associated with the philosopher John Locke (and exploited by Thomas Kuhn in his famous arguments for the incommensurability of scientific theories), faces serious logical problems; we discuss these in detail in Appendices 5.1–5.2 at the end of this chapter. For this reason most contemporary philosophers reject the view that theoretical identity statements are stipulative definitions. Some radical changes in the concept of an old word are the result of discovering that we were wrong about the familiar phenomenon that the word designates; for more detail, see Appendix 5.3. Put more concretely, we know something about water that Aristotle and Anaximenes didn't know: water is not a primitive element, but a molecular compound.

5.5 What is 'life'?

Let us return to the definition of 'life.' If (as seems likely, but not certain) life is a natural kind, then attempts to define 'life' are fundamentally misguided. Definitions serve only to explain the concepts that we currently associate with terms. As human mental entities, concepts cannot reveal the objective underlying nature (or lack thereof) of the categories designated by natural kind terms. Yet when we use a natural kind term, it is this underlying nature (not the concepts in our heads) that we are interested in. 'Water' means whatever the stuff in streams, lakes, oceans, and *everything* else that is water has in common. We currently believe that this stuff is H_2O, and our belief is based on a well-confirmed, general scientific theory of matter. We cannot, of course, be absolutely positive that molecular theory is the final word on the nature of matter; *conclusive* proof is just not possible in science. Nevertheless, our current scientific concept of water as H_2O represents a vast improvement over earlier concepts based on superficial sensory experience. If we someday discover that molecular theory is wrong, we will change the concept that we associate with 'water,' but we will still be talking about the same thing.

Analogously to 'water,' 'life' means whatever cyanobacteria, hyperthermophilic archaeobacteria, amoebae, mushrooms, palm trees, sea turtles, elephants, humans, and *everything* else that is alive (on Earth or elsewhere) has in common. No purported definition of 'life' can provide a scientifically satisfying answer to the question "What is life?" because no mere analysis using human concepts can reveal the nature of a world that lies beyond them. The best we can do is to construct and empirically test scientific theories about the general nature of living systems, theories that settle our classificatory dilemmas by explaining puzzling cases – why things that are alive sometimes lack features that we associate with life and why things that are non-living sometimes have features that we associate with life. No

scientific theory can be conclusive, but someday we may have a well-confirmed, adequately general theory of life that will allow us to formulate a theoretical identity statement providing a scientifically satisfying answer to the question "What is life?"

5.5.1 Dreams of a general theory of life

In order to formulate a convincing theoretical identity statement for life we need a general theory of living systems. The problem is that we are currently limited to only one sample of life, namely, terrestrial life. Although the morphological diversity of terrestrial life is enormous, all known life on Earth is extraordinarily similar in its biochemistry. With the exception of some viruses, the hereditary material of all known life on Earth is DNA of the same right-handed chirality. Furthermore, life on Earth utilizes 20 amino acids to construct proteins, and these amino acids are typically of left-handed chirality. These biochemical similarities lead to the conclusion that life on Earth had a single origin. Darwinian evolution then explains how this common biochemical framework yielded such an amazing diversity of life. But because the biochemical similarities of all life on Earth can be explained in terms of a single origin, it is difficult to decide which features of terrestrial life are common to *all* life, wherever it may be found. Many biochemical features that currently strike us as important (because all terrestrial life shares them) may derive from mere chemical or physical contingencies present at the time life originated on Earth (Sagan, 1974). In the absence of a general theory of living systems, how can we discriminate the contingent from the essential? It is a bit like trying to come up with a theory of mammals when one can observe only zebras. What features of zebras should one focus upon – their stripes, common to all, or their mammary glands, characteristic only of the females? In fact, the mammary glands, although present in only some zebras, tell us more about what it means to be a mammal than do the ubiquitous stripes. Without access to living things having a different historical origin, it is difficult and perhaps ultimately impossible to formulate an adequately general theory of the nature of living systems.

This problem is not unique to life. It reflects a simple logical point. One cannot generalize from a single example. What makes the case of life seem different is the amazing diversity of life on Earth today. We risk being tricked into thinking that terrestrial life provides us with a variety of different examples. But biochemical analyses coupled with knowledge of evolution reveals that much of this diversity is a historical accident. Had the history of the Earth been different, life on Earth today would certainly be different. "*How* different?" is a crucial question for astrobiology. In the absence of a general theory of living systems, one simply cannot decide. In essence, the common origin of contemporary terrestrial life blinds us to the possibilities for life in general.

A look at some popular definitions of 'life' illustrates the problem of trying to identify the nature of life in the absence of an adequately general theoretical framework for living things. Many definitions (e.g., Conrad and Nealson, 2001; Koshland, 2002) cite sensible properties of terrestrial life – features such as metabolism, reproduction, complex hierarchical structure, and self-regulation. But defining 'life' in terms of sensible properties is analogous to defining 'water' as being wet, transparent, tasteless, odorless, thirst quenching, and a good solvent. As we have discussed, reference to sensible properties is unable both to exclude things that are not water (e.g., nitric acid) and to include everything that is water (e.g., ice). Similarly, this approach will be unsuccessful for defining 'life.'[4]

Definitions of 'life' that do not make reference to sensible properties typically suffer from being too general. Definitions of 'life' based on thermodynamics provide good examples. As discussed in Section 5.2, it is difficult to exclude systems (e.g., crystals) that are clearly non-living without introducing ad hoc devices (Chyba and McDonald, 1995). Similarly, the "chemical Darwinian" definition discussed earlier (Section 5.2.2) excludes problematic cases (such as artificial or computer life) by simply stipulating that something must be a chemical system in order to qualify as living. If we had an adequate theoretical framework for understanding life, we could avoid the problem of being too general without resorting to ad hoc devices.

New scientific theories change old classifications, for example by uniting mass and energy under mass–energy, or, less profoundly, by splitting jade into the two minerals jadeite and nephrite. A general theory of living systems might well change our current classifications of living and non-living. These changes in classification will be convincing only if an empirically tested, general theory of living systems can explain, for example, why a system that we once viewed as

non-living is really living, or vice versa.[5] But to be in a position to formulate such a theory will require a wider diversity of examples of life. Current laboratory investigations (e.g., research on the hypothesized prebiotic "RNA World" on Earth) and empirical searches for extraterrestrial life are important steps in supplying these examples. Until the formulation of such a theory, we will not know whether such a theoretical identity statement for life exists.

5.5.2 How to search for extraterrestrial life

There remains the problem of how to hunt for extraterrestrial life without either a definition of 'life' or a general theory of living systems. One approach is to treat the features that we currently use to recognize terrestrial life as *tentative criteria* for life (as opposed to defining criteria). These features will then necessarily be inconclusive; their absence cannot be taken as sufficient for concluding that something is not alive. Therefore they cannot be viewed as providing operational definitions of 'life' (in the strict sense of that term). The purpose of using tentative criteria is not to definitively settle the issue of whether something is alive, but rather to focus attention on possible candidates, namely, physical systems whose status as living or non-living is genuinely unclear. Accordingly, the criteria should include a wide diversity of the features of terrestrial life. Indeed, diversity is absolutely crucial (Cleland, 2001; 2002) when one is looking for evidence of long past extraterrestrial life, e.g., in the martian meteorite ALH84001 (Section 18.5.3), or with instrument packages delivered to ancient martian flood plains or to europan frozen ice "ponds" (Section 19.7). Some features for shaping searches for extraterrestrial life (whether extant or extinct) may not even be universal to terrestrial life. For example, features that are common only to life found in certain terrestrial environments may prove more useful for searching for life in analogous extraterrestrial environments than features that are universal to terrestrial life. Similarly, features that are uncommon or non-existent among non-living terrestrial systems may make good criteria for present or past life, even if they are not universal to living systems, because they stand out against a background of non-living processes. The chains of chemically pure, single-domain magnetite crystals found in ALH84001 provide a potential example

---

[4] For further discussion of the relation between the concept of life and the features that we use to recognize it, see Lange (1996).

[5] Other possibilities include *three* distinct categories of life, or *no* distinct categories, but rather a continuum.

(Section 18.5.3). If (as is still quite controversial) it turns out that these chains can only be produced biogenically (except perhaps under circumstances that are exceedingly unlikely to occur in nature), then they will provide a good biosignature for life, despite the fact that most terrestrial bacteria do not produce them.

The basic idea behind our strategy for searching for extraterrestrial life is to employ empirically well-founded, albeit provisional, criteria that increase the probability of recognizing extraterrestrial life while minimizing the chances of being misled by inadequate definitions. This is similar in spirit (though with greater care given to the limitations of 'definition') to suggestions that *in situ* searches for extraterrestrial life should rely when possible on contrasting definitions of life (Chyba and Phillips, 2001; 2002). Unlike efforts that focus on a favored definition, our suggestions are perhaps closest to the strategy proposed by Nealson and his colleagues, who (despite their liberal use of the word "definition") emphasize the use of a number of widely diverse biosignatures (atmospheric, hydrospheric, and lithospheric) (Conrad and Nealson, 2001; Storrie-Lombardi *et al.* 2001; Chapter 23). The important point, however, is that our strategy is deliberately designed to probe the boundaries of our current concept of life. It is only in this way that we can move beyond our Earth-centric ideas and recognize genuinely weird extraterrestrial life, should we be fortunate enough to encounter it. And it is only by keeping the boundaries of our concept of life adaptable and open to unanticipated possibilities that we can accrue the empirical evidence required for formulating a truly general theory of living systems.

Appendix 5.1 Locke's theory of meaning

The idea that theoretical identity statements represent stipulative definitions receives support from a problematic theory of meaning associated with seventeenth-century philosopher John Locke (1689; see Schwartz, 1977, for a review). According to this theory, the meaning of *any* term in a language is completely exhausted by the concepts associated with it, and concepts are identified with descriptions. On some versions of the theory, concepts are analyzed as clusters (rather than logical conjunctions) of descriptions. Wittgenstein's oft-cited analysis of the meaning of the word 'game' provides a good illustration (see Wittgenstein, 1953; also Schwartz, 1977). The items (e.g., chess, solitaire, water polo, charades) that we call "games" are

extraordinarily diverse, so diverse that it seems highly improbable that any conjunction of descriptions could distinguish everything that is a game from everything that is not a game. Wittgenstein concludes that there are no necessary and sufficient conditions for being a game. According to Wittgenstein, what distinguishes games from things that are not games is family resemblance: if an item has enough of the pertinent properties, then it is a game. But whether concepts are identified with clusters or conjunctions of descriptions, the question of whether an item falls into the extension of a term is taken to be completely settled by whether it fits the descriptions that we happen to associate with the term. The upshot is that anything that fits our current concept of water qualifies as "water." If our concept of water were completely founded on sensible properties and the sensible properties that we deemed to be most important failed to exclude nitric acid, then not only would we call nitric acid "water" (which, historically speaking, we once did), but on the Lockean view, nitric acid would actually *be* water. On this view, there is no possibility of discovering that we are wrong – that our descriptions are too inclusive or exclusive – since the only thing that qualifies an item as a member of the extension of a term is whether it happens to fit the descriptions that we associate with the term. If we change our concept of water by stipulating, in the context of a new theory, that water is H_2O, then we are no longer talking about the same thing. Thus Aristotle, who held that water is an indivisible element, cannot be interpreted as talking about the same thing that we are talking about when we use the word 'water' because, for us, water is a composite of hydrogen and oxygen atoms.

Locke's theory is unable to distinguish natural kind terms from non-natural kind terms. Locke was fully aware of this; his solution was to bite the bullet, and reject the distinction. In a revealing discussion Locke (1689: Book III, Chapter XI, Section 7) argues that the seventeenth-century debate over whether bats are birds has little scientific merit since the (seventeenth-century) concepts of bat and bird are compatible with either position; for Locke, the debate is merely verbal. Yet in hindsight this seems wrong. The question of whether bats are birds is not merely verbal – a matter of what description we decide to associate with 'bat' and 'bird.' Indeed, we have *discovered* that the things we call "bats" are far more like mammals than birds. It is instructive to compare this situation with an analogous argument over whether bachelors could be married. No one can discover that bachelors are married. Any one who claims that they have done so either does not understand the meaning of 'bachelor' or, alternatively, is simply stipulating (vs. discovering) a new meaning for 'bachelor.' In other words, unlike the debate over whether bats are birds, the question of whether bachelors can be married *is* purely verbal. An adequate theory of meaning should be able to explain the difference between common nouns like 'bat' and 'bachelor.'

Appendix 5.2 John Locke and Thomas Kuhn

The Lockean view of meaning underlies Thomas Kuhn's famous argument for the incommensurability of scientific theories (Kuhn, 1962). When the defining descriptions associated with a term drastically change, as happens in scientific revolutions, the Lockean theory says that the meaning of the term also drastically changes. Thus the term 'mass' means something drastically different in Newtonian mechanics (where mass is conserved) than it does in the special theory of relativity (where only mass–energy is conserved). The upshot is that we can't say that the special theory of relativity tells us something new about the thing referred to by the old term 'mass.' Rather than expanding our knowledge of the natural world, on Kuhn's account, new scientific theories only alter our conceptual framework. Yet this conclusion seems wrong. Surely we know more about the natural world than we did a hundred years ago!

The inadequacy of the Lockean framework for meaning cannot, in our view, be overstated. A successful theory of meaning must account for indisputable facts about language and thought; after all, language and thought *are* the subject matter of a theory of meaning. It is undeniable that we speak and think differently about natural kinds than we do about conventional kinds. Because it treats the meaning of *every* term as *just* a matter of convention – as depending only upon the concepts that we happen to associate with it – the Lockean view cannot accommodate this difference; it lacks the resources to explain it.

Appendix 5.3 A new theory of meaning

In contemporary philosophy, the Lockean view has been challenged by a new theory of meaning (Schwartz, 1977 gives a review). This new theory solves the problems of the old theory by dispensing with the whole project of identifying meanings with concepts, whether construed as conjunctions or as clusters of descriptions.

There are a number of different versions of the new theory of meaning. All of them, however, agree that meaning involves reference, and reference is not determined by concepts. The word 'water' *means* whatever has the same intrinsic nature as the stuff that we typically call 'water' regardless of the descriptions that we happen to associate with it. While it is undeniable that we use descriptions (derived from our sensible experiences with paradigmatic examples) to recognize things as water, these descriptions do not (as in the old Lockean view) determine what it is for something *to be* water. Thus something can fit descriptions that we associate with 'water' and yet fail to qualify as water by virtue of having the wrong intrinsic nature.

This point is illustrated by Hilary Putnam (1973; 1975), a founder of the new theory, in a well-known thought experiment. Putnam asks us to suppose that there existed a fantastic planet called "Twin Earth." Twin Earth is like Earth, but the liquid called "water" on Twin Earth is not H_2O but a different liquid whose chemical formula is abbreviated as "XYZ." XYZ and H_2O have the same sensible properties; XYZ is wet, transparent, odorless, tasteless, and a good solvent. In Putnam's thought experiment, Twin Earthers from the seventeenth century (before molecular theory appeared) and seventeenth-century Earthlings have the same concept of water. A seventeenth-century Earthling might well believe that there is water on Twin Earth. But that conclusion would be wrong. The stuff on Twin Earth that looks like water is not water because it is not H_2O, even though Twin Earthers and Earthlings might not understand this until the end of the eighteenth century.

It is important to understand the point of Putnam's thought experiment. The fact that it makes little scientific sense to speak of Twin Earth being just like Earth *except* for the chemical composition of water is not relevant to his argument. Putnam is making a point about language and concepts. Language is used to describe many kinds of situations, from actual to hypothetical (e.g., what if Al Gore had been the US President in 2003?), to fantastic (e.g., the adventures of the young wizard Harry Potter). An adequate theory of meaning must do justice to hypothetical and fantastic situations as well as factual ones. Putnam's thought experiment about Twin Earth demonstrates that the meaning of a natural kind term is not fully captured by the descriptions that we associate with it. If it were, we would have to conclude that our seventeenth century Earthling is correct about there being water on Twin Earth.

We now have the tools to evaluate the proposal that theoretical identity statements (the theoretical identity statement "water is H_2O") are stipulative definitions. On either the old or the new theory of meaning, definitions are concerned only with language and concepts. If statements such as "water is H_2O" are stipulative definitions, then (à la Kuhn) they don't tell us anything new about the world of nature. They represent nothing more than linguistic decisions to attach concepts (H_2O) derived from theoretical frameworks (molecular theory) to familiar old terms ('water'). But this interpretation does not do justice to the way we think and speak about theoretical identity statements: we take them to be making defeasible (capable of being invalidated) claims about the old familiar world of experience. On the problematic old theory of meaning, this aspect of our conceptual structure and linguistic behavior could not be explained. The upshot was that scientific debates over the underlying nature of natural kinds had to be interpreted (e.g., as in Locke's analysis of whether bats are birds) as merely verbal. The new theory of meaning restores the connection between language and the world, and allows us to make good sense of our intuitions about the contingent empirical status of theoretical identities.

Rather than viewing theoretical identity statements as stipulative definitions, it is more accurate to construe them as empirical conjectures, situated within the context of a well-confirmed scientific theory, about a category of items treated in common discourse as a natural kind. Theoretical identities are contingent in the sense that (unlikely as it now seems in some cases) we might someday discover that they are false. Moreover, it is important to keep in mind that there is no guarantee that our best scientific theories will carve up the world in exactly the same way as natural language. As an example, jadeite and nephrite were once included under the common term 'jade,' but it is now clear from chemical analysis and microscopic examination that they are different (Bauer, 1968; Putnam, 1975); the term 'jade' does not designate a (single) natural kind after all. Similarly, in the context of the right theoretical framework, we may discover that what we thought were different natural kinds are actually part of the same natural kind. To cite another example from mineralogy, we now know that rubies and sapphires, despite their striking sensible differences, are members of the natural kind corundum (Al_2O_3) (Bauer, 1968). In short, when an old theory is replaced by a new theory, we do not simply change the subject and began talking about something entirely different. We learn something new about an old familiar subject, and this may include discovering that our language and concepts have badly misled us.

References

Adami, C. (1998). *Introduction to Artificial Life*. New York: Springer-Verlag.

Aristotle. *De Anima*. Trans. by J. A. Smith (1941). In *The Basic Works of Aristotle*, ed. R. McKeon. New York: Random House.

Audi, R. (ed.) (1995). *The Cambridge Dictionary of Philosophy*. Cambridge: Cambridge University Press.

Bada, J. (2001). State-of-the-art instruments for detecting extraterrestrial life. *Proc. Natl. Acad. Sci. USA*, **98**, 797–800.

Banfield, J. F., Moreau, J. W., Chan, C. S., Welch, S. A., and Little, B. (2001). Mineralogical biosignatures and the search for life on Mars. *Astrobiology*, **1**(4), 447–465.

Bauer, M. (1968). *Precious Stones Vol. II*. New York: Dover Publications.

Bedau, M. (1996). The nature of life. In *The Philosophy of Artificial Life*, ed. Margaret Boden. Oxford: Oxford University Press, 332–357.

Benner, S., Devine, K., Matueeva, L., and Powell, D. (2000). The missing molecules on Mars. *Proc. Natl. Acad. Sci. USA*, **97**, 2425–2430.

Biemann, K., Oro, J., Toulmin, P. III, *et al.* (1977). The search for organic substances and inorganic volatile compounds in the surface of Mars. *J. Geophys. Res.*, **82**, 4641–4658.

Chao, L. (2000). The meaning of life. *BioScience*, **50**, 245–250.

Chyba, C. F. and McDonald, G. D. (1995). The origin of life in the solar system: current issues. *Ann. Rev. Earth Planet. Sci.*, **23**, 215–249.

Chyba, C. F. and Phillips, C. B. (2001). Possible ecosystems and the search for life on Europa. *Proc. Natl. Acad. Sci. USA*, **98**, 801–804.

Chyba, C. F. and Phillips, C. B. (2002). Europa as an abode of life. *Orig. Life Evol. Biosph.*, **32**, 47–68.

Cleland, C. E. (2001). Historical science, experimental science, and the scientific method. *Geology*, **29**, 978–90.

Cleland, C. E. (2002). Methodological and epistemic differences between historical science and experimental science. *Philosophy of Science*, **69**, 474–496.

Cleland, C. E. and Chyba, C. F. (2002). Defining 'life'. *Orig. Life Evol. Biosph.*, **32**, 387–393.

Conrad, P. G. and Nealson, K. H. (2001). A non-Earth-centric approach to life detection. *Astrobiology*, **1**, 15–24.

Da Vinci, Leonardo (1513). Quoted in Whitcombe, C., Leonardo da Vinci and Water, at witcombe.sbc.edu/water/artleonardo.html.

Trans. by MacCurdy (2003: 734) from *Il Codice Arundel*, No. 263, fol. 57r.

Dawkins, R. (1983). Universal Darwinism. In *Evolution from Molecules to Men*, ed. D. S. Bendall. Cambridge: Cambridge University Press, 403–425.

Dennett, D. C. (1995). *Darwin's Dangerous Idea*. New York: Simon and Schuster.

Dyson, F. (1985). *Origins of Life*. Cambridge: Cambridge University Press.

Feinberg, G. and Shapiro, R. (1980). *Life Beyond Earth: Intelligent Earthlings, Guide to the Universe*. New York: William Morrow.

Fleischaker, G. R. (1990). Origins of life: an operational definition. *Orig. Life Evol. Biosph.*, **20**, 127–137.

Gilbert, W. (1986). The RNA world. *Nature*, **319**, 618.

Glavin, D., Schubert, M., Botta, O., Kminek, G., and Bada, J. (2001). Detecting pyrolysis products from bacteria on Mars. *Earth Planet. Sci. Lett.*, **185**, 1–5.

Harold, F. M. (2001). Postscript to Schrodinger: so what is life? *Am. Soc. Microbio. News*, **67**, 611–616.

Joyce, G. F. (1994). Forword. In *Origins of Life: the Central Concepts*, eds. Deamer, D. and Fleischaker, G., pp. xi–xii. Boston: Jones & Bartlett.

Joyce, G. F. (1995). The RNA world: life before DNA and protein. In: *Extraterrestrials – Where Are They? II.* eds. Zuckerman, B. and Hart, M., 139–151. Cambridge: Cambridge University Press.

Kamminga, H. (1988). Historical perspective: the problem of the origin of life in the context of developments in biology. *Orig. Life Evol. Biosph.*, **18**, 1–11.

Kauffman, S. (2000). *Investigations*. Oxford: Oxford University Press.

Klein, H. P. (1978). The Viking biological experiments on Mars. *Icarus*, **34**, 666–674.

Klein, H. P. (1979). Simulation of the Viking biology experiments: an overview. *J. Mol. Evol.*, **14**, 161–165.

Klein, H. P. (1999). Did Viking discover life on Mars? *Orig. Life Evol. Biosph.*, **29**, 625–631.

Koshland, D. E. (2002). The seven pillars of life. *Science*, **295**, 2215–2216.

Kuhn, Thomas S. (1962). *The Structure of Scientific Revolutions*. Chicago: The University of Chicago Press.

Lahav, N. (1999). *Biogenesis: Theories of Life's Origins*. New York: Oxford University Press.

Lange, M. (1996). Life, "artificial life," and scientific explanation. *Philosophy of Science*, **63**, 225–244.

Lavoisier, A. L. (1783). On the nature of water and on experiments which appear to prove that this substance is not strictly speaking an element but that it is susceptible of decomposition and recomposition. *Observations sur la Physique*, **23**, 452–455

(trans. Carmen Giunta. Available online at webserver.lemoyne.edu/faculty/giunta/laveau.html).

Levin, G. V. and Levin, R. L. (1998). Liquid water and life on Mars. *Proc. SPIE – The International Society for Optical Engineering*, **3441**, 30–41.

Levin, G. V. and Straat, P. A. (1979). Completion of the Viking labeled release experiment on Mars. *J. Mol. Evol.*, **14**, 167–183.

Lloyd, G. E. R. (1982). *Early Greek Science: Thales to Aristotle*. London: Chatto & Windus.

Locke, J. (1689). *An Essay Concerning Human Understanding*. Oxford: Oxford University Press.

MacCurdy, Edward (2003). *The Notebooks of Leonardo da Vinci, Definitive Edition in One Volume*. Old Saybrook: Konecky and Konecky.

Matthews, G. B. (1977). Consciousness and life. *Philosophy*, **52**, 13–26.

Mayr, E. (1982), *The Growth of Biological Thought*. Cambridge, MA: Belknap Press.

McKay, C. P. (1994). Origins of Life. In *Van Nostrand Reinhold Encyclopedia of Planetary Sciences and Astrogeology*, eds. Shirley, J. and Fairbridge, R. New York: Van Nostrand.

Monod, J. (1971). *Chance and Necessity: an Essay on the Natural Philosophy of Modern Biology*. London: Alfred A. Knopf.

Nagel, E. (1961). *The Structure of Science: Problems in the Logic of Scientific Explanation*. New York: Harcourt, Brace & World.

Nealson, K. H. and Conrad, P. G. (1999). Life: past, present and future. *Philos. Trans. R. Soc. Lond. B Biol. Sci.*, **354**, 1923–1939.

New, M. and Pohorille, A. (2000). An inherited efficiencies model of non-genomic evolution. *Simulation Practice and Theory*, **8**, 99–108.

Pohorille, A. and New, M. (2000). Models of protocellular structures, functions, and evolution. In *Frontiers of Life*, eds. Palyi, G., Zucchi, C., and Caglioti, L. New York: Elsevier, 37–42.

Putnam, H. (1973). Meaning and reference. *J. Philos.*, **70**, 699–711.

Putnam, H. (1975). The meaning of meaning. In: *Mind, Language and Reality: Philosophical Papers, Volume 2*. Cambridge: Cambridge University Press, 215–271.

Rasmussen, S. (1992). Aspects of information, life, reality, and physics. *Artificial Life*, **2**, 767–74.

Ray, T. S. (1992). An approach to the synthesis of life. *Artificial Life*, **2**, 371–408.

Roberts, G. (1994). *The Mirror of Alchemy: Alchemical Ideas in Images, Manuscripts and Books*. Toronto: University of Toronto Press.

Rode, B. M. (1999). Peptides and the origin of life. *Peptides*, **20**, 773–786.

Rizzotti, M. (ed.) (1996). *Defining Life*. Padova: Padova University Press.

Sagan, C. (1970). Life. In: *Encyclopaedia Britannica* (fifteenth edn.), **22**, 985–1002.

Sagan, C. (1974). The origin of life in a cosmic context. *Orig. Life Evol. Biosph.*, **5**, 497–505.

Schrödinger, E. (1945). *What is Life? The Physical Aspect of the Living Cell*. Cambridge: Cambridge University Press.

Schulze-Makuch, D., Guan, H., Irwin, L., and Vega, E. (2002). Redefining life: an ecological, thermodynamic, and bioinformatic approach. In *Fundamentals of Life*, eds. Palyi, G., Zucchi, C., and Caglioti, L. New York: Elsevier, 169–179.

Schwartz, S. P. (1977). Introduction. In *Naming, Necessity, and Natural Kinds*, ed. Schwartz, S. Ithaca: Cornell University Press, 13–41.

Shapiro, R. and Feinberg, G. (1995). Possible forms of life in environments very different from the Earth. In: *Extraterrestrials: Where Are They?* eds. Zuckerman, B. and Hart, M. Cambridge: Cambridge University Press, 165–172.

Shelley, M. (1818). *Frankenstein: or the Modern Prometheus*. London: Lackington, Hughe Harding, Mavor & Jones.

Shields, C. (1999). *Order in Multiplicity*. Oxford: Oxford University Press.

Sober, E. (1992). Learning from functionalism – prospects for strong artificial life. *Artificial Life*, **2**, 749–765.

Storrie-Lombardi, M., Hug, W., McDonald, G., Tsapin, A., and Nealson, K. (2001). Hollow cathode ion laser for deep ultraviolet Raman spectroscopy and fluorescence imaging. *Rev. Sci. Instrum.*, **72**, 4452–4459.

Wittgenstein, L. (1953). *Philosophical Investigations*. Trans. by G. E. M. Anscombe. New York: Macmillan.

Further reading and surfing

Boyd, R., Gasper, P. and Trout, J. D. (1997). *The Philosophy of Science*. Cambridge, MA: MIT Press.

Kim, J. and Sosa, E. (1995). *A Companion to Metaphysics*. Oxford: Blackwell Publications.

Newton-Smith, W. H. (2001). *A Companion to the Philosophy of Science*. Oxford: Blackwell Publications.

Rosenberg, A. (2000). *Philosophy of Science*. New York: Routledge.

Schwartz, S. P. (ed.) (1977). *Naming, Necessity, and Natural Kinds*. Ithaca, NY: Cornell University Press.

Stanford Encyclopedia of Philosophy at plato.stanford.edu. A marvellous and authoritative source about all aspects of philosophy and philosophers.

6 Origin of life: the crucial issues

Robert Shapiro
New York University

6.1 Why is this question important?

In observing our vast Universe thus far, we have encountered life only on or near the surface of our home planet. Yet life in its properties and behavior is so different from the barren realms that we have surveyed elsewhere, that we cannot help but wonder how it first took root here, and whether things that we would consider alive exist elsewhere. The fossil record on Earth appears to extend to 3.5 Gyr (Schopf *et al.*, 2002) and isotopic evidence suggests the presence of life several hundred million years earlier than that. Recently, however, this evidence has come into question (Brasier *et al.*, 2002), so caution should be used in relying on these conclusions (Section 12.2.1). No hard evidence exists at all, however, concerning the mechanism by which life first began here.

Every human culture has felt the need to address this question, considering its importance in defining our place in the cosmos. In the absence of firm evidence, the door has been left open to a variety of answers from science, mythology, and religion, each defining our place in the Universe in different ways. I will follow a scheme put forward by the scientist and philosopher Paul Davies (1995: 21) and separate the competing points of view into three groups, called Biblical–Creationist, Improbable Event, and Cosmic Evolution.

6.1.1 Group I: Biblical–Creationist

Life began through the actions of a Creator, using supernatural means. This point of view is accepted by a large fraction of our population, but I will not consider it further here as it falls outside the boundaries of science.

A modified version of this theory, Intelligent Design, states that life on Earth arose through the actions of an intelligent being, but does not identify the creator or specify the means that were used to create life. If we presume that the Creator worked in a supernatural manner, then we are back in the position described above. On the other hand, Francis Crick has suggested that life began here when intelligent beings from a far-off planet seeded the Earth by releasing bacteria from a spaceship (Crick, 1981). Such ideas explain only the first arrival of life on Earth, and are unsatisfactory in that the origin of the living intelligent beings is not addressed.

6.1.2 Group II: improbable event

Life began through natural causes, but through events that were extremely improbable. It follows that life beyond this planet may be rare; the remainder of the Universe may be barren. The essence of this position was captured by George Wald (1979: 48, 50) in an article in *Scientific American* where he considered the chances for the sudden and spontaneous generation of life from non-living matter by the random shuffling of simple chemicals.

> One has only to contemplate the magnitude of this task to concede that spontaneous generation is impossible. Yet we are here – as a result, I believe, of spontaneous generation.
>
> Time is in fact the hero of the plot. The time with which we have to deal is of the order of two billion years. Given so much time, the 'impossible' becomes possible, the possible probable, and the probable virtually certain. One has only to wait: time itself performs the miracles.

In our current thinking, the amount of time available has contracted from two billion years to perhaps a few hundred million, though the maximum amount of space that can be used for random trials is presumably

Planets and Life: The Emerging Science of Astrobiology, eds. Woodruff T. Sullivan, III and John A. Baross. Published by Cambridge University Press. © Cambridge University Press 2007.

the same: the surface of the Earth and the areas that exchange material with it. Both could be expanded greatly if the possibility of *panspermia* were taken into account. This theory presumes that once life has started on a particular world, it may then be spread to others by spores transported through space, perhaps within protective rocky material. Much of the Universe and much of the time since its creation would then be available for trials on the origin of life.

Unfortunately, it is easy to be misled here. The concepts of billions and billions of stars, and the great age of our Universe come readily to mind, but there is less appreciation of how rapidly small improbabilities can multiply. For example, consider the English word "at." If it is defined by two characters, how long would it take to generate it on my computer keyboard by random strokes? Assume that I limited my strokes to the fifty keys that produce characters, and could carry out one trial every second. My chances of producing "at" would be 1 in 2,500 and I could confidently expect this word to appear within an hour or so.

But suppose we raise the level of difficulty: *What are the chances that I could type this line at random?* The italicized sentence has 48 characters (we will ignore the spaces – the spacebar is large and it is harder to calculate the chances of hitting it). The odds of getting their order completely correct would be 1 in 50^{48}, about 1 in 4×10^{81}! To appreciate the size of that number, consider that the age of the Universe is about 4×10^{17} s, the number of stars in the Galaxy and the number of galaxies in the observable Universe is each about 10^{11}, and the number of humans on this planet is about 6×10^9. Thus, if every star in the Universe harbored an Earth-like planet, and on each planet the present number of humans had been sitting at word processors since the formation of the Universe, and grinding out one line every second, the number of trials run so far would be about 10^{49} (the product of the above four quantities). If, as an approximation, we assume that the chances for a successful trial become favorable when the number of trials is roughly the same as the odds against success on a single trial, then we see that the chances for getting the italicized line right in a Universe full of typists are still less than 1 in 10^{30}.

The incredibly slim chances for success would be masked if one broke up that sentence into small fragments which were considered individually. The chances for getting the first two characters right in a finite time would seem reasonable, as we have seen above, as would those for the next two, and so on. If we assume, wrongly, that the whole is the sum (rather than the

product) of its parts, the entire sequence appears reasonable. This type of reasoning, while seldom stated explicitly, underlies much of the approach used by researchers for what is called *prebiotic synthesis*. A number of steps is demonstrated separately in the laboratory, each under conditions that may have prevailed in one location or another on the early Earth. The products of the reaction are then termed *prebiotic*, a distinction that then allows the experimenter to use any of them, in pure form and enhanced concentration, in a subsequent prebiotic transformation. This next transformation may be conducted under very different conditions, compatible with some other location on the early Earth. Little consideration is generally given to the adverse probabilities generated by the numerous purification, concentration, and transport processes which are implied by the suggested sequence of steps.

For example, consider part of a proposed prebiotic synthesis of adenine (one of the four bases present in DNA) and hypoxanthine (a hydrolysis product of adenine and a minor component of RNA) described in a recent origin of life text intended for the college level (Zubay, 2000: 240) (the initials AICN, AICI, and DAMN stand for chemical intermediates):

> Formation of hydrogen cyanide in the atmosphere would probably have proceeded best in a region of high volcanic activity. ... The hydrogen cyanide formed in the atmosphere would be expected to rain down because of its high solubility in water. A freshwater pond on a mountainside might have served as a convenient catch basin for the hydrogen cyanide; here the compound could have become concentrated over the winter months by partial freezing. In a cold concentrated solution, DAMN would be expected to form slowly over a period of many months. The conversion of DAMN to AICN might occur in the spring after a thaw as a stream containing DAMN irradiated by the sunlight flows down the mountainside to a second location, where a fraction of the AICN might be converted to AICA by contact with a clay at somewhat elevated temperatures (e.g., 75 °C). Following this, the stream now containing both AICN and AICA could continue to flow until it reached warmer waters containing ammonium formate. The slow evaporation of this pool over the summer months should result in the efficient conversion of the two imidazoles into the corresponding purines.

This sequence of steps was invoked to rationalize the production of a single batch of products which would be of little use on their own, but require further extended processing to place them within functional biological macromolecules. None of the individual steps violates any law of physics, but their combination into an extended series generates odds that makes the overall process seem near miraculous. With some very good fortune, however, such an event might have occurred within the history of the Universe, and given rise to us. A consequence of this theory, however, would be that separate origins of life would be very rare in the Universe. Panspermia would still be a possibility, but any life that we encountered elsewhere would then be related to our own. Astrobiology would represent a simple extension of Earth biology.

6.1.3 Group III: cosmic evolution

In this viewpoint, life began through the inevitable consequences of the working of the laws of nature. There exists a principle of self-organization that has guided the development of the Universe from the Big Bang to the present. Life is common in the Universe. It will arise whenever an appropriate set of materials interacts with a suitable energy source. In this viewpoint, life's origin is not linked to the improbable formation of a particular functioning large biomolecule, and it requires a much simpler set of ingredients: (i) a suitable source of free energy, (ii) a set of chemicals that can utilize the energy to increase its own ability to survive (through enhanced stability, for example, or improved access to the energy source), and (iii) an environment that is sufficiently stable so that the evolving system is not disrupted by extreme changes in conditions.

This theory encourages the expectation that life is likely to arise on other worlds that have enjoyed Earth-like conditions during some part of their history. Further, no recipe for a specific set of ingredients for life is offered. The possibility remains open that alternative forms of life may arise in environments that are very different from Earth but still meet the necessary, limited set of requirements (Feinberg and Shapiro, 1980; Chapter 14).

Unfortunately, those requirements have been explored only in computer simulations so far, by theoreticians such as Stuart Kauffman and Doron Lancet. No specific prescription for a set of chemicals whose evolution could be followed in the laboratory has appeared. Kauffman (1995: 8, 47, 75, 24, 72, 69), however, has eloquently pictured this scenario in his book *At Home in the Universe*:

Order is not at all accidental. ... Vast veins of spontaneous order lie at hand. Laws of complexity spontaneously generate much of the order of the natural world.

Life is a natural property of complex chemical systems. ... When the number of different molecules in a chemical soup passes a certain threshold, a self-sustaining network of reactions – an autocatalytic metabolism – will suddenly appear.

Autocatalytic metabolisms arose in the primal waters spontaneously, built from a random conglomeration of whatever happened to be around. One would think that such a haphazard collection of thousands of molecular species would most likely behave in a manner that was disorderly and unstable. In fact, the opposite is true: order arises spontaneously, order for free.

The collective system does possess a stunning property not possessed by any of its parts. It is able to reproduce and to evolve. The collective system is alive. Its parts are just chemicals.

Once an autocatalytic set is enclosed in a spatial compartment of some sort ... the self-sustaining metabolic processes can actually increase the number of copies of each type of molecule in the system. In principle, when the total has doubled, the compartmentalized system can 'divide' into two daughters. Self-reproduction can occur.

If all this is true, life is vastly more probable than we have supposed. Not only are we at home in the Universe, but we are far more likely to share it with unknown companions.

This stance towards the origin of life is the most optimistic one for the field of astrobiology, but it remains to be demonstrated that it is correct.

6.2 Historical background

Here we present a brief overview of the historical development of our ideas on the origin of life; a more detailed account is found in Sections 1.10 through 1.18.

6.2.1 Spontaneous generation

The conflict between the differing philosophical points of view described in Section 6.1 arose only within recent centuries. For millennia before that, a consensus existed in favor of *spontaneous generation*, which holds that life may arise quickly by chance from non-living matter, independent of the action of parents (Farley,

1977). This concept was supported by many observations in which living creatures seemed to appear suddenly in the midst of inanimate matter. Aristotle and his followers claimed that fireflies emerged from morning dew. Many kinds of small animals seemed to arise from the mud at the bottom of ponds and streams. In Shakespeare's *Antony and Cleopatra*, Lepidus comments to Antony: "Your serpent of Egypt is bred now of your mud, by the action of your sun; so is your crocodile." The seventeenth-century philospopher René Descartes suggested that the process was driven by heat agitating the subtle and dense particles of putrefying matter.

When later workers applied careful experimental tests to these speculative ideas, however, they got negative results for a whole array of lower life forms. By the mid-nineteenth century the spontaneous generation question had contracted to the realm of microorganisms. As stated by biologist and historian John Farley (1977: 71):

> The cell theory was modified in the 1860s so that protoplasm came to be considered the structural unit of life. Moreover, this protoplasm was viewed as an essentially simple substance, theoretically capable of being produced directly from inorganic matter. That the simplest organisms were generally regarded merely as naked lumps of protoplasm added credence to the belief that they, too, could be produced spontaneously.

The paradigm collapsed in the late nineteenth century as a result of the experiments of Louis Pasteur (Section 1.13). He succeeded so well that, in an 1862 lecture at the Sorbonne, he was able to proclaim "never will the doctrine of spontaneous generation recover from the mortal blow of this simple experiment." In one of the best-known and simplest of his experiments, Pasteur heated and thus sterilized a broth of sugared yeast water. In earlier disputes, supporters of spontaneous generation maintained that such a step would destroy the capacity of the broth to generate life, and the failure of new microbes to appear subsequently had little meaning. In Pasteur's work, however, the flask was not sealed, but remained in contact with the outside air through a long S-shaped neck. The shape of the flask protected the contents from the entry of dust, which remained trapped within the neck. The bacteria carried by the dust had no chance to enter the flask, and its contents remained sterile indefinitely. When the neck was removed, however, bacteria soon reappeared within the broth, showing that no life-supporting force had been destroyed by the heat.

Experiments of this type demonstrated that all living things arise today from existing life. The results were comforting to those who believed that the action of a Creator was necessary in order to bring life into this world for the first time. Pasteur himself apparently belonged to this camp, and in a later (1864) Sorbonne lecture declared:

> What a victory for materialism if it could be affirmed that it rests on the established fact that matter organizes itself, takes on life itself, matter which has in it already all known forces! ... Of what good would it be then to have recourse to the idea of a primordial creation, before which mystery it is necessary to bow? What good then would be the idea of a creator God? (Farley, 1977: 108)

Those who wished a scientific rather than theological origin for life were forced to push the origin back from the present time to an earlier era when conditions on the Earth were very different, and when timespans greater than a few days could be invoked. In addition, it was recognized that the generation of a fully equipped modern functioning cell might not be necessary to get life going. Some simpler system which utilized only a portion of a cell might suffice. These ideas were placed in a definite context in the 1920s by Soviet biologist Alexander I. Oparin and British geneticist J. B. S. Haldane, and their proposals were subsequently extended by the American chemist Harold Urey.

6.2.2 The Oparin–Haldane hypothesis

The theory, in its mature form, proposed that life arose slowly on the early Earth in a "prebiotic soup" of chemicals that covered the planet. There as a succession of steps. (1) The early Earth had an oxygen-free atmosphere, with ammonia, hydrogen gas, water, and carbon present in its reduced (bound to hydrogen) form, as methane. (2) This atmosphere was exposed to a variety of energy sources: solar radiation, meteorite and comet infall, volcanic eruptions, lightning, and others, which led to the formation of organic compounds. (3) These substances, in the words of Haldane, "must have accumulated until the primitive oceans reached the consistency of hot dilute soup." Urey suggested that the soup components "would remain for long periods of time in the primitive oceans. ... This would provide a very favorable situation for the origin of life." (4) By further (unspecified) transformations, life developed in this soup. Oparin and Haldane differed in that the former held that

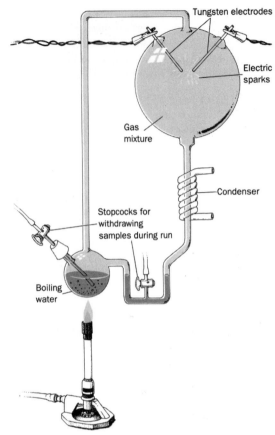

FIGURE 6.1 A Miller–Urey spark-discharge apparatus, first used in 1953 to simulate production of organic compounds as might have occurred on the early Earth. From Voet, D. and Voet, J. G., *Biochemistry*, p. 21. © 1995 John Wiley & Sons, Inc.

the carbon source, methane, was consumed. What was produced in its place?

The results apparently depended upon the arrangement of the components and the nature of the spark. The pictured apparatus afforded a mixture containing amino acids, the building blocks of proteins. In an earlier run with a modified design, a hydrocarbon layer was observed, which was not fully analyzed, but no amino acids were detected (Miller, 1974). In a later experiment, an intermittent spark that better simulated the effect of actual lightning was employed, but few organic compounds were produced. Further studies were carried out under the conditions that afforded amino acids.

In these studies, the prominent product was a tarry, insoluble substance that coated the walls of the vessel. Fifteen percent of the methane had been converted to a mixture of simple organic compounds. Fourteen of them were present in yields over 0.1%. All of these were carboxylic acids (organic acids containing the group COOH), which may have been preserved from destruction by their formation of non-volatile ammonium salts. The salts would have remained in the liquid phase and evaded continuous recycling through the spark. Two of the group were members of the set of amino acids used in life's proteins; the remainder were not.

The experiment did confirm one tenet of the hypothesis: that an atmosphere of methane, hydrogen, and ammonia, when exposed to certain types of energy, would afford a mixture of organic compounds. It revealed nothing, however, about the processes that could produce a living organism from such a mixture. Indeed, as our understanding of the complexity of life has increased, that problem has become more and more difficult over the ensuing half-century.

6.2.3 A difficulty: even the simplest forms of life are extremely complex

In the mid-nineteenth century, at the time of Pasteur's experiments, a common scientific idea was that the building material of life was a pulpy, gel-like, structure-less material called protoplasm. For example, in 1857, biologist Thomas Henry Huxley dredged a "transparent, gelatinous matter" from the sea bottom and concluded that it was protoplasm. Zoologist Ernst Haeckel provided an alternative name, *Urschleim*, for the substance (Farley, 1977). If the construction of life was relatively simple, as such observations and others implied, then it would not be surprising if life could

metabolism came first while the latter supported the primacy of the gene. For more on the history of the development of these ideas, see Shapiro (1986) and Section 1.14.

Some elements of the hypothesis were demonstrated in a famous 1953 experiment by Urey and his student Stanley Miller (Miller, 1953). They set up a flask of boiling water (Fig. 6.1) to represent the oceans of the early Earth. The water vapor passed through a compartment that held two electrodes and supported a spark discharge. The spark served as an energy source and represented lightning. After the vapors passed through the spark, they were condensed to water droplets that returned to the original flask. The entire system was enclosed and filled with a mixture of methane, ammonia, and hydrogen gases, to mimic the primitive atmosphere as it was then understood. The experiment was run for a week, during which all of

be prepared from simple broths of nutrients. After a century and a half of careful study, however, a very different picture has emerged: a bacterium is far more complex than the most intricate machine constructed by humans.

6.3 The architecture of life

6.3.1 Higher organisms' (eukaryotic) cells have complex internal structure

The complexity of life over a number of levels of magnification is illustrated in Figure 6.2. A number of distinct features is seen at the human level (limbs, eyes, nose, etc). Magnification to the organ or tissue level apparently produces simplification, but at the cellular level a wealth of complex structures (called *organelles*) emerges once more. If one of them, the mitochondrion, is enlarged further, it reveals considerable intricacy in its construction. Further magnification produces additional details, until at the molecular level, we encounter an individual *protein*. Chemical analysis reveals that the protein is a chain molecule whose links are amino acids. A set of twenty related amino acids (Fig. 7.6) is used in biology on Earth to construct all proteins. To get a protein with a particular shape and function, it is necessary to connect the amino acids in a specific order – in a similar way we endow English sentences with a variety of meanings by arranging a limited set of 26 letters in different sequences.

6.3.2 Bacterial (prokaryotic) cells are simpler, but still intricate

For most of the history of this planet, single-celled organisms were the only form of life. In seeking an origin for life, the appearance of such forms must be explained. In a superficial examination, a bacterial cell reveals fewer details than a eukaryotic cell, but a deeper inspection at higher magnification uncovers many complexities (Fig. 6.3). A wealth of intricate structures comes into view. They are crafted of large molecules, the same ones used in the construction of a eukaryotic cell. We have mentioned the proteins; another class of large (or macro-) molecules, the *nucleic acids*, is assembled in a similar manner. Individual units called *nucleotides*[1] are strung together in a linear array to

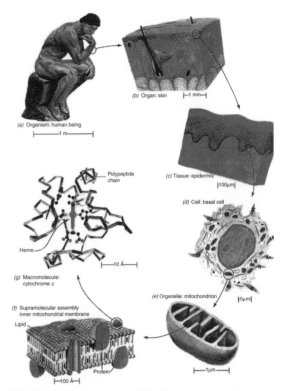

FIGURE 6.2 The complexity of life. The structure of a human being is depicted at different levels of magnification from the organism down to a protein. From Voet, D. and Voet, J. G., *Biochemistry*, p. 15. © 1995 John Wiley & Sons, Inc.

construct long chains. Only four different nucleotide units are combined to construct a nucleic acid, in contrast to the set of twenty amino acids employed in the construction of proteins. But what nucleic acids lack in variety, they make up for in size. In the case of *DNA* (*deoxyribonucleic acid*), one of the two nucleic acid subclasses, several million nucleotides are linked in a huge circle to form the bacterial chromosome (by comparison, proteins generally have a few hundred to a thousand amino acid units). Another nucleic acid subclass, called *RNA* (*ribonucleic acid*), does not attain such lengths, but appears in a greater variety of cellular contexts than DNA, and performs many more functions.

An additional class of macromolecule, the polysaccharides, can be seen in the diagram as branched chains projecting from the cell surface. These chains serve as recognition elements in cellular contacts. Another major type of molecule important for life are the carbohydrates, used for energy storage (as in starch) and as structural materials (as in cellulose). Finally, another group of biomolecules called *lipids* do not form long

---

[1] Each nucleotide consists of a sugar (deoxyribose in DNA and ribose in RNA), a phosphate group, and a base; the bases are discussed further in Section 6.3.3.2.

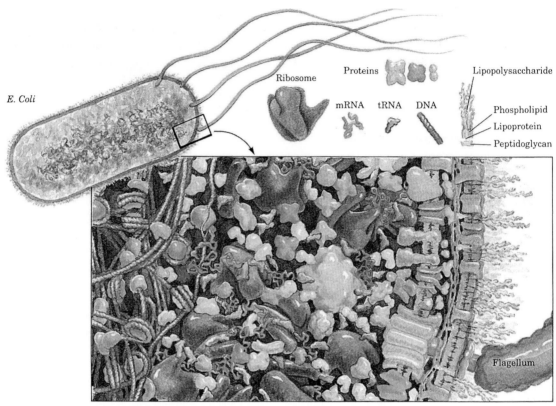

FIGURE 6.3 Cross section of a bacterial cell. An artist's rendition of a portion of a bacterial cell at a magnification of one millionfold. The structures observed are large molecules such as DNA, ribosomes, and proteins. From Voet, D. and Voet, J. G., *Biochemistry*, p. 14. © 1995 John Wiley & Sons, Inc.

chains, but the components instead aggregate together through weaker attractions (ones not involving covalent bond formation) to form the membrane of the cell. We will now discuss these categories of biomolecules in more detail.

6.3.3 Life remains very organized at the molecular level

Historically, biologists have used a "top-down" approach to understand life. Anatomical studies of the larger creatures first drew their interest. On the other hand chemists traditionally work "bottom-up," and have studied the small molecule components of life before attempting to explore the larger structures. If we examine the composition of a bacterial cell using the latter approach, we find that 70% (by weight) is water, H_2O, which serves as a solvent. Hundreds of small organic molecules are present, with molecular weights ranging to the hundreds, but together they comprise only a few percent of the weight of the cell.

Inorganic ions of many kinds provide only about 1% of the cellular mass. The remainder of a bacterial cell is made up of biopolymers or macromolecules, particularly the proteins (15%) and nucleic acids (about 7%). Proteins and nucleic acids comprise the bulk of the organic components, and play a number of roles in vital cellular processes. H, O, C, and N provide well over 99% of all the atoms in a bacterium (Table 3.1).

6.3.3.1 *Proteins*

Proteins control most of the functions of a cell. By contracting, they move our muscles; their interaction permits the rotation of flagella (also protein structures) in bacteria, which allows them to swim. In both bacteria and humans, proteins form channels in cellular membranes, allowing specific substances to enter and leave, while excluding others. Proteins form brigades that transport electrons within cellular organelles, allowing energy production from reactions with nutrients. They can function as biological weapons and poisons, or in the form of antibodies, as defenses against such weapons. As structural materials, they can form an interior scaffolding for cells, or more visibly, form the cartilage that shapes our noses. Above all, they act as *enzymes*, catalysts that permit only a few of the many possible reactions among our cellular components to take place. In this manner they allow the reactions vital to life to proceed, while the many irrelevant or destructive ones are limited to low rates that can be tolerated.

All proteins are manufactured by cellular machinery that joins amino acids in a linear array, head to tail. As mentioned before, only twenty of a vast number of possible amino acids are selected by this machinery. Further, all but the simplest amino acid (glycine) can occur in two mirror-image forms. Only one of these forms, arbitrarily called left-handed, is utilized to construct proteins for all life on Earth today. Similar considerations apply to the building blocks used to construct nucleic acids (nucleotides) and polysaccharides (carbohydrates). The reasons for this selectivity remain unclear, despite determined efforts by researchers. We do not know whether the selection was influenced by physical processes that established an environmental bias, before life began, or whether it was arbitrary, and governed by some accident. The choice may have been made when life began, or at some subsequent stage in evolution. Because of our ignorance, we cannot anticipate what we would find if we encountered life beyond Earth. If this life used left-handed amino acids to construct proteins, and in particular if it employed the same set of twenty, it would provide a strong signal for common ancestry through panspermia. On the other hand, if the alien life used right-handed amino acids, or none at all, this would signal a separate origin.

The linear sequence of amino acids in a protein governs the way that it folds in three-dimensional space. The proteins produced by billions of years of evolution fold into distinct shapes, often with charges at particular locations and reactive groups in proximity to one another. These properties in turn determine how a protein biologically functions: whether it will bind certain organic molecules and catalyze their reactions, for example, or form a regular structure such as a helix and act as a building material. Sometimes, a protein in its form as originally synthesized in the cell must be modified in order to perform a useful function. In such a case the protein may be altered by other proteins, or may recruit an additional needed factor such as a small organic molecule or an inorganic ion.

6.3.3.2 *Nucleic acids*

Nucleic acids store hereditary information and reproduce themselves (with the help of proteins). They also play a crucial role in the synthesis of proteins. The synthesis of proteins from amino acids takes place in an intricate structure, an organelle within the cell called a *ribosome*. The bacterial ribosome is made of over fifty different proteins and three RNA molecules. As two of the RNA molecules have more than 1,000 nucleotides strung together, and because nucleotide units are larger than amino acids, the RNA provides about two-thirds of the mass of the ribosome. The RNA also catalyzes (in enzyme-like fashion) the central step of the protein synthesis process, the linking of each new amino acid to the growing chain of linked amino acids, called a *polypeptide* chain. Individual amino acids are transported to the ribosome for manufacturing purposes by a smaller specialized type of RNA called *transfer RNA*.

A bacterial cell may contain as as many as 20,000 ribosomes able to produce several thousand different proteins. Ribosomes are generalists, however; each of them can manufacture any of the proteins. The protein that a particular ribosome manufactures is determined by an information-bearing "tape" that threads through the ribosome. This tape is also made of a large RNA, one fittingly called *messenger RNA*. The information is carried in the sequence of *bases* (a subunit of a nucleotide) that project from the backbone of the RNA. The bases are adenine, cytosine, guanine, and uracil (A, C, G, and U). As four different bases must specify twenty different amino acids, a set of three successive bases are in fact used to specify which of the twenty amino acids should next be placed by the protein-synthesizing machinery into the growing polypeptide chain. The recognition scheme is termed the genetic code, and the overall process is called *translation*.

Messenger RNA is not the original source of the information, but it has been carried from DNA, the master information storehouse of the cell. A portion of the information stored in DNA is read into a messenger in a process called *transcription*. DNA exists normally as the famous double helix, whose structure was elucidated by James Watson and Francis Crick in 1953. In this structure, two DNA chains (polymers) wrap around one another, linked by an affinity of A for T (thymine, a close relative of the uracil in RNA) and C for G, the well-known Watson–Crick pairs. Each chain thus contains all of the information present in the double helix; given the sequence of one chain, the sequence of the other is known and the two chains are said to be complementary. One chain can thus be used as a template for the construction of the other from appropriate nucleotide subunits. This process requires the assistance of a specialized protein enzyme called a polymerase, as well as the participation of other helpful proteins. When a bacterial cell divides to give two daughters, the two DNA strands of its chromosome come apart, and each is copied to provide a complement. This process, called replication, provides a complete chromosome for each daughter cell.

As RNA is structurally a close cousin of DNA (RNA has the replacement of T by U and an additional oxygen bound to the sugar subunit that is part of each nucleotide), a messenger RNA chain can be produced by copying a portion of one chain of DNA in the chromosome, again with the help of a protein polymerase. The information that determines the function of a protein thus originates in DNA and is transmitted by RNA; a simple expression of this relation is "DNA makes RNA makes protein."

6.3.3.3 *Lipids*

Lipids form a boundary, separating the contents of the cell from the external world. This function appears essential, as it prevents the contents of a cell from being dispersed into the broader environment in which it operates. This boundary must not be impermeable however, as the processes of metabolism within a cell require that nutrients be admitted selectively, wastes be allowed to leave, and undesired substances be ejected. The recipe for basic membrane construction is far less rigorous and information-rich than those used for proteins and nucleic acids. Certain small organic molecules, called amphiphiles, contain water-loving (hydrophilic) parts, often the salts of carboxylic acids or phosphates, and water-shunning (hydrophobic) parts, often long chains of hydrocarbon (molecules made only of hydrogen and carbon). When amphiphiles are placed in water, they arrange themselves so that the hydrophilic parts are exposed to the solvent, and the hydrophobic parts cluster inside. One arrangement that achieves this is a *micelle* (Fig. 6.4), a sphere with the hydrophilic groups on the surface and the hydrophobic ones within. Another is a *bilayer* (Fig. 6.4), with two aligned layers of amphiphiles arranged in a tail-to-tail structure that resembles a membrane. A bilayer can also curve to produce an enclosed structure that resembles the membrane of a cell. Biological cellular membranes are modified, however, by embedded proteins, which control the access of small molecules into the cell.

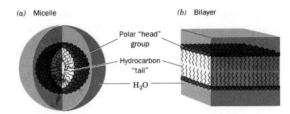

FIGURE 6.4 Structures of a micelle and a bilayer. Both concentrate hydrophobic groups in a central position, away from water. The bilayer (right), however, may curve to form a closed spheroid enclosing an aqueous solution. From Voet, D. and Voet, J. G., *Biochemistry*, p. 33. © 1995 John Wiley & Sons, Inc.

6.4 Central ideas about the origin of life

6.4.1 The Second Law of Thermodynamics

This law specifies that spontaneous processes in a closed system are characterized by the conversion of order to disorder. Only those events can take place in the Universe (or in any closed-off part of it) that increase the total amount of disorder. The entropy of the system, a measure of the disorder, must increase. As entropy is also associated with probability, the formation of a organized system from a disordered one is not forbidden, but is highly improbable. Living systems "pay" for their continuing state of order by creating even more disorder elsewhere in the system. For example, they may convert nutrients to simpler chemicals, with release of heat to the general environment.

6.4.2 It is extremely unlikely that highly organized structures arise by chance

The spontaneous generation of a living bacterium from small molecules would be near-miraculous, because of the intricacy of bacterial construction. Physicist Harold Morowitz has calculated the odds that the component atoms, present as a hot gas, would assemble into a bacterium as 1 in $10^{100,000,000,000}$, i.e., a 1 followed by 100 billion zeroes (Morowitz, 1968). Matters might be improved by thousands of orders of magnitude if we started with an appropriate mixture of small molecular species, but the overall prospect would still be hopeless.

6.4.3 Life probably started with a simple chemical system

To reduce the improbability, scientists prefer to believe that life began with only a very limited subset of the components of a bacterial cell. They disagree, however, on the identity of this initial state. The heart of the origin-of-life problem lies in finding the way in which this initial functioning system was formed from disorganized chemical mixtures produced by the normal processes of abiotic chemistry.

6.4.4 Origin of life on Earth and the origin of Earth's organics are separate problems

In the early nineteenth century, many chemists felt that the distinction between life and materials that had never been part of life lay in the very chemicals used to construct them. Living systems contained carbon, and the chemicals in them were called "organic," while substances considered to be inorganic (with the exception, perhaps, of some minerals and gases in the atmosphere) had no carbon. This distinction was erased in 1828 when the German chemist Friedrich Wöhler prepared urea, a component of urine, from another substance that was classified as inorganic. "I must tell you that I can prepare urea without requiring a kidney or an animal, either man or dog," Wöhler wrote to a colleague.

We now recognize that mixtures of simple organic chemicals are produced in many locations off this planet. They occur, for example, in certain meteorites called carbonaceous chondrites, in interstellar dust clouds, in the tails of comets, and in the atmosphere of Saturn's large satellite Titan. Mixes of varying composition have also been prepared in laboratory experiments that simulate processes in possible planetary atmospheres, such as the Miller–Urey experiment that we have discussed.

The most informative studies have been those performed on meteorites, as they represent the products of authentic extraterrestrial abiotic processes, and yet are amenable to detailed analysis in laboratories on Earth. The Murchison meteorite, in particular, has been subjected to extensive analysis (Cronin et al., 1988; Cronin and Chang, 1993); the results from others are generally similar, though they may differ in minor details.

Overall, the contents of the Murchison meteorite are quite complex, with many chemical classes represented; Tables 7.1 through 7.3 summarize the contents and Section 7.4.1 discusses these results. The bulk of the carbon present is in the form of insoluble macromolecules, while soluble organic compounds comprise 10–20%. The most predominant class (abundance greater than 1000 ppm) is organic sulfonic acids, followed by polar aromatic hydrocarbons and carboxylic

acids (100 to 1000 ppm). Amino acids occur at levels between 10 and 100 ppm, and the base components of the nucleic acids at about 1 ppm. Sugars that are subunits of nucleotides or building blocks of polysaccharides are absent, as are phosphate esters of all kinds.

Within a particular class, all isomers[2] of a given carbon number can be detected, if only a few carbons are present. The amount of material of a given carbon number declines logarithmically with the number of carbon atoms, however. Within a particular class, there is no striking resemblance between the substances present and those important in Earth biology. More than seventy amino acids have been detected in Murchison, but only eight of the twenty needed for Earth life were among them. The abundant macromolecular material bears no resemblance to the highly ordered biopolymers of Earth. To quote the review of Cronin and Chang (1993: 212):

> The bulk of the organic matter is retained in the residual insoluble fraction along with the exotic carbon phases and carbonate minerals. It is composed of a poorly characterized, structurally heterogeneous, macromolecular material, which has been variously named, but commonly referred to as either meteorite "polymer" or "kerogen-like" material. ... [Studies show] that the material is comprised of condensed aromatic, heteroaromatic and hydroaromatic ring systems in up to four-ring clusters, cross linked by short methylene chains, ethers, sulfides and biphenyl groups.

If we presume that life on Earth began amidst a chaotic chemical mixture such as that present in the Murchison meteorite, we must look for a pathway by which the smaller components could organize themselves, as it seems unlikely that amorphous, tarry mixtures of the type described above could play any role. But before we consider this problem, another one arises: how did the components of that mixture come to be present on the early Earth?

6.4.5 Possibilities for the origin of reduced organics on Earth

6.4.5.1 Synthesis through atmospheric chemistry
Underlying the Miller–Urey experiment was the assumption that the primitive atmosphere of the

---

[2] *Isomers* are molecules that have in common the same numbers and types of atoms, but have different physical structures and therefore varying chemical properties.

Earth was derived from the cosmic dust clouds from which our planet was formed, and was therefore rich in hydrogen (Miller and Urey, 1959). An analogy was made to planets in the outer Solar System, which have reducing atmospheres where H is present as H_2 and other compounds, but not as H_2O. Such an atmosphere, when subjected to energy from various sources, was expected to give rise to a global "prebiotic soup." More recently, however, a consensus has arisen that any primary atmosphere of that type was lost quickly, and replaced by a secondary one derived by outgassing (e.g., volcanism) from within the planet (Kasting, 1993). While a reducing atmosphere for the early Earth may still be favored by some prebiotic chemists, the weight of geochemical opinion favors an early atmosphere dominated by CO_2, H_2O, and N_2, with sulfur as SO_2 (Chapter 4; Kasting, 1993; Delano, 2001). This composition is more in accord with current observations on volcanic emissions. In this view, the primary difference between the early atmosphere of Earth and our present one would be the absence of oxygen in the past. But spark discharge experiments run in non-reducing atmospheres, such as those produced by outgassing from a planet, afford meager yields of organic substances.

6.4.5.2 *Meteor and comet input*

An alternative source has been evoked to provide the oceans of the early Earth with a supply of reduced organic materials (C and H bound in various compounds), even in the absence of a reduced atmosphere: infall by comets and meteorites (Chyba and Sagan, 1992; Delsemme, 2001). Uncertainties have existed concerning the total quantity of material that would be available from such sources, and whether the organic components would survive the impact, although some survival seems likely.

6.4.5.3 *Local syntheses in specialized environments*

It is not necessary, of course, to supply the entire globe with organics in order to furnish raw materials for the origin of life. Some very specialized and favorable local environments may have been sufficient for the purpose. This thought was expressed as early as 1871 by Charles Darwin in a letter to botanist Joseph Hooker (Wills and Bada, 2001: 35):

> If (and oh! what a big if!) we conceive in some warm little pond, with all sorts of ammonia and phosphoric salts, light, heat, electricity present, that

protein compound was chemically formed, ready to undergo still more complex changes. At the present day, such matter would be instantly devoured or absorbed, which would not have been the case before living creatures were formed.

More recently, Darwin's "warm little pond" has been joined by a number of other suggested suitable locations for life's origin, for example, submarine hydrothermal vents (Baross and Hoffman, 1985; Holm, 1992). Vent conditions have been demonstrated to be suitable for mineral-catalyzed abiotic synthesis of a number of small organic molecules relevant to our biochemistry (Chapter 8; Huber and Wächtershäuser, 1997; Cody *et al.*, 2000). Abiotic hydrocarbon reservoirs within mineral enclosures deep within Earth's crust represent another proposed site (Gold, 1992). Igneous mineral matrices have been demonstrated to be suitable locations for the assembly of carboxylic acids from CO and CO_2 (Freund *et al.*, 2001).

6.4.5.4 *Biological fixation*

In our current ignorance about the varieties of life that may be present in the Universe, we cannot exclude another alternative; that life may have started on this planet without the assistance of organic molecules. In the theory of chemist Graham Cairns-Smith, layered clay minerals such as kaolinite provide all of the necessary qualities (see Fig. 6.5) needed to support life: information storage, reproductive ability, catalysis, and the capacity to evolve (Cairns-Smith, 1982). In this picture organic molecules were introduced as an evolutionary improvement only after mineral life was well under way; eventually they displaced the mineral components entirely.

6.4.6 The mechanism is the solution

The above theories, with the exception of that of Cairns-Smith, do not address the central problem of the origin of life: that of the mechanism by which an abiotic chemical mixture absorbs energy, increases in complexity, and evolves. The availability of a suitable energy source, necessary raw materials, and a suitable locale are prerequisites, but not the answer to the riddle of self-organization. In the search for that answer, investigators have taken two fundamentally different tacks. Which was first to appear on Earth – replicating molecules or metabolic processes? Section 8.2 also discusses this issue from a different viewpoint.

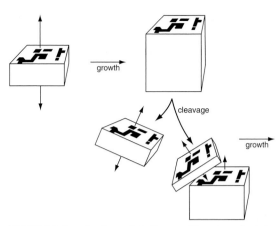

FIGURE 6.5 A replication cycle is illustrated for a hypothetical genetic crystal. The black pattern represents information storage in defects in the crystal lattice. Growth takes place on the crystal faces, with retention of the information, and reproduction occurs when the crystal fractures. Cairns-Smith (1982: 158). © 1982 Cambridge University Press.

6.5 Replicator-first theories: life began with the random formation of a self-copying molecule

The publication of the Watson–Crick theory of the structure of DNA in 1953 marked a new era of biology. Before then, proteins had been considered to be the central substances of life, undoubtedly the hereditary material as well; afterwards, the replicator DNA was at the center of our understanding of life. The demonstration by O. T. Avery and his colleagues in 1944 that the transforming principle (a fragment of DNA that could be absorbed and change the heredity of a bacterium) was in fact DNA and not protein had not been readily accepted. The Miller–Urey experiment, which was published only a few weeks after the first paper of Watson and Crick, undoubtedly gained some of its significance because its synthesis of amino acids tied in with many scientists' view that genes were made of protein. The beauty of the structure of the double helix, which demonstrated how both reproduction and information storage could be carried out at the molecular level, nevertheless convinced many in the field that nucleic acids were at the center of life. Among them was the geneticist H. J. Muller (1966: 512), who expressed the following thoughts in *American Naturalist*.

> The 'stripped down' definition of a living thing offered here may be paraphrased: that which possesses the potentiality of evolving by natural selection. ... The gene material also, of natural materials, possesses these faculties and it is legitimate to call it living material, the present-day representative of the first life. ... Primitive conditions afforded it enough means of exercising them to allow it to evolve the protoplasm that served it. ... Thus the gene material itself has the properties of life.

6.5.1 Advantages of the replicator theory

Charles Darwin's theory of natural selection provided a path for the evolution of simple cells to humans. It could now be extended so that a replicator could also evolve by natural selection, but in a milieu where molecules (rather than species) reproduced, mutated, and survived according to their ability to further reproduce in the environment. Thus, to explain the origin of life, one only had to account for the origin of the first replicator. The gap in organization between simple chemicals and the first cells would then be closed.

Considerable support for these concepts was provided by the experiments of biochemist Sol Spiegelman and his group in the 1960s and 1970s, using the RNA of a virus called Qβ. These studies were extended and furnished with a mathematical basis by Manfred Eigen and his colleagues. Qβ uses RNA rather than DNA as its genetic material, with replication carried out in the normal Watson–Crick manner. In the presence of the appropriate polymerase enzyme (Qβ RNA replicase) and the four necessary building blocks (nucleoside triphosphates), this RNA can be replicated in a test tube. The newly synthesized RNA can in turn act as a template for further synthesis, so the amount of Qβ RNA increases exponentially, as long as the supplies of building blocks hold up. By using appropriate dilution techniques, Spiegelman's group was able to follow this process for a time equivalent to over seventy RNA "generations." In one thought-provoking experiment, they added a drug that bound to certain sites on the RNA, greatly slowing down the rate of copying. After multiple generations of replication, however, they found that an altered RNA now dominated their colony. The favored binding place of the drug had been destroyed by three changes (or mutations) in the product RNA. The product RNA thus replicated more rapidly than the parent, and its descendents soon became an overwhelming majority in the colony. This process was considered to represent Darwinian evolution at the molecular level.

6.5.2 One early problem: a naked replicator could carry information, but could not carry out tasks today performed by proteins

The Spiegelman–Eigen concept could not be extended immediately to the origin of life because, in addition to the replicator, a protein enzyme was needed for the copying process to take place. Chemist Leslie Orgel (1968) described the paradox that came to be known as "the chicken or the egg" problem. A nucleic acid could be copied, and evolve through mutation, but could not by itself carry out the necessary catalytic functions. On the other hand protein enzymes could catalyze all manner of processes, but could not undergo any type of replication. Each seemed necessary for the conduct of life, but as it is difficult to account for the appearance of even *one* such substance on the early Earth, the abiotic formation of *both* together seemed out of the question.

6.5.3 A solution: life began with an "RNA World"

A solution to this dilemma appeared with the discovery by chemist Thomas R. Cech and biochemist Sidney Altman that RNA molecules could carry out some of the functions of proteins. RNA could thus carry heredity *and* catalyze reactions. The term *ribozyme* was coined to describe the newly discovered class of RNA enzymes. Walter Gilbert, a physicist who had converted to biology, applied the name "RNA World" to his vision of a biosphere in which RNA alone performed all of the key functions of life, before proteins entered the scene. In Gilbert's (1986) words: "The first stage of evolution proceeds, then, by RNA molecules performing the catalytic activities to assemble themselves out of a nucleotide soup."

The elegant way in which catalytic RNA appeared to solve the "chicken-or-the-egg" problem led many scientists to the conviction that an RNA World indeed existed in the earliest days of life on Earth. Such beliefs were strengthened by the observations that RNA plays many diverse roles in modern metabolism. Further, many present-day biological cofactors (substances that assist enzymes in catalysis, such as nicotinamide adenine dinucleotide, Coenzyme A, and adenosine triphosphate (ATP)) contain an attached RNA nucleotide that does not participate directly in the enzyme-catalyzed reaction. Such attached nucleotides are considered to be the molecular equivalent of fossils,

"vestiges of an ancient metabolism based on RNA catalysis" (Benner *et al.*, 1989).

Such indicators were sufficient to give RNA World a stamp of authenticity. Lubert Stryer's widely employed biochemistry text, for example, carries a chapter section titled: "RNA probably came before DNA and proteins in evolution" (Stryer, 1995: 115). This viewpoint was echoed in many other sources, and an entire monograph has been devoted to the ramifications of this idea (Gesteland *et al.*, 1999). For a brief summation, we quote Bartell and Unruh (1999: M9):

> The appeal of RNA-based life is that catalytic RNAs, which could have served as their own genes, would have been much simpler to duplicate than proteins. According to this theory, RNA first promoted the reactions required for life with the help of minerals, pyridines, and other small-molecule cofactors. Then, as metabolism became more complex, RNA developed the ability to synthesize coded polypeptides that served as more sophisticated cofactors. DNA eventually replaced RNA as the genetic polymer and protein replaced RNA as the prominent biocatalyst.

The concept requires that ribozymes be capable of catalyzing a wide variety of metabolic reactions. Seven naturally occuring classes of catalytic RNA have in fact been isolated from contemporary cells. All of these, however, focus on one task: catalyzing the cleavage or joining together of RNA backbones (Doherty and Doudna, 2000). Supplementing this list is the striking discovery that the key catalytic step in protein synthesis, the one that joins an additional amino acid to the growing peptide chain, is catalyzed by a ribozyme component of the ribosome (Cech, 2000).

To demonstrate that RNA can exhibit an even broader repertoire of catalysis, many ribozymes with additional capabilities have been artificially prepared in the laboratory. The process has employed many of the advanced techniques of modern molecular biology. Large libraries of ribozyme variants that contain many randomized sequences are synthesized. The successful candidates are isolated, sequenced, and then subjected to additional rounds of mutation and selection. In this manner, new ribozymes have been prepared that can catalyze such exotic processes as the Diels–Alder reaction and the isomerization of bridged biphenyls (Bartel and Unrau, 1999).

One triumph of these techniques has been the preparation of a ribozyme polymerase, 189 units of RNA in length, that was capable of carrying out a limited amount of RNA-catalyzed replication (Johnston *et al.*, 2001). When provided with a partly copied RNA chain (a primer-template complex) as a substrate, and suitably activated RNA building blocks, the new ribozyme catalyzed the addition of up to 11 units to the shorter chain. The added units were the ones expected from the Watson–Crick bonding rules, with an error rate of 3% or less. Johnston *et al.* (2001) stated that "The new polymerase ribozyme was isolated from a pool of over 10^{15} different RNA sequences." Note that the starting pool was not randomly chosen, but was itself the product of an equally extensive mutation and selection procedure.

The authors termed their achievement "a testament both to the catalytic abilities of RNA, as well as to modern combinatorial and engineering methodology." With further engineered improvements, an RNA ribozyme might conceivably replace $Q\beta$ replicase in the Spiegelman and Eigen type of experiments described above, allowing new insights into the possibilities of molecular self replication. Speculations have even been made about the possibility of constructing a primitive cell in the laboratory, based on two ribozymes and a suitable membrane (Szostak *et al.*, 2001). One ribozyme would convert membrane precursors, provided by the environment, into a functioning component. The other one, a polymerase, would use nucleotide building blocks (again from the environment) to copy itself and the first ribozyme. (The membrane components and RNA building blocks would be placed in the environment as "food" by the attending scientists.)

Such an achievement would represent a triumph for experimental molecular biology, and a non-theological demonstration of Intelligent Design. It would say very little about the actual origin of life on this planet, however, due to the absence of chemists and high-tech labs on the early Earth, and the extreme improbability of the abiotic synthesis of even a single functioning RNA molecule under realistic natural conditions (see the next section).

6.5.4 A problem: RNA is too complicated to form spontaneously – adenosine synthesis as an example

Gilbert's seminal paper on RNA World spoke of RNA molecules catalyzing their own assembly from a "nucleotide soup." Many other authors have presumed that nucleotides, the building blocks of nucleic acids, were readily available on the early Earth. For example, Eigen and Schuster (1978: 346) stated: "Here we simply

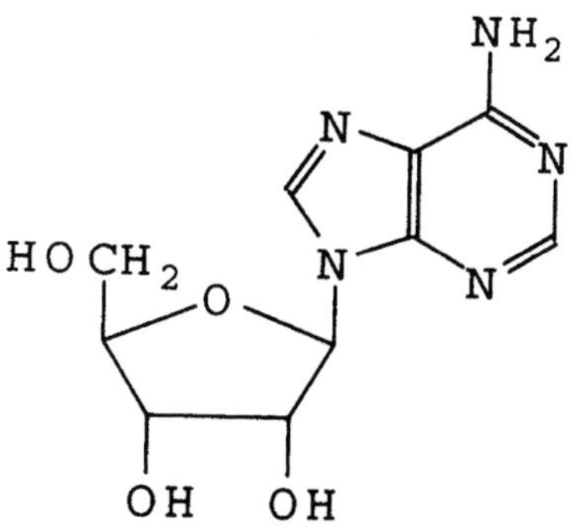

FIGURE 6.6 The chemical structure of adenosine. The base adenine (upper rings, with nitrogen) is attached in a very specific way to a form of the sugar, ribose (lower five-membered ring, with oxygen). With an attached phosphate, this structure constitutes a subunit of RNA.

start from the assumption that when self-organization began, all kinds of energy-rich materials were ubiquitous, including in particular: amino acids in varying degrees of abundance, nucleotides involving the four bases A, U, G, C, polymers of both preceding classes." The available literature, however, offers no support for the ideas that nucleotides are prominent products of prebiotic chemistry, nor that, as minor components of a complex abiotic mixture, they would have any tendency to combine in a highly regular manner to form RNA. This author has been one of the primary contributors to this viewpoint, and for a lengthy exposition, my articles should be consulted (Shapiro, 1995; 1999; 2000; 2006).

We will illustrate the significant difficulties in synthesizing RNA by considering the likely abiotic availability of one of the basic components of RNA, the nucleoside (sugar-base combination) adenosine (Fig. 6.6). This substance has never been detected in any quantity in a Miller–Urey-type experiment or in a meteorite. Advocates of an RNA World generally presume that its two components, the base adenine and the sugar ribose, were formed independently on the early Earth, brought together, and combined to form adenosine. How likely are these events?

Ribose formation

The sugar ribose is the essential backbone component of RNA. The formose reaction, a polymerization of the simple chemical formaldehyde (CH_2O) has generally been cited as a plausible source of ribose on the early Earth. However, this procedure affords an extremely complex mixture and the ribose soon decomposes (Shapiro, 1988). The half-life for decomposition at pH 7 has been measured as 73 minutes at $100\,^{\circ}$C and 44 years at $0\,^{\circ}$C (Larralde *et al.*, 1995). Ribose yields can be improved by the addition of specific minerals and metals selected for that purpose, by careful control of experimental conditions, and by the input of pure reagents rather than typical abiotic mixtures (Pitsch *et al.*, 1995; Zubay, 1998; Ricardo *et al.*, 2004), but each such specification escalates the improbability that such an event would occur under natural conditions.

Adenine formation

(Shapiro, 1995) One of the early triumphs of laboratory prebiotic chemistry was the discovery that adenine can be prepared in yields of up to 0.5% by the simple oligomerization (combining with itself to make larger molecules) of HCN (Oró and Kimball, 1961). However, yields fell drastically when the concentrations of HCN dropped below 0.1 M, due to competing hydrolysis reactions. One estimate for a maximal plausible HCN concentration in a prebiotic ocean of Earth has been 4×10^{-5} M. A remedy proposed for this situation was that the adenine was produced in lakes that froze seasonally, concentrating the HCN. Another problem was the need for ammonia at one stage of the reaction, which need could be avoided, however, by introducing a photochemical step. The need for a high-temperature hydrolysis step, preferably in acid, provided another difficulty. Without such treatment, most of the adenine would be tied up in another form. The introduction of formaldehyde (a starting point for sugar synthesis) was also unwelcome here. Formaldehyde and HCN react, almost irreversibly, to form glyconitrile, consuming both. To quote Schlesinger and Miller (1973):

> the great stability of glyconitrile shows that the two types of compounds (sugars and bases) could not have been synthesized at the same time on the primitive earth unless there was a mechanism to concentrate the formaldehyde and hydrogen cyanide in different areas. It is more plausible to think that the adenine was synthesized during one period and the sugars during another period.

The scenario quoted at length in Section 6.1.2 above represented an effort to satisfy all of these requirements at one imaginable geological site.

Combination of adenine and ribose to form adenosine

If we assume that adenine and ribose, formed separately, were somehow brought together, would they then combine in the configuration required for RNA?

Surprisingly little work has been devoted to this question. In one systematic study (Fuller *et al.*, 1972) it was found that the two pure components, heated together to dryness, would indeed react. The concentrations used were above those reasonable for a natural setting, and no other components of a plausible prebiotic mixture were present. The product of this experiment, however, was not the one used in RNA. The ribose was present for the most part in a six-membered ring (Maurel and Convert, 1990), with the adenine connected through the amino group, not at the desired 9-position. When additional components (0.25 M magnesium chloride and 0.25 M sodium trimetaphosphate) were specified for the reaction, however, a mixture containing some adenosine was produced, but still in the presence of other isomers that would be expected to interfere with RNA synthesis.

The above considerations apply to the partial preparation of one RNA building block. To proceed further, phosphorylation of the adenosine on a particular hydroxyl group would be needed, and the adenosine monophosphate would have to be brought together with the three other nucleotide building blocks of RNA in a specific location. We would then have created Gilbert's "nucleotide soup." A creation of this type would violate no natural law, but would obviously fall within the category of an extremely improbable event. In the words of Graham Cairns-Smith (1982: 56), "these experiments allow us to see, in much greater detail than would otherwise have been possible, just why pre-vital nucleic acids are highly implausible."

Some scientists feel nevertheless that the current state of prebiotic chemistry simply reflects the presence of unanswered problems, which can be cured by the application of additional ingenuity. For example, see the article by Mojzsis *et al.* (1999) in a volume devoted to RNA World. In the same volume, however, Gerald Joyce and Orgel (1999: 68) provide this summary:

> Scientists interested in the origin of life seem to divide neatly into two classes. The first, usually but not always molecular biologists, believe that RNA must have been the first replicating molecule and that chemists are exaggerating the difficulty of nucleotide synthesis. ... The second group of scientists are much more pessimistic. They believe that the *de novo* appearance of oligonucleotides [small nucleotide chains] on the primitive earth would have been a near miracle. (The authors subscribe to the latter view.)

6.6 Pre-RNA World: another replicator preceded RNA as the genetic material

A number of scientists who felt that the formation of the first replicator was coincident with the origin of life, but were unhappy with the prebiotic synthesis of RNA, have sought another alternative. As expressed by Bartell and Unruh (1999: M9):

> Most professional advocates of an RNA world are doubtful that life began with RNA *per se*. Instead, they propose that life began with an RNA-like polymer, yet to be identified, which possessed the catalytic and templating features of RNA but mirac-ulously lacked RNA's undesirable traits, most nota-bly, its intractable prebiotic synthesis. The era of this RNA-like polymer is the 'pre-RNA world,' which presumably gave way to the RNA world in a manner analogous to that in which the RNA world gave rise to the protein-nucleic-acid world of today.

The task of the pre-RNA World advocates then was to identify another polymeric substance that possessed the catalytic and self-replication capabilities of RNA, but which was a more plausible product of abiotic synthesis. Ideally, this substance would in addition form a stable double helix with an RNA strand, in order to provide a mechanism by which information could be transferred smoothly from pre-RNA to RNA during the course of evolution. While a number of candidate structures have been suggested, we shall mention only three: p-RNA, TNA, and PNA.

Two of these were synthesized as part of a compre-hensive effort to prepare analogues of DNA and RNA (Eschenmoser, 1999). These substances carry the same Watson–Crick base pairs but employ a different sugar in the alternating sugar–phosphate backbone. In some instances, inter-strand pairing was weak or not detect-able. In other cases, however, the base pairing observed was stronger than that of RNA, and also more accurate in its fidelity. One extensively investigated example of this type was p-RNA (technically a pentoribopyrano-syl-(2'→4') linked oligonucleotide) (Fig. 6.7). p-RNA retained the sugar ribose, but replaced the five-membered ring with a six-membered (pyranosyl) ring. As free ribose in solution prefers a six-membered ring to a five-membered one, the prospects for the abiotic synthesis of p-RNA would appear marginally better than those of RNA. However, all the difficulties con-nected with ribose synthesis and its instability would still be present.

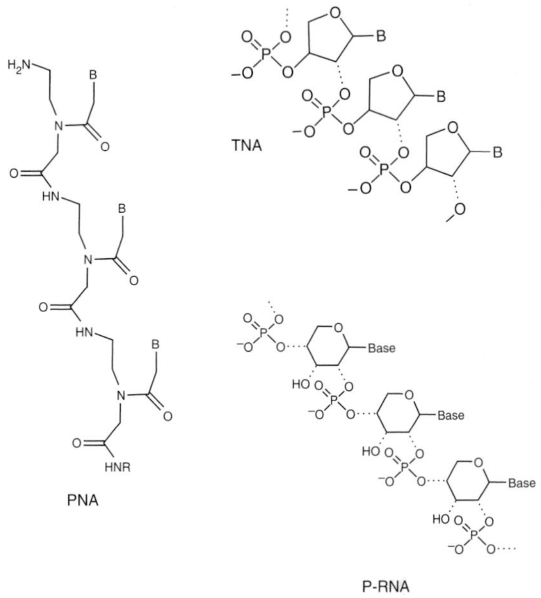

FIGURE 6.7 The structures of three possible genetic materials for a pre-RNA World: p-RNA, TNA, and PNA. p-RNA contains the sugar ribose in a six-membered ring form. TNA uses the simpler sugar threose in place of ribose. PNA substitutes a peptide-like backbone for the sugar–phosphate backbone of RNA.

Physical chemical studies revealed no reason why p-RNA should not be able to perform the same func-tions as does RNA in forming complex structures and serving as a template for replication. However, it failed to cross-pair with RNA, so no obvious mechanism for information transfer was available. Further, it was not clear why p-RNA, once established as a genetic mate-rial, should have yielded that function to RNA.

Another candidate RNA precursor that arose from the efforts of the Eschenmoser group is TNA (Fig. 6.7), with a threofuranosyl (3'→2') phosphodiester back-bone (Schöning *et al.*, 2000). In this case, the sugar contains only four carbons, and may be more accessi-ble to abiotic synthesis than ribose. A four-carbon sugar can form five-membered rings, but not six-mem-bered rings. TNA double helices exhibit satisfactory Watson–Crick base pairing, and a TNA strand will indeed cross-pair with both RNA and DNA. Despite these advantages, there are problems of sugar instabil-ity and other difficulties that apply to the abiotic syn-thesis of any sugar-based replicator.

Sugars contain multiple chiral centers (groups that can take on only right- and left-handed forms; see Section 6.3.3.1), and so give rise to many isomers. Any abiotic process that produces a particular sugar would be expected to produce a host of related ones as well. Any further processing would increase the diversity of the mixture. No driving force exists that would link the sugars to (the desired) information-bearing side chains, rather than elsewhere. If the resulting zoo of structures were then polymerized and condensed, amorphous substances would be the most likely result. One process of this type is the so-called Maillard reaction, in which amino acids and sugars, heated together, produce polymeric pigments (Shapiro, 1988).

Our third example of an RNA precursor, PNA (peptidyl nucleic acid) (Fig. 6.7), escapes the complications of carbohydrate chemistry (Neilsen, 1999). The structure contains two alternating units and has no chiral center. The amide links resemble those that hold proteins together, but the familiar bases of DNA are attached to side chains. PNA can form stable double helices with itself, and it can cross-pair with RNA and DNA. Although PNA was first prepared for potential medicinal uses, its favorable properties have called attention to it as a possible RNA precursor in evolution (Nelson et al., 2000).

The evidence presented in support of PNA includes the determination that traces of the fundamental repeating unit of the backbone (an ethylene diamine unit linked to the simplest of the amino acids, glycine) were formed in a Miller–Urey spark discharge experiment. The amount produced, however, was <0.01% of the glycine. Although this ratio could undoubtedly be improved by an experimenter with that specific purpose in mind, it seems unlikely that the amount of PNA backbone component on the early Earth would ever approach that of its much simpler constituent, glycine, let alone that of the combined mass of other amino acids, amines, and acids. In any undirected abiotic polymerization, PNA subunits would not be expected to seek each other out, but would simply be incorporated as isolated ingredients in high molecular weight materials dominated by other substances.

Considerations of this type have been extended to abiotic polymers in general (Shapiro, 2000). Processes such as chain termination, branching, the incorporation of irregular subunits, and cross-linking reactions would be expected to interfere with the abiotic formation of any linear polymer of uniform structure, let alone one that could function effectively in self-replication.

6.7 Metabolism-first theories

If replicator theories are set aside as extremely improbable, though not impossible, scenarios for the origin of life, then a surprising conclusion emerges: although proteins, nucleic acids, and other macromolecules dominate the activities of present cells, in the beginning life functioned without them. Small molecules carried out the processes of catalysis, information storage and reproduction using energy-driven cycles of chemical reactions. I would call this scenario "monomer world" to contrast it with the ones that require functioning biopolymers, but it has been more commonly called "metabolism-first." Oparin (Section 6.2.2) is commonly considered to be the founding father of this set of ideas; for a summary of his thought, see Oparin (1964). In a wider sense, such theories consider the origin of life to be one episode of the broader processes described under the "cosmic evolution" approach discussed in Section 6.1.3. In exploring them, we first have to consider how the basic life processes could be carried out by small molecules.

6.7.1 The compositional genome: information is stored in the mixture

We are so used to information storage in DNA, and in lists of all kinds, that the alternative is best explained by an analogy. My wife sometimes sends me to the supermarket with a list of the provisions we need. When I return, however, I present her with paper bags containing the desired items, but in no particular order. Unless I have made some error, the list can be thrown away. The bags contain the same information that was present in the list, but in a more directly usable form.

In the same way, a set of molecules within a primitive cellular compartment on the early Earth would, by their mere presence, carry their own heredity. An information-storing system of this type has been called a "compositional genome" (Segré and Lancet, 2001).

6.7.2 Reproduction is carried out by splitting the compartment

Using the above analogy, let us suppose that a neighbor wanted the identical set of groceries that I had brought home, and I offered to collect it. In the absence of a list, I could carry this out by returning to the supermarket with my collection of groceries and matching each item with a similar one from the shelves.[3] When I returned home, I would simply separate my goods into two identical sets: one for me and one for the neighbor.

---

[3] We will ignore the problems this might cause at checkout!

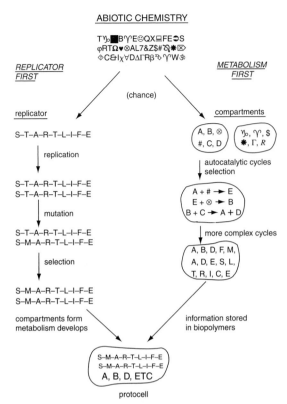

FIGURE 6.8 A comparison of the replicator and compositional genome proposals for the origin of life. Letters and other symbols represent small molecules; the ones of use to life are those of the English alphabet. In the replicator-first path, an information-rich replicator is formed by chance to start life. In the compositional genome (metabolism-first) path, life begins when an appropriate selection of small molecules within a compartment interacts with an energy source in a way that promotes the evolution of the mixture.

In a similar fashion, a primitive replicator-free cell could reproduce by collecting a duplicate set of its contents from the environment, and then splitting in two. Of course, improved ingredients could also be collected from the environment, which would be the equivalent of a mutation. The process is illustrated in Fig. 6.8, which compares the evolution of a compositional genome with that of a replicator. At the top, the mix of keyboard symbols represents an abiotic mixture, such as that present in a meteorite. Letters of the English alphabet represent substances of use to life, and the other characters stand for irrelevant or interfering chemicals.

In the replicator-first scenario (on the left), a replicator forms by chance; this is likely to be a very improbable event. Subsequent replication, mutation, and selection afford an improved replicator, which acquires a supporting metabolism.

In the alternative metabolism-first path, compartments form from amphiphilic substances in the soup, and collect different sets of components at random. When the right combination appears, a network of autocatalytic cycles develops in an evolutionary manner. Better ingredients are recruited from the environment or prepared internally, and undesirable ones are eliminated. Eventually, when the information carried has increased significantly, storage within biopolymers is "invented." This type of origin has been vividly described by Freeman Dyson (1999: 37): "Life began with little bags, the precursors of the cell, enclosing small volumes of dirty water containing miscellaneous garbage."

6.7.3 Enzymatic catalysis is carried out by monomers

Small molecules cannot approach the speed or specificity of highly evolved enzymes, but it is not clear how much of these abilities would be needed to drive primitive autocatalytic cycles. Bar-Nun et al. (1994) observed that mixtures of simple amino acids could achieve activities that varied from 10^{-3} to 10^{-7} of that demonstrated by enzymes. In the case of β-galactosidase activity, three of the four most active amino acids were ones present at the enzyme's active site. They suggested that weak interactions with the substrate might cause the monomers to assemble in a catalytic array. Such processes, perhaps aided by alignment along a mineral surface or within a membrane, might have served as the catalysts needed in primitive life. As evolution progressed, monomers may have yielded to dimers and trimers, gradually ascending the ladder of complexity (Bar-Nun et al., 1994; Kochavi et al., 1997).

6.7.4 Energy-driven catalytic cycles perform the essential processes of the cell

Some possible functions of importance would be the synthesis of improved catalysts and membrane components, and especially of the ingredients that permit improved interaction with the energy supply. Possible energy sources include light, redox reactions involving organic substances, volatile inorganics and minerals, pH gradients, and ionic potentials across membranes (Deamer, 1997). It is not clear, at present, which of these was most suitable on the early Earth, or which particular interactions allowed this energy source to be tapped to power self-organizing cycles.

An equal amount of uncertainty also exists about the specific chemicals involved in the cycles. As already mentioned, a number of mathematical treatments have demonstrated the viability of such cycles and attempted to constrain them in terms of the number of interacting variables; see for example Kauffman (1993), Dyson (1999), and Segré *et al.* (2001). Such efforts help demonstrate the physical feasibility of autocatalytic cycles, but do not specify the detailed chemical combinations that will sustain them. Many workers have suggested particular transformations or sets of reactions as likely participants in the entire interactive network of interactions, without attempting to specify the entire set. Morowitz *et al.* (2000), for example, have nominated the reductive citric acid cycle, which converts carbon dioxide to small organic molecules, as a likely contributor to the original metabolism. Finally, a few instances exist where fairly complete pictures of early metabolism-driven organisms have been provided. Two of them are discussed below; the lipid world and the iron–sulfur world. A more complete description of possible early metabolisms is provided in Chapters 8 and 10.

6.7.5 The lipid world

This scenario closely resembles the descriptions in Sections 6.7.1 through 6.7.4. A self-assembling lipid micelle (Fig. 6.4) would of itself catalyze necessary reactions (Segré *et al.*, 2001). The name "lipozyme" has been given to such an assembly. A multi-component catalytic assembly would be established, grow, divide by splitting, and evolve. Eventually dimers and higher oligomers would form, leading to a nucleic acid "takeover" of the hereditary function.

6.7.6 The iron–sulfur world

In this metabolism-first theory, the evolving system is segregated from the environment not by a membrane, but by being constrained to the surface of a positively charged mineral (Section 8.4; Wächtershäuser, 1992). The candidate of choice is an iron sulfide called pyrite, whose shiny luster has led others to call it "fool's gold." The metabolites are of necessity negatively charged, and their proximity and orientation on the essentially two-dimensional surface facilitates chemical reactions. The formation of pyrites from hydrogen sulfide and ferrous ion provides a source of chemical energy to drive a variety of processes including the fixation of carbon monoxide and carbon dioxide into small

organic molecules. No pre-formed chemical "soup" is necessary. Hydrothermal vents are the preferred location for these transformations. Several of the proposed reactions have in fact been individually demonstrated. In this theory, as in the lipid world, metabolic cycles (which are described in great detail) enter evolution first, and hereditary biopolymers come later.

6.8 Prospects for a solution

New approaches are clearly needed for progress in this field. I will make two suggestions for the collection of experimental data which may move the field closer to consensus.

6.8.1 Realistic experiments on prebiotic simulation

Many origin-of-life experiments have identified themselves as studies in prebiotic synthesis. In such experiments reactions of choice are studied, using a limited number of pure reagents under carefully controlled conditions, though usually carried out in water. If a particular chemical is successfully prepared by any such procedure, it is then declared to be "prebiotic," and the assumption is made that it would be available in unlimited amounts for any other transformations that need to have taken place on the early Earth. Strings of such reactions are then logically linked together as proof that some complicated material, usually a replicator, could originate through chemical processes before life began.

To my view, these demonstrations resemble the act of a skilled lion tamer, who can induce his beasts to sit on stools, leap through rings, and refrain from biting off his head when placed in their mouths. The act is very entertaining, but does not represent what the beasts would do on their own in the jungle. For this purpose, we would do better to plant hidden cameras in the wilderness.

Similarly, to learn of the origin of life, we should bring a likely set of chemicals and energy sources together, let them interact, and watch what happens. Such prebiotic simulations would be unlikely to capture a rare, once-in-a-Universe occurrence, such as the formation of a complex replicator by chance. They might, however, permit the establishment of a set of autocatalytic cycles that would continue to change and evolve as long as we continued to watch. Although we would probably not have the time to follow events until the formation of anything that we would recognize as a living organism, one can nevertheless learn something about the process of hiking up a mountain just by taking the first few uphill steps.

6.8.2 The exploration of other worlds: Titan as an example

To explore the results of chemical self-organization experiments conducted for longer periods of time, we must turn to nature, and in particular to sites off the Earth. There are many places that could instruct us, but one has special relevance for questions concerning the origin of life: Titan, the largest moon of Saturn. This world has an atmosphere that is denser than that of Earth, with a composition that is closer to the original Urey–Miller specification than our current concept of the early Earth's atmosphere. Nitrogen is the major component, but hydrogen is present, as well as hydrocarbons. An extensive photochemistry has shrouded Titan's surface in a perennial haze, and there has been extensive speculation concerning bodies of liquid hydrocarbon on its surface (e.g., Lunine, 1994) – the Cassini Mission and Huygens probe should tell us a great deal (Chapter 20).

The low temperature at the surface of Titan (about $-180\,°C$) would appear to preclude any chemistry relevant to the origin of life on Earth. However, volcanic eruptions of liquid water or asteroid impacts on the icy crust of Titan may have created temporary aqueous environments, in contact with organic mixtures, that may have persisted for thousands of years (Lorenz *et al.*, 2001). After refreezing, these sites represent preserved long-term experiments in chemical self-organization, awaiting our inspection. The possibility that Titan contains an internal ammonia–water ocean whose inhabitants, if any exist, may be contributing substances of biogenic origin to the atmosphere adds another dimension to the investigation (Fortes, 2000). For a full discussion of Titan, see Chapter 20.

6.8.3 Afterword

In the past century and a half, we have learned much about the structure and function of life and its evolution, but almost nothing about its origin. The field has been hindered, perhaps because of its connection to great philosophical problems, and perhaps by a tendency of some workers to leap ahead to sweeping conclusions, which they then attempt to buttress by selective collection of data. We can hope that the newly named field of astrobiology will give first priority to less biased observation of the cosmos, and thus produce more abundant insights in the future.

REFERENCES

Bar-Nun, A., Kochavi, E., and Bar-Nun, S. (1994). Assemblies of free amino acids as possible prebiotic catalysts. *J. Mol. Evol.*, **39**, 116–122.

Baross, J. A. and Hoffman, S. E. (1985). Submarine hydrothermal vents and associated gradient environments as sites for the origin and evolution of life. *Orig. Life*, **15**, 327–345.

Bartell, D. P. and Unruh, P. J. (1999). Constructing an RNA World. *TIBS*, **24**, M9–M13.

Benner, S. A., Ellington, A. D., and Tauer, A. (1989). Modern metabolism as a palimpsest of the RNA World. *Proc. Natl. Acad. Sci. USA*, **86**, 7054–7058.

Brasier, M. D., Green, O. R., Jephcoat, A. P., *et al.* (2002). Questioning the evidence of earth's oldest fossils. *Nature*, **416**, 76–81.

Cairns-Smith, A. G. (1982). *Genetic Takeover and the Mineral Origins of Life*. Cambridge: Cambridge University Press.

Cech, T. R. (2000). The ribosome is a ribozyme. *Science*, **289**, 878–880.

Chyba, C. and Sagan, C. (1992). Endogenous production, exogenous delivery and impact-shock synthesis of organic molecules: an inventory for the origin of life. *Nature*, **355**, 125–132.

Cody, G. D., Boctor, N. Z., Filley, T. R., *et al.* (2000). Primordial carbonylated iron-sulfur compounds and the synthesis of pyruvate. *Science*, **289**, 1337–1340.

Crick, F. H. C. (1981). *Life Itself*. New York: Simon and Schuster.

Cronin, J. R., Pizzarello, S., and Cruikshank, D. P. (1988). Organic matter in carbonaceous chondrites, planetary satellites, asteroids and comets. In *Meteorites and the Early Solar System*, eds. Kerridge, J. S. and Matthews, M. S., pp. 819–857. Tucson: University of Arizona Press.

Cronin, J. R. and Chang, S. (1993). Organic matter in meteorites: molecular and isotopic analyses of the Murchison meteorite. In *The Chemistry of Life's Origins*, eds. Greenberg, J. M., Mendoza-Gómez, C. X., and Piranello, V., pp. 209–258. Dordrecht: Kluwer Academic Publishers.

Davies, P. (1995). *Are We Alone?* New York: Basic Books.

Deamer, D. W. (1997). The first living systems: a bioenergetic perspective. *Microbiol. Mol. Biol. Rev.*, **61**, 239–261.

Delano, J. W. (2001). Redox history of the Earth's interior since $\sim$3900 Ma: implications for prebiotic molecules. *Orig. Life Evol. Biosphere*, **31**, 311–341.

Delsemme, A. H. (2001). An argument for the cometary origin of the biosphere. *Amer. Scientist*, **89**, 432–442.

Doherty, E. A. and Doudna, J. A. (2000). Ribozyme structures and mechanisms. *Ann. Rev. Biochem.*, **69**, 597–615.

Dyson, F. (1999). *Origins of Life*, second edn. Cambridge: Cambridge University Press.

Eigen, M. and Schuster, P. (1978). The hypercycle. A principle of natural self-organization. Part C: the realistic hypercycle. *Naturwissenschaften*, **65**, 341–365.

Eschenmoser, A. (1999). Chemical etiology of nucleic acid structure. *Science*, **284**, 2118–2123.

Farley, J. (1977). *The Spontaneous Generation Controversy from Descartes to Oparin*. Baltimore: Johns Hopkins University Press.

Feinberg, G. and Shapiro, R. (1980). *Life Beyond Earth. The Intelligent Earthling's Guide to Life in the Universe*. New York: Morrow.

Fortes, A. D. (2000). Exobiological implications of a possible ammonia-water ocean inside Titan. *Icarus*, **146**, 444–452.

Freund, F., Staple, A., and Scoville, J. (2001). Organic protomolecule assembly in igneous minerals. *Proc. Natl. Acad. Sci. USA*, **98**, 2142–2147.

Fuller, W. D., Sanchez, R. A., and Orgel, L. E. (1972). Studies in prebiotic synthesis. VI. Synthesis of purine nucleosides. *J. Mol. Biol.*, **67**, 25–33.

Gesteland, R. F., Cech, T. R., and Atkins, J. F., eds. (1999). *The RNA World*, second edn. Cold Spring Harbor: Cold Spring Harbor Laboratory Press.

Gilbert, W. (1986). The RNA World. *Nature*, **319**, 618.

Gold, T. (1992). The deep hot biosphere. *Proc. Natl. Acad. Sci. USA*, **89**, 6045–6049.

Holm, N. G. (1992). Why are hydrothermal systems proposed as plausible environments for the origin of life? *Orig. Life Evol. Biosphere*, **22**, 5–14.

Huber, C. and Wächtershäuser, G. (1997). Activated acetic acid by carbon fixation on (Fe, Ni)S under primordial conditions. *Science*, **276**, 245–247.

Johnston, W. K., Unrau, P. J., Lawrence, M. S., Glasner, M. E., and Bartel, D. P. (2001). RNA-catalyzed RNA polymerization: accurate and general RNA-templated primer extension. *Science*, **292**, 1319–1325.

Joyce, G. F. and Orgel, L. E. (1999). Prospects for understanding the origin of the RNA World. In *The RNA World*, eds. R. F. Gesteland, T. R. Cech, and J. F. Atkins, second edn., pp. 49–77. Cold Spring Harbor: Cold Spring Harbor Laboratory Press.

Kasting, J. F. (1993). Earth's early atmosphere. *Science*, **259**, 920–926.

Kauffman, S. (1993). *The Origins of Order. Self-Organization and Selection in Evolution*. New York: Oxford University Press.

Kauffman, S. (1995). *At Home in the Universe*. New York: Oxford University Press.

Kochavi, E., Bar-Nun, A., and Fleminger, G. (1997). Substrate-directed formation of small biocatalysts under prebiotic conditions. *J. Mol. Evol.*, **45**, 342–351.

Larralde, R., Robertson, M. P., and Miller, S. L. (1995). Rates of decomposition of ribose and other sugars: implications for chemical evolution. *Proc. Natl. Acad. Sci. USA*, **92**, 8158–8160.

Lorenz, R. D., Junine, J. I., and Mckay, C. P. (2001). Geologic settings for aqueous organic synthesis on Titan revisited. *Enantiomer*, **6**, 83–96.

Lunine, J. (1994). Does Titan have oceans? *Amer. Scientist*, **82**, 134–143.

Maurel, M.-C. and Convert, O. (1990). Chemical structure of a prebiotic analogue of adenosine. *Orig. Life Evol. Biosphere*, **20**, 43–48.

Miller, S. L. (1953). A production of amino acids under possible primitive Earth conditions. *Science*, **117**, 528–529.

Miller, S. (1974). The first laboratory synthesis of organic compounds under primitive Earth conditions. In *The Heritage of Copernicus: Theories Pleasing to the Mind*, ed. J. Neyman, pp. 228–242. Cambridge, MA: MIT Press.

Miller, S. L. and Urey, H. C. (1959). Organic compound synthesis on the primitive Earth. *Science*, **130**, 245–251.

Mojzsis, S. J., Krishnamurthy, R., and Arrhenius, G. (1999). Before RNA and after: geophysical and geochemical constraints on molecular evolution. In *The RNA World*, eds. R. F. Gesteland, T. R. Cech, and J. F. Atkins, pp. 1–47. Cold Spring Harbor: Cold Spring Harbor Laboratory Press.

Morowitz, H. (1968). *Energy Flow in Biology*. New York: Academic Press.

Morowitz, H. J., Kostelnik, J. D., Yang, J., and Cody, G. D. (2000). The origin of intermediary metabolism. *Proc. Natl. Acad. USA*, **97**, 7704–7708.

Muller, H. J. (1966). The gene material as the initiator and the organizing basis of life. *American Naturalist*, **100**, 493–517.

Neilsen, P. E. (1999). Peptide nucleic acid. A molecule with two identities. *Accounts Chem. Res.*, **32**, 624–630.

Nelson, K. E., Levy, M., and Miller, S. L. (2000). Peptide nucleic acids rather than RNA may have been the first genetic material. *Proc. Natl. Acad. Sci. USA*, **97**, 3868–3871.

Oparin, A. I. (1964). *Life, Its Nature, Origin and Development*, translated from the Russian by A. Synge. New York: Academic Press.

Orgel, L. E. (1968). Evolution of the genetic apparatus. *J. Mol. Biol.*, **38**, 381–393.

Oró, J. and Kimball, A. P. (1961). Synthesis of purines under possible primitive Earth conditions. I. Adenine from hydrogen cyanide. *Arch. Biochem. Biophys.*, **94**, 217–227.

Pitsch, S., Eschenmoser, A., Gedulin, B., Hui, S., and Arrhenius, G. (1995). Mineral induced formation of sugar phosphates. *Orig. Life Evol. Biosphere*, **25**, 297–334.

Ricardo, A., Carrigan, M. A., Olcott, A. N., and Benner, S. A. (2004). Borate minerals stabilize ribose. *Science*, **303**, 196–197.

Schlesinger, G. and Miller, S. L. (1973). Equilibrium and kinetics of glyconitrile formation in aqueous solution. *J. Amer. Chem. Soc.*, **95**, 3729–3735.

Schöning, K.-U., Scholz, P., Guntha, S., Wu, X., Krishnamurthy, R., and Eschenmoser, E. (2000). Chemical etiology of nucleic acid structure: the α-threofuranosyl-($3'\rightarrow2'$) oligonucleotide system. *Science*, **290**, 1347–1351.

Schopf, J. W., Kudryavtsev, A. B., Agresti, D. G., Wdowiak, T. J., and Czaja, A. D. (2002). Laser-raman imagery of Earth's earliest fossils. *Nature*, **416**, 73–76.

Segré, D., Ben-Eli, D., Deamer, D. W., and Lancet, D. (2001). The lipid world. *Orig. Life Evol. Biosphere*, **31**, 119–145.

Segré D. and Lancet, D. (2001). Composing life. *EMBO Reports*, **1**, 217–222.

Shapiro, R. (1986). *Origins. A Skeptic's Guide to the Creation of Life on Earth*. New York: Summit.

Shapiro, R. (1988). Prebiotic ribose synthesis: a critical analysis. *Orig. Life*, **18**, 71–85.

Shapiro, R. (1995). The prebiotic role of adenine: a critical analysis. *Orig. Life Evol. Biosphere*, **25**, 83–98.

Shapiro, R. (1999). Prebiotic cytosine synthesis: a critical analysis. Implications for the origin of life. *Proc. Nat. Acad. Sci. USA*, **96**, 4396–4401.

Shapiro, R. (2000). A replicator was not involved in the origin of life. *IUBMB Life*, **49**, 173–176.

Shapiro, R. (2006). Small molecule interactions were central to the origin of life. *Quarterly Rev. Biol.*, **81**, 105–125.

Stryer, L. (1995). *Biochemistry*, fourth edn. New York: Freeman.

Szostak, J. W., Bartel, D. P., and Luisi, P. L. (2001). Synthesizing life. *Nature*, **409**, 387–390.

Wächtershäuser, G. (1992). Groundworks for an evolutionary biochemistry: the iron-sulfur world. *Prog. Biophys. Molec. Biol.*, **58**, 85–201.

Wald, G. (1979). The origin of life. In *Life, Origin and Evolution*, introduction by C. E. Folsome, pp. 47–56. San Francisco: Freeman. Reprinted from *Sci. Am.*, August 1954.

Wills, C. and Bada, J. (2001). *The Spark of Life. Darwin and the Primeval Soup*. Cambridge: Perseus.

Zubay, G. (1998). Studies on the lead-catalyzed synthesis of aldopentoses. *Orig. Life Evol. Biosphere*, **25**, 297–324.

Zubay, G. (2000). *Origins of Life on the Earth and in the Cosmos*, second edn. San Diego: Academic Press.

FURTHER READING AND SURFING

Cairns-Smith, A. G. (1985). *Seven Clues to the Origin of Life*. Cambridge: Cambridge University Press. The minerals-first theory of the origin of life is presented in a format suitable for the layman.

Campbell, N. A. (1996). *Biology*, fourth edn. Menlo Park, CA: Benjamin Cummings. Introductory, comprehensive, well-illustrated textbook for science students.

Davies, P. C. W. (1999). *The Fifth Miracle: the Search for the Origin and Meaning of Life*. New York: Simon & Schuster. A noted physicist and philosopher places the topic in a cosmic context.

Deamer, D. W. and Fleischaker, G. R. (1994). *Origins of Life: the Central Concepts*. Boston: Jones and Bartlett. A collection of important papers in the field organized by topic and accompanied by explanatory comments.

DeDuve, C. (2002). *Life Evolving – Molecules, Mind and Meaning*. Oxford: Oxford University Press. An engaging discourse in which a distinguished Nobel laureate speculates on the origin, evolution, and future of life, including issues of philosophy.

Fry, I. (2000). *The Emergence of Life on Earth – a Historical and Scientific Overview*. New Brunswick: Rutgers University Press. A well-written and up-to-date survey of ideas in the field, also including historical and philosophical aspects.

Hazen, R. M. (2005). *Genesis: The Scientific Quest for Life's Origin*. Washington, DC: Joseph Henry Press. An excellent, engaging account of the science and scientists involved in research on the origin of life.

Lahav, N. (1999). *Biogenesis: Theories of Life's Origin*. Oxford: Oxford University Press. Tutorial treatment of the basics of life, its history, and ideas on its origin.

Shapiro, R. (1986). *Origins. A Skeptic's Guide to the Creation of Life on Earth*. New York: Summit. A survey of ideas on the origin of life from the point of view of a questioning skeptic. The use and misuse of scientific method receives special emphasis.

www.issol.org/ The International Society for the Study of the Origin of Life.

cmex.ihmc.us/VikingCD/Puzzle/Prebiot.htm A lushly illustrated site sponsored by NASA that attempts to integrate the origin of life into the greater concept of Cosmic Evolution.

www.us.net/life/ The Origin of Life Foundation offers a US$1 million prize for a solution to the origin-of-life problem; site also contains an extensive list of readings.

7 The origin of proteins and nucleic acids

Alonso Ricardo, *University of Florida (currently at Massachusetts General Hospital)*
Steven A. Benner, *University of Florida*

Astrobiologists hope to understand the origin, history, extent, and future of life in the Universe. This is a huge task, considering that two of the terms in this mission statement are difficult to define. The definition of "life" is itself worth a chapter in this book (Chapter 5). Here in this chapter, we must not only concern ourselves with a definition of life, but also with the concept of "understanding." What does it mean to say that we "understand life"?

Any attempt to understand life soon engages organic chemistry. Biology today is increasingly focused on the molecular scale. Indeed, it is difficult to find a biologist today who is *not* attempting to put a molecular structure on the phenomenon that they are studying, so much so that biology can be (provocatively!) viewed as the subfield of chemistry dealing with chemical systems capable of Darwinian evolution.

Some illustrations make this point. The human genome is nothing more (and nothing less) than a collection of chemical structures, recording how carbon, oxygen, nitrogen, hydrogen, and phosphorus atoms are bonded in the natural products directly responsible for heritance. Molecular evolution uses organic chemistry to describe the Darwinian evolution of species, the process that drives biology. Neurobiologists are attempting to describe the inventory of molecules, including messenger RNA, that allow neurons to learn and remember.

Nowhere is this more evident than in the segment of astrobiology that investigates the origin of life. At some point, either on Earth or elsewhere in the Cosmos, a collection of inanimate organic molecules found themselves assembled in a way that supported Darwinian evolution. This chemical system must have had three properties. First, it must have been able to direct the synthesis of more of itself, an ability that is uncommon in chemistry, although not unknown.

But simple replication is not sufficient to support Darwinian evolution. To support evolution, the replicates generated by the original Darwinian system must have been imperfect; the replicated chemical structures must have been different from their parents. Lastly, the imperfections themselves must have been heritable. Any system that has these properties (in correct measure) will improve itself via two Darwinian mechanisms: mutation and natural selection.

Unfortunately, we have little idea of what chemical structures support these properties, let alone what structures actually supported them in the first living systems on Earth. No examples are known of Darwinian systems other than the ones that Nature herself presents to us. We do not have an artificial life form working in the laboratory, something that might better define models for the first Darwinian molecular systems. Although many chemists have attempted to create artificial Darwinian molecular systems in the laboratory, they have failed so far because too little is known about chemistry.

These facts create challenges for future astrobiologists, challenges that go far beyond the description of planetary environments or the characterization of unusual forms of life that are found on Earth. The first goal of chemical astrobiology is to define molecular structures that support Darwinian evolution. The second will be to create, in the laboratory, artificial chemical systems that evolve as they exploit these structures. Only by creating life in the laboratory will we demonstrate that we truly understand life.

The purpose of this chapter is to lay the grounds for this future. We begin by describing some general tools for analyzing the reactivity of organic molecules, to allow the non-chemist to better understand the ways that molecules can become other molecules. We then apply these tools to ask simple questions about how

Planets and Life: The Emerging Science of Astrobiology, eds. Woodruff T. Sullivan, III and John A. Baross. Published by Cambridge University Press. © Cambridge University Press 2007.

building blocks for proteins and nucleic acids might be created abiologically. Finally, we apply these tools to constrain the historical question about origins: what chemistry might *actually have occurred* four billion years ago that led to the emergence of the life on Earth that we know today.

7.1 Carbon and biochemistry

The carbon produced by stars and the hydrogen that presumably was present near the origin of the Universe can combine to form an unlimited number of organic molecules. Carbon atoms form four bonds. A single bond joining two carbon atoms is strong, on the order of 400 kJ ($\sim$100 kcal) per mole (or $\sim$4 eV per bond). This means that a typical pair of carbon atoms joined by a single bond will remain joined for many thousands of years at the temperatures for which water is a liquid at sea level on Earth.[1] Bonds between carbon and hydrogen, carbon and oxygen, and carbon and nitrogen are similarly strong.

No other element forms so many single bonds that are so strong. Nitrogen–nitrogen single bonds, like oxygen–oxygen single bonds, sulfur–sulfur single bonds, and other single bonds between two identical atoms, are weak enough to easily fall apart at liquid-water temperatures. Nitrogen can use all three of its valences to bond to another nitrogen atom; two nitrogens held together by this triple bond are indeed hard to pull apart ($\sim$800 kJ/mole). But by using all of their three valences simply to hold themselves together, two nitrogen atoms have no valences remaining to form extended chains, or to carry substituents (an atom or group of atoms that can vary in a series of analogous molecules).

As a consequence of their strength, the bonds between two carbon atoms break only when an energetically favorable path exists that allows them to do so. At temperatures accessible to liquid water, such a path generally requires that the carbon form a bond to *another* atom just as, or soon after, the original bond breaks. Understanding ways that a new bond can form as the old bond is breaking is a key to understanding reactivity.

7.2 Reactivity

Reactivity in organic molecules is based on *Structure Theory*, which holds that the behavior of all organic

molecules, including those in living matter, can be understood in terms of the structures of collections of atoms held together by bonds.

Practitioners of Structure Theory sometimes seem to some to have an almost mystical understanding of the relationship between chemical structure and reactivity. In fact, there is nothing mystical to it. Chemists know thousands of reactions that organic molecules undergo, the conditions under which they go, and how forming bonds can rush in and snap up energetic intermediates arising when an old bond breaks. From this, they acquire an intuition as to the plausibility that any given chemical transformation, perhaps one that they have never seen before, might occur. They are similarly able to propose new chemical transformations.

This understanding is critical to evaluate any model relating to the origins of proteins, nucleic acids, fats, or sugars of contemporary *terran* (Earth) life. The difficulty, of course, is that this intuition cannot be gained in a single chapter, or even a semester of study. Thus, we can provide here only the basic skills that are needed to assess the likelihood of chemical pathways that might have led to the components of early life.

7.2.1 Pairs of electrons form bonds between atoms

We begin by noting that covalent bonds between two atoms are formed by the sharing of a pair of electrons. Thus, in dihydrogen (often written as H-H, or H_2) the line joining the two hydrogen atoms represents a pair of electrons, or a single bond. In water (H-O-H), the lines between the hydrogen atoms and the oxygen atom each represent a pair of electrons that form a single bond holding the hydrogens to the oxygen. In formaldehyde ($H_2C=O$), the double line between the carbon and the oxygen represents two pairs of electrons, four electrons in total (Fig. 7.1).

To simplify the diagrams that represent organic molecules, chemists frequently omit some of the atoms and electrons. Therefore, the first skill that is required in analyzing reactivity in organic chemistry requires that one completes the structure of the organic molecule as it is written on a sheet of paper. This ensures that we know the locations of all of the molecule's electrons.

A completed structure is known as a *Lewis structure*, and must include the electrons present that are *not* involved in bonding. For example, in the Lewis structure of water, oxygen carries two pairs of unshared electrons from the outer valence shell. We represent

---

[1] Liquid water will be regarded here as "standard," even though other solvents might conceivably be used for some type of "weird life" – see Chapter 27 for a discussion of the possibilities.

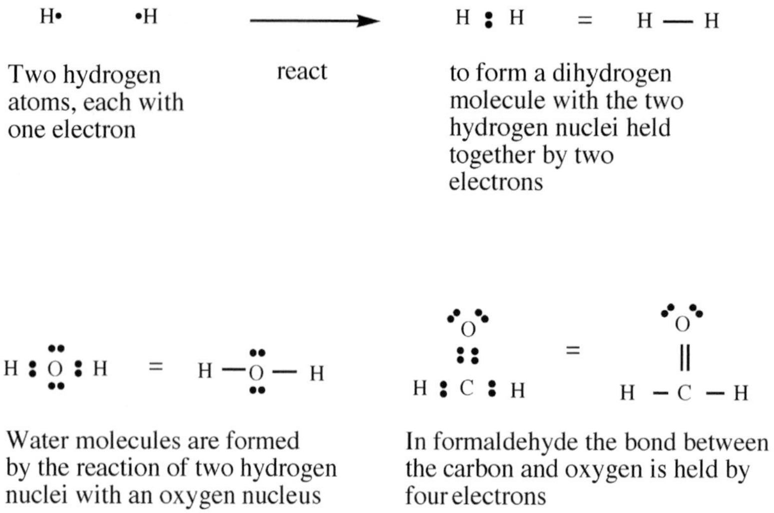

FIGURE 7.1 Lewis representation of chemical bonds for dihydrogen (H_2), water (H_2O), and formaldehyde (H_2CO).

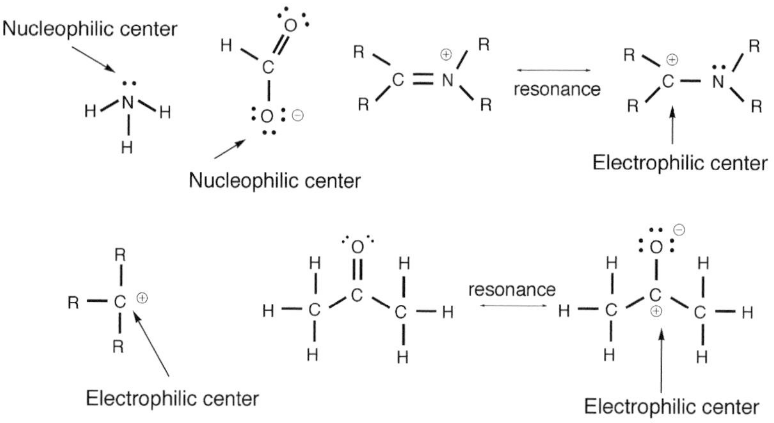

Two hydrogen atoms, each with one electron react to form a dihydrogen molecule with the two hydrogen nuclei held together by two electrons

Water molecules are formed by the reaction of two hydrogen nuclei with an oxygen nucleus

In formaldehyde the bond between the carbon and oxygen is held by four electrons

FIGURE 7.2 Nucleophilic centers have an unshared pair of electrons that can form a new bond, or can get one (via resonance, for example). Electrophilic centers have a vacant orbital (or can get one via resonance, for example) that can accept an unshared pair of electrons from a nucleophilic center to form a new bond. In a *resonance* form a pair of electrons moves between adjacent atoms (electrons are dynamic entities), creating a new representation for the same molecule. The resulting resonance forms are joined by a double headed arrow to indicate equivalence.

each of these valence electrons not involved in a bond by a dot. Hence, the oxygen in the H-O-H structure has four dots, representing electrons on the oxygen that are not involved in bonding. Likewise, the oxygen in formaldehyde carries two pairs of unshared electrons, represented again by four dots on the oxygen.

A simple electrophile, for example, is a proton (H^+). H^+ is not bonded to anything, and can intrinsically form one bond. H^+ is therefore looking for a single partner with which to bond. But since H^+ itself has no electrons that it can use to form a bond, it must find a nucleophilic center as that partner.

7.2.2 A nucleophilic center brings a pair of electrons to form a new bond

As a pair of electrons is needed to form a chemical bond, any unshared pair of electrons is available in principle to form a new bond. Atoms that contain pairs of electrons available to form a new bond are called *nucleophilic centers* (Fig. 7.2). To form a bond, the electron pair on the nucleophile must find an atom that lacks a bond, or will soon lack a bond through breakage. This atom is called an *electrophilic center*.

7.2.3 Curved arrows describe the movement of pairs of electrons in reactions that form and break bonds between atoms

Organic chemists use curved arrows to describe reactions between nucleophilic and electrophilic centers that produce a new bond. The curved arrow begins with an unshared pair of electrons on the nucleophile, the pair that will form the new bond in the product. The arrow is drawn to end at a position (on the structures of the reactants) where the electron pair *will be*

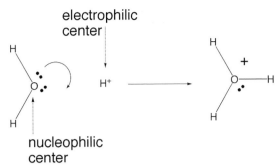

FIGURE 7.3 Reaction of the nucleophilic center on the oxygen of water with an electrophilic center, H^+. The movement of a pair of electrons in the reaction is illustrated using a curved arrow. The result is H_3O^+, the hydronium ion.

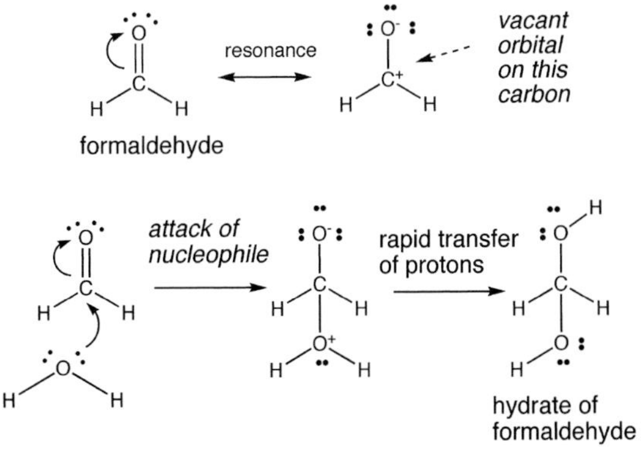

formaldehyde

FIGURE 7.4 Reaction of the nucleophilic center on the oxygen of water with an electrophilic center, the carbon atom of formaldehyde. The movement of a pair of electrons in the reaction is illustrated using a curved arrow.

after the bond is formed. Figure 7.3 shows the reaction of the unshared pair of electrons on the oxygen of water (the nucleophilic center) with H^+ (the electrophilic center) to give H_3O^+ (the hydronium ion).

While nucleophilic centers are often easy to spot (they can bear the relevant electron pair prominently in a correctly drawn Lewis structure), electrophilic centers are frequently less so. This is especially true when the electrophilic center is a carbon atom. For example, the carbon of formaldehyde (H_2CO) has all of its four valences filled. That carbon does not seem to have a valence available to form a new bond with anything.

If, however, one of the two bonds between carbon and oxygen breaks, with the electron pair moving from a position between the carbon and the oxygen to a new position on the oxygen, then the carbon center has a valence free. It then welcomes nucleophilic attack from an oxygen atom, such as from any nearby H_2O molecule.

This process is shown using curved arrows in Fig. 7.4. Here, a bond between carbon and oxygen is broken at the same time as the carbon forms a new bond to an incoming oxygen atom. As the energy in the second C-O bond is lost through breakage, the energy of a new C-O bond is gained. The resulting product is known as the hydrate of formaldehyde (H_4CO_2).

7.3 Curved arrow mechanisms for the synthesis of biological molecules

The curved arrow tool can be used to describe most reactions of organic molecules under standard conditions (water at room temperature). This includes transformations that might have converted formaldehyde, present in a prebiotic world, into molecules that are characteristic of contemporary terran life, as we now show.

7.3.1 Curved arrow mechanisms for the formation of amino acids

Let us use the curved arrow formalism to describe the synthesis from formaldehyde, ammonia, cyanide, and water of a simple amino acid, glycine (NH_2CH_2COOH), one of the basic building blocks of proteins. The following numbered steps are shown in Fig. 7.5.

Ammonia (1) has three hydrogen atoms and one nitrogen atom. A Lewis structure shows that the nitrogen in ammonia also carries an unshared pair of electrons. The nitrogen atom is therefore a nucleophilic center. Ammonia should therefore react with formaldehyde (2) for the same reason that water does. In this reaction, the unshared pair of electrons on ammonia forms a new bond between its nitrogen and the carbon of formaldehyde, just as the pair of electrons forming the second carbon–oxygen bond leaves to form a new bond to H^+. This generates an "amino alcohol" (3).

After the transfer of various hydrogens, the nitrogen of the amino alcohol again has an unshared pair of electrons, and is able to form a second bond with carbon atom. The resulting compound is known as an imine (4), which contains a (C=N) unit having a carbon atom bonded twice to a nitrogen atom.

The carbon of the C=N unit is also an electrophilic center. This sets the stage for the next reaction, where the carbon of the cyanide anion (5) attacks the imine (4) carbon to form an aminonitrile (6). The nitrile has a C≡N unit, where the nitrogen is bonded three times to the carbon. This carbon is again an electrophilic center. If a pair of electrons forming one of the bonds between

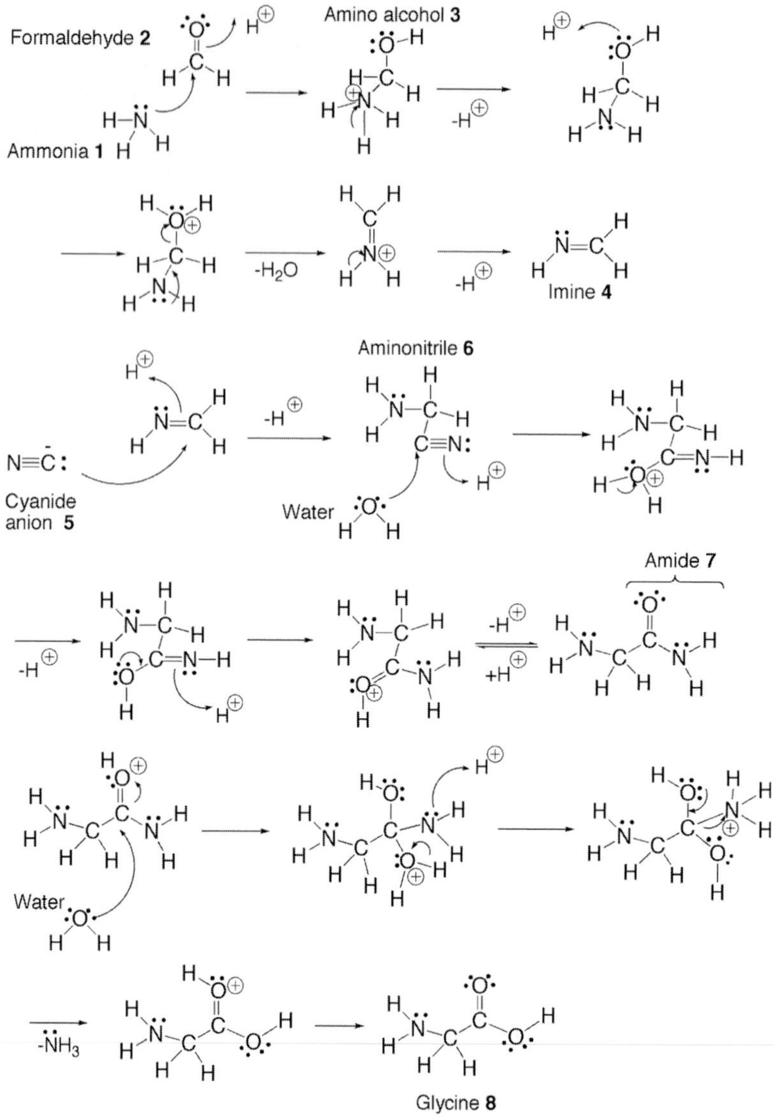

FIGURE 7.5 The Strecker synthesis of glycine. Reaction of the nucleophilic centers on the nitrogen of ammonia, the carbon of the cyanide anion, and the oxygen of water with electrophilic centers on formaldehyde and key intermediates. The movements of pairs of electrons in the reactions are illustrated using curved arrows.

carbon and nitrogen leaves to form a bond with H^+, then the carbon has a free valence. It is therefore available to form a bond with a nucleophilic oxygen atom from water.

The product, again after H^+ atoms are transferred, has another C=O group in a unit known as an *amide* (7). The carbon of the amide is again an electrophilic center, and can therefore now be attacked by the nucleophilic oxygen of another water molecule. This leads to the *hydrolysis* (taking on a water molecule) of the amide and the formation of the amino acid glycine, together with an ammonia molecule.

The net process is the reaction of one molecule of formaldehyde, one molecule of hydrogen cyanide, and

one molecule of water to give one molecule of glycine (8). Ammonia is used in the first step, and is released in the last step. Therefore, ammonia is a catalyst for the reaction, being consumed and formed in equal amounts in the reaction cycle.

This sequence of reactions is known as the *Strecker synthesis of amino acids*, named after the chemist who developed it in the 1860s. The Strecker synthesis is driven by the innate reactivity of nucleophiles and electrophiles, and proceeds spontaneously and in reasonable yield. Further, the Strecker synthesis is quite general. It can be used to prepare any amino acid for which the corresponding aldehyde is available, not just formaldehyde; many of these amino acids are among

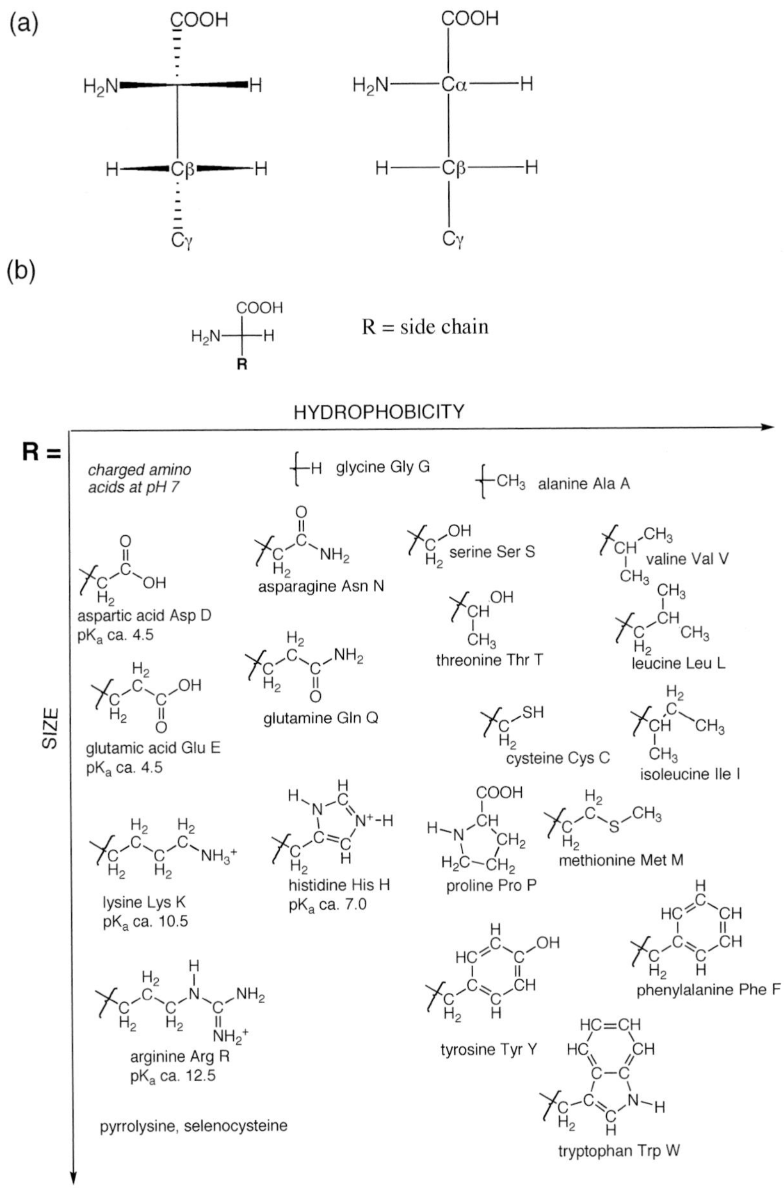

(a)

(b)

R = side chain

HYDROPHOBICITY

R =

FIGURE 7.6 (a) The general structure of an α-amino acid. The α refers to the position on the carbon chain that carries the amino group H$_2$N; other possibilities are the β and γ carbons. The four atoms attached to each carbon are arranged in space above and below the plane of the paper, as indicated by the wedged lines. On the right is a Fischer projection, which places bonds that project above the paper horizontally, and those below vertically.

(b) The side chains of the 20 predominant amino acids for Earth life. In general, organic chemists do not represent all of the carbon atoms in a molecular structure with the letter "C", or all of the hydrogen atoms in the structure with a letter "H". Rather, carbon atoms are understood to be present at any unlabelled vertex. These carbon atoms are assumed to form four bonds. If four lines are not shown explicitly on the structure, the missing lines are assumed to be bonds to hydrogen atoms, which are not explicitly shown. Recent discoveries have increased the number of encoded amino acids to 22, with the inclusion of the "minor" amino acids selenocysteine and pyrrolysine (Rother et al., 2000; Berry et al., 2001; Hao et al., 2002; Srinivasan, James, and Krzycki, 2002).

those that link together to form the proteins of terran organisms (structures are shown in Fig. 7.6). For example, if we start with acetaldehyde (CH$_3$CHO) rather than formaldehyde, the amino acid alanine (CH$_3$CH(NH$_2$)COOH) is formed (Fig. 7.7).

7.3.2 Curved arrow mechanisms for forming nucleobases for RNA and DNA

Curved arrow mechanisms can be used to generate nonbiological routes for the synthesis of many molecules in biology. For example, the Oró–Orgel synthesis

exploits the reactivity of HCN to make adenine (C$_5$H$_5$N$_5$), one of the five *nucleobases* (also called *bases*, symbolized as A, C, T, G, and U) used to store information in DNA and RNA (Fig. 7.8). The cyanide anion again reacts as a nucleophile, this time with HCN, whose carbon atom serves as an electrophilic center.

7.3.3 Curved arrow mechanisms for forming sugars for RNA

Eschenmoser has considered the special reactivity of hydrogen cyanide to generate the sugar ribose, also a

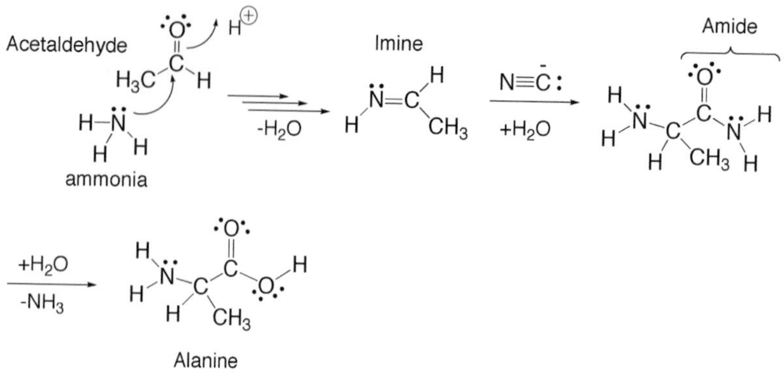

FIGURE 7.7 Strecker synthesis of alanine, starting from acetaldehyde.

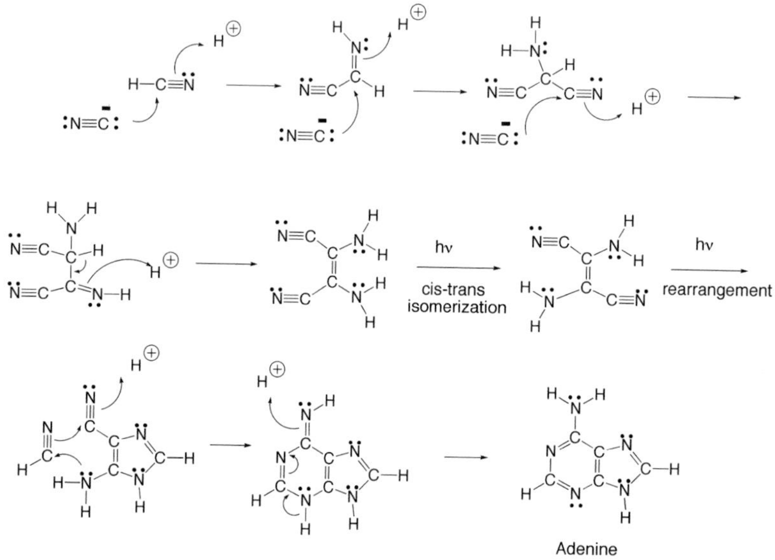

FIGURE 7.8 HCN yields adenine via the Oró–Orgel synthesis.

part of RNA. An intriguing sequence of reactions from an interesting starting material (derived from HCN) is shown in Fig. 7.9. This process occurs in the laboratory (Mueller *et al.*, 1990). It is not known, however, to occur in nature, and the starting material has not been detected naturally in the cosmos.

7.4 Extraterrestrial sources of organic building blocks

What starting materials might have been available in the cosmos or on Earth to support the origin of life? What kinds of molecules might have been available as precursors for prebiological synthesis? The Cosmos offers several places to look directly. One is the meteorites that fall to Earth continuously. These come in all sizes, including submicron particles that are too small to analyze easily, but may contribute most of the carbon to a forming planet. Another place to look is in the

interstellar dust in clouds surrounding stars that are forming.

7.4.1 Organics from meteorites

Many meteorites that fall to Earth are believed to originate in or near the asteroid belt. Asteroids most likely arose as the consequence of a failed planet formation (Section 3.9.2). Particularly interesting are carbonaceous chondrites, which have evidently suffered exposure to water when within their original parent body. These have frequently been reported to contain amino acids, principally glycine and alanine. Early reports may, however, have been partly unreliable, reflecting contamination of the meteorite after it arrived on Earth.

Some of the most reliable work exploited a meteorite that fell in 1969 near Murchison, Australia. This meteorite was quickly recovered, and presumably

FIGURE 7.9 The Eschenmoser synthesis of ribose-2,4-diphosphate from a proposed starting material derived from HCN.

FIGURE 7.10 A fragment of the carbonaceous chondrite Murchison meteorite.

therefore suffered little contamination (Fig. 7.10). The Murchison meteorite therefore has become the principal sample from which conclusions are drawn about organic molecules that might exist in the Solar System outside of Earth.

Murchison contains a large amount of organic material, summarized in Tables 7.1 and 7.2. The amino acids found in Murchison have been classified

into two general groups (Fig. 7.11). Within these groups, some 70 specific amino acids have been reported, some of which are shown in Fig. 7.12. Many of these are not known to occur in contemporary Earth life, diminishing the likelihood that they arose in the meteorite as a result of contamination from a terran source.

Several of these amino acids contain *stereogenic centers*. These are, in these cases, atoms in which carbon atoms are bonded to four different substituents. Compounds having stereogenic centers can give rise to *enantiomers*. A pair of enantiomers have mirror image structures, i.e., related to each other like the left hand is related to the right hand.

Enantiomers are isomers, meaning the left-handed and right-handed forms are different. In natural processes, left- and right-handed molecules are often formed in equal amounts. To form regular structures when these monomers are assembled, however, it is generally more efficient to have the same handedness in all molecules. Thus, 100% of amino acids in terran proteins are left-handed, and all sugars in terran nucleic acids are right-handed. Enantiomeric excess,[2] the predominance of one enantiomer over the other for a given assembly of atoms, is therefore believed to be a

---

[2] This property is often also called *chirality*, or handedness.

TABLE 7.1 Carbon in the Murchison meteorite

Total carbon	2.12% (Jarosevich, 1971), 1.96% (Fuchs, Olsen, and Jensen, 1973)
Carbon as interstellar grains	
Diamond	400 ppm (Lewis et al., 1987)
Silicon carbide	7 ppm (Tang et al., 1989)
Graphite	<2 ppm (Amari et al., 1990)
Carbonate minerals	2–10 % of total carbon (Grady et al., 1988)
Macromolecular carbon	70–80 % of total carbon

TABLE 7.2 Organic compounds in the Murchison meteorite (Cronin and Pizzarello, 1988)

Amino acids 60 ppm	Purines and pyrimidines 1.3 ppm
Aliphatic hydrocarbons >35 ppm	Basic N-heterocycles 7 ppm
Aromatic hydrocarbons 15–28 ppm	Amines 8 ppm
Carboxylic acids >300 ppm	Amines 55–70 ppm
Dicarboxylic acids >30 ppm	Alcohols 11 ppm
Hydroxycarboxylic acids 15 ppm	Aldehydes and ketones 27 ppm

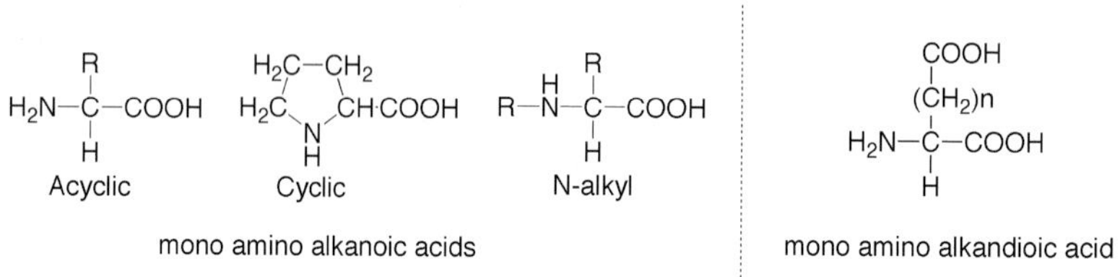

mono amino alkanoic acids

mono amino alkandioic acid

FIGURE 7.11 Typical types of amino acids in the Murchison meteorite. Adapted from Cronin and Pizzarello (1988).

universal feature of chemistry derived from living systems.

In this light, researchers were surprised to find that several of the amino acids found in the Murchison meteorite are in fact slightly enantiomerically enriched (Table 7.3). Especially notable were enantiomeric enrichments of 7–9% in three specific amino acids not known in forms of life presently living on Earth. This makes contamination from terrestrial sources unlikely.

If we assume that the amino acids in Murchison were *not* generated by living processes, this suggests that enantiomeric enrichment can be achieved by processes that are independent of life. Several of these are known in the laboratory, but not in nature. This weighs against the usual thinking that enantiomeric enrichment is a unique signature of life. However, the data available are also consistent with a biotic origin. This would be the case if the Murchison meteorite originated in an asteroid that supported life based on α-methyl amino acids that preferred one enantiomer to the other. While few would argue so, this result is consistent with life having emerged in the asteroid belt.

We do not know the extent to which the Murchison organics reflect what was available on early Earth before life emerged. This rich inventory of amino acids does not appear to be universal in carbonaceous chondrites (although the number of these that have

been examined in detail is very small). For example, only a few amino acids (glycine, alanine, α-aminoisobutyric acid, α-amino-*n*-butyric acid, γ-aminobutyric acid) are found in the meteorite that fell in 2000 on Tagish Lake, Canada (Pizzarello *et al.*, 2001; Table 7.4). The near absence of complex amino acids is significant, as the meteorite was captured in a pristine condition soon after it fell.

Also discouraging is the fact that not even two joined amino acids have ever been found in meteorites. Joining two amino acids is the first step towards the synthesis of proteins, such as those found in contemporary terran life. If the meteorite organics analyzed to date are representative of planetary processing of primitive organic compounds, the problem of assembling amino acids into *polypeptides* (short strings) remains to be solved.

Some of the chemical fragments of DNA and RNA likewise can be found in meteorites. For example, some meteorites have been reported to contain small amounts of adenine, one of the nucleobases found in RNA and DNA. The current view is that Murchison contained adenine, guanine, and their hydrolysis products hypoxanthine and xanthine, as well as uracil. The reported concentration of adenine, however, is low, ~1.3 ppm. Murchison and other meteorites may also contain ribitol and ribonic acid, respectively the reduced and oxidized forms of ribose (Cooper *et al.*, 2001).

TABLE 7.3 Enantiomeric enrichments (EE) for amino acids in two meteorites (Pizzarello and Cronin, 2000). Column 2 refers to the natural abundance of the amino acids in Earth's biosphere

Meteorite sample compound	on Earth	Murchison EE (%)	Murray EE (%)
2-amino-2,3-dimethyl-pentanoic acid			
2 S,3S/2R,3R	unknown	7.6	1.0
2 S,3S/2R,3S	unknown	9.2	2.2
α-methylnorleucine	unknown	4.4	1.8
α-methylnorvaline	unknown	2.8	1.4
α-methylvaline	unknown	2.8	1.0
Isovaline	rare	8.4	6.0
Norvaline	rare	0.4	0.8
α-amino-n-butyric acid	common	0.4	−0.4
Valine	ubiquitous	2.2	−0.4
Alanine	ubiquitous	1.2	0.4

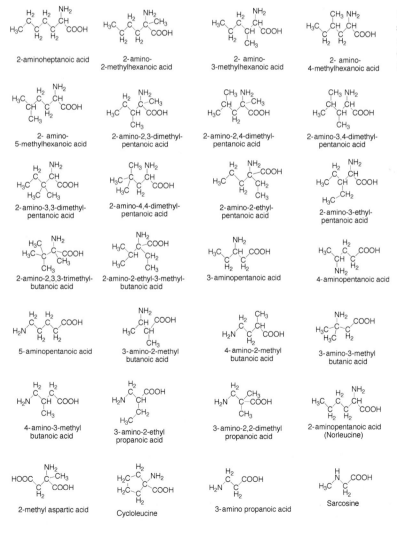

2-aminoheptanoic acid

2-amino-2-methylhexanoic acid

2- amino-3-methylhexanoic acid

2- amino-4-methylhexanoic acid

2- amino-5-methylhexanoic acid

2-amino-2,3-dimethyl-pentanoic acid

2-amino-2,4-dimethyl-pentanoic acid

2-amino-3,4-dimethyl-pentanoic acid

2-amino-3,3-dimethyl-pentanoic acid

2-amino-4,4-dimethyl-pentanoic acid

2-amino-2-ethyl-pentanoic acid

2-amino-3-ethyl-pentanoic acid

2-amino-2,3,3-trimethyl-butanoic acid

2-amino-2-ethyl-3-methyl-butanoic acid

3-aminopentanoic acid

4-aminopentanoic acid

5-aminopentanoic acid

3-amino-2-methyl butanoic acid

4- amino-2-methyl butanoic acid

3-amino-3-methyl butanic acid

4-amino-3-methyl butanoic acid

3-amino-2-ethyl propanoic acid

3-amino-2,2-dimethyl propanoic acid

2-aminopentanoic acid (Norleucine)

2-methyl aspartic acid

Cycloleucine

3-amino propanoic acid

Sarcosine

FIGURE 7.12 Some of the 70 amino acids reported in the Murchison meteorite (Kvenvold *et al.*, 1970; Cronin and Moore, 1971; Cronin and Pizzarello, 1986).

TABLE 7.4 The organic content of the Tagish Lake meteorite (Pizzarello *et al.*, 2001)

Aliphatic hydrocarbons 5 ppm	Dicarboximides 5.5 ppm
Aromatic hydrocarbons ≥ 1 ppm	Sulfonic acids ≥ 20.0 ppm
Dicarboxylic acids 17.5 ppm	Amino acids < 0.1 ppm
Carboxylic acids 40 ppm	Amines < 0.1 ppm
Pyridine carboxylic acids 7.5 ppm	Amides < 0.1 ppm

TABLE 7.5 Organic compounds identified in tholin mixtures (Sagan *et al.*, 1978; Sagan and Khare, 1979; Pietrogrande *et al.*, 2001)

Hydrogen sulfide	Hexene	Formamide
Hydrogen cyanide	Heptene	Pyridine
Ammonia	Butadiene	Styrene
Ethane	Benzene	2,3-pentadiene
Propane	Toluene	2-methylpyrimidine
Butane	Thiophene	4-methylpyrimidine
Ethene	2-methyltiophene	3-butenenitrile
Propene	Methylmercaptan	Butyne
Butene	Ethylmercaptan	Acetonitrile
Pentene	Propylmercaptan	Carbon dioxide
Carbon disulfide	Methylisocyanate	Acetamide

Phosphorus is an important component of terran life. It is abundant on the Earth, both as an element (the eleventh most abundant atom in the Earth's crust) and as phosphate. Meteorites hold a variety of phosphate-containing minerals, and some phosphide minerals (Moore, 1971). Unfortunately, a clear prebiotic pathway for the chemical incorporation of phosphate into RNA or DNA has not yet been found.

No *nucleosides* (a nucleobase joined to a sugar), have been reported from meteorites. Nor has evidence been found in any meteorite for the joining of *nucleotides* (a nucleoside attached to a phosphate), as is necessary for RNA or DNA.

7.4.2 Organics from the interstellar medium

The interstellar medium is an alternative source of organic molecules that may have been delivered to the primitive Earth. Organic molecules are assembled from the elements in interstellar space, without the need for a planet (Section 3.8.1 discusses this chemistry). The inventory of organic molecules found in interstellar space is now at 120 species and is illustrated in part in Fig. 7.13. These molecules include formaldehyde, cyanide, acetaldehyde, water, and ammonia, all of which we found useful above when forming amino acids. Glycine, the simplest amino acid, has also been detected in the interstellar medium.

7.4.3 Laboratory simulations

In a number of cases, laboratory environments designed to model conditions on early Earth have generated amino acids and/or components of nucleosides from species known or suspected to be present in the cosmos. For nucleic acids, work by Oró and Orgel showed that the key nucleobase adenine can

be prepared from hydrogen cyanide, as discussed above (Fig. 7.8) (Oró, 1960; Sanchez, Ferris, and Orgel, 1967).

Stanley Miller, working in Harold Urey's laboratory in 1953, made the first conscious attempt to model organic synthesis under prebiotic conditions. Amino acids were found to be generated after electrical discharges from electrodes were passed through an atmosphere of hydrogen, methane, and ammonia over water. These amino acids presumably arose from compounds generated in the discharge that later self-assembled in the water (Miller, 1955). The electrical discharge was necessary simply to generate the formaldehyde and cyanide needed as starting materials for the synthesis. Once these precursors were formed, the synthesis of amino acids occurred without the need for any energy at all, presumably via the Strecker synthesis.

The simulated atmosphere chosen by Miller for his laboratory experiments was considered at the time to approximate the atmosphere of early Earth, but today many models hold that the amount of methane on early Earth was much smaller than that used in the Miller experiments. Instead, the carbon inventory of the early Earth is today modeled as being present largely as carbon dioxide (Kasting, 1993; Section 4.4).

Accordingly, much effort has been devoted to seeking nonbiological syntheses of biomolecules under conditions presumed to better fit those of early Earth. Some success has been achieved, reflecting the general reactivity of organic molecules. In general, when simple carbon-containing compounds are treated with energy

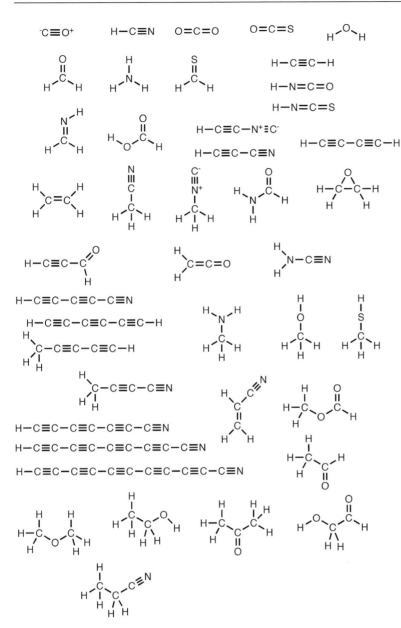

FIGURE 7.13 Some of the 120 organic compounds detected in the interstellar medium, mostly by microwave spectroscopy. (Courtesy Lucy Ziurys.)

sources, they form more complex organic molecules. These energy sources include electrical discharge, ultraviolet light, impact, and ionizing radiation. One example of this is "tholin," a red–brown collection of organic molecules made by irradiating simple molecules (where carbon is present largely as carbon dioxide) (Sagan *et al.*, 1978). A partial inventory of molecules comprising tholins is shown in Table 7.5. Tholins have been proposed as an important part of the atmosphere of Saturn's moon Titan (Chapter 20), and considerable effort is now being made to determine the ratio of carbon to hydrogen to oxygen in an atmosphere that will optimize the yield of organic species when energy is applied.

Analogous work has been extended to simulate interstellar environments (also see Section 3.8.1). For example, Bernstein, Allamandola, and their colleagues recently synthesized the amino acids glycine, alanine, and serine in their laboratory model of icy interstellar grains irradiated by ultraviolet light as an energy source (Bernstein *et al.*, 2002). This suggests that at least some amino acids may have arrived on early Earth as a product of interstellar photochemistry, rather than through formation in liquid water on an early Solar System body.

7.5 Thermodynamic equilibria

Given a source of organic precursors, the question remains: which reactions will occur and in what yields? For this, we must consider thermodynamic properties of molecules.

First, we consider the concept of the *reduction/ oxidation* (or *redox*) state, frequently used to describe organic and other molecules. The ratio of the number of hydrogen atoms in a molecule to the number of non-hydrogen atoms, excluding carbon, determines the redox state of the molecule.

Carbon dioxide, because its carbon is bonded to two oxygen atoms and no hydrogen atoms, is the most oxidized that a carbon atom can be. Methane, where carbon is bonded only to hydrogen, is the most reduced that a carbon atom can be. Formaldehyde is at the same "oxidation level" as elemental carbon (because it has an equal number of bonds to H and O). Viewed alternatively, the ratio of hydrogen atoms (2) to oxygen atoms (1) in formaldehyde is the same as in water (Fig. 7.14). Thus, compounds of the formula $C_n(H_2O)_n$ can be converted to elemental carbon by heating, which extrudes water without a net change in the redox state of the carbons.

At one level, understanding the thermodynamics of carbon-containing molecules with respect to oxidation or reduction is as simple as asking whether hydrogen or oxygen is more abundant in the environment. In the modern terran atmosphere, which contains abundant dioxygen, essentially all compounds containing reduced carbon are thermodynamically unstable with respect to oxidation to carbon dioxide. From a thermodynamic perspective, virtually all organic matter placed in today's atmosphere will eventually "burn" to give carbon dioxide and water. The *rate* of the burning, however, can be very slow in the 20°–40 °C temperature range and at today's amospheric oxygen partial pressure.

In the absence of oxygen and in the presence of H_2, reduced carbon is thermodynamically preferred. This is certainly true deep in the ocean, near hydrothermal vents for example. Here, the synthesis of reduced organic compounds is thermodynamically favored. Shock, Cody, and others have exploited this fact to propose net synthesis of organic molecules in anoxic environments (Chapter 8).

A reaction that is thermodynamically "uphill" (not energetically favored) in one direction can become

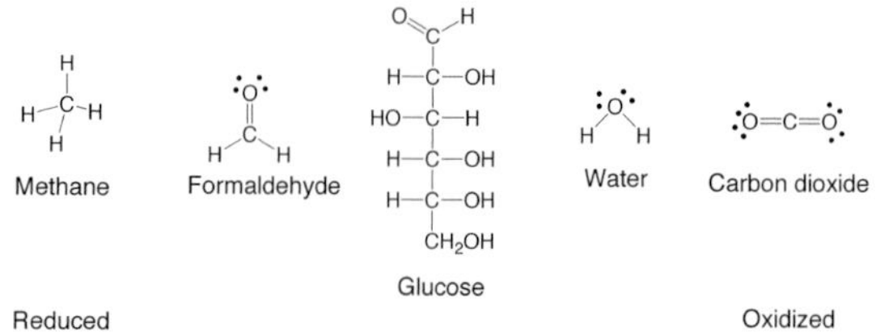

FIGURE 7.14 Redox states. Formaldehyde and glucose have the same oxidation state as elemental carbon, because the ratio of hydrogen to oxygen atoms in both is 2:1, the same as in water. Methane is more reduced (the ratio of hydrogen to oxygen atoms is 4:0). Carbon dioxide is more oxidized (the ratio of hydrogen to oxygen atoms is 0:2).

"downhill" in the same direction if the environmental conditions are changed. If $A + B \Leftrightarrow C + D$, the reaction can be pulled to the right if D is removed, converting all of $A + B$ to C. Conversely, if excess D is added, C will be driven to $A + B$. This behavior of equilibria often appears in textbooks as Le Chatelier's Principle.

It is important to note that no biological compound can ever be said to be universally "high in energy." Students (and sometimes even famous scientists) can easily misunderstand this fact.

Each reaction has a free energy, or ΔG^0, which is defined as $-RT \ln [\text{product}]_{eq}/[\text{reactant}]_{eq}$, where [product]$_{eq}$ and [reactant]$_{eq}$ are the concentrations of product and reactant *at equilibrium*, R is the universal gas constant, and T is the absolute temperature. The value for ΔG^0 can be either positive or negative, but this value does *not* determine whether the reaction runs in the forward or reverse direction under any set of conditions.

Nor is it useful to speak of the "energy" of any particular compound. Rather, the free energy ΔG of a system, which makes a statement about whether it can do thermodynamic work, is determined by the degree to which the system is *out of* equilibrium. This, in turn, is defined by the equation: $\Delta G = \Delta G^0 + RT \ln [\text{product}]/[\text{reactant}]$. Systems where ΔG is less than zero will spontaneously yield more product, until the amount of product rises (and the term $RT \ln [\text{product}]/[\text{reactant}]$) rises, ΔG approaches zero, and the net flux ceases. Conversely, systems where ΔG is greater than zero will spontaneously yield more reactant.

In this context, adenosine triphosphate (ATP), the "currency of energy" in all cells (Chapter 10), is viewed as "high energy" only because at equilibrium the reaction: ATP + water $\Leftrightarrow$ ADP + inorganic phosphate contains more ADP and inorganic phosphate than ATP. If, however, the initial state contains ADP + inorganic phosphate and *no* ATP, the process spontaneously

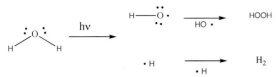

FIGURE 7.15 Possible photodissociation of water on the surface of Mars. Products are hydrogen peroxide (a strong oxidant) and hydrogen gas (which then escapes the planet).

proceeds to the left, i.e., in the direction of the *synthesis* of ATP from ADP and inorganic phosphate. Under these initial conditions, ADP + inorganic phosphate are the "high energy" compounds.

Other generalizations concerning reactivity are based on the principles of thermodynamics. For example, organic molecules contain hydrogen atoms that, given an appropriate catalyst or source of energy (ultraviolet light, for example), might generate H_2. Because H_2 molecules have a low mass compared to other molecules, they move faster on average and therefore preferentially escape from planetary bodies, especially those with low mass and, consequentially, weak gravitational attraction. Although both the formation and loss of H_2 may be slow, cosmic processes have time. A collection of organic molecules thus slowly becomes more oxidized through loss of H_2 (Section 4.2.2 discusses this in detail for the Earth's early atmosphere).

This is presumably what is occurring today on the surface of Mars, whose relatively weak gravity means that the escape of light molecules is easier than on Earth. Above Mars, water is dissociated by ultraviolet light to give H· and ·OH, the hydrogen radical and the hydroxy radical. As illustrated in Fig. 7.15, two H· units can combine to give H_2. The H_2 then escapes from Mars, leaving behind HOOH, hydrogen peroxide. Under typical conditions on Earth, hydrogen peroxide might be viewed as "high energy," but on Mars escape of its reaction partner leads to its formation over time.

For the same reasons, carbon is likely to congeal to high molecular weight polymers as H_2 distills off. In extraterrestrial environments, we thus expect lower hydrocarbons to eventually transform into pure carbon – either diamond (where each carbon is singly bonded to all neighboring carbons), fullerenes and graphite (where each carbon has on average 1.5 bonds to other carbons), or carbon bonded to other elements that cannot be converted to a volatile form.

Polycyclic aromatic hydrocarbons (PAHs) can be viewed as "carbon on the way to forming graphite." These are common in extraterrestrial environments

(Section 3.8.1 and Fig. 3.4). Their central structures are fragments of graphite with bonding to hydrogen atoms at the edges of the structures. These become larger and larger, and more and more like graphite, as more hydrogen distills away.

7.6 Problems in origins and their partial solution

These examples illustrate how concepts of nucleophilicity, electrophilicity, oxidation, and reduction unify processes that might turn prebiotic mixtures into key components of life such as amino acids, nucleobases, and sugars. The Miller experiments are cited in many texts as providing laboratory support to these ideas.

These and other examples over the years have generated a view that the spontaneous emergence of life from inanimate matter in nonbiological environments is easy, in particular in environments that may have been present on the early Earth. In this view, life is a natural consequence of organic chemistry. Organic molecules, if provided with the right kind of energy in the right amounts under the right conditions, will self-organize to give chemical systems capable of Darwinian evolution.

This view, as well as a hint that it might have problems, is reflected by a fictional exchange recorded during a television program between a computer Mnemosyne and the journalist John Hockenberry:[3]

Mnemosyne: Well, chemists think that if you could recreate the conditions of the earth about four and a half billion years ago, you'd see life happen spontaneously. You'd just see DNA ... just pop out of the mix.

John Hockenberry: Can they do that in a laboratory?

Mnemosyne: Well, no. Actually, they've tried, but so far they can't seem to pull it off. In fact chemists have a little joke about that, you know: they say that life is impossible. Experience shows that it can't happen. That we're just imagining it. Ha ha ha ha. . .

Hockenberry: Right. Those chemists. . .

As Mnemosyne noted, the chemists' objection to the notion that life is a natural consequence of organic reactivity is simple, and comes from experience that many individuals have had, perhaps in an organic chemistry laboratory, but more commonly in the kitchen. When one bakes a cake too long, it chars, forming a complex mixture of components with the more volatile species lost. With each additional minute

---

[3] *The DNA Files: Astrobiology* (2001).

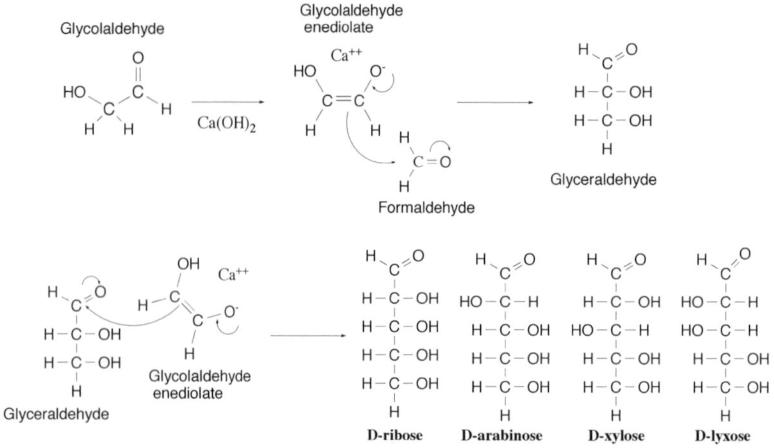

FIGURE 7.16 The formose reaction yields D, L-pentoses (only D-pentoses shown) by combination of the enediolate of glycolaldehyde and glyceraldehyde.

at 350 °F (175 °C), the material resembles less and less something that we might call "living." Similarly, when one applies random energy to mixtures of organic molecules, one gets *tar*, a complex, ill-defined mixture of organic molecules. Robert Shapiro has provided a thoughtful and detailed discussion of these difficulties (Shapiro, 1987; Chapter 6).

From this perspective, existing prebiotic chemistry experiments are not particularly compelling. In the complex chemical mixtures generated under prebiotic conditions, one may indeed be able to find trace amounts of amino acids and perhaps nucleobases. Some of these might indeed catalyze reactions that might have some utility. But other compounds in these types of mixtures may well inhibit this catalysis, or catalyze undesired reactions. For example, Joyce and Orgel (1999) pointed out that the clay-catalyzed condensation of nucleotides to give small chains must have only one enantiomer; if both are present, the desired reactions are inhibited.

Even crystallization, a well-documented method to obtain order through self-organization, is not a particularly powerful way of separating mixtures of organic chemicals into their constituents. Normally, an organic compound must be relatively pure before crystallization occurs. Salts crystallize better, which may explain why crystals are more common in the mineral world than the organic world. But even organic salts can have problems crystallizing from an impure mixture.

These facts generate the central paradox in prebiotic chemistry. Spontaneous self-organization of organic matter is not known to be an intrinsic property of most organic matter, at least as we observe it in the laboratory. Hockenberry asked the correct question of Mnemosyne. Obtaining the chemical order that is

(presumed to be) necessary for life from complex mixtures of organic compounds (such as those presumed to arise from planetary organic chemistry, the type that is occurring on Titan, for example) seems as unlikely to a chemist as a ball rolling uphill seems to a child. It opposes the apparent natural tendency of organic molecules to become tar.

7.6.1 Nucleophilic and electrophilic reactions can destroy as well as create

The same inherent reactivities that generate organic molecules can also convert them into complex mixtures. This is well exemplified by processes that might have generated the sugar ribose, a key component of RNA and DNA, under prebiotic conditions. A reaction known as the *formose reaction* is known to produce ribose by converting formaldehyde in the presence of calcium hydroxide into several sugars, including ribose (Butlerow, 1861; Breslow, 1959; Zubay, 1998).

The formose reaction exploits the natural electrophilicity of formaldehyde and the natural nucleophilicity of the enediolate of glycolaldehyde,[4] a carbohydrate glycoaldehyde that has been detected in interstellar clouds (Hollis, Lovas, and Jewell, 2000). This species reacts as a nucleophile with formaldehyde (acting as an electrophile) to give glyceraldehyde. Reaction of glyceraldehyde with a second equivalent of the enediolate generates a pentose sugar (ribose, arabinose, xylose, or lyxose, depending on stereochemistry). A curved arrow mechanism describes this process as well (Fig. 7.16).

---

[4] The enediolate of glycolaldehyde is a closely related molecule to glycolaldehyde, as illustrated in Fig. 7.16.

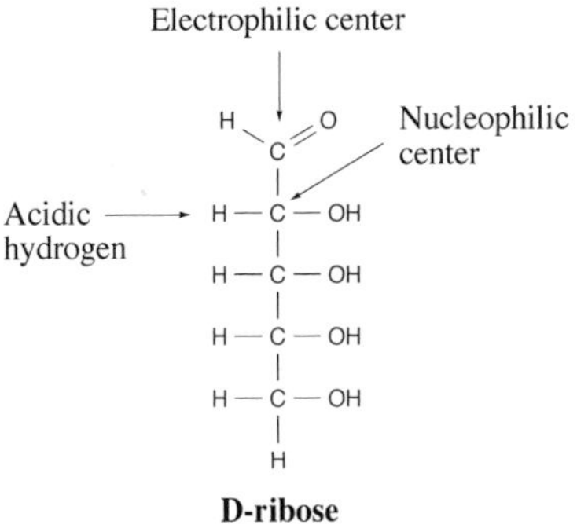

FIGURE 7.17 The open form of ribose contains reactive electrophilic and nucleophilic centers.

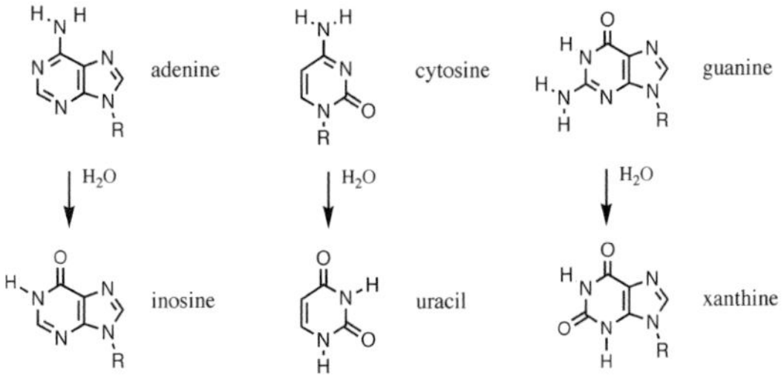

FIGURE 7.18 Degradation pathways for nucleosides through the action of water.

Despite this reactivity inherent to glycolaldehyde and formaldehyde, the formose reaction does not offer a compelling source of prebiotic ribose. Under typical formose conditions, ribose not only forms, but also decomposes. In the presence of calcium hydroxide, ribose is rapidly converted to a brown, complex mixture of organic species. This mixture has never been thoroughly characterized, but does not appear to contain much ribose, and is not an auspicious precursor for life.

The further reaction of ribose in the presence of calcium hydroxide arises because ribose itself has both electrophilic and nucleophilic sites, respectively at the aldehyde carbon and at the carbon directly bonded to the aldehyde (following enolization, Fig. 7.17). Molecules having both reactivities are, as expected, prone to polymerization as the nucleophilic sites and electrophilic sites react with each other, or with more formaldehyde, or with water, or with other electrophiles in the increasingly complex mixture.

These reactivities undoubtedly cause the rapid destruction of the ribose formed under formose conditions. Based on this reactivity, Larralde, Robertson, and Miller (1995) concluded that "ribose and other sugars were not components of the first genetic material."

For these reasons, some have suggested that life may have begun with an alternative organic compound as a genetic material, not RNA, but based on molecules that are less fragile (Schöning et al. 2000; Nielsen, 1999). Underlying this concept is the notion of a "genetic takeover," where delicate RNA and/or DNA molecules arose rather late in the development of life, supplanting a hardier genetic molecule that founded life (Cairns-Smith, 1982).

In short, the reactivity of nucleophilic and electrophilic centers can convert molecules that were plausibly present on early Earth into biologically interesting products. But these products themselves often have nucleophilic and electrophilic centers, and will therefore further react to give uninteresting products. This is the central paradox associated with the origin of life, even given plausible mechanisms to create its components.

7.6.2 Water as a substance antithetical to life

Water creates its own set of problems for prebiotic synthesis. As noted above, water is an essential ingredient in the formation of amino acids from aldehydes and cyanide. Yet many biological molecules, once they are formed in water, are unstable in the water where they formed. For example, the nucleobase adenine (A in the genetic code) spontaneously hydrolyzes in water to give inosine; cytosine (C) hydrolyzes to give uracil (U); and guanine (G) hydrolyzes to give xanthine (Fig. 7.18).

This is also true for polypeptide chains of amino acids. Two amino acids do not spontaneously join in water. Rather, the *opposite* reaction is thermodynamically favored at any plausible concentrations: polypeptide chains spontaneously hydrolyze in water, yielding their constituent amino acids.

The same is true for RNA and DNA. Small chains spontaneously hydrolyze in water to generate individual nucleotides. Thus, even if the building blocks for proteins and nucleic acids are obtained, water and environments rich in water mean that their assembly is thermodynamically uphill.

In this regard, it is remarkable that water is viewed as essential for life. We believe that some liquid is needed for life, as a solvent within which to hold chemical reactions. Certainly, in modern terran life, water is a very useful solvent. It is also clear, however, that water is inimical to the stability of many key biological

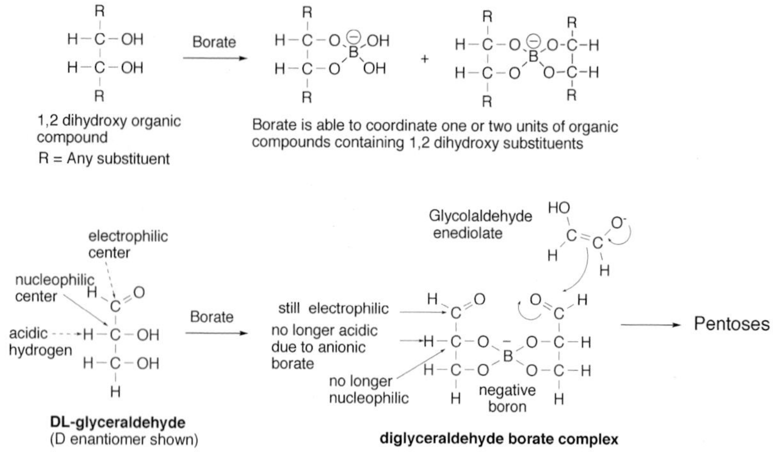

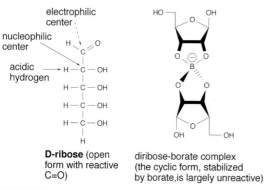

FIGURE 7.19 Borate removes the nucleophilicity of glyceraldehyde (a 1,2-dihydroxy compound) by making an anionic complex. The electrophilic center remains, allowing the addition of the enediol of glycolaldehyde to yield pentoses.

FIGURE 7.20 The cyclic form of ribose forms a stable complex with borate.

polymers. For example, every minute, perhaps ten cytosines in the genome of each human cell suffer spontaneous hydrolysis. These must be repaired to maintain information stored in DNA. It is not easy to see how such repair would occur in the first forms of life.

7.6.3 Minerals as a possible solution to the instability of ribose

To find solutions to individual prebiotic problems, we need to search the full range of chemical reactivity. For example, to make the formose reaction work as a source of ribose that is stable, a prebiotic way is needed to remove the nucleophilicity of glyceraldehyde, and to remove both the nucleophilicity and electrophilicity of ribose.

One promising approach exploits the fact that borate forms complexes with 1,2-dihydroxy units in organic molecules (Fig. 7.19). The borate complex carries a negative charge. This anionic nature of the complex should prevent glyceraldehyde from losing a

proton to create a nucleophilic enolate, but not prevent glyceraldehyde from reacting as an electrophile with the enediolate of glycolaldehyde to generate pentoses.

Further, the 1,2-dihydroxy unit of the cyclic form of ribose should form a stable complex with borate. This will stabilize the cyclic form of ribose at the expense of the aldehyde form. This should render ribose largely unreactive as either a nucleophile or an electrophile, as the cyclic form lacks a C=O carbonyl group, which is the center of this molecule's electrophilicity (Fig. 7.20). Experiments confirm this reasoning. In the presence of $Ca(OH)_2$ under formose conditions at a range of temperatures from 25 to 85 °C, a solution of glycolaldehyde and glyceraldehyde rapidly turns brown, and holds little ribose. However, when the same incubation is done in the presence of the borate-containing mineral colemanite ($Ca_2B_6O_{11} \cdot 5H_2O$), the solution does not turn brown, and one instead finds ribose, together with other five-carbon carbohydrates (Ricardo *et al.*, 2004). Borate has evidently constrained the intrinsic reactivity of glyceraldehyde to form tar, causing it to enter a productive reaction with glycolaldehyde.

Boron is not abundant in the Solar System, although it is known in carbonaceous chondrites, where it is almost certainly present as borate (Zhai and Shaw, 1994). Borate is, however, excluded from many mineral forming processes, appearing mostly in tourmalines, minerals best known as gemstones. Tourmaline weathers to generate borates, which have appreciable solubility in water. As a consequence, colemanite and other borate-containing minerals are delivered to deserts by runoff from the weathering mountains, and then crystallize from water as it evaporates to give *evaporites*. Such conditions are close to what is needed to generate ribose and related sugars.

Curiously, such conditions are also found in the most recent explorations on Mars. Although the instrument package delivered to Mars was not configured to detect either boron or ribose, the ratio of chloride and bromide salts found by the Mars Exploration Rover *Opportunity* is consistent with the formation of evaporite minerals on Mars (Section 18.8). Given the appropriate source rocks, these should contain alkaline borate minerals, analogous to those found in (for example) Death Valley. These are exactly the conditions described above that generate ribose. Indeed, noting that such conditions may not have been present on early Earth, Joe Kirschvink has suggested that terran life originated from ribose formed on Mars.

This synthesis of these carbohydrates in the presence of the mineral colemanite is therefore a plausible prebiotic pathway. Indeed, in the presence of borate, the formation of ribose appears to be a natural consequence of the intrinsic chemical reactivity of compounds available from the interstellar medium under alkaline, calcium-rich conditions.

This example of how minerals can productively control organic reactivity reminds us of the fact that prebiotic chemistry is occurring on a planet, in the context of a larger geology. Minerals must be considered as we constrain models for the origin of life. For example, Orgel and his co-workers have prepared short chains of nucleotides by template-directed polymerization, where clay acts as a catalyst (Ferris *et al.*, 1996). In addition, Szostak and his group have recently shown how clays might have helped the isolation of catalytically interesting biological macromolecules (Hanczyc, Fujikawa, and Szostak, 2003). Going further, Cairns-Smith (1982) has suggested that minerals themselves may have provided the genetic material for early forms of life.

7.6.4 Formamide as a possible solvent

The notion that ribose can be stabilized by minerals found in deserts also suggests an approach to the water problem. Given an activity of water that is sufficiently low, the synthesis of polypeptides from amino acids becomes thermodynamically favorable. But then the absence of water prompts the search for an alternative solvent. Formamide ($HCONH_2$), present in the interstellar medium, is an excellent candidate solvent, being able to dissolve RNA, DNA, and proteins, as well as their precursors. It is created by the reaction of hydrogen cyanide with water following simple rules of nucleophilicity and electrophilicity (Fig. 7.21).

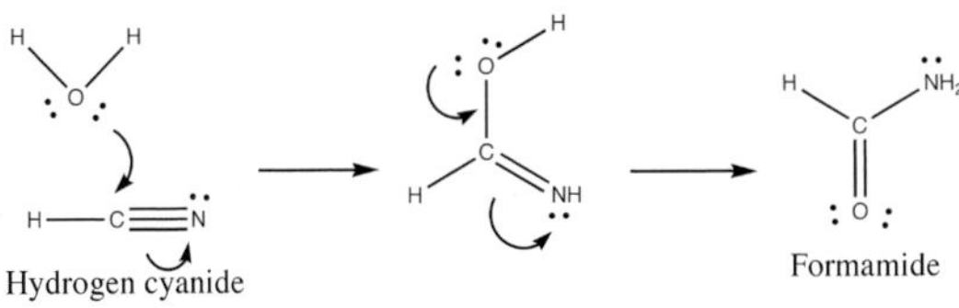

FIGURE 7.21 Formamide is obtained by the hydrolysis of hydrogen cyanide.

In formamide, the synthesis of products through dehydration is thermodynamically favored. Since formamide boils at $\sim$220 °C, a mixture of formamide and water, if placed in the desert, would lose its water over time and end up as a pool of formamide. Within this pool, many syntheses are thermodynamically favorable: polypeptides from amino acids, nucleosides from sugars and bases, nucleotides from nucleosides and inorganic phosphate, and RNA from nucleotides.

7.7 Final thoughts and future directions

The literature contains many theoretical papers that outline possible syntheses of life's organic molecules from possible precursors based on a general knowledge of organic chemical reactivity. A second literature covers experimental work that is presumed to model prebiotic conditions. A third literature objects to the other two because they do not meet a standard of proof that allows one to conclude that such processes actually did generate life, or because such experiments do not generate biological compounds of the correct structure, or chirality, or purity.

This illustrates a defining problem in the field. We are unlikely ever to know the actual events that historically led to the formation of life on Earth. At best, we can simulate conditions in the laboratory believed to have prevailed on early Earth, and try to generate Darwinian molecular systems. Such models will always be open to doubt. There is no reason to insist that the *average* conditions on early Earth were those where life emerged; life could have emerged in a niche with a special, and perhaps unusual, local environment. The environment, or the specific conditions, may have been highly contingent on accident. This has certainly been the case for more recent innovations in the biosphere.

We can, of course, gather more information to constrain such models. The future holds many opportunities to do so. For example, in early 2006 NASA's Stardust mission gathered dust particles from a comet and returned them to Earth for analysis. Comets are believed to contain substantial reservoirs of organic compounds, and are black in color because of this. Analysis of organics in a comet may generate an "Aha!" experience if discovered compounds clarify issues of origins.

Next, the Cassini–Huygens mission is currently studying Titan. This moon of Saturn contains a large reservoir of organic matter, evidently undergoing chemical transformations on a large scale (Chapter 20). Experiments on board Cassini may generate insight into the fate of organic molecules undergoing transformation on a planetary scale.

In addition to going elsewhere with spacecraft, information from Earth is relevant to the opinion that life is a natural consequence of organic reactivity. If structures in 3.5 Ga rocks from Australia and at least some of the carbon-containing materials in 3.8 Ga rocks from Greenland are indeed biogenic, these would indicate that life emerged on Earth "soon" (that is, within a few hundred million years) after the surface of the Earth cooled (see Chapter 12). Rapid emergence of life is consistent with an easy emergence of life, which in turn suggests that one might see life happen spontaneously if one could only reproduce the conditions on early Earth. If, however, these structures and materials are not biogenic, then this argument is weakened.

Another outcome that may emerge from exploratory missions is a better understanding of the nature of early oceans on Earth. It is not yet clear whether the concentrations of salt in Earth's early oceans was high or low (Knauth, 1998). The activity of water is greatly different at the extremes of salinity (Chapter 15), and this has an impact on the constraints imposed on prebiotic chemistry by the nature of water.

Further progress is also needed in understanding the potential on the early Earth for mineral catalysis in the transformation of organic molecules. We are far from being able to construct an inventory of the minerals present then, and know very little about what kinds of reactions minerals catalyze. Current models for the accretion of continents speak of an evolving mineralogy, and it may be that life emerged only when the minerals arose that could catalyze key prebiotic reactions.

References

Amari, S., Anders, E., Virag, A., and Ziner, E. (1990). Interstellar graphite in meteorites. *Nature*, **345**, 238–240.

Bernstein, M. P., Dworkin, J., Sandford, S. A., Cooper, G. W., and Allamandola, L. J. (2002). Racemic amino acids from the ultraviolet photolysis of interstellar ice analogues. *Nature*, **416**, 401–403.

Berry, M. J., Tujebajeva, R. M., Copeland, P. R., *et al.* (2001). Selenocysteine incorporation directed from the 3′UTR: characterization of eukaryotic EFsec and mechanistic implications. *Biofactors*, **14**, 17.

Breslow, R. (1959). On the mechanism of the formose reaction. *Tetrahedron Lett.*, **21**, 22–26.

Butlerow, A. (1861). Bildung einer zuckerartigen Substanz durch Synthese. *Annalen Chemie*, **120**, 295–298.

Cairns-Smith, Alexander (1982). *Genetic Takeover and the Mineral Origins of Life*. Cambridge: Cambridge University Press.

Cooper, G., Novelle, K., Belisle, W., Sarinana, J., Brabham, K., and Garrel, L. (2001). Carbonaceous meteorites as a source of sugar-related organic compounds for the early earth. *Nature*, **414**, 879–884.

Cronin, J. R. and Moore, C. B. (1971). Amino acid analyses of the Murchison, Murray, and Allende carbonaceous chondrites. *Science*, **172**, 1329.

Cronin, J. R. and Pizzarello S. (1986). Amino-acids of the murchison meteorite. III. Seven carbon acyclic primary alpha-amino alkanoic acids. *Geochim. Cosmochim. Acta*, **50** (11), 2419–27.

Cronin, J. R. and Pizzarello, S. (1988). Organic matter in carbonaceous chondrites, planetary satellites, asteroids and comets. In *Meteorites and the Early Solar System*, eds. J. Kerridge and M. S. Matthews, pp. 819–857. Tucson, AZ: The University of Arizona Press.

Ferris, J. P., Hill, A. R. Jr., Liu, R., and Orgel, L. (1996). Synthesis of long prebiotic oligomers on mineral surfaces. *Nature*, **381**, 59–61.

Fuchs, L. H., Olsen, E., and Jensen, K. J. (1973). Mineralogy, crystal chemistry and composition of the Murchison (C2) meteorite. *Smithsonian Contrib. Earth Sci.*, **10**, 1–39.

Grady, M. M., Wright, I. P., Swart, P. K., and Pillinger, C. T. (1988). The carbon and oxygen isotopic composition of meteorite carbonates. *Geochim. Cosmochim. Acta*, **52**, 2855–2866.

Hanczyc, M. M., Fujikawa, S. M., and Szostak, J. W. (2003). Experimental models of primitive cellular compartments: encapsulation, growth, and division. *Science*, **302**, 618–622.

Hao, B., Gong, W. M., Ferguson, T. K., James, C. M., Krzycki, J. A., and Chan, M. K. (2002). A new UAG-encoded residue in the structure of a methanogen methyltransferase. *Science*, **296**, 1462.

Hollis, J. M., Lovas, F. J., and Jewell, P. R. (2000). Interstellar glycolaldehyde: the first sugar. *Astrophys. J.*, **540**, L107–L110.

Jarosevich E. (1971). Chemical analysis of the Murchison meteorite. *Meteoritics*, **6**, 49–52.

Joyce, G. F. and Orgel, L. E. (1999). Prospects for understanding the origin of the RNA world. In *The RNA World*, second edn., eds. R. Gestland, J. Atkins, and T. Cech. Cold Spring Harbor, NY: Cold Spring Harbor Press.

Kasting, J. F. (1993). Earth's early atmosphere. *Science*, **259**, 920–925.

Knauth, L. P. (1998). Salinity history of the Earth's early ocean. *Nature*, **395**, 554–555.

Kvenvold, K., Lawless, J., Pering, K., *et al.* (1970). Evidence for extraterrestrial amino-acids and hydrocarbons in Murchison meteorite. *Nature*, **228**, 923.

Larralde, R., Robertson, M. P., and Miller, S. L. (1995). Rates of decomposition of ribose and other sugars. Implications for chemical evolution. *Proc. Natl. Acad. Sci. USA*, **92**, 8158–8160.

Lewis, R. S., Tang, M., Wacker, J. F., Anders, E., and Steel, E. (1987). Interstellar diamonds in meteorites. *Nature*, **326**, 160–162.

Miller, L. Stanley (1955). Production of some organic compounds under possible primitive earth conditions. *J. Am. Chem. Soc.*, **77**(9), 2351.

Moore, B. Carleton. (1971). Phosphorus. In *Handbook of Elemental Abundances in Meteorites*, ed. Mason, Brian, pp. 131–135. New York, NY: Gordon and Breach, Science Publishers, Inc.

Mueller, D., Pitsch, S., Kittaka, A., Wagner, E., Wintner, C. E., and Eschenmoser, A. (1990). Chemistry of a-aminonitriles. Aldomerization of glycolaldehyde phosphate to rac-hexose 2, 4, 6-triphosphates and (in presence of formaldehyde) rac-pentose 2, 4-diphosphates: rac-allose 2, 4, 6-triphosphate and rac-ribose 2, 4-diphosphate are the main reaction products. *Helv. Chim. Acta*, **73**, 1410–68.

Nielsen, P. E. (1999). Peptide nucleic acid. A molecule with two identities. *Acc. Chem. Res.*, **32**, 624–630.

Oró, J. (1960). Synthesis of adenine from ammonium cyanide. *Biochem. Biophys. Res. Commun.*, **2**, 407–412.

Pietrogrande, M. C., Coll, P., Sternberg, R., Szopa, C., Navarro-Gonzalez, R. Vidal-Madjar, C., and Dondi, F. (2001). Analysis of complex mixtures recovered from space missions. Statistical approach to the study of Titan atmosphere analogues (tholins). *Journal of Chromatography A*, **939**, 69–77.

Pizzarello, S. and Cronin J. R. (2000). Non-racemic amino acids in the Murray and Murchison meteorites. *Geochim. Cosmochim. Acta*, **64** (2), 329–338.

Pizzarello, S., Huang, Y. S., Becker, L., Poreda, R. J., Nieman, R. A., Cooper, G., and Williams, M. (2001). The organic content of the Tagish Lake meteorite. *Science*, **293**, 2236.

Ricardo, A., Carrigan, M. A., Olcott, A. N., and Benner, S. A. (2004). Borate minerals stabilize ribose. *Science*, **303**, 196.

Rother, M., Wilting, R., Commans, S., and Böck, A. (2000). Identification and characterisation of the Selenocysteine-specific Translation Factor SelB from the Archaeon Methanococcus jannaschii. *J. Mol. Biol.*, **299**, 351.

Sagan, C. and Khare, N. B. (1979). Tholins: organic chemistry of interstellar grains and gas. *Nature*, **277**, 102–107.

Sagan, C., Khare, N. B., Bandurski, L. E., and Bartholomew, N. (1978). Ultraviolet-photoproduced organic solids synthesized under simulated jovian conditions: molecular analysis. *Science*, **199**, 1199–1201.

Sanchez, R. A., Ferris, J. P., and Orgel, L. E. (1967). Studies in prebiotic synthesis. II. Synthesis of purine precursors and amino acids from aqueous hydrogen cyanide. *J. Mol. Biol.*, **30**, 223–253.

Schöning, K. U., Scholz, P., Guntha, S., Wu, X., Krishnamurthy, R., and Eschenmoser, A. (2000). Chemical etiology of nucleic acid structure: the α-Threofuranosyl-(3'→2') Oligonucleotide System. *Science*, **290**, 1347–1351.

Shapiro, R. (1987). *Origins: a skeptic's guide to the creation of life on earth.* New York: Bantam Books.

Srinivasan, G., James, C. M., Krzycki, J. A. (2002). Pyrrolysine encoded by UAG in Archaea: charging of a UAG-decoding specialized tRNA. *Science*, **296**, 1459.

Tang, M., Anders, E., Hoppe, P., and Zinner, E. (1989). Meteoritic silicon carbide and its stellar sources: implications for galactical chemical evolution. *Nature*, **339**, 351–354.

Zhai, M. and Shaw, D. M. (1994). Boron cosmochemistry. Part I: Boron in meteorites. *Meteoritics*, **29**, 607–615.

Zubay, G. (1998). Studies on the lead-catalyzed synthesis of aldopentoses. *Orig. Life Evol. Biosphere*, **28**, 13–26.

8 The roots of metabolism

George D. Cody and James H. Scott
Carnegie Institution of Washington

8.1 Introduction to cellular metabolism

While it may be impossible to derive a satisfactory
definition of life in a global or astrobiological context,
we can circumvent such metaphysical questions if we
define life not by what it *is*, but rather by what it *does*.
Every function of life requires *metabolism*, the coupled
chemical processes that generate and exploit biochemical energy. This chapter discusses what can be deduced
from examination of extant organisms about the origin
and early history of life. First, however, in this section
we lay out the basic principles of cellular metabolism.

It is constructive to divide metabolism into functional categories and then consider the various biochemical details. The primary distinction between
metabolic functions is whether they generate biologically useful energy or, instead, use this energy.
Processes that use energy from the environment for
the production of biological energy are referred to as
catabolic metabolic processes. *Anabolic* metabolic processes, on the other hand, use stored biological energy
to do *biosynthesis*, i.e., to synthesize the required building blocks for cellular structure. In any organism, the
coordination (or regulation) of anabolic and catabolic
processes is the essence of cellular metabolism
(Fig. 8.1). At the core of metabolism lies the flow of
carbon within an organism – its chemical pathways
govern all essential cellular function and are the basis
of what is called *intermediary* metabolism.

For biological energy conversion and structural biosynthesis, it is fundamental that life exploits chemical
gradients and thermodynamic potentials that naturally
exist in the external environment. Life does this by
developing sophisticated strategies for accelerating reactions that extract energy from the surrounding chemical
environment faster than the environment naturally
drifts towards thermodynamic equilibrium. For example, if the physiological goal of an organism is to

accumulate molecules having highly energetic triphosphates, then the biochemical synthesis of triphosphates
must occur at a rate far in excess of the natural tendency
of the environment to break them down through combination with water. In order to succeed at this and other
crucial biochemical reactions, life has evolved complex
and highly efficient protein enzymes to gain catalytic
advantages over the surrounding abiotic chemistry.

All life uses energy available in its environment for
growth, replication, and in most cases movement.
Evidence of life may therefore be best sought in observing signs of energy utilization. An excellent example of
this is the chemically stratified water in aquatic ecosystems and sediments, wherein steep gradients of dissolved O_2, nitrate, sulfide, Mn^{2+}, and Fe^{2+} only exist
because of a corresponding stratification of organisms
that use these chemicals. Such stratification is created
by the presence of life and opposes the natural environmental tendency to dissipate such gradients by molecular diffusion (Hutchinson *et al.*, 1939).

All organisms oxidize chemical "fuels" for catabolism.[1] Furthermore, the energetic requirements that
govern catabolic processes are the same as for any
chemical reaction, i.e., the reactions must be spontaneous, meaning that the free energy of the reaction
(ΔG_r, where G is the Gibbs free energy) is less than
zero. Such reactions are referred to as *exergonic* (producing energy), whereas those with ΔG_r greater than
zero are referred to as *endergonic* (requiring energy
to go). Catabolic processes convert environmental
energy into biological energy, often in the form of

---

[1] The *oxidation* of one substance (the oxidant) by another (the fuel or
reductant) refers not just to combining reduced compounds with
oxygen, but to any reaction in which electrons are transferred. The
oxidant is thus also called an *electron acceptor* and the reductant an
electron donor. The opposite reaction to oxidation is called *reduction*,
in which electrons are added. A given pair of oxidation and reduction
reactions is called a *redox couple*.

Planets and Life: The Emerging Science of Astrobiology, eds. Woodruff T. Sullivan, III and John A. Baross. Published by Cambridge University
Press. © Cambridge University Press 2007.

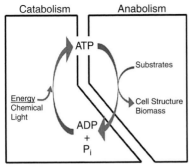

FIGURE 8.1 The catabolic and anabolic metabolism branches. Catabolic processes use external environmental energy to create useful biological energy, stored in the form of ATP (adenosine 5' triphosphate), the essential energetic currency of biochemistry. Anabolic processes use ATP to promote biosynthesis of most of the cellular material necessary for an organism's viability.

adenosine-5'-triphosphate (ATP). ATP is an essential molecule capable of driving multitudes of anabolic synthetic reactions, e.g., combining various monomers to synthesize large molecules such as peptides, polysaccharides, and polynucleotides. Under physiological conditions the reaction of adenosine-5'-diphosphate (ADP) with inorganic phosphate P_i, forming ATP, is highly endergonic. Therefore it is a thermodynamic requirement of any catabolic reaction that the overall change in the free energy resulting from the oxidation of a given fuel be large enough to offset the energy cost of synthesizing ATP.

The essential requirement for ATP synthesis is environmentally available fuels and oxidants. The key to obtaining the most energy is to combine the highest potential oxidant with the lowest potential fuel. Oxygen is the highest potential natural oxidant and organic carbon comprises some of the lowest potential fuels, and thus coupling oxygen with virtually any organic fuel produces significant biological energy. Many organisms, however, are able to thrive utilizing considerably less favorable redox couples, e.g., CO_2 and SO_4 as oxidants and H_2 and organic molecules as fuels.

Organisms are classified according to the types of fuels and oxidants they use. *Heterotrophs* are organisms that use organic compounds as sources of carbon for biosynthesis. Depending on whether O_2 is used as the oxidant further defines the heterotroph as being either *aerobic* or *anaerobic*. Anaerobic heterotrophs often use *fermentation*[2] for energy production and

<hr />

[2] Fermentation refers to metabolic processes that occur in the absence of light and that do not involve respiratory chains that use either oxygen or nitrate as terminal oxidants.

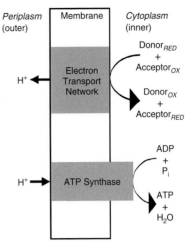

FIGURE 8.2 A schematic depiction of electron transport-coupled phosphorylation (ETP). Electrons generated from the oxidation of a donor and the reduction of an acceptor are coupled with proton movement to the outside, setting up the proton gradient. The proton motive force related to this gradient drives the synthesis of ATP using a membrane-bound enzyme called ATP synthase.

subsequent biosynthesis. *Autotrophs*, on the other hand, are capable of obtaining energy and performing biosynthesis utilizing single-carbon compounds such as CO_2, CO, and formate ($HCOO^-$). Autotrophs may also be either aerobic or anaerobic. Finally, *chemoautotrophs* are organisms whose sole source of energy is through chemical reactions, while *photoautotrophs* use light energy.

Life has developed two fundamental strategies for the production of ATP, depending on the available energy. In environments where high-energy reduced molecules, such as sugars (e.g., glucose), and strong oxidants such as O_2 are available, organisms use a process called substrate level phosphorylation (SLP). In SLP, ATP is synthesized when a phosphoryl group ($-PO_3^-$) is transferred to ADP from high-energy compounds such as phosphenol pyruvate, PEP. For the glycolysis pathway (glucose catabolism), for example, 1 mole of glucose is broken down into 2 moles of PEP, which then produces 2 moles of ATP.

A second means of ATP synthesis by respiration processes involves electron transport-coupled phosphorylation (ETP) (Fig. 8.2). In the ETP process electrons generated from a redox reaction (e.g., O_2 and a reduced organic compound such as succinic acid) are used to generate an electrical potential across the cell membrane, which then drives the phosphorylation reaction. An essential difference between ETP and SLP is that in ETP the molecular machinery responsible for electron transfer and ATP synthesis is located within the cell membrane.

ETP-directed ATP synthesis involves coupling respiration (the transfer of electrons from a donor to an acceptor) through a membrane-bound electron transport chain along a redox gradient. In cells (and mitochondria[3]) the development of an electron gradient enables the movement of protons from the inside of the membrane to the outside, creating a proton gradient across the membrane. The magnitude of the difference in the proton activity from one side of the membrane to the other defines a *proton motive force (pmf)*; alternatively, other cations such as Na^+ or K^+ may be used for this purpose. In prokaryotes, as well as in mitochondria, it is the protomotive force that generates ATP by driving protons back across the membrane through a membrane-bound enzyme called ATP synthase.

From an astrobiological perspective the ETP process is particularly interesting as it provides a source of biological energy (ATP) capable of supporting the complex metabolic strategies of chemoautotrophic anaerobic organisms. Since aerobic and multicellular life may have been derived from initially anaerobic lineages, this suggests that ATP synthesis originated as an ETP (rather than SLP) process.

8.2 The TCA cycle and its evolution

The core of anabolic synthesis in most aerobic organisms and all multicellular life is the oxidative tricarboxylic acid (TCA) cycle (Fig. 8.3) (Kornberg, 1970). Key nodes or branches exist within the cycle that lead outwards toward the synthesis of important biochemicals such as amino acids. Such branch points include oxalacetic acid (to form aspartate and other amino acids), alpha-ketoglutaric acid (to form glutamate), and pyruvic acid (to form alanine, sugars, and many other compounds). In fact, the synthesis of *all* cellular materials can be traced back to reactions originating within the oxidative TCA cycle.

Chemoautotrophic anaerobic prokaryotes face the daunting task of having to synthesize all of their cellular constituents from purely inorganic sources of carbon, nitrogen, and sulfur. Their primary route for *carbon fixation* (acquiring inorganic carbon for biosynthesis) follows one of two paths; the acetyl-CoA pathway (used by a wide array of anaerobic

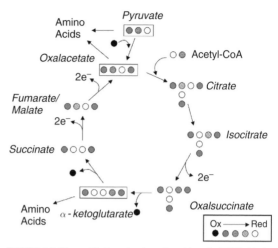

FIGURE 8.3 The oxidative tricarboxylic acid cycle (TCA). Each circle corresponds to a carbon atom contained within one of the metabolic molecules. The relative oxidation state of carbon is depicted by the shade of the circle (with the convention that hydrogen is always +1): black corresponds to the most oxidized form of carbon (C^{4+} as found in CO_2) while white is the most reduced, with carbon's valence charge being −4 for CH_4 and about −2 for CH_2. The arrows depict the clockwise direction of the cycle's reactions. Note that when two electrons are removed, one of the carbons becomes more oxidized; when one CO_2 is removed, a given molecule is reduced by one carbon atom. Boxes highlight key anabolic branch points, i.e., intermediate molecules used for the synthesis of amino acids and other cellular constituents.

chemoautotrophic prokaryotes; also see Section 11.6) and the reductive tricarboxylic acid cycle (rTCA) (used by relatively few). The reductive TCA cycle (Fig. 8.4) superficially looks like the oxidative TCA cycle run backwards, i.e., the rTCA cycle uses biological energy to fix CO_2. But while the metabolic intermediates are identical to those in the oxidative TCA cycle, the specific reactions that occur in the rTCA cycle are more than just the reverse of oxidative TCA reactions and require a specialized set of enzymes. Many of these enzymes contain [Fe_4S_4] clusters (ferredoxins) that serve to store and transport electrons. They thus can operate at very low redox potential and drive reductive anabolic chemistry towards the key anabolic nodes.

There is some evidence that today's ubiquitous TCA cycle may be an evolutionary amalgamation of two separate reaction pathways (the reductive and oxidative branches). Some methanogenic organisms, for example, only use a reductive branch of the pathway, stopping at alpha ketoglutarate synthesis (Fig. 8.5), in which case citric acid is not required for either catabolism or anabolism. Note that this pathway

[3] A mitochondrion is a specialized cellular component, or organelle, that is the site of ATP synthesis in most eukaryotes.

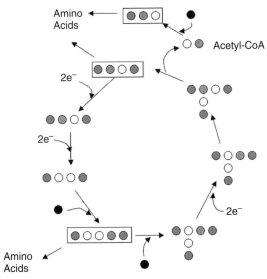

FIGURE 8.4 The reverse or reductive TCA cycle. Note that the direction of reactions is opposite to that of the oxidative TCA cycle in Fig. 8.3. The rTCA cycle fixes CO_2 at the expense of reduction potential.

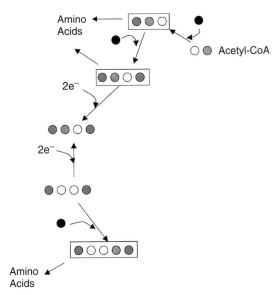

FIGURE 8.5 The acetyl Co-A metabolic path, as used by some methanogenic organisms, follows an incomplete version of the TCA cycle. Carbon fixation again occurs at the expense of reduction potential. Note that not all of the intermediates, e.g., citrate and isocitrate, are required to achieve a metabolic path to the critical anabolic branch points.

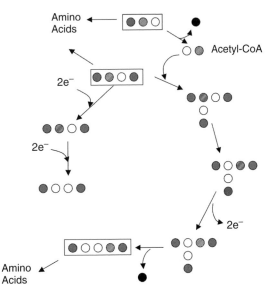

FIGURE 8.6 Fermentive and obligate chemolithoautotrophic organisms such as E. coli use both oxidative (clockwise) and reductive (counterclockwise) branches of an incomplete TCA cycle. In this case one of the anabolic branch points (α-ketoglutarate) is reached via an oxidative path from citrate.

still conserves the critical anabolic branch points. Of the many anabolic synthesis pathways, this acetyl CoA pathway (Fig. 8.5) provides the most direct method to fix CO_2 and CO. The formation of acetate from CO_2 (and/or CO) is carried out by a multi-enzyme complex called the acetyl CoA synthetase (see Section 8.3). Like the TCA cycle, the acetyl CoA pathway is used as both a catabolic or energy-generating pathway and as a route into anabolic synthesis. Considering the paucity of environmental energy available for these chemoautotrophic organisms, the evolutionary development of an abridged TCA cycle for intermediary metabolism is perhaps not surprising.

Other organisms also use an incomplete TCA cycle for intermediary metabolism. E. coli, for example, grown under anaerobic (fermenting) conditions uses both reductive and oxidative chemical strategies to reach key anabolic branch points (Fig. 8.6). A similar strategy is also employed by many chemolithoautotrophic organisms such as sulfate- and nitrate-reducing prokaryotes. The use of both oxidative and reductive branches is often referred to as the "horseshoe," as seen in Fig. 8.6.

In order for the TCA cycle (Fig. 8.3) to have evolved from the "horseshoe" (Fig. 8.6), a critical enzyme, 2-oxoglutarate dehydrogenase, must have evolved and allowed for the connection of the reductive and oxidative branches. The reverse TCA cycle may have similarly originated, but required the development of a completely different set of enzymes.

The apparent similarity and modularity of the intermediary metabolisms manifested by different organisms (Figs. 8.3–8.6) under vastly different environmental conditions is truly remarkable. Notwithstanding the similarity in the various chemical intermediates, however, substantially different enzymes are used to move the chemistry in either an oxidative or reductive sense. From an astrobiological perspective, it is interesting to speculate how the TCA cycle may have evolved. Did it begin in pieces (such as shown in Figs. 8.5 and 8.6) and only close later as an evolutionary response to a changing environment? Did it perhaps happen during the rise of atmospheric oxygen levels about two billion years ago? The evolution of the TCA cycle may provide key insights into how metabolic strategies have evolved in response to changes in the planetary environment. It may also provide insight into how and where the Earth's first organisms arose.

8.3 Two approaches to the emergence of the first metabolic pathways

Although it is not known precisely when life emerged on Earth, it is obvious that it was not present from Earth's beginnings. Consequently, life was either imported to Earth or natural processes intrinsic to the Earth's initial state led to its development. As one could not have proteins without first having amino acids, nor RNA without first having nucleotides, it is clear that the roots of biochemistry lie in abiotic inorganic and organic chemistry that preceded it. The point at which life may be said to begin is when the formation of complex molecules such as proteins and RNA becomes coordinated with the extraction of energy from the surrounding environment.

Researchers have proposed two prominent and opposing hypotheses regarding the stage at which metabolism first appeared along the path to the emergence of life. The first theory was proposed by A. Oparin in 1938 (see Oparin, 1953), wherein metabolic chemistry arose out of the natural chemistry of the primitive Earth (Section 1.14). This hypothesis has been largely superseded by the RNA World hypothesis (Gilbert, 1986) wherein metabolism followed the development of RNA synthesis. The apparent conflict between these theories is outlined in Freeman Dyson's elegant 1985 treatise *Origins of Life*.

8.3.1 Replication preceded metabolism (RNA World)

The notion that replication preceded metabolism grew out of experiments by Manfred Eigen, Leslie Orgel, and numerous colleagues. Eigen and colleagues first showed that RNA could be synthesized without an initial template, although it still required a polymerase enzyme (e.g., Eigen *et al.*, 1981). Inoue and Orgel (1983) demonstrated that RNA could be synthesized using chemically activated nucleotide monomers and an oligonucleotide (a few nucleotides) template *without* the aid of a polymerase enzyme. The essential attraction of RNA, and polynucleotides in general, is derived from their capacity to base-pair with complementary nucleotides with high fidelity. From these experiments and many that followed, the *RNA World hypothesis* for the origin of life developed. Starting from a natural source rich in nucleotides, it is proposed that random polymerization led to the formation of a vast array of initial oligonucleotide templates. These templates had the ability to selectively base-pair with complementary monomers in solution, thus creating a new complementary RNA strand.

But was the new strand capable of base-pairing with monomers in solution to produce a copy of the initial template? Such a reaction scheme is generalized in Fig. 8.7, wherein an RNA strand ultimately produces a copy of itself. In such an *autocatalytic reaction* the product of the reaction is also a reactant; e.g. $A + B \rightarrow C + A$. If such reactions were competing against non-autocatalytic reactions for the same reactant B, they could eventually outcompete non-catalytic reactions by virtue of progressively increasing the concentration of A. Although such chemistry has not yet been demonstrated with monomers, several experiments have shown that small RNA strands can promote autocatalytic replication (Achilles and von Kiedrowski, 1993).

Within the RNA World scenario, RNA catalysis came first and led to metabolism. Once RNA template molecules evolved beyond simple replicators, they became capable of promoting a range of useful chemical reactions. RNA molecules with catalytic qualities are called *ribozymes*, analogous to catalysts for peptides (small chains of amino acids that are building blocks for proteins) being called enzymes. While today most catalytic reactions are promoted by protein enzymes, it is proposed that in the RNA world, RNA must have served as the primary catalyst. Naturally occurring ribozymes are well known (Doudna and Cech, 2001). Elegant experiments (e.g., Bartel and Szostak, 1993) have led to the synthesis of numerous unnatural ribozymes capable of performing a broad range of catalytic functions mimicking enzyme reactions, in some cases performing comparably to similar protein enzymes.

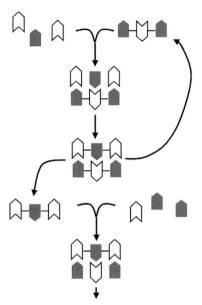

FIGURE 8.7 A schematic representation of a small and simple replicator. The two symbols represent nucleotides that base-pair (through hydrogen bonds) in complementarity (e.g., guanine and cytosine). In the first step, a trimer (three linked nucleotides) base-pairs with complementary nucleotides in solution (indicated by a lack of connecting links). The trimer in this regard acts as a catalytic template: when in close proximity the templated nucleotide monomers will link forming a new, complementary trimer. The new trimer then repeats the first step by base-pairing with additional nucleotides in solution, promoting formation of the initial trimer. At this point the net reaction is autocatalytic in the sense that the final product is a primary reactant.

Within the RNA World two key developments had to occur. First, cellular encapsulation was required for the development of a genetic line, i.e., mutations must have been kept spatially separated from the pool. The second critical development was the emergence of ribozymes used to synthesize proteins at first and later protein enzymes. Once this stage occurred, the threshold to "modern" metabolism would have been crossed and life as we know it would have started.

The key issue is that in the RNA World hypothesis, metabolic chemistry is slave to RNA (and later DNA) evolution. Mutation of a given ribozyme to promote the synthesis of a more efficient protein enzyme imparts to the RNA World proto-organism a greater fitness, i.e., increases its chance for survival and its probability of passing its primitive genetic code to a successive generation. Therein lies the primary attraction of the RNA World hypothesis: in a coarse sense it mimics the essential evolutionary principles found in today's life.

A variation to the RNA World hypothesis is the mineral-based first genetic code (Cairns-Smith, 1985).

However, as this situation would have eventually bridged into the initial stages of RNA World, we will not explore it further here.

8.3.2 Metabolism preceded replication

The second approach to the origin of life hypothesizes that abiotic organic chemistry first led to the development of macromolecules such as peptides and polysaccharides (Oparin, 1953; Section 1.14). The peptides then developed catalytic qualities similar to those of today's protein enzymes, initiating the Earth's first metabolic chemistry. Following encapsulation in a primitive membrane (Section 9.6), an early cellular metabolism evolved. Oparin recognized that individuality based on cellularity was critical to evolution, i.e., Darwinian selection for the fittest metabolic entities. Replication under these circumstances would be governed purely by the physical chemistry dictating the stability of the membrane; if the cell got too large, it split, presumably sharing its chemistry fairly among its progeny.

Dyson (1985) amended the Oparin theory to include the possibility that RNA replication arose from the early use of ATP. Dyson imagined that if the first protein-based organisms consumed ATP for biosynthesis, over time the concentration of the byproduct AMP might reach high levels in the primitive cells. At some point the AMP monomers may have linked up to form self-replicating RNA strands. Cells "infected" by these early replicators may have learned to live with and ultimately benefit from the presence of this useful macromolecule. Such a symbiosis may have then led to the emergence of the genetic code. The essence of Oparin's hypothesis, and of Dyson's modification, is that metabolic chemistry provided for the emergence of the first organisms and that genetic information need not have been initially stored within specific large molecules. Upon the advent of an effective molecule-based genetic code, however, any organism so endowed evidently gained a competitive advantage over strictly physico-chemical organisms. Such advantage presumably resulted from greater fidelity of reproduction of metabolic function, as well as from a more rapid means of evolving improved metabolic function.

An intriguing variation to the "metabolism first" hypothesis has been proposed by Gunther Wächtershäuser (1988a), a patent lawyer by profession, but with an intellectual passion for understanding the roots of life. Wächtershäuser has formulated a detailed theory of how metabolic biochemistry may have begun from purely chemical origins. Specifically, he proposed that the primordial energy source for prebiotic fixation of CO_2 involved the oxidation of the mineral pyrrhotite (FeS) to form pyrite (FeS_2), with the origins of anabolic metabolism arising from this thermodynamic push. Under STP conditions the reduction of CO_2 by H_2 to make formic acid (HCOOH) is energetically unfavorable, while the reaction of FeS with H_2S to form FeS_2 is favorable. Coupling these reactions leads to what Wächtershäuser calls "pyrite pulled" metabolism (Wächtershäuser, 1988a, 1988b):

$$FeS + H_2S \rightarrow FeS_2 + H_2,$$

$$CO_2 + H_2 \rightarrow HCOOH \text{ (formic acid)},$$

and

$$FeS + CO_2 + H_2S \rightarrow FeS_2 + HCOOH.$$

These equations define a very simple anabolic reaction wherein environmental energy in the form of a redox reaction provides the energy for the synthesis of a simple organic molecule (formic acid). In principle this source of energy could be used for a wide range of organosynthetic reactions. Wächtershäuser's theory has come to be referred to as the *iron–sulfur world hypothesis* (Wächtershäuser, 1988a). Central to this scenario is the role of iron sulfide as both a catalyst and source of energy for the earliest life. Gases such as CO_2, CO, COS, H_2S, N_2, and NH_3 could have come from volcanic outgassing.

The attractive aspect of Wächtershäuser's hypothesis is the identification of a reasonable geochemical energy source. Both FeS and FeS_2 are found in abundance in hot spring regions, volcanic vents, and deep-sea hydrothermal vents. One of the crucial questions regarding the first stages of prebiotic chemistry is how to jump-start the anabolic metabolic pathways in the absence of a powerful agent like ATP. Wächtershäuser recognized that within the prebiotic world there needed to be both an initiator, e.g., FeS oxidation, as well as development of chemical "vitalizers," i.e., molecules that operate similarly to ATP in facilitating the synthesis of a broad range of critical biomolecules. Wächtershäuser particularly favors compounds such as thioacids and thioesters as "vitalizers", a point of view shared by de Duve (1991) in his treatise on a "thioester world."

Another interesting aspect of Wächtershäuser's hypothesis is the central role that iron sulfur minerals play in catalyzing the first biochemistry (Wächtershäuser, 1988a). In the iron–sulfur world hypothesis FeS_2 is the catalytic substrate upon which the first metabolic reactions occurred. Recognizing that within a cell most biomolecules are anions (negatively charged), whereas FeS_2 is thought to have a largely positively charged surface, FeS_2 should be able to scavenge dilute solutions for organic anions to promote bimolecular reactions. This is a particularly desirable quality that mimics the action of protein enzymes. The mineral surface increases the effective concentration of potentially reactive substrates and thus drives reactions forward.

Perhaps the most inventive aspect of Wächtershäuser's theory is the proposal that the first life would have consisted of organisms tethered to mineral surfaces by the need for mineral-based catalysis for promotion of metabolic chemistry and separated from the larger environment by a single membrane (Wächtershäuser, 1988a), in essence a two-dimensional proto-organism. The evolutionary step from this organism to simple prokaryotic organisms occurred at the stage that RNA synthesis began, initially using FeS_2 surfaces as templates. Such a development preceded the emergence of a genetic code and of, ultimately, translational synthesis of protein enzymes. The final evolutionary step would involve the two-dimensional organism leaving the mineral surface, taking with it all of the catalytic strategies it had "learned" during the nascent surface-bound stages of its development.

8.3.3 Reconciling metabolism and replication theories

There has been a tendency to consider the RNA world and the Oparin-type theories as mutually exclusive. This view led to a long-standing debate between the so-called "metabolist" and "replication" intellectual camps as to which best described the origins of life. This debate, however, hinged on what one allowed under the rubric of metabolism. Specifically, if the definition of metabolism is limited to translation-directed protein synthesis and DNA- and RNA-based regulation, then some form of the RNA World must be the more correct hypothesis. On the other hand, if metabolism is defined by what is accomplished in a purely chemical sense rather than precisely how this chemistry occurs, then the apparent conflict between the two theories is removed.

Analysis of the Oparin and RNA World hypotheses reveals more a difference in emphasis than a fundamental incompatibility. The RNA World focuses on replication and, in general, pays minimal attention to how the first nucleotides were synthesized. The Oparin hypothesis focuses on chemistry that mimics extant metabolic chemistry and, although acknowledging the eventual emergence of RNA synthesis and the genetic code, places minimal attention on how the RNA world emerged from such chemistry.

De Duve (1991) has provided the most satisfying reconciliation to this metabolist–replication debate with his introduction of the term *proto-metabolism* for a process distinct from metabolism. Proto-metabolism refers to a stage wherein primordial synthetic organic chemistry led to the synthesis of the fundamental monomers used by life, e.g., nucleic acids, sugars, and amino acids. A viable source of such compounds provided an initiation point for an RNA World. Proto-metabolism is not constrained by mechanism, whereas de Duve's *metabolism proper* is. Metabolism proper is defined as today's biochemistry (and any earlier, now extinct manifestations) that exclusively uses protein-based enzymes synthesized in ribosomal factories. With de Duve's general picture, there does not exist any incompatibility between the Oparin view and the RNA World – both address critical stages in the development of life.

8.4 Proto-metabolism and acetyl CoA synthesis[4]

From an astrobiological perspective the probability of a given planet harboring and sustaining life is subject to the probability of a prior stage of proto-metabolic synthesis. The requirements for both life and proto-metabolism are essentially the same: (1) a sustained source of energy driven by chemical disequilibrium originating from either geological processes or sunlight, and (2) the presence of certain elements, in particular transition metals, to provide catalytic potential for synthesis of organics.

It is worthwhile to consider an example where minerals appear to provide catalytic qualities very similar to those of complex protein enzymes. Many of today's microorganisms have evolved efficient catalytic methods to extract energy and synthesize biomolecules exploiting the natural disequilibrium of coexisting CO_2 and H_2 within fluids from the Earth's crust. Certain microorganisms, e.g., methanogenic archaea[5]

and acetogenic bacteria, are true autotrophs in the sense that they synthesize their "food" directly from the reduction of CO_2 by H_2. In order to survive in an environment devoid of sources of food, e.g. sugars, methanogenic archaea use the acetyl-CoA metabolic pathway (Fig. 8.5) both to fix carbon and to synthesize ATP. The first critical step in this metabolic pathway involves the production of a very useful biochemical intermediate, the thioester acetyl-CoA. The synthesis of acetyl-CoA constitutes the first step from which all subsequent anabolic syntheses occur.

The acetyl-CoA synthase enzyme complex has been extensively studied in a variety of organisms, in particular in the acetogenic bacterium *Morella thermoaceticum* (Lindahl *et al.*, 1990; Qiu *et al.*, 1994). Note that, although acetogenic bacteria and methanogenic archaea lie within different domains of the prokaryotic phylogenetic tree, considerable similarity has been preserved in the enzyme complexes with which they synthesize acetyl-CoA.

The reaction pathways within the overall acetyl-CoA synthase enzyme complex for the synthesis of acetyl CoA are sketched in Fig. 8.8. The essential feature of the enzyme complex is a pair of reaction branches, one leading to the progressive reduction of CO_2 (ultimately to a transferable methyl group CH_3), and a second branch that promotes the reduction of CO_2. The two reaction branches join at the point where a methyl group is transferred, first to a cobalt atom in a cofactor (cobalamin; an enzymatic cofactor similar to vitamin B_{12}) and ultimately to a nickel atom in an Ni-X-Fe$_4$S$_4$ cluster (where X is believed to be sulfur). The key reaction involves a carbonyl (CO) insertion reaction that occurs on the Ni-X-Fe$_4$S$_4$ cluster buried within the enzyme, yielding the transferable acyl (CH_3CO) group on the nickel atom. Transfer of this acyl group first to a neighboring cysteine (a thiol (SH)-bearing amino acid) and then to the thiol end-group of acetyl CoA-SH completes the catalytic pathway.

There exist various lines of evidence (Peretó *et al.*, 1999) that suggest that the acetyl-CoA pathway (also known as the Wood–Ljundahl pathway) may be more ancient than other extant autotrophic carbon fixation pathways such as the reverse TCA cycle (Fig. 8.4). It is interesting then to consider whether such metabolic chemistry retains any connections to its proto-metabolic roots, i.e., geochemical reactions promoted without peptide enzymes. Certainly, prior to the development of enzyme catalysts, nonbiological, perhaps mineral, catalysts must have played a primary role in carbon fixation. The potential of transition

---

[4] Readers without a biochemistry background may wish to skip this section.

[5] Archaea are single-celled organisms that represent one of the three primary divisions of the Tree of Life.

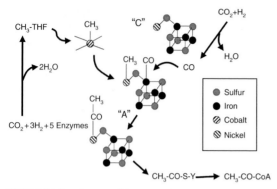

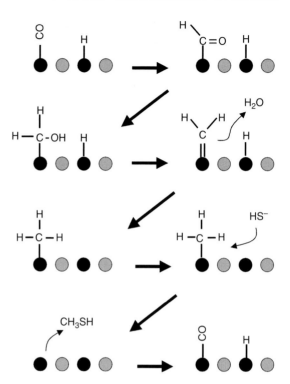

FIGURE 8.8 Initial stages of the acetyl-CoA pathway, highlighting the extensive use of transition metals and sulfur. The left branch sequentially reduces CO_2 to form a methyl (CH_3) group. The methyl group is transferred to a cobalt atom and then to a nickel atom in a $Ni-X-Fe_4S_4$ cluster ("A"). The right reaction branch reduces CO_2 to form carbon monoxide that is transferred to an iron atom in the Fe_4S_4 cluster. Migration and insertion of the carbon monoxide between the methyl group and the nickel atom yields the acetyl group. This acetyl group is ultimately transferred to a biochemical co-factor yielding acetyl-CoA (lower right), providing the initiation point to the acetyl-CoA metabolic pathway (Fig. 8.5).

FIGURE 8.9 The stepwise reduction of carbon monoxide on the surface of a transition metal sulfide mineral. Here the black disks correspond to iron atoms and the gray disks correspond to sulfur atoms. Progressive hydrogen addition to the metal-bound CO group leads ultimately to a methyl group. Attack by bisulfide (HS^-) yields methane thiol, regenerating the catalytic surface for further reaction.

metal sulfides as useful proto-metabolic catalysts has been shown, for example, in experiments performed by Heinen and Lauwers (1996) wherein the reduction of CO led to the formation of methane thiol (CH_3SH) in the presence of the mineral pyrrhotite ($Fe_{1-x}S$) and H_2S at temperatures of $100\,°C$ (Fig. 8.9). In essence they were able to show that transition metal sulfides were capable of promoting chemistry similar to that of the tetrahydrofolate branch of the acetyl-CoA synthesis pathway (Fig. 8.8). In their experiments, the conversion of pyrrhotite to pyrite was presumed to be the source of reducing power, as predicted by Wächtershäuser (1988a) in his iron–sulfur world theory. Methane thiol is a very useful molecule as it provides a facile source of transferable methyl cations.

When comparing proto-metabolic and extant metabolic reactions it is intriguing to note that while acetogenic bacteria extensively use iron–sulfur clusters for various metabolic tasks, they do not directly use transition metals to promote the formation of transferable methyl groups as in Fig. 8.8. Rather, these organisms use a coenzyme called tetrahydrofolate (THF). Methanogenic archaea, on the other hand, use a very similar coenzyme, tetrahydromethanopterin (H_4MPT). Both THF and H_4MPT have the capacity to efficiently reduce oxidized C_1 compounds such as formaldehyde to methylene and methyl groups via a sequential reduction of carbon. During the proto-metabolic stages, chemistry

such as shown in Fig. 8.9 utilizing transition metal sulfides presumably sufficed to produce transferable methyl groups. Extant life, however, evidently required a much higher degree of chemical fidelity, necessitating the development of H_4MPT or THF.

With a source of transferable methyl groups, a major hurdle in a proto-metabolic path mimicking acetyl-CoA was surmounted. In a landmark experiment Huber and Wächtershäuser (1997) converted methane thiol, in a high yield, to acetic acid (also see Section 11.6.3 for discussion of this). Methyl thioacetate (a simple thioester, but a "vitalizer" in the sense that it has the capacity to drive key anabolic reactions; de Duve, 1991) was also formed in the presence of CO, and a Ni- and Fe-containing sulfide precipitate formed from aqueous reactions of Na_2S with $FeSO_4 \cdot 7H_2O$ and $NiSO_4 \cdot 6H_2O$. They proposed that the reaction was catalyzed on the surface of an ordered (Ni,Fe)S precipitate, following a mechanism similar to the methanol carbonylation to acetic acid from CO in the presence of a cobalt catalyst (Cotton and Wilkinson,

1988). It has been subsequently shown that many different transition metal (Fe, Co, Ni, and Cu) sulfides are capable of promoting the essential chemistry catalyzed by the acetyl-CoA synthase enzyme (Cody *et al.*, 2004).

In addition to the acetyl-CoA synthase enzyme complex, transition metals such as Fe, Co, and Ni and others play a critical role in a broad range of metabolic strategies used by chemoautotrophic prokaryotes. For example, molybdenum is involved in enzymes that serve to convert N_2 to ammonia and tungsten is required by some hyperthermophilic prokaryotes for important redox reactions (Howard and Rees, 1996; Kletzin and Adams, 1996). The reason that transition metals play such a prominent role in acetyl-CoA synthase and work so well in a protometabolic sense is due to the high bonding affinity of CO for, particularly, group VIII transition metals. The electronic configurations of CO and transition metals allow for considerable sharing of charge, resulting in a weak bond.

8.5 Proto-metabolism and mineral catalysts on the early Earth

Many of the transition metals used in the active centers of various prokaryotic enzymes have low average abundance in the Earth's crust. If prebiotic carbon fixation (proto-metabolism) arose using minerals with a relatively high abundance of such elements, this provides a clue as to where life may have begun. Surveying likely ancient geological environments, one finds that massive ore bodies provide a favorable source of base metals and, more specifically, magmatic and deep-sea hydrothermal hot springs are sources of both transition metal sulfides and fluids containing H_2 and CO_2.

Both proto-metabolic synthesis and extant chemoautotrophy likely benefited from an essentially continuous source of CO_2 and H_2 derived from the natural chemistry of the fluids emitted from Earth's crust. The source of this bounty lies in iron's natural oxidation state within the lower crust and mantle, which buffers the activity of hydrogen in fluids migrating from the hot depths to the Earth's surface. Such oxidation buffering is governed by mineral–fluid reactions (Huebner, 1971), e.g., the equilibrium of the mineral fayalite (Fe_2SiO_4, containing exclusively ferrous iron, Fe^{2+}) with quartz (SiO_2) and magnetite (Fe_3O_4, containing ferrous and ferric [Fe^{3+}] iron). The activity of hydrogen is fixed via the equilibrium

$$3Fe_2SiO_4 + H_2O \Leftrightarrow 2Fe_3O_4 + 3SiO_2 + H_2.$$

In the natural world fayalite is rarely found, and the buffering capacity is instead provided by a fayalitic component present in the abundant mineral olivine [$(Mg,Fe)_2SiO_4$]. The second critical component, CO_2, is a common volatile constituent in present day volcanic gases and magma. In the early 1990s Everett Shock and colleagues (Shock, 1992) made a series of calculations that demonstrate a strong thermodynamic driving force for protometabolic chemistry arising from the natural thermal evolution of crustally derived fluids. Shock's argument hinged on the concept that below $\sim 400\,°C$ thermodynamic equilibrium would be kinetically inhibited. Thus, although the reaction $CO_2 + 4H_2 \rightarrow CH_4 + 2H_2O$ would be most favored energetically, the reduction of CO_2 by H_2 to form virtually any hydrocarbon in fact significantly reduces the overall free energy of the system.

Another significant source of hydrogen is derived from the aqueous alteration of basalt at relatively low temperatures (e.g., Berndt *et al.*, 1996). Basalt is the principal rock type of the floors of deep oceanic basins and forms at the spreading ridges where upwelling magma is derived from the partial melting of Earth's mantle. Large scale mantle convection drives the oceanic crustal plates apart while magma rises and cools, forming new crust at the spreading center. With continued convection, the newly formed basalt moves away from the spreading center and cools. This new crust is still hot enough to set up hydrothermal circulation cells, wherein cold sea water is drawn down through fractures into the basaltic crust, heated up, and expelled at deep-sea hydrothermal vents. The hot hydrothermal fluids react with the basalt and alter its chemistry. One reaction, the alteration of anhydrous olivine [$(Mg,Fe)_2SiO_4$] to form the hydrous mineral serpentine [$Mg_3Si_2O_5(OH)_4$] generates hydrogen as a byproduct:

$$6[(Mg_{1.5}Fe_{0.5})SiO_4] + 7H_2O \Rightarrow 3[Mg_3Si_2O_5(OH)_4] + Fe_3O_4 + H_2.$$

The chemistry outlined above has been demonstrated experimentally by following long-term, high-temperature reactions using rocks of the appropriate composition and synthetic sea water. Intriguingly, when bicarbonate was added to these experiments, methane, ethane, and propane were synthesized, indicating the catalytic reduction of CO_2 due to the presence of H_2.

In addition to these experiments the above ideas are also confirmed by natural occurrences of abiotic hydrocarbon synthesis detected in fluids associated with ore bodies and deep-sea hydrothermal vents.

Methane and larger hydrocarbons are derived from fluids passing through ancient massive sulfide deposits (e.g., the Kidd Creek mine in Canada); careful analysis of the isotopic composition of these gases indicates a purely abiotic origin (Sherwood Lollar *et al.*, 2002). Similarly, certain deep-sea hydrothermal vents emit large quantities of methane, derived from the reduction of CO_2 during the aqueous alteration of basalt and related rocks (Kelley *et al.*, 2001).

Natural ocean waters also contain SO_4^{2-} that in the presence of H_2 can be readily reduced to form bisulfide, HS^-. High activities of dissolved transition metals and bisulfide lead to the widespread precipitation of transition metal sulfides. Spectacular examples of such sulfide deposits include the black smoker chimneys ubiquitous along the axis and flanks of deep-sea spreading centers.

The natural environment clearly can and does provide both the chemical energy and the catalytic mineral resources to promote abiotic organosynthesis. Even in the absence of life, proto-metabolic synthesis of organics will most certainly occur in all environments, terrestrial or non-terrestrial, that are endowed with reduced carbon-bearing fluids and transition metal sulfides.

8.6 Has primitive metabolic chemistry left us a "chemical fossil"?

Based on the above discussions it is reasonable to consider whether aspects of extant intermediate metabolism constitute a "chemical fossil" of the first organism, as proposed in Wächtershäuser's hypothesis and considered by other theorists (e.g., Morowitz, 1992). The apparent ubiquity of certain TCA cycle intermediates is certainly intriguing. From a purely chemical perspective there is no obvious reason why a greater variety of compounds could not have appeared in the TCA or other metabolic cycles of different organisms; for example, substitution of methyl succinate for succinate. An explanation for the fidelity of the TCA intermediates across species lies in the structure of the catalytic enzymes that drive the anabolic chemistry. Evolution in enzyme function must of course occur in a way that maintains a viable organism; consequently, the emergence of new or improved catalytic functionality must not compromise primary functionality. One of the reasons that protein enzymes operate so efficiently as catalysts is that the catalytic sites are highly shape specific. Such a restriction likely limits the extent to which new molecules may enter intermediary

metabolism. New pathways may develop among a conservative group of molecular species by virtue of evolution in catalytic functionality, but the addition of new organic compounds in intermediary metabolism will be severely inhibited by the topology of the catalytic sites in enzymes. The TCA cycle and its variants owe their similarity, therefore, to shape-selective catalysts derived from some primary set of enzymes, perhaps those of the last common ancestor. Finally, if the first enzymes owe their origins to an RNA World, then intermediary metabolism cannot be traceable back to a proto-metabolic world, rather only as far back as the root ribozymes.

We have seen that there exist intriguing similarities between the chemistry promoted naturally by transition metal sulfides and those directed by the acetyl-CoA synthesis pathway. However, this is a superficial similarity likely based on chemical necessity. In a highly reduced environment with minimal or no available organic substrates, the only chemistry available for creating a cell's carbon is metal-catalyzed reductive carbonylation reactions (Fig. 8.8). Thus, it should not surprise us that abiotic organic chemistry under such circumstances crudely mimics extant metabolism. Although such abiotic chemistry appears similar to that manifested by methanogenic archaea, there remain enormous differences, for example the choice of using THF or H_4MPT for the formation of transferable methyl groups, as opposed to metallic reaction centers.

With extant life one cannot escape the fact that with the onset of the genetic code, organisms developed the ability to adapt to their environment. The chemical similarity between proto-metabolic pathways and extant metabolism lies in the fact that organisms adapt chemical strategies that work best in their environments. The metabolic strategies of organisms within a given environment, therefore, highlight the possible proto-metabolic opportunities that might have existed in that environment prior to the emergence of life.

8.7 Summary

Regardless of one's definition of life, metabolism is the most robust signature of a living system. Throughout life's existence on Earth the natural environment has provided the fuel and oxidants to promote sustained biosynthesis and provide the essential building blocks for the cell's function and reproduction. When considering the prebiotic Earth, whether employing polynucleotides (RNA) or polypeptides (proteins) as the first biocatalysts, the natural environment had to first supply the energy and mineral catalysts to produce the critical organic precursors during the nascent proto-metabolic stages of life. And just as on Earth we see this intricate chemical interplay between the environment and life, in astrobiology we must remember that the same must hold in any extraterrestrial environment where life is to arise and thrive.

All extant organisms exhibit gross similarities in their general pathways within intermediary metabolism. For example, they all use the same anabolic branch points for the synthesis of critical amino acids. Molecular evolution has primarily modified the enzyme catalysts used to promote intermediate metabolic reactions; improvement or broadening of enzyme function has not led to the introduction of new molecules participating in intermediary metabolism. Such consistency in the core of metabolism is expected from the situation where all life evolved from an ancient common ancestor.

In the case of life arising in some extraterrestrial environment, it would not be unreasonable to assume that different molecular constituents might be present in a biochemical equivalent of central metabolism. An early choice of different anabolic nodes would necessarily lead to the biosynthesis of a different suite of amino acids. Such potential differences aside, however, it would not be likely that any form of extraterrrestrial life would utilize non-organic chemical intermediates, such as silicon-based life. All metabolic strategies exploit an essential characteristic of carbon, namely the relative stability of the many allowed oxidation states of carbon atoms within a given molecule. It is this capacity of carbon that allows for the enormous range of chemical structures enabling a proto-organism to develop the range of biosynthetic pathways necessary for biological energy conversion, replication, and motility. If life arose spontaneously in non-terrestrial environments such as Mars, the specific molecular details might well differ from those of extant Earth life, but carbon-based molecules would nevertheless be required to support a sufficiently broad range of catabolic and anabolic strategies to capitalize on that environment's energy resources.[6]

References

Achilles, T. and von Kiedrowski, G. (1993). A self-replicating system from three starting materials. *Angew. Chem. Int. Ed. Engl.*, **32**, 1198–1201.

Bartel, D. P. and Szostak, J. W. (1993). Isolation of new ribozymes from a large pool of random sequences. *Science*, **261**, 1411–1418.

Berndt, M. E., Allen, D. E., and Seyfried Jr., W. E. (1996). Reduction of CO_2 during serpentinizatioin of olivine at 300 °C and 500 Bar. *Geology*, **24**, 351–354.

Cairns-Smith, A. G. (1985). *Seven Clues to the Origin of Life*. Cambridge: Cambridge University Press.

Cody, G. D., Boctor, N. Z., Brandes, J. A., Filley, T. R., Hazen, R. M., and Yoder, H. S. Jr. (2004). Assaying the catalytic potential of transition metal sulfides for abiotic carbon fixation. *Geochim. Cosmochim. Acta.*, **68**, 2185–2196.

Cotton, F. A. and Wilkinson, G. (1988). *Advanced Inorganic Chemistry*. New York: John Wiley & Sons.

De Duve, C. (1991). *Blueprint for a Cell: the Nature and Origin of Life*. Burlington: Portland Press Ltd.

Doudna, J. A. and Cech, T. R. (2001). The chemical repertoire of natural ribozymes. *Nature*, **418**, 222–228.

Dyson, F. (1985). *Origins of Life*. Cambridge: Cambridge University Press.

Eigen, M., Gardiner, W., Schuster, P., and Winckler-Oswatitch, R. (1981). The origin of genetic information. *Sci. Am.*, **244**, 88–118.

Gilbert, W. (1986). The RNA World. *Nature*, **319**, 618.

Heinen, W. and Lauers, A. M. (1996). Organic sulfur compounds resulting from the interaction of iron sulfide, hydrogen sulfide and carbon dioxide in an aerobic aqueous environment. *Orig. Life Evol. Biosphere*, **26**, 131–150.

Huebner, J. S. (1971). Buffering techniques for hydrostatic systems at elevated pressures. In *Research Techniques for High Pressure and High Temperature*, ed. G. C. Ulmer. New York: Springer-Verlag.

Howard, J. B. and Rees, D. C. (1996). Structural basis for nitrogen fixation. *Chem. Rev.*, **96**, 2952–2982.

Huber, C. and Wächtershäuser, G. (1997). Activated acetic acid by carbon fixation on (Fe, Ni)S under primordial conditions. *Science*, **276**, 245–247.

Hutchinson, G. E., Deevey, E. S., and Wollack, A. (1939). The oxidation-reduction potential of lake waters and their ecological significance. *Proc. Nat. Acad. Sci.*, **25**, 87–90.

Inoue, T. and Orgel, L. E. (1983). A nonenzymatic RNA polymerase model. *Science*, **219**, 859–862.

Kelley, D. S., Karson, J. A., Blackman, D. K., *et al.* (2001). An off-axis hydrothermal vent field near the Mid Atlantic ridge at 30° N. *Nature*, **412**, 145–149.

Kletzin, A. and Adams, M. W. W. (1996). Tungsten in biology. *FEMS Microbiol. Rev.*, **18**, 5–64.

Kornberg, H. L. (1970). The role of maintenance of the tricarboxylic acid cycle in *Escherichia coli*. *Biochem. Soc. Symp.*, **30**, 155–171.

[6] See Chapter 27 for a full discussion of possible alien biochemistries.

Lau, N.C., Lim, L.P., Weinstein, E.G., and Bartel, D.P. (2001). An abundant class of tiny RNAs with probable regulatory roles in *Caenorhabditis elegans. Science*, **294**, 858–862.

Lindahl, P.A., Münck, E., and Ragsdale, S.W. (1990). CO dehydrogenase from *Clostridium thermoaceticum. J. Biol. Chem.*, **265**, 3873–3879.

Morowitz, H. (1992). *Beginnings of Cellular Life*. New Haven: Yale Press.

Oparin, A.I. (1953). *Origin of Life*, translated from Russian. New York: Dover.

Peretó, J., Velasco, A.M., Becerra, A., and Lazcano, A. (1999). Comparative biochemistry of CO_2 fixation and the evolution of autotrophy. *Internatl. Microbiol.*, **2**, 3–10.

Qiu, D., Kumar M., Ragsdale, S.W., and Spiro, T.G. (1994). Nature's carbonylation catalyst: Raman spectroscopic evidence that carbon monoxide binds to iron, not nickel, in CO dehydrogenase. *Science*, **264**, 817–819.

Lollar, S., Westgate, B., Ward, T.D., Slater, J.A., and Lacrampe-Couloume, G.F. (2002). Abiogenic formation of alkanes in the Earth's crust as a minor source for global hydrocarbon reservoirs. *Nature*, **416**, 522–524.

Shock, E. (1992). Chemical environments of submarine hydrothermal systems. *Orig. Life Evol. Biosphere*, **22**, 67–107.

Wächtershäuser, G. (1988a). Before enzymes and templates: theory of surface metabolism. *Microbiol. Rev.*, **52**, 452–484.

Wächtershäuser, G. (1988b). Pyrite formation, the first energy source for life: a hypothesis. *System. Appl. Microbiol.*, **10**, 207–210.

FURTHER READING

Abeles, R.H, Frey, P.A., and Jencks, W.P. (1992). *Biochemistry*. Boston: Jones and Bartlett.

Brack, A. (1998). *The Molecular Origins of Life: Assembling the Pieces of the Puzzle*. Cambridge: Cambridge University Press. A broad overview of current ideas on the origin of life.

Fenchel, T., King, G.M., and Blackburn, T.H. (2000). *Bacterial Biogeochemistry: the Ecophysiology of Mineral Cycling*, second edn. London: Academic Press. An extensive overview of the ecology and physiology of key microbial communities that drive elemental cycles on Earth.

Lengeler, J.W., Drews, G., and Schlegel, H.G. (1999). *Biology of Prokaryotes*. Stuttgart: Blackwell Science. A comprehensive, albeit technical, discussion of the biochemistry and ecology of prokaryotes.

Murphy, M.P. and O'Neill, L.A.J. (eds.) (1995). *What is Life?: the Next Fifty Years: Speculations on the Future of Biology*. Cambridge: Cambridge University Press. A superb collection of essays that put in perspective the power of Erwin Schrödinger's lectures (see next entry) and speculate about the future of origin of life studies.

Schrödinger, E. (1992). *What is Life?: the Physical Aspect of the Living Cell*. Cambridge: Cambridge University Press. Later edition of the 1944 classic *What is Life?* (plus other writings by Schrödinger). This slim volume formed the framework for identification of the chemical unit (DNA) responsible for molecular evolution.

9 The origin of cellular life

David W. Deamer
University of California, Santa Cruz

9.1 Introduction

How does life begin? Can life arise elsewhere than the Earth? These questions are among the most fundamental and challenging in all of biology. Charles Darwin once wrote to a friend, "It is mere rubbish, thinking at present of the origin of life; one might as well think of the origin of matter." (Letter to J. D. Hooker, March 29, 1863.) Darwin made this comment when the knowledge required to think about the origin of life and matter simply did not exist. Now, 150 years later, we understand much more. We know that new elements are constantly being synthesized by nuclear fusion of hydrogen and helium in the interiors of stars, then expelled into interstellar space when stars reach the ends of their lives. This matter is the source of new stars and planetary systems, and it is literally true that planets like the Earth and the biogenic elements that give rise to life are composed of "star dust" (Chapter 3). We also know that liquid water once existed on Mars, and perhaps still does beneath the martian surface, suggesting that microbial life may exist elsewhere than on the Earth. Probably most important is that we understand living cells in unprecedented detail, even to the point of knowing the entire sequence of three billion nucleotide bases in the human genome, and we have begun to manipulate the genetic blueprint of life.

Recent scientific advances have provided a more coherent picture of the events leading up to the origin of life. This integrated vision has played an important role in the new field of astrobiology. The basic premise of astrobiology is that life on the Earth is probably only one example of a universal process, so that living systems are expected to arise wherever there are habitable Earth-like planets. The main point to be made in this chapter is that certain kinds of molecules have physical and chemical properties that allow them to self-assemble into orderly structures, and these are the molecules and structures that are likely to be used by life processes anywhere. The self-assembly process seems to defy our intuitive expectation from the laws of physics that everything on average becomes more disordered, or that entropy increases as described by the second law of thermodynamics. An ordinary soap bubble is an example of a self-assembled molecular system. It is easy to calculate that a single soap bubble could never be produced by chance in the lifetime of the Universe, yet they are a common occurrence in accord with the principles of self-assembly.

The origin of life must first be understood in the context of our knowledge of the prebiotic (before life began) environment. For this reason, we will begin with a brief description of the history of our planet and other Earth-like planets in terms of astronomy and planetary science. Other chapters in this book deal with this question in detail, so here we will summarize those properties of the early Earth 4 Ga that are most closely related to life's beginnings. Second, a fundamental characteristic of life is the ability to capture and use free energy available in the environment. Therefore we will discuss in detail sources of energy on the prebiotic Earth and self-assembly of simple membrane structures that can capture that energy. Finally, the origin of cellular life occurred when a membrane boundary structure encapsulated catalytic and replicating macromolecules (large molecules). We will describe laboratory studies of such systems that have one or more functions related to cellular life. Section 9.9 lists the key steps, many already demonstrated, required to achieve a "second origin of life" in the twenty-first century laboratory.

9.2 The early Earth environment

The origin of life on the prebiotic Earth required three essential conditions to be met: liquid water, organic compounds, and free energy. By *free energy* we mean

Planets and Life: The Emerging Science of Astrobiology, eds. Woodruff T. Sullivan, III and John A. Baross. Published by Cambridge University Press. © Cambridge University Press 2007.

energy that is available to do work in a chemical system. For example, a mixture of hydrogen and oxygen gas has a certain amount of free energy, which is explosively released if the mixture is ignited. On the early Earth, sources of free energy included sunlight, geothermal energy (heat), and various forms of chemical energy.

The origin of life is investigated by first proposing a hypothetical scenario that incorporates water, organic compounds, and free energy, then setting up a laboratory experiment to test the hypothesis. But how can we choose which scenarios are worth testing? Because the early Earth was a highly complex geochemical and geophysical system that existed over 4 Ga, such scenarios are often contradictory. Here we must use our scientific judgment. In fact, conflicting ideas are characteristic of any emerging scientific field, and testing the ideas is part of the excitement. Prebiotic scenarios for the origin of life rely on plausibility arguments, very much as circumstantial evidence is argued in a court of law. For the purposes of this chapter, we will judge plausibility according to three basic principles. The first is the principle of *continuity*: those models that most clearly demonstrate a continuous evolutionary pathway leading from proposed origins to extant life forms will be considered to be more plausible than models requiring a discontinuity. The second principle concerns *ubiquity*. In general, ubiquitous conditions will be considered more plausible than special cases, unless the special case is necessary in some fundamental way. The third might be called the principle of *robustness*. A robust model does not depend on precisely defined environmental parameters. Plausible simulations of the origin of life should therefore incorporate relatively common organic compounds that were likely to be present in the prebiotic environment over periods of several hundred million years.

9.2.1 Chronology of the early Earth

The Solar System formed approximately 4.6 Ga, a date taken from radioactive decay rates and isotope ratios measured in meteorites (Stevenson, 1983). The current consensus is that the Earth–Moon system was the result of a collision between a Mars-sized object and the Earth about 4.4 Ga, after which the Moon accreted from the mass of material ejected into orbit during the collision. An important point is that the energy of the collision melted the Earth's crust, so that the lava-like temperatures degraded any organic carbon compounds that were present at that time. By organic carbon we mean chemical compounds of carbon such

as hydrocarbons and amino acids. These compounds are known to be present in comets and meteorites, so it is reasonable to assume that there were similar sources of organic carbon on the early Earth. When organic carbon is degraded by heat, it reverts to inorganic forms of carbon, such as soot, graphite, and carbon dioxide.

The fact that the Earth was molten in its early history means that all of the organic material associated with the origin of life must have accumulated *after* the Moon-forming event. Furthermore, from the lunar cratering record, it is clear that the Moon and the early Earth were bombarded by comets and asteroid-sized objects until about 3.9 Ga. The energy released by the largest impacts would have been sufficient to partially or completely vaporize the ocean and sterilize the Earth's surface (Maher and Stevenson, 1988; Sleep *et al.*, 1989).

After the Earth cooled sufficiently to allow permanent oceans to form, organic molecules could begin to accumulate and produce a variety of more complex self-assembled structures against dispersive thermal effects. Only then could life begin. The oldest geological formations on the Earth are in Canada, about 4.0 Gyr old, and some studies suggest that biological fractionation of stable carbon isotopes was occurring ~3.8 Ga (Mojzsis *et al.*, 1996). The first claimed microfossils appear in sedimentary deposits ~3.5 Ga (Schopf, 1993),[1] suggesting that cellular life with a complete genetic system and translation apparatus for protein synthesis existed at that time.

9.2.2 The atmosphere of the early Earth

The first scientists to consider the gas composition of the early Earth's atmosphere reasoned that the Solar System is mostly hydrogen (Miller and Urey, 1959), and that the atmosphere at that time would be strongly reducing, rather than strongly oxidizing as it is today. That is, the most common biogenic elements – carbon, oxygen, and nitrogen – would be in highly reduced states as CH_4 (methane), H_2O (water), and NH_3 (ammonia).

More recent evidence does not support a strongly reducing atmosphere. Instead, the most abundant components of the early atmosphere were probably CO_2 (carbon dioxide) and N_2 (Holland, 1984; Kasting and Brown, 1998). A carbon dioxide atmosphere is

---

[1] The validity of these microfossils has recently been seriously questioned; see Chapter 12 for a full discussion of this and other evidence for early life.

TABLE 9.1 Mineral salts in seawater relevant to the origin of life

Name	Formula	Concentration	Properties
Sodium chloride	NaCl	0.5 M	Produces osmotic gradients[1]
Calcium and magnesium ions	Ca^{2+}, Mg^{2+}	$\sim$10–20 mM	Interact with anions[2]
Ferrous iron	Fe^{2+}	$\sim$10 mM	Source of reducing power Precipitates anions[3]
Phosphate	HPO_4^{--}	Micromolar	Essential for metabolism[4]
Bicarbonate	HCO_3^-	Millimolar	Buffering agent, forms calcite minerals[5]

[1] Sodium chloride is the most highly concentrated ionic solute in sea water.
[2] Calcium, magnesium, and iron ions can bind to anions such as carboxyl groups and phosphate, thereby precipitating them. This reaction poses a problem for scenarios in which cellular life begins in a marine environment.
[3] Ferrous iron in solution can donate an electron to other solutes if they have a greater oxidation potential. Such reactions are referred to as electron transfer, and are a source of chemical free energy.
[4] Phosphate in contemporary seawater is a limiting nutrient in the micromolar concentration range. A source of phosphate for early cellular life is not yet established.
[5] Carbon dioxide was removed from the early atmosphere by precipitation as calcium carbonate mineral (calcite) to form ocean sediments.

indicated by the fact that the carbonate now present at the Earth's surface as calcium carbonate sediments must once have been present as carbon dioxide gas. If all of this carbonate mineral could somehow be transformed back into carbon dioxide, it would be equivalent to a pressure of $\sim$60 atm (where 1 atm is the pressure of today's Earth atmosphere). A significant fraction of this would presumably have contributed to the Earth's early atmosphere. This idea is given additional support by comparing Earth and Venus. The two planets are similar in mass, and probably underwent similar processes of early planetary evolution. Venus, however, has retained its original carbon dioxide atmosphere, equivalent to $\sim$100 atm. Venus is closer to the Sun, and is relatively hot because of a solar greenhouse effect. Therefore water on Venus remained in the vapor phase, rather than condensing as oceans, and carbon dioxide could not be removed from the atmosphere as calcium carbonate minerals in marine sediments.

Nitrogen on the early Earth would be approximately at the partial pressure of today's atmosphere, and oxygen would have been a rare, short-lived species produced by high-energy reactions and quickly consumed by reduced solutes in the ocean such as ferrous iron. Trace amounts of hydrogen, methane, hydrogen cyanide, formaldehyde, and hydrogen sulfide would have been continuously added by outgassing from the Earth's interior and by photochemical reactions in the atmosphere. This would have produced a modest level of reducing power in localized environments such as hydrothermal vents, geothermal regions, and volcanic

regions. See Chapter 4 for a full discussion of the history of Earth's atmosphere.

9.2.3 Composition of the early ocean

Liquid water is essential for all life today, and the first forms of life must have begun in physically and chemically complex aqueous environments. Possible sites include intertidal zones, lakes and ponds, geothermal springs on volcanic land masses, or submarine hydrothermal vents. A marine site might at first seem plausible because oceans contain most of the Earth's water. Furthermore, the earliest fossil stromatolites apparently developed in marine environments (Schopf, 1993). The ionic solutes in the early ocean were probably similar to those of contemporary seas (see Table 9.1), with dissolved salts present as sodium chloride (at concentrations of $\sim$0.5–1.0 M) and Mg^{++} and Ca^{++} ions ($\sim$10–20 mM). Ferrous iron would have been at much higher concentrations than contemporary values because oxygen had not yet accumulated in the atmosphere. Much of the mineral iron ore today was originally present in the ocean as soluble ferrous chloride, which much later reacted with photosynthetic oxygen to produce ferric oxide (Fe_2O_3). Dissolved carbon dioxide and bicarbonate would also have been relatively abundant because of the high levels of atmospheric CO_2.

Phosphate is a primary limiting nutrient in the present biosphere, and plays a central role in metabolic processes. Phosphate must have been involved in the origin of life, yet a plausible prebiotic source of soluble phosphate has not yet been established. Because of the limited solubility of calcium phosphate salts at the pH of seawater, only a small fraction of crustal phosphate is available for biological reactions. Seawater presently contains about 10^{-6} M phosphate, which suggests that phosphate availability probably constrained chemical evolution, just as it limits biomass production today. (See Gedulin and Arrhenius (1994) for a review of primitive phosphate chemistry.)

At some point the first land masses rose above sea level as volcanic islands resembling today's Hawaii and Iceland, and fresh water became available from precipitation. Fresh water today (lakes, rivers, and aquifers) represents less than 1% of the total water of Earth's hydrosphere, even with extensive continental land masses. It follows that sources of fresh water on the early Earth, with its limited land area, were rare. However, it is significant that some of the self-assembly processes to be discussed later in this chapter are inhibited by the salt content of seawater, particularly by high concentrations of sodium chloride and by cations such as Mg^{++}, Ca^{++} and Fe^{++} in the millimolar range. For this reason we should keep open the possibility that the first forms of life arose in a fresh water environment, then adapted later to seawater by evolving mechanisms to deal with high salt and ion concentrations.

9.2.4 Mineral surfaces, temperature, and organic geochemistry

Mineral surfaces have several important properties related to the origin of life (Table 9.2). First, they provide an interface that can absorb thin films of organic solutes from aqueous solutions and thereby concentrate them. Certain minerals also have catalytic properties which can activate chemical reactions that are otherwise unlikely to occur. Minerals composed of iron and sulfur have the potential to act as oxidants and reductants and thereby provide a source of chemical free energy for primitive metabolic pathways. Assuming that volcanic land masses had emerged from the early ocean, mineral surfaces could then serve as a site for heat-driven reactions that produce polymers.[2] Even today lava surfaces approach 80 °C by solar heating, and geothermal heat can produce much higher temperatures. Dry films of organic material on

TABLE 9.2 Mineral surfaces relevant to the origin of life

Mineral surface	Elemental composition	Properties
Lava minerals	Si, O, Fe	Major mineral surface on early Earth
Apatite	Ca, PO_4	Primary phosphate mineral
Clays	Si, Al, O	Can organize organics in films, catalyze polymerization reactions
Pyrite	FeS_2	Source of reducing power
Calcite	$CaCO_3$	Chiral surfaces
Quartz and glass	SiO_2	Chiral surfaces

such mineral surfaces would be subjected to temperatures sufficient to form covalent chemical bonds through condensation reactions in which water is driven off to produce ester[3] bonds, peptide bonds (bonds between amino acids), and other chemical bonds that link monomers into polymers.

An important property of all life today is that the basic monomers have the property of *chirality*, or "handedness." Just as the right and left hand are mirror images of each other, the molecular structures of carbon compounds with four different chemical groups attached to a central carbon also can occur as mirror images called L (levo-) and D (dextro-) forms. By a yet unknown process, at some point life began to use only one of the two mirror image structures, so that today all amino acids in a living cell are L form and all sugars are D form. The question of how life became chiral is still unanswered, and represents a deep clue related to life's origin. Surprisingly, certain mineral surfaces such as quartz and calcite have chiral surfaces that can preferentially adsorb L or D molecules from racemic (equally L and D forms) mixtures, thereby introducing chiral selection into a system. For instance, Hazen *et al.* (2001) recently demonstrated that a calcite mineral surface can selectively bind L forms of an amino acid. When the amino acid is released from the calcite surface back into solution, it is thus enhanced in the number of L

---

[2] A *polymer* (such as proteins and DNA) is a molecular chain of identical or nearly identical molecules called *monomers*. *Polymerization* is the process of chemically assembling a polymer.

[3] An *ester* is a compound produced by combining an acid with an alcohol (a compound containing one or more OH groups).

molecules. This result suggests that certain minerals may have contributed to the origin of chirality, since the L enhancement would be expected to have become entrained in whatever process led to the first protein-like polymers.

9.2.5 The origin of life at different scales

The scales to be considered in thinking about the origin of life range from global to local to microscopic scales. Each of the three scales has corollaries that must be taken into account. For instance, at the global scale, the ocean becomes a kind of giant test tube and concentrations of organic solutes turn out to be very dilute, in the nanomolar to micromolar range. At such concentrations, self-assembly and polymerization reactions are highly unfavorable. For this reason most investigators assume that a concentrating mechanism was available at a local or microscopic scale.

Virtually any environment can be available at the local scale, many of which provide sources of energy for concentrating solutes and driving chemical reactions. Examples of diverse local environments on the Earth today include Antarctic deserts, tropical tide pools, geothermal subsurface sites, and hydrothermal vents. It is reasonable to assume that a similar variety of sites as available at the time of life's origin.

Microscopic scales are most useful for laboratory models of the processes leading to the origin of life. Micro-environments such as mineral surfaces with adsorbed films of organics and membrane-encapsulated volumes have been used in such work, as will be discussed later. Micro-environments are able to significantly concentrate reactants in the molecular films of organic substances or within the membranes of protocell structures.

9.3 The site of life's origin

We can now use our knowledge of the early Earth to consider possible sites in which life may have begun (Table 9.3). For over a hundred years, beginning with Darwin's "warm little pond" and then followed by Oparin's conjectures (Oparin, 1924; Chapter 1) and Stanley Miller's pioneering experiments (Miller, 1953), it has been thought that life most probably began at the Earth's surface. Assuming that synthesis of organic compounds occurred through chemical reactions in the atmosphere, the resulting organic compounds would have tended to accumulate in bodies of water at the Earth's surface. Furthermore, fluctuating

TABLE 9.3 Geological sites relevant to the origin of life

Site	Properties
Intertidal zone, tidepools, beaches, sand	Fluctuating environment can concentrate organic solutes
Hydrothermal vents	Temperature can range from moderate (40–60 °C) to very hot (>300 °C). Reducing power available.
Subterranean geothermal regions	Temperature can range from moderate (40–60 °C) to >500 °C. Reducing power available.
Fresh water ponds, lakes	Moderate temperature ranges. Low mineral content can be conducive to self-assembly processes.
Ice fields	Organics can be concentrated in eutectics within ice matrix. Low temperature preserves organic compounds.

environments such as ponds and tide pools would provide a mechanism for concentrating dilute solutions of organics, and further drying and heating would create polymers if monomers like amino acids were present.

A second possible site is under the ocean or underground. As previously mentioned, the lunar cratering record suggests that the Earth was subjected to giant impacts of comets and asteroid-sized objects just before the time that the first living organisms appeared. The heat energy associated with such events would have vaporized some or all of the early ocean, in effect sterilizing the entire upper portion of the Earth's surface. The origin of life on the surface could only have occurred after the last such event, a concept referred to as "impact frustration" (Sleep *et al.*, 1989). For this reason, sites for the origin of life have been extended to include other environments that may have served as refuges, in particular submarine hydrothermal vents. Shortly after the discovery of hydrothermal vents, Corliss *et al.* (1981) and Baross and Hoffman (1985) proposed vents as a potential site for the first life on Earth. Stevens and McKinley (1995) reported that hydrothermal regions surprisingly deep in the Earth's crust also support extensive microbial populations. Both hydrothermal vents and deep geothermal regions may have provided a refuge from giant impacts that sterilized the surface of the early Earth. This idea is supported by evidence from molecular phylogeny that strongly suggests that the last common ancestor was likely to have been a thermophilic microorganism (Stetter, 1996; Chapter 14). Hydrothermal vents also provide a potential source of energy in the form of reduced minerals and dissolved gases such as hydrogen sulfide.

A third alternative site was proposed by Bada *et al.* (1994), who noted that models of the Sun's evolution predict the luminosity of the young Sun to have been 20–30% less than today. Unless there was significant greenhouse warming (and many feel that this was the case; see Chapter 4), the early oceans may have frozen to a depth of approximately 300 m, albeit with periodic thaws from the heat of impact events. This proposal addresses the problem that organic solutes in the prebiotic ocean would have been unstable and subject to rapid turnover as a result of ultraviolet radiation and thermal degradation. That is, organic compounds such as amino acids have finite lifetimes in solution, and their lifetimes become very short as temperatures increase. At the highest temperatures associated with hydrothermal vents, amino acids cannot survive at all (Bada *et al.*, 1995). Colder temperatures and a global ice cover would clearly afford significant protection from such degradation reactions. It also seems reasonable that the concentrated mixture of solutes available during thaws could have undergone a burst of chemical reactions leading to more complex molecules, a few of which would be on the evolutionary pathway to life.

Of the three alternative sites proposed for the accumulation of organic compounds pertinent to the origin of life, two are at the Earth's surface, and the third is a subsurface hydrothermal region associated with relatively high temperatures. The surface sites have access to light energy, chemical energy, and concentrating mechanisms, while the subsurface site has access to chemical energy in the form of certain mineral surfaces (pyrite and clays), dissolved gases like hydrogen and methane, and solutes such as ferrous iron. We do not have enough knowledge yet to choose among the three sites proposed for the origin of life. They can never be tested by direct experiment, because the origin of life occurred in an unknown environment over 3.5 Ga in conditions that cannot be reproduced with certainty today. However, if a system of molecules could be assembled in the laboratory that used energy and monomers to grow through polymerization, as well as incorporating some simple catalysts together with an information transfer mechanism resembling today's

genetic code, it would at least provide a plausible scenario for how life began. Furthermore, if the system functioned optimally under conditions associated with one of the sites described above, it would permit an informed choice among the three alternatives. The discovery of such a system is a primary goal of research on the origin of life.

9.4 Organic carbon compounds on the early Earth

The origin of life must have taken place after oceans formed but prior to the appearance of stromatolites, the earliest fossil indicators of cellular activity (Section 12.2.2). During this time, membrane-bounded structures appeared (later to become cell boundaries), simple replicating molecules began to interact with catalytic molecules, and mechanisms evolved to capture light and chemical energy, making it available to drive polymerization reactions. An essential point to understand is that the organic complexity required for life to begin can only occur under conditions in which reduced carbon compounds are available. For this reason we will often refer to *reducing power*, which simply means that a given chemical system has sufficient free energy to add electrons to any carbon compounds that have the ability to accept electrons (called *oxidized* compounds). An example on today's Earth is the reducing power of photosynthesis, in which light energy is captured and used to reduce carbon dioxide to carbohydrates. This free energy is then used to drive virtually all of the chemical reactions that occur in the biosphere. We assume that on the early Earth a source of reduced carbon compounds was available, either from synthesis of organic carbon using free energy in the environment, or from delivery of organic carbon compounds by meteorites and comets.

Table 9.4 summarizes the organic substances that were likely to be available on the prebiotic Earth, both from synthesis on our planet and delivery from outside.

9.4.1 Delivery of extraterrestrial organics

An extraterrestrial source of organic compounds at first might seem unlikely, but in fact all biogenic elements on the Earth (C, H, O, N, P, S) had an extraterrestrial origin, in that they were present in the interstellar medium and the solar nebula that gave rise to the Solar System, then delivered to the Earth's surface in molecular form during accretion. So the question instead concerns how much chemical processing

TABLE 9.4 Summary of organic compounds on the early Earth

Compound	Elemental composition	Properties
Polar compounds (water soluble)		
Cyanide	HCN	Polymerizes to form purines.
Formaldehyde	HCHO	Undergoes formose reaction to produce simple carbohydrates (sugars)
Amino acids	H, C, N, O (S)	Synthesized from HCN and HCHO by Strecker synthesis; monomer of peptides, proteins
Purines	H, C, N, O	Synthesized from HCN; monomer of nucleic acids, with sugar and phosphate
Sugars	H, C, O	Synthesized from formaldehyde; potential source of chemical energy
Nonpolar compounds (water insoluble)		
Amphiphiles	H, C, O	Self-assembly into membranous vesicles
PAH	H, C	Abundant feedstock of organic carbon; potential pigment system

occurred, and where, before the elements were incorporated into the first life forms.

It has long been known that certain stony meteorites called carbonaceous chondrites contain a surprising amount of organic carbon, ranging up to several percent of their mass. Analysis of one of these, the Murchison meteorite, supports the concept that organic material can readily be synthesized in prebiotic conditions. It contains over 1% of its mass as organic material, including amino acids, hydrocarbons, and even traces of purines, one of the basic components of nucleic acids (see Tables 7.1 and 7.2 and their discussion for details). Because carbonaceous meteorites represent samples of the primitive Solar System, it is reasonable to assume that a similar

suite of organic chemicals was also present on the Earth's surface.

As discussed earlier, the consensus today is that the early atmosphere was not a mixture of reduced gases, but instead was composed of carbon dioxide and nitrogen. Carbon dioxide does not support a rich array of synthetic pathways leading to possible monomers, and alternative sources of organic material are therefore being considered. One possibility is that extraterrestrial infall in the form of comets and micrometeorites (interplanetary dust particles) provided significant amounts of organic carbon to the Earth's surface. This was first suggested by Oró (1961) and Delsemme (1984) and more recently elaborated by Anders (1989) and Chyba and Sagan (1992). Estimates of accumulated organic compounds during a late accretion period of 0.1 Gyr ending $\sim$3.9 Ga are in the range of 10^{16}–10^{18} kg (Chyba and Sagan, 1992). Although this is much less than the total organic carbon stored as oil shales, coal, and other fossil deposits on the Earth (10^{21} kg), it is several orders of magnitude greater than the organic carbon now circulating in the biosphere, estimated to be 6×10^{14} kg. See Fig. 3.11 and its discussion for more information on delivery of organics.

9.4.2 Synthesis of organic compounds on Earth

The primary synthetic reactions that have been considered important in the origin of life are summarized in Table 9.5; further details regarding these reactions are in Chapter 7. The classic experiments of Stanley Miller and Harold Urey (1953, 1959) showed that impressive yields of certain amino acids can be obtained when a mixture of gases (hydrogen, methane, ammonia, and water vapor) is exposed to the energy of an electrical discharge; the mixture was assumed to be a simulation of the original terrestrial atmosphere, but we now know that the early atmosphere was not like this (Section 9.2.2). This discovery represented a major breakthrough, since amino acids are the monomers that compose all proteins. At sufficiently high energy fluxes, such reducing systems of gases generate hydrogen cyanide (HCN) and formaldehyde (HCHO), which in turn react to produce amino acids, purines, sugars, and other water-soluble organic compounds.

Although we tend to think of life's chemistry in terms of water-soluble reactants and products, cellular life also requires a source of water-*insoluble* hydrocarbon derivatives that compose the membranes surrounding all cells. Hydrocarbons can be produced

TABLE 9.5 Synthetic organic reactions potentially occurring on the early Earth

===

Gas phase reactions
Reduced gases (hydrogen, methane, ammonia, water) energy (electric discharge, UV light) → cyanide (HCN) and formaldehyde

Reactions producing water soluble products
HCN → purines (example: adenine)
HCHO → simple sugars (glyceraldehyde, ribose, glucose)
HCN + HCHO → amino acids (Strecker synthesis)

Reactions producing water insoluble products (hydro-carbon derivatives)
CO, hydrogen + heat, iron catalyst → hydrocarbons and amphiphiles (long chain acids, alcohols) (Fischer–Tropsch type reaction)

Polymerization reactions
Amino acids + dry heat → peptide bonds (protein-like polymers)
Glyceraldehyde → polyglyceric acid
Purines, pyrimidines, sugars, phosphate (unknown pathway) → nucleic acids

===

under simulated prebiotic conditions by so-called Fischer–Tropsch type reactions, in which a gas mixture like carbon monoxide and hydrogen is passed over a hot metallic catalyst such as metallic iron. Under these conditions numerous hydrocarbon compounds are synthesized, including alkanes, long-chain monocarboxylic acids, and alcohols (McCollom *et al.*, 1999). As will be discussed later, such compounds are essential components of membranes and must have been available on the prebiotic Earth to support the origin of cellular life.

Overall, it seems reasonable that the most basic organic compounds came both from synthesis on Earth and from infall. For instance, amino acids are present only in trace amounts in carbonaceous meteorites, and furthermore are relatively unstable in water. It follows that amino acids and other water soluble organic compounds were probably continuously synthesized at the Earth's surface. On the other hand, hydrocarbon derivatives are relatively stable and compose several percent of the mass of carbonaceous meteorites, yet are not major products of Miller–Urey type reactions. It may be that hydrocarbons required by early life were primarily delivered by meteoritic infall, rather than being synthesized by terrestrial chemistry.

The most important synthetic reaction for the origin of life is polymerization. All life today is based on a few kinds of polymers, and the origin of life required polymer synthesis by a nonbiological reaction mechanism. We can speculate that a variety of energy-dependent processes could produce an assortment of polymers (examples are discussed in Chapter 7), a few of which had the potential to be incorporated into what would become a living system.

9.5 Sources of energy on the early Earth

Energy flow through molecular systems is integral to life's origin since life is an energy-consuming phenomenon in which free energy is used to produce order from disorder. Certain aspects should be clarified before going on, particularly in defining the levels of organization related to energy sources involved in prebiotic evolution. Energy sources on the early Earth can be understood in terms of four levels that are related to the complexity of the system (Fig. 9.1).

The first level is the energy required to drive synthesis of reactive molecules like formaldehyde and hydrogen cyanide, which then have the chemical potential to produce more complex molecules such as amino acids, sugars, and purines. These are relatively high energies, modeled in the laboratory by electrical discharge and ultraviolet light. The second level is the energy required for concentrating dilute solutions of potentially reactive solutes. Concentration could occur by adsorption to mineral surfaces, drying at intertidal zones, or self-assembly processes producing molecular complexes in the form of films and membrane-bound structures. The third level is the input of external energy that would drive polymerization reactions in a non-living assembly of molecules, ultimately increasing the chemical complexity of the system. The final level is the metabolic energy made available by internalized chemical reactions in a cellular micro-environment.

9.5.1 Global energy sources

In considering the sources of energy available in the prebiotic environment, we first ask what energy fluxes at the global scale could have driven the chemical evolution leading to increasingly complex organic molecules. Estimates of such energy sources make it clear that light energy is the most abundant by 2–3 orders of magnitude. However, the *efficiency* with which energy sources can be used for chemical synthesis must also be

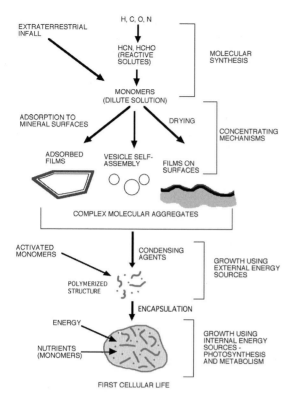

FIGURE 9.1 Four levels of energy sources were available on the early Earth (from top to bottom, labeled on the right-hand side of the diagram). (1) High-energy reactions driven by ultraviolet radiation and electric discharge can synthesize reactive compounds like cyanide and formaldehyde, which later undergo lower energy reactions to produce water soluble substances (amino acids, sugars, purines, pyrimidines). Energy was also required to (2) concentrate dilute solutions for more efficient reactions, and (3) to drive ever-increasing levels of chemical complexity, especially using polymers. (4) At the point that cellular life begins, energy flow becomes internalized in the form of metabolic pathways.

considered (Stribling and Miller, 1987). Other forms of energy, even though less abundant than light, can nevertheless be significant in driving synthetic reactions. These global energy inputs have been incorporated into a variety of schemes that lead to the synthesis of simple organic compounds.

An important example of a primary photochemical pathway is the atmospheric synthesis of formaldehyde, which has been calculated to have been produced at the rate of 10^{11} moles per year (Pinto *et al.*, 1980). To give a perspective on this value, if we assume that synthesis proceeded for 10 Myr and that no degradative reactions occurred, this rate would result in the accumulation of 10^{-3} M formaldehyde in the ocean. But because formaldehyde is highly reactive and unstable, the actual concentrations would likely have been lower.

9.5.2 Energy sources in localized environments

Which energy sources were available for prebiological systems at the local scale and in micro-environments? Possible prebiotic energy sources can be understood in terms of contemporary metabolic pathways, in that they would have driven a network of chemical reactions that could later evolve into metabolic pathways associated with the living state. These hypothetical chemical processes are potentially testable, in that remnants of the network might be found in the metabolic pathways of extant organisms (Morowitz, 1992).

One possibility is that the earliest forms of life had the ability to capture energy by electron transport from donors like hydrogen or hydrogen sulfide to electron acceptors such as iron, either in solution or present on a mineral surface. This would provide an energy source for life existing in an environment such as submarine hydrothermal regions.

Energy may also have been captured by macromolecular structures rather than chemical reactions in solution. This idea can be traced back to Oparin's proposal that colloidal gels (called coacervates) could potentially incorporate chemical reactions that allowed them to grow and reproduce (Oparin *et al.*, 1976). A contemporary version is that membrane compartments were available on the early Earth, which self-assembled from compounds in the environment and later provided a suitable micro-environment for life processes (Deamer, 1997; Deamer *et al.*, 2002). If we assume that such membranes were sufficiently impermeable to ions that electrochemical ion gradients were able to develop, chemiosmotic energy sources would become available even to the earliest forms of life (Koch, 1991). Furthermore, if pigments were available to capture light energy by combining electron transport with a proton pump in the membrane, such systems could become an early form of life.

In the following sections we discuss specific forms of energy that were available to early forms of life, then critically assess whether each could plausibly have been incorporated into a primitive microorganism to drive metabolic processes. The primary forms are easily enumerated for local and micro-environments: heat energy, chemical energy (including oxidation and reduction reactions), and light energy (Table 9.6). We then consider possible sources of chemiosmotic energy across membranes, which raises the question of whether primitive membrane structures were plausibly available on the early Earth.

TABLE 9.6 Summary of energy sources on the early Earth

Energy source	Properties
Electrical discharge	High energy can activate gases, leading to synthesis of reactive compounds such as cyanide and formaldehyde
Light	Ultraviolet light can disrupt chemical bonds; visible light can provide energy to a molecular system if a pigment is available to absorb it
Anhydrous heating	Ready source of energy for condensation reactions that can link monomers into polymers
Chemical bond energy	Basic energy source of metabolic pathways
Oxidation- reduction	Typically requires a membrane to capture energy
Chemiosmotic energy	Requires closed membranes, electron and proton transport

9.5.3 Heating and drying as an energy source

Heating and drying are robust ways to drive the condensation reactions required to produce polymers of simple organic molecules. Dry (*anhydrous*) heating can activate dehydration reactions in which water is lost and a variety of covalent bonds are produced. The chemical potential is provided by the anhydrous condition, so that condensation reactions become favorable as water leaves the reactants. Fox and Harada (1958) first demonstrated that simple heating of amino acid mixtures to temperatures of ~160 °C produced polymeric substances they termed "proteinoids." Under some conditions, the polymers could be induced to form spherical structures referred to as protocells, and Rohlfing (1976) has shown that, given time, similar polymers can form at lower temperatures as well. Even though heating and drying remain an attractive way to introduce polymers of amino acids into the prebiotic environment, it is not easy to imagine how these could evolve into true replicating molecular systems. Nor has it been convincingly demonstrated that in simulated early Earth environments very dilute solutions of amino acids mixed with other organic solutes can assemble into polymeric structures.

Heating and drying have also been used to drive polymerization of nucleotides, the monomers of nucleic acids. For example, Usher (1977) has discussed the ubiquitous nature of heating and drying as a condensation

mechanism, and developed a model based on temperatures found on today's Earth. He demonstrated experimentally that heating could drive phosphodiester bond formation (as in DNA) in which a 12-base nucleotide polymer was produced from 6-base polymers in the presence of a complementary template (Usher and McHale, 1976). It was also shown in this study that the $3'$–$5'$ bonds characteristic of modern nucleic acids are thermodynamically more stable than $2'$–$5'$ under such conditions. There have been few attempts to carry this work further, probably because of the success of other experimental approaches to be discussed later.

Heat can also degrade organic substances, and it has been proposed that volcanic and other high-temperature activity continuously removed organic compounds from the early environment. For example, using reasonable assumptions one finds that the equivalent of the entire global ocean volume passes through deep sea hydrothermal vents every 10 Myr (Bada *et al.*, 1995). The highest temperatures associated with some hydrothermal vents (>300 °C) would presumably degrade all but the most robust organic solutes.

9.5.4 Chemical energy

Many abundant sources of chemical energy were possible in the prebiotic environment. In contemporary cells chemical energy is derived from nutrients and released through metabolism. The argument that early life was similar springs from this line of thinking. For instance, Wächtershäuser (1988a) and Morowitz (1992) proposed that metabolism could have developed from specific reactions that were spontaneous on the early Earth, and only later became entrenched in today's life in a defined series of enzyme-catalyzed reactions. All metabolic pathways today are catalyzed by enzymes that not only speed up the reactions by enormous factors, but also regulate the flow of reactants through the living system. Could a given energy source plausibly drive specific kinds of reactions in a general prebiotic environment containing complex mixtures of dissolved organic substances? Or could life processes only have been initiated in a *specialized* micro-environment with reduced complexity, but also fewer potential reaction pathways?

One example of a model prebiotic system driven by chemical energy involves the reactions of glyceraldehyde. Weber (1987, 1997) has proposed that glyceraldehyde and glyceric acid esters could have been central molecules in the chemical evolution leading up to metabolism. Glyceraldehyde is a relatively simple

TABLE 9.7 Activated monomers

Monomer	Activated form	Example	Reaction	Polymer
Amino acids	Thioamino acids	Thioglutamic acid	Peptide bond formation	Proteins
Nucleotides	Di- and triphosphate	ADP, ATP	Phosphodiester bonds	DNA, RNA
Nucleotides	Imidoesters	Methylimidazolide	Phosphodiester bonds	RNA (synthetic)
Fatty acids	Anhydrides, thioesters	Acyl-coenzyme A	Lipid synthesis	Lipid bilayers[*]

*Not a true polymer, but instead a self-assembled structure of individual molecules.

molecule that may have been synthesized from formaldehyde produced in atmospheric reactions, thus providing a continuous source. Glyceraldehyde, once synthesized, can generate other useful reactants such as glycerol and energy-rich glyceroyl thioesters. Weber has argued that such a pathway could plausibly evolve into modern metabolism. A limitation of this argument is that there is no obvious way to link glyceraldehyde reactions to the primary reactions of life that involve nitrogen and phosphate, such as protein synthesis and replication of macromolecules.

Another form of prebiotic chemical energy includes chemically activated monomers. In this scenario, it is supposed that relatively high energy biomolecules were available to chemically activate monomers in a way that was energetically favorable ("downhill") to produce the key polymers of life. De Duve and Miller (1991) summarized this as follows:

> The pathway to life must have been *downhill* all the way, with at most a few rare humps that could be negotiated with the help of the acquired momentum. One would expect such a roadway to be readily visible. Yet, so far, like some artfully hidden jungle trail, it has eluded every search, despite extensive experimentation and much imaginative theorization and speculation.

Several activated monomers related to the origin of life are summarized in Table 9.7. As one example of a prebiotic synthetic reaction, De Duve suggested that possible activated monomers were the thioesters of amino acids. Synthesis of short chains (peptides) of thio-activated amino acids occurs spontaneously (Maurel and Orgel, 2000) and a pathway to thioesters has been demonstrated by Miller and Schlesinger (1993). It is clear that this process, even in the absence of a catalyst, can produce peptides with a random assortment of sequences. A few of these could have had catalytic properties which then would be preferentially selected if they became incorporated into a more

complex system of molecules involved in primitive metabolic pathways.

The imidoesters of nucleotides are another version of activated monomers. This reaction was pioneered by Lohrmann and Orgel and co-workers (1973, 1976), and has been extensively investigated (Inove and Orgel, 1983; Joyce and Orgel, 1986). Examples of chemically activated monomers include imidazole esters of mononucleotides, with the imidazole acting as a leaving group in a polymerization reaction. In a typical reaction, chemically activated nucleotides are allowed to interact with a template that has favorable base-pairing possibilities. The monomers line up along the template, and energy is released as phosphodiester bonds are synthesized.

Although no plausible pathway to the prebiotic synthesis of imidoesters has been established, this system has provided significant insight into what might be required of a true self-replicating system of molecules. For instance, the polymerization reaction was markedly inhibited if a racemic mixture of activated nucleotides was used, rather than pure D-ribonucleotides (Joyce et al., 1984). This sensitivity offers a strong clue to how chirality originally developed, in that polymerization of chiral compounds would be more efficient than the same reaction with a racemic mixture.

Perhaps the most significant limitation of activated imidoesters is that purine nucleotides readily polymerize on a pyrimidine template, but there has been no experimental success in closing the cycle by demonstrating that pyrimidine nucleotides can polymerize on a purine template (see Orgel, 1987, 2004 for reviews). In other words, the template-dependent polymerization as it stands is a one-step reaction, and could not be incorporated into a full system capable of continuous template-directed growth.

9.5.5 Pyrophosphate bond energy

Adenosine triphosphate (ATP) is the energy currency for all life today (Chapter 10). The chemical energy present in ATP is in the form of *anhydride bonds* between two of the three phosphates. Anhydride is a term used to describe high-energy bonds that are formed when a molecule of water is removed from between two acidic groups. When the acidic groups are phosphoric acid, the bonds are called *pyrophosphate bonds*, from the Greek word for fire. In contemporary cells the pyrophosphate bonds in ATP are synthesized by a complex enzymatic process embedded in the membranes of bacteria, mitochondria, and chloroplasts.

It seems unlikely that a molecule as complex and unstable as ATP was a component of the prebiotic environment. However, phosphate itself must have been present, both in dilute solutions and in the form of minerals such as apatite (calcium phosphate). It is well known that when phosphate dries at temperatures of $>100\,°C$, water is driven off to form anhydride bonds:

Phosphate (PO_4) + dry heat $\rightarrow$ pyrophosphate $(PO_3 - O - PO_3)$ + H_2O.

Pyrophosphate contains the same anhydride bond energy as ATP, but in a much simpler molecule. In fact, at least one pyrophosphate mineral has been described, supporting the possibility that pyrophosphate may have been available in the early Earth environment. Baltscheffsky (1994) has proposed that pyrophosphate bond energy could have been used by primitive forms of life. According to this hypothesis, at some stage this pyrophosphate chemistry became linked with cells we would call living, and the earliest microorganisms thus acquired metabolic pathways involving pyrophosphate bonds as an energy source.

Pyrophosphate as an energy source for early cells also has significant limitations. As noted earlier, phosphate is a trace nutrient in the biosphere, and geological conditions that could have provided a concentrated source of phosphate have not been identified. But even given sufficient phosphate in a local environment, and assuming it was in some fashion continuously formed into pyrophosphate to balance the hydrolytic breakdown that would otherwise remove it, how could the first cellular forms of life have taken advantage of pyrophosphate bond energy? Most phosphate chemistry in cellular metabolism is based on enzyme-catalyzed group transfer or enzyme-catalyzed hydrolysis. In the absence of strong catalysts there is no obvious way that such reactions could be incorporated into primitive forms of life.

9.5.6 Mineral surfaces as sources of chemical energy and molecular order

Research has indicated that mineral surfaces on the early Earth are excellent candidates to have organized organic compounds and catalyzed specific reactions related to life processes. In a typical scenario it is proposed that life began as an organic film on mineral surfaces in subsurface geothermally active sites (Pace, 1991). The film would provide a micro-environment of low water activity so that hydrolytic back reactions would not continuously degrade more complex molecules such as polymers formed by condensation reactions. As an energy source, either dissolved hydrogen gas or the mineral surface itself would provide a source of reducing power. Membrane encapsulation and a system of information transfer would evolve at some later time.

An example of one such mineral surface is that of pyrite ("fool's gold") composed of iron and sulfur (FeS_2). Wächtershäuser (1988a, 1988b) proposed that the first metabolic pathways took place in films of simple organic compounds coating pyrite minerals. In this proposal, such films are expected to undergo a variety of energetically favorable reactions; examples include polymerization reactions, lipid chain synthesis from isoprene derivatives, and fixation of carbon dioxide to formic acid. Pyrite has a number of features that make it pertinent to the origin of metabolic pathways. For instance, it has a cationic surface (positively charged) so that negatively charged organic compounds would presumably be absorbed to its surface through electrostatic interactions. Organic films could then undergo a kind of surface metabolism driven by thermodynamically favorable reactions. For instance, pyrite formation can serve as a source of reducing power in which ferrous iron reacts with hydrogen sulfide to produce pyrite and free electrons. The reaction is energetically highly favorable due to the fact that pyrite is removed from the reaction as a virtually insoluble solid.

An experimental test of this idea was reported by Huber and Wächtershäuser (1997), who showed that a slurry of nickel and iron sulfide was able to promote the formation of acetic acid from carbon monoxide and methyl mercaptan (CH_3SH). The conditions were considered to represent a primordial geothermal system in which metal sulfides at high temperatures $(\sim100\,°C)$ provided a reaction pathway for the initial steps of a primitive autrophic metabolism. This is yet another source of free energy that can drive the formation of simple organic molecules. The chemical potential is contained in the original reactants, rather than being provided by electrical discharge as in the Miller–Urey synthesis, and the process is catalyzed by a mineral surface.

Although this proposal has a number of ingenious features, De Duve and Miller (1991) have discussed a number of shortcomings. One is that the proposed reactions are highly speculative and have only minimal experimental support. They also point out that the thermodynamic aspects are improbable. For instance, only two forms of free energy are proposed to drive the reactions: sulfur oxidation and bonding of anions to charged surfaces. However, no coupling reactions are proposed to link sulfur oxidation to the other reactions. Furthermore, if ionic bonding to pyrite surfaces is an absolute requirement, then even if polymers are synthesized on the surface, one eventually ends up with no more than a strongly bound coat of polymer, an evolutionary dead end. A more detailed discussion of Wächtershäuser's proposed reactions is presented in Chapter 8.

9.5.7 Light as an energy source for primitive cells

After chemical energy, the two most common energy sources used by contemporary microorganisms are (a) light energy and (b) ionic potentials commonly referred to as chemiosmotic energy (discussed in the next section). For light energy to be used, a pigment must exist to capture it. Chemiosmotic energy requires a membrane with sufficient electrical capacitance to maintain electrochemical gradients for a useful amount of time. We first ask whether plausible pigment systems might have been available for photochemistry to be of use.

Light energy was presumably the most abundant source of energy on the prebiotic Earth, as it is today. However, to capture this energy the light must be absorbed by a photochemical process, then transduced into other usable forms of energy rather than being transformed into (unusable) heat or fluorescence. Contemporary photosynthetic pigment systems (such as chlorophyll) use light energy to transfer electrons from water to carbon dioxide, thereby making chemical energy available to the biosphere. As the electrons travel through a complex chain of reactions, energy is conserved and exchanged between electrochemical proton potentials, pyrophosphate bond energy, and "reducing power." This highly evolved reaction

sequence is only just now being understood at the molecular level, and turns out to have very precise structural and chemical requirements. It is unlikely that such a system could have sprung full-blown from the prebiotic mix. Therefore we break the sequence into steps and ask whether plausible prebiotic assemblies of organic compounds might have been capable of a given step. Here is a list of the main steps.

1. Transfer of electrons from a pigment donor molecule to an acceptor molecule.
2. Proton uptake and release, coupled to production of proton gradients across membranes.
3. Hydrolysis of water to protons and oxygen.
4. Generation of an electrical potential (a voltage) across a membrane.
5. Pyrophosphate bond formation, catalyzed by an enzyme.

A variety of models have been proposed for the photochemical reactions pertinent to a primitive photosynthetic system. For example, polycyclic aromatic hydrocarbons (PAHs) are among the most abundant organic components of carbonaceous meteorites, and such compounds absorb light in the blue and near-ultraviolet region of the spectrum. The photochemical reactions that follow this absorption are good examples of what is possible in a simple chemical system plausible for the early Earth. For instance, if pyrene is present in phospholipid bilayers, it can release a hydrated electron upon illumination and reduce benzophenone (Escabi-Perez et al., 1979). This photochemical pathway has clear relevance to possible light energy capture by a primitive pigment system. A photon generates an excited state in a membrane-bound pigment, an electron is released, and an acceptor molecule is reduced. In another relevant PAH reaction, when 1-naphthol is illuminated and shifts to an excited state, its pK shifts dramatically from 9 to 0.5, so that a proton is transiently released (Harris and Selinger, 1983). A third experimental system was established by Warman et al. (1986), who first synthesized a dimethoxynaphthalene compound attached to a hydrocarbon chain with two cyanide groups at the other end. Upon illumination, the excited state of this molecule became a "giant dipole" and generated a voltage across a membrane, similar to what occurs in the initial steps of today's photosynthesis.

To summarize, even though chlorophyll itself could not have been among the pigments available to primitive cellular life, it is not difficult to conceive of other, simpler organic molecules with photochemical properties. Only a few investigators have begun to work on this problem, which will be a very fruitful area for future exploration.

9.5.8 Ionic potentials across membranes

The *chemiosmotic theory*, originally developed by Peter Mitchell (1961), proposes that proton gradients across the membranes of bacteria, mitochondria, and chloroplasts are today the primary source of free energy used to synthesize ATP. The proton gradients are coupled to electron transport enzymes in the membrane, and the energy of the gradient drives other enzymes that produce ATP from ADP and inorganic phosphate.

There are several ways in which early cells could have developed ionic potentials from which energy in the form of a chemiosmotic potential could be derived. Koch (1991) has proposed a detailed hypothetical mechanism by which cells may have generated a chemiosmotic potential and synthesized high-energy phosphate bonds as well. In this scheme, hydrogen is used as a source of electrons, which are donated to an internal acceptor so that charge is separated across the bilayer. In a second reaction, external inorganic phosphate forms a bond with a compound analogous to creatinine which is linked to a hydrocarbon moiety that makes its membrane soluble. The phosphate then picks up protons so that its charges are neutralized, and the positively charged phosphorylated compound is actively transported down the electrochemical gradient to the cell interior. At this point the phosphate is transferred to an acceptor and the neutral molecule diffuses back across the bilayer to begin a new cycle. The phosphorylated acceptor on the interior now serves as a source of high-energy phosphate to drive metabolic reactions and other energy-requiring processes. This pathway has a plausible energy source in the form of molecular hydrogen, which could have been available from a number of geochemical sources. The other components are somewhat less plausible in terms of what we understand about prebiotic chemistry.

9.6 The first cells

The origin of cellular life must have involved chemical and physical mechanisms that permitted systems of molecules to self-assemble from available organic compounds. The life process probably did not begin in a single moment when a tiny portion of inanimate matter suddenly took on the properties of the living state. Instead, life more plausibly arose from countless natural experiments in which mixtures of organic molecules were subjected to energy sources such as light, heat, and electric discharge. Various combinations of the molecules were mixed and recombined into complex interacting systems. Through this process, an imperceptible molecular evolution took place over a period of several hundred million years after the Earth had cooled sufficiently for water vapor to condense into oceans. At some point, membrane-bound systems of molecules appeared that could grow and reproduce by using energy and nutrients from their environment. An observer seeing this end-product would conclude that such systems were alive, but would be unable to point to the exact time when the complex mixtures took on the property of life. This first cell had to have four key properties: (1) encapsulated materials within a membrane-bound structure; (2) the ability to capture energy and nutrients from the local environment; (3) growth through catalyzed polymerization; and (4) replication using an information-storage molecule.

Self-assembly refers to the ability of certain kinds of molecules to form aggregates with novel properties that emerge only in the resulting structure. Two macroscopic examples are the self-assembly of liquid water into highly structured ice crystals during freezing, and the membranous bubbles that are produced by soap molecules in solution. At the molecular level, examples of self-assembly include the double helix formed by complementary strands of DNA, and the manner in which proteins fold into functional structures after synthesis on ribosomes (Table 9.8).

Self-assembly of molecular systems occurs through hydrogen bonding and nonpolar forces that stabilize orderly arrangements of small and large molecules. A central event in the origin of life was the self-assembly of a molecular system in which catalytic polymers could interact with a second class of polymers having the capacity to store information in a sequence of monomers. That sequence in turn would in some manner determine the sequence of monomers in the catalyst, so that the resulting catalytic-information cycle was able to direct growth. In the contemporary cell, the cycle involves protein enzymes and nucleic acids. However, in the protocell, both catalytic and information-containing functions could have been present in the same molecule, as suggested by recent studies of RNA ribozymes (the "RNA World"; see Chapter 8).

Another important self-assembly process involves certain compounds that can form closed, membrane-bounded micro-environments (Section 9.6.1). Such boundary structures can make energy available in the

TABLE 9.8 Examples of self-assembled structures

Compound	Structure	Example
Water	Ice	Snow crystals
Protein strands	Secondary structure	Alpha helix, beta sheet
	Tertiary structure	Protein folding
Nucleic acid strands	Complementary base pairing	Double helix, DNA hairpins
Amphiphilic molecules	Micelles, monolayers, bilayers	Phospholipid vesicles

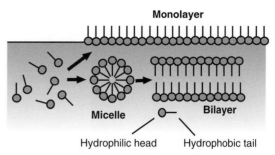

FIGURE 9.2 Self-assembled structures of amphiphiles. Amphiphilic molecules having carbon chain lengths greater than six carbons form micelles and monolayers as concentrations increase above a critical value. At chain lengths of eight or more carbons, bilayers begin to appear in the form of closed-membrane vesicles, which become the dominant structure as concentration and chain lengths increase further.

form of ion gradients, and can provide a selective inward transport of nutrients. Furthermore, enclosed membranes in principle are capable of growth and division, and there exist several possible mechanisms through which the molecular constituents of membranes could have been synthesized from "nutrient" precursors.

If a macromolecular replicating system could be encapsulated within a membrane, the components of the system would share the same micro-environment, and the result would be a major step toward true cellular function. The encapsulation process is not difficult to imagine, since macromolecules have been shown to be readily encapsulated by certain molecules under simulated prebiotic conditions (see the review by Deamer (1997) and Section 9.6.1). An energy-yielding process is also necessary to provide for growth of the encapsulating membrane concurrently with growth of the catalytic/information system.

9.6.1 Self-assembly of closed-membrane vesicles

A hypothetical scenario arising from our knowledge of self-assembly processes is that the origin of cellular life involved the formation of cell-sized compartments, called *vesicles*, from organic compounds available in the prebiotic environment. To understand this approach we need a short tutorial on the self-assembly of amphiphilic molecules. An *amphiphile* is a molecule that has both hydrophilic ("water-loving") and hydrophobic ("water-hating") groups in its structure. The simplest amphiphile is a fatty acid, or soap molecule,

and everyone is familiar with the formation of soap bubbles from such molecules. Other examples of amphiphiles include biological lipids such as phospholipids, cholesterol, and triglyceride (fat).

A universal property of amphiphiles is that they form monomolecular films when present at an air–water interface. It is easy to demonstrate this fact by sprinkling pepper or talcum powder onto a clean water surface, then touching the surface with a bit of soap. The particles of pepper or talcum powder rapidly move away from the soap, driven by the formation of a soap monolayer at the surface. Certain lipids can also assemble into micelles and layers in the aqueous phase, as shown in Fig. 9.2. In a *lipid bilayer*, two layers of molecules are arranged such that hydrophobic tails of the lipid in each layer are directed into the interior of the membrane, and the hydrophilic heads line the two surfaces. All biological membranes consist of lipid bilayers composed of a substance called *phospholipid*, providing a permeability barrier essential for cellular life. If phospholipid is isolated from membranes and dispersed in water, it readily self-assembles into lipid bilayer vesicles referred to as *liposomes*.

In order for cellular life to have begun on the early Earth, is seems inescapable that membranous vesicles resembling liposomes were involved. Such vesicles have several properties that would permit systems of organic molecules to assemble and interact in a manner conducive to prebiotic evolution. The components of lipid bilayer structures are highly concentrated, thereby overcoming the dilution that occurs with water soluble organic species. Vesicles are also capable of maintaining specific groups of macromolecular species within a compartment, which facilitates their interaction and

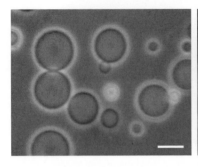 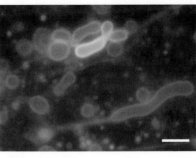

FIGURE 9.3 Membrane formation from nonbiological amphiphiles. Left panel: membranous vesicles self-assemble from amphiphilic compounds present in the Murchison carbonaceous meteorite. Right panel: vesicles are also formed from mixtures of 10-carbon amphiphiles (decanoic acid + decanol) in simulations of meteoritic organic mixtures. Bars show 20 µm. (Apel *et al.*, 2002.)

provides a form of speciation that is lacking in the general environment. The membranes of self-assembled compartments also have the potential to maintain concentration gradients of ions, thus providing a source of free energy that can drive otherwise energetically unfavorable processes. Finally, if certain components of the prebiotic organic inventory happened to be nonpolar pigments, they would be incorporated into the hydrophobic phase of a membrane and have the potential to capture light energy. This set of biophysical properties can only arise from amphiphilic molecules that have the capacity to self-assemble into more complex vesicular structures.

Some of the amphiphilic molecules required for early membrane structure may have been delivered by carbonaceous meteorites. This possibility has been investigated by studying the physical properties of organic compounds present in the Murchison meteorite. A variety of organic acids represent the most abundant fraction, some with chain lengths up to 12 carbons long. Aliphatic and aromatic hydrocarbons, ureas, ketones, alcohols, aldehydes, and purines are also present (Cronin *et al.*, 1988; Section 7.4.1). Samples of the Murchison meteorite have been extracted in an organic solvent commonly used to extract membrane lipids from biological tissues (Deamer and Pashley, 1989). When this material was allowed to interact with water, one class of compounds with acidic properties was clearly capable of forming membrane-bounded vesicles (left panel of Fig. 9.3). This fraction contains monocarboxylic acids such as nonanoic and decanoic acid (nonanoic acid is $H_3C–(CH_2)_7–COOH$ with nine carbon atoms in the chain) and such compounds are the simplest amphiphilic molecules capable of assembling into a true membrane structure (right panel of Fig. 9.3).

9.7 Membrane permeability

Although membranes define all living cells, the membrane also limits access to nutrients and energy sources.

The first living cellular systems were unlikely to have evolved specialized membrane transport systems, and it is interesting to consider how early cells might have coped with the negative aspects of their barriers. To give a perspective on permeability and transport rates, we can compare the fluxes of relatively permeable and relatively impermeable solutes across contemporary lipid bilayers. The permeabilities of lipid bilayers to small, uncharged molecules such as water, oxygen, and carbon dioxide are $\sim 10^9$ greater than the permeabilty to ions (Table 9.9). Measurements show that half the water in a liposome exchanges in milliseconds, while potassium ions have exchange half-times measured in days.

We can now consider some typical nutrient solutes like amino acids and phosphate. Such molecules are ionized, which means that they would not readily cross the permeability barrier of a lipid bilayer. One can estimate that if a primitive microorganism depended on passive transport of phosphate across a phospholipid bilayer, it would require several years to accumulate phosphate sufficient to double its DNA content, or pass through one cell cycle!

If lipid bilayers are so impermeable to solutes like amino acids and phosphate, how could primitive cells have had access to these essential nutrients? The answer lies in the fact that modern lipids are highly evolved products of several billion years of evolution, and typically contain hydrocarbon chains 16 to 18 carbons in length. These chains provide an interior "oily" portion of the lipid bilayer that represents a nearly impermeable barrier to the free diffusion of ions such as sodium, potassium, and protons. If an ion dissolved in water attempts to leave the water and dissolve in the oil phase, it faces a very high energy barrier (called the Born energy), which is associated with the difference in energy for an ion in a high-dielectric medium (water with its dielectric constant of 80) compared to the same ion in a low-dielectric medium (hydrocarbon with a dielectric constant of 2).

TABLE 9.9 Permeability coefficients[*] of solutes in cm s^{-1}

Water	Urea	Glycerol	Glucose	Cl-	K+	Glycine	Phosphate
10^{-3}	10^{-5}	10^{-6}	10^{-7}	10^{-9}	10^{-11}	10^{-11}	10^{-12}

*These values are proportional to the rate at which solutes can diffuse across membranes. They vary markedly depending on lipid composition; the approximate values given here are typical for lipid bilayers composed of phospholipids from biological membranes.

This energy barrier is immense, up to 40 kcal per mole. For comparison, the energy released by ATP hydrolysis is only 7 kcal per mole.

Studies have shown, however, that permeability is strongly dependent on chain length (Paula *et al.*, 1996). For instance, shortening phospholipid chains from 18 to 14 carbons increases permeability to ions by a thousand-fold. The reason is that thinner membranes have increasing numbers of transient defects that open and close on nanosecond timescales, allowing ionic solutes to move across the membrane without dissolving in the oily interior phase of the bilayer. That is, shortening the chains of a given membrane lipid dramatically increases permeation rates of ionic solutes. On the early Earth, shorter hydrocarbon chains would have been much more common than longer chain amphiphiles (Table 9.10), suggesting that the first cell membranes were sufficiently leaky that ionic and polar nutrients could enter, while still maintaining larger polymers in the encapsulated volume.

9.8 Encapsulation of macromolecules

It is reasonable to think that the origin of cellular life took place in a concentrated mixture of chemical components that had already established simple metabolic pathways, and perhaps even a catalyzed polymerization of primitive genetic material. In order for lipid bilayer vesicles to capture the large molecules involved and thereby produce the first cells, there must have been a robust encapsulation mechanism. Encapsulation entails the ability of the bilayer barrier first to open, allowing entry of large molecules, then to reseal. One encapsulation mechanism that has the potential to function in the prebiotic environment depends on the fact that when lipid vesicles are dried in the presence of macromolecules, they tend to fuse into multilayered structures that "sandwich" the solutes, as shown in Fig. 9.4 (Shew and Deamer, 1983). The macromolecules are then captured upon rehydration when the lipid layers reseal into vesicles. Such

TABLE 9.10 Comparison of solutions, films, and vesicles of organic amphiphiles

System	Kind of amphiphile	Examples
Solutions	Short chain lengths 4–6 carbons	Hexanoic acid (6 carbons)
Micelles	Medium chain lengths 6–8 carbons	Octanoic acid (8 carbons)
Monomolecular films	Longer chain lengths 10–20 or more carbons	Stearic acid (18 carbons)
Lipid bilayers	Phospholipids, sterols, fatty acids	Phosphatidylcholine, cholesterol

hydration–dehydration cycles leading to encapsulation may well have occurred in ponds or intertidal zones.

From these observations, it seems plausible that lipid-like molecules were available on the early Earth and could self-assemble into membrane-bound vesicles that captured large molecules. In order to investigate this conjecture, several research groups have developed laboratory models of simple cellular systems containing enzymes encapsulated by lipid bilayers. The first such experiments incorporated an RNA polymerase called polynucleotide phosphorylase (Chakrabarti *et al.*, 1994; Walde *et al.*, 1994a. See Fig. 9.5a). This enzyme does not depend on a template to synthesize RNA. Instead, it can utilize nucleotide diphosphates such as ADP as both an energy source and a monomer. In a typical experiment, the enzyme was captured in liposomes by a simulated tide pool cycle in which a mixture of the enzyme and lipids was first dried, then rehydrated in the reaction medium (see Fig. 9.4). Under these conditions about half of the original enzyme can

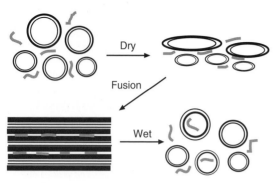

FIGURE 9.4 Encapsulation of macromolecules by lipid vesicles. When a mixture of lipid vesicles and macromolecules such as nucleic acids or proteins is dried, the vesicles fuse and capture the macromolecules in a "sandwich" structure. Upon rehydration, the vesicles reform and approximately half of the macromolecules are encapsulated. (Shew and Deamer, 1983.)

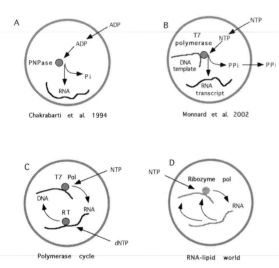

FIGURE 9.5 Models of protocells. (a) In this laboratory model, RNA is synthesized inside lipid vesicles containing an encapsulated polynucleotide phosphorylase (PNPase) using externally added substrate (ADP) (Chakrabarti et al., 1994). (b) In this laboratory model an encapsulated RNA polymerase (T7 polymerase) transcribes a DNA template into an RNA transcript using externally added nucleoside triphosphates (ATP, UTP, GTP, and CTP) as substrates (Monnard, 2003). (c) If polymerases such as the T7 polymerase (T7 pol) and reverse transcriptase (RT) could both be encapsulated in a lipid vesicle, the system would have the capacity to undergo molecular evolution (see text for details). (d) In a hypothetical RNA World, encapsulated RNA in the form of a ribozyme could function as both template and polymerase.

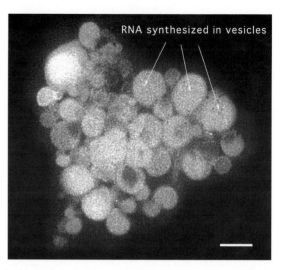

FIGURE 9.6 An image of lipid vesicles with encapsulated T7 RNA polymerase and DNA template, corresponding to Fig. 9.5b. A mixture of four nucleoside triphosphates was added, and these diffused into the vesicles and were used by the polymerase to synthesize RNA from the DNA template. The RNA was stained with ethidium bromide and appears as fluorescent material within the vesicles. Note that some of the vesicles do not contain fluorescent RNA, presumably because they lacked sufficient enzyme or template. The scale bar shows 20 micrometers.

This demonstrated that the lipid bilayer protected the reaction mixture in the vesicle interior.

The next experimental step toward a laboratory model of early cellular life will be to develop an encapsulated, template-directed replicating system as a model of genetic information transfer between two molecules. Fig. 9.5b shows one such system that is currently being investigated (Monnard et al., 2007). T7 RNA polymerase is an enzyme purified from bacteriophages, which can synthesize RNA from a DNA template. In the process, a base sequence in the DNA is precisely transcribed into the complementary base sequence in RNA. This enzyme is so efficient that it is available as a commercial kit for making large amounts of RNA with defined base sequences. In preliminary studies, it has been possible to capture T7 RNA polymerase together with template DNA in liposomes and demonstrate synthesis of RNA with specific transcribed sequences. Figure 9.6 shows lipid vesicles containing newly synthesized RNA that has been stained with a fluorescent dye.

The next step in this progression is to capture both a reverse transcriptase and the T7 RNA polymerase in liposomes (Fig. 9.5c). This system was first developed to study molecular evolution in RNA and ribozymes (Beaudry and Joyce, 1992; Wilson and Szostak, 1995).

be encapsulated. ADP was added to the external medium, and after an incubation period RNA synthesis was detected in the vesicles. RNA was synthesized even if an enzyme called RNase, which readily attacks and breaks down RNA, was present in the medium.

To carry out an experiment, a base sequence in DNA is first transcribed to RNA by the RNA polymerase, then back to DNA by the reverse transcriptase. This cycle of polymerization can occur dozens of times in an hour, and the result is a 10,000-fold amplification of the original base sequence. The most important property of the system is that the catalytic properties of the RNA can be made to evolve by repeating the amplification ten times and imposing a selective barrier. For instance, Beaudry and Joyce (1992) were able to evolve a ribozyme that could catalyze the hydrolysis of DNA.

Encapsulating such a system in lipid vesicles would represent a significant experimental hurdle because it requires two functional enzymes, two primers, and a template to be present in the same vesicle. It also requires that four different ribonucleotide triphosphates and four deoxyribonucleotide triphosphates be able to cross the membrane barrier at a rate sufficient to supply the encapsulated enzymes with substrates. Although this system works very well in a test tube environment, it has not yet been demonstrated in lipid vesicles.

The next step in the progression is illustrated in Fig. 9.5d, which shows ribozymes incorporated into a hypothetical protocellular system in an early "RNA World" (Section 8.2.1). Here a ribozyme serves both as a polymerase and as a store of sequence information that allows it to guide its own replication. Such a system is entirely hypothetical, but it is interesting to think about its properties. First, there is no metabolism, so that the system is entirely dependent on an external source of nutrient monomers and energy. Another point is that the boundary membrane is not growing, so even if the ribozyme could grow and replicate itself, at some point it would fill the internal volume and stop growing. On the other hand, the self-replicating ribozyme clearly would have the capacity to undergo evolutionary change, in that errors made in replication would be preserved and undergo selective pressure. At first, the selection would simply be a matter of which of the systems were more efficient as a polymerase.

It is interesting to note the possibility that one ribozyme species in the system described above might be able to catalyze the synthesis of amphiphilic molecules that could be incorporated into the boundary membrane. Such a system would have a selective advantage over a similar system lacking this capacity, in that the membrane boundary could enlarge to accommodate growth of the encapsulated catalysts and genetic material. Growth of primitive boundary membranes by lipid addition has been discussed by Morowitz et al. (1988),

Walde et al. (1994b), Segré et al. (2001), and Szostak et al. (2001). An experimental system demonstrating both growth and division has also been reported by Hanczyc et al. (2003).

As a final step in this logical progression, protein synthesis in lipid vesicles has recently been demonstrated. The first attempt to assemble a translation system in a lipid vesicle system was made by Oberholzer et al. (1999). However, only very small amounts of peptides were synthesized, largely because the lipid bilayer was impermeable to amino acids, so that ribosomal translation was limited to the small number of amino acids that were encapsulated within the vesicles. Yu et al. (2001) and Nomura et al. (2003) substantially improved the yield by using larger vesicles, demonstrating that green fluorescent protein (GFP) can be synthesized by an encapsulated translation system. Most recently, Noireaux and Libchaber (2004) published an elegant solution to the dilemma of the impermeable vesicle membrane. They broke open bacteria and captured samples of the bacterial cytoplasm in lipid vesicles. The samples included ribosomes, transfer RNAs and the hundred or so other components required for protein synthesis. The researchers then carefully chose two genes to translate, one for GFP and a second for a pore-forming protein called alpha hemolysin. If the system worked, the GFP would accumulate in the vesicles as a visual marker for protein synthesis, and the hemolysin would allow externally added "nutrients" in the form of amino acids and ATP, the universal energy source required for protein synthesis, to cross the membrane barrier and supply the translation process with energy and monomers. The system worked, and the vesicles began to glow with the classic green fluorescence of GFP. Nutrient transport through the hemolysin pore allowed synthesis to continue for as long as four days. Can both a gene transcription system and protein synthesis be encapsulated in lipid vesicles? This was in fact achieved by Ishikawa et al. (2004), who managed to assemble a two-stage genetic network in liposomes, in which the gene for an RNA polymerase was expressed first, and the polymerase then used to produce mRNA required for GFP synthesis.

9.9 A second origin of life in the laboratory?

Can we now synthesize life? All that remains to be done, it seems, is to add up the individual processes and integrate them into a complete system. But would the system really be alive? What would such a system do? We can answer this question by listing the steps that would be required for a microorganism to emerge as the first cellular life on the early Earth, or as artificial life in the laboratory.

- Boundary membranes self-assemble from soap-like molecules to form microscopic cell-like compartments.
- Energy is captured by the membranes, either from chemical energy or from light via a pigment system, or both.
- Ion concentration gradients are maintained across the membranes and serve as a major source of metabolic energy.
- Macromolecules are encapsulated in the compartments, but smaller molecules can cross the membrane barrier to provide nutrients and chemical energy for a primitive metabolism.
- The macromolecules grow by polymerization of the nutrient molecules.
- Macromolecular catalysts evolve and speed the growth process.
- The macromolecular catalysts themselves are reproduced during growth.
- Information is captured in the sequence of monomers in one set of polymers.
- The information is used to direct the growth of catalytic polymers.
- The membrane-bounded system of macromolecules divides into smaller structures that continue to grow.
- Genetic information is passed between generations by duplicating the gene sequences and sharing them among daughter cells.
- Occasional mistakes (mutations) are made during replication or transmission of information, allowing the system to evolve through natural selection.

Examining this list, one is struck by the complexity of even the simplest form of life. This is one reason why it has been so difficult to "define" life in the usual sense of a definition, that is, boiled down to a few sentences in a dictionary. Life is a complex system that cannot be captured in a few sentences, so perhaps a list of its observed properties is the best we can ever hope to do. One is also struck by the fact that *all but one of the functions* in the list have now been reproduced in the laboratory. The one exception is that everything in the system grows and reproduces *except* the catalytic macromolecules themselves (the polymerase enzymes or ribosomes), which get left behind.

This is the final challenge: to encapsulate a system of macromolecules that can make more of themselves. Bartel and co-workers (Johnston *et al.*, 2001) have produced a ribozyme that can grow by copying a sequence of bases in its own structure. So far, the polymerization has only been able to copy a short string of fourteen nucleotides, but this is a good start. If a ribozyme system can be found that catalyzes its own complete synthesis using genetic information encoded in its structure, it could rightly be claimed to have the essential property of self-replication. Given such a ribozyme, it is not difficult to imagine its incorporation into lipid vesicles in order to produce a very simple cell that would have the basic properties of the living state.

9.10 Summary

This chapter has provided a perspective on the most primitive forms of cellular life. In the early Earth environment, there must have been a variety of amphiphilic hydrocarbon derivatives that could self-assemble into bilayer boundary structures and encapsulate polymers that were being synthesized by a separate process. The vesicle membranes would have been sufficiently permeable to allow passage of smaller ionic substrates required for metabolism and biosynthesis, yet maintain larger molecules within. Encapsulated catalysts and information-bearing molecules would thus have had access to nutrients required for growth. Furthermore, specific groupings of macromolecules would be maintained, rather than drifting apart. This would allow true Darwinian-type selection of such groupings to occur, a process that could not take place in mixtures of molecules free in solution. A small number of the encapsulated molecular systems were likely to have the specific set of properties that allowed them to capture free energy and nutrients from their environment and undergo growth by polymerization. At some point, the growth would become catalyzed by the encapsulated polymers, and then begin to be directed by a primitive genetic process. Such structures would be on the evolutionary path to the first forms of cellular life.

References

Anders, E. (1989). Pre-biotic organic matter from comets and asteroids. *Nature*, **342**, 255–256.

Apel, C. L., Deamer D. W., and Mautner, M. (2002). Self-assembled vesicles of monocarboxylic acids and alcohols: conditions for stability and for the encapsulation of biopolymers. *Biochim. Biophys. Acta*, **1559**, 1–10.

Bada, J. L., Bigham, C., and Miller, S. L. (1994). Impact melting of frozen oceans on the early Earth – implications for the origin of life. *Proc. Natl. Acad. Sci. USA*, **91**, 1248–1250.

Bada, J. L., Miller, S. L., and Zhao, M. (1995). The stability of amino acids at submarine hydrothermal vent temperatures. *Orig. Life Evol. Biosphere*, **25**, 111–118.

Baltscheffsky, H. and Baltscheffsky, M. (1994). Molecular origin and evolution of early biological energy conversion. In *Early Life on Earth. Nobel Symposium No. 84*, S. Bengston, ed. New York: Columbia University Press, pp. 81–90.

Baross, J. A. and Hoffman, S. E. (1985). Submarine hydrothermal vents and associated gradient environments as sites for the origin and evolution of life. *Orig. Life*, **15**, 327.

Beaudry, A. A. and Joyce, G. F. (1992). Directed evolution of an RNA enzyme. *Science*, **342**, 255–257.

Chakrabarti, A., Breaker, R. R., Joyce, G. F., and Deamer, D. W. (1994). Production of RNA by a polymerase protein encapsulated within phospholipid vesicles. *J. Mol. Evol.*, **39**, 555–559.

Chyba, C. F. and Sagan, C. (1992). Endogenous production, exogenous delivery and impact-shock synthesis of organic molecules: an inventory for the origin of life. *Nature*, **355**, 125–13.

Corliss, J. B., Baross, J. A., and Hoffman, S. E. (1981). An hypothesis concerning the relationship between submarine hot springs and the origin of life on Earth. *Oceanol. Acta*, **4** (Suppl.) 59–69. Proc. 26th Intl. Geolog. Congress, Geology of the Oceans symposium, Paris.

Cronin, J. R., Pizzarello, S., and Cruikshank, D. P. (1988). In *Meteorites and the Early Solar System*, J. F. Kerridge and M. S. Matthews, eds. Tucson AZ: University of Arizona Press, pp. 819–857.

Deamer, D. W. (1997). The first living systems: a bioenergetic perspective. *Microbiol. Mol. Biol. Rev.*, **61**, 239–261.

Deamer D. W. and Pashley, R. M. (1989). Amphiphilic components of the Murchison carbonaceous chondrite: surface properties and membrane formation. *Orig. Life Evol. Biosphere*, **19**, 21–38.

Deamer, D.W., Dworkin, J. P., Sandford, S. A., Bernstein, Max P., and Allamandola, L. J. (2002). The first cell membranes. *Astrobiology*, **2**, 371–382.

De Duve, C. and Miller, S. L. (1991). Two-dimensional life? *Proc. Natl. Acad. Sci. USA*, **88**, 10014–10017.

Delsemme, A. (1984). The cometary connection with prebiotic chemistry. *Orig. Life*, **14**, 51–60.

Escabi-Perez, J. P., Romero, A., Lukak, S., and Fendler, J. H. (1979). Aspects of artificial photosynthesis. Photoionization and electron transfer in dihexadecyl phosphate vesicles. *J. Am. Chem. Soc.*, **101**, 2231–2233.

Fox, S. W. and Harada, K. (1958). Thermal copolymerization of amino acids to a product resembling protein. *Science*, **128**, 1214.

Gedulin, B. and Arrhenius, G. (1994). Sources and geochemical evolution of RNA precurosor molecules: the role of phosphate. In *Early Life on Earth. Nobel Symposium No. 84*, S. Bengston, ed. New York: Columbia University Press, pp. 91–106.

Hanczyc, M. M., Fujikawa, S. M., and Szostak, J. W. (2003). Experimental models of primitive cellular compartments: encapsulation, growth, and division. *Science*, **302**, 618–22.

Harris, C. M. and Selinger, B. K. (1983). Excited state processes of naphthols in aqueous surfactant solution. *Z. Phys. Chem.*, **134**, 65–92.

Hazen, R. M, Filley, T. R., and Goodfriend, G. A. (2001). Selective adsorption of L- and D-amino acids on calcite: implications for biochemical homochirality. *Proc. Natl. Acad. Sci. USA*, **98**, 5487–90.

Holland, H. D. (1984). *The Chemical Evolution of the Atmosphere and Oceans*. Princeton, NJ: Princeton University Press.

Huber, C. and Wächtershäuser, G. (1997). Activated acetic acid by carbon fixation on (Fe,Ni)S under primordial conditions. *Science*, **276**, 245.

Inove, T. and Orgel, L. E. (1983). A nonenzymatic RNA polymerase model. *Science*, **219**, 859–862.

Ishikawa, K., Sato, K., Shima, Y., Urabe, I., and Yomo, T. (2004). Expression of a cascading genetic network within liposomes. *FEBS Lett.*, **576**, 387–390.

Johnston, W. K., Unrau, P. J., Lawrence, M. S., Glasner, M. E., and Bartel, D. L. (2001). RNA catalyzed RNA polymerization: accurate and general RNA-templated primer extension. *Science*, **292**, 1319–1325.

Joyce, G. F. and Orgel, L. E. (1986). Non-enzymic template-directed synthesis on RNA random copolymers – poly (C,G) templates. *J. Mol. Biol.*, **188**, 433–441.

Joyce, G. F., Visser, G. M., van Boeckel, C. A. A., van Boom, J. H., Orgel, L. E., and van Westrenen, J. (1984). Chiral selection in poly(C)-directed synthesis of oligo(G). *Nature*, **310**, 602–604.

Kasting, J. F. and Brown, L. L. (1998). The early atmosphere as a source of biogenic compounds. In *The Molecular Origins of Life*, A. Brack, ed. Cambridge: Cambridge University Press.

Koch, A. L. and Schmidt, T. M. (1991). The first cellular bioenergetic process: primitive generation of a proton motive force. *J. Mol. Biol.*, **33**, 297–304.

Lohrmann, R. and Orgel, L. E. (1973). Prebiotic activation processes. *Nature*, **244**, 418–420.

Lohrmann, R. and Orgel, L. E. (1976). Template directed synthesis of high molecular weight polynucleotide analogues. *Nature*, **261**, 342–344.

McCollom, T. M., Ritter, G., and Simoneit, B. R. T. (1999). Lipid synthesis under hydrothermal conditions by Fischer-Tropsch-type reactions. *Orig. Life Evol. Biosphere*, **29**, 153–166.

Maher, K. A. and Stevenson, D. J. (1988). Impact frustration of the origin of life. *Nature*, **331**, 612–614.

Maurel, M.-C. and Orgel, L. E. (2000). Oligomerization of alpha-thioglutamic acid. *Orig. Life Evol. Biosphere*, **30**, 423–430.

Miller, S. L. (1953). Production of amino acids under possible primitive Earth conditions. *Science*, **117**, 528–529.

Miller, S. L. and Urey, H. C. (1959). Organic compound synthesis on the primitive Earth. *Science*, **130**, 245–251.

Miller, S. and Schlesinger, G. (1993). Prebiotic syntheses of vitamin coenzymes: I. Cysteamine and 2-mercaptoethanesulfonic acid (Coenzyme M). *J. Mol. Evol.*, **36**, 302–7.

Mitchell, P. (1961). The chemiosmotic hypothesis. *Nature*, **191**, 144–148.

Mojzsis, S. J., Arrhenius, G., McKeegan, K. D., and Harrison, T. M. (1996). Evidence for life on Earth before 3,800 million years ago. *Nature*, **384**, 55–59.

Monnard, P.-A. (2003). Liposome-entrapped polymerases as models for microscale/nanoscale bioreactors. *J. Membrane Biol.*, **191**, 87–97.

Monnard, P.-A., Luptak, A., and Deamer, D. W. (2007). Models of primitive cellular life: polymerases and templates in liposomes. *Phil. Trans. Roy. Soc. B*, in press.

Morowitz, H. J. (1992). *Beginnings of Cellular Life*. New Haven, CT: Yale University Press.

Morowitz, H. J., Heinz, B., and Deamer, D. W. (1988). The chemical logic of a minimum protocell. *Orig. Life Evol. Biosph.*, **18**, 281–7.

Noireaux, V. and Libchaber, A. (2004). A vesicle bioreactor as a step toward an artificial cell assembly. *Proc. Natl. Acad. Sci. USA*, **101**, 17669–74.

Nomura, S., Tsumoto, K., Hamada, T., Akiyoshi, K., Nakatani, Y., and Yoshikawa, K. (2003). Gene expression within cell-sized lipid vesicles. *Chem. Biochem.*, **4**, 1172–1175.

Oberholzer, T., Nierhaus, K. H., and Luisi, P. L. (1999). Protein expression in liposomes. *Biochem. Biophys. Res. Commun.*, **261**, 238–241.

Oparin, A. I. (1924). *The Origin of Life*. Moscow: Izd. Moskovshii Rabochii. English translation in J. D. Bernal. (1967). *The Origin of Life*. London: Weidenfeld and Nicolson, p. 199–234.

Oparin, A. I., Orlovskii, A. F., Bukhlaeva, V. Ya., and Gladilin, K. L. (1976). Influence of the enzymatic synthesis of polyadenylic acid on a coacervate system. *Dokl. Akad. Nauk. SSSR*, **226**, 972–974.

Orgel, L. E. (1987). Evolution of the genetic apparatus: a review. *Cold Spring Harbor Symp. Quant. Biol.*, **52**, 9–16.

Orgel, L. E. (2004). Prebiotic chemistry and the origin of the RNA world. *Crit. Rev. Biochem. Mol. Biol.*, **39**, 99–123.

Oró, J. (1961). Comets and the formation of biochemical compounds on the primitive Earth. *Nature*, **190**, 389–390.

Pace, N. R. (1991). The origin of life: facing up to the physical setting. *Cell*, **65**, 531–533.

Paula, S., Volkov, A. G., Van Hoek, A. N., Haines, T. H., and Deamer, D. W. (1966). Permeation of protons, potassium ions and small polar molecules through phospholipid bilayers as a function of membrane thickness. *Biophys. J.*, **70**, 339–348.

Pinto, J. P., Gladstone, G. R., and Yung, Y. L. (1980). Photochemical production of formaldehyde in Earth's primitive atmosphere. *Science*, **210**, 183–185.

Rohlfing, D. L. (1976). Thermal polyamino acids: synthesis at less than 100 °C. *Science*, **193**, 68–70.

Schopf, J. W. (1993). Microfossils of the Early Archean Apex Chert: new evidence of the antiquity of life. *Science*, **260**, 640–646.

Segré, D., Deamer, D. W., and Lancet, D. (2001). The lipid world. *Orig. Life Evol. Biosphere*, **31**, 119–145.

Shew, R. and Deamer, D. (1983). A novel method for encapsulating macromolecules in liposomes. *Biochim. Biophys. Acta*, **816**, 1–8.

Sleep, N. H., Zahnle, K., Kasting, J. F., and Morowitz, H. J. (1989). Annihilation of ecosystems by large asteroid impacts on the early Earth. *Nature*, **342**, 139–142.

Stevens, T. O. and McKinley, J. P. (1995). Lithoautotrophic microbial ecosystems in deep basalt aquifers. *Science*, **270**, 450–455.

Stetter, K. O. (1996). Hyperthermophiles in the history of life. *Ciba Found. Symp.*, **202**, 1–10.

Stevenson, D. J. (1983). The nature of the Earth prior to the oldest known rock record: the Hadean Earth. In *Earth's Earliest Biosphere, Its Origin and Evolution*, J. W. Schopf, ed. NJ Princeton: Princeton University Press.

Stribling, R. and Miller S. L. (1987). Energy yields for hydrogen cyanide and formaldehyde synthesis: the HCN and amino acid concentration in the primitive ocean. *Orig. Life Evol. Biosphere*, **17**, 261–73.

Szostak, J. W., Bartel, D. P., and Luisi. P. L. (2001). Synthesizing life. *Nature*, **409**, 387–390.

Usher, D. A. (1977). Early chemical evolution of nucleic acids: a theoretical model. *Science*, **196**, 311–313.

Usher, D. A. and McHale, A. H. (1976). Nonenzymic joining of oligoadenylates on a polyuridylic acid template. *Science*, **192**, 53–54.

Wächtershäuser, G. (1988a). Before enzymes and templates: theory of surface metabolism. *Microbiol. Rev.*, **52**, 452–484.

Wächtershäuser, G. (1988b). Pyrite formation, the first energy source for life: a hypothesis. *Syst. Appl. Microbiol.*, **10**, 207–210.

Walde, P., Goto, A., Monnard P.-A., Wessicken, M., and Luisi, P. L. (1994a). Oparin's reactions revisited: enzymatic synthesis of poly(adenylic acid) in micelles and self-reproducing vesicles. *J. Am. Chem Soc.*, **116**, 7541–7547.

Walde, P., Wick, R., Fresta, M., Mangone, A., and Luisi, P. L. (1994b). Autopoietic self-reproduction of fatty acid vesicles. *J. Am. Chem. Soc.*, **116**, 11649–11654.

Warman, J. M., de Haas, M. P., Paddon-Row, M. N., *et al.* (1986). Light-induced giant dipoles in simple model compounds for photosynthesis. *Science*, **320**, 615–616.

Weber, A. L. (1987). The triose model: glyceraldehyde as a source of energy and monomers for prebiotic condensation reactions. *Orig. Life*, **17**, 107–119.

Weber, A. L. (1997). Energy disproportionation of sugar carbon drives biotic and abiotic synthesis. *J. Molec. Evol.*, **44**, 354–360.

Wilson, C. and Szostak, J. W. (1995). In vitro evolution of a self-alkylating ribozyme. *Nature*, **374**, 777–782.

Yu, W., Sato, K., Wakabayashi, M., *et al.* (2001). Synthesis of functional protein in liposome. *J. Biosci. Bioengin.*, **92**, 590–593.

FURTHER READING

Hazen, R. M. (2001). Life's rocky start. *Sci. Amer.* (April), **284**, 76–85.

Deamer, D., Dworkin, J. P., Sandford, S. A., Bernstein, Max P., and Allamandola, L. J. (2002). The first cell membranes. *Astrobiology*, **2**, 371–82.

Pohorille, A. and Deamer, D. W. (2002). Artificial cells: prospects for biotechnology. *Trends in Biotechnology*, **20**, 123–128.

Rasmussen, S., Chen, L., Deamer, D., Krakauer, D. C., Packard, N. H., Stadler, P. F., and Bedau, M. A. (2004). Transitions from nonliving to living matter. *Science*, **303**, 963–5.

Szostak, J. W., Bartel, D. P., and Luisi, P. L. (2001). Synthesizing life. *Nature*, **409**, 387–90.

Part IV
Life on Earth

10 Evolution: a defining feature of life

John A. Baross
University of Washington

In biology nothing makes sense except in the light of evolution. It is possible to describe living beings without asking questions about their origins. [But] the descriptions acquire meaning and coherence only when viewed in the perspective of evolutionary development.

Theodosius Dobzhansky (1970: 6)

10.1 From Lamarck to Darwin to the central dogma

The basic notion of evolution is that inherited changes in populations of organisms result in expressed differences over time – these differences are at the gene level (the *genotype*) and/or expression of the gene into an identifiable characteristic (the *phenotype*). The important underlying fact of evolution is that all organisms share a common inheritance, or, put more dramatically, all extant organisms on Earth evolved from a common ancestor. We see this in the universal nature of the genetic code and in the *unity of biochemistry*: (a) all organisms share the same biochemical tools to translate the universal information code from genes to proteins, (b) all proteins are composed of the same twenty essential amino acids, and (c) all organisms derive energy for metabolic, catalytic, and biosynthetic processes from the same high-energy organic compounds such as adenosine triphosphate (ATP).

In *On the Origin of Species* Charles Darwin (1859) (Fig. 10.1) built his theory of evolution using evidence that included an ancient Earth thought at the time by many geologists to have an age in millions of years. He also took the extinction of species to be a real phenomenon since fossils existed that were without living representatives. Since different species showed close phenotypic similarities, he argued that existing organisms descended from other organisms including extinct groups. The key to his evolutionary theory therefore was that inherited characteristics of organisms can change through time and that these changes occur gradually and without discontinuities. Jean Baptiste Lamarck (1809) had earlier recognized a similar principle of evolution and offered an explanation generally

referred to as "inheritance of acquired characters." By this Lamarck meant that the variations in characteristics or adaptations seen in organisms were acquired in response to the environment. Classic examples include the long neck of the giraffe as an adaptation for foraging tender foliage on treetops, or the use of long legs by some aquatic fowl to venture into deep waters in search of prey. While this is certainly how these phenotypic characteristics are utilized to the advantage of the organism, under Darwinian principles they were not acquired as a response to the needs of the environment.

One of Darwin's major contributions was his explanation of how and why organisms change over time and how they acquire characteristics useful for living in different environments. Darwin referred to the mechanism for character changes through time as *natural selection*. Natural selection is based on the idea of the struggle for existence (survival of the fittest) in populations where there are more individuals of each species than can survive. A variation in any characteristic of an individual that gives an advantage in surviving (and therefore reproducing) will be "naturally selected" since the new trait will be preferentially inherited by subsequent generations. Darwin differed from Lamarck by recognizing that character changes that offer a survival advantage and are "naturally selected" originate from a pool of randomly generated character changes that are not directed by environmental conditions. Note that Darwin proposed his theory without knowing the mechanism for the inheritance of acquired traits – not until forty years later would the field of genetics begin with the recognition that Gregor Mendel's principle of discrete units of inheritance (genes) was correct.

Planets and Life: The Emerging Science of Astrobiology, eds. Woodruff T. Sullivan, III and John A. Baross. Published by Cambridge University Press. © Cambridge University Press 2007.

Ch. Darwin

FIGURE 10.1 Charles Darwin (1809–1882) in his later years at Down House, his combined home, office, and laboratory (see Appendix D). When the German biologist and philosopher Ernst Haeckel came to visit Darwin, he saw:

> The great naturalist himself, a tall and venerable figure, with the broad shoulders of an Atlas supporting a world of thought, his Jupiter-like forehead, highly and broadly arched ... and deeply furrowed with the plough of mental labour; his kindly, mild eyes looking forth under the shadow of prominent brows. (Photo courtesy John van Wyhe.)

Mendelian genetics established that phenotypes are transmitted from one generation to another following statistical principles and that these phenotypes reside in simple heritable "characters." The nature of these heritable characters were unknown to Mendel, but their location was confined to chromosomes by 1910 and then to DNA as the genetic material by Hershey and Chase (1952). This immediately led to the discovery by Watson and Crick (1953) of the double-helix structure of DNA and shortly afterwards to the elucidation of the genetic code and to the understanding that a gene is primarily a sequence along a section of DNA that codes for a protein using an alphabet composed of the four bases that constitute DNA. The steps leading from DNA to a specific protein are referred to as the *central dogma*: a DNA gene is transcribed to make messenger RNA (mRNA), followed by translation of mRNA into a protein. The exceptions to the central dogma are those genes that specify not proteins, but instead the various classes of RNA

that are involved in both transcription and translation, such as ribosomal RNA (rRNA) and transfer RNA (tRNA).

10.2 Evolution at the molecular level

The Watson and Crick discovery also opened the doors to studies of evolution at the molecular level and helped develop classification schemes that allow for the evolutionary comparisons of all groups of extant organisms, as well as the construction of models for inferring the nature of Earth's earliest microbial communities and the emergence of multicelled organisms (Hedges, 2002). Usually, ribosomal RNA genes and ribosomal protein genes are used for evolutionary studies because they have *highly conserved* sequences, meaning sequences that are found across all domains of life (Woese *et al.*, 1990). Most functional RNA molecules have secondary structures that are associated with their function. The base sequence determines to a great extent the functional secondary structure. Mutations that change the secondary structure of RNA molecules will frequently render them inactive. These conserved sequences must have originated a long time ago in a common ancestor and be of fundamental importance to all species. They are especially associated with the cell's ribosomes, where proteins are assembled, because this process is so fundamental to the functioning of all cells.

A central concept in evolutionary theory is that a gene coding for a characteristic is subject to *mutation* (change) in a random fashion, which in some cases can lead to variability in that characteristic in the next generation. Mutations come about due to mistakes made during DNA replication, or through external factors such as ionizing radiation or toxic chemicals. Most mutations are moot, i.e., they have little or no effect on the protein product of the gene or (for ribosomal RNA genes) the function of the RNA. Others, particularly those involving deletions and insertions that can result in structural changes in the transcribed protein or in a ribosomal RNA, can render it inactive. The most lethal mutations are those that damage the genes involved in DNA replication, transcription of DNA into mRNA, or translation of mRNA into a protein – in particular, mutations involving deletions or insertions of bases can change the structure of the transcribed mRNA and protein.

Changing environmental conditions can negatively affect growth and survival (inducing *stresses*) and, depending on the degree and kind of stress, can result in death of the organism. To survive such a lethal stress the organism must have mutation rates sufficiently high to handle the stress, but not so high as to cause lethal damage to the genome (the entire set of genes defining the species). Moreover, all extant organisms have a set of conserved genes for repairing mutations or proteins affected by environmental stresses such as starvation, heat, radiation, changes in pH, etc. While these "stress" genes are not 100% effective, they greatly reduce the number of deleterious mutations. The same genes can also target other specific genes for an *increased* mutation rate under stress conditions – these are called "stress-directed adaptive mutations" (Wright, 2004). For example, this mechanism can be observed when bacteria are starving from lack of their usual nutrient, but then undergo increased evolutionary rates of specific genes involved in the metabolism of alternative nutrients for growth.

There is debate about how random mutations lead to useful characters, and particularly about the mechanisms involved in adaptation that eventually results in useful complex structures such as enzymes, bacterial flagella motors, eyes, and brains. In evolution, adaptation means more than simply being well suited to the environment; it also involves in any generation the selection of one particular genetic change (over many other possibilities) that results in maximum reproductive success. But since *many* incremental steps are involved in evolving complex structures and processes, it would seem that adaptation involves a sequence of coordinated (not random) steps. Until recently, there was no satisfactory mechanism that could account for the evolution of complex structures.

Two kinds of evolutionary change are recognized. *Microevolution* results in changes at the species level and accounts for the short-term variability observed in populations. The second process of *macroevolution* involves the more substantial changes that over long times result in the development of new higher taxa such as genera, families, orders, etc. Macroevolution affects the genotypes of individuals within populations and thus also involves microevolution. Macroevolution is also invoked as the mechanism that results in the gradual formation of novel complex structures that involve multiple genes.

Development of the eye has provided a classic illustration for gradualism producing increasing complexity and function. There are more than forty different eye structures found in both invertebrates and vertebrates with a range of complexity from light sensitive patches to compound eyes (Parker, 2003). It was once thought that these photosensitive organs developed

independently along several different branches of the Tree of Life, a classic example of *convergent evolution* (independent evolution of morphologically and/or functionally similar structures). Recent molecular data, however, show that in many cases macroevolution is not totally a gradual set of changes based on mutation and natural selection. There appears to be a common set of genes that instigated the evolution of the eye in as diverse a group as fruit fly, squid, and humans. These genes are called "tool box" genes (Carroll, 2005), and are common to many diverse organisms, implying that they are inherited from a common ancestor. For example, one group of genes called *HOX genes*[1] accounts for the incredibly high diversity found in animal body plans. The three principal anatomical plans for wings exemplified in birds, bats, and pterosaurs were also thought to be the products of convergent evolution. The bird wing developed from the entire arm, the bat wing from a hand, and the pterosaur wing from a single finger. Similar HOX genes, acting on different sets of genes in birds, bats, and pterosaurs, resulted in the evolution of different kinds of wings (Carroll, 2005). The profoundly important point is that the origin of diverse body forms of animals and their organs may have more to do with the way multiple genes are expressed and less to do with the number of different kinds of genes. We are learning that a basic set of genes is used in animals in different ways to produce the myriad different body forms, appendages and organs. "Genetic switches," specific gene sequences that "instruct tool kit genes where to act and what to do" (Carroll, 2005), select which specific genes get expressed.

This new combination of evolution with developmental biology is called *Evo-Devo* and is revolutionizing our understanding of macroevolution and embryology. Evo-Devo also offers an explanation for the rapid macroevolutionary changes (termed *punctuated equilibrium* by Eldredge and Gould (1972)) that appear in the fossil record and that cannot be explained by gradualism. An example of punctuated equilibrium is the sudden appearance of diverse animal forms during the Cambrian explosion 550 Ma (Section 16.3.1). Evo-Devo studies indicate that this sudden emergence of highly diverse animal forms was due to the evolution of key regulatory HOX genes in the common ancestor to all Cambrian animals (Carroll, 2005).

---

[1] *HOX* comes from "homeo-" (*like*), and "box," from the fact that the DNA sequence is short enough to fit into a box drawn on paper.

10.3 Mechanisms for acquiring new genes

Besides mutation, other mechanisms can effect changes in genes that coordinate cell structures, metabolism, or physiological traits, whether for sudden acquisition of new genes or incremental changes of individual genes or groups of genes. These mechanisms are:

- fusion of different cells, sometimes called *endosymbiosis*;
- coevolution;
- lateral gene transfer.

Symbiosis is any interaction between two organisms (occasionally more than two organisms are involved) in which at least one of the organisms benefits from the relationship. This broad definition includes *parasitic associations* in which the parasite benefits at the expense of the host, or *mutualistic associations* in which both organisms benefit. Some symbiotic associations are obligatory where either the host or symbiont (or both) is unable to live independently. For instance, a recent model based on whole genome sequences indicates that the first eukaryote cell may have formed by the *fusion* of Bacteria and Archaea, and that Bacteria contributed the operational genes while Archaea contributed the informational genes (Rivera and Lake, 2004).[2] While there are other models for the origin of eukaryotes (Gupta, 1998; Martin and Müller, 1998), there is agreement about the bacterial and archaeal origin of informational and key operational genes in eukaryotes. Such a fusion would fall into the category of mutualistic symbiosis since both cells benefited from this association. Furthermore, we have evidence for ancient symbioses in eukaryote cells in that their mitochondria (involved in oxygen respiration) and chloroplasts (involved in photosynthesis) both first occurred in specific groups of bacteria (Sapp, 2005; Wakeford, 2001).

The proposed fusion of an archaeum with a bacterium somehow resulted in conditions favorable for evolution to greater complexity, multicellularity, and sexual reproduction. Similarly, the later acquisition by early eukaryotes of the mitochondria and chloroplast from bacteria must have had a profound effect on eukaryote evolution and particularly on their adaptation into habitats bathed in light and oxygen. Unfortunately, most of the evolutionary steps from the proposed fusion-based "proto-eukaryote" to single-cell eukaryotes (Chapter 13) are unknown since no known extant organism retains characteristics that can definitely be interpreted as intermediate to those of modern-day eukaryotes. For example, we do not know the intermediate steps/structures involved in the transition from the generally circular, double-stranded DNA chromosome of bacteria and archaea to the complex linear DNA-protein chromosomes of eukaryotes.

While **the fusion of two cells** is not believed to have been a common occurrence in the early life history of organisms, there are many examples of other forms of symbiosis that are widely distributed in eukaryotes, allowing them to live under conditions that otherwise would not be possible (Sapp, 1994). One of the first cases to be identified was the symbiosis of an alga and a fungus to form a lichen, researched in detail and recognized in the late nineteenth century by Beatrix Potter (better known as the author of *Peter Rabbit*) well before symbiosis was accepted by the British scientific community.[3] Other examples of symbiosis include hydrothermal vent tubeworms and clams that utilize inorganic chemical energy sources, plants that assimilate nitrogen via nitrogen fixation by root hair bacteria, and the microbial communities in the guts of ruminants and insects that anaerobically digest complex polysaccharides such as cellulose (Sapp, 1994; Wakeford, 2001). Parasitism, another form of symbiosis, can result in radical changes in the physiology of the host that include mating and feeding behavior, and morphological changes. It has been suggested that other ancient symbiotic events involving bacteria and eukaryotes occurred that are not as obviously visible in the cell as are mitochondria and chloroplasts, but are nevertheless important in the early evolution of eukaryotes (Margulis, 1993; Margulis *et al.*, 2000).

Coevolution is a special kind of symbiosis in which two kinds of organisms interact in such a way that each exerts a selective pressure on the other. Classical examples include flowering plants and their insect pollinators, and predators and their prey. Less obvious examples may include whole ecosystems in which all trophic levels from bacteria to animals have coevolved. Understanding the nature of coevolving ecosystems is one of the most difficult and important challenges in ecology.

Lateral gene transfer (also referred to as genetic exchange and horizontal gene transfer) is the transfer

---

[2] The Tree of Life divides all species into three Domains called Archaea (containing the archaea), Bacteria (containing the bacteria), and Eukarya (containing the eukaryotes). See Fig. 11.1.

[3] Wakeford (2001) has an interesting account of Potter's futile attempt to convince the British scientific community of the importance of her observations.

of DNA from one organism to another such that it effects a "permanent" change in the genetic composition of the recipient. Genetic exchange can be mediated by cell–cell contact (conjugation), by viral infection (transduction), or by incorporation of DNA from the environment (transformation).

The recent accumulation of complete genome sequences from representatives of all the domains of life has revealed a universal pattern of lateral gene transfer for acquiring genes or parts of genes. Woese *et al.* (1990) and Woese (2002) speculated that this mechanism was widespread in the early evolutionary stages of life and vital to producing the diversity reflected in the present three domains of life. Furthermore, we now know that *viruses*[4] have played and continue to play a significant role in the evolution of life through lateral gene transfer (Canchaya *et al.*, 2003). This is illustrated by the high abundance of bacterial viruses (bacteriophages or phages) in marine environments, exceeding the bacterial population by an order of magnitude. Viruses are the most abundant biological entity on Earth, yet are poorly understood. It is presumed that the primary role of viruses in the environment is causing death in bacteria (or producing disease in eukaryotes), but their significance as vehicles for transmitting new genes to bacteria *in situ* is not well understood, although likely extensive. Jiand and Paul (1998) calculated that at the low rate of infection of 10^{-8} per infected bacterial population, viral-mediated gene transfer takes place in Earth's oceans at the rate of $\sim 2 \times 10^{16}$ per second.

The characters transmitted by phage in the environment, their rate of transmission, and the environmental factors involved in the transfer of genes are generally unknown. However, recent evidence shows the presence of bacterial genes in marine phages, including genes that code for proteins necessary for photosynthesis (Hambly and Suttle, 2005). Similarly, a significant portion of eukaryote chromosomes (approximately 45% for humans and a much higher percentage for some plants and an amoeba species) is composed of remnants from RNA viruses (called retrotransposons, or mobile genetic elements that replicate by reverse transcription: RNA to DNA rather than DNA to RNA) (Bushman, 2001). Most of these viral sequences in eukaryotes are not transcribed into proteins. But do they nevertheless serve some important function to the organism? And if not, why do organisms retain them anyway? Some retroviral sequences have been implicated in the evolution of vertebrate genes including the development of the human placenta and the regulation of the gene for starch hydrolysis, but most have no apparent function and along with other non-protein coding regions on eukaryotic chromosomes, have been called "selfish DNA" (Bushman, 2001).

How important is lateral gene transfer in evolution? Results from whole genome sequences of bacteria and archaea indicate that lateral gene transfer may be the most important mechanism for acquiring new genes, including those involved in complex and coordinated phenotypes. For example, $\sim 16\%$ of the genome of *Escherichia coli* K12 is viral genes. Microorganisms have evolved elaborate mechanisms for incorporating acquired genes into their chromosome at specific sites. These sites can serve as "pathogenicity islands" if all of the acquired genes are involved in disease production, such as for the cholera-producing bacterium *Vibrio cholerae* (Faruque and Mekalanos, 2003), or they may be "genetic islands," which align acquired genes involved in key physiological activities such as magnetotaxis (Grünberg *et al.*, 2001) or the dissimilatory reduction of sulfate (Mussmann *et al.*, 2005).

It is very unlikely that the formation of a genome with sufficient information to lead to free-living (self-sufficient) cells could have originated without a mechanism for acquiring "functional" genes from other early cells or communities of interdependent cells or "precells" (Baross and Hoffman, 1985). This is certainly consistent with the fact that all life on Earth is derived from a common ancestral pool of genes based on a universal genetic code (Woese, 1998). Darwinian evolution would have played an important role in these early stages and selection would have favored specific biochemical and molecular structures and mechanisms over others. Could this imply that if we started over again by resetting the clock to 4 Ga, the resultant life would have the same biochemical and molecular properties, including the same genetic code, as present-day Earth life? If environmental conditions and the starting pool of organic compounds were the same, it is probable that a second genesis would result in biochemistry that would resemble or possibly be indistinguishable from present-day Earth life. The strong link between specific nucleotide bases and specific amino acids is one more example verifying that there are "rules of organic chemistry" that favor specific reactions or macromolecular structures (Copley *et al.*, 2005). In such a second

---

[4] A *virus* is defined as an intracellular parasite and is incapable of living without a host cell. While it shares many of the biochemical characteristics of a living cell including nucleic acids and proteins (although much smaller and simpler), it cannot reproduce independently, only by infecting a normal cell.

genesis, however, contingency in evolution could result in the selection of organisms and ecosystems significantly different from those found on Earth. Yet, compared to present-day organisms, they would share a similar biochemistry and evolve many or all of the same phenotypes (both structural and functional), albeit possibly with different genotypes. Further discussion of these points is found in Chapter 27.

10.4 Could there be life without evolution?

Many of the definitions of life include the phrase "undergoes Darwinian evolution" (Chapter 5). The implication is that phenotypic changes and adaptation are necessary to exploit unstable environmental conditions, to function more optimally in the environment, and to provide a mechanism to increase biological complexity. Evolutionary changes have even been suggested for hypothesized "clay crystal life" of Cairns-Smith (1982), referring to randomly occurring errors in crystal structure during crystal growth as analogous to mutations (Section 27.4.2). Would a self-replicating chemical system capable of chemical transformations in the environment be considered life? If self-replicating chemical compounds are not life, then replication by itself is not sufficient as a defining characteristic of life. Likewise, the ability to undergo Darwinian evolution, that is, a process that results in heritable changes in a population, is also not sufficient to define life if we consider minerals that are capable of reproducing errors in their crystal structure to be equivalent to evolution. Although this property of clays may have been vital in the origin of life and particularly in the prebiotic synthesis of organic macromolecules and as catalysts for metabolic reactions, can the perpetuation of "mistakes" in crystal structure result in the selection of a "more fit" crystal structure? It is important to emphasize that evolution is not simply reproducing mutations (mistakes in clays), but selecting those variants that are functionally more fit.

The canonical characteristics of life are an inherent capacity to adapt to changing environmental conditions and to increase in complexity by multiple mechanisms, but particularly by interactions with other living organisms (and, at least on Earth, also with viruses). Natural selection is the key to evolution and the main reason why Darwinian evolution persists as a characteristic of many definitions of life. Clays could never evolve an eye or a nose, or adapt behavioral strategies to exclude clays with other crystal characteristics. Hmmm – would Michael Crichton's *Andromeda Strain* (1969), a carbon-based crystal capable of using chemical and physical energy sources, be considered life? (Incidentally, the *Andromeda Strain* could also mutate, which was probably a necessity to reach a happy ending to the story.) The only alternative to evolution for producing diversity would be to have environmental conditions that continuously create different life forms, or similar life forms with random and frequent "mistakes" made in the synthesis of chemical templates used for replication or metabolism. These mistakes would be equivalent to mutations and could lead to traits that gave some selective advantage in an existing community or in exploiting new habitats. This could lead to life forms that undergo a form of evolution without a master information macromolecule such as DNA or RNA. It is difficult, however, to imagine such life forms being able to "evolve" into complex structures unless other mechanisms such as symbiosis or cell–cell fusion are available.

10.5 Evolution and extraterrestrial life

We have seen that evolution is much more than mutation and natural selection. It is the key mechanism for heritable changes to occur in a population. Mutation is not the only mechanism for acquiring new genes. Lateral gene transfer appears to be one of the most important mechanisms and clearly one of the earliest mechanisms for creating diversity and possibly for building genomes with the requisite information to result in free-living cells, as opposed to codependent communities of "precells" with insufficient genetic information to escape communal life (Baross and Hoffman, 1985). Lateral gene transfer is also one of the mechanisms to align genes from different sources into complex functional activities such as magnetotaxis and dissimilatory sulfate reduction. It is possible that this mechanism was important in the evolution of metabolic and biosynthetic pathways and other physiological traits that may have evolved only once even though they are present in a wide diversity of organisms. The coevolution between two or more species is also a hallmark of evolution manifested in many ways from insect–plant interactions to the hundreds of species of bacteria involved in the nutrition of ruminant animals. The organisms and the environment also coevolve depending on the dominant characteristics of the environment and the availability of carbon and energy sources. Even some of the most extreme environments on Earth, such as hydrothermal vent sulfide chimneys and the very acidic Rio Tinto River in Spain, have a remarkably high diversity of organisms (Kelley *et al.*, 2002; Zettler *et al.*, 2003). Diversity drops off, however, in environments with *combinations* of stressors such as high temperature and high pH, or high salt and high or low temperature (Chapter 14).

If the ability to undergo Darwinian evolution is a canonical trait of life no matter how different that life form is from Earth life, then are there properties of evolving extraterrestrial organisms that would be detectable as positive signs of life? Evolution provides organisms the opportunity to exploit new and changing environments, and one piece of evidence for the probable cosmic ubiquity of evolution is that on Earth life occupies all available habitats and even creates new habitats as a consequence of its metabolisms. Another hallmark of evolution is the ability of organisms to coevolve with other organisms and to form permanent and obligatory associations. Also, it is highly probable that an inevitable consequence of evolution is the elimination of radically different biochemical lineages of life that may have formed during the earliest period of evolution of life. Extant Earth life is the result of either selection of the most fit lineage or homogenization of some or all of the different lineages into a common ancestral community that developed into the present three major lineages (domains). All have a common biochemistry based on presumably the most "fit" molecular information strategies and energy yielding pathways among a potpourri of possibilities. One caveat and perhaps a verification of the above statement is that genetic remnants of other lineages may still exist in some of the deeply rooted archaea, as evidenced from the unique 16 S rRNA found in *Nanoarchaeum equitans* and novel and presently undelineated metabolic pathways in some hyperthermophilic Crenarchaeota (Huber *et al.*, 2003; Hügler *et al.*, 2003).

Thus, one of the apparent generalizations that can be made from extant Earth life, and the explanation for the development of a "unity of biochemistry" in all organisms, is that lateral gene transfer is both an ancient and an efficient mechanism for rapidly creating diversity and complexity. Lateral gene transfer is also an efficient mechanism for selecting the genes that are most "fit" for specific proteins and transferring them into diverse groups of organisms. The result is both the addition of new genes and the replacement of less-fit genes having a similar function. Natural selection based solely on mutation is not likely an adequate mechanism for evolving complexity. More importantly, lateral gene transfer and endosymbiosis are probably the most obvious mechanisms for creating complex genomes that can lead to free-living cells and complex cellular communities in the short geological time available from life's origin to the establishment of microbial communities more than 3.8 billion years ago (Section 12.3). An important implication of the existence of viruses or virus-like entities during the early evolution of cellular organisms is that their genomes may have been the source of most genetic innovations, due to their rapid replication rates, high rates of mutation from replication errors, and gene insertions from diverse host cells. It is interesting that Darwin perceived evolution as random changes in individual species that could lead to selection of more fit traits, but he could not have known that some of these fit traits could be transferred to species that were not only not sexually compatible but belonged to separate domains.

It is clear that both the individual organism and its community coevolve. In a sense, evolution is evolving, allowing cells to control their own evolution – accept or reject changes in genotypes from newly acquired foreign genes. While this has already been demonstrated in bacteria, the source of foreign genes and the kinds of genes most likely to be selected for permanence are largely not known. However, it is clear that the evolution of a useful trait by one organism frequently means that it is likely to be acquired by other organisms. It appears that the field of biology is beginning to break out of its molecular-reductionist "egg" and emerging more focused on what Carl Woese (2004) terms "holistic biology," where the emphasis, rather than just on genes, is on the cell, communities and ecosystems. This would also take evolution to a new level of inquiry with emphasis on coevolution, cellular complexity, and the re-examination of the concept of ecosystems as "super-organisms." The new science of Evo-Devo integrates well with this new holistic approach while offering another lesson about evolution: chance mutations or microevolution create the panoply of gene variation, but it is key genes and combinations of key genes that "better meet the imperatives of ecological necessity, and they arise and are selected for repeatedly" (Carroll, 2005).

Finally, what are the limits of evolution for Earth life? This is a complex question with many different components. On the one hand, it involves the different possible biochemistries from carbon chemistry that are not found in extant Earth organisms but could be better suited for environmental conditions that exist on other planets and moons (Chapter 27). The technology exists to design genes and groups of genes that could lead to novel phenotypes suited to exploit new habitats and novel energy sources. These kinds of studies would be important and perhaps essential in our quest to search for life elsewhere. Another component to the question of the limits of evolution is where *we* are going and what will *Homo sapiens* or its successors be like if it continues to evolve for tens of thousands or millions of years? This is an integral part of our search for advanced extraterrestrial intelligence, which requires us to imagine our future portrait (Chapter 26). We cannot imagine all of the possible changes that will occur after millions of years of evolution, but based on just the tens of thousands of years of primate evolution, it is likely that one possible outcome will be an increasing ability to control our environment and all that is evolving. It is also likely that we will someday know if we are alone in the Universe.

10.6 Summary

It is evident that cells are more than the information encoded in their genomes. They are part of a highly integrated biological and geochemical system in whose creation and maintenance they have participated. The unity of biochemistry among all of Earth's organisms emphasizes the ability of organisms to interact with other organisms to form coevolving communities, to acquire and transmit new genes, to use old genes in new ways, to exploit new habitats, and most important to evolve mechanisms to help control their own evolution. It is expected that these characteristics are likely to be present in extraterrestrial life even if it has had a separate origin and a very different unified biochemistry from that of Earth life.

Since evolution is an essential feature of Earth life and probably all life, the search for life elsewhere should include a search for evidence of evolution.

REFERENCES

Baross, J. A. and Hoffman, S. (1985). Submarine hydrothermal vents and associated gradient environments as sites for the origin and evolution of life. *Orig. Life*, **15**, 327–345.

Bushman, F. (2001). *Lateral DNA Transfer, Mechanisms and Consequences*. New York: Cold Spring Harbor Laboratory Press.

Cairns-Smith, A. G. (1982). *Genetic Takeover and the Mineral Origins of Life*. Cambridge: Cambridge University Press.

Canchaya, C., Fournous, G., Chibani-Chennoufi, S., Dillmann, M.-L., and Brüssow, H. (2003). Phage as agents of lateral gene transfer. *Curr. Opin. Microbiol.*, **6**, 417–424.

Carroll, S. B. (2005). *Endless Forms Most Beautiful: the New Science of Evo-Devo*. New York: W. W. Norton & Co.

Copley, S. D., Smith, E., and Morowitz, H. J. (2005). A mechanism for the association of amino acids with their codons and the origin of the genetic code. *Proc. Natl. Acad. Sci. USA*, **102**, 4442–4447.

Crichton, M. (1969). *The Andromeda Strain*. New York: Alfred A. Knopf.

Darwin, C. (1859). *On the Origin of Species*. London: John Murray.

Dobzhansky, T. (1970). *Genetics of the Evolutionary Process*. New York: Columbia University Press.

Eldredge, N. and Gould, S. J. (1972). Punctuated equilibria: an alternative to phyletic gradualism. In *Models in Paleobiology*, T. J. M. Schopf and J. M. Thomas (eds.), pp. 82–115. San Francisco: Freeman.

Faruque, S. M. and Mekalanos, J. J. (2003). Pathogenicity islands and phage in *Vibrio cholerae* evolution. *Trends in Microbiol.*, **11**, 505–510.

Grünberg, K., Wawer, C., Tebo, B. M., and Schüler, D. (2001). A large gene cluster encoding several magnetosome proteins is conserved in different species of magnetotactic bacteria. *Appl. Environ. Microbiol.*, **67**, 4573–4582.

Gupta, R. S. (1998). Protein phylogenies and signature sequences: a reappraisal of evolutionary relationships among Archaebacteria, Eubacteria and Eukaryotes. *Microbiol. Mol. Biol. Rev.*, **62**, 1435–1491.

Hambly, E. and Suttle, C. A. (2005). The virosphere, diversity, and genetic exchange within phage communities. *Curr. Opin. Microbiol.*, **8**, 444–450.

Hedges, S. B. (2002). The origin and evolution of model organisms. *Nature Reviews Genetics*, **3**, 838–849.

Hershey, A. D. and Chase, M. (1952). Independent functions of viral protein and nucleic acid in growth of bacteriophage. *J. Gen. Physiol.*, **36**, 39–56.

Huber, H., Hohn, H. J., Stetter, K. O., and Rachel, R. (2003). The phylum Nanoarchaeota: present knowledge and future perspectives of a unique form of life. *Res. Microbiol.*, **154**, 165–171.

Hügler, M., Huber, H., Stetter, K. O., and Fuchs, F. (2003). Autotrophic CO_2 fixation pathways in archaea (Crenarchaeota). *Arch. Microbiol.*, **179**, 160–173.

Jiang, S. C. and Paul, J. H. (1998). Gene transfer by transduction in the marine environment. *Appl. Environ. Microbiol.*, **64**, 2780–2787.

Kelley, D. S., Baross, J. A., and Delaney, J. R. (2002). Volcanoes, fluids, and life at mid-ocean ridge spreading centers. *Ann. Rev. Earth Planet. Sci.*, **30**, 385–491.

Lamarck, J. B. (1809). *Zoological Philosophy*. Translated into English by H. Elliott, 1914. New York: Macmillan.

Margulis, L. (1993). *Symbiosis in Cell Evolution*, second edn. New York: W. H. Freeman.

Margulis, L., Dolan, M. F., and Guerrero, R. (2000). The chimeric eukaryote: origin of the nucleus from the

Karyomastigont in amitochondriate protists. *Proceed. Natl. Acad. Sci. USA*, **97**, 6954–6959.

Martin, W. and Müller, M. (1998). The hydrogen hypothesis for the first eukaryote. *Nature*, **392**, 37–41.

Mussmann, M., Richter, M., Lombardot, T., *et al.* (2005). Clustered genes related to sulfate respiration in uncultured prokaryotes support the theory of their concomitant horizontal transfer. *J. Bacteriol.*, **187**, 7126–7137.

Parker, A. (2003). *In the Blink of an Eye*. Cambridge, MD: Perseus Publishing.

Rivera, M. C. and Lake, J. A. (2004). The ring of life provides evidence for a genome fusion origin of eukaryotes. *Nature*, **431**, 152–155.

Sapp, J. (1994). *Evolution by Association. A History of Symbiosis*. New York: Oxford University Press.

Sapp, J. (2005). The bacterium's place in nature. In *Microbial Phylogeny and Evolution*, J. Sapp (ed.) pp. 3–52. New York: Oxford University Press, Inc.

Wakeford, T. (2001). *Liaisons of Life*. New York: John Wiley & Sons, Inc.

Watson, J. D. and Crick, F. H. C. (1953). A structure for deoxyribose nucleic acid. *Nature*, **171**, 737–738.

Woese, C. R. (1998). The universal ancestor. *Proc. Natl. Acad. Sci. USA*, **95**, 6854–6859.

Woese, C. R. (2002). On the evolution of cells. *Proc. Natl. Acad. Sci. USA*, **99**, 8742–8747.

Woese, C. R. (2004). A new biology for a new century. *Microbiol. Mol. Biol. Rev.*, **68**, 173–186.

Woese, C. R., Kandler, O., and Wheelis, M. L. (1990). Towards a natural system of organisms: proposal for the domains Archaea, Bacteria and Eukarya. *Proc. Natl. Acad. Sci. USA*, **87**, 4576–4579.

Wright, B. E. (2004). Stress-directed adaptive mutations and evolution. *Mol. Microbiol.*, **52**, 643–650.

Zettler, C. A. A., Messerli, M. A., Laatsch, A. D., Smith, P. J. S., and Sogin, M. L. (2003). From genes to genomes: beyond biodiversity in Spain's Rio Tinto. *Biol. Bull.*, **204**, 205–209.

FURTHER READING AND SURFING

Desmond, A. and Moore, J. (1991). *Darwin: the Life of a Tormented Evolutionist*. New York: Warner. Excellent and thoroughly researched biography of Darwin. Darwin struggled as to how to present the theory of evolution in a way acceptable to a community shackled by Victorian mores.

Judson, H. W. (1979). *The Eighth Day of Creation: the Makers of Revolution in Biology*. New York: Simon and Schuster. The definitive historical study of the mid-twentieth-century birth of molecular biology, and a must-read for anyone interested in how revolutions in science get started. Now reprinted with an updated preface (New York: Cold Spring Harbor Laboratory Press, 1996).

Knoll, A. H. (2003). *Life on a Young Planet: the First Three Billion Years of Evolution on Earth*. Princeton: Princeton University Press. A "tour de force" portrait of life from Earth's beginning to the Cambrian explosion. Knoll is masterful in blending geology, geochemistry, and biology in the context of Earth history.

Lovelock, J. E. (1979). *Gaia: a New Look at Life on Earth*. Oxford: Oxford University Press. James Lovelock and Lynn Margulis first put forth the proposition that the composition and temperature of the atmosphere is an evolutionary product of interrelated activities in the biosphere, especially those of microorganisms, and that the entire biosphere behaves as a single self-regulating organism. The Gaia Hypothesis has been every bit as influential as it is controversial (see Section 10.5 for related thinking). www.mendelweb.org. An excellent website for learning the details of what Mendel actually did.

Conway Morris, S. (2003). *Life's Solution: Inevitable Humans in a Lonely Universe*. Cambridge: Cambridge University Press. Conway Morris has a different perspective on evolution. He argues for determinism rather than contingency, i.e., that evolution has predictable and inevitable outcomes. His metaphysical arguments are interesting and thought-provoking and somewhat reminiscent of Chapter 31 in Christian de Duve's excellent book, *Vital Dust: Life as a Cosmic Imperative* (New York: Basic Books, 1995).

Ptashne, M. (1992, second edn.). *A Genetic Switch*. Cambridge, MA: Blackwell Science. This classic work describes the basic molecular reactions underlying the regulation of gene transcription in all organisms and how the genes involved in these reactions, when combined, produce complex regulatory circuits. The regulatory circuits found in bacteria are the forerunners to the evolution of the more complex regulatory circuits involved in macroevolution and the emergence of Evo-Devo (Carroll, 2005).

Ridley, M. (ed.) (1997). *Evolution*. Oxford: Oxford University Press. Excellent compilation of many of the classic papers on evolution. The list of authors is the "Who's Who" of great evolution thinkers and includes Charles Darwin, Stephen J. Gould, Ernst Mayr, George Gaylord Simpson, Richard Dawkins, and Francis Crick. Topics include adaptation, macroevolution, molecular evolution, biodiversity, human evolution, and evolution and philosophy.

11 Evolution of metabolism and early microbial communities

John A. Leigh, David A. Stahl, and James T. Staley
University of Washington

11.1 Introduction

The process of metabolism, in which cells carry out biochemical reactions, is a hallmark of all living organisms. *Catabolic* reactions generate energy for the organism while *anabolic* reactions are used for the synthesis of cell material. Metabolic pathways in today's living organisms have been evolving for more than 3.5 Gyr. In fact, since metabolism would have been necessary even for the earliest organisms, its evolution cannot be separated from the origin of life. Contemporary metabolic pathways are presumed to be much more elaborate and sophisticated than those that first evolved. Indeed, metabolism today is extraordinarily rich and diverse, ranging from the use of various inorganic chemicals such as hydrogen or sulfur for nutrients and energy, to several forms of photosynthesis, to the metabolism of hundreds of organic compounds. It is impossible for us, at least at this time, to know which pathways originated first and how they evolved. Nonetheless, because metabolism is essential to life, understanding how metabolism evolved is of considerable importance. Furthermore, we have good grounds to speculate on which of life's diverse metabolisms evolved earliest and which could only have come later. Microorganisms, most likely resembling present day Archaea and Bacteria, were the first organisms, so it is their metabolism that is of relevance. Indeed, all basic metabolic pathways on Earth today can be traced to microorganisms.

The goal of this chapter is to describe, insofar as possible, the evolution of metabolism. Although there are several principles that guide our considerations, two are predominant. First, we want to understand the nature of the chemical milieu (energy sources and biogenic compounds) in which life's metabolisms originated and evolved.[1] Our second guiding principle is that of evolving complexity. Life on Earth has been

continuous; any pair of organisms on this planet are distant, albeit very distant, cousins – as revealed by the genetic relationship between humankind and microorganisms. In looking for clues about the character of early life, we must incorporate this continuity. Implicit in this perspective is the hypothesis that early life used a simpler form of a contemporary metabolism. We emphasize those metabolisms that are represented in the contemporary biosphere that could have existed in earlier and simpler forms.

This chapter begins with the setting on early Earth in which the origin of life and its attendant metabolism occurred. We next discuss the rather limited categories of information that give us any direct evidence for the nature of early metabolism. We then present the principles and mechanisms of contemporary metabolism as a basis for considering the simplicity and feasibility of possible early metabolisms. This background enables us to argue for specific metabolic types as the earliest to evolve and others as later innovations. We end by describing microbial communities in which different metabolic types interact to form entire microbial ecosystems probably similar to those on early Earth.

11.2 The setting: conditions on early Earth

In order to discuss early metabolism, it is important to briefly consider the conditions on early Earth in which metabolic processes originated and evolved. Let us consider *physical* conditions and *chemical* conditions. One factor that influenced physical conditions was

---

[1] For present purposes we describe metabolism as if it evolved on Earth. Nevertheless, we do not exclude the possibility that life arrived here from an extraterrestrial source. In either case, certainly the chemical environment would have had to be suitable on Earth for life to thrive.

Planets and Life: The Emerging Science of Astrobiology, eds. Woodruff T. Sullivan, III and John A. Baross. Published by Cambridge University Press. © Cambridge University Press 2007.

instability. It is noteworthy, as discussed elsewhere in several chapters, that conditions on Earth, though continuously stable for many millions of years, can be dramatically affected by cosmic events such as asteroid and comet impacts and climatic events such as "Snowball Earth" episodes. Indeed, an impact of an asteroid only 100 km in diameter can sterilize all surface life on our planet and evaporate a 3-km-deep global ocean. Only microbial life at a depth of several km could then survive. Thus, these cosmic events have had major and largely unknown effects resulting in extinction of many species and re-direction of the evolution of Earth's biota. It is therefore possible that catastrophic events during the early history of Earth may have also affected the evolution of metabolism, constraining it throughout the Archaean period (before 2.5 Ga) to simple forms that could have evolved in relatively short intervals of time.

Chemical conditions on early Earth were marked by a relative dearth of nutrients that make life so rich today. Nevertheless, a steady supply of certain nutrients was available, and some of the most critical are listed in Table 11.1. First of all, life could not have evolved unless organic materials were already present. As discussed in Sections 3.8 and 7.4, organic and other volatile materials are synthesized in space and became incorporated into the Earth through impacts by large bodies (as part of the formation process) and by comets and asteroids (after Earth's formation) in the so-called early heavy bombardment. These impacts may have disrupted evolution, but they also provided nutrients for life. In addition, the abiotic synthesis of organic materials occurred *on* Earth. Acetic acid (CH_3COOH) synthesis has been shown to occur under primordial conditions, and it has been suggested that this laid the groundwork for an important biochemical pathway (discussed below). Therefore, although on the early Earth organic materials were less abundant than today, we may reasonably assume that they were present in sufficient quantities and varieties for life to originate and evolve. But as discussed below, organic materials alone would not likely have been sufficient to sustain life. Other nutrients were necessary to allow organisms to carry out the early equivalent of eating and breathing, in short, to utilize electron donors and electron acceptors (explained below) to generate the metabolic energy needed for life. Perhaps most notably, early Earth had no free oxygen (Chapter 4) and thus the first metabolisms that arose were carried out *anaerobically*. Fortunately, we have some idea of what nutrients were available when life originated and where

TABLE 11.1 Building blocks of life and replenishable sources of energy on early Earth

Nutrient	Sources on primitive Earth
Fixed carbon[a]	
Simple amino acids, sugars, other organics	Meteors and asteroids
Acetic acid	Chemosynthesis on Earth
Electron donors	
Hydrogen (H_2)	Volcanic outgassing
Ferrous iron (Fe^{2+})	Mineral dissolution
Electron acceptors	
Carbon dioxide and carbon monoxide (CO_2 and CO)	Volcanism
Sulfite ($SO_3^=$)	Volcanism; hydration of sulfur dioxide (SO_2)
Elemental sulfur (S^o)	Mineral dissolution

[a] Most of these carbon compounds can also be synthesized by electric discharge, heat, and impacts.

they came from (Table 11.1). In this chapter we discuss how these nutrients constrained early life and how they were used metabolically.

11.3 Evidence for the nature of early metabolisms

Although much of our discussion rests on considerations of complexity and Earth's early chemistry, a few categories of more direct evidence can be brought to bear on the nature and timing of early metabolisms. This evidence comes from three primary sources: biomarkers, the geochemical record, and phylogenetic evidence.

11.3.1 Biomarkers

Unique biomarkers, organic compounds thought to be characteristic of certain organisms, have been useful for understanding early metabolisms. For example, hopanoids are sterol-like compounds that are found only in cell membranes of the bacterial group, cyanobacteria, which carry out oxygenic (oxygen-producing) photosynthesis. The presence of hopanoids has been used to date the evolution of these organisms to about 2.5 Ga

(Summons *et al.*, 1999; Section 12.3.3). This date agrees well with the geochemical record for oxygen production on Earth and supports the view that the first oxygenic photosynthetic organisms were cyanobacteria.

11.3.2 Geochemical record

The geochemical record has been useful in establishing evidence for certain types of metabolic activity. Strong evidence indicates that carbon dioxide fixation was an early metabolic event on Earth; this is called *primary production*, in which environmental CO_2 is taken up and the carbon incorporated into organic matter. Carbon dioxide-fixing organisms, including Bacteria and Archaea, also cause isotopic fractionation. They preferentially fix the lighter form, $^{12}CO_2$, into organic material in comparison with the heavier isotopic form, $^{13}CO_2$ (see Section 11.3 for details). As a result, in bands from the Isua deposits in Greenland that date to 3.5 Ga, kerogen (deposited organic carbon that has been chemically altered in sedimentary rocks) has a lighter isotopic signature than carbonate (Schidlowski, 1988). This is excellent evidence that primary production was an early process on Earth. However, there is no definitive evidence about the nature of the earliest primary producing organisms. Were they chemosynthetic (using chemical energy) or photosynthetic (using light energy), and which organisms were first responsible for this process? Interestingly, the anaerobic methane oxidizing microbes are known to produce the lightest organic carbon, which may explain the preponderance of lighter organic carbon isotopes in certain layers of the Isua sediments (Schidlowski, 1988).

The other isotope for which isotopic fractionation has provided information on early metabolism is sulfur. The process of sulfate reduction (described below) has been recently dated to about 3.5 Ga based upon the fractionation of ^{32}S versus ^{34}S (Shen *et al.*, 2001; Section 12.3).

11.3.3 Phylogenetic analysis

Phylogenetic analysis is another useful approach for ascertaining which bacterial groups were involved in early metabolism. By knowing which group of organisms evolved first, it is possible to hypothesize which metabolic activities arose first. Indeed, the Tree of Life based upon analysis of the gene for the RNA of the ribosomal small subunit (16 S rDNA) is of interest in this regard (Fig. 11.1). The deepest branches in the Tree

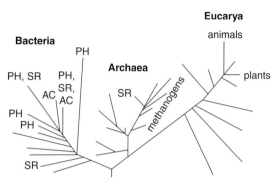

FIGURE 11.1 The Tree of Life derived from comparison of each organism's small subunit ribosomal RNA gene (16 S rDNA). Life is divided into three domains, Bacteria, Archaea, and Eucarya (Woese *et al.*, 1990). The phylogenetic positions of organisms having the major types of metabolism discussed here are indicated: methanogens, acetogens (AC), sulfate reducers (SR), and phototrophs (PH).

are of thermophilic Archaea and Bacteria, implying that these are the earliest ancestors of any organisms still alive today. Some have argued that members of these phyla, many of which grow anaerobically, were likely the first organisms on Earth.

However, in this regard, two issues are important: (1) discernible detail regarding the major groups of microorganisms is inadequate, so there remains a question about which organisms might have been the earliest; and (2) horizontal gene transfer, which involves the exchange of genes among different lineages (Section 10.3). Most genes are transferred by vertical inheritance from one generation to another within a species. However, in bacteria, genes can be transferred from one species to another via horizontal gene transfer. Thus some of the genes responsible for carrying out methanogenesis (methane production) have also been found in bacteria that are methane oxidizers (methane consumers) (Chistoserdova *et al.*, 1998). Methanogens are found in the domain Archaea whereas methane oxidizers are found in the domain Bacteria. This sort of horizontal gene transfer makes it very difficult to trace the evolution of metabolic functions and other characteristics of organisms.

Analyses of complete genome sequences should help resolve the issue of which organisms and which metabolisms occurred early. One of the hopes is that a sufficient number of different organisms representing all of the diverse phyla of the Bacteria and Archaea will have been sequenced during the next decade. Through such an effort, a more robust Tree of Life and tree of metabolic functions will result.

11.4 Contemporary metabolism: principles an astrobiologist needs to know

The evolution of life has changed the Earth, even as the Earth has provided the habitats for life. For example, the evolution of oxygenic photosynthesis gave rise to oxygen on Earth and enabled the later evolution of aerobic (oxygen-utilizing) metabolism. On the other hand, it is unlikely that any habitat that once existed has disappeared completely (for example, anaerobic habitats still exist), and the vast metabolic diversity on Earth today likely contains all the major metabolic types that ever evolved. Therefore, to understand the evolution of metabolism, we study the essential metabolic features of present-day organisms.

11.4.1 Anabolism

One of the two major parts of metabolism is *anabolism*, "constructive" metabolism or *biosynthesis*, the building of new cell material for growth and reproduction. Some organisms, the *heterotrophs* (Greek, "nourished from other"), synthesize cell material from organic molecules that are present in their environment. In contrast, other organisms, the *autotrophs* ("self-nourished"), make organic molecules from carbon dioxide in a process called *carbon dioxide fixation*. The autotrophs are of special interest because they are the *primary producers* that ultimately provide the entire biosphere with organic material. Depending on the organism, carbon dioxide fixation can take place through a variety of pathways. The pathway that appears to be most advanced is the Calvin–Benson cycle, used by plants and some bacteria. Other pathways for carbon dioxide fixation include the acetyl coenzyme A pathway, the reverse tricarboxylic acid pathway, and the hydroxypropionate pathway. In this chapter we emphasize the acetyl coenzyme A pathway (Section 11.6) because of its possible early evolution. See Chapter 8 for a discussion of the reverse tricarboxylic acid pathway and its possible significance in the origin of life.

11.4.2 Catabolism

The chemical reactions that constitute metabolism must be thermodynamically feasible. Anabolic reactions require energy to create order, and thus organisms must also have *catabolic* pathways, "destructive" metabolism, to generate the energy needed for anabolism.

Catabolic pathways are remarkably diverse. The organisms that are most like humans "burn" organic food using oxygen. But this type of catabolism, termed *aerobic organotrophy*, is only one specialization. Before oxygen, organisms "breathed" substances such as carbon dioxide, sulfate, iron, and other substances, and some organisms still do. In addition, many organisms, the *lithotrophs*, "eat" hydrogen, sulfur, ammonia, and other unlikely-seeming foods. Still other organisms, the *phototrophs*, harvest the energy of light.

11.4.3 Oxidation, reduction, and electron flow

Catabolism follows the chemical principles of oxidation and reduction. These principles are best illustrated using the form of catabolism called respiration. For respirers, oxidation corresponds to "eating," and reduction to "breathing." Between the two types of reactions, electrons flow. (Note that oxidation does not necessarily involve oxygen, only the removal of electrons!) The work done using catabolic energy is analogous to the work done as electrons flow between the terminals of a battery. The only requirement is that the reaction has the potential to release energy, which condition is realized if the substance oxidized has a more negative electrical potential than the substance reduced. Therein lies the constraint that determines which catabolisms are possible. To generate energy the organism oxidizes, or "eats," the *electron donor*, while it reduces, or "breathes," the *electron acceptor*.

Table 11.2 lists the major types of metabolism and illustrates that the electron acceptor is a major differentiating characteristic. Thus, respirers range from methanogens and acetogens, which utilize carbon dioxide and produce methane and acetic acid, respectively, as waste products, to aerobes, which use oxygen and produce water. The electron acceptor has also been a major constraint in the evolution of catabolism because many electron acceptors (e.g., oxygen) became abundant relatively late (note that many of the substances listed in Table 11.2 are absent from Table 11.1). On the other hand, the early presence of electron acceptors such as carbon dioxide is one reason why organisms such as methanogens will be discussed below as possible early catabolic types.

11.4.4 Respiration, photosynthesis, and fermentation

The three forms of catabolism – respiration, photosynthesis, and fermentation – are distinguished largely by

TABLE 11.2 Major types of catabolism

Organism	Electron donor	Electron acceptor
Respirers		
Methanogen	H_2, organic	carbon dioxide
Acetogen	H_2, organic	carbon dioxide
Sulfate reducer	H_2, organic	sulfate ($SO_4^=$)
Sulfur reducer	H_2, organic	sulfur ($S°$)
Iron reducer	H_2, organic	iron (Fe^{3+})
Denitrifier	various	nitrate ($NO_3^=$)
Aerobe	various	oxygen
Photosynthesizers	various	carbon dioxide[a]
Fermenters	organic	no external acceptor

[a] In photosynthesis, carbon dioxide is a biosynthetic, not a catabolic, substrate.

their patterns of electron flow (Fig. 11.2). *Respiration* (described above) is the most straightforward because separate *substrates* (substances in the environment used for metabolism) are used by the organism as electron donor and electron acceptor. In photosynthesis and fermentation, electrons still flow from a donor to an acceptor, but the pattern is less intuitive to respiring organisms such as you, the reader. *Photosynthesizers* take advantage of the key property of chlorophyll. This marvelous molecule functions well as an electron acceptor but also, when energized by light, is an effective electron donor. As a result, electrons flow from chlorophyll and back to chlorophyll in a cyclic pattern, doing work in the process. In addition, superimposed on this chlorophyll-driven cyclic electron flow, electrons from an electron-donating substrate are used to reduce carbon dioxide in carbon dioxide fixation. *Fermenters* have yet another pattern. An electron-donating substrate is oxidized, but the electron acceptor is an internal metabolic intermediate rather than a substrate from the environment. It is worth noting that although electron flow in photosynthesizers and fermenters is seemingly more complex, these organisms are free from the constraint imposed by the use of an electron acceptor from the environment.

11.4.5 ATP and the proton-motive force

So far we have described catabolism as consisting of metabolic pathways involving oxidation and reduction. But how is the energy actually stored and transferred? ATP (adenosine triphosphate) is the energy currency of biology. ATP contains phosphoanhydride bonds that yield energy upon hydrolysis (breakdown through combination with water). Certain other compounds have energy-rich phosphoanhydride bonds too, including ADP (adenosine diphosphate), PP (pyrophosphate), and other metabolic intermediates, but ATP predominates in present-day organisms. Consequently, the formation of ATP and the hydrolysis of ATP provide the energy link between catabolism and anabolism; catabolism generates ATP and anabolism uses it. The mechanism of ATP generation is usually simplest in the case of *fermentation* (Fig. 11.2). Here, ATP is directly generated by reactions that are part of the catabolic pathway in a process termed *substrate-level phosphorylation* – a metabolic intermediate containing an energy-rich phosphoanhydride bond transfers its phosphate to ADP, forming ATP.

For *respiration* and *photosynthesis*, however, ATP is generated in a two-stage process in which the cell membrane plays an essential role. Indeed, just as a membrane is necessary to separate the contents of a cell from the surroundings, so is a membrane necessary for energy generation and storage. The principle is akin to charge separation. First the flow of electrons through a membrane-bound *electron transport chain* creates a potential between the interior and exterior of the cell called a *proton-motive force*. The proton-motive force consists of gradients across the membrane of proton concentration (pH) and ionic charge. The energy supplied by the proton-motive force is then used to make ATP, via a membrane-bound enzyme[2] called ATP synthase that couples the synthesis of ATP to the import of protons. This process is termed *electron transport phosphorylation* or *oxidative phosphorylation*. Hence, the cell membrane plays a central role not only in the cell's structural integrity, but also in its ability to generate energy.

11.4.6 Energy yields of catabolism

As described above, all catabolic mechanisms lead to the generation of ATP. But which forms of catabolism generate the most ATP and are the most advantageous for the organism? A quantity used by chemists to evaluate the energy yield of a reaction (or metabolic pathway) is the change in Gibbs free energy, or ΔG, expressed

---

[2] An *enzyme* is a protein that acts as a catalyst for a biochemical reaction.

Respiration

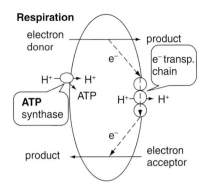

Photosynthesis

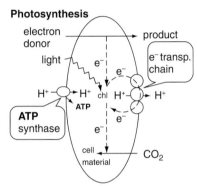

Fermentation

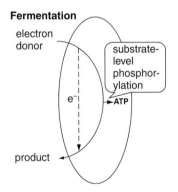

FIGURE 11.2 The three types of catabolism: respiration, photosynthesis, and fermentation. In each diagram, the bacterial cell membrane (wall layer) is represented as an oval. The conversion of substrates (electron donors and electron acceptors) to products is shown as solid arrows, and the flow of electrons is shown as dashed arrows.

In *respiration*, separate substrates donate and accept electrons. The flow of electrons through a membrane-bound electron transport chain generates a proton-motive force, usually by exporting protons (H^+) from the inside to the outside of the cell. A membrane-bound enzyme called ATP synthase uses the proton motive force, coupling ATP synthesis to proton import.

Photosynthesis differs from respiration in the use of light to stimulate electron flow between reduced and oxidized chlorophyll, resulting in a cyclic pattern. An electron donor supplies electrons needed for carbon dioxide fixation. These electrons receive a "boost" to a more negative electrical potential through the chlorophyll. In the case of oxygenic photosynthesis (not shown), electrons from water receive two boosts from two different chlorophylls.

In *fermentation*, no electron transport chain or membrane-bound ATP synthase is involved. Oxidation, ATP synthesis (through substrate-level phosphorylation), and reduction occur from a single electron-donating substrate.

in kilojoules (kJ). A large negative ΔG indicates a high yield of energy, while a positive ΔG indicates that energy is consumed. Hence, all catabolic pathways must have a negative ΔG, and the more negative the better. Since catabolism is based on electron flow, the ΔG depends on the relative electrical potentials (called redox potentials) of the electron donors and acceptors.

Some examples of catabolic pathways are given below (from Gottschalk, 1986). Aerobic respiration using glucose as electron donor and oxygen as electron acceptor is one of the most effective in producing ATP:

Aerobic respiration (ATP yield = 38)
glucose ($C_6H_{12}O_6$) + 6O_2 → 6CO_2 + 6 H_2O,
$\Delta G = -2870$ kJ.

ΔG has a large negative value because the oxidation of glucose occurs at a very negative redox potential while the reduction of oxygen occurs at a very positive redox potential. Organisms that carry out aerobic respiration can therefore grow fast – these include highly evolved forms such as animals.

Photosynthesis is also very effective. The effectiveness of light-driven electron flow via chlorophyll can be appreciated when one realizes that oxygenic photosynthesis (see below) is the reverse of aerobic respiration. Since the reverse of an energy-yielding process is an energy-requiring one, the use of light via chlorophyll must be very effective. Like animals, plants are also highly evolved and successful.

In contrast, other forms of respiration, such as methanogenesis or sulfate reduction, yield less energy:

Methanogenesis (respiration)
4H_2 + CO_2 → CH_4 + 2 H_2O, $\Delta G = -131$ kJ.

Sulfate reduction (respiration)
4H_2 + SO_4^{2-} + H^+ → HS^- + 4H_2O,
$\Delta G = -152$ kJ.

Although methanogens and sulfate reducers use electron donors with large negative redox potentials such as hydrogen or organic substances, the electron acceptors (carbon dioxide and sulfate) have redox potentials that are also in the negative range. The ΔGs have small negative values and the ATP yields are low (the exact ATP yields are unknown).

Fermenters are also limited by small redox potential differences between the electron donor and the internal electron acceptor:

Lactic acid fermentation (ATP yield = 2)
glucose $\rightarrow 2$ lactic acid $(CH_3CH(OH)COOH)$, $\Delta G = -196\,kJ$.

Compounding this limitation, fermenters also lack the opportunity to fully oxidize their substrates.

11.5 Early metabolic mechanisms

We now speculate on the order in which different forms of metabolism evolved. The above discussion has served to illustrate that metabolism today is complex and diverse. One of our tenets is that life, and metabolism, began simpler and more uniform. The continuity of evolution, in which all present day organisms arose from a common ancestor, means that all contemporary metabolism evolved from these modest beginnings. Our other tenet is that early life of course had to make do with the nutrients that were available.

Oparin (1938) originally hypothesized that the first metabolisms were not photosynthetic but heterotrophic (Section 1.14). In other words, these organisms obtained organic carbon from their surroundings rather than synthesizing it photosynthetically from carbon dioxide. The basis for this view is that photosynthetic organisms are much more complex than heterotrophic organisms because they need pathways to carry out photosynthesis and carbon dioxide fixation in addition to the anabolic pathways for synthesis of amino acids and other cell materials and structures. In contrast, heterotrophic metabolism requires relatively few enzymes to produce cell material from organic precursors.

As a corollary to this simple anabolism, a simple catabolism was also thought possible. A simple type of fermentative catabolism is the Stickland reaction that occurs in certain bacteria such as the anaerobic genus *Clostridium* and involves the fermentation of amino acid pairs. One example of the Stickland reaction is:

alanine $+2$ glycine $+P_i$ (phosphate) $+ADP \rightarrow$ 3 acetate $+CO_2 + 3NH_3 + ATP$.

In this reaction, only two amino acids are needed and very few enzymes (and hence genes in the organism's genome to store the blueprints for these enzymes)

are required for catabolism and generation of ATP. Also, it is interesting to note that alanine and glycine in this particular example of the Stickland reaction are two amino acids that are very stable and commonly found in carbonaceous chondrite meteorites (Kvenvolden *et al.*, 1970; Schroeder and Bada, 1976). Therefore, these substrates may have been available for metabolism early on.

However, the hypothesis that heterotrophic metabolism such as the Stickland reaction arose first ignores the importance of primary production, i.e., the production of organic material either by chemosynthesis or photosynthesis. Heterotrophic organisms *require* preformed organic material. Although organic substrates may have been abundant on early Earth, within a short time they would have been depleted by any heterotrophic organisms. Therefore, some process in which organic material is produced biologically on Earth is of fundamental importance in early metabolism. That is why we emphasize primary production in this chapter. We regard the most important question about early metabolism as: what were the most likely metabolism(s) of the initial primary producers?

We now develop in more detail the arguments in favor of two hypothetical metabolisms being important in early primary production: (1) methanogenesis and acetogenesis, and (2) sulfate and sulfite reduction.

11.6 Evolution of methanogenesis and acetogenesis: the first metabolisms?

11.6.1 Methanogens and acetogens: today and on the early Earth

Though not phylogenetically related (Fig. 11.1), methanogens and acetogens partly share analogous metabolic pathways. Methanogens are widespread within the sub-domain Euryarchaeota (in the domain Archaea), whereas acetogens are mostly in the kingdom known as Gram positive Bacteria (in the domain Bacteria). Methanogens and acetogens are both anaerobes that specialize in metabolism using one-carbon intermediates. Hydrogen is a common electron donor and carbon dioxide a common electron acceptor for both types of organisms (Table 11.2), and methane (CH_4) and acetic acid (CH_3COOH) are the catabolic products, respectively. In addition, both organisms fix carbon dioxide by the so-called acetyl coenzyme A pathway, discussed below. Today methanogens and acetogens play essential roles in anaerobic ecosystems and live in a variety of habitats ranging from hydrothermal vent chimneys to pond sediments to animal digestive tracts. For example, in sediment methanogens team up with a variety of fermenters to convert cellulose-type materials to methane and carbon dioxide. However, before photosynthesis, the Earth and its biosphere were quite different. Organic materials were limited to those formed by abiotic processes and perhaps by a relatively feeble primary production. Autotrophy may have been preferred over heterotrophy, necessitating a means of carbon dioxide fixation. In addition, not only was the early Earth without free oxygen, but electron donors and acceptors for catabolism may have been rare other than hydrogen, carbon dioxide, and a few others (Table 11.1). In this most ancient of biological worlds, methanogens or acetogens may have been the *only* metabolic types.

Analogues of such a world may exist today: microbial communities in the Earth's subsurface are apparently dominated by methanogens that obtain hydrogen from reduced minerals in rock (Stevens and McKinley, 1995; Chapelle *et al.*, 2002). Deep basalt aquifers, hot springs, and submarine hydrothermal vents are examples of sites where microbial ecosystems may be supported not by organic nutrients of photosynthetic origin, but by hydrogen of geological origin. These environments, which are poor in organic material but potentially rich in mineral nutrients, could resemble ecosystems that dominated on the early Earth or that may exist on other planetary bodies.

11.6.2 Metabolic simplicity and the acetyl CoA pathway

It is reasonable to assume that the earliest organisms were also the simplest. Moreover, frequent bombardment of the early Earth might have limited the time available for complexity to evolve. From this standpoint, methanogenesis and acetogenesis are appealing as the earliest metabolisms. In its simplest form, the metabolisms of methanogens and acetogens each consist of three essential pathways (Fig. 11.3). One pathway reduces carbon dioxide (CO_2) to a methyl (CH_3) group, a process that occurs in analogous series of reduction steps in both organisms. The second pathway, called the acetyl coenzyme A (acetyl CoA)[3] pathway, uses the methyl group to form acetyl coenzyme A

[3] A *coenzyme* is any non-protein molecule required for the functioning of an enzyme. Coenzyme A is a specific sulfur-containing organic coenzyme.

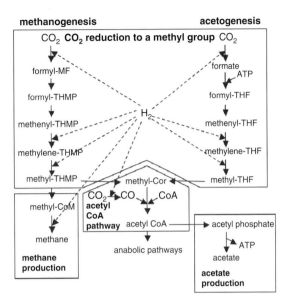

FIGURE 11.3 Pathways of methanogenesis and acetogenesis from hydrogen and carbon dioxide, and carbon dioxide fixation by the acetyl coenzyme A pathway. One-carbon intermediates are shown bound to various coenzymes: MF, methanofuran; THMP, tetrahydromethanopterin; THF, tetrahydrofolate; CoM, coenzyme M; Cor, corrinoid. Electron flow from H_2 is indicated with dashed arrows.

(CH_3CO-CoA), a process that is essentially the same in the two organisms. Note that these first two pathways can contribute to both catabolism and carbon dioxide fixation. The third pathway consists of a single energy-yielding step that differs between methanogens and acetogens. Methanogens reduce additional methyl groups, also obtained by reduction of carbon dioxide, to methane (CH_4). Acetogens convert some of their acetyl CoA to acetic acid, gaining an ATP in the process. Of all these pathways, the acetyl CoA pathway may have been the crux of an ancient autotrophic mode of life. Notice its simplicity. The methyl group joins with carbon monoxide, obtained by the reduction of another carbon dioxide), forming an acetyl (CH_3CO) group. The acetyl then joins with coenzyme A to form acetyl CoA. Other pathways of carbon dioxide fixation are more complex, involving various numbers of multicarbon intermediates. Because of its simplicity, the acetyl CoA pathway may have been the easiest to evolve.

11.6.3 A prebiotic precursor for the acetyl CoA pathway

Further support for the ancient nature of the acetyl CoA pathway comes from evidence that the crucial

reaction could have occurred prebiotically. Chemist Günther Wächtershäuser has propounded the theory that metabolism evolved from prebiotic chemistry (Wächtershäuser, 1990). Furthermore, he has shown in the laboratory that under primordial conditions a methyl group can react with carbon monoxide to yield an acetyl group (Huber and Wächtershäuser, 1997). Specifically, methyl mercaptan (CH_3SH) reacted with CO to form CH_3-CO-SCH, chemically speaking a thioester between an acetyl group and methyl mercaptan. Particularly interesting was the observation that the reaction required nickel sulfide and iron sulfide as catalysts. The enzyme that catalyzes the reaction in present-day organisms also contains a nickel-iron-sulfur center. Furthermore, the thioester bond formed in the Wächtershäuser reaction chemically resembles the thioester bond that acetyl forms with coenzyme A in today's enzyme-catalyzed reaction. Therefore, a core step in the metabolism of methanogens and acetogens may have derived from a prebiotic chemical reaction. Wächtershäuser has even suggested that the combined presence of methyl mercaptan, iron sulfide, and nickel sulfide could constitute important components of a "marker" for primitive habitats on Earth or Mars. Section 8.3 discusses this issue further.

11.6.4 Hydrogen oxidation and the proton-motive force

Another significant feature of methanogenic and acetogenic metabolism is the use of H_2 as the catabolic electron donor. Recall that a proton-motive force across the cell membrane is an essential component of respiratory catabolism. The electron transport chain of many organisms is a complex multi-component assembly that moves proteins from the inside to the outside of the cell as electrons flow from one electron carrier to another. Thus, the transport of both electrons and protons occurs. These complex electron transport chains are characteristic of organotrophs and phototrophs as well as aerobic lithotrophs that oxidize inorganic compounds other than hydrogen. But the use of hydrogen offers the possibility of a simpler mechanism in which a proton-motive force is generated without transport of protons. The products of hydrogen oxidation are simply protons and electrons. If the oxidation of hydrogen occurs on the outside of the cell membrane, then protons can accumulate there. A simple electron carrier that spans the membrane can then deliver electrons to the inside. The electrons can then be used on the inside in a proton-consuming reaction to

reduce carbon dioxide to methane or acetic acid. The result is a proton-motive force. This simple mechanism, requiring only an extracellular hydrogen-oxidizing enzyme (hydrogenase) and a membrane-bound electron carrier (such as an iron-sulfur protein), has not yet actually been demonstrated in the methanogens or acetogens, but neither is it excluded as a primordial mechanism or as one that operates today.

11.6.5 Summary

We have considered arguments that methanogenesis, acetogenesis, and carbon dioxide fixation via the ACoA pathway were likely early metabolisms on Earth. The possibility that hydrogen and carbon dioxide were the only abundant electron donor and acceptor produces obvious constraints. Simplicity of mechanism, both in terms of metabolism with one-carbon compounds and hydrogen oxidation to generate a proton-motive force, is significant as well. Finally, the resemblance of reaction mechanisms to possible prebiotic chemistry is striking. Because substrate availability on Mars and Europa may parallel the early Earth, and because any life there could be quite primitive, methanogenesis or acetogenesis may be the most likely of any extraterrestrial metabolism.

11.7 Counterpoint: sulfate respiration very early?

11.7.1 Evidence for an early origin of sulfate respiration

Sulfate reduction is also a simple form of metabolism that could have evolved early. Sulfate reducing bacteria respire sulfate ($SO_4^=$), reducing it to sulfide (HS^-) via the addition of eight electrons. This reduction takes place in two steps, an initial energy-consuming two-electron reduction to sulfite ($SO_3^=$) and a second energy-liberating six-electron reduction to sulfide. Energy is recovered in the second step via a simple respiratory chain, as previously discussed. The key point with respect to early evolution is that sulfate-reducing bacteria prefer to respire sulfite, since the reduction of sulfite does not require an energy investment. As noted in Table 11.1, sulfite was almost certainly abundant on the very early Earth, formed via reaction of SO_2 from abundant volcanos with water ($SO_2 + H_2O \rightarrow HSO_3^- + H^+$). Many contemporary sulfate (and sulfite) respiring microorganisms have the capacity to use H_2 as an electron donor for energy generation and for fixing

CO_2. Thus a very early presence of this type of metabolism is plausible. The ability to use hydrogen also allows for a simple mechanism for the generation of the proton-motive force, as discussed above. Finally, many sulfate reducing bacteria use the acetyl-coenzyme A pathway for CO_2 fixation, another relatively simple mechanism shared with the methanogens and acetogens.

These considerations are bolstered by the geochemical record. Stable isotopes in the early rock record support an origin of sulfate respiration as early as 3.47 Ga (Shen *et al.*, 2001; Section 12.3.8). Although supporting an early origin, these data of course do not tell us how much earlier organisms may have been respiring sulfite. We also note that the interpretation of the early isotopic record is subject to a number of caveats (Farquhar *et al.*, 2001).

Early sulfate reduction is also inferred from studies of dissimilatory sulfite reductase (Dsr), the enzyme catalyzing the second, energy-liberating step of sulfate respiration. The genes encoding Dsr reveal a remarkably high degree of sequence similarity, suggesting either high evolutionary constraints on structural change or relatively recent horizontal transfer of these genes between lineages. The presence of this enzyme in at least five highly divergent prokaryotic lineages (both bacterial and archaeal) and overall phylogenetic congruence of this gene tree with that of 16S rRNA suggests that the Dsrs of extant sulfate-reducing prokaryotes evolved vertically from common ancestral genes (Wagner *et al.*, 1998). The basis for inferring horizontal transfer (or not) is the assumption that the 16S rRNA phylogeny reflects the organismal phylogeny (Woese, 1987). More recently, evidence for horizontal gene exchange has been obtained (Klein *et al.*, 2001). However, transfer appears to have been episodic and has not been so pervasive as to erode overall phylogenetic relationships. Within-lineage relationships inferred by Dsr gene sequence analysis are generally consistent with the 16S rRNA phylogeny (Klein *et al.*, 2001). Thus, sulfate respiration is a strong contender for a very early mode of chemosynthetic metabolism.

11.8 Evolution of photosynthesis

The evolution of photosynthesis greatly enhanced the capability of the planet to carry out primary production. Organisms have evolved to make efficient use of light energy, which has always been available and abundant on the surface of the planet. Therefore, although the evolution of photosynthesis most likely followed the evolution of simpler primary production

mechanisms as discussed above, when it evolved it had a major impact on all life.

Photosynthesis has been found in two domains, the Bacteria and the Eucarya (although its presence in the latter, in plants, arose by "endosymbiotic" transfer from ancestral cyanobacteria, an event akin to a major horizontal gene transfer). A remarkable variety of photosynthetic bacterial groups exists. Within the Bacteria, five of the major phyla contain photosynthetic members (Fig. 11.1). These include the Proteobacteria, Firmicutes, Chlorobi, Chloroflexi, and Cyanobacteria.

Based on his investigations of photosynthetic bacteria, van Niel (1941) proposed the following general reaction for photosynthesis:

$$CO_2 + 2 H_2A + light \rightarrow (CH_2O) + H_2O + 2 A, \quad (11.1)$$

where H_2A can stand for H_2S, H_2, H_2O (in cyanobacteria), an organic compound, or even ferrous iron (Fe^{2+}). (CH_2O) represents an organic product of the photosynthesis.

Many of the photosynthetic bacteria, those referred to as the purple and green sulfur photosynthetic bacteria (found in the Proteobacteria and Chlorobium, respectively) are restricted to anoxic habitats and use H_2S as a source of electrons. Thus, the reaction for these bacteria is:

$$CO_2 + 2 H_2S + light \rightarrow (CH_2O) + H_2O + 2 S. \quad (11.2)$$

These phototrophs can often further oxidize the elemental sulfur to produce sulfate anaerobically during photosynthesis:

$$3 CO_2 + 2 S + 5 H_2O + light \rightarrow 3 (CH_2O) + H_2SO_4.$$

It is interesting to note that some cyanobacteria can carry out reaction (11.2) in their photosynthesis. However, they are best known as the first organisms to use water to carry out *oxygenic* photosynthesis, in which oxygen is produced:

$$CO_2 + H_2O + light \rightarrow (CH_2O) + O_2.$$

Note that the O_2 in this reaction does not come from CO_2 but from H_2O, a conclusion inferred from the use of "A" in general reaction (11.1) for photosynthesis. For this reason oxygenic photosynthesis is called the "water splitting" reaction. (The reaction has been confirmed by experiments with ^{18}O-labeled H_2O, in which the labeled atoms end up in O_2.) Apart from the cyanobacteria, however, all photosynthetic bacteria carry out photosynthesis anaerobically and oxygen is *not* produced. This anaerobic process is referred to as *anoxygenic* photosynthesis.

When did photosynthesis evolve? The availability of suitable substrates is not a constraint, since CO_2, H_2S, H_2, Fe^{2+}, and certainly water were all available, as was light. The main constraint has to do with complexity. To generate a proton-motive force photosynthetically requires the addition of chlorophyll (Fig. 11.2), as well as a relatively complex electron transport chain involving transport of protons across the membrane. This mechanism is more likely to have followed the relatively simple H_2-dependent chemosynthetic mechanisms described above.

Although the time of emergence of different photosynthetic types is unresolved in the geological record, available evidence strongly suggests that anoxic phototrophs predated the cyanobacteria. At that time the entire planet was anoxic, since the cyanobacteria had yet to begin to oxygenate the atmosphere and oceans. The evolutionary emergence of the cyanobacteria was the result of a remarkable biochemical innovation – the development of a light-driven electron transport system that could fix CO_2 with electrons extracted from water, liberating its oxygen. This photosystem, composed of two photosynthetic reaction centers, was likely built upon the machinery of earlier simpler microorganisms (with only one reaction center) that were restricted to anoxic habitats and that used reduced sulfur as a source of electrons (Blankenship and Hartman, 1998). The time of cyanobacterial emergence has been dated to 2.1–2.5 Ga on the basis of protein sequences (Feng *et al.*, 1997), which is consistent with the geochemical estimate for the era of oxygenation of the Earth's atmosphere and oceans. Geochemical imprints of the first major oxygenation event date to ca. 2.2–1.9 Ga (Holland, 1984). An early Proterozoic time of 2.5 Ga is also supported by the abundance of specific cyanobacterial biomarkers, 2-methylhopanoids, in organic-rich marine sediments of this age (Summons *et al.*, 1999). Chapter 4 and Section 12.3 discuss the geological record in more detail.

The evolution of oxygenic photosynthesis liberated microbial life from dependence on geochemically produced, reduced substrates such as H_2S, resulting in an explosive increase in biomass on Earth. By using water as a reductant, life was free to fully colonize the planet, now requiring only water, light, CO_2, and inorganic substrates for growth and division.

The innovation of oxygenic photosynthesis began a remarkable evolutionary progression, culminating in the development of complex animal and plant species (note that the chloroplasts of today's plants are descendents of an ancient symbiosis between a cyanobacterium and a eukaryote). The development of an oxygen-rich atmosphere prepared the biosphere for the emergence of higher life forms. Only by burning reduced carbon with oxygen, via an electron transport chain, could sufficient energy be released to fuel large multicellular organisms. As an additional effect, the resultant atmospheric oxygen produced an ozone layer in the stratosphere, strongly absorbent of ultraviolet radiation (which severely damages biological macromolecules). When the ozone layer became significant, the flux of ultraviolet radiation falling on land was greatly reduced, and this opened up land to colonization by plants and animals who no longer needed water for protection.

11.9 Aerobic metabolism

Aerobic metabolism is the culmination of metabolic evolution on Earth. One tremendous innovation made possible due to the superior energy yield from using oxygen as an electron acceptor has already been mentioned: the evolution of multicellular organisms. But the microbial world expanded too. The aerobes that evolved included not only organotrophs but new kinds of chemolithotrophs: oxidizers of nitrogen, sulfur, and iron compounds that play important roles in today's nutrient cycles. As for earlier metabolisms, aerobes benefited not only from the addition of a new nutrient, but also from metabolic mechanisms that had previously evolved. Thus, the innovations in the electron transport chain that came with photosynthesis allowed aerobes to take advantage of this new source of energy. Indeed, many of the same components of the electron transport chain that evolved for photosynthesis are used in aerobic respiration.

We have presented a speculative scenario for the evolution of metabolism, progressing from hydrogen-utilizing anaerobic respirers (methanogens, acetogens, and sulfate/sulfite reducers), to phototrophs (anoxygenic then oxygenic), and finally to aerobic respirers. Clearly, other scenarios are possible. For example, iron reducers ($Fe^{+3} \rightarrow Fe^{+2}$) could have been another kind of early respirer. The periodic action of iron-reducing respirers, alternating with iron-oxidizing phototrophs, has been suggested as one possible origin of banded iron geological formations that preceded the oxygen-rich atmosphere of Earth (Chapter 4).

11.10 Earth's earliest communities: microbial mats

If you were to visit planet Earth during any average point in its history, the scene would be other-worldly.

For most of Earth's history it was the planet of the microbes. The land would be mostly barren of visible life. Only in shallow freshwater and intertidal marine basins would there be visible accumulation of life, in the form of microbial mat communities. Microbial mats are macroscopically visible microbial ecosystems in which microbes build communities that in many ways are analogous to rain forests. They are layered communities that today develop as carpets, often many centimeters thick, in certain types of shallow aquatic settings. The structural coherence of the community is provided by extracellular polymers produced by the mat microorganisms and by the filamentous forms of some dominant populations. The top layer, the "canopy," primarily consists of oxygen-producing bacteria, the cyanobacteria. The cyanobacteria, together with anoxygenic phototrophs sustain a remarkably diverse "understory" of other microorganisms. In fact, virtually all major physiological types of microbes are resident in contemporary microbial mats, and they sustain all central biogeochemical cycles of C, N, and S, where each element is alternately oxidized and reduced by the metabolism of different microorganisms. When examined under the microscope each mat is revealed to be highly stratified by different populations of microorganisms. The activities of these highly organized populations shape the chemical structure of the mat, forming sharp gradients of oxygen, pH, sulfide, and other chemical species that are either produced or consumed by microbial metabolism at different depths. High productivity by photo- or chemosynthesis, combined with internal nutrient recycling among closely interdependent populations, allows each mat to sustain significant biomass on little more than water, sunlight, and available inorganic nutrients.

The most impressive mat communities today occur in rather extreme habitats (extreme, that is, as viewed by multicellular life), since significant biomass accumulation can only occur in the absence of grazing. For example, today's mats of thermophilic cyanobacteria are common in hot springs worldwide (Fig. 11.4). Mats also thrive in hypersaline lagoons (Fig. 11.5), as well as in intermittently dry intertidal regions. Mats also build calcified stromatolites in warm seas, form desert soil crust, and grow within the frozen rocks of Antarctica as endolithic mats. They even thrive embedded in the ice cover of Antarctic dry valley lakes (Pearl *et al.*, 2000).

Despite their restricted distribution today, the fossil record suggests that mats were once widely distributed on early Earth. The decline of extensive microbial mat systems in the rock record is generally attributed to the

FIGURE 11.4 Transparent core recovered from the effluent channel of Mushroom Spring in Yellowstone National Park, USA. The source water of this hot spring is approximately 92 °C, but conspicuous microbial mat development does not occur until the channel water has cooled to ~60 °C, a temperature that allows growth of thermophilic cyanobacteria and associated mat community members, yet limits grazing by eukaryotes.

emergence of multicelluar grazers and water plants that ultimately forced them to a much more limited habitat range. The most conspicuous and at the same time often enigmatic vestiges of the early microbial biosphere and microbial mats are stromatolites, finely laminated sedimentary structures formed by trapping of particles or precipitation of dissolved minerals in the marine environment (Sections 12.2.2 and 16.2.3; Fig. 16.2). It is generally accepted, but not always proven, that the sediment trapping or mineral precipitation processes that formed stromatolites were mediated by ancient microbial communities. The filamentous morphology of microfossils within stromatolites and the apparently phototactic growth patterns of stromatolite-building microorganisms support the view that phototrophs were involved. The existence of fossil stromatolites beginning in the early and middle Archean is consistent with isotopic evidence for autotrophic carbon fixation in the early Archean, and very strongly suggests that photoautotrophy existed.

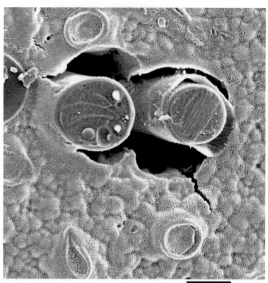

10 μm

FIGURE 11.5 Scanning electron micrograph of a cross section of an ensheathed bundle of the filamentous cyanobacterium *Microcoleus* embedded in the exopolymer matrix of a microbial mat. This specimen is from a mat at Solar Lake, a hypersaline pond in the Sinai (Egypt).

However, fossil stromatolites show considerable diversity in morphology and environmental setting over time, which could indicate different microbial communities forming them (Walter, 1994). The physiological tolerance and adaptability of cyanobacteria makes them today's premier mat-building organism. With increasing age, however, stromatolites become harder to interpret with respect to how exactly microbes contributed to their formation, and which microbes were most likely involved. A direct association of stromatolites with mat-building microorganisms is only possible in a few cases where stromatolites harbor exceptionally well-preserved and identifiable microfossils. The first generally accepted examples of stromatolites built by cyanobacteria come from the Paleoproterozoic era (~2 Ga) (Golubic and Hofmann, 1976). It is therefore not clear whether anoxygenic or oxygenic phototrophs dominated earlier mat communities.

Prior to the emergence of cyanobacteria, any photosynthetic mats would have been restricted to aquatic locations where light and reduced sulfur were available. Examples of these types of organisms today include *Chloroflexus*, *Chromatium*, and *Chlorobium*, all species that inhabit hot sulfidic springs (Ward *et al.*, 1989). An iron cycle has also been implicated in early photosynthetic systems (Widdel *et al.*, 1993; Vargas *et al.*, 1998).

The source of reduced sulfur (or iron) used by anoxygenic mat communities was either geothermal or self-generated by the biological reduction of sulfate/sulfite to sulfide. Such a source is tenable if sulfite respiration was an ancient mechanism for energy recovery, predating photosynthesis. The enzyme that anoxic phototrophs use to oxidize sulfide is evolutionarily related to the sulfite reductase used by sulfate/sulfite-reducing microorganisms (Hipp *et al.*, 1997), but modified to run efficiently in reverse. That is, both contemporary enzymes are derived from a common ancestral enzyme that we hypothesize functioned in a reductive capacity. If the sulfate/sulfite reducing bacteria were among the earliest organisms, an attractive progression of metabolic innovation involving sulfur compounds would be sulfite respiration first (using sulfite reductase), followed by the emergence of anoxic phototrophs using a reverse sulfite reductase to strip electrons from sulfide. By coupling sulfite reduction with sulfide oxidation (Fig. 11.6) an internal sulfur cycle could be established in a mat community, requiring little or no input of reduced sulfur (e.g., HS^-, SO_2). However, such systems would still be dependent on geochemical input of reduced substrates (i.e., a source of electrons) for growth.

Today there is increasing recognition that microorganisms are extremely promiscuous. As discussed earlier in this chapter, horizontal DNA exchange, the transfer of genetic information between different lineages, is not restricted to closely related organisms. Exchange within and between domains is now well documented and increasingly recognized to be a major force in shaping metabolic innovation. We speculate that the high density and close associations among microorganisms in microbial mats provided, and still provides, a superb environment for genetic exchange between populations. If this is true, some of the most significant events in the history of our planet may have occurred in microbial mats. For example, the cyanobacteria may have emerged within this context as a result of melding the photosystems of anoxic phototrophs.

11.11 Concluding thoughts

We have presented a story about early Earth and early life. It is a story inferred from a spotty (bio)geochemical record, uncertain phylogeny reconstruction, and an extrapolation from contemporary metabolic modes to versions that might have been present in the distant past. The connection between today's

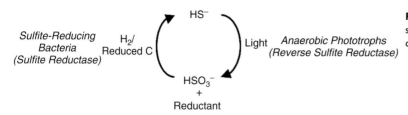

FIGURE 11.6 Possible internal sulfur cycle of early microbial mat communities.

organisms and the first organisms spans some 3.5 Gyr of evolution. It is a tenuous connection. Today's relatively primitive metabolisms, defined as having simple biochemical pathways and compatibility with substrates available on early Earth, are nonetheless highly evolved biochemical and genetic systems. We anticipate that a more persuasive story will in part require a more complete (bio)geochemical data set and a better census of the diversity of microorganisms now living on our planet.

Today there is general agreement that only a few percent of the microbial diversity on our planet has been described (Staley and Reysenbach, 2002). With incomplete understanding of contemporary microbial diversity, the puzzle of life's origins and the nature of biochemical innovation will be incomplete, and possibly misleading. As we look to other planets for signs of life, it is equally important that we look closely at life on Earth. Microorganisms represent the most ancient and most biochemically diverse group of organisms on Earth. They "invented' the biosphere and provided a foundation for the emergence of more complex life, the plants and animals. However, the mechanisms of biochemical invention are not well understood. We are only now beginning to appreciate the role that horizontal gene transfer has played in this process. We need to know more about the mechanisms of evolution, how they affected the evolutionary trajectory on Earth, and how they might affect it on another planet. For example, what is the likelihood that oxygenic photosynthesis would emerge on another planet having initial conditions comparable to early Earth?

The most difficult problems associated with reconstructing life's origin(s) and its diversification are the many apparent discontinuities. The evolution of increasingly complex biological systems was not gradual and continuous, but punctuated by biochemical innovations. Horizontal gene exchange was certainly an important contributor to innovation, but only with continued analysis of whole microbial genomes will we be able to better constrain its relative contribution. The greatest discontinuity of all was the transition from abiotic chemical systems to self-sustaining metabolic systems that we now call biology. What were the early non-protein catalysts that were improved upon by the development of protein-based catalysts? Just as environments that provided abiotic catalysts may help narrow down possible early metabolic pathways, understanding the first metabolic processes may shed light on the steps involved in crossing the boundary from chemistry to biology.

REFERENCES

Blankenship, R. E. and Hartman, H. (1998). The origin and evolution of oxygenic photosynthesis. *Trends Biochem. Sci.*, **23**, 94–97.

Chapelle, F. H., O'Neill, K., Bradley, P. M., Methe, B. A., Ciufo, S. A., Knobel, L. L., and Lovley, D. R. (2002). A hydrogen-based subsurface microbial community dominated by methanogens. *Nature*, **415**, 312–315.

Chistoserdova, L., Vorholt, J. A., Thauer, R. K., and Lidstrom, M. E. (1998). C1 transfer enzymes and coenzymes linking methylotrophic bacteria and methanogenic Archaea. *Science*, **281**, 99–102.

Farquhar, J., Bao, H., Thiemens, M. H. (2001). Questions regarding Precambrian sulfur isotope fractionation. *Science*, **292**, 1959a–1959b.

Feng, D.-F., Cho, G., and Doolittle, R. F. (1997). Determining divergence times with a protein clock: update and reevaluation. *Proc. Natl. Acad. Sci. USA*, **94**, 13028–13033.

Golubic, S. and Hofmann, H. J. (1976). Comparison of modern and mid-precambrian Entophysalidaceae (Cyanophyta) in stromatolitic algal mats: cell division and degradation. *J. Palaeontol.*, **50**, 1074–1082.

Gottschalk, G. (1986). *Bacterial Metabolism*, second edn. New York: Springer-Verlag.

Hipp, W. M., Pott, A. S., ThumSchmitz, N., Faath, I., Dahl, C., and Truper, H. G. (1997). Towards the phylogeny of APS reductases and sirohaem sulfite reductases in sulfate-reducing and sulfur-oxidizing prokaryotes. *Microbiology-UK*, **143**, 2891–2902.

Holland, H. (1984). *The Chemical Evolution of the Atmosphere and the Oceans*. Princeton, NJ: Princeton University Press.

Huber, C. and Wächtershäuser, G. (1997). Activated acetic acid by carbon fixation on (Fe, Ni)S under primordial conditions. *Science*, **276**, 245–247.

Klein, M., Friedrich, M., Roger, A. J., *et al.* (2001). Multiple lateral transfer events of dissimilatory sulfite reductase genes between major lineages of bacteria. *J. Bacteriol.*, **183**, 6028–6034.

Kvenvolden, K. A., Lawless, J., Pering, K., *et al.* (1970). Evidence for extraterrestrial amino-acids and hydrocarbons in the Murchison Meteorite. *Nature*, **228**, 923–926.

Oparin, A. I. (1938). *The Origin of Life*. London: The Macmillan Co.

Pearl, H. W., Pinckney, J. L., and Steppe, T. F. (2000). Cyanobacterial-bacterial mat consortia: examining the functional unit of microbial survival and growth in extreme environments. *Environ. Microbiol.*, **2**, 11–26.

Schidlowski, M. (1988). A 3800 million-year isotopic record of life from carbon in sedimentary rocks. *Nature*, **333**, 313–318.

Schroeder, R. A. and Bada, J. L. (1976). A review of the geochemical applications of the amino acid racemization reaction. *Earth-Science Rev.*, **12**, 347–391.

Shen, Y., Buick, R., and Canfield, D. E. (2001). Isotopic evidence for microbial sulphate reduction in the early Archaean era. *Nature*, **410**, 77–81.

Staley, J. T. and Reysenbach, A.-L. (eds) (2002). *Biodiversity of Microbial Life: Foundation of Earth's Biosphere*. New York: John Wiley & Sons.

Stevens, T. O. and McKinley, J. P. (1995). Lithoautototrophic microbial ecosystems in deep basalt quifers. *Science*, **270**, 450–454.

Summons, R. E., Jahnke, L. L., Hope, J. M., and Logan, G. A. (1999). 2-methyl-hopanoids as biomarkers for cyanobacterial oxygenic photosynthesis. *Nature*, **400**, 554–557.

Vargas, M., Kashefi, K., Blunt-Harris, E. L., and Lovley, D. R. (1998). Microbial evidence for Fe(III) reduction on early Earth. *Nature*, **395**, 65–67.

Van Niel, C. B. (1941). The bacterial photosyntheses and their importance for the general problem of photosynthesis. *Adv. Enzymol.*, **1**, 263–328.

Wächtershäuser, G. (1990). Evolution of the first metabolic cycles. *Proc. Natl. Acad. Sci. USA*, **87**, 200–204.

Wagner, M., Roger, A. J., Flax, J. L., Brussean, G. A., and Stahl, D. A. (1998). Phylogeny of dissimilatory sulfite reductases supports an early origin of sulfite respiration. *J. Bacteriology*, **180**, 2975–2982.

Walter, M. R. (1994). Stromatolites: the main geological source of information on the evolution of the early benthos. In *Early Life on Earth*. Nobel Symposium No. 84, Stefan Bengtson, ed. New York: Columbia University Press, pp. 270–286.

Ward, D. M., Weller, R., Shiea, J., Castenholz, R. W., and Cohen, Y. (1989). Hotspring microbial mats: anoxygenic and oxygenic mats of possible evolutionary significance. In *Microbial Mats: Physiological Ecology of Benthic Microbial Communities*. Y. Cohen and E. Rosenberg, eds. Washington DC: American Society of Microbiology, pp. 3–15.

Widdel, F., Schnell, S., Heising, S., Ehrenreich, A., Assmus, B., and Schink, B. (1993). Ferrous iron oxidation by anoxygenic phototrophic bacteria. *Nature*, **362**, 834–836.

Woese, C. R. (1987). Bacterial evolution. *Microbiol Rev.*, **51**, 221–271.

Woese, C. R., Kandler, O., and Wheelis, M. L. (1990). Towards a natural system of organisms: proposal for the domains Archaea, Bacteria, and Eucarya. *Proc. Natl. Acad. Sci. USA*, **87**, 4576–4579.

12 The earliest records of life on Earth

Roger Buick
University of Washington

12.1 Problems with the record

Astrobiology has only a single successful experiment in planetary life available to investigate: that on the Earth. Hence, the history of terrestrial life must act as the archetype, albeit an ever more contingent and unique one, for astrobiological models of the appearance and radiation of life anywhere in the Universe. Indeed, it could be argued that all habitable planets would have had similar environmental constraints and pathways of physical and chemical development, so the process of biological initiation elsewhere should be broadly reminiscent of Earth's experience of the phenomenon. If so, astrobiology is saved from the challenges of studying things far away, but is instead faced with the difficulties of examining events here long ago.

Unfortunately, and perhaps surprisingly, the origin and early evolutionary history of terrestrial life is poorly known, as is the corresponding record of environmental conditions on the early Earth. There are many reasons for this. Firstly, like all old things, ancient rocks are rare (Fig. 12.1). Almost all potential information about the first half of Earth's history[1] is contained in geological materials. But rocks of such great antiquity have mostly been hidden or destroyed by geological processes like burial, erosion, or subduction back into the mantle via plate-tectonic[2] recycling of crust along ocean trenches. Even ejection into space by catastrophic meteorite impacts, of which there were plenty during the heavy bombardment that occurred over the first billion years

of Earth history (Chapter 3), is a viable mechanism for destruction of the earliest crust.

Secondly, for those rocks that survived obliteration, their information content often has been perverted by metamorphism (mineralogical changes in rocks due to pressure and heat) and/or deformation. Biological signatures are fragile, usually surviving only mild metamorphism and moderate deformation. But on a tectonically active planet like Earth, the cumulative probability of deep burial or intense folding steadily increases over time. Radioactive heating was assuredly greater on the Archean Earth, and tectonic activity may also have been more intense, with *post-depositional*[3] modification of rocks correspondingly more likely. For the earliest crust, the effects of shock metamorphism during meteorite impact should also have been severe, judging by the amount of cratering on the Moon and Mars. For most chemical biosignatures, heating to much above 300 °C is significantly destructive. For most physical signatures, recrystallization into platey or elongate minerals is sufficient to mask visible biological remains. So, the rare rocks surviving from early in the geological record are even more rarely informative on issues of astrobiological importance.

Thirdly, those few well-preserved rocks that have survived from the Archean are often located in unfortunate places (Fig. 12.1). Many occur on continents where recent environmental conditions have been inimical to pristine preservation. Some are situated on old flat Gondwana continents (e.g., Australia, Africa, India, South America) where prolonged weathering under subtropical climates has transformed most surface rocks into varieties of soil. Others occur in subpolar settings in northern Asia, Europe, Greenland,

---

[1] This chapter will focus on the *Archean* eon, defined as before 2.5 Ga, but will also consider the early part of the *Proterozoic* eon (0.55–2.5 Ga), called the *Paleoproterozoic* era (1.6–2.5 Ga). Chapter 16 covers the later fossil record.

[2] *Tectonics* refers to the structure, motions, and dynamics of a planet's crust.

[3] *Deposition* is the process of laying down a sedimentary layer. *Post-depositional processes* can later modify this layer in many ways and thus complicate its interpretation.

Planets and Life: The Emerging Science of Astrobiology, eds. Woodruff T. Sullivan, III and John A. Baross. Published by Cambridge University Press. © Cambridge University Press 2007.

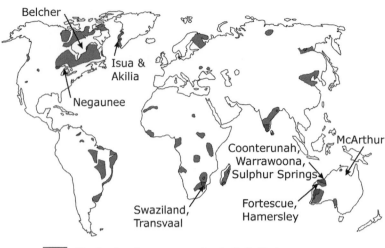

FIGURE 12.1 Global distribution of Archaean rocks.

Early Archaean rocks (>3.0 Ga)

Late Archaean rocks (2.5-3.0 Ga)

and North America where glaciation over the past million years has covered them with till, shattered them by frost action, swamped them in lakes, or encouraged the overgrowth of luxuriant lichens and mosses. In such circumstances, the best way to get fresh samples is by drilling. However, most core boring of this type has been concentrated in the vicinity of mines, where the rocks are unrepresentative. Mines are always situated where there is an anomalous concentration of some particular element, usually requiring that the rocks underwent marked post-depositional modification to induce this anomalous abundance. So, the opportunity for obtaining fresh samples of well-preserved rocks of great age is restricted to a few isolated locations.

Fourth, not all such rocks were formed in settings likely to yield evidence of life. Sedimentary rocks are the only ones from which paleobiological[4] data can be readily obtained, and even then, only particular types are suitable. For the early Earth, these are commonly shales (sedimentary rocks composed of finely layered clay) rich in kerogen,[5] cherts (rocks composed almost entirely of microcrystalline quartz), banded iron formations (sediments chemically precipitated from the ocean consisting of alternating layers of iron minerals, mostly oxides, and chert), and carbonates such as limestone and dolomite. A few other rock types can yield

important environmental information and, in exceptional instances, biological relics – these include sandstones, evaporites (sediments precipitated by evaporating seawater), hydrothermal deposits (rocks precipitated from geothermally heated fluids), paleosols (fossil soils), tuffs (sediments composed of volcanic ash), and hyaloclastites (rocks composed of volcanic glass fragments). All of these, however, constitute only a minority of the primordial geological record, which mostly consists of granites and other intrusive igneous rocks, basalts and other volcanic rocks, schists, gneisses, and other highly metamorphosed and deformed materials.

Finally, the very few well-preserved remnants of early Earth have mostly not yet been studied in a systematic fashion and from first principles. Their sedimentary setting, stratigraphic relationships, post-depositional evolution, and geochronological framework all must be understood before paleobiological and paleo-environmental interpretations can be drawn. Moreover, because the environment in which life originated and radiated was very different from today's, the initial organisms would have been very different from those living now and so primordial rocks and relics of early life might well have no modern analogues. Disregard of these issues can lead to serious problems in interpretation with respect to contamination and biogenicity. *Contamination*, the introduction of younger material into older rocks, has plagued Archean paleontology because the opportunities for such adulteration of the fossil record during the long

---

[4] The prefix *paleo-* means "ancient" (Greek) and is attached to many scientific terms to refer to the distant past.

[5] *Kerogen* refers to insoluble sedimentary carbonaceous solids.

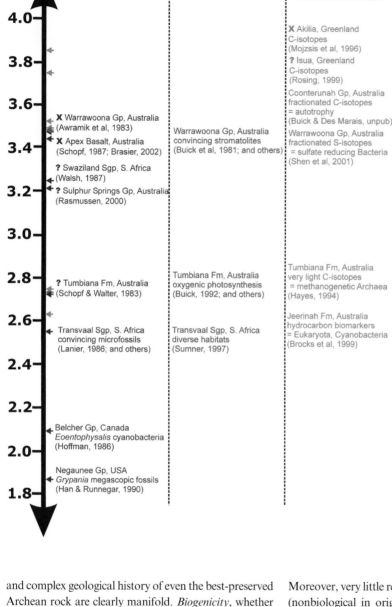

FIGURE 12.2 Timeline (in Ga) of key evolutionary events in the geological record. The left column refers to microfossil evidence, the middle column to stromatolites, and the right column to biomarkers and isotopic evidence. Names and dates refer to initial discoveries, not necessarily to cited papers. X = doubtful; ? = possible.

Microfossils

X Warrawoona Gp, Australia
(Awramik et al, 1983)
X Apex Basalt, Australia
(Schopf, 1987; Brasier, 2002)
? Swaziland Sgp, S. Africa
(Walsh, 1987)
? Sulphur Springs Gp, Australia
(Rasmussen, 2000)

? Tumbiana Fm, Australia
(Schopf & Walter, 1983)

Transvaal Sgp, S. Africa
convincing microfossils
(Lanier, 1986; and others)

Belcher Gp, Canada
Eoentophysalis cyanobacteria
(Hoffman, 1986)

Negaunee Gp, USA
Grypania megascopic fossils
(Han & Runnegar, 1990)

Stromatolites

Warrawoona Gp, Australia
convincing stromatolites
(Buick et al, 1981; and others)

Tumbiana Fm, Australia
oxygenic photosynthesis
(Buick, 1992; and others)

Transvaal Sgp, S. Africa
diverse habitats
(Sumner, 1997)

Chemical Fossils

X Akilia, Greenland
C-isotopes
(Mojzsis et al, 1996)
? Isua, Greenland
C-isotopes
(Rosing, 1999)
Coonterunah Gp, Australia
fractionated C-isotopes
= autotrophy
(Buick & Des Marais, unpub)
Warrawoona Gp, Australia
fractionated S-isotopes
= sulfate reducing Bacteria
(Shen et al, 2001)

Tumbiana Fm, Australia
very light C-isotopes
= methanogenetic Archaea
(Hayes, 1994)

Jeerinah Fm, Australia
hydrocarbon biomarkers
= Eukaryota, Cyanobacteria
(Brocks et al, 1999)

and complex geological history of even the best-preserved Archean rock are clearly manifold. *Biogenicity*, whether or not a specimen is of biological origin, is similarly contentious for many putatively primordial fossils, largely because early organisms are expected to be morphologically simple, chemically unsophisticated, and environmentally benign. But many inorganic objects also have such characteristics and so distinguishing them from primitive fossils is difficult, especially as the nature of early life forms is unpredictable. Modern analogues can help, but even for today's microbes little is known about their diversity, the range of their biogeochemical activities, and their environmental impact.

Moreover, very little research has been done on abiogenic (nonbiological in origin) mimics of possible early biosignatures. Because of all these shortcomings, every claimed Archean fossil must be subjected to strict scrutiny, if only because of the extreme improbability of their survival.

Despite all these caveats, accessible outcrops of relatively well-preserved Archean rocks of suitable composition for paleobiological and paleo-environmental study still do exist. In the following sections, these will be assessed to compile a picture of what early life on Earth was like, where it lived, and how it influenced its host planet.

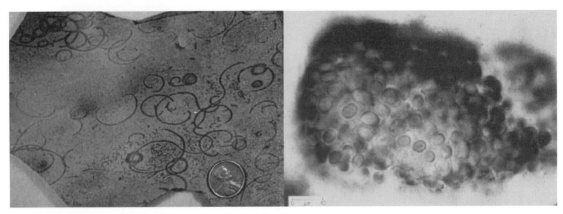

FIGURE 12.3 Early Proterozoic fossils.

(a) The macrofossil *Grypania* (spiral seaweed) from the ∼1.87 Ga Negaunee Formation, USA (Han and Runnegar, 1992).
(b) The cyanobacterium *Eoentophysalis* from the ∼2.1 Ga Belcher Group, Canada (Hofmann, 1976).

12.2 Types of evidence

12.2.1 Microfossils

Microfossils are the preserved remains of microbial organisms. In the case of the early Earth, those that have been found are exclusively of organic-walled *prokaryotes* (microbes lacking chromosomes, a nucleus, and other organelles, as opposed to *eukaryotes*, which have these). They are a few tens of microns in size at maximum – large life-forms are only evident in the second half of Earth's geological record. The oldest fossil visible to the unaided eye, *Grypania*, a carbonaceous compression (a layer of carbon tracing out the original organism's shape) of a spiral seaweed about a centimeter across, appears at ∼1.85 Ga (Han and Runnegar, 1992; see Fig. 12.3a). Despite a reasonable amount of searching, no older microfossils have been found. Figure 12.2 gives a timeline for the major evidences of life >1.8 Ga, all of which are discussed in this chapter.

It is not surprising that the preserved remains of any Archean organisms should be microscopic, as large multicellular organisms appear only on peripheral branches of the "Tree of Life" (the universal phylogeny of living organisms based on sequence similarities in ribosomal RNA; see Chapter 13). Moreover, the earlier appearance of microorganisms might also be expected from the overall trend within individual evolutionary lineages towards larger sizes over time. Unfortunately, however, remnants of very small organisms are most prone to destruction by metamorphic and deformational recrystallization. Moreover, the vagaries of *taphonomy* (the processes involved in fossil preservation) also vary on a microscopic scale, depending on the degree of oxidative degradation, biological reprocessing, and compactional dismemberment. To counter these destructive taphonomic processes, microbial fossils are generally only preserved where lithification (entombment in rock) was almost instantaneous upon death, where the host sediment is extremely fine in grain size and where the mineralizing material is inimical to microbial activity. The two main rock types that satisfy these requirements are chert and shale. Cherts readily precipitated in evaporative carbonate environments during the Proterozoic and Archean before silica-shelled organisms had evolved and thereby lowered dissolved silica concentrations in the oceans to levels far below saturation. Shales are widespread in Archean terrains, but are generally recrystallized or sheared along bedding planes, to the great detriment of fine-scale fossil preservation. Thus, most searches for Archean microbial fossils have concentrated on black kerogenous chert.

However, not all ancient cherts were passively precipitated from silica-saturated seawater. In Archean terrains in particular, hydrothermal processes have also formed microcrystalline siliceous rocks by replacement of pre-existing permeable rocks or by fracture-filling in subsurface settings. Such hydrothermal cherts are precipitated rapidly from hot, acidic, reduced fluids, often crystallizing initially as rather coarse-grained chalcedony or quartz. Though organic matter can survive such conditions, it is not yet clear that morphologically intact microfossils can. Moreover, such fluids apparently often exceeded the thermal limit of life (the current record is ∼120 °C – see Chapter 14)) and so were unlikely to contain any organisms. These hydrothermal cherts are likely to preserve only organisms or fossils that were inadvertently engulfed by the lithifying fluid.

Because of these various difficulties, the search for Archean microfossils has been fraught with controversy. The two key issues that have emerged from the many vigorous debates are those of biogenicity and syngenicity. *Biogenicity* arises as a problem because of the extreme morphological simplicity of prokaryotic organisms and their likely remains. For virtually all microbial groups, excepting some cyanobacteria, their shapes can be categorized as "balls or sticks"; either simple spheroids or uncomplicated tubular filaments. In addition, they are minute in size (near the limits of optical microscopic resolution), lack any discernible internal structure, and have minimal surface ornamentation. All this makes it very difficult to distinguish true prokaryotic microfossils from the many abiogenic structures that can mimic such simple shapes. Several schemes have been proposed for differentiating biogenic from abiogenic microstructures (e.g., Schopf and Walter, 1983; Buick, 1990), none of which have proved infallible. In the light of this, perhaps the best way of discriminating fossilized simple lifeforms from non-life is to find evidence of past biological *behavior* not explicable in terms of physical or chemical processes. Features such as varying responses to different substrates, varying orientations with respect to environmental gradients, and cellular distribution indicating reproduction, coloniality, or matting habit could be considered relics of biological behaviors.

Syngenicity (whether a presumed microfossil indeed has an ancient, indigenous origin) arises as a problem because of the long and complex geological history of even the best-preserved Archean rock. This provides many potential opportunities for younger biological contaminants to infiltrate, evidence of which can then be obscured by later metamorphism and deformation. Again, several schemes have been proposed for recognizing secondary microfossils (e.g., Schopf and Walter, 1983; Buick, 1990), all of which seem to work if strictly followed. However, these require detailed field mapping, petrological examination, and geochemical analysis in order to succeed – contamination can sometimes only be distinguished by features like slightly discordant host rocks, lack of fossil–sediment interaction, incongruous kerogen color, or anomalous geochemistry compared to metamorphic grade.[6]

The most ancient microfossils that can be *confidently* assigned to an extant group of organisms are variably pigmented colonies of spheroidal cells showing a distinct division pattern from the ~1.85 Ga Belcher Group of arctic Canada (Hofmann, 1976). These are so similar in structure and behavior to an extant genus of morphologically complex cyanobacteria (blue-green bacteria performing oxygenic photosynthesis) that they have been given an almost identical name: *Eoentophysalis* as opposed to *Entophysalis* (Fig. 12.3b). The degree of resemblance is striking in so many ways that there can be little doubt that the Proterozoic fossils are also cyanobacteria.

Before this time, however, the microfossil record diminishes markedly in quantity and quality. The only absolutely convincing Archean microfossils, from various ~2.55 Ga cherty carbonate formations in the Transvaal Supergroup of South Africa (Fig. 12.4), barely fall into that eon. These are assemblages of solitary or paired ellipsoids 0.2–2.5 μm in diameter, solitary, paired, or clustered spheroids 1.5–20 μm in diameter, solitary tubular filaments 0.5–3 μm in diameter, and interwoven mats of tubular filaments 10–30 μm in diameter (Lanier, 1986; Klein *et al.*, 1987; Altermann and Schopf, 1995). They are composed of kerogen (or replacive iron oxides) that has similar isotopic values to younger biogenic carbon. Their complex divisional and matting habits are strongly reminiscent of biological behavior, and their irregular distribution aligned along their host sediments' layering indicates contemporaneous deposition. Thus these objects are clearly syngenetic and evidently biogenic, and hence are incontestable Archean microfossils.

All older assemblages are more controversial. Two types of rare, vaguely visible, and poorly preserved filaments ~1 μm and ~10 μm in diameter have been described from the ~2.7 Ga Fortescue Group of Australia (Schopf and Walter, 1983). Though they are associated with stromatolites (laminated structures with convex-upward flexures, presumably deposited by microbial sediment accretion; see the following section), their relationships with these structures are not evident from the published photographs, which show secondary veins cross-cutting the stromatolitic fabric. The narrow filaments have opaque "cells" and translucent "septa" (partitions across the filament), the converse of most younger fossils of septate microbial filaments. Moreover, it is not clear that the filaments have an organic composition, a matting habit, or a

---

[6] *Metamorphic grade* is a qualitative measure of the degree of heating and pressure to which a rock has been exposed; low-grade means less heat and pressure, high-grade means more.

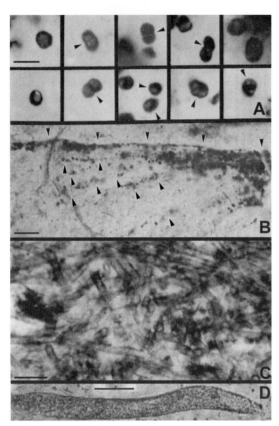

FIGURE 12.4 Archean microfossils from the Transvaal Supergroup, South Africa (2.55 Ga).

(a) Spheroidal microfossils showing cell division (Lanier, 1986; scale = 5 μm).
(b) Spheroidal microfossils aligned along the growth layering of stromatolites (from Lanier, 1986; scale = 25 μm).
(c) Filamentous microfossils interwoven into a mat (Klein et al., 1987; scale = 100 μm).
(d) Large filamentous microfossil (Klein et al., 1987; scale = 50 μm).

particular orientation with respect to sedimentary particles or structures. Without this information, it is still debatable whether these filaments are biogenic and syngenetic or if they are contaminants or pseudofossils.

Some intriguingly life-like filaments have been discovered in the unlikely setting of a ~3.23 Ga hydrothermal base-metal deposit from the Sulphur Springs Group of Australia (Rasmussen, 2000). The filaments are 0.5–2 μm wide and up to 300 μm long, sinuous and composed of solid pyrite (Fig. 12.5a). Interestingly, they are often intertwined and show differing orientations when in planar-layered versus diffusely layered substrate (e.g., Fig. 12.5b), strongly suggestive of biological behavior. They are clearly indigenous because they are overprinted by sulfides precipitated during the later stages of hydrothermal mineralization. As mineralization took place in an anaerobic subsurface setting at high temperatures, the discovery raises the possibility that these filaments represent ancient analogues of the modern microbiota of submarine "black smokers," the sulfide-belching hydrothermal vents along mid-ocean ridges (Chapter 14). However, it is hard to be absolutely sure of their biological origin as they have no detectable organic content from which to measure geochemical signatures of life. This is somewhat baffling because there are preserved organics elsewhere in the deposit, in the form of bitumen and oil (Rasmussen and Buick, 2000), so the cautious assessment of these filaments as "probable microfossils" in their original description is justified.

Many assemblages of spheroidal and filamentous objects have been reported from rocks ~3.23–3.47 Ga in the Swaziland Supergroup of South Africa. Though many are now interpreted as non-fossils (see Schopf and Walter, 1983, for a review), some are still plausibly biogenic. Of the spheroidal objects (Fig. 12.5c), several reports (Muir and Grant, 1976; Knoll and Barghoorn, 1977) illustrate kerogenous objects that could be cells in the process of division by binary fission. However, they do not display any particular relationship with respect to sedimentary grains or structures, and they have a markedly granular and degraded microstructure, so it's not clear that they formed during deposition. Of the filaments (Fig. 12.5d), the most life-like assemblage consists of hollow kerogenous tubes 1.5–5 μm in diameter and up to 100 μm long, apparently oriented nearly parallel to lamination (Walsh and Lowe, 1985; Walsh, 1992). As they are composed of kerogen of much finer grain size than in the surrounding matrix, it is hard to envisage a secondary inorganic process that could have produced them. However, before they can be considered assuredly biogenic, some evidence of behavior would be desirable.

Remarkably life-like filaments have also been reported from North Pole, northwest Australia, in the ~3.47 Ga Warrawoona Group (Fig. 12.5e) (Awramik et al., 1983). Though evidently kerogenous, diverse (four morphotypes), sinuous, and septate, none has a clear syngenetic relationship with sedimentary grains or structures, nor do they show a matting habit or other signs of behavior. Indeed, there are good reasons for believing that they are in fact relatively youthful contamination, perhaps more than half a billion years younger than the surrounding Warrawoona Group rocks. Their kerogen is dark brown in color, which

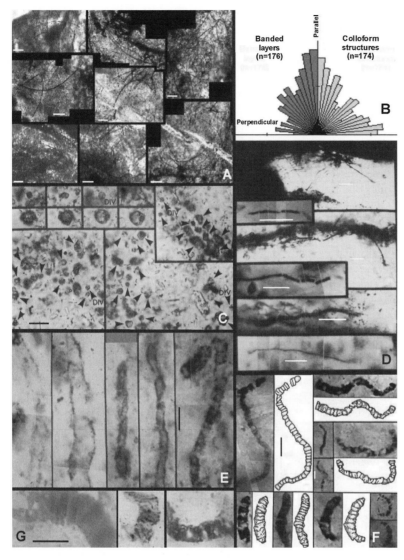

FIGURE 12.5 Microfossil-like objects older than 3.0 Ga.

(a) Pyritic filaments from the ~3.2 Ga Sulphur Springs Group, Australia (Rasmussen, 2000; scales = 10 μm).

(b) Orientation of Sulphur Springs pyritic filaments in two different substrates (from Rasmussen, 2000).

(c) Kerogenous spheroids from ~3.3 Ga rocks in the Swaziland Supergroup, South Africa (Muir and Grant, 1976; scale = 10 μm).

(d) Kerogenous filaments from ~3.35 Ga rocks in the Swaziland Supergroup, South Africa (Walsh and Lowe, 1985; scales = 10 μm).

(e) Filaments in ~3.47 Ga rocks from the Warrawoona Group, Australia (Awramik et al., 1983; scale = 10 μm).

(f) Filaments from the ~3.45 Ga Apex Basalt, Australia (Schopf, 1993; scale = 10 μm).

(g) Filaments from the ~3.45 Ga Apex Basalt, Australia (Brasier et al., 2002; scale = 20μm).

suggests that they have undergone less intense metamorphism ($< \sim300\,°C$) than the totally black indigenous Warrawoona organic matter. Moreover, the filament-bearing sample was apparently collected from a site that subsequent detailed mapping indicates is an exhumed unconformity (erosion) surface between the early Archean Warrawoona Group and the late Archean (~2.75 Ga) Fortescue Group. Hence, the brown filaments could have been emplaced in exposed crevices within the former during deposition of the latter. As the exact sample site unfortunately can no longer be located with greater accuracy than 100 m, it is even possible that the filaments came from the overlying younger rocks.

Recently, controversy has exploded over another set of Warrawoona Group filaments, perhaps the most widely accepted Archean microfossils, from the Apex Basalt near Marble Bar (Fig. 12.5f). The host rocks for these are, at ~3.46 Ga, slightly younger than the North Pole rocks and are ~5 km higher in the regional stratigraphy. A diverse set (eleven named species) of dark brown to black, septate, curved filaments has been described from a body of chert breccia (a composite rock of broken fragments) within the basalt and interpreted as "probably cyanobacterial" microfossils in several influential publications (e.g., Schopf and Packer, 1987; Schopf, 1993). However, doubts about the biogenicity of the filaments have recently been raised (Brasier *et al.*, 2002) and rebutted (Schopf *et al.*, 2002). Specifically, the doubters claim that the filaments are abiogenic artifacts constructed by crystal growth from amorphous organic matter, based on their three-dimensional morphology (in many cases branching) and their angular septation (resembling crystal outlines; Fig. 12.5g). Furthermore, their synsedimentary origin is also questioned, with a hydrothermal breccia suggested as the actual host rock of the filaments. The rebuttal consists of a demonstration that the filaments are composed of moderately metamorphosed kerogen, as shown by laser-Raman spectroscopy. However, neither the color nor the Raman spectra are consistent with rocks that have suffered temperatures above 400 °C, as inferred from the surrounding Warrawoona rocks. At such temperatures, organic matter should be completely black and Raman spectra should show narrow peaks. Hence, it can be concluded that, regardless of their origin, these particular filaments are not particularly old and are thus contaminants. Their widely touted status as the world's oldest fossils is therefore invalid.

So, the microfossil record thus leaves us with no certain members of modern microbial lineages before ∼2.1 Ga and no certain preserved cells before ∼2.55 Ga, but plenty of possible and probable microbial remains from older rocks dating back as far as ∼3.45 Ga. However, the record also shows how easy it is to mistake abiogenic artifacts and younger contaminants for genuine Archean microfossils and illustrates that careful geology and converging lines of evidence are needed in order to be confident of biogenicity and syngenicity. In particular, signs of biological behavior and indications of metamorphic grade have proved to be the most powerful tools for distinguishing reliable results from the misleading.

12.2.2 Stromatolites

Stromatolites are laminated sedimentary structures accreted as a result of microbial growth, movement, or metabolism. As such, they are trace fossils of microbial activity and are thus less direct evidence of life than microfossils. Their shapes vary, but they often show a predominance of convex-upward flexures forming domes or columns, although some conical forms show the converse. Several scales of flexuring and layering give them a complex internal structure, and a pronounced irregularity of flexuring and layering makes them look wrinkly. If accreted by photosynthetic or otherwise sun-loving microbes, the layers show

thickening over flexure crests because the constructing microbes grow more successfully on topographic highs where there is more light. The constructing microbes accrete sediment by three distinct processes. *Trapping* occurs when erect microbial filaments baffle passing water currents, causing entrained sediment to be deposited, just as a carpet traps dirt. *Binding* happens when passively deposited sediment is caught up in and overgrown by a microbial mat, either by lodging in irregularities in the mat or getting stuck in the extracellular mucilage secreted by the microbes. *Precipitation* results from microbial photosynthesis removing CO_2 from the surrounding water, causing calcium carbonate to be deposited as the equilibrium of the following reaction is forced towards the right side because of product depletion:

$$Ca^{2+} + 2HCO_3^- \leftrightarrow CaCO_3 + CO_2^\uparrow + H_2O.$$

There are several classification schemes for stromatolites, with the principal ones being a binomial, quasi-biological taxonomic system like that used for animal trace fossils and a morphological system like that used for sedimentary structures. Proponents of the binomial system emphasize its usefulness for biostratigraphy, with characteristic stromatolites classified for various Proterozoic time intervals on several continents, principally north Asia and Australia. Proponents of the morphological system emphasize the involvement of inorganic processes in stromatolite construction and thus regard them as better environmental indicators than time markers. Disputes have thus ensued about the relative importance of biotic versus sedimentary controls on their structure and even about the definition of the term "stromatolite." Hence, there has also been discussion about their utility for monitoring the early evolution of life, especially since Grotzinger and Rothman (1996) showed that it was possible, using only sedimentary parameters, to model growth of stromatolite-like structures with fractal geometry and self-similar lamination. But as stromatolites are generally large and obvious structures with a better preservation potential than microfossils, a considerable amount of effort has been devoted to their study in Archean terrains. Clearly, however, before using stromatolites as evidence for early biology, multiple lines of supporting evidence should be sought, preferably including fabrics indicative of microbe-sediment interaction.

Many assemblages of stromatolites are now known from rocks older then 2.5 Ga.[7] Those of late Archean

---

[7] See Section 16.2.3 for discussion of younger stromatolites.

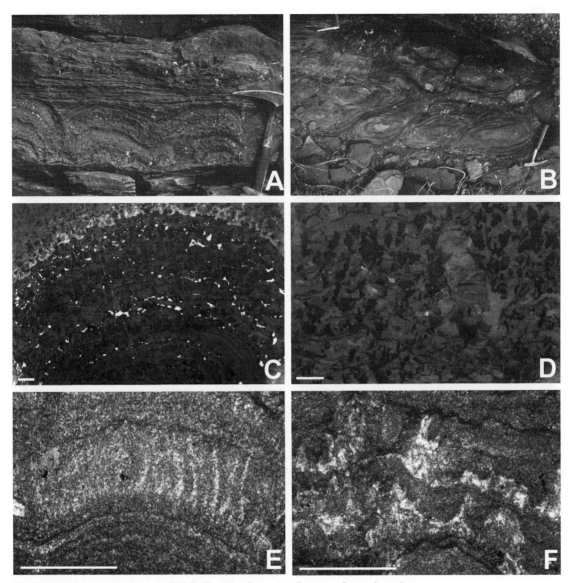

FIGURE 12.6 Stromatolites from the 2.72 Ga Tumbiana Formation, Fortescue Group, Australia.

(a) Pseudo-columnar domes in cross section, showing wrinkly lamination.
(b) Same structures in plan view.
(c) Same structures in a close-up view of a cut slab, showing fenestral fabric of subspherical voids filled by diagenetic carbonate (scale = 10 mm).
(d) Finger-like microstructure (scale = 10 mm).
(e) Pale palimpsests of vertically oriented filaments (scale = 100 μm).
(f) Pale palimpsests of filament tufts (scale = 100 μm).

age (<3.0 Ga) are widely regarded as showing strong evidence of biological activity during their formation. For instance, the ~2.7 Ga Fortescue Group from Australia (Fig. 12.6) has a diverse assemblage of stromatolites (at least ten distinct morphologies) in both tuffaceous and carbonate sediments that were apparently deposited in evaporative, fault-bounded lakes (Buick, 1992). As well as the large-scale structural features typical of biogenic accretion, they also have internal fabrics such as fenestrae (precipitate-filled gas

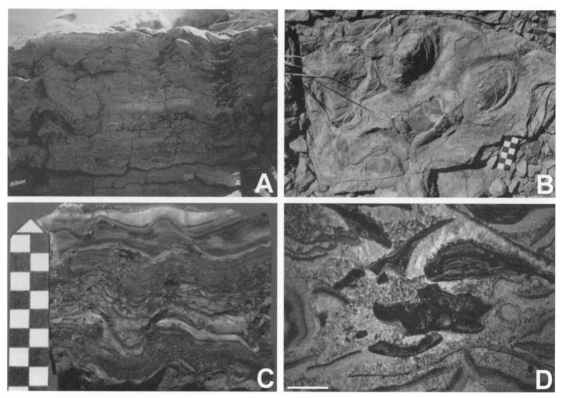

FIGURE 12.7 Stromatolites from the 3.47 Ga Warrawoona Group at North Pole, Australia.

(a) Pseudo-columnar domes in cross section showing wrinkly lamination.
(b) Same structures in plan view.
(c) Same structures in a close-up view of a cut slab, showing lamination thickening over flexure crests (scale in centimetres).
(d) Desiccated flakes eroded from flexure crests, showing kerogenous microlamination (scale = 500 μm).

bubbles produced by metabolic activity) (Fig. 12.6c), cuspate lamination (tufted layers formed by gliding filaments congregating on topographic high-points; Fig. 12.6f), and filament palisades (carpet-like palimpsests[8] of vertically oriented filaments thickening on substrate highs (Fig. 12.6e). There can be little doubt that these structures were accreted by a community of microbes that responded in some way to sunlight.

The origin of older Archean stromatolite-like structures is more controversial. Indeed, it has been proposed (Lowe, 1994) that all such structures older than 3.2 Ga are in fact abiogenic in origin, either evaporative precipitates, hydrothermal deposits, or deformational features. But this analysis was based on only some of the known assemblages and neglected perhaps the most complex, from the ~3.47 Ga Warrawoona Group at

North Pole, Australia (Fig. 12.7). These show a range of large-scale features also seen in younger, clearly biogenic stromatolites, such as pseudo-columnar structure (stacked, domically flexed layers that are laterally continuous), semi-columnar structure (stacked domical flexures that are only continuous in one direction), and complex wrinkly lamination that thickens over flexure crests (Buick et al., 1981). Moreover, they defy categorization as evaporative, hydrothermal, or deformational structures. They are not simple chemical precipitates, because they show lateral variation in lamina thickness, merely moderate inheritance of lamina irregularities, and several orders of flexures that are not self-similar. Moreover, the presence of lenses filled by desiccated fragments of wrinkly laminated stromatolitic intraclasts in troughs between domes indicates that the structures are not deformational but had primary relief above the seafloor. Lastly, the persistence of kerogenous microlaminae in less-weathered parts of

---

[8] A palimpsest is an original texture preserved in a rock of altered mineral composition.

the outcrop implies microbial inhabitants. So, as a result of the many biotic features of these North Pole structures, it has been argued that Lowe's (1994) dismissal of all early Archean stromatolites as abiogenic is perhaps unjustified (Buick *et al.*, 1995).

In support of this reasoning, Hofmann *et al.* (1999) have added another stromatolite assemblage from ~5 km higher in the ~3.46 Ga Warrawoona Group to the list of potentially biogenic structures from the early Archean. These stromatolites have more complex large-scale structures than any others known from such ancient rocks, consisting of linked conical pseudo-columns with parasitic domical pseudo-columns branching from them, similar to some assuredly biogenic Proterozoic stromatolite forms. Thus it seems that some early Archean stromatolites do indeed represent the activities of primordial microbial communities.

12.2.3 Carbon isotopes

Both the microfossil and stromatolite records in the Archean are very patchy. But a more continuous record of Archean biology is preserved in a less direct way: in isotopic chemofossils, particularly the carbon isotope ratios of sedimentary rocks. *Autotrophic* metabolisms (using CO_2 for manufacturing cellular carbon compounds) preferentially incorporate the light stable isotope of carbon (^{12}C) over the heavy (^{13}C) into their synthesized organic matter,[9] so any sedimentary kerogen derived from such organisms will inherit this characteristic isotopic fractionation. Reflecting this biological concentration of ^{12}C, sedimentary carbonate that is deposited from water bodies inhabited by autotrophs will consequently be depleted in ^{12}C, i.e., relatively enriched in ^{13}C. Hence, on the modern Earth, organic carbon has a carbon isotopic value (called $\delta^{13}C_{org}$) that averages around −22‰ (i.e., depleted in ^{13}C by 22 parts per thousand relative to an arbitrary standard), whereas marine carbonate has a $\delta^{13}C_{carb}$ of about 0‰ and the entire Earth is near −6‰. If autotrophy has persisted through time, similar fractionations should be evident in the isotopic ratios of ancient sedimentary carbonates and kerogens.[10]

Unfortunately, there are some complications to this simple picture. Fluctuations in the fractionation record (Fig. 12.8) might be expected to appear at times when the burial ratio of sedimentary kerogen to carbonate (f_{org}) varies or when the mean autotrophic fractionation between organic and inorganic carbon (Δ_C) varies. Indeed, it appears that now is a somewhat unusual time in Earth history with very low Δ_C. This may have resulted from declining atmospheric pCO_2 associated with the onset and growth of Antarctic glaciation over the past 30 Myr. In contrast, a general survey going back to ~1.0 Ga (Hayes *et al.*, 1999) shows a persistent range for Δ_C between −28‰ and −32‰ and for f_{org} between 0.15 and 0.3, except for bizarre intervals around the Neoproterozoic glacial events at 0.6 Ga and 0.75 Ga. Overall, it seems reasonable to assume that these values represent steady-state operation of a global carbon cycle (Chapter 11) based on autotrophy of both independent microbes and endosymbiotic microbes such as chloroplasts in plants and algae.

There are, however, several other variables that need to be considered before extrapolating back to more ancient rocks. Firstly, older rocks tend to be more metamorphosed and, at metamorphic grades above greenschist facies (i.e., $T > \sim 550\,°C$), $\delta^{13}C_{org}$ and $\delta^{13}C_{carb}$ begin to re-equilibrate and approach each other in value. Hence, highly metamorphosed rocks can yield low Δ_C indistinguishable from non-biologic fractionations. Secondly, $\delta^{13}C_{org}$ can shift to heavier values at lower metamorphic grades because of loss of light hydrocarbons during thermal maturation. Thirdly, isotopically light hydrocarbons can migrate into ancient sedimentary rocks and become solidified, sometimes appearing almost indistinguishable from primary kerogen and yet having a very different isotopic composition. Fourthly, metasomatism and metamorphism can introduce secondary carbonate or graphite into ancient rocks by precipitation from or reaction with migrating aqueous fluids rich in dissolved carbonic species. All of these mechanisms are apt to confound the ancient Δ_C record, particularly in deformed and metamorphosed terrains. It is thus necessary to have a good understanding of post-depositional history, $\delta^{13}C_{org} - \delta^{13}C_{carb}$ pairs, and organic hydrogen/carbon ratios (a measure of metamorphic grade) in order to be certain of obtaining a valid isotopic record from ancient sediments.

The most noticeable feature of the pre-1.0 Ga carbon isotope record (Fig. 12.8) is that the fractionations of sedimentary kerogen and of carbonate have

---

[9] The light isotope becomes relatively more abundant because compounds containing it have a lower zero-point energy and therefore reactions using light-isotope molecules are very slightly energetically favored.

[10] See Section 4.6.2.1 for more introduction to the concepts in this paragraph.

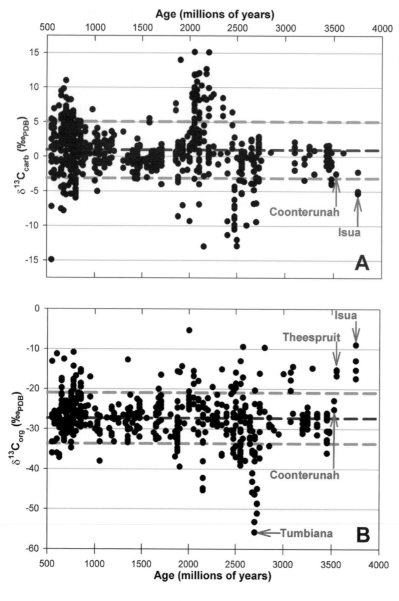

FIGURE 12.8 Sedimentary carbon isotope values through time (middle dashed line is the mean and the outer two lines are one standard deviation excursions).

(a) Carbonate carbon.
(b) Organic carbon.

(Modified and updated from the database of G.A. Shields and J. Veizer, www.science.uottawa.ca/geology/isotope_data/.)

generally remained fairly constant back to ∼3.5 Ga. There is a weak trend towards lighter (lower) values of $\delta^{13}C_{org}$ between ∼1.0 and 2.5 Ga (e.g., Des Marais *et al.*, 1992), perhaps the result of decreasing atmospheric pCO_2, increasing crustal storage of C_{org}, or a greater input of organic matter derived from chemolithotrophic microbes (Chapter 11). There is also a marked positive excursion in $\delta^{13}C_{carb}$ between ∼2.4 and 2.0 Ga resembling that associated with the Neoproterozoic glacial events ("Snowball Earths"; Section 4.24), and similarly marked by low-latitude glaciations. However, the most anomalous excursion is during the late Archean between ∼2.8 and 2.5 Ga,

when $\delta^{13}C_{org}$ reaches exceptionally light values, as low as −60‰. This has been explained as a period of enhanced microbial methanogenesis and methanotrophy (Hayes, 1994); methane-generating microbes produced methane much more depleted in ^{13}C than standard autotrophic processes, and methane-consuming microbes incorporated the light isotopic signature into their cellular biomass.

Apart from these weak trends and temporary excursions, the overall pre-1.0 Ga carbon isotope record is similar to that of more modern times. $\delta^{13}C_{carb}$ is generally around 0‰ and $\delta^{13}C_{org}$ is ∼−30‰ in low-grade metasedimentary rocks, implying a pre-metamorphic

Δ_C of about $-28‰$ (hydrocarbons expelled from kerogen during diagenesis[11] are generally somewhat heavier than the original organic matter). Shallow marine carbonates in particular are very close to 0‰ as far back in time as the undisputed sedimentary record goes. Organic carbon isotopic values are similarly consistent, though with greater variability about their mean value. Hence, autotrophic carbon fixation has almost certainly been the predominant process of primary biological productivity on the Earth since at least 3.5 Ga. Before that, the record is sparse, confused, and controversial (see Section 12.3).

12.2.4 Sulfur isotopes

Some microbial metabolic processes also fractionate the stable isotopes of sulfur, ^{32}S and ^{34}S (discussed more fully in Chapter 11). In particular, dissimilatory sulfate reduction (sulfate reduction for metabolic purposes rather than for biosynthesis), which is performed by several groups of Bacteria and one of Archaea, can impart a large ($-10‰$ to $-45‰$) fractionation in favor of the light isotope. This reaction, which can use either organic carbon:

$$SO_4^{2-} + 2H^+ + 2CH_2O \rightarrow H_2S + 2CO_2 + 2H_2O$$

or molecular hydrogen:

$$SO_4^{2-} + 2H^+ + 4H_2 \rightarrow H_2S + 4H_2O$$

as an electron donor during the reduction of sulfate, only takes place in anaerobic settings. Isotopic fractionation only occurs if reduction is performed under conditions of moderate to high (>1 mM) sulfate abundance (Canfield and Raiswell, 1999; Canfield et al., 2000). The fractionation is preserved in the geologic record when the metabolic product hydrogen sulfide (H_2S) reacts with dissolved ferrous ion during early diagenesis to form sedimentary pyrite.

Though not particularly abundant or widely distributed in sedimentary rocks, such pyrite is still useful as a tracer of ancient biological activity because it is relatively immune to metamorphic resetting of isotopes. Its low solubility hinders isotopic re-equilibration and the scarcity of mobile sulfur species in most geological environments limits post-depositional alteration. However, evidence of the isotopic composition of seawater sulfate, the parent reactant, is only rarely preserved in rocks, principally in evaporitic sulfate

minerals such as gypsum and anhydrite, or in trace quantities in the crystal lattice of carbonate minerals. Thus, it is usually impossible to determine directly the Δ_S (analogous to Δ_C in the last section) of ancient sediments, weakening the biological inferences that can be drawn from this isotopic system.

The sulfur isotope record (Fig. 12.9) shows three phases to its evolution. Most recently, since about 0.8 Ga, $\delta^{34}S_{sulfate}$ has fluctuated around $+20‰$, reflecting the removal of the light sulfur isotope from seawater by sedimentary pyrite. $\delta^{34}S_{sulfide}$ shows a wide spread of values, from as light as $-50‰$ to as heavy as $+40‰$. This apparently reflects the superimposition of several processes. Firstly, standard microbial sulfate reduction in sulfate-rich waters has imparted a fractionation of $-10‰$ to $-45‰$ on most sedimentary pyrites. Secondly, because the environment over the past 0.8 Gyr has been highly oxygenated, inimical to sulfate-reducing microbes, most sulfate reduction has been forced into semi-closed anoxic systems such as diagenetic pore fluids or restricted euxinic basins. This has led to Rayleigh distillation (selective loss of a particular isotope in a closed system, thereby enriching the residue in the other isotope) during continuing reduction, with removal of the light isotope forcing the residual sulfate heavier and so producing isotopically heavy pyrite, sometimes with highly positive values, as sulfate depletion proceeds. Thirdly, also as a result of environmental oxygenation, up to 90% of microbially reduced H_2S has been re-oxidized to sulfur species of intermediate redox state such as elemental sulfur (S^0), sulfite (SO_3^{2-}), and thiosulfate ($S_2O_3^{2-}$), which can then be microbially disproportionated into H_2S and SO_4^2 with further moderate isotopic fractionations.

Between $\sim$2.3–2.7 and 0.8 Ga, a second stage of $\delta^{34}S$ evolution is evident when $\delta^{34}S_{sulfate}$ rises gradually from about $+10‰$ to $+25‰$. $\delta^{34}S_{sulfide}$ suggests only a single reduction step, with values below $-20‰$ only rarely recorded. Some pyrite heavier than seawater sulfate is evident, indicating that at least some reduction occurred in diagenetic or euxinic settings. Such a pattern might signify a partially oxidized environment, in which sulfate was abundant enough (>1 mM) in the upper ocean for significant isotopic fractionation to occur upon microbial reduction, but where H_2S reoxidation only occurred rarely with concomitantly negligible disproportionation. This would imply that the deeper ocean was anoxic and sulfidic (Canfield, 1998).

During the earlier Archean, before 2.7 Ga, $\delta^{34}S_{sulfate}$ was close to 0‰, implying negligible burial of light

---

[11] *Diagenesis* is the process of lithification of sediments into sedimentary rocks.

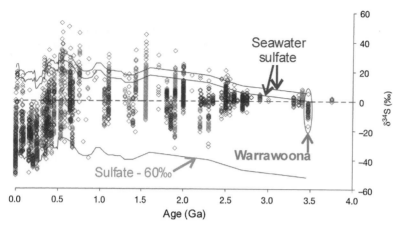

FIGURE 12.9 Sedimentary sulfur isotope values through time (from Shen *et al.*, 2001). Diamonds mark individual measurements of sedimentary sulfide isotopic values, upper parallel lines indicate seawater sulfate values, lower line indicates a $-60‰$ sulfide fractionation from seawater values. Note the Warrawoona Group values lying below the general trend of Archean results.

sedimentary pyrite. Almost everywhere, $\delta^{34}S_{sulfide}$ was similarly close to 0‰. This could indicate that the biological capability for dissimilatory sulfate reduction had not yet evolved (Lambert *et al.*, 1978), but evidence from an early Archean sulfate oasis suggests the contrary. ~3.47 Ga sedimentary rocks in the Warrawoona Group at North Pole, Australia, contain beds of barite crystals that originally formed as evaporative gypsum, indicated by their interfacial angles and crystallographic habit (Buick and Dunlop, 1990). Lining the growth faces of the original crystals are microscopic pyrite crystals that show exclusively negative $\delta^{34}S_{sulfide}$ values with some as light as $-17‰$ (Shen *et al.*, 2001). $\delta^{34}S_{sulfate}$ is consistently around $+4‰$ with the mean Δ_S about $-12‰$, clearly in the biological range. As the surrounding rocks are basalts that have never been heated to much more than 300 °C, inorganic isotopic fractionations caused by hydrothermal or magmatic processes are inconsistent with both the geological setting and isotopic spectrum. Hence, these results indicate that sulfate reducing microbes had evolved and were metabolically active in the early Archean in the rare sites where sulfate was concentrated by evaporation. Nevertheless, the open Archean ocean was evidently depleted in sulfate to such an extent that any microbial sulfate reduction imparted minimal isotopic fractionation. It has been argued (Ohmoto *et al.*, 1993) that the Archean ocean was hot, promoting high rates of sulfate reduction and thus minimal isotopic fractionation, implying high oceanic sulfate concentrations and thus atmospheric oxygenation. However, the required temperature increase (30–40 °C above present levels) is incompatible with the paleotemperatures derived from the Archean evaporite record, where the former presence of gypsum implies much cooler conditions

(Section 12.3.2). Moreover, microbial sulfate reduction always imparts large fractionations regardless of temperature or reduction rate (Canfield *et al.*, 2000).

12.2.5 Nitrogen isotopes

Nitrogen is an essential nutrient for all living things, as it is a key constituent of the amino acids of proteins and the bases of nucleic acids. It has two stable isotopes: ^{14}N and ^{15}N, and predominantly resides in a single reservoir, the atmosphere, as N_2 gas. However, its biogeochemical cycle is more complex than those of either carbon or sulfur and is consequently more difficult to decode from sedimentary isotopic ratios. Moreover, because nitrogen is never more than a trace component of sedimentary rocks, with no long-lived minerals dominated by nitrogenous species, the geological record of nitrogen abundance and isotopic fractionation is very sparse. As a result, comparatively little is known about the impact of post-depositional processes upon nitrogen isotopes. However, because of the scarcity of mobile nitrogenous species in most geologic environments, isotopic exchange and re-equilibration reactions should be minimal. Hence, isotopic compositions in the two main crustal repositories of nitrogen, organic matter and ammonium ions in clays and micas, probably fairly closely reflect their biological parents, particularly at lower metamorphic grades.

There are four principal steps in the biogeochemical nitrogen cycle (Fig. 12.10a), but only two of them influence nitrogen isotopic ratios. Atmospheric N_2 gas (with a δ^{15} value of 0‰) is relatively unreactive and can only be incorporated into organic matter after fixation into ammonia NH_3. This reaction,

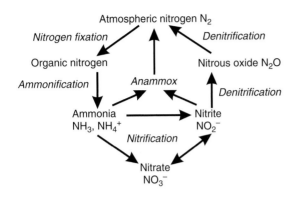

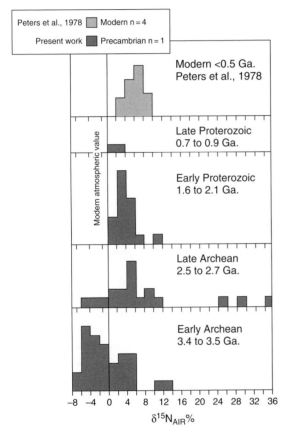

FIGURE 12.10 (a) The biogeochemical nitrogen cycle. (b) Nitrogen isotopes in chert through time (Beaumont and Robert, 1999).

$$N_2 + 3\,H_2 \rightarrow 2NH_3,$$

can occur only in a reducing environment. On the early, abiotic Earth, it could have occurred around hydrothermal vents using dissolved N_2. It is now performed by a wide range of microorganisms including many cyanobacteria, methanogenic Archaea, the aerobic *Azotobacter* group, some members of the anaerobic

Clostridium group, and the symbiotic *Rhizobium* in plant roots. These all possess varieties of the enzyme nitrogenase, containing one or more of the transition metals Fe, Mo, or V, which catalyze the breaking of the molecular nitrogen triple bond and the reduction of the resulting nitrogen. This reaction in simplest form is

$$2N_2 + 6H_2O \rightarrow 4NH_3 + 3O_2.$$

It causes little isotopic fractionation and all ammonia produced intracellularly is rapidly incorporated into organic matter.

Organic nitrogen is broken down by ammonification, performed by heterotrophic bacteria and producing either ammonium ion or ammonia depending on whether the environment is aqueous or not. In anaerobic settings (lacking O_2), these products can then be directly assimilated by other organisms, with the whole process of ammonification and assimilation involving minimal isotopic fractionation. Alternatively, in aerobic settings, ammonium is oxidized to nitrate by nitrifying microbes. This is done in two steps, from ammonium to nitrite, and then to nitrate:

$$2NH_4^+ + 3O_2 \rightarrow 2NO_2^- + 2H_2O + 4H^+,$$

$$2NO_2^- + O_2 \rightarrow 2NO_3^-.$$

Different bacteria undertake the two reactions, with *Nitrosomonas* principally responsible for the first and *Nitrobacter* dominant for the second. Recently, indications that microbes can also oxidize ammonia anaerobically have been found, whereby manganese rather than molecular oxygen is utilized to produce nitrate. Ammonia is also oxidized by the "anammox organism", a planctomycete bacterium that reacts ammonia with nitrite to produce nitrogen gas. Neither aerobic nor anaerobic nitrification induces a significant isotopic effect. Nor does the process whereby organisms, particularly plants, assimilate and reduce nitrate as their principal source of nitrogen.

The breakdown of dissolved nitrate to N_2 gas (denitrification) removes nitrogen from the biosphere and returns it to the atmosphere, thus completing the biogeochemical nitrogen cycle. The process occurs in several steps involving intermediate compounds such as NO_2^-, NO, and N_2O. The resulting N_2 is enriched in the light isotope ^{14}N by up to 30‰, depending on the degree of denitrification. This light N_2 diffuses out of aqueous settings, leaving the residual dissolved nitrate isotopically heavy. Because dissolved nitrate is the dominant source of organic nitrogen on the modern oxygenated Earth, living and sedimentary organic matter strongly reflects the effects of this isotopic fractionation during denitrification and is thus heavy like dissolved nitrate (average $+7‰$). The process is an anaerobic respiration that can be expressed in abbreviated form as:

$$5CH_2O + 4NO_3^- + 4H^+ \rightarrow 5CO_2 + 2N_2 + 7H_2O.$$

The various reactions in the pathway are performed by a range of bacteria, most notably by some members of the genera *Pseudomonas* and *Thiobacillus*. Under conditions of low pH, high O_2, or low reaction rates, denitrification does not always proceed to completion, instead stopping at N_2O.

The secular record of sedimentary $\delta^{15}N$ values shows little change from modern ranges back to the end of the Archean (Schidlowski *et al.*, 1983). This implies that an active biological nitrogen cycle with all of its modern complexity was operating as early as 2.5 Ga. Before this time, however, the data are contradictory. In a study of Archean and Proterozoic cherts (Fig. 12.10b), Beaumont and Robert (1999) found that the older samples (>2.5 Ga) had markedly light $\delta^{15}N$ values, with 3.4–3.5 Ga cherts yielding values as low as $-6.2‰$. Some late Archean cherts were also as light as $-4.1‰$. Proterozoic cherts, on the other hand, all yielded positive $\delta^{15}N$ values. Beaumont and Robert (1999) interpreted these data as indicating that a curtailed nitrogen cycle, consisting only of fixation, ammonification, and assimilation, operated in the early Archean. They attributed this to low levels of O_2, since nitrification occurs only in the presence of O_2. The scatter of data in the late Archean was, they considered, indicative of a transitional environment in which aerobic and anoxic micro-environments coexisted.

More recent work challenges these interpretations. Pinti and Hashizume (2001) suggested that the cherts studied by Beaumont and Robert (1999) represent an environmentally biased sample of the potential Archean biota, in that many were probably formed by precipitation from hydrothermal fluids. In such settings, conditions were almost certainly anoxic, necessarily preventing the activities of nitrifiers and denitrifiers. This contention is evidently borne out by $\delta^{15}N$ values obtained from Archean shales (Yamaguchi, 2002) and other ancient sediments (Jia and Kerrich, 2001). These are heavier and more varied in their isotopic composition than those recorded from contemporary cherts, regardless of age. This suggests that the full suite of microbial reactions that constitute the modern nitrogen cycle reactions were in operation

early in Earth history. Moreover, just as now, differing Archean environments show varying nitrogen isotope values, indicating that this complex biogeochemical cycle functioned in a similarly diverse geographic pattern.

12.2.6 Molecular biomarkers

Certain hydrocarbon molecules found in kerogen or oil are recognizable derivatives of biological molecules. In mature sedimentary rocks, these have had their functional groups removed and multiple bonds broken, but their carbon skeleton is intact and clearly indicates their parent biomolecule. When these molecules have a specific biological source such as the Domain Bacteria or the Phylum Porifera, they are called *biomarkers*. There are many such biomarker molecules, but most that survive in mature rocks are derived from cellular lipids, in particular those of the cell membrane and pigments (Fig. 12.11). When temperatures pass into the metamorphic range, however, the abundance and range of preserved hydrocarbon molecules decreases markedly because long chains break down into methane or form linked aromatic ring structures. It was once thought that such destruction was largely complete by $\sim 200\,°C$ and would occur in less than a billion years, but it is now clear that this is not so. Hence, hydrocarbon biomarkers can also be used to inform us about life's early diversity (Brocks *et al.*, 1999).

Though potentially powerful, organic geochemical analysis of particularly ancient rocks is fraught with difficulties. First, in thermally altered rocks, the concentrations of the relevant compounds diminishes to extremely low levels, making it very difficult to distinguish indigenous molecules from laboratory contamination. Usually only differences in isomer ratios and chain lengths (maturity parameters) serve to differentiate very old from very young molecules. Second, old samples can be contaminated during the recent geological past by groundwater, subsurface microbes or, if from the surface, lichens and endolithic (living within rocks) bacteria. Such youthful contaminants can be distinguished, however, because they often show preferences for handedness in compounds with stereo-isomers such as amino acids, as well as detectable levels of radioactive ^{14}C, an isotope with a geologically short half-life. Third, during the long and complex geological history of all Archean rocks, there have generally been many opportunities for post-Archean introduction of organic molecules from younger

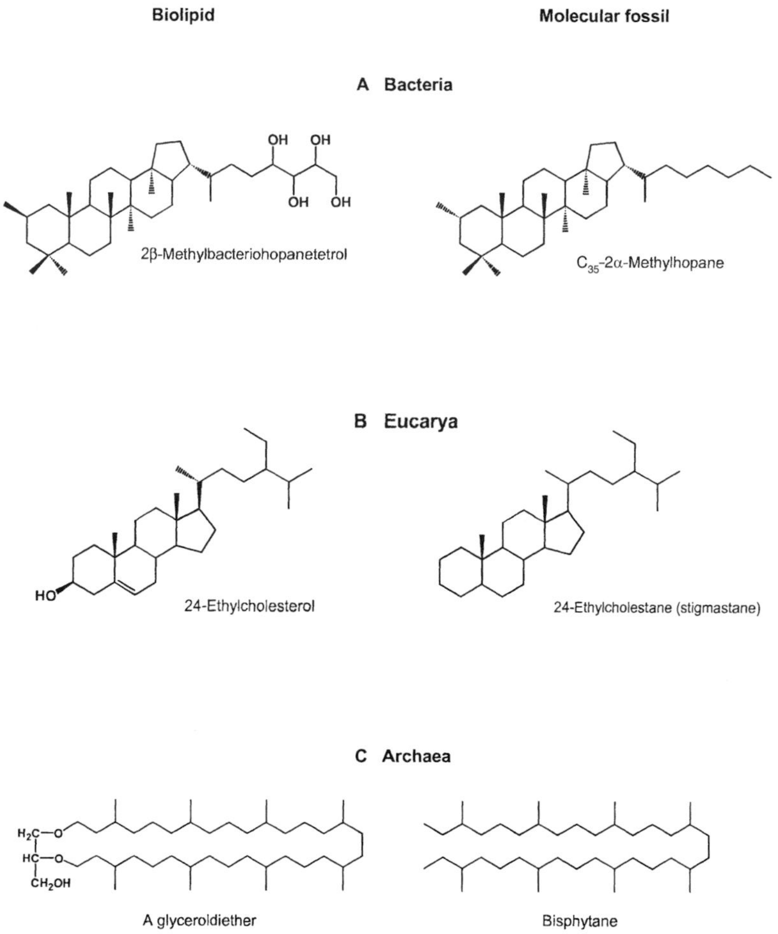

FIGURE 12.11 Molecular structures of hydrocarbon biomarker molecules and their biological precursors. (a) Hopanes from Bacteria. (b) Steranes from Eukaryota. (c) Acyclic isoprenoids from Archaea.

geological fluids. These can be particularly difficult to recognize as they may have had similar thermal histories to indigenous molecules and may have been emplaced from migrating hydrocarbons. However, subtle differences in carbon isotope ratios, aromatic contents, or in evolutionarily inappropriate biomarker distributions (e.g., plant biomarkers in Archean rocks) can sometimes be used to discriminate between older and younger molecules. As a result of these problems, many studies before 1990 are now regarded as dubious, especially those of porous rocks allowing ingress of soluble contaminants (Hayes *et al.*, 1983). Even pyrolysis studies, in which kerogen is combusted at high temperatures to drive off bound biomarkers, are now viewed with some suspicion because kerogen can be irreversibly contaminated by younger hydrocarbons (Oehler, 1977).

Several recent discoveries, however, have re-invigorated the field. Since the oldest known biomarkers came from the oldest oil-bearing rocks in the ~1.7 Ga McArthur Group from northern Australia (e.g., Summons *et al.*, 1988), it is thus reasonable to suppose that where oil survives, so will biomarkers. Hence, the demonstration that oil exists in fluid inclusions in many Archean sandstones, that it was clearly generated before peak metamorphism of the host rocks, and that this maximum thermal episode occurred in Archean or early Proterozoic time indicated that complex organic compounds may survive for much longer than previously suspected (Dutkiewicz *et al.*, 1998). Moreover, it became apparent that not all Archean rocks have been highly metamorphosed, with the recognition of rocks as old as ~3.47 Ga that have seen temperatures no higher than ~200–300 °C (Buick *et al.*, 1995). Lastly, as instrumentation and techniques improved, it became possible to analyze more complex molecules at lower concentrations, in some instances with compound-specific isotopic data as well.

As a result of these advances, Brocks *et al.* (1999, 2003) discovered a diverse suite of apparently indigenous biomarker molecules in 2.7 – 2.5 Ga kerogenous shales in the Archean Fortescue and Hamersley Groups in north-western Australia. The biomarkers possessed most of the distinctive features of Proterozoic hydrocarbons, but lacked any evidence of higher plants and animals or modern bacteria, thus ruling out more recent contamination. Most samples showed thermal maturity parameters consistent with their geological setting, indicating that the biomarkers were older than peak metamorphism, which in this area was ~2.1 Ga. Hence, the molecules were apparently not introduced into their host rocks by ancient fluid migration.

The molecular assemblages consist predominantly of low molecular weight n-alkanes, cycloalkanes, and monomethylated alkanes that are not very informative about specific biology because they have widely distributed biochemical precursors. However, they also contain abundant $C_{\leq 20}$ acyclic isoprenoids (Fig. 12.11c), molecules usually derived from the phytyl side-chain of chlorophyll and thus indicating the former presence of photosynthetic organisms. They also contain a variety of hopanes, which are polycyclic hydrocarbons with five carbon rings (Fig. 12.11a) that are derived from bacterial cell membranes. These include $C_{>31}$ 2α-methylhopanes that are predominantly produced by cyanobacteria (Summons et al., 1999) and 3β-methylhopanes which are principally derived from methylotrophic bacteria (Zundel and Rohmer, 1985). In addition, they contain significant quantities of steranes (Fig. 12.11b), four-ring polycyclic hydrocarbons derived from the steroids acting as membrane-rigidifying molecules in eukaryote cells (Summons and Walter, 1990). A wide range of steranes was detected, including many cholesterol derivatives, but none were necessarily indicative of any organism more complex than unicellular algae.

12.3 The oldest evidence of life

The most ancient rocks that could possibly contain evidence of life, i.e., the oldest deposited on the Earth's surface, are within the >3.7 Ga supracrustal successions at Isua and Akilia in western Greenland. These controversial outcrops have been both highly metamorphosed and severely deformed to such an extent that their exact age is now difficult to determine because most geochronologically useful isotopic systems have been reset. Though microfossils have been reported from Isua (e.g., Pflug and Jaeschke-Boyer, 1979), these have now been dismissed as either fluid inclusions (Bridgwater et al., 1981) or weathered

carbonate crystals (Roedder, 1981) and, indeed, it is hard to imagine any kerogenous microfossil surviving such rigorous conditions with their morphology intact and recognizable. Biomarker molecules have also been reported from these rocks (Walters et al., 1981), but the preservation of such complex chemicals under these severe conditions is even less likely than microfossils, and permeability and racemization experiments have shown that it is unlikely that they are indigenous (Nagy et al., 1981). It seems that, at such high grades, only isotopic geochemical data are robust enough to provide satisfactory evidence of life and even then, their interpretation can be fraught with problems. Moreover, in particularly ancient terrains such as the early Archean of Greenland, several abiogenic sources could have contributed isotopically light carbon to the metamorphosed sedimentary rocks. These include carbonaceous chondritic meteorites, prebiotic organic synthesis, and reduced igneous rocks, all of which can contain organic carbon with light $\delta^{13}C_{org}$ values resembling those of biogenic carbon. As the Greenland rocks were deposited close to the end of the primordial period of intense meteoritic bombardment, cosmic carbon in particular may have been incorporated into the sedimentary precursors of the schists and gneisses now present. Upon metamorphism, this would have been rendered isotopically indistinguishable from authentic biogenic carbon.

The Akilia rocks have undergone granulite facies metamorphism at temperatures >600 °C and have been intensely deformed (Fedo and Whitehouse, 2002). In rocks that have been interpreted as metasedimentary banded iron formation >3.85 Ga (Mojzsis et al., 1996) are apatite crystals containing graphite inclusions a few microns across, which yield a bimodal distribution of $\delta^{13}C_{org}$ values (mean of $-37‰$, modes at $-44‰$ and $-27‰$). These light $\delta^{13}C_{org}$ ratios have been cited as evidence that the original carbon was of biological origin, the oldest sign of life (Mojzsis et al., 1996). However, these values were determined by ion-microprobe analysis, a new technique that at least in one instance has yielded isotopic results inconsistent with conventional whole-rock analyses (for example, for Isua, ion-microprobe analysis gives a mean of $-30‰$, whereas conventional analysis gives values closer to $-10‰$). Moreover, the very high metamorphic grade suggests that if the organic carbon were originally biogenic, then light hydrocarbon loss and carbonate re-equilibration should have made its initial $\delta^{13}C_{org}$ value lighter by as much as 25‰. Such extremely light values are only biologically produced by methanotrophic and methylotrophic metabolisms that require free oxygen or high sulfate, for which there is no evidence so early in Earth's history. Furthermore, Rayleigh distillation upon reduction of metamorphic CO_2 could conceivably have produced organic carbon with the isotopic distribution observed (Eiler *et al.*, 1997). Such inorganic scenarios seem more likely now that it has been shown that the host rock is not a metamorphosed banded iron formation, but is apparently a highly deformed and metasomatized ultramafic igneous rock (Fedo and Whitehouse, 2002). Finally, field relations indicate that the rocks may not be quite as old as previously believed. Earlier age determinations were based on dates from a supposedly intrusive sheet of tonalitic gneiss, which now appears to be tectonically interleaved rather than cross-cutting (Myers and Crowley, 2000), making its age irrelevant to that of the graphite. Moreover, the Akilia supracrustal rocks themselves yield ages of 3.65–3.7 Ga (Schiøtte and Compston, 1990; Kamber and Moorbath, 1998), but the apatite grains containing the actual graphite inclusions return U-Pb dates of just ~1.5 Ga (Sano *et al.*, 1999). This all suggests that it is perhaps premature to regard the Akilia gneisses as the hosts of the oldest evidence of life on Earth.

The Isua evidence is somewhat less controversial but still not fully convincing. These rocks are generally regarded as 3.7–3.8 Ga and they have been metamorphosed to temperatures of ~550 °C, but some still display original depositional structures and fabrics (Appel *et al.*, 1998). Initial carbon isotope studies were performed on rocks believed to be sedimentary carbonates and banded iron formations. These yielded whole-rock $\delta^{13}C_{org}$ values from −5 to −28‰ (mean of −15.5‰) and $\delta^{13}C_{carb}$ values from −7‰ to +5‰ (mean of −2.5‰), which have been interpreted as biological, representing metamorphically reset carbon reservoirs modulated by photoautotrophic fractionation (Schidlowski, 1993). However, it is not clear that this explanation is necessarily valid. Rose *et al.* (1996) have mounted a convincing case that carbonates in the Isua succession are not sedimentary but were introduced after deposition when metasomatic fluids altered the chemical composition of the rocks, thereby making it impossible to extrapolate back through metamorphism to the original sedimentary carbon isotope fractionation. Although Rosing (1999) has reported light $\delta^{13}C_{org}$ values (−20‰ to −11‰) from organic particles in definite metasediments, in the absence of any definite sedimentary carbonate for comparison it is not certain

that these values indeed represent metamorphosed biogenic carbon. So, though permissive of a biogenic interpretation, judgment on their origin should be suspended until there is additional supporting evidence.

If neither of the early Archean Greenland successions provides compelling isotopic evidence for the existence of life, what then? The next oldest rocks that might yield biological information are the ~3.52 Ga Coonterunah Group in the Pilbara Craton of northwestern Australia and the similarly ancient Theespruit Formation in the Swaziland Supergroup of South Africa. These rocks have experienced metamorphic temperatures above 450 °C (the Theespruit higher), about the level at which substantial resetting of carbon isotope values begins. They also contain organic carbon in quantities sufficiently large to analyze by conventional whole-rock methods. The Theespruit Formation yields $\delta^{13}C_{org}$ values between −23‰ and −15‰ (Strauss and Moore, 1992), certainly within the range expected for biogenic carbon at these metamorphic grades. However, there are no analyses of coeval sedimentary carbonate with which to compare them, so the same caveats apply as to the Isua data. On the other hand, the Coonterunah Group contains both sedimentary carbonate and kerogenous chert (Green *et al.*, 2000), the former with $\delta^{13}C_{carb}$ averaging about −2‰ and the latter with $\delta^{13}C_{org}$ averaging about −24‰ (Buick *et al.*, unpublished data). As the metamorphic grade should not have markedly modified these values from their original state, their $\Delta^{13}C$ of 22‰ can be compared with that of Proterozoic banded iron formations, which evidently formed in a similar sedimentary environment. These typically show somewhat light carbonate values, around −2‰ (Kaufman *et al.*, 1990), with similar $\Delta^{13}C$ (Strauss and Moore, 1992). Given the unambiguous association of banded iron formations with microbial fossils, e.g., in the Gunflint Iron Formation, such $\Delta^{13}C$ can reasonably be regarded as representing a biological isotopic effect. Hence, the Coonterunah isotopic data are fully consistent with photoautotrophic fractionation and thus perhaps provide the oldest compelling evidence for life on Earth.[12]

12.3.1 Physical characteristics

Though there is little preserved evidence, it all suggests that the earliest organisms were morphologically

---

[12] In reading this section, consult the summarizing timeline in Fig. 12.2.

simple. All genuine Archean microfossils are of spheroidal or ellipsoidal single cells that reproduced by binary fission, or of cylindrical sheaths that surrounded filamentous chains of linked cells (Fig. 12.4). Usually the cells are aggregated into clusters with many representatives of each morphological type present. Filaments may be interwoven into mats in which all individuals have a similar orientation and size (Fig. 12.4c), implying some degree of environmental sensitivity. Though only a limited range of morphological and size diversity is evident in the Archean microfossil record, this does not necessarily mean taxonomic diversity was similarly restricted. Microbes are in general architecturally conservative, so beneath the disguise of a simple spheroid or filament may have lurked a wide range of metabolic styles and phylogenetic diversity. Amongst prokaryotes today, only certain rather recently evolved groups of cyanobacteria show much distinctive morphological elaboration. It is possible, then, that the Archean microbiota was almost as diverse as its modern counterpart but did not reveal this in its appearance.

Community organization was also apparently simple. Assemblages of microfossils are locally homogeneous, as are stromatolite microstructures (Fig. 12.6). However, the gregarious habits of the organisms provided certain ecological advantages. Benthic filaments formed interwoven mats, and spheroids grew in dense clusters that provided protection for individuals against radiation, erosion, and desiccation. Some microbial aggregates clearly modulated sedimentation, as indicated by the complex stromatolites that were widespread by the late Archean. Some microbial communities were clearly very robust, e.g., many stromatolites in the ~2.7 Ga Fortescue Group of northwestern Australia were able to accommodate frequent deluges of coarse volcanic ash into their structures. Others could withstand complete desiccation, as shown by mud-cracks in some Fortescue stromatolites that cut across microbial laminations.

Regardless of their morphological complexity or community structure, primordial organisms were apparently small. The largest late Archean filamentous microfossils are no more than a few hundred microns in length. Individual cells must have been much smaller, with the greatest possible diameter allowed by the filaments being ~20 µm. Preserved unicells are mostly less than 10 µm across. The layering of stromatolites also provides a proxy gauge for the size of the constructing organisms. The finest lamination in cherty stromatolites in which lithification occurred before compaction

is 20 µm, which constrains the maximum cell diameter. Evidently, then, the Archean biota was entirely microbial.

12.3.2 Environmental preferences

Though it used to be dogma in geology textbooks that the Archean environment was exclusively hot, volcanic, and deep-water, this is not so. A wide range of environmental settings are represented in the Archean geological record and, by the end of the eon, life was present in many of them. From this, some conclusions about the ecological tolerances of primordial organisms can be drawn.

Benthic organisms (living on the sea floor) clearly flourished in the photic zone (where sunlight is present), as shown by the abundance of stromatolites and other microbially laminated sediments (Figs. 12.6 and 12.7). Such structures have been widely reported from intertidal, shallow subtidal, and shelf settings, and in the Transvaal Supergroup of South Africa, filamentous microbial mats formed at depths of up to 50 m (Klein et al., 1987). This suggests that sediment-accreting microbial mats formed wherever light was available and other conditions were suitable. Periodic exposure was not a problem for intertidal mats, as early Archean stromatolites overgrow desiccation fragments sloughed from their crests (Buick et al., 1981). Neither were strong currents much of a restriction, as Archean stromatolites are frequently elongated, constructed of coarse sand, and have asymmetric current ripples and cross-beds in the interspaces between columns (Sumner and Grotzinger, 2000). But current activity was not a necessity, because some stromatolites in the Transvaal Supergroup evidently formed at depths where water movement was negligible (Sumner, 1997). Mat-forming benthic microbes could also tolerate a wide range of salinity conditions, from normal marine through carbonate precipitation (Sumner and Grotzinger, 2000) to sulfate saturation (Buick et al., 1981), indicated by the evaporitic minerals deposited during stromatolite growth.

Marine plankton (organisms that float) must also have been abundant, because kerogenous shales deposited well below the photic zone are widespread in Archean deep-water environments. These have such high organic contents (up to ~10% organic carbon), even after diagenetic and metamorphic loss of hydrocarbons, that chemoautotrophic productivity alone is very unlikely to have been responsible for so much carbon fixation at such depths; the organic material must have been largely produced by organisms living at shallow depths. Moreover, some late Archean black shales deposited at considerable depths contain hydrocarbon biomarkers for photosynthetic cyanobacteria (Brocks *et al.*, 1999), which can only have lived in the photic zone of the overlying water column as plankton. But since no certain microfossils representing the Archean plankton have ever been found, it is unknown if they were, as today, mostly minute spheroidal microbes.

Both benthic and planktonic microbes also inhabited non-marine aqueous settings such as lakes. A wide range of stromatolite morphologies occur in lacustrine (deposited in lakes) rocks of the ~2.75 Fortescue Group in northwestern Australia, occupying similar depth and salinity ranges as comparable marine stromatolites. However, some are interbedded with sediments containing molds of halite crystals, indicating that the constructing microbes could withstand hypersaline evaporative conditions (Buick, 1992). Others contain thick laminae and lenses of coarse volcanic ash, indicating that they could overcome extremely high sedimentation rates and that the constructing microbes could move very fast, grow very rapidly, or reproduce very quickly. Planktonic organisms apparently also inhabited these lakes, because associated black shales have post-metamorphic organic carbon contents of up to 2.5% and contain cyanobacterial biomarker molecules (Brocks *et al.*, 2003).

Despite much debate, it is still unclear whether land surfaces were inhabited during the Archean. There has been a long-held belief that before the advent of an oxygenated atmosphere with its concomitant ozone shield, life on land would have been impossible because the high solar ultraviolet flux would have destroyed the DNA of any organism unfortunate enough to be exposed. However, endolithic and soil microbes could have thrived where ultraviolet light could not penetrate. The oldest uncontroversial record of life on land dates from only 1.2 Ga (Horodyski and Knauth, 1994), but colonization of the land was undoubtedly earlier because land settings are prone to erosion and thus unlikely to preserve relics of life. Indeed, several recent studies of fossil soils have suggested that organisms may have inhabited dry land long before this time. Rye (1998) described organic matter in a ~2.77 Ga soil from the Fortescue Group, northwestern Australia, with very light $\delta^{13}C_{org}$ values, which was initially interpreted as representing the remains of soil-dwelling methanotrophic microbes. However, re-examination of the material (Rye and Holland, 2000) showed that the organic matter may have been derived from

ephemeral ponds. Watanabe *et al.* (2000), studying a 2.6–2.7 Ga soil from South Africa, also reported organic matter, this time isotopically heavy (−14‰ to −17‰), which they interpreted as representing microbial mats that developed on the soil surface. However, the immediately overlying rocks of the Black Reef Formation contain abundant hydrocarbon residues (Dutkiewicz *et al.*, 1998) and so it is not inconceivable that the organic matter in the paleosol was derived from downwards percolating oil.

Another widespread belief about early organisms is that they were thermophilic, based on the heat-loving tendencies of many modern microbes that appear to be ancient based on their basal branching on ribosomal RNA phylogenetic trees (see Chapter 14). It is difficult, however, to determine from the geological and paleontological records whether the Archean biota in fact reflected this tendency. The oxygen isotope ($\delta^{18}O$) geothermometer, which is the most reliable measure of aquatic paleotemperatures over the last 30 Myr, is unreliable if seawater $\delta^{18}O$ has evolved through time, which it clearly has. Furthermore, most Archean cherts and carbonates have been hydrothermally altered to such an extent that the $\delta^{18}O$ of the minerals now present almost certainly does not record ambient seawater temperatures. A better constraint on temperature is provided by the sequence of sulfate minerals precipitated from evaporating seawater. At temperatures above ~57 °C, calcium sulfate crystallizes as the blocky orthorhombic mineral anhydrite, but in cooler water it precipitates as bladed to prismatic crystals of monoclinic gypsum. This transition temperature is lowered to near 20 °C if the evaporating brine is saturated with NaCl. Interbedded with the ~3.47 Ga stromatolites in the Warrawoona Group at North Pole, northwestern Australia, are voluminous sulfate evaporites, now barite ($BaSO_4$) but once gypsum ($CaSO_4 \cdot 2H_2O$) as shown by the interfacial angles of the crystals (Lambert *et al.*, 1978; Buick and Dunlop, 1990). As relics of halite (NaCl) are preserved in similarly old marine sediments elsewhere (Boulter and Glover, 1986; Westall *et al.*, 2001), the evaporite brine was probably rich in chloride. This implies that the water in which the stromatolites formed was cool and thus that the earliest microbial community for which we have tangible evidence was not dominated by thermophiles.

12.3.3 Metabolic activity

Only a few of the many microbial metabolic strategies leave detectable signatures in the geological record.

However, several of those metabolic pathways that control the workings of biogeochemical cycles had already evolved by the end of the Archean. Moreover, some organisms preferentially utilize particular metabolisms in particular circumstances, so records of those organisms in those conditions can be used to infer the existence of that metabolism.

Autotrophy is the process whereby organisms transform inorganic carbon in the form of CO_2 into cellular organic carbon (carbon fixation). As this requires input of energy, either photons from light or electrons from inorganic chemical compounds, to form the newly synthesized organic molecules, autotrophy thus modulates the flow of energy through the biosphere. This metabolism imposes a negative isotopic fractionation between the source CO_2 and the organic product, varying somewhat according to the particular metabolic pathway employed, the supply of source carbon, the availability of nutrients, and the ambient temperature. In semi-closed systems like the ocean where CO_2 supply is somewhat restricted by diffusion, residual carbonate becomes heavier as a result of autotrophic depletion of the light isotope, usually with a $\Delta^{13}C$ of 20–30‰. A $\Delta^{13}C$ fractionation of this magnitude is apparent in the 3.52 Ga rocks of the Coonterunah Group, Australia. When considered along with the evidence for phototropic (responding to light) microbes in the form of stromatolites in the ~3.47 Ga Warrawoona Group, then it strongly implies that photoautotrophy (autotrophy using light) had indeed arisen by then. Though there are isotopic data compatible with autotrophy having appeared even earlier, in the >3.7 Ga Isua supracrustals, the degree of metamorphism and absence of sedimentary carbonate precludes an unequivocal interpretation.

By far the most important process of carbon fixation on the modern Earth is oxygenic photosynthesis:

$$CO_2 + H_2O \rightarrow CH_2O + O_2.$$

Though there is debate about the evolutionary history of the five photosynthetic bacterial groups, each with a distinct type of chlorophyll and a differing style of photosynthesis, oxygenic photosynthesis is thought to be the most recently evolved, based on molecular phylogenies derived from multiple photosynthetic genes (Xiong *et al.*, 2000). The only microbes to use this are the cyanobacteria, some of which have distinctive structures. Though some microfossil-like filaments from the ~3.46 Ga Apex Basalt have been interpreted as "probable cyanobacteria ... essentially indistinguishable from specific oscillatoriaceans" (Schopf, 1993), it

is now likely that these are not biological and not old (Section 12.2.1). None of the well-preserved late Archean microfossils show distinctively cyanobacterial morphologies; indeed all could be the remains of non-photosynthetic bacteria because none show behavior patterns suggestive of photoautotrophy.

However, biological activity in late Archean environments lacking external sources of reducing power provides indirect evidence for the early advent of oxygenic photosynthesis (Buick, 1992). In the 2.72 Ga Tumbiana Formation in northwest Australia, abundant stromatolites with palimpsests of phototropic microbial filaments (Fig. 12.6) occur in sulfur-deficient lakes that formed on flood basalts. In such a setting, the lakes and their biota would have been deprived of inflowing organic matter or upwelling hydrocarbons, precluding an ecosystem based exclusively on heterotrophy. As the carbonate mineral precipitating in the lakes was calcite rather than siderite, conditions were not so highly reducing that hydrogen gas would have been available for chemoautotrophy. As sulfides and sulfates are scarce, H_2S was evidently not present in sufficient abundance to power an ecosystem either by chemoautotrophy or anaerobic photosynthesis. Hence, we are left with oxygenic photosynthesis, which must have been the primary process of biological production.

Corroborative evidence in the form of molecular biomarkers has been recently obtained from this and slightly younger formations in the same area. Brocks *et al.* (1999) discovered high relative concentrations of syngenetic 2α-methylhopanes with chain lengths greater than C_{31}, indicative of cyanobacteria (Summons *et al.*, 1999). Though some cyanobacteria can temporarily metabolize by anoxygenic photosynthesis under reducing sulfidic conditions, these particular 2α-methylhopane biomarkers show evidence of oxidative degradation (Brocks *et al.*, 2003), implying that the source organisms were indeed metabolizing by oxygenic photosynthesis.

These same sediments contain organic carbon with extremely light $\delta^{13}C$ values (-40‰ to -60‰), implying that microbial methanogenesis and methanotrophy were important metabolisms during the recycling of organic matter produced by cyanobacteria (Hayes, 1994). The former process produces methane even more depleted in ^{13}C than the source organic matter, whereas the latter consumes it, incorporating the extremely light isotopic signature of the methane into cellular material. Though methanogenesis can also be an autotrophic metabolism,

$$CO_2 + 4H_2 \rightarrow CH_4 + 2H_2O,$$

in the case of the Tumbiana sediments it was almost certainly the heterotrophic form

$$2CH_2O \rightarrow CH_4 + CO_2$$

that was dominant. As mentioned above, calcite and not siderite was the dominant carbonate precipitated, indicating limited hydrogen availability and therefore little autotrophic methanogenesis. Because methanotrophs usually consume oxygen:

$$CH_4 + O_2 \rightarrow CH_2O + H_2O$$

(some anaerobic methanotrophs form symbioses with sulfate reducers, but there is little evidence for sulfate in these particular sediments), this very light isotopic signature provides further support for the existence of oxygenic photosynthesis. These particular recycling pathways probably prevailed in these settings because the scarcity of sulfate and nitrate inhibited other forms of respiratory recycling. Hence, even though oxygenic photosynthesis was locally active, oxygen levels had not built up to levels where large amounts of oxygen or oxidized substrates were available.

In modern marine sediments, organic recycling is usually dominated by dissimilatory sulfate reduction:

$$2CH_2O + SO_4^{2-} + 2H^+ \rightarrow H_2S + CO_2 + 2H_2O.$$

Large $\Delta^{34}S$ fractionations ($\sim$20‰) between coexisting sulfates and sulfides in the $\sim$3.47 Ga barite beds in the Warrawoona Group at North Pole indicate that this metabolism arose very early in Earth's history (Section 12.2.4). However, it seems that it was restricted to localized sulfate oases until after $\sim$2.7 Ga, when large isotopic fractionations became widespread in marine sediments as dissolved sulfate concentrations rose above 1 mmol (Canfield and Raiswell, 1999; Canfield et al., 2000). Despite this late rise to ecological prominence, its early appearance marks the first evidence for a specific metabolic pathway in the geological record (Fig. 12.2). As dissimilatory sulfate reduction is a complex metabolism, requiring enzyme catalysts, energy regulation and genetic control, it seems that by $\sim$3.47 Ga microbes had all the biochemical and cellular sophistication of their modern descendants.

12.3.4 Phylogenetic relationships

To understand how Archean organisms were related to their more modern descendants, we have to use the "Tree of Life," the family tree of all species living today, derived from the comparative sequences of their ribosomal RNA molecules. This phylogeny works on the assumption that slowly evolving and universally shared molecules will reveal evolutionary relationships throughout the history of life (Chapter 13). Though it is a powerful device for exploring early evolution, it has significant drawbacks. First, it is necessarily constructed from data derived from modern organisms; nothing extinct appears anywhere on the tree. In reality, the tree should be a lot bushier than it is. Second, it assumes linear descent of characters, but it is now known that significant lateral gene transfer has occurred, particularly in prokaryotic organisms. This means that the simple branching tree structure is somewhat unrealistic, as it should show lower branches with many cross-connections. Third, it has no temporal scale, only an ordering; the length of branches is related only to evolutionary distance, not geological time. But despite these shortcomings, the RNA Tree is still the best guide we have to the pattern of early evolution. In the following section, the Archean evidence will be integrated with the Tree to enrich it by adding time markers based on Archean evidence, with the key assumption that distinctive metabolisms and biomarkers have arisen only once and were not also present in unknown extinct groups.

The most obvious feature of the RNA Tree of Life is that it has three main divisions, termed *Domains* (Fig. 12.12). The *Bacteria* represent most known *prokaryotes* (the simplest one-celled organisms, without a nucleus), the *Eukarya* (*eukaryotes*) are organisms with complex cells containing organelles, and the *Archaea* (not to be confused with the *Archean* Eon!) are newly recognized prokaryotes with very different cell wall composition, membrane stiffening molecules, gene structure, DNA binding, and RNA transcription. The only evidence for the existence of the Archaea before 0.5 Ga comes from extreme depletions in $\delta^{13}C_{org}$ values imparted by methanogenesis, a metabolism restricted to members of the Archaea. $\delta^{13}C_{org}$ values lighter than -40‰ are common in sediments aged between 2.8–2.6 Ga, implying that Archaeal methanogens were indeed extant by the late Archean (Section 12.2.3). In Fig. 12.12 this point is labeled "2.8 Ga." But as Archaea (including methanogens) occupy some of the lowest branches on the RNA phylogenetic tree, it is somewhat paradoxical that the first geological signal for methanogenesis appears so late in the record.

Clear evidence for Bacteria in the Archean is provided by the presence in sediments 2.72 Ga and younger of hopanes (Section 12.2.6), a diagnostic

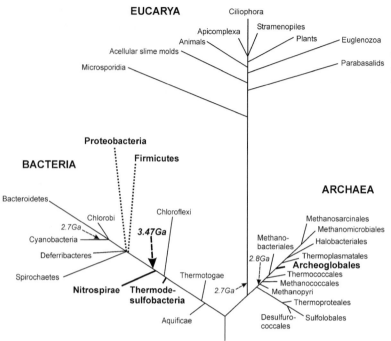

FIGURE 12.12 The "Tree of Life" (short-subunit ribosomal RNA phylogenetic tree) with ages shown for four key points, as estimated from the fossil record. See discussion in the text.

biomarker for the domain (Brocks *et al.*, 2003). A biomarker for cyanobacteria, $C_{>31}$ 2α-methylhopanes (Summons *et al.*, 1999), is also present in these sediments (Brocks *et al.*, 1999). In Fig. 12.12 this point is labeled "2.7 Ga." But an earlier date for the appearance of Bacteria can be obtained from the sulfur isotope systematics of the ~3.47 Ga baritic sediments in the Warrawoona Group (discussed in Section 12.2.4). From these, it is apparent that dissimilatory sulfate reduction had already evolved. This metabolism is fairly widely distributed through both the Archaea and Bacteria, but all known Archaeal sulfate reducers are hyperthermophiles (with optimal growth temperatures of $>80\,^\circ$C) of the genus *Archaeoglobus* (Stetter, 1996). Moreover, the most deeply branching Bacterial sulfate reducers, belonging to the *Thermodesulfobacterium* group, are also hyperthermophiles. As the former presence of gypsum indicates that temperatures were below 60 °C, the Warrawoona sulfate reducing microbes were not hyperthermophiles, but at best moderate thermophiles or mesophiles. Those fitting these criteria are all Bacteria: members of the δ-subdivision of the proteobacteria (purple bacteria), the low $G+C$, Gram-positive Firmicute bacteria, and the *Nitrospira* group. So, given current knowledge of the phylogenetic distribution of thermal adaptations among sulfate reducers, the Warrawoona δ^{34}S data suggest a minimum age of 3.47 Ga for the appearance of Bacteria.

Recent genetic data (Klein *et al.*, 2001) indicate that the gene for dissimilatory sulfite reductase (*dsr*), the enzyme catalyzing the key step in the microbial sulfate reduction metabolic pathway, has undergone significant lateral transfer between lineages. As the *dsr* phylogenetic tree is evidently rooted in the thermophilic *Thermodesulfovibrio* lineage of the *Nitrospira* group, it would appear that the origin of sulfate reduction thus resides within this peripherally branching division. If so, then the dated node provided by the North Pole data can be moved up the Tree of Life to a position (as shown in Fig. 12.12) at the divergence point between the *Nitrospira* division and the mesophilic sulfate-reducers (δ-Proteobacteria and low $G+C$, Gram-positive bacteria). This suggests that a great deal of microbial evolution must have occurred during the first billion years of Earth history.

Surprisingly, it now appears that the origin of the domain Eucarya can also be traced back to the Archean. The diversity and abundance of indigenous sterane biomarkers in sediments of the 2.8–2.5 Ga Fortescue and Hamersley Groups of northwest Australia indicates the presence of eukaryotes in the Archean ecosystem (in Fig. 12.12 this point at the base of the Eucarya Domain is labeled "2.7 Ga"), as the few prokaryotes that produce steroids (sterane precursors) synthesize only a few specific varieties (Brocks *et al.*, 1999). Steroids, which act as rigidifying molecules within the lipid bilayers that make up cell membranes, provide eukaryotes with the ability to engulf large particles (phagocytosis) and thus allow for the possibility of endosymbiosis of organelles. They are therefore intimately involved with the evolution of microstructural complexity within cells, and are necessary precursors to the development of multicellularity, tissue differentiation, and sex.

Though some steranes are characteristic of particular subdivisions of eukaryotes, none of those are abundant in the Archean biomarker suites. Hence, it is impossible to determine what types of eukaryotic organisms were present on the early Earth. They may well have been merely single-celled algae, but there is nothing in the biomarker data that precludes large multicellular organisms as complex as rhodophytes (red algae) or even animals. In the light of recent evidence suggesting that large animal-like organisms were present on the Earth as early as 1.65 Ga (Rasmussen *et al.*, 2002), perhaps it would be wise to suspend judgment on just how complex Archean eukaryotes were.

12.4 Astrobiological implications

It is clear that this planet was inhabited within a billion years of its formation. Not only that, but the organisms present 3.5 Ga were as complex and competent in their cellular and biochemical functioning as the bulk of the modern biota. If it is indeed true that cataclysmic meteoritic bombardment ended as recently as ~3.9 Ga, then life obviously radiated rapidly thereafter and quickly attained considerable sophistication. This suggests that other planets such as Mars that had short habitable windows may also have potentially evolved biotas of some diversity and complexity.

It is also evident that all three main lineages of life, including our own, were present by the end of the Archean at 2.5 Ga. Just what controls how fast intelligent life evolves on a planet is unknown, but it now seems unlikely that it is acquisition of a complex cell structure. Eukaryotes have been around for at least 60% of Earth's history, but intelligent technological eukaryotes evolved only in the latest 0.1% of that time. Accordingly, it seems that attempts to detect extraterrestrial life directed towards dumb germs and worms are more likely to be successful in the near-Earth region of space.

Any such investigations of planetary bodies in the Solar System will require many of the techniques and strategies used in the search for Archean life here on Earth. Exploring for signs of extinct life on Mars should be quite analogous to searching for biosignatures in old Earth rocks. Perhaps it would make sense to include an expert in Archean paleobiology amongst the first Martian exploration party. Indeed, as suitable Archean rocks for paleobiology tend to occur in remote places and in inhospitable environments, such a scientist should be particularly well trained for the rigors of the mission. This should be borne in mind by all budding astrobiological astronauts when making future career choices.

REFERENCES

Altermann, W. and Schopf, J. W. (1995). Microfossils from the Neoarchean Campbell Group, Griqualand West Sequence of the Transvaal Supergroup, and their paleoenvironmental and evolutionary significance. *Precambrian Res.*, **75**, 65–90.

Appel, P. W. U., Fedo, C. M., Moorbath, S., and Myers, J. S. (1998). Recognizable primary volcanic and sedimentary features in a low strain domain of the highly deformed, oldest known (3.7–3.8 Gyr) greenstone belt, Isua, west Greenland. *Terra Nova*, **10**, 57–62.

Awramik, S. M., Schopf, J. W., and Walter, M. R. (1983). Filamentous fossil bacteria from the Archean of Western Australia. *Precambrian Res.*, **20**, 357–374.

Beaumont, V. and Robert, F. (1999). Nitrogen isotope ratios of kerogens in Precambrian cherts: a record of the evolution of atmosphere chemistry? *Precambrian Res.*, **96**, 63–82.

Boulter, C. A. and Glover, J. E. (1986). Chert with relict hopper moulds from Rocklea Dome, Pilbara craton, Western Australia: an Archaean halite-bearing evaporite. *Geology*, **14**, 128–131.

Brasier, M. D., Green, O. R., Jephcoat, A. P., *et al.* (2002). Questioning the evidence for Earth's oldest fossils. *Nature*, **416**, 76–81.

Bridgwater, D., Allaart, J. H., Schopf, J. W., *et al.* (1981). Microfossil-like objects from the Archaean of Greenland: a cautionary note. *Nature*, **289**, 51–53.

Brocks, J. J., Logan, G. A., Buick, R., and Summons, R. E. (1999). Archean molecular fossils and the early rise of eukaryotes. *Science*, **285**, 74–77.

Brocks, J. J., Buick, R., Summons, R. E., and Logan, G. A. (2003). A reconstruction of Archean biological diversity based on molecular fossils from the 2.78 to 2.45 billion year old Mount Bruce Supergroup, Hamersley Basin, Western Australia. *Geochim. Cosmochim. Acta*, **67**, 4321–4335.

Buick, R. (1990). Microfossil recognition in Archean rocks; an appraisal of spheroids and filaments from a 3500 m.y. old chert-barite unit at North Pole, Western Australia. *Palaios*, **5**, 441–459.

Buick, R. (1992). The antiquity of oxygenic photosynthesis: evidence from stromatolites in sulphate-deficient Archaean lakes. *Science*, **255**, 74–77.

Buick, R. and Dunlop, J. S. R. (1990). Evaporitic sediments of early Archaean age from the Warrawoona Group, North Pole, Western Australia. *Sedimentology*, **37**, 247–277.

Buick, R. Dunlop, J. S. R., and Groves, D. I. (1981). Stromatolite recognition in ancient rocks: an appraisal of irregularly laminated structures in an early Archaean chert-barite unit from North Pole, Western Australia. *Alcheringa*, **5**, 161–181.

Buick, R., Dunlop, J. S. R., and Groves, D. I. (1995). Abiological origin of described stromatolites older than 3.2 Ga: discussion. *Geology*, **23**, 191.

Canfield, D. E. (1998). A new model for Proterozoic ocean chemistry. *Nature*, **396**, 450–453.

Canfield, D. E. and Raiswell, R. (1999). The evolution of the sulfur cycle. *Am. J. Sci.*, **299**, 697–723.

Canfield, D. E., Habicht, K. S., and Thamdrup, B. (2000). The Archean sulfur cycle and the early history of atmospheric oxygen. *Science*, **288**, 658–661.

Des Marais, D. J., Strauss, H., Summons, R. E., and Hayes, J. M. (1992). Carbon isotope evidence for the stepwise oxidation of the Proterozoic environment. *Nature*, **359**, 605–609.

Dutkiewicz, A., Rasmussen, B., and Buick, R. (1998). Oil preserved in fluid inclusions in Archaean sandstones. *Nature*, **395**, 885–888.

Eiler, J. M., Mojzsis, S. J., and Arrhenius, G. (1997). Carbon isotope evidence for early life. *Nature*, **386**, 665.

Fedo, C. M. and Whitehouse, M. J. (2002). Metasomatic origin of quartz-pyroxene rock, Akilia, Greenland, and implications for Earth's earliest life. *Science*, **296**, 1448–1452.

Green, M. G., Sylvester, P. J., and Buick, R. (2000). Growth and recycling of early Archaean continental crust: geochemical evidence from the Coonterunah and Warrawoona Groups, Pilbara Craton, Australia. *Tectonophysics*, **322**, 69–88.

Grotzinger, J. P. and Rothman, D. H. (1996). An abiotic model for stromatolite morphogenesis. *Nature*, **383**, 423–425.

Han, T.-M. and Runnegar, B. (1992). Megascopic eukaryotic algae from the 2.1-billion-year-old Negaunee Iron-Formation, Michigan. *Science*, **257**, 232–235.

Hayes, J. M. (1994). Global methanotrophy at the Archean–Proterozoic transition. In *Early Life on Earth*, ed. S. Bengtson, pp. 220–236. New York: Columbia University Press.

Hayes, J. M., Strauss, H., and Kaufman, A. J. (1999). The abundance of ^{13}C in marine organic matter and isotopic fractionation in the global biogeochemical cycle of carbon during the past 800 Ma. *Chem. Geol.*, **161**, 103–125.

Hofmann, H. J. (1976). Precambrian microflora, Belcher Islands, Canada: significance and systematics. *J. Paleontol.*, **50**, 1040–1073.

Hofmann, H. J., Grey, K., Hickman, A. H., and Thorpe, R. I. (1999). Origin of 3.45 Ga coniform stromatolites in Warrawoona Group, Western Australia. *Geol. Soc. Am. Bull.*, **111**, 1256–1262.

Horodyski, R. J. and Knauth, L. P. (1994). Life on land in the Precambrian. *Science*, **263**, 494–498.

Jia, Y. and Kerrich, R. (2001). Nitrogen recycling in the atmosphere-crust-mantle systems: evidence from secular variation of crustal N abundances and $\delta^{15}N$ values, Archean to present. *Eos*, **82**(47), F695.

Kamber, B. S. and Moorbath, S. (1998). Initial Pb of the Amîtsoq gneiss revisited: implication for the timing of early Archaean crustal evolution in West Greenland. *Chem. Geol.*, **150**, 19–41.

Kaufman, A. J., Hayes, J. M., and Klein, C. (1990). Primary and diagenetic controls of isotopic compositions of iron-formation carbonates. *Geochim. Cosmochim. Acta*, **54**, 3461–3473.

Klein, C., Beukes, N. J., and Schopf, J. W. (1987). Filamentous microfossils in the early Proterozoic Transvaal Supergroup: their morphology, significance, and paleoenvironmental setting. *Precambrian Res.*, **36**, 81–94.

Klein, M., Friedrich, M., Roger, A. J., et al. (2001). Multiple lateral transfers of dissimilatory sulfite reductase genes between major lineages of sulfate-reducing prokaryotes. *J. Bacteriol.*, **183**, 6028–6035.

Knoll, A. H. and Barghoorn, E. S. (1977). Archean microfossils showing cell division from the Swaziland System of South Africa. *Science*, **198**, 396–398.

Lambert, I. B., Donnelly, T. H., Dunlop, J. S. R., and Groves, D. I. (1978). Stable isotope studies of early Archaean sulphate deposits of probable evaporitic and volcanogenic origins. *Nature*, **276**, 808–811.

Lanier, W. P. (1986). Approximate growth rates of Early Proterozoic microstromatolites as deduced by biomass productivity. *Palaios*, **1**, 525–542.

Lowe, D. R. (1994). Abiological origin of described stromatolites older than 3.2 Ga. *Geology*, **22**, 387–390.

Mojzsis, S. J., Arrhenius, G., McKeegan, K. D., Harrison, T. M., Nutman, A. P., and Friend, C. R. L. (1996). Evidence for life on Earth before 3800 million years ago. *Nature*, **384**, 55–59.

Muir, M. D. and Grant, P. R. (1976). Micropalaeontological evidence from the Onverwacht Group, South Africa. In *The Early History of the Earth*, ed. B. F. Windley, pp. 595–604. London: Wiley.

Myers, J. S. and Crowley, J. L. (2000). Vestiges of life in the oldest Greenland rocks? A review of early Archean geology in the Godthåbsfjord region, and reappraisal of field evidence for >3850 Ma life on Akilia. *Precambrian Res.*, **103**, 101–124.

Nagy, B., Engel, M. H., Zumberge, J. E., Ogino, H., and Chang, S. Y. (1981). Amino acids and hydrocarbons 3,800-Myr-old in the Isua rocks, southwestern Greenland. *Nature*, **289**, 53–56.

Oehler, J. H. (1977). Irreversible contamination of Precambrian kerogen by 14C-labeled organic compounds. *Precambrian Res.*, **4**, 221–227.

Ohmoto, H., Kakegawa, T., and Lowe, D. R. (1993). 3.4-billion-year-old biogenic pyrites from Barberton, South Africa: sulfur isotope evidence. *Science*, **262**, 555–557.

Pflug, H. D. and Jaeschke-Boyer, H. (1979). Combined structural and chemical analysis of 3,800-Myr-old microfossils. *Nature*, **280**, 483–486.

Pinti, D. L. and Hashizume, K. (2001). ^{15}N-depleted nitrogen in early Archean kerogens: clues on ancient marine chemosynthetic-based ecosystems? A comment to Beaumont, V., Robert, F., 1999 in *Precambrian Res.*, **96**, 62–82. *Precambrian Res.*, **105**, 85–88.

Rasmussen, B. (2000). Filamentous microfossils in a 3,235-million-year-old volcanogenic massive sulphide deposit. *Nature*, **405**, 676–679.

Rasmussen, B. and Buick, R. (2000). Oily old ores: evidence for hydrothermal petroleum generation in an Archean volcanogenic massive sulfide deposit. *Geology*, **28**, 731–734.

Rasmussen, B., Bengtson, S., Fletcher, I. R., and McNaughton, N. J. (2002). Discoidal impressions and trace-like fossils more than 1200 million years old. *Science*, **296**, 1112–1115.

Roedder, E. (1981). Are the 3,800-Myr-old Isua objects microfossils, limonite-stained fluid inclusions, or neither? *Nature*, **293**, 459–462.

Rose, N. M., Rosing, M. T., and Bridgwater, D. (1996). The origin of metacarbonate in the Isua Archean Supracrustal Belt, west Greenland. *Am. J. Sci.*, **296**, 1004–1044.

Rosing, M. T. (1999). ^{13}C-depleted carbon microparticles in >3700-Ma sea-floor sedimentary from west Greenland. *Science*, **283**, 674–676.

Rye, R. (1998). Highly negative δ^{13}C values in organic carbon in the Mt. Roe #2 paleosol: terrestrial life at 2.765 Ga? *Mineralogical Mag.*, **62A**, 1308–1309.

Rye, R. and Holland, H. D. (2000). Life associated with a 2.76 Ga ephemeral pond? Evidence from Mount Roe paleosol. *Geology*, **28**, 483–486.

Sano, Y., Terada, K., Takahashi, Y., and Nutman, A. P. (1999). Origin of life from apatite dating? *Nature*, **400**, 127.

Schidlowski, M. (1993). The initiation of biological processes on Earth: summary of empirical evidence. In *Organic Geochemistry*, eds. M. H. Engel and S. A. Macko, pp. 639–655. New York: Plenum.

Schidlowski, M., Hayes, J. M., and Kaplan, I. R. (1983). Isotopic inferences of ancient biochemistries: carbon, sulfur, hydrogen, and nitrogen. In *Earth's Earliest Biosphere: its Origin and Evolution*, ed. J. W. Schopf, pp. 149–186. Princeton: Princeton University Press.

Schiøtte, L. and Compston, W. (1990). U-Pb age pattern for single zircon from the early Archaean Akilia association south of Ameralik fjord, southern West Greenland. *Chem. Geol.*, **80**, 147–157.

Schopf, J. W. (1993). Microfossils of the Early Archean Apex chert: new evidence of the antiquity of life. *Science*, **260**, 640–646.

Schopf, J. W. and Packer, B. M. (1987). Early Archean (3.3-billion to 3.5-billion-year-old) microfossils from Warrawoona Group, Australia. *Science*, **237**, 70–73.

Schopf, J. W. and Walter, M. R. (1983). Archean microfossils: new evidence of ancient microbes. In *Earth's Earliest Biosphere: its Origin and Evolution*, ed. J. W. Schopf, pp. 214–239. Princeton: Princeton University Press.

Schopf, J. W., Kudryavtsev, A. B., Agresti, D. G., Wdowiak, T. J., and Czaja, A. D. (2002). Laser-Raman imagery of Earth's earliest fossils. *Nature*, **416**, 73–76.

Shen, Y., Buick, R., and Canfield, D. E. (2001). Isotopic evidence for microbial sulphate reduction in the early Archaean era. *Nature*, **410**, 77–81.

Stetter, K. O. (1996). Hyperthermophiles in the history of life. In *Evolution of Hydrothermal Systems on Earth (and Mars?)*, eds. G. R. Bock and J. A. Goode, pp. 1–10. New York: Wiley.

Strauss, H. and Moore, T. B. (1992). Abundances and isotopic compositions of carbon and sulfur species in whole rock and kerogen samples. In *The Proterozoic Biosphere: a Multidisciplinary Study*, eds. Schopf, J. W. and Klein, C. Cambridge: Cambridge University Press, 709–798.

Summons, R. E. and Walter, M. R. (1990). Molecular fossils and microfossils of prokaryotes and protists from Proterozoic sediments. *Am. J. Sci.*, **290A**, 212–244.

Summons, R. E., Jahnke, L. L., Hope, J. M., and Logan, G. A. (1999). 2-Methylhopanoids as biomarkers for cyanobacterial oxygenic photosynthesis. *Nature*, **400**, 554–557.

Summons, R. E., Powell, T. G., and Boreham, C. J. (1988). Petroleum geology and geochemistry of the Middle Proterozoic McArthur Basin, northern Australia. III Composition of extractable hydrocarbons. *Geochim. Cosmochim. Acta*, **52**, 1747–1763.

Sumner, D. Y. (1997). Late Archean calcite-microbe interactions: two morphologically distinct microbial communities that affected calcite nucleation differently. *Palaios*, **12**, 302–318.

Sumner, D. Y. and Grotzinger, J. P. (2000). Late Archean aragonite precipitation; petrography, facies associations, and environmental significance. *Soc. Sed. Geol. Spec. Publ.*, **67**, 123–144.

Walsh, M. M. (1992). Microfossils and possible microfossils from the Early Archean Onverwacht Group, Barberton Mountain Land, South Africa. *Precambrian Res.*, **54**, 271–293.

Walsh, M. M. and Lowe, D. R. (1985). Filamentous microfossils from the 3,500-Myr-old Onverwacht Group, Barberton Mountain Land, South Africa. *Nature*, **314**, 530–532.

Walters, C., Shimoyama, A., and Ponnamperuma, C. (1981). Organic geochemistry of the Isua Supracrustals. In *Origin of Life*, ed. Y. Wolman, pp. 473–479. Amsterdam: Reidel.

Watanabe, Y., Martini, J. E., and Ohmoto, H. (2000). Geochemical evidence for terrestrial ecosystems 2.6 billion years ago. *Nature*, **408**, 574–578.

Westall, F., de Wit, M. J., Dann, J., van der Gaast, S., de Ronde, C. E. J., and Gernekee, D. (2001). Early Archean fossil bacteria and biofilms in hydrothermally-influenced sediments from the Barberton greenstone belt, South Africa. *Precambrian Res.*, **106**, 93–116.

Xiong, J., Fischer, W. M., Inoue, K., Nakahara, M., and Bauer, C. E. (2000). Molecular evidence for the early evolution of photosynthesis. *Science*, **289**, 1724–1730.

Yamaguchi, K. (2002). *Geochemistry of Archean-Paleoproterozoic black shales: the early evolution of the atmosphere, oceans, and biosphere*. Ph. D. thesis, Pennsylvania State University.

Zundel, M. and Rohmer, M. (1985). Prokaryotic triterpenoids 3. The biosynthesis of 2β-methylhopanoids and 3β-methylhopanoids of *Methylbacterium organophilum* and *Acetobacter pasteurianus* spp. *Pasteurianus. Eur. J. Biochem.*, **150**, 35–39.

FURTHER READING

Hoefs, J. (2004). *Stable Isotope Geochemistry*. Berlin: Springer.

Knoll, A. H. (2003). *Life on a Young Planet: the First Three Billion Years of Evolution on Earth*. Princeton, NJ: Princeton University Press. An authoritative and sweeping popular science treatment.

Nisbet, E. G. (1987). *The Young Earth: an Introduction to Archaean Geology*. Boston: Allen & Unwin.

Peters, K. E. (2005). *The Biomarker Guide*. Cambridge: Cambridge University Press.

Schopf, J. W. (ed.) (1983). *Earth's Earliest Biosphere: Its Origin and Evolution*. Princeton NJ: Princeton University Press. The "bible" of its time, and still very useful.

Schopf, J. W. (1999). *Cradle of Life: the Discovery of Earth's Earliest Fossils*. Princeton, NJ: Princeton University Press. A personal account of the author's career and what he's found.

Walter, M. R. (ed.) (1976). *Stromatolites*. Amsterdam: Elsevier Scientific.

13 The origin and diversification of eukaryotes

Mitchell L. Sogin, David J. Patterson, and Andrew McArthur
Marine Biological Laboratory at Woods Hole

13.1 Introduction

Prokaryotic microorganisms[1] were the only form of life for at least 80 percent of our evolutionary history (Schopf and Packer, 1987). Multicellular organisms including plants, animals and fungi evolved a mere 0.5–1.0 Ga from single-celled eukaryotic[2] ancestors. Geologists and paleontologists debate the age of life on this planet and when the major microbial lineages first diverged (see Chapter 12 for details). Cyanobacterium-like fossils suggest that life emerged at least 3.45 Ga (Schopf *et al.*, 2002; Schopf and Packer, 1987), but the biogenic origins of these structures are contested (Brasier *et al.*, 2002; Section 12.2.1). The chemical record documents prokaryotic metabolisms that may have existed 3.47–3.85 Ga (Mojzsis *et al.*, 1996) and eukaryotic biosignatures that may be as old as 2.7 Gyr (Brocks *et al.*, 1999). Yet, these are still imprecise interpretations (some might be more recent microbial contamination) and do not set absolute limits on the possible origins of life on Earth. Early periods of heavy bombardment between 4.1 and 3.8 Ga might constrain when life first appeared on Earth, although microorganisms living off chemical energy at kilometer depths could have survived even the largest impact events.

By the standards of multicellular plants and animals, single-cell organisms look relatively simple (Patterson and Sogin, 1993), yet they transformed the atmosphere, the waters, the surface, and the subsurface of the Earth. Even today, single-celled eukaryotes, most of which are members of the group *Protista* (which includes all eukaryotes excepting plants, animals, and fungi), and prokaryotes continue to dominate our biosphere in terms of genetic diversity, total biomass, and metabolic activity (Chapters 4 and 11). Microbes are usually members of complex communities and are directly or indirectly responsible for all of Earth's biogeochemical cycles. Because single-cell organisms dramatically alter planetary environments, the multicellular world is completely dependent upon diverse microbial species for continued survival. If life exists on a world beyond Earth, it would be microbial, either now or in its earliest days. Understanding Earth's evolutionary history or the kinds of life most likely to be encountered beyond Earth requires that we study the history, habitats, ecology, and full diversity of microorganisms that shape our biosphere.

In this chapter we outline the systems of classification for protists, and develop in detail their evolutionary history as revealed through genomic trees. We discuss various possibilities for the origin of eukaryotes and their subsequent evolutionary pathways on a changing Earth.

13.2 Microbial diversity

Our understanding of how multicellular organisms relate to each other has traditionally surpassed what we know about microbial evolution. Shared morphological features within the plants, the animals, and the fungi provide a robust basis for defining evolutionary relationships and these characteristics are frequently preserved in the rich fossil record (Chapters 16 and 17) (Barnes, 1984). Biodiversity experts claim that the number of different plant, animal, and fungal species ranges from 1 million to 10 million. Contrast this with our current census, based upon traditional methods

---

[1] *Prokaryote*s are microorganisms whose cells have no nucleus, organelles (organ-like parts such as a flagellum or chloroplast), or tubulin-based cytoskeleton (a network of protein filaments and tubules in the cytoplasm that gives the cell shape and coherence).

[2] *Eukaryote*s are single-cell microorganisms with flexible cytoskeletal structures, various organelles, and chromosomes bound within a nuclear membrane.

Planets and Life: The Emerging Science of Astrobiology, eds. Woodruff T. Sullivan, III and John A. Baross. Published by Cambridge University Press. © Cambridge University Press 2007.

using morphology (the visible shape and form of an organism), of only ∼5,000 species of prokaryotes (www.dsmz.de/bactnom/bactname.htm) and ∼200,000 species of protists (Margulis *et al.*, 1990; Patterson and Sogin, 1993). But this census is argued to understate the number of microbial species by orders of magnitude (Pace, 1997). It is very difficult to identify *phenotypic* traits – morphological features or physiological and biochemical capabilities of an organism – that can inform us about how to distinguish microbial species and study their interrelationships. There are no widely accepted standards for interpreting the *phylogenetic* (evolutionary) significance of similarities in microbial cell morphology and other traits over large evolutionary distances (Sleigh, 1989; Patterson and Sogin, 1993) and the geological record is silent for most microbial groups (Lipps, 1993). Even if there were consensus about an effective set of phylogenetic markers, changes in the genomes of species (*genotypic* variation) are not necessarily closely linked to phenotypic variation. "Natural systems" based upon resemblance are useful for the taxonomist but this method alone is not sufficient to reconstruct the evolutionary history of distantly related organisms.

13.3 Molecular phylogeny

Deciphering phylogenetic relationships among microorganisms requires an objective measure of relatedness. Physical appearance is not sufficient, but most of these creatures have small genomes that are more amenable to DNA sequence analyses than are the much larger genomes of plants and animals. Over the past twenty years, molecular biology tools have transformed studies of how all organisms are related to each other. In a landmark paper, Zuckerkandl and Pauling (1965) first described the use of protein sequences as "documents of evolutionary history." Today, molecular data provide a practical metric for assessing biodiversity in the context of 3.5–3.8 Gyr of evolutionary history. The comparison of DNA sequences that share a common ancestry makes possible the measurement of genetic differences that are common to members of populations, species, or kingdoms.

Advances in DNA sequencing technology and well-established protocols for isolating particular genes from diverse organisms have fueled an explosive growth of molecular databases. In 1994, GenBank, a public repository of genetic sequences maintained by the US National Institutes of Health, contained 173 million basepairs from a total database of ∼5,000 organisms. By 2007 those numbers reached over 100 billion basepairs from more than 100,000 organisms (ftp://ftp.ncbi.nih.gov/genbank/gbrel.txt). A single automated DNA-sequencing machine can now determine ∼1.5 million basepairs per day.

The ideal strategy for reconstructing evolutionary history would be to sequence the entire genomes of all species that might inform us about microbial phylogeny. But so far fewer than 50 protist genomes have been sequenced, versus ∼500 for prokaryotes. Protists can present technological challenges because their chromosomes can be large and a single genome can contain many millions or even billions of basepairs. With rare exception, eukaryotic microbial genome projects are only being conducted for medically important parasites (such as the hiker's companion *Giardia lamblia*). But these do not at all adequately represent the diversity of microorganisms on this planet. Not only are we dealing with a tiny fraction of the species present in natural settings, but parasitic protists are very specialized organisms that almost certainly appeared late in the history of Earth. Truly primitive taxa[3] are most likely free-living in anaerobic niches that have been present on Earth for billions of years. If we sequenced *entire* genomes, adequate taxon sampling for molecular evolution studies would require impractically large resources. Consequently, most molecular evolution studies focus upon comparisons of a limited number of genes for a large number of species. The ultimate goal is to determine how the major lines of descent are related to each other, and to understand how major protist innovations (in organization within cells, morphology, and life styles) contributed to evolution before the appearance of the multicellular world.

In order to construct evolutionary trees, specific genes are isolated from the genome. Because different genes, domains within a gene, or specific nucleotide positions can display dissimilar rates of evolutionary change, the selection of a particular gene or suite of genes for phylogenetic analysis depends upon the evolutionary distances separating the taxa under study. For example, for closely related species slowly evolving genes may show an insufficient number of differences to resolve evolutionary patterns. At the opposite end of the spectrum, sequences for rapidly evolving genes are of little value for inferring phylogenies that span the major domains of life.

---

[3] *Taxon* (plural *taxa*) is a general term for any grouping of organisms, such as by species, genus, family, etc.

Other criteria must also be considered when constructing phylogenetic frameworks from molecular data. The compared genes must have a common evolutionary history and perform the same role in all organisms in the sample. The compared genes must change slowly enough to allow us to unambiguously identify the nucleotide positions that share a common ancestry. To infer evolutionary relationships, compared sequences must contain a statistically significant number of variable sites. Ideally the sequences should also span several functional domains (regions of the genome that serve different purposes). This condition allows recognition of *convergent evolution*, which is when structures or molecules come to resemble each other even though they have independent evolutionary sources. If, instead, identical trees are found in two sequence domains that can vary independently, we have divergent evolution from a common ancestor. Finally, the genes defining the macromolecules must not be transferred between species. If such *lateral transfer* (Section 10.3) has occurred, the inferred phylogeny will be that of the genes, not of the organisms.

Most of what we have learned about the evolution of microbes comes from studies of the small subunit ribosomal RNA (rRNA) genes (Sogin and Silberman, 1998), large subunit rRNA genes (Ben Ali *et al.*, 2001), or from combined analysis of both (Medina *et al.*, 2001). These genes are common to the protein synthesis machinery of all cells. Comparisons of small subunit rRNA in many organisms led to a new understanding of the *Universal Tree of Life*, which has three primary lineages (Fig. 13.1) with a common origin in a population of primordial cells referred to as the Last Universal Common Ancestor (Woese, 1987). These primary domains are called Archaea (also referred to as Archaebacteria), Bacteria (also referred to as Eubacteria) and Eukarya (the eukaryotes). The root of the Universal Tree of Life may lie within the Bacteria, while the Eukarya and Archaea shared a more recent common ancestor (Woese *et al.*, 1990). The line segment lengths in this Universal Tree of Life are proportional to the evolutionary distances (amount of differences in the compared gene sequences) that separate organisms and shared nodes, but they do not necessarily correspond to real time. The so-called molecular clock inferred from evolutionary distances is imprecise since the rate of evolution can be quite variable in different evolutionary lineages. Furthermore, the tree is based on evolutionary relationships for only a single gene (rRNA), a very small portion of the complete genomes.

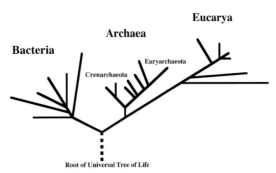

FIGURE 13.1 The phylogenetic Universal Tree of Life in rooted form, showing the three domains. Branching order and branch lengths are based upon comparisons of ribosomal RNA sequences. The position of the root was determined by comparing early divergences in sequences of pairs of paralogous genes (genetic elements that were duplicated in an ancestral organism). (Woese *et al.*, 1990.)

Understanding how the primary lineages are interrelated is a question of great interest. A simplistic interpretation of the Universal Tree of Life is that the history of life on Earth was a series of bifurcating events leading to millions of terminal nodes. The competing view is that Earth's extant biodiversity was strongly shaped by frequent lateral gene transfer, which would make it difficult if not impossible to identify the earliest diverging lineages within each of the primary lines of descent. Through the exploration of evolutionary history, we can also address other issues that are important for astrobiologists. For example, combining studies of the Tree of Life with research on today's microbial communities, we can begin to understand how complex ecosystems evolved on Earth, in particular how the emergence of the biosphere influenced planetary evolution (Chapter 4).

13.4 Molecular ecology

Microbes of untold diversity are directly or indirectly responsible for all biogeochemical cycles on Earth. For instance, photosynthesis by cyanobacteria in the oceans produces more oxygen than do all higher plants. Microorganisms also maintain the global carbon cycle primarily by producing and consuming CO_2. Single-celled organisms control global utilization of nitrogen through nitrogen fixation, nitrification, and nitrate reduction. Microbes impact all ecosystems on Earth, yet we still have only sparse information about their true diversity and know very little about microbially mediated mechanisms that control key biogeochemical cycles.

Molecular evolution studies have provided new frameworks and tools for deciphering the complexity of microbial ecology. It is now possible to identify all microorganisms within a sample (say, of water or soil) without cultivating them species by species in the laboratory (Pace, 1997). Such surveys targeting natural populations of prokaryotes have revealed largely unexplored microbial diversity in geothermal environments, cold oceans, temperate aquatic and marine environments, soils, and animal digestive tracts. Whereas some argue that most existing eukaryotic microorganisms have now been discovered (Patterson, 1999), a plethora of novel prokaryotic microorganisms, sometimes living in seemingly inhospitable environments (Chapter 14), has been revealed by molecular inventories of places such as hydrothermal anoxic sediments, the general ocean, and various acidic and alkaline drainages. Species numbers represented in existing culture collections do not at all reflect the true microbial diversity of planet Earth.

The next advances in microbial ecology will come from studies that use genomes to measure the interplay between microbial populations and diverse ecosystems on planet Earth. We refer to this interdisciplinary approach as *ecogenomics*. The power of ecogenomic technology for understanding how organisms in complex communities interact cannot be overstated – it will redefine how we understand the interactions between the living world and the physical world. A rich vista is opening up in which molecular tools will move beyond their present role in establishing relationships and identities, and go on to define units of diversity, then catalogue that diversity, map the distributions of organisms, measure their abundances, and clarify their interactions.

This chapter presents an integrated view of protist phylogeny based upon genome analyses, as well as *conservation* (degree of commonality) of ultrastructure features. *Ultrastructure* refers to the features made visible by electron microscopy. These include mitochondrial and chloroplast morphology, structure of the flagella and basal bodies, and the deployment of microtubular and cytoskeletal arrays within the nucleus or under the cell surface. This historical perspective on molecular phylogenetics of protists does not attempt to present a definitive view of protist evolution. Just as molecular phylogenetics of protists is a dynamic field, tools for molecular ecology are just coming to fruition. To interpret molecular trees and molecular ecology, we must design and capitalize on internet resources that can integrate rapidly growing

sequence databases with what we know about the biology, ecology, and diversity of microorganisms. Such integration has led to the creation of micro*scope (www.mbl.edu/microscope; described in Appendix E), an internet knowledge network being developed to link molecular approaches with traditional knowledge.

13.5 Morphology and ultrastructure studies of protists

Protists are an eclectic assemblage of predominantly unicellular eukaryotes. Free-living protists thrive in diverse environments, while some species parasitize other protists, fungi, plants, or animals (Margulis *et al.*, 1990; Patterson, 1999). They are morphologically diverse and their phenotypic and genotypic variation far exceeds that seen in other eukaryotic kingdoms. Morphology and ultrastructure alone have not provided enough information to reach conclusions on the relationships within the protist species. There is no trait that unifies protists to the exclusion of all other eukaryotes and there is no agreement about the relative importance of distinct characters for tests of evolutionary hypotheses. Rather than delineating a cohesive evolutionary group, protist species are more fruitfully organized by their degree of cellular organization (ultrastructure) and their paraphyletic lines of descent (evolutionary lineages that contain the ancestor but not all of its descendants).

Progress in understanding the evolution of protists has been advanced by ultrastructure studies. Prior to these studies, characteristics visible by *light* microscopy were mostly used to define ancestral groups. Our current evaluation of the major evolutionary groups within the protists is now largely based on *electron* microscope studies that can distinguish types based on members sharing a common organization, called an "ultrastructural identity" (Patterson and Brugerolle, 1988).

An example of the differing amount of information provided by light- and electron-microscope studies is illustrated in Fig. 13.2. Ultrastructure has been such an influential tool because many protists diversified by inventing and modifying their organelles, details only detectable with the power of an electron microscope. This stands in sharp contrast to prokaryotes and even many fungal eukaryotes where diversification occurred at the molecular and metabolic levels.

It is possible to infer phylogenetic trees based upon shared characteristics such as ultrastructural traits, morphology as viewed through the light microscope,

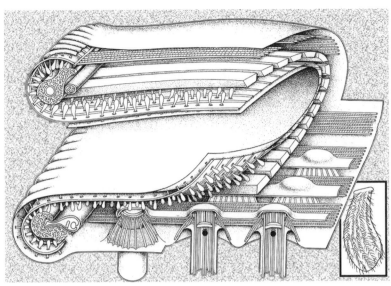

FIGURE 13.2 *Stephanopogon apogon*, a medium-sized protist that lives in marine sediments, is illustrated with observations made by a light microscope (insert at lower right) and a detail of the anterior mouth (large drawing) as determined from ultrastructure studies using an electron microscope. The much greater information provided by electron microscopy has led to wholly different ways to classify protists. The lower surface of the cell shows the bases of the flagella which allow the cell to move, as well as the arrays of microtubules and other structures which hold the flagella in place. The cell is about 20 μm long.

and biochemical capabilities. An important underlying assumption is that the character states that are being compared are indeed linked by evolutionary history. About 60 different patterns of ultrastructural organization have been recognized, although ∼200 genera of protists have yet to be examined (Patterson, 1999). Since we cannot always be certain about the common evolutionary origins for any given phenotypic character, it is necessary to confirm new hypotheses with independent data such as molecular sequences. An example of this process is given by our changing perceptions of the affinities of the pelobionts, a group of flagellates with an unusually limited array of organelles. The first few species to be described by electron microscopy had an aberrant organization of flagella unlike that of other eukaryotes (they lacked mitochondria, dictyosomes, and many familiar cytoskeletal structures). It was therefore hypothesized that the pelobionts represented a very early step in the evolution of eukaryotes. As more ultrastructural evidence emerged, however, the situation became ambiguous, and in the end it was molecular data which showed that pelobionts were not primitive but had a sister relationship with certain amoebae (the entamoebae). The process of combining molecular and phenotypic data is leading to a comprehensive evolutionary tree that provides detailed insights about how the eukaryotic cell was assembled.

13.6 Molecular studies of protists

Success in doing molecular phylogeny requires that the DNA sequences selected for analysis reflect the

historical evolution of the relevant organisms. Before any nucleotide sequence data were available, Woese (1987) recognized the extraordinary potential of ribosomal RNA (rRNA) for inferring phylogenies that include all cellular life forms. Ribosomal RNA is ideal because it evolves very slowly (is conserved) as compared to most protein-coding genes. Ribosomal RNA is the functional core of the information translation machinery found in a cell's ribosomes (protein synthesis organelles). Given their mandatory involvement in the synthesis of every protein in a cell, the extraordinary conservation of ribosomal RNA genes is not surprising. Yet rRNA sequences can display rapidly evolving regions interrupted by highly conserved elements. The rapidly changing sequences are useful for measuring evolutionary distance between closely related species in the same genus, while the highly conserved sequences are used to infer relationships that span large evolutionary distances.

Early rRNA phylogenies for protists claimed that protists without mitochondria (*amitochondriates*) were ancestral to all other eukaryotes, as can be seen in Fig. 13.3 (Sogin *et al.*, 1989). These organisms lack many of the organelles found in other eukaryotes. The molecular trees based upon rRNAs also described new complex evolutionary assemblages (Gajadhar *et al.*, 1991) and confirmed relationships previously inferred from morphology and ultrastructure (Leipe *et al.*, 1994). Measured in terms of lifestyle and phenotypic variation, some major protist clades appear to be as complex as plants, animals, or fungi (Fig. 13.4). But the several radiations in the eukaryotic rRNA

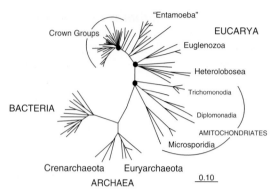

FIGURE 13.3 An unrooted tree of life showing the three domains and deep branching eukaryotic lineages, based upon comparisons of ribosomal RNA gene sequences. It can be seen that the amitochondriates are older than all other eukaryotes. The bar labeled 0.10 indicates 10 nucleotide substitutions per 100 positions. (Sogin and Silberman, 1998.)

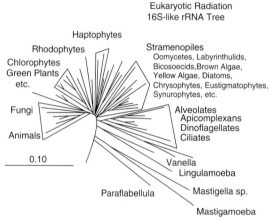

FIGURE 13.4 Phylogentic tree for "crown group" eukaryotes, based upon comparisons of ribosomal RNA gene sequences. (Sogin and Silberman, 1998.)

trees make it difficult to define the major protist clades. More importantly, we still have no clear understanding about which extant protist lineages might represent the earliest branches in the eukaryotic line of descent.

13.6.1 Multi-gene studies and the Archezoan hypothesis

One of the striking features of these early molecular "trees" was the basal location of amitochondriates such as microsporidia and the diplomonad *Giardia* (Fig. 13.3). As mentioned above, this led to hypotheses that the first eukaryotes were organisms without mitochondria. In this scenario, the transition more

than 2.4 Ga from a reducing atmosphere to one that contained oxygen would *not* have been a prerequisite for the formation of eukaryotic cells. Cavalier-Smith (1983) introduced the term *Archezoa* to refer to these apparently most primitive eukaryotes. In one model of how the major eukaryotic organelles were assembled (Knoll, 1992), those lineages appearing before the acquisition of mitochondria occurred just prior to the largely unresolved radiation of many eukaryotic lineages (called the "crown groups"; Fig. 13.3).

To address questions of uncertainty about the identity of the earliest eukaryotes and to test the rRNA scheme, several groups of researchers shifted their attention to gene families other than rRNA (Baldauf *et al.*, 2000). In some cases these multi-gene phylogenies confirmed what we have learned from rRNA comparisons, while in others, consistent disagreements suggest alternative scenarios for the evolution of protists (Embley and Hirt, 1998). Combined phylogenies of multi-gene families have identified potential super-clades of protists (Fig. 13.4), but different studies disagree and the number of taxa in most of these analyses is sparse. It is now evident, however, that the simple division of the eukaryotes into two broad territories is incorrect. Many of the taxa presumed to have been originally amitochondriate have genes which might be derived from mitochondria. The "primitively amito-chondriate" taxa are now mostly regarded as having derived from "crown" taxa by *loss* of mitochondria. Of the amitochondriate taxa, *Giardia* and the other diplomonads remain the best candidates for early eukaryotes, although not without debate.

Conflicting molecular trees suggest that apparently deep-branching protists might be artifacts of "long branch attraction" between rapidly evolving, basal eukaryotic branches and distantly related archaeal and bacterial out-groups. This conflict among genealogies has led to the "Big Bang Hypothesis" (Philippe *et al.*, 2000), which states that all extant eukaryotes are descendants of a sudden evolutionary radiation that occurred about ~0.6–1.0 Ga (Fig. 13.5). This interpretation is not without problems. The paleontological record offers evidence of 1.8–2.1 Ga eukaryotic microfossils (Knoll, 1992; Section 12.3) and 2.7 Ga archaean molecular signatures in the form of steranes that are attributed to eukaryotes (Brocks *et al.*, 1999; Section 12.2.6). Moreover, the Big Bang Hypothesis does not account for consistent hierarchical relationships evident in combined, multi-gene studies.

A big bang in eukaryotic evolution?

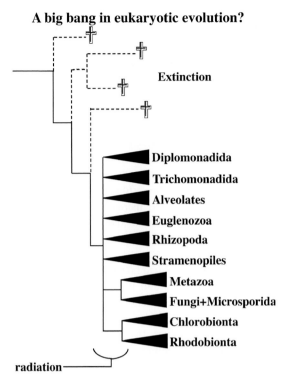

Extinction

Diplomonadida

Trichomonadida

Alveolates

Euglenozoa

Rhizopoda

Stramenopiles

Metazoa

Fungi+Microsporida

Chlorobionta

Rhodobionta

radiation

FIGURE 13.5 "Big Bang Hypothesis" for the evolution of eukaryotes. A grand radiation that occurred between 1 Ga and 0.6 Ga is hypothesized to be responsible for the origins of contemporary eukaryotic biodiversity (Philippe et al., 2000). This radiation was followed by the subsequent divergence of animals (Metazoa) from fungi, and of red algae (Rhodobionta) from green algae (Chlorobionta).

13.6.2 Evolving ideas about the origin of protists

To explain the disparity between the Big Bang Hypothesis and the paleontological record, Philippe et al. (2000) hypothesize that only a single early eukaryotic lineage survived extinction events that preceded the ~1.0 Ga evolutionary radiation of plants, animals, fungi, and all other protists. If this scenario is correct, the thread of life leading to contemporary eukaryotes is incredibly thin and difficult to explain. In molecular phylogenies, long unbroken basal branches are characteristic of extinction events. This phenomenon is well described for mammals, birds, and angiosperms, but not protists. Eukaryotic microbial communities consist of associations of thousands of species separated by large evolutionary distances relative to those separating angiosperms, birds, or mammals. Their mutual extinction would require global events encompassing an enormous variety of the Earth's environments and ecological diversity. A Snowball Earth scenario

(Hoffman et al., 1998; Section 4.2.4) might account for a global extinction pattern for protists, but we must then explain how complex communities capable of supporting growth of diverse protists were re-established. Finally, analyses of protein clocks do not support global extinctions or the Big Bang Hypothesis. Comparisons of 57 sets of amino acid sequences suggest a 2.0–3.5 Ga divergence between eukaryotes and prokaryotes followed by a ~1.5 Ga divergence of protists and a ~1.2 Ga separation of plants from animals plus fungi (Feng et al., 1997). Another study using refined methods of sequence alignment, site selection and time estimation using 87 protein sequences concluded that there were two primary horizontal transfer events in the origin of eukaryotes, one at 2.7 Ga and the other at 1.8 Ga (Hedges et al., 2001).

13.6.3 Sampling bias and other problems

Our models for understanding how complexity evolved in the eukaryotic cell are in flux. For example, interpretations of early molecular trees argued that the most basal protist lineages lacked mitochondria and introns.[4] The absence of introns in early diverging eukaryotes and most prokaryotes suggested their relatively late introduction into microbial genomes. And if mitochondria were not present or required for eukaryotic life, protists could have appeared prior to the advent of significant quantities of oxygen in Earth's atmosphere. However, the discovery of bacterial-like molecular chaperones in putatively deep-branching amitochondriate protists (Roger et al., 1998) suggests that symbionts ancestral to mitochondria could have been present in the early stages of eukaryote evolution. Molecular chaperones, particularly cpn60 genes, provide robust phylogenetic descriptions of alphaproteobacteria, which shared a recent common ancestry with symbionts that gave rise to mitochondria. Although cpn60 genes are encoded by the nucleus, the proteobacterial-like cpn60 proteins of eukaryotes localize or target the mitochondria. Their presence in the genomes of protists lacking mitochondria is prima facie evidence that mitochondria were either lost or reduced to form organelles (hydrogenosomes) where anaerobic energy metabolism occurs. Thus the absence of mitochondria may be a secondary adaptation to anaerobic environments.

These discoveries emphasize that our molecular perspective of early eukaryote evolution is strongly

---

[4] An intron is a non-coding segment of a gene.

biased by the limited selection of taxa and gene families available for molecular studies. Broad-scale genomic sampling from taxa that might represent basal lineages in the eukaryotic line of descent is very limited. The *Giardia lamblia* (diplomonad) genome project (www.mbl.edu/Giardia) provides our first glimpse of a genome from an early diverging, amitochondriate protist. Surprisingly, this genome contains introns that are removed by splicesomal machinery (Nixon *et al.*, 2002) and a mitochondrial-like cpn60 gene (Roger *et al.*, 1998), but there are no other credible examples of coding regions derived from the symbiont that was ancestral to mitochondria. Although there are many amitochondriate flagellates that lack dictyosomes (membrane vesicles that play a role in the modification and transport of proteins in most eukaryotic cells) and complex cytoskeletal systems, few of these lineages have been included in molecular trees. These protists occur in anoxic environments and are only now being cultured and characterized in detail at the ultrastructural level. Since some are likely to represent even deeper diverging eukaryote lineages, their phenotypic properties will be important in understanding the evolution and assembly of the first eukaryotes.

In addition to limited taxon representation, alternative mechanisms of genome evolution might explain discrepancies between molecular trees inferred from different protist gene families. Conflicting phylogenies could be due to (a) paraphyletic relationships inferred from comparisons of genes that were duplicated in an ancient ancestor, (b) horizontal gene transfer mediated by fusion of two or more genomes, (c) endosymbioses, or (d) viral-mediated, cross-species transfers (Section 10.3). Single-gene phylogenies cannot resolve which of these mechanisms might be major factors in the evolution of eukaryotic genomes. To understand phylogenetic patterns for protists, we must combine from as many taxa as possible more extensive molecular and phenotpypic data such as morphological features, ultrastructure, and biochemical capabilities.

The first models for early evolution of protists were misled by (a) structural simplicity that can reflect either ancestral or highly reduced states, (b) the use of highly derived (parasitic) taxa and therefore failure to adequately sample the full range of protist diversity, and (c) algorithms that did not consistently handle both fast- and slow-evolving sequences. Regarding the last point, data sets sometimes include sequences that continue to evolve at very slow rates after their divergence from a common ancestor. In contrast, the same data sets may include sequences that are more recently diverged but display accelerated rates of evolutionary change. Algorithms that inadequately compensate for different rates of evolutionary change can incorrectly cluster together the slowly evolving sequences and describe the recently diverged, rapidly evolving sequences as deep phylogenetic branches. Genes that share a common ancestry might display differential rates of evolution in distinct evolutionary lineages for many reasons: (a) environmental shifts requiring a large number of changes at the amino acid level in order to maintain function, (b) altered mutation rates occurring in response to increased errors by DNA polymerases, or (c) reduced ability to repair DNA replication errors leading to an increase in nucleotide substitution patterns.

We must rethink how to take advantage of both the huge rRNA databases and the high-speed molecular tools at our disposal. A major advance would be the broader application of comparative genomics for inferring evolutionary history. Instead of the serendipitous selection of taxa, comprehensive molecular studies should include at least two representatives from each major taxonomic assemblage. Investigation of earlier diverging eukaryotic lineages from rarely studied environments might provide even greater insights into early eukaryotic evolution.

13.7 Future directions of ecogenomics

Understanding how microbial ecosystems function and learning how they evolve over geological timescales requires far more than enumerating microbial populations and attendant shifts in concentrations of key molecular components characterizing biogeochemical cycles. We must integrate our rapidly expanding information about microbial diversity in natural settings with information about the metabolic activities and gene expression patterns that coordinate biogeochemical processes. If we are ever to understand how microorganisms shape our environment on a global scale, we must develop an experimental paradigm that links measurement of biogeochemical processes with temporal and spatial distributions of microbial populations and their metabolic properties. To date, gene expression studies have been restricted to a handful of genes, usually from closely related populations (Wawer *et al.*, 1997). Investigators have monitored one or a few metabolic functions, encoded by a specific gene or set of genes, and ignored all other parts of the microbial system. But technological advances promise to change this gene-by-gene approach to molecular ecology. Microbial genomes can be sequenced in weeks, bioinformatics is revolutionizing the storage and manipulation of biological data, and DNA micro-arrays permit massive, parallel expression analyses.

It will soon become possible to apply these powerful tools to unravel the mysteries of microbially-mediated processes that drive our biosphere. We will be able to measure how microbial gene expression patterns for entire ecosystems respond to altered chemical, physical, and biological parameters. The integration of molecular studies of microbial diversity, ecosystem-wide patterns of gene expression, and biogeochemistry will provide a new foundation for interpreting paleon-tological and geological aspects of Earth's early history, as well as data gathered by robotic missions to other Solar System bodies. Ecogenomic studies will continue to provide new insights about the kinds of organisms, metabolisms, and biosignatures that might be present on other planets or satellites. Finally, know-ing the distribution of microbes and their biochemical properties in extreme environments will profoundly shape the design of life detection missions to other Solar System bodies.

REFERENCES

Baldauf, S. L., Roger, A. J., Wenk-Siefert, I., and Doolittle, W. F. (2000). A Kingdom-level phylogeny of eukaryotes based on combined protein data. *Science*, **290**, 972–977.

Barnes, R. S. K. (1984). *A Synoptic Classification of Living Organisms*. Oxford: Blackwell Scientific Publications.

Ben Ali, A., De Baere, R., Van der Auwera, G., De Wachter, R., and Van de Peer, Y. (2001). Phylogenetic relationships among algae based on complete large-subunit rRNA sequences. *Int. J. Syst. Evol. Microbiol.*, **51**, 737–749.

Brasier, M. D., Green, O. R., Jephcoat, A. P., *et al.* (2002). Questioning the evidence for Earth's oldest fossils. *Nature*, **416**, 76–81.

Brocks, J. J., Logan, G. A., Buick, R., and Summons, R. E. (1999). Archean molecular fossils and the early rise of eukar-yotes. *Science*, **285**, 1033–1036.

Cavalier-Smith, T. (1983). A 6-kingdom classification and unified phylogeny. In *Endocytobiology II, Intracellular Space as an Oligogenetic Ecosystem*, Schwemmler, W. and Schenk, H. E. A. (eds.). Berlin: De Gruyter, 1027–1034.

Embley, T. M. and Hirt, R. P. (1998). Early branching eukar-yotes? *Curr. Opin. Genet. Dev.*, **8**, 624–629.

Feng, D. F., Cho, G., and Doolittle, R. F. (1997). Determining divergence times with a protein clock: update and reevalua-tion. *Proc. Nat. Acad. Sci. USA.*, **94**, 13028–13033.

Gajadhar, A. A., Marquardt, W. C., Hall, R., Gunderson, J., Ariztia-Carmona, E. V., and Sogin, M. L. (1991). Ribosomal RNA sequences of *Sarcocystis muris*, *Theileria annulata* and *Crypthecodinium cohnii* reveal evolutionary relationships among apicomplexans, dinoflagellates, and ciliates. *Mol. Biochem. Parasitol.*, **45**, 147–154.

Hedges, S. B., Chen, H., Kumar, S., Wang, D. V.-C., and Watanabe, H. (2001). A genomic timescale for the origin of eukaryotes. *BMC Evol. Biol.*, **1**, 4.

Hoffman, P. F., Kaufman, A. J., Halverson, G. P., and Schrag, D. P. (1998). A neoproterozoic snowball earth. *Science*, **281**, 1342–1346.

Knoll, A. H. (1992). The early evolution of eukaryotes: a geo-logical perspective. *Science*, **256**, 622–627.

Leipe, D. D., Wainright, P. O., Gunderson, J. H., Porter, D., Patterson, D. J., Valois, F., Himmerich, S., and Sogin, M. L. (1994). The stamenopiles from a molecular perspective: 16S-like rRNA sequences from *Labyrinthuloides minuta* and *Cafeteria roenbergensis*. *Phycologia*, **33**, 369–377.

Lipps, J. H. (1993). *Fossil Prokaryotes and Protists*. Boston: Blackwell Scientific Publications.

Margulis, L., Corliss, J. O., Melkonian, M., and Chapman, D. J. (1990). *Handbook of the Protoctista*. Boston: Jones and Bartlett Publishers.

Medina, M., Collins, A. G., Silberman, J. D., and Sogin, M. L. (2001). Evaluating hypotheses of basal animal phylogeny using complete sequences of large and small subunit rRNA. *Proc. Natl. Acad. Sci. USA*, **98**, 9707–9712.

Mojzsis, S. J., Arrhenius, G., McKeegan, K. D., Harrison, T. M., Nutman, A. P., and Friend, C. R. (1996). Evidence for life on Earth before 3,800 million years ago. *Nature*, **384**, 55–59.

Nixon, J. E., Wang, A., Morrison, H. G., McArthur, A. G., Sogin, M. L., Loftus, B. J., and Samuelson, J. (2002). A spli-ceosomal intron in Giardia lamblia. *Proc. Natl. Acad. Sci. USA*, **99**, 3701–3705.

Pace, N. R. (1997). A molecular view of microbial diversity and the biosphere. *Science*, **276**, 734–740.

Patterson, D. J. (1999). The diversity of eukaryotes. *American Naturalist*, **154**, 96–124.

Patterson, D. J. and Brugerolle, G. (1988). The ultrastructural identity of *Stephanopogon apogon* and the relatedness of the genus to other kinds of protists. *Eur. J. Protistol.*, **23**, 279–290.

Patterson, D. J. and Sogin, M. L. (1993). Eukaryote origins and protistan diversity. In *The Origin and Evolution of Prokaryotic and Eukaryotic Cells*, Hartman, H. and Matsun, K. (eds.). New Jersey: World Scientific Publishing Co., 13–46.

Philippe, H., Germot, A., and Moreira, D. (2000). The new phylogeny of eukaryotes. *Curr. Opin. Genet. Dev.*, **10**, 596–601.

Roger, A. J., Svard, S. G., Tovar, J., Clark, C. G., Smith, M. W., Gillin, F. D., and Sogin, M. L. (1998). A mitochondrial-like chaperonin 60 gene in *Giardia lamblia*: evidence that diplomonads once harbored an endosymbiont related to the progenitor of mitochondria. *Proc. Natl. Acad. Sci. USA*, **95**, 229–234.

Schopf, J. W. and Packer, B. M. (1987). Early Archaean (3.3-billion to 3.5-billion-year-old) microfossils from warrawoona Group, Australia. *Science*, **237**, 70–73.

Schopf, J. W., Kudryavtsev, A. B., Agresti, D. G., Wdowiak, T. J., and Czaja, A. D. (2002). Laser–Raman imagery of Earth's earliest fossils. *Nature*, **416**, 73–76.

Sleigh, M. (1989). *Protozoa and Other Protists*. London: Edward Arnold.

Sogin, M. L. and Silberman, J. D. (1998). Evolution of the protists and protistan parasites from the perspective of molecular systematics. *Int. J. Parasitol.*, **28**, 11–20.

Sogin, M. L., Gunderson, J. H., Elwood, H. J., Alonso, R. A., and Peattie, D. A. (1989). Phylogenetic meaning of the kingdom concept: an unusual ribosomal RNA from *Giardia lamblia*. *Science*, **243**, 75–77.

Wawer, C., Jetten, M. S. M., and Muyzer, G. (1997). Genetic diversity and expression of the [NiFe] hydrogenase large subunit gene of *Desulfovibrio* spp. in environmental samples. *Appl. Environ. Microbiol.*, **63**, 4360–4369.

Woese, C. R. (1987). Bacterial evolution. *Microbiol. Rev.*, **51**, 221–271.

Woese, C. R., Kandler, O., and Wheelis, M. L. (1990). Towards a natural system of organisms: proposal for the domains Archaea, Bacteria and Eucarya. *Proc. Natl. Acad. Sci. USA*. **87**, 4576–4579.

Zuckerkandl, E. and Pauling, L. (1965). Molecules as documents of evolutionary history. *J. Theor. Biol.*, **8**, 357–366.

14 Limits of carbon life on Earth and elsewhere

John A. Baross,
Matthew O. Schrenk*, and
Julie A. Huber*, *University of Washington*

14.1 Introduction

The search for extraterrestrial life is intimately linked with our understanding of the distributions, activities, and physiologies of Earth-life.[1] This is not to say that only Earth-life could exist on other planets and moons but it is important to know the extent of environmental conditions that can support terrestrial organisms as a first-order set of criteria for the identification of potential extraterrestrial habitats. Even though the life forms may have different biochemistries and in fact may have had different origins, the limits of life on Earth may help define the potential for habitability elsewhere. It is also likely that many of the limits of Earth-life could extend out of the bounds of extreme conditions found on modern-day Earth. This is the case for the bacterium *Deinococcus radiodurans* that can tolerate levels of radiation beyond those found naturally on Earth, and also for the apparent tolerance by *Escherichia coli* to hydrostatic pressures that exceed by more than ten times the pressures in the deepest ocean trenches (Cox and Battista, 2005; Sharma *et al.*, 2002).

Since Earth is the only planet that unequivocally supports modern, living ecosystems, it is logical to first look for life elsewhere that resembles Earth-life. Earth-life requires either light or a chemical energy source, and other nutrients including nitrogen, phosphorus, sulfur, iron, and a large number of elements in trace concentration; 70 elements in all are either required or are targets of interaction by various species of Earth-life (Wackett *et al.*, 2004). Earth-life also requires liquid water as a medium for both energy transduction and biosynthesis. Additionally, chemical disequilibria are required to fuel the maintenance and growth of organisms. Thus the search for extraterrestrial life is focused on planets and moons that currently have or have had liquid water; that have a history of geological and geophysical properties that favor the synthesis of organic compounds and their polymerization; and that provide the energy sources and nutrients needed to sustain life.

However, inasmuch as we can use Earth-life as a point of comparison, we are also limited by what we currently know about life on this planet. In this regard there are two main issues being addressed by the astrobiology community: (1) our incomplete understanding of the physiological diversity of Earth-life, and (2) our almost complete lack of data about possible alternative biochemistries. Regarding the first issue, little or nothing is known about the physiological and metabolic diversity of the majority of microorganisms in most Earth environments; for understudied environments such as deep marine sediments, we have cultured far less than 1% (0.01%?) of the microbial population that is detected by molecular methods, and in seawater the figure is $\sim 1\%$ (note that molecular methods provide only an index of genetic diversity but tell us nothing about physiological diversity). Sleuthing the physiological characteristics of some of the organisms represented in the "unknown majority" has provided new insights into their adaptations to "extreme" environments, those at the fringes of habitability as we know it. For example, a novel marine photosynthetic microorganism, recently isolated from deep-sea hydrothermal vents, may be utilizing the blackbody radiation from hot sulfides for photosynthesis (Beatty *et al.*, 2005). In another case,

---

* Matthew Schrenk is currently at the Carnegie Institution of Washington and Julie Huber at the Marine Biological Laboratory at Woods Hole.

[1] The term *Earth-life* is used in this chapter for life as we know it on Earth.

newly discovered microorganisms have extended the upper temperature for growth to 121 °C and lowered the pH "record" to below 0. One hyperthermophilic microbe lacks consensus sequences (strings of nucleotide bases that are common to all other known organisms) in its 16S rRNA and, incidentally, is symbiotic to another archaeal species (Huber *et al.*, 2002). Discoveries such as these only emphasize how little we know.

Regarding alternative biochemistries, we begin any extraterrestrial search by assuming carbon-based life. The key arguments in favor of carbon-based life are the ubiquity of organic compounds in the Universe and the ability of carbon to form stable compounds with a high number of different inorganic elements, thus creating the wide variety of structural, catalytic, and informational macromolecules[2] that make up Earth-life (Chapter 27). But how versatile and adequate is the carbon-based life model to environmental conditions that have either not been adequately explored on Earth, or that extend beyond the bounds found on Earth? Are there alternative carbon-based biochemistries that would allow organisms to exist under more extreme conditions than can Earth-life? Embedded in this question are two others: What are the limitations to evolutionary innovations in carbon-based life? How do environmental parameters or "extrinsic" factors, such as hydrostatic pressure and solute concentrations, affect these limitations?

Powerful new molecular methods now allow for the construction of novel enzymes and, eventually, novel organisms. A viral genome has already been constructed from synthetic oligonucleotides (short strings of nucleotides) (Smith *et al.*, 2003). These "directed evolution" methods have the potential to explore whether it is possible to design life that uses different information macromolecules or new metabolic pathways exploiting energy sources not known in extant Earth organisms – energy sources such as ultraviolet radiation, gravity, and heat. Furthermore, explorations of deep subsurface environments have uncovered microbial communities potentially decoupled from photosynthetic reactions occurring at the Earth's surface. The communities use magmatic degassing, radiolysis of water, or serpentinization reactions (aqueous alterations of mantle material) to drive their metabolisms. These ecosystems persist in environments that have remained essentially unchanged for millions, if not billions, of years, and that overlap with conditions favorable for abiotic organic synthesis. Do organisms in these ecosystems contain relict biochemistries that have eluded evolutionary pressures?

There are thus two questions associated with understanding the limits of carbon-based life: (1) what are the limits of extant Earth-life?, and (2) what are the limits for carbon-based life? To answer the second question, we must do more than study Earth's habitats and Earth-life. It is a common tendency to extrapolate the physiologies of Earth organisms that might be best suited to live under the extreme conditions found on other planets to the extraterrestrial milieu. However, this assumes that an extraterrestrial organism would have the same physiological characteristics of an Earth organism if the environmental conditions were the same. In thinking this way, we assume that life originated on that planet and then evolved physiologies so as to take advantage of the available habitat conditions. But any planet or moon that has or has had environmental conditions that could support Earth-life or an Earth-like organism may not ever have had conditions suitable for a *separate* origin of life and thus could be sterile. This may be particularly true for icy planets with liquid water even if they have the cache of chemicals necessary to support life. Could life originate *de novo*, or would it have to be seeded from a neighboring planet or moon, one that during its early history had more suitable conditions for spawning life? How many types of environments can lead to the origin of life?

For the case of Earth, various models postulate geophysical conditions that may have favored early organic and biochemical stages leading to life. Some models rely upon subsurface hydrothermal settings because they can provide all the chemical precursors and catalysts essential to generating complex carbon chemistry (Section 8.5). Others exploit alternative settings such as meteoritic input, reduced atmospheres, or freshwater ponds (Chapters 6–9). The origin of life is likely to involve multiple environmental conditions that span both spatial and temporal dimensions. Since we still do not understand how Earth-life started, it is impossible to answer the general question of whether there could be multiple settings that independently give rise to different carbon-based life forms. Would a separate origin of life under different settings than those that produced life on Earth create a carbon-based life form capable of different evolutionary innovations, or are there rules of organic chemistry that limit carbon-based life to only the physiological diversity that we observe today?

---

[2] A *macromolecule* is a very large molecule such as DNA, a protein, etc.

14.2 Looking for Earth-like environments elsewhere

Do the limits of life on Earth set the bounds for *all* carbon-based life? As mentioned above, this question is difficult to approach because we do not yet even know the limits of Earth-life. There are many environments on Earth that are not adequately explored, such as the deep sub-seafloor crust or most submarine hydrothermal environments (particularly those in slow-spreading ridge axis environments). Also, we have a very incomplete knowledge of the physiological diversity of the majority of microorganisms in extreme environments since so few have been isolated and studied in pure culture. Even those organisms that have been sustained in the laboratory have been studied only under near-optimum conditions, far from what we would expect to find in natural environments. Furthermore, there are also *combinations* of extreme conditions not found or yet sampled on Earth that may be relevant to extraterrestrial situations – examples include high temperature and high salinity, low temperature and acid pH, and, relevant to Titan, extreme cold and non-aqueous organic solvents or liquid ammonia.

But even if we did know the full range of conditions suitable for Earth-life, we have a very limited knowledge of the possible biochemical structural and catalytic macromolecules and metabolic pathways that are possible from carbon chemistry (Chapter 27). On Earth evolution long ago selected for one kind of biochemistry – surely there were alternative biochemistries that existed during the early stages of life's origin and "lost out" and can now only be imagined. Many of these might well be able to sustain life in very unEarth-like conditions.

Our present search for extraterrestrial microbial life has been focused on Mars and Europa (Chapters 18 and 19), but it is likely that Saturn's small moon Enceladus will be added to this list, based on recent Cassini Mission observations that it is spewing a plume containing water, methane, carbon dioxide, and propane (Porco *et al.*, 2006). Thinking beyond the constraints of Earth-life and its requirement for liquid water, we should also not forget the possibility that Titan is swimming in complex biochemical compounds (Chapter 20), or that on Venus some form of life could exist in its sulfidic "greenhouse" atmosphere. And in the not too distant future we will be debating the merits for life of Earth-like planets in *other* planetary systems (Chapter 21).

The Earth-life model limits biological energy sources to light and chemicals. Mars and Europa, however, presently lack viable habitats for Earth-life that are bathed in light; for life to persist on these worlds, there must be a source of chemical energy or a novel means of photosynthesis, possibly using blackbody radiation from hotspots. A possible chemical energy source could be organic material left over from a bygone era when either Mars or Europa had an atmosphere and possibly a short-lived photosynthetic microbial community. The organic carbon would eventually dissipate as volatile products of microbial metabolism or chemical reactions. On Earth, active geophysical processes such as volcanism and particularly hydrothermal systems can support life without photosynthesis. In deep-sea hydrothermal systems, hyperthermophiles ("heat-loving" microbes thriving at temperatures $> 80 \,^{\circ}C$), including methanogens and some sulfur reducers, use hydrogen generated by sub-seafloor water–rock reactions or magmatic degassing as energy sources. The "primary producers" in hydrothermal vent environments derive their carbon from CO_2 and possibly abiotically produced organic compounds of low molecular weight. Several different metabolic pathways are used to fix CO_2 in these systems and include at least two that are completely unlike any well-characterized pathways (Hügler *et al.*, 2003). In the hot subsurface associated with hydrothermal systems, hyperthermophiles that fix CO_2, including methanogens and some sulfur reducers, also use hydrogen as an energy source.

Hydrothermal systems can result in hot salty fluids, acid or alkaline. These fluids are saturated in volatiles including carbon dioxide, methane, hydrogen, and hydrogen sulfide, as well as a wide range of metals. Moreover, hydrothermal systems can be a source of abiotically produced organic compounds and ammonia. While there is evidence for tectonics and volcanism on Mars during its past history, very little is known about present subsurface heat sources and hot water/rock interactions. The detection of methane on Mars (Section 18.8) does suggest a hydrothermal source. Likewise, models of heat production within Europa from tidal flexing and radiogenic decay suggest that this jovian satellite may be hydrothermally active. Clearly, any evidence for hydrothermal activity on Mars or Europa would greatly increase the chance for the existence of thriving microbial ecosystems. By this same rationale, Enceladus, which shows the most compelling evidence for active hydrothermal hot springs, should be a focus for future life-detection missions.

14.3 Extremophiles and the limits of life

The extreme conditions that limit growth or prove lethal to most organisms are "Garden of Eden" conditions for others. Extremes of high temperature, high and low pH, high salt concentration, toxic metals, toxic organic chemical compounds, and high levels of radiation kill the overwhelming majority of Earth's organisms. However, there are organisms from all three domains of life that have adapted to many of the terrestrial extremes. High temperature, low pH, and high salinity environments are likely to have persisted throughout Earth history, and this also may be the case for icy environments. Note that these extreme environments are not rare on modern-day Earth: most of the ocean is cold ($\sim$2–4 °C) and a vast portion of the subsurface is hot ($>$50 °C).

There are very few natural environments on Earth where life is absent – life is the rule rather than the exception. Microbial life on Earth has proliferated into habitats that span nearly every imaginable physico-chemical variable. Only the very highest temperatures or low water activities[3] render terrestrial environments unsuitable for growth. However, even these conditions do not necessarily render the environment sterile since many organisms have adapted mechanisms for long-term survival at temperatures more than 100 °C above their maximum growth temperature, or in a desiccated state. There are few environments on Earth entirely free from surviving life. Viable microorganisms were even detected, albeit in low numbers, from Chile's Atacama Desert, perhaps the driest environment on Earth and thought to be an analogue of sterile Martian soil (Navarro-Gonzáles et al., 2003). On the other hand, although rare, there are some environments with liquid water that do not appear to support life. These include the $>$400 °C water at submarine hydrothermal vents that is kept liquid due to hydrostatic pressure (Kelley et al., 2002) and the high-brine liquid water found in sea-ice inclusions at $-$30 °C (Section 15.3). But even in these extreme cases, there is evidence for viable microorganisms that apparently survive exposure to temperature extremes well outside their growth range (Baross and Deming, 1995; Straube et al., 1990).

There are several recent review articles that discuss the limits of life, the characteristics of extremophiles ("extreme (condition)-loving" microbes) and astrobiology implications (López-García, 2005; Rothschild and Mancinelli, 2001; Holland and Baross, 2003). Most discussions of the limits of life focus on the extreme range of single physical or chemical conditions such as temperature, salinity, heavy metal concentrations, desiccation, and pH. There are also many excellent reviews of single classes of extremophiles with information on their ecology, physiology, and biochemistry (Bartlett, 2002; Horikoshi, 1999; Silver and Phung, 1996; Wiegel and Kevbrin, 2003; Ventosa et al., 1998). The terms used to define organisms are based on their ability to grow at different degrees of extreme condition; Table 14.1 defines these terms. Implied in these definitions is that some organisms have adapted to grow best (fastest growth rates) at discrete levels of extreme conditions. For example, obligately psychrophilic[4] ("cold-loving") bacteria are incapable of growth above 15 °C, and obligately piezophilic ("pressure-loving") bacteria can grow only under large hydrostatic pressure and usually grow best at the in situ pressure of their natural environments. Not included in Table 14.1 are separate categories or terms to describe organisms that grow in environments where there may be multiple extreme parameters. For example, organisms that grow at high temperatures and high alkalinity are called alkali-thermophiles (Wiegel and Kevbrin, 2003). There is also evidence for alkali-hyperthermophilic, methane-metabolizing Archaea (which carry out both methane production and consumption) at 90 °C and pH 10 from the Lost City hydrothermal vent environment in the Atlantic Ocean (Schrenk et al., 2004). Another example includes the majority of characterized obligate piezophilic bacteria, which are also obligate psychrophiles.

Individual organisms are frequently highlighted for their ability to lead the pack in tolerance or ability to grow under the most extreme conditions – some record-holders (not all of them microbes) are listed in Table 14.2. In some cases, such as high pH, high hydrostatic pressure, and high metal content, the limits for life reflect the limits found in natural environments and not necessarily the actual limits for life. There does appear to be some absolute maximum temperature and minimum concentration of water that will prevent cellular growth (Table 14.2). There are two distinct classes of extreme environmental conditions based on how they affect cells. Extremes in pressure and temperature extend their effect into the cytoplasm. Intracellular biosynthesis, metabolism, and macromolecular

---

[3] *Water activity* is a measure of the extent to which water molecules are available to participate in chemical reactions (see Section 15.2.1.2 for further details).

[4] *Obligately* psychrophilic means that the condition of cold is *necessary* for the organism to exhibit growth.

TABLE 14.1 Terminology used to classify extremophilic Bacteria and Archaea and examples of organisms and habitats

Parameter	Term	Definition	Example organism	Habitats
Temperature	Obligately psychrophilic	No growth >15 °C	*Colwellia demingi*	Polar sediments, sea ice
	Psychrophilic	Optimum growth <20 °C	*Moritella marina*	Sediments, sea ice
	Psychrotolerant	Growth to <0 °C; optimum growth 20–40 °C	*Halomonas neptunia*	Deep and surface seawater
	Mesophilic	Optimum growth 20–45 °C	*Pseudomonas putida*	Surface seawater
	Thermophilic	Optimum growth 45–80 °C	*Thermus aquaticus*	Geothermal sites
	Hyperthermophilic	Optimum growth >80 °C	*Pyrolobus fumarii*	Hydrothermal vents
Hydrostatic pressure	Piezotolerant	Optimum growth at 0.1 MPa; (pressure decreases growth rate)	*Halomonas pacificus*	Water column (marine)
	Piezophilic	Optimum growth >0.1 MPa; growth at 0.1 MPa	*Photobacterium profundum*	Marine sediments
	Obligately piezophilic	Optimum growth >0.1 Mpa; no growth at 0.1 MPa	*Colwellia hadaliensis*	Deep-sea trench
pH	Acidophile	Optimal growth at pH < 2–3	*Picophilus torridus*	Solfataric fields
	Alkaliphile	Optimal growth at pH > 9–10	*Natronobacterium sp.*	Soda lakes
Salinity	Nonhalophilic	Optimum growth rate 0% salt; growth rate decreases with increasing salinity	*Escherichia coli*	Human gut
	Slightly halophilic	Optimum growth rate 0–2% salt	*Vibrio cholerae*	Estuaries
	Halophilic	Optimum growth rate at seawater salinity	*Vibrio splindidus*	Seawater
	Moderately halophilic	Optimum growth rate 5–10% salt	*Halomonas elongata*	Hypersaline lagoon
	Extremely halophilic	Optimum growth >10–15% salt	*Halobacterium salinarum*	Dead Sea
	Euryhaline	Growth over a wide salt range	*Halomonas elongata*	Solar saltern
Metals	Metal-sensitive	Sensitive to μM metal conc.	*Escherichia coli*	Human gut
	Metal-resistant	Growth at metal concentrations above those which *E. coli* can tolerate	*Ralstonia metallidurans*	Polluted sediment
Low water activity A_w	Xerophile	Any organisms that can grow with low water content; includes yeasts, molds, and extremely halophilic bacteria and Archaea	*Zygosaccharomyces rouxi* (yeast); *Halobacterium salinarum* (Archaea)	Dried fruits Dead Sea
Radiation	Radiotolerant	Survive > 5,000 Gray (gamma rays)	*Deinococcus radiodurans*	Soil

Source: modified from Kaye (2003) and Holland and Baross (2003).

TABLE 14.2 Limits to microbial growth and survival at the extreme environmental conditions on Earth

Parameter	Most extreme value		Environments
	for growth	for tolerance	
High temperature	121 °C; strain 121 113 °C; *Pyrolobus fumarii*	>120–200 °C in hydrothermal sulfides; brief exposure to 300 °C fluids	Submarine hydrothermal systems; geothermal hot springs (e.g., Yellowstone)
Low temperature	<−12 °C; pure cultures ~−15 °C; growth detected (see Chapter 15)	~−20 °C; respiration/ protein synthesis in sea ice brine channels; preservation by freezing at −196 °C	Brine pockets in sea ice at ~−30 °C
Acid pH	pH 0; acidophilic Archaea, e.g. *Picrophilus* sp. & *Ferroplasma* sp.	not definitively explored	acid mine drainge, solfataric sites
Alkaline pH	pH 13; *Plectonema* pH 10.5; *Natrobacterium*	not definitively explored	soda lakes, serpentinization associated systems (e.g., Lost City Field, mud volcanoes)
High pressure (hydrostatic)	>102 MPa; activity at 1.06 GPa	survival at 1.6 GPa	11,000 m water depth; Challenger Deep, Marianas Trench
Low water activity	$a_w = 0.75$; ~35% NaCl halophilic Archaea and Eubacteria $a_w = 0.62$; yeast	dessication; preservation by freeze-drying	soda lakes, evaporite ponds, deep-sea brines, dry soils/ rocks, foods with high solute content (e.g., jams, honey, dried fruit)
Ionizing radiation	60 Gray h^{-1} *Deinococcus radiodurans*	15,000 Gray gamma-ray irradiation; *Deinococcus radiodurans* & *Chroococcidiopsis* (cyanobacterium) (see Table 14.3)	nuclear waste (no natural sources at this level)
Solar radiation (ultraviolet, visible)	phytoplankton cyanobacteria	100 J m^{-2} *Deinococcus radiodurans*	high altitudes low latitudes
Toxic heavy metals	millimolar concentrations of As, Cd, Zn, Ni, and Co tolerated by many species of Archaea and Eubacteria	not definitively explored	submarine hydrothermal vents; acid mine drainage

Sources: Holland and Baross (2003); Rothschild and Mancinelli (2001).

structures in extremophiles are adapted to function under such conditions. In contrast, organisms capable of growing in extremes of pH, salinity, and irradiation, and in the presence of high levels of organic solvents and toxic metals are adapted to either (a) maintain intracellular conditions that are typical for non-extremophiles, or (b) compensate for the extreme condition. There are some exceptions. While most acidophiles ("low pH-loving" microbes) maintain an internal pH near neutrality, a recently described acidophilic archaeon, *Picrophilus torridus*, grows optimally at pH 0.7 (equivalent to 1.2 M sulfuric acid) and maintains an intracellular pH value of 4.6 (Ciaramella *et al.*, 2004). Among extremely halophilic ("salt-loving") Archaea, some have an absolute requirement for salt and grow best at salt concentrations of 3.5 to 4.5 M, but can also grow in saturated NaCl (5.2 M); their intracellular functional and structural components are adapted to high salt concentration (up to 5 M, mainly KCl) and their enzymes require high salt to maintain their active structure. On the other hand, there are moderately halophilic bacteria that grow over a wide range of salt concentrations that maintain internal salt levels typical of non-halophilic organisms (∼0.85% NaCl).

There are combinations of extreme conditions that apparently prevent cells from growing. For example, so far, no organisms have been characterized that are capable of growing in high salt concentrations at the upper and lower limits of temperature and pH (Figure 14.1). It is not known whether this combination effect is due to an insurmountable barrier posed by these combinations of extreme conditions, insufficient sampling, or lack of assessable habitats with these combinations of extremes. For example, a marine environment having the combination of high temperature (up to 90 °C) and high pH (up to 11) was only recently discovered and is teeming with microorganisms (see Kelley *et al.*, 2005). Some other combinations known to exist in natural environments, such as high pressure, high salt, and low temperature, have rarely been studied (Kaye and Baross, 2004). There are also combinations of extreme conditions that have a synergistic effect on the growth or survival of cells not adapted to either of the specific extreme conditions. This is the case for hydrostatic pressure and temperature, or salt and temperature. Low temperature and high hydrostatic pressures affect cell processes in the same way, with the result that the minimum growth temperature of non-piezophilic microbes is increased with increasing pressure. Similarly, high salt concentrations increase the minimum or decrease the maximum growth temperatures of non-halophilic microbes.

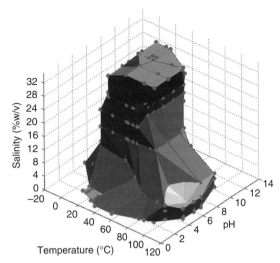

FIGURE 14.1 The growth range of cultured microorganisms as a function of temperature, salinity, and pH. The "mountain-peak" shape indicates an absence of known organisms capable of growing in high salt concentrations at the upper and lower limits of both temperature and pH. (Courtesy Julian Wimpenny; presented at UK Astrobiology Forum meeting, Girton College, Cambridge, March 2003.)

Organisms have also evolved a wide range of physiological and metabolic strategies to grow in environments where nutrients are limited. Carbon and energy sources are extremely scarce (*oligotrophic*) in most of the open ocean, and particularly in deep waters and deep sediments. Yet, active microorganisms are present although their growth rates may be extremely slow. Recently, *Pelagibacter ubique*, a representative of one of the most cosmopolitan microorganisms in oligotrophic oceans, was found to grow only in its *in situ* micromolar concentrations of organic carbon. *P. ubique* has one of the smallest genomes, yet it includes all of the essential genes to exist without help from other organisms) (Giovannoni *et al.*, 2005). *P. ubique* and related marine oligotrophs may help set the limits on concentration of organic compounds that can support the growth of heterotrophs and thus could serve as model organisms for designing strategies to detect similar organisms in Europa's ocean or in Antarctica's Lake Vostok.

14.4 Water, desiccation, and life in non-aqueous solvents

The absence of available water and extremes of temperature are the only single variables known to prevent growth and survival of organisms. The other physical and chemical factors that are thought of as extreme conditions, such as pH, pressure, damaging radiation, and toxic metals, are life-prohibiting factors for most organisms but not for all. Life has adapted to the entire terrestrial ranges of these variables (Table 14.2). However, there are some *combinations* of physical and chemical conditions for which no known organisms have been found to grow (Fig. 14.1). These include environments that have both high salt (>30% NaCl by weight) and low temperatures (<0 °C), such as in sea-ice inclusions, and high salt (>30% NaCl) and high temperatures (>90 °C), known to exist in brine pools beneath the Red Sea and the Mediterranean Sea. Is life not capable of growing under these combinations of stressors or have we not looked hard enough in the appropriate environments? If the last 30 years of extremophile research are any indication of future discoveries, then it is highly probable that novel organisms will be discovered that extend the limits of life to include combinations of stressors.

Earth-life can be described as a web of aqueous (water-based) chemical reactions. There is a point where the intracellular water activity decreases so far that most cells will die. Desiccation causes DNA to break, lipids to undergo permanent phase changes, and proteins to crystallize, denature, and undergo condensation reactions (Billi and Potts, 2002; Potts, 1994). Saturated brine pools (35% salt, water activity 0.75) are environments of low water activity that are inhabited by bacteria, Archaea, eukaryotic algae, and brine shrimp. Organisms that grow or survive in dry environments or in solutions with low water activity such as brines and syrups match their internal water activity with that of their surroundings. This is accomplished by accumulating compatible solutes which can be either ions such as K^+, or low molecular weight soluble organic compounds that do not interfere with the normal physiological functions of the cell (Müller *et al.*, 2005). The obligately halophilic Archaea accumulate K^+ and Cl^+ ions in their cytoplasm to a concentration similar to the surrounding environment. This strategy required evolutionary adaptations of intracellular macromolecules, and of metabolic and biosynthetic processes to operate despite high salt concentrations. Most other microorganisms and eukaryotes deal with desiccation by accumulating compatible organic compounds that include organophosphate compounds, β-amino acids, and glycerol. Some organisms survive desiccation by forming spores or cysts, while others such as the bacterium *Deinococcus radiodurans* have mechanisms to repair any damage to their DNA caused by desiccation. Both bacteria and eukaryotes have been found to grow in Antarctic rocks that have liquid water for only short thaw periods. So far, nothing is known about the mechanisms that are used by these organisms to survive long periods of desiccation at cold temperatures.

An important issue regarding water and astrobiology is the degree to which water is required by carbon-based life, and whether or not an organic solvent could replace water as the primary solvent (see Section 27.3). Another issue is the ability of organisms to survive environmental conditions that are outside their limits for growth. These usually involve decreasing the internal water content of the cell, as in the case of bacterial and fungal spores and some animals such as tartigrades (water bears). These issues are important in assessing whether or not carbon-based life could exist in liquid methane or ethane pools on Titan, survive the harsh physical conditions that would be encountered during transport from one planet to another, or survive long periods in a completely desiccated state and still retain the ability to grow if water is eventually introduced.

Is it possible for carbon-based life to exist in solvents other than liquid water? Many organic solvents, including alcohols, phenols, and toluene, are extremely toxic to microorganisms. The degree of antimicrobial action of a solvent depends on its hydrophobicity. The more hydrophobic a solvent, the more readily it can accumulate in cellular membranes. The toxicity of the solvents to cells is due to their ability to make the membrane permeable, resulting in the leakage of macromolecules including RNA and proteins (Isken and de Bont, 1998). While organic solvents kill most microorganisms, there are some bacteria that can tolerate relatively high concentrations. Two mechanisms have been identified for solvent tolerance: membranes that limit the diffusion of solvents into the cell, and specialized mechanisms that remove any solvents that have diffused into the cell. Another key issue is whether the carbon-based biochemistry of life as we know it can function in organic solvents. Certainly, many enzymes function in organic solvents and many organic reactions fundamental to biochemistry can occur in non-aqueous solvents (Benner *et al.*, 2004). However, even those enzymes active in organic solvents still have some bound water necessary to maintain their active structure. Moreover, water is important in other vital biochemical reactions during metabolism and biosynthesis; water is also a product of metabolic reactions. It appears that carbon-based life is unlikely to be able to adapt to a pure solvent milieu unless (a) it has mechanisms to form water from solvents such as alcohols, or

(b) it can produce all the necessary water *de novo* from biochemical reactions.

Are these mechanisms possible and is there evidence that bacteria could grow in some organic solvent with or without low levels of water? Until recently, it has been generally believed that external water passively diffuses through cell membranes. Recent work, however, has shown that this is not necessarily true and that there exist specific channels in bacterial membranes that function as pores for transport of water. Named *aquaporins*, these channels can facilitate rapid water transport during osmotic stress (Stroud *et al.*, 2003). The presence of specific water-transport pores makes it imaginable that there could be organisms with membranes that are resistant to organic solvents and yet with their specialized aquaporins extract and selectively transport low concentrations of water from the solvent/water mixture. Moreover, up to 70% of the intracellular water in actively growing *Escherichia coli* cells is generated by metabolic processes and not derived from the external environment (Kreuzer-Martin *et al.*, 2005). Is it possible for organisms to grow if their intracellular water is exclusively generated *de novo*? There are two obvious problems that a cell would have to overcome. If the cell is a heterotroph and requires transport of organic nutrients from an organic solvent milieu, it will require a membrane that can differentially transport the organic nutrients and prevent solvent toxicity. On the other hand, if the organism is a methanogen or some other chemolithoautotroph that requires carbon and energy sources that are gases, then only a passive diffusion mechanism through a solvent resistant membrane would be necessary. Do such organisms exist or can carbon-based organisms be engineered to grow under these conditions? It would be worthwhile to address this question in the context of life in organic ponds on Titan, or life in a gaseous milieu of CO_2, H_2, N_2, etc.

14.5 Temperature extremes

Temperature is a fundamental thermodynamic parameter that affects all biochemical reactions, in particular setting limits on life because temperature and pressure determine whether H_2O is in the liquid phase (Fig. 15.3). Given that liquid water exists in the liquid state, however, what then is the allowable range of temperatures for life? Microorganisms have been cultured with growth observed at temperatures as high as 121 °C (Kashefi and Lovley, 2003) or as low as -15 °C

(Chapter 15 and Fig. 15.1). There is even evidence for intact microorganisms with DNA and RNA in hydrothermal vent sulfides at temperatures exceeding 200 °C (Schrenk *et al.*, 2003). Eukaryotic organisms, however, are not known to live above 60 °C. Despite this wide gap in maximum growth temperatures between prokaryotic and eukaryotic cells, eukaryotes share all other extremes of life with bacteria and Archaea, including growth in environments of great acidity, high salt concentration, high pressure, and high concentrations of toxic metals.

Freezing temperatures can kill cells if internal ice crystals are formed. Cells survive freezing if quickly frozen – viable cells are even preserved for long periods if quickly frozen in liquid nitrogen (-196 °C). On the other hand, slow freezing or slow thawing of cells favors ice crystal formation that can damage macromolecules and structural polymers and lead to death. Some cellular adaptations for preventing internal ice crystal formation during slow freezing include increasing intracellular solute concentrations, similar to low water activity adaptation, production of exopolysaccharides, and modification of lipids and proteins to increase the fluidity of membranes and mobility of enzymes.

Microorganisms that grow best at temperatures above 80 °C are called *hyperthermophiles*. The maximum growth temperature of cultured hyperthermophiles varies from 80 to 121 °C, and the minimum growth temperature varies from ~40 °C to ~80 °C, depending on the organism (Holden and Daniel, 2004; Kashefi and Lovley, 2003; Table 14.2). Hyperthermophiles include bacteria and Archaea, aerobes and anaerobes, and heterotrophs and autotrophs. Some acidophiles, alkalophiles ("high pH-loving" microbes), and radiation-resistant organisms are also hyperthermophiles.

Hyperthermophiles have protein and lipid structures that are adapted to high temperature. While there are no generalizations that can be made about how enzymes and other proteins are thermally stable, there are some recurrent characteristics (Charlier and Droogmans, 2005). Protein structures are stabilized at high temperature through amino acid substitutions and most importantly through increased use of disulfide bonds for structural stabilization (Beeby *et al.*, 2005). Heat-stable, ether-linked lipids are universal in hyperthermophilic Archaea and in some hyperthermophilic bacteria. Fundamental changes in protein and lipid structure compensate for the increased mobility and fluidity at high temperatures. All hyperthermophiles studied have a reverse gyrase that positively supercoils DNA, which along with cationic proteins increases the thermal stability of the DNA (Daniel *et al.*, 2004).

Is 121 °C the highest possible temperature for growth of life? The chemical properties of water and biological macromolecules are affected by high temperature. Theoretically, as temperatures increase, it should be possible to engineer biological macromolecules to maintain their three-dimensional structure by compensating temperature effects with either higher pressure or with increasing salt concentrations. This appears to be the case with hyperthermophilic proteins (Hei and Clark, 1994; Sterner and Liebl, 2001; Summit *et al.*, 1998). Important cofactors such as adenosine triphosphate (ATP) and nicotine-adenine dinucleotide (NAD) are thermally labile (subject to chemical modifications in order to cope with temperature stress). Hyperthermophiles stabilize ATP and NAD by increasing their turnover rates (rate of synthesis), by extrinsic factors such as high ion concentrations, or by using intrinsically more stable replacements (Daniel *et al.*, 2004). There is also evidence that attachment and biofilm formation on minerals increases the thermal stability of hyperthemophilic Archaea (Schrenk, 2005). The labilities of proteins, DNA, and cofactors do not preclude life at temperatures higher than now known – the upper temperature for life is still to be determined.

At the other extreme, is there a *low*-temperature limit for life? Microbial activity has been measured at −20 °C in ice (Chapter 15), and photosynthesis has been observed in Antarctic cryptoendolithic ("hidden within rock") lichens at −20 °C (Friedmann and Sun, 2005). Water can remain liquid at temperatures lower than −30° in the presence of salts or other solutes and at even lower temperatures in combination with soluble organic solvents (Chapter 15). Enzyme activity has been measured at −100 °C in a mixture of methanol, ethylene glycol, and water (Bragger *et al.*, 2000). There is also evidence for the transfer of electrons and enzyme activity at −80 °C in a marine psychrophilic bacterium (Junge *et al.*, 2006). The process of vitrification (liquid water moving directly into a glassy state without ice-crystal formation), facilitated by the presence of salts and exopolymers (insoluble high molecular weight organic compounds, usually carbohydrates) in the starting solution, may be essential to such activity (Chapter 15). It is even possible that there may be no lower temperature limit for enzyme activity, nor even for cell growth, if a suitable solvent or solvent mixture is available (Price and Sowers, 2004).

14.6 Survival while traveling through space

The ability of microorganisms to survive the most extreme environmental conditions, such as high and low temperature, desiccation, radiation, and extremes in pressures and pH, increases the probability that they could survive transport to other planets and moons and thereby effect *panspermia*, the transfer of life from one world to another (Section 18.4.4). Panspermia is an important issue in the context of the "limits" of life since the presence of life in any extraterrestrial setting requires an origin – *de novo* or via panspermia. Two groups of microorganisms have received most of the attention as possible successful space travelers. These are the spore-forming bacteria and the radiation-resistant microorganisms. Spore formation by bacteria and fungi is usually in response to stress conditions including limiting nutrients, drying, and heat-shock. Spores are capable of long-term survival and have been recovered from environmental samples that are more than a million years old. There is even a report of viable spores from the gut contents of a bee entombed in amber estimated to be more than 25 million years old (Cano and Borucki, 1995). Spores are also known to remain stable against heat and radiation; this combined with their small size allows them to travel on the winds to distant locations, or slowly to sediment down into deep ocean trenches. Some spores can survive more than an hour of exposure to dry heat at 150 °C, although survival is greatly reduced in moist heat (Nicholson *et al.*, 2000). Spores from thermophilic bacteria are more resistant to heat than mesophilic spores. Similarly, spores are significantly more resistant to ultraviolet radiation than cells that are actively growing (Nicholson *et al.*, 2000).

Any possibility that microorganisms can be transported from one planet to another necessitates that they have the ability to resist the lethal effects of radiation. High-energy electromagnetic radiation (ultraviolet, X-ray, and gamma) and high-energy alpha and beta particles damage DNA, resulting in cytotoxic and mutagenic effects to the cell. Ultraviolet radiation is the most abundant form of damaging radiation and probably the most common natural mutagen. Ionizing radiation kills cells in general by causing multiple breaks in double-stranded DNA, although ultraviolet light can also kill cells as a result of dimerization of thymidine residues in the DNA, which prevents replication. Most organisms protect themselves from damaging radiation with

TABLE 14.3 Bacteria and Archaea resistant to ionizing radiation

Species	Phylum	Representative value[*] (kGy = 10^3 gray)
Rubrobacter radiotolerans	Actinobacteria	11
Deinococcus radiodurans R1	Deinococcus-Thermus	10
Thermococcus gammatolerans[1]	Euryarchaeota (Archaea)	8
Rubrobacter xylanophilus	Actinobacteria	5.5
Chroococcidiopsis species	Cyanobacteria	4
Hymanobacter actinoscierus	Flexibacter-Cytophaga-Bacteroides	3.5
Kineococcus radiotolerans	Actinobacteria	2
Acinetobacter radioresistens	γ-Proteobacteria	2
Kocuria rosea	Actinobacteria	2
Methylobacterium radiotolerans	α-Proteobacteria	1

*D_{10} is the dose of ionizing radiation that kills 90% of the population. The D_{10} values listed are for actively growing cells. (Modified from Cox and Battista, 2005.)
[1] Jolivet *et al.* (2004).

measures such as radiation absorbing pigments and DNA repair mechanisms.

The most radiation-resistant microorganisms include both bacteria and Archaea (Table 14.3). *Deinococcus radiodurans* is one of the most studied of these organisms and can survive radiation levels up to 10 kGy.[5] This level of radiation is much higher than is naturally found anywhere on Earth. Radiation resistance in *D. radiodurans* is the result of extremely efficient DNA repair mechanisms that with high fidelity can reassemble DNA that has been sheared into multiple double-stranded pieces (Cox and Battista, 2005). It is believed that these DNA repair mechanisms evolved in response to DNA damage due to desiccation rather than radiation. Recently, a hyperthermophilic archaeon, from a submarine hydrothermal vent environment, was found to be resistant to 8 kGy (Table 14.3) (Jolivet *et al.*, 2004).[6]

Besides being desiccated and irradiated, microorganisms traveling in space will also be exposed to the interplanetary vacuum of order 10^{-14} Pa (Nicholson *et al.*, 2000). Exposure to this level of vacuum causes extreme dehydration of cells, such that naked spores can survive for only days. Survival of spores is increased if they are associated with various chemicals such as sugars, or are embedded in salt crystals. Nicholson *et al.* (2000) discuss the various stresses that a microbial cell or spore would have to endure in order to survive interplanetary travel. These include the ability to survive the process that transports them out of the Earth's atmosphere, such as volcanic eruptions and bolide impacts, long periods of transit in the cold of space, and the entry into a new planetary home. Spores have been demonstrated to survive the shock conditions of a meteorite impact and the ultraviolet and cold temperature in space (Horneck *et al.*, 2002). It is clear that panspermia is possible and even probable if bacterial spores become embedded in rocks that get ejected from one planet and eventually enter the atmosphere of another.

While bacterial spores have received most of the attention for panspermia they are not the only good candidates for long-term survival in space. There is evidence that microorganisms that are attached to surfaces such as minerals and organic polymers have enhanced survival to a variety of stress conditions. These *biofilms* are physically and physiologically complex (Section 14.7; Fig. 14.2c). The niche within a mature biofilm allows both physiological and genetic transformations of the encapsulated organisms (Wimpenny, 2000). The immediate environment of cells within a biofilm can affect the delivery of nutrients and the availability of energy sources, leading to a range of activities, even amongst monoclonal

[5] kGy = kilogray, where a *gray* is a unit of absorbed dose of ionizing radiation corresponding to the absorption of 1 J per kg of absorbing material. 1 gray = 100 rads. On Earth a typical mid-latitude, sea-level natural background level is ~0.3 mGy per year, whereas on the surface of Europa the level is almost 10^{10} times higher, enough to kill humans in one minute of exposure (Section 19.4).

[6] The German cockroach (*Blattella germanica*) is the most radiation resistant metazoan; the tardigrade (*Milnesium tardigradum*) in its desiccation-resistant "tun" state can also survive extremely high X-ray exposure.

A. Physiological Effects

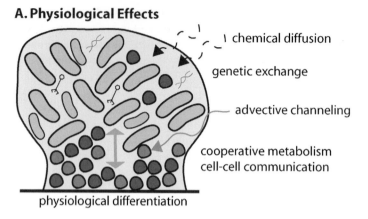

chemical diffusion

genetic exchange

advective channeling

cooperative metabolism
cell-cell communication

physiological differentiation

B. Geochemical Effects

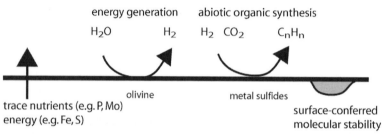

energy generation   abiotic organic synthesis

H_2O H_2 H_2 CO_2 C_nH_n

olivine   metal sulfides

trace nutrients (e.g. P, Mo)
energy (e.g. Fe, S)

surface-conferred
molecular stability

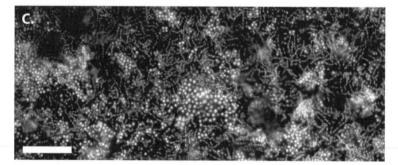

C.

FIGURE 14.2 (a) Physiological and geochemical effects of microbial growth on surfaces. The development of complex microbial biofilms on surfaces can have physical, chemical, and biological consequences upon the local environment. Biofilm structures can mitigate transport of energy, nutrients, and stress factors. Byproducts of these activities include cooperative behavior amongst microbial consortia and/or differentiation into various physiological states. The close association of cells within biofilms can also influence genetic change between organisms. (b) Geochemical effects pertinent to microbial growth at mineral surfaces include the presence of trace nutrients or energy in the solid phase, generation of redox-active compounds, abiotic synthesis of organic molecules, or stabilization of bio-active compounds on surfaces. (c) Photomicrograph of a complex microbial biofilm formed on a pyrite surface at a deep-sea hydrothermal vent. A DNA-specific stain was used so that individual cells and chains of cells in the biofilm could be observed and enumerated by epifluorescence microscopy. The biofilm is both morphologically diverse and physiologically complex, extending in three dimensions. Scale bar is 20 μm.

populations. Heterogeneity within biofilms has also been shown to play a role in resistance to environmental stress. Stress resistance has commonly been attributed to two features of the biofilm microbial communities: their ability to limit rates of diffusion, and the presence of multiple physiological states, including "persister" cells with recalcitrant physiologies. Explanations for enhanced resistance of biofilms to antimicrobial compounds and toxic metals are primarily based on diffusion effects (Wimpenny, 2000). However, these physical phenomena are commonly coupled to physiological changes such as the diversity of growth stages, including slow-growing, stationary phase cells – these too may contribute to the observed resilience of attached microbial communities to environmental stress. Studies of gene expression within attached communities have demonstrated a remarkable similarity between biofilm-regulated genes and generalized stress responses. Perhaps adaptation to the biofilm milieu predisposes a fraction of the surface-associated cells to tolerate environmental change. Could such biofilms show greater survival than spores during space travel to another planet?

14.7 Extremophiles, hydrothermal vents, and Earth's earliest microbes

There has been much research on possibilities for the earliest ecosystems and the metabolic groups of organisms that could exploit available carbon and energy sources. Models are based on geological, geochemical, and isotopic evidence of biosignatures in the earliest rocks, inferences from global phylogenetic trees, and the biochemical characteristics of extant organisms. The deep-sea and the sub-seafloor are two of the modeled sites for the earliest microbial ecosystems. These are safe havens from killer impact events (which were much more common in Earth's early history – Chapter 3), as well as sites of ubiquitous and active geophysical processes, including hydrothermal activity that can generate carbon and chemical energy sources, along with other elements required for life. There are models that support the hypothesis that hydrothermal systems were involved in the origin of life and that microorganisms related to extant hydrothermal vent extremophiles were the earliest organisms (Baross and Hoffman, 1985; Holm and Andersson, 2005; Nisbet and Sleep, 2001; Russell and Martin, 2004). The implication is that submarine hydrothermal systems can both generate and support life without the need for oxidants produced by photosynthesis. While these remain hypotheses, there is convincing evidence that microorganisms capable of exploiting the chemical energy sources at vents, particularly H_2, preceded photosynthesizing microorganisms (Nisbet and Sleep, 2001). Biochemical evidence indicates that the first photosynthetic organisms were anaerobic and coupled the reduction of CO_2 with the oxidation of H_2S rather than H_2O.

To date on Earth we know of two primary types of hydrothermal processes that drive warm- to high-temperature fluid flow in the sub-seafloor. The first of these is the most well studied: ridge-axis, hydrothermal circulation driven by upwelling magma or recently solidified hot rock (Fig. 14.3a), often producing spectacular hydrothermal chimneys or "black smokers." Water–rock reactions commensurate with high-temperature-driven fluid flow result in fluid temperatures of up to 407 °C and produce distinct chemistries containing high concentrations of metals and low pH (Kelley et al., 2002). The second process, discovered only in 2000 at the Lost City field in the mid-Atlantic, is the exothermic serpentinization of ultramafic rocks, which yields moderate temperature fluids (<150 °C) in comparison to ridge-axis systems. These fluids are highly alkaline (up to pH 12), and contain elevated levels of volatile compounds (e.g., CH_4, H_2) relative to ridge-axis systems (Kelley et al., 2005). A third source of hydrothermal activity is tidal heating that occurs when differential gravitational forces from a planet "flex" one of its moons, usually in concert with gravitational effects from other moons. Tidal heating is

the best explanation for geysers on Enceladus and is also believed to be important in maintaining a liquid ocean on Europa (Section 19.3).

Hydrothermal fluids emanating from cracks and fractures in the seafloor have been investigated and used to infer biological and chemical properties of their source regions. While this approach has provided considerable insight, much of the sub-seafloor remains an enigma. The gradients between hot, highly reduced fluids and cold, oxygenated seawater also exist in a compressed fashion within the walls of hydrothermal chimneys found at the seafloor and provide a more direct means to access this environmental system (Fig. 14.3b). These environments are similar in many ways, but also differ with regard to the scales of the gradients (drastic in chimneys, expansive in the subsurface), and the residence time of fluids within pore spaces of the rocks.

Strong hydrothermally driven fluid circulation is a fundamental physical–chemical process that was probably present on Earth as soon as it had appreciable amounts of liquid water. In fact, hydrothermal activity on the early Earth may have been even more pervasive than today due to higher global heat fluxes (Baross, 1998). A hallmark feature of hydrothermal systems is dynamic mixing between fluids of varying compositions. These mixing processes occur at multiple spatial and temporal scales, ranging from the instantaneous process of sulfide mineral precipitation seen at black smokers, to the modest thermal and chemical gradients in the sub-seafloor. A vast majority of the mixing occurs within the fractures and pore spaces of rocks, where the fluids interact with and alter the host materials. Hydrothermal systems also experience periodic perturbations linked to the tides, and are intertwined with episodic processes such as earthquake activity, opening and sealing of fractures, and magma intrusion events.

As mentioned above, microbes are commonly found in a physiological form known as a biofilm (Fig. 14.2). In natural environments, a majority of any microbial population is associated with surfaces, whether inorganic, biological, or particulate. Biofilms may have a number of ecological roles, including consortial interactions, utilization of surface-bound energy and nutrient sources, and as an adaptation to environmental stress. This is especially true at the deep-sea vents, where fluid flow maintains favorable habitats by providing distinct thermal–chemical niches (Schrenk et al., 2003). As the vents are in continual flux, the biofilms may help with transient exposure

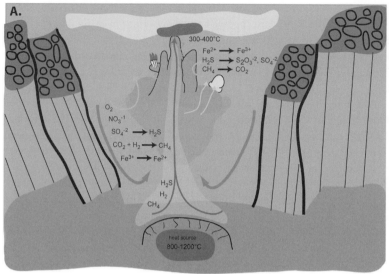

FIGURE 14.3 (a) Schematic of the environmental setting for high-temperature hydrothermal venting for sulfide chimneys along a mid-ocean ridge axis. Crustal magma chambers typically form at depths of 1–4 km below the seafloor. Figure not to scale. (b) Generalized mineralogical and thermal-chemical gradients present within the walls of mature hydrothermal chimney structures. The widths of sulfide chimneys vary from centimeters to meters and the gradients are dependent on degree of seawater intrusion, presence of cracks, and the shapes of the hot fluid flow channels.

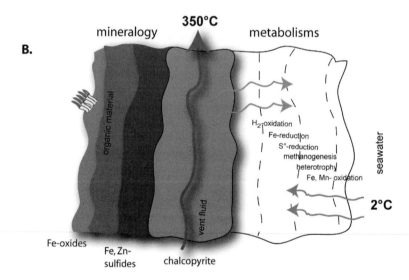

to oxygen, storage of nutrients in oligotrophic circumstances, and resistance to other stresses. Biofilms are also important for microbial communities that live on rocks in dry situations.

The surface-associated microbiology of present-day hydrothermal vent environments may also be relevant to the origin and early evolution of Earth-life. Abiotic reactions facilitated by mineral surfaces catalyze a number of reactions that may have substituted for early metabolic processes (Chapter 8). This idea has been expanded to where mineral catalysis could have preceded all enzyme catalysis during the inferred early, non-protein stage in the origin of life. The ubiquity and importance of metal–sulfur cluster proteins in all

Earth-life supports this idea. Mineral surfaces also selectively absorb a number of organic molecules, including nucleic acids, and under-hydrating and dehydrating conditions lead to polymer formation. Furthermore, the aggregated communities found on mineral surfaces may have been important to the early evolution of life in that their diversity could have facilitated a high rate of genetic exchange and overcome the evolutionary limitations of any single lineage. Hydrothermal systems of the present-day Earth have not changed significantly from those present millions of years ago. Such systems may harbor relict microorganisms and biochemistries that can aid in our understanding of the origin of Earth-life.

14.8 Summary and future directions

Carbon-based life is extremely versatile on Earth and has adapted to grow in niches representing multifarious physical and chemical conditions. On Earth, only high temperatures and low water activities limit the growth of organisms. Even in these growth-limiting conditions, some organisms have adapted strategies for long-term survival. Moreover, the ability to grow in the most extreme conditions on Earth is not restricted to bacteria and Archaea. There are examples of eukaryotes that can grow at most of the extreme conditions, with the exception of high temperatures.

There is still much to be understood about the limits of carbon-based life. There are extreme terrestrial environments that have not been adequately explored for novel microbes or for evidence of novel microbial interactions. These include the hot sub-seafloor crust, hot brine sediments and crustal aquifers, and rock-hosted microbial communities, particularly those affected by hydrothermal activity or subjected to changing environmental extremes such as hydrating and dehydrating conditions. Even within relatively well-studied extreme environments, very little is known about the microbial inhabitants other than their identity based on molecular signatures. There is also a need to understand how different physiological characteristics of organisms, such as biofilm formation and production of exopolysaccharides, affect their growth and survival in extreme environments. It is also probable that the limits of extant Earth-life might not be indicative of limits possible for organisms with Earth-life biochemistry. Life occupies most known habitats and in some cases can tolerate extreme conditions such as irradiation and hydrostatic pressure that exceed levels found in natural environments. What is the maximum growth temperature for life – are there extrinsic or intrinsic factors that facilitate growth above 120 °C? Above 200 °C? Can organisms adapt to grow in much lower levels of available water or higher concentrations of organic solvents than are currently accepted limits? The ability of cells to grow in high concentrations of organic solvents mixed with low levels of water would also change significantly our perception of the lower and upper temperature limits for life.

In the absence of a visible-light driven ecosystem, other planets and moons would have to have a source of chemical energy and other inorganic nutrients to maintain a microbial ecosystem. The presence of hydrothermal systems would not only provide a constant chemical energy source, but would also provide carbon sources and other elements required by life as we know it. All source reactions for hydrothermal systems produce high temperatures and anoxic conditions. This underscores the need to better understand microbial communities that grow at high temperatures on hydrothermal nutrients, and particularly the communities associated with sub-seafloor minerals. Similarly, if hydrothermal systems exist on Europa and other icy moons, or in the deep subsurface of Mars, than the resultant carbon and energy sources could reach into the cold ocean of Europa or into the Martian regolith. So far, little is known about anaerobic life in ice and particularly anaerobic psychrophilic microbial communities that use hydrogen as their primary energy source.

In summary, the two key issues that relate astrobiology to the limits for carbon-based life are the needs to (1) gain a better understanding of the actual limits of extant life by searching understudied extreme environments for novel organisms, and (2) explore experimentally the actual limits for carbon-based life, and whether or not these limits extend beyond conditions found on Earth, but not beyond those at other Solar System locations.

REFERENCES

Baross, J. A. (1998). Do the geological records of the early Earth support the prediction from global phylogenetic models of a thermophilic cenancestor? In *Thermophiles: the Keys to Molecular Evolution and the Origin of Life?* J. Wiegel and M. W. W. Adams (eds.), pp. 1–18. London: Taylor & Francis.

Baross, J. A. and Deming, J. W. (1995). Growth at high temperatures: isolation and taxonomy, physiology, and ecology. In *The Microbiology of Deep-Sea Hydrothermal Vents*, D. M. Karl (ed.), pp. 169–217. Boca Raton: CRC Press.

Baross, J. A. and Hoffman, S. (1985). Submarine hydrothermal vents and associated gradient environments as sites for the origin and evolution of life. *Origins of Life*, **15**, 327–45.

Bartlett, D. H. (2002). Pressure effects on in vivo microbial processes. *Biochemica et Biophysica Acta*, **1595**, 367–381.

Beatty, J. T., Overmann, J., Lince, M. T., *et al.* (2005). An obligately photosynthetic bacterial anaerobe from a deep-sea hydrothermal vent. *Proc. Natl. Acad. Sci. USA*, **102**, 9306–9310.

Beeby, M., O'Connor, B. D., Ryttersgaard, C., *et al.* (2005). The genomics of disulfide bonding and protein stabilization in thermophiles. *Biology Public Library of Science*, **3**, 1549–1558.

Benner, S. A., Ricardo, A., and Carrigan, M. A. (2004). Is there a common chemical model for life in the universe? *Curr. Opin. Chem. Biol.*, **8**, 672–689.

Billi, D. and Potts, M. (2002). Life and death of dried prokaryotes. *Res. Microbiol.*, **153**, 7–12.

Bragger, J. M., Dunn, R. V., and Daniel, R. M. (2000). Enzyme activity down to −100 °C. *Biochim. Biophys. Acta.*, **1480**, 278–282.

Cano, R. J. and Borucki, M. K. (1995). Revival and identification of bacterial spores in 25- to 40-million-year-old Dominican amber. *Science*, **268**, 1060–1064.

Charlier, D. and Droogmans, L. (2005). Microbial life at high temperatures, the challenges, the strategies. *Cell. Mol. Life Sci.*, **62**, 2974–2984.

Ciaramella, M., Napoli, A., and Rossi, M. (2004). Another extreme genome: how to live at pH 0. *Trends Microbiol.*, **13**, 49–51.

Cox, M. M. and Battista, J. R. (2005). *Deinococcus radiodurans* – the consummate survivor. *Nature Rev. Microbiol.*, **3**, 882–892.

Daniel, R. M., van Eckert, R., Holden, J. F., *et al.* (2004). The stability of biomolecules and the implications for life at high temperatures. In *The Sub-seafloor Biosphere at Mid-Ocean Ridges*, W. S. D. Wilcock, E. F. DeLong, D. S. Kelley, J. A. Baross, and S. C. Cary (eds.), Geophysical Monograph 144, pp. 25–39. Washington, DC: American Geophysical Union.

Friedmann, E. I. and Sun, H. J. (2005). Communities adjust their temperature optima by shifting producer-to-consumer ratio, shown in lichens as models: 1. Hypothesis. *Microbial Ecol.*, **49**, 523–527.

Giovannoni, S. J., Tripp, H. J., Givan, S., *et al.* (2005). Genome streamlining in a cosmopolitan oceanic bacterium. *Science*, **309**, 1242–1245.

Hei, D. J. and Clark, D. S. (1994). Pressure stabilization of proteins from extreme thermophiles. *Appl. Environ. Microbiol.*, **60**, 932–939.

Holden, J. F. and Daniel, R. M. (2004). The upper temperature for life based on hyperthermophile culture experiments and field observations. In *The Sub-seafloor Biosphere at Mid-Ocean Ridges*, W. S. D. Wilcock, E. F. DeLong, D. S. Kelley, J. A. Baross, and S. C. Cary (eds.), Geophysical Monograph 144, pp. 13–24. Washington, DC: American Geophysical Union.

Holland, M. and Baross, J. A. (2003). Limits of life in hydrothermal systems. In *Energy and Mass Transfer in Marine Hydrothermal Systems*, P. E. Halbach, V. Tunnicliffe, and J. Hein (eds.), pp. 235–250. Proc. 89th Dahlem Conference. Berlin: Springer-Verlag.

Holm, N. G. and Andersson, E. (2005). Hydrothermal simulation experiments as a tool for studies of the origin of life on Earth and other terrestrial planets: a review. *Astrobiology*, **5**, 444–460.

Horikoshi, K. (1999). Alkaliphiles: some applications of their products for biotechnology. *Microbiol. Mol. Biol. Rev.*, **63**, 735–750.

Horneck, G., Mileikowsky, C., Melosh, H. J., *et al.* (2002). Viable transfer of microorganisms in the solar system and beyond. In *Astrobiology: the Quest for the Conditions of Life*, eds. G. Horneck and C. Baumstark-Khan, pp. 57–76. Berlin: Springer-Verlag.

Huber, H., Hohn, M. J., Rachel, R., *et al.* (2002). A new phylum of Archaea represented by a nanosized hyperthermophilic symbiont. *Nature*, **417**, 63–67.

Hügler, M., Huber, H., Stetter, K. O., and Fuchs, G. (2003). Autotrophic CO_2 fixation pathways in archaea (Crenarchaeota). *Arch. Microbiol.*, **179**, 160–173.

Isken, S. and de Bont, J. A. M. (1998). Bacteria tolerant to organic solvents. *Extremophiles*, **2**, 229–238.

Jolivet, E., Corre, E., L'Haridon, S., *et al.* (2004). *Thermococcus marinus* sp. Nov. and *Thermococcus radiotolerans* sp. Nov., two hyperthermophilic archaea from deep-sea hydrothermal vents that resist ionizing radiation. *Extremophiles*, **8**, 219–227.

Junge, K, Eicken, H., Swanson, B. D., and Deming, J. W. (2006). Bacterial incorporation of leucine into protein down to −20 °C with evidence for potential activity in subeutectic saline ice formations. *Cryobiology*, **52**, 417–429.

Kashefi, K. and Lovley, D. R. (2003). Extending the upper temperature limit for life. *Science*, **301**, 934.

Kaye, J. Z. (2003). Ecology, phylogeny and physiological adaptations of euryhaline and moderately halophilic bacteria from deep-sea and hydrothermal-vent environments. Ph.D. thesis, University of Washington.

Kaye, J. Z. and Baross, J. A. (2004). Synchronous effects of temperature, pressure and salinity on growth, phospholipid profiles, and protein patterns of four *Halomonas* species isolated from deep-sea hydrothermal-vent and sea surface environments. *Appl. Environ. Microbiol.*, **70**, 6220–6229.

Kelley, D. S., Baross, J. A., and Delaney, J. R. (2002). Volcanoes, fluids, and life at mid-ocean ridge spreading centers. *Ann. Rev. Earth Planet. Sci.*, **30**, 385–491.

Kelley, D. S., Karson, J. A., Früh-Green, G. L., *et al.* (2005). A serpentinite-hosted ecosystem: the Lost City hydrothermal field. *Science*, **307**, 1428–1434.

Kreuzer-Martin, H. W., Ehleringer, J. R., and Hegg, E. L. (2005). Oxygen isotopes indicate most intracellular water in log-phase *Escherichia coli* is derived from metabolism. *Proc. Natl. Acad. Sci. USA*, **102**, 17337–17341.

López-García, P. (2005). Extremophiles. In *Lectures in Astrobiology*, Volume 1, M. Garguad, B. Barbier, H. Martin, and J. Reisse (eds.). Berlin: Springer-Verlag.

Müller, V., Spanheimer, R., and Santos, H. (2005). Stress response by solute accumulation in archaea. *Curr. Opin. Microbiol.*, **8**, 729–736.

Navarro-González, R., Rainey, F. A., Molina, P., *et al.* (2003). Mars-like soils in the Atacama Desert, Chile, and the dry limit of microbial life. *Science*, **302**, 1018–1021.

Nicholson, W. L., Munakata, N., Horneck, G. *et al.* (2000). Resistance of *Bacillus* endospores to extreme terrestrial and extraterrestrial environments. *Microbiol. Mol. Biol. Rev.*, **64**, 548–572.

Nisbet, E. G. and Sleep, N. H. (2001). The habitat and nature of early life. *Nature*, **409**, 1083–1091.

Porco, C. C., Helfebstein, P., Thomas, P. C., *et al.* (2006). Cassini observes the active south pole of Enceladus. *Science*, **311**, 1393–1401.

Potts, M. (1994). Desiccation tolerance of prokaryotes. *Microbiol. Rev.*, **58**, 755–805.

Price, B. and Sowers, T. (2004). Temperature dependence of metabolism rates for microbial growth, maintenance, and survival. *Proc. Natl. Acad. Sci. USA*, **101**, 4631–4636.

Rothschild, L. J. and Mancinelli, R. L. (2001). Life in extreme environments. *Nature*, **409**, 1092–1101.

Russell, M. J. and Martin, W. (2004). The rocky roots of the acetyl-CoA pathway. *Trends Biochem. Sci.*, **29**, 358–363.

Schrenk, M. O. (2005). Exploring the diversity and physiological significance of attached microorganisms in rock-hosted deep-sea hydrothermal environments. Ph.D. thesis, University of Washington.

Schrenk, M. O., Kelley, D. S., Delaney, J. R., and Baross, J. A. (2003). Incidence and diversity of microorganisms within the walls of an active deep-sea sulfide chimney. *Appl. Environ. Microbiol.*, **69**, 3580–3592.

Schrenk, M. O., Kelley, D. S., Bolton, S., and Baross, J. A. (2004). Low archaeal diversity linked to sub-seafloor geochemical processes at the Lost City Hydrothermal Field, Mid-Atlantic Ridge. *Environ. Microbiol.*, **6**, 1096–1095.

Sharma, A., Scott, J. H., Cody, G. D., Fogel, M. L., Hazen, R. M., Hemley, R. J., and Huntress, W. T. (2002). Microbial activity at gigapascal pressures. *Science*, **295**, 1514–1516.

Silver, S. and Phung, L. T. (1996). Bacterial heavy metal resistance: new surprises. *Ann. Rev. Microbiol.*, **50**, 753–789.

Smith, H. O., Hutchison III, C. A., Pfannkoch, C., and Venter, J. C. (2003). Generating a synthetic genome by whole genome assembly: ØX174 bacteriophage from synthetic oligonucleotides. *Proc. Natl. Acad. Sci. USA*, **100**, 15440–15445.

Sterner, R. and Liebl, W. (2001). Thermophilic adaption of proteins. *Crit. Rev. Biochem. Mol. Biol.*, **36**, 39–106.

Straube, W. L., Deming, J. W., Somerville, C. C., *et al.* (1990). Particulate DNA in smoker fluids: evidence for existence of microbial populations in hot hydrothermal systems. *Appl. Environ. Microbiol.*, **56**, 1440–1447.

Stroud, R. M., Miercke, L. J. W., O'Connell, J., *et al.* (2003). Glycerol facilitator GlpF and the associated aquaporin family of channels. *Curr. Opin. Struct. Biol.*, **13**, 424–431.

Summit, M., Scott, B., Nielson, K., Mathur, E., and Baross, J. A. (1998). Pressure enhances thermal stability of DNA polymerase from three thermophilic organisms. *Extremophiles*, **2**, 339–345.

Ventosa, A., Nieto, J. J., and Oren, A. (1998). Biology of moderately halophilic aerobic bacteria. *Microbiol. Mol. Biol. Rev.*, **62**, 504–544.

Wackett, L. P., Dodge, A. G., and Ellis, L. B. M. (2004). Microbial genomics and the periodic table. *Appl. Environ. Microbiol.*, **70**, 647–655.

Wiegel, J. and Kevbrin, V. V. (2003). Alkalithermophiles. *Biochem. Soc. Trans.*, **32**, 193–198.

Wimpenny, J. (2000). Heterogeneity in biofilms. *FEMS Microbiol. Rev.*, **24**, 661–671.

FURTHER READING

Bains, W. (2004). Many chemistries could be used to build living systems. *Astrobiology*, **4**, 137–167. Excellent review of the complex carbon- and silicon-based biochemistry that might be possible in non-aqueous solvents.

Gaidos, E., Deschienes, B., Dundon, K., *et al.* (2005). Beyond the principle of plenitude: a review of terrestrial planet habitability. *Astrobiology*, **5**, 100–126. Comprehensive review of the important factors responsible for the habitability of terrestrial rocky planets based on the Earth model.

Hochachka, P. W. and Somero, G. N. (2002). *Biochemical Adaptation: Mechanism and Process in Physiological Evolution.* New York: Oxford University Press. Much more than a description of animal physiology, this volume provides the details of how animals have adapted to extreme conditions of temperature, high salt, hydrostatic pressure, pH, low levels of oxygen, etc.

Russell, M. (2006). First life. *American Scientist*, **94**, 32–39. An excellent synthesis of Russell's ideas on how life could originate in alkaline hydrothermal vents, and particularly on the critical role that minerals would play in this scenario.

Schulze-Makuch, D. and Irwin, L. N. (2004). *Life in the Universe.* Berlin: Springer-Verlag. Excellent summary of life as we know it and as it might be in the different environmental conditions found in the Universe.

15 Life in ice

Jody W. Deming, *University of Washington, Seattle*
Hajo Eicken, *University of Alaska, Fairbanks*

White men think of ice as frozen water, but Inuit think of water as melted ice. To us, ice is the natural state.
Nuka Pinguaq, nineteenth-century native Arctic hunter (Kobalenko, 2002)

15.1 Introduction

Ice is the "natural state" or predominant form of water in our Solar System. The surfaces of most planets and moons are currently at temperatures well below the freezing point of pure water, including two of the more promising sites in the search for traces of extraterrestrial life, Mars and Europa. The amount of water-ice on Europa exceeds the volume of liquid water on Earth. Comets, considered potential vectors for precursors or early stages of life, are also icy bodies (Chapter 3). Even Earth may have undergone a series of complete (or near-complete) glaciations in its recent history, earning the title "Snowball Earth" (Section 4.2.4).

The presence of the liquid phase of water, however, is essential to the prospering of life as we know it. In order to study where liquid water can occur, we must understand scales ranging from the structure of the water molecule to the temperatures possible on a planetary surface or subsurface. In our Solar System, only Earth allows for a planetary *surface* with abundant liquid water (Chapter 4). Mars, however, may well have some liquid water in permafrost (perennially frozen soil) beneath its surface today (Section 18.4.1), and the evidence is strong that Europa and perhaps other moons of Jupiter have water oceans below their icy crusts (Chapter 19).

Dismissing the surfaces of other Solar System bodies as currently lifeless for lack of a significant body of liquid H_2O might seem reasonable if only planetary-scale parameters are considered. However, on the scale of micrometers relevant to microbial life, the "natural state of water" – ice – does not in fact automatically exclude the simultaneous presence of the liquid phase. Consider that natural ice is rarely

if ever formed from pure water. The impurities of ice on Earth – for example, the salts in ice derived from seawater or the impurities in permafrost – allow for the presence of liquid water as a significant fraction of the ice volume (Table 15.1) even though, on a larger bulk scale, the ice appears to be solid. Like virtually all natural waters on Earth, this remaining liquid within the ice is inhabited by microorganisms. The impurity effect allowing for liquid water holds true even at temperatures approaching the average surface temperature of Mars today ($-55\,°C$). Indeed, recent evidence suggests the potential existence of unfrozen water in subsurface permafrost on Mars (Sec. 18.4.1).

Water is bound in the solid phase on Earth in five major types of so-called *ice formations* (Table 15.1). Listed by global volume, they are the terrestrial ice sheets (mainly those covering Antarctica and Greenland), mountain glaciers and polar marine ice shelves, sea ice (encircling Antarctica and covering the Arctic Ocean and surrounding seas), and permafrost (as in Siberia and Alaska). Although seasonal snow covers the largest fraction of Earth's surface area compared to the other major ice formations (9% versus 0.1–5%), snow represents only a minor form of frozen water volumetrically ($< 0.007\%$). The most prevalent type of ice on Earth is the terrestrial ice sheet ("terrestrial" meaning "on land", as opposed to "marine"), which is formed from water vapor deposited onto snow crystals with limited impurities; such ice contains the lowest fraction by volume of liquid water (0.0001, a generous upper limit). Underscoring the importance of salt impurities for the presence of a significant liquid phase, the average fraction of liquid present in sea ice (0.08) greatly exceeds that of all other ice forms on Earth (Table 15.1). For perspective, the total amount

TABLE 15.1 Bodies of ice on Earth and unfrozen (liquid) water within them

Type of ice formation	Global surface area covered (%)	Volume of ice $(10^6 \, km^3)$	Fraction of liquid water within the ice	Volume of liquid water $(10^3 \, km^3)$
Ice sheets	3	30	0.0001	3
Mountain glaciers	0.1	0.1	0.001	0.1
Marine ice shelves	0.1	0.1	0.001	0.1
Sea ice	5	0.04	0.08	3
Permafrost[a]	3	0.025	0.01	0.25
Seasonal snow	9	0.002	0.01	0.02

Note:
[a] excluding sub-glacial permafrost in Antarctica and Greenland and sub-seabed permafrost under the Arctic Ocean (extent is not well known) (from Zhang et al., 2000).

of liquid water present at sub-freezing temperatures, as a result of salt and other impurities, within all ice formations on Earth exceeds the volume of freshwater flowing in all rivers. From the microbial perspective, even a fraction of a microliter of water contained within a block of ice represents a luxurious water world. Critical to that water being supportive of ongoing metabolism and growth, however, is the presence of connections between numerous liquid niches:[1] an open system allows for the essential exchange of nutrients and waste products by diffusion or *advection* (carrying of materials or molecules via (small-scale) currents). At the very lowest temperatures of ice on Earth, and thus for our best analogues to the frozen environments on Mars and Europa, salt or organic impurities are essential to the presence of any liquid water within the ice. Thermal gradients across an ice formation can even allow fluids to flow within the ice on a scale relevant to microorganisms.

In this chapter, we present the underlying physical and chemical reasons for the occurrence and characteristics of liquid water in ice. We then discuss emerging relationships between the amount of water available in various ice formations and the abundance and activity of microbial life forms found therein. Figure 15.1 presents the temperature ranges for the occurrence of liquid water on Earth and for laboratory-confirmed microbial growth and enzyme activity. We focus on microbial life (single-celled organisms, typical dimension of 1 μm or less) in the domains of Bacteria and Archaea (Chapter 10), here generically called bacteria (with lower case b). Bacteria, themselves composed of 70–90% water, must conduct all of their basic metabolic functions within an aqueous medium,

[1] A *niche* is a habitat providing the factors necessary for growth of a particular organism or species.

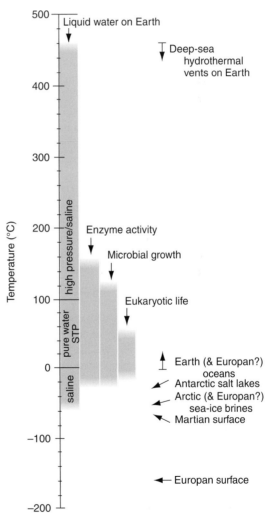

FIGURE 15.1 Liquid water and life on planets. Temperature scale for the presence of liquid water on Earth and for observed enzyme activity and growth of microorganisms (Bacteria and Archaea). Note that more complex organisms (Eukarya) occupy a more restrictive thermal range. Average surface temperatures on Mars ($-55\,°C$) and Europa ($-160\,°C$) are also shown.

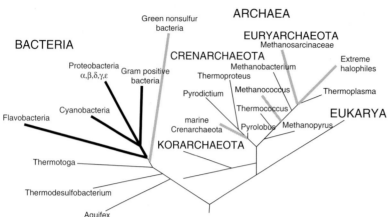

FIGURE 15.2 Universal phylogenetic Tree of Life based on 16 S rRNA sequences, showing the three domains of Bacteria, Archaea, and Eukarya. Originally featuring hyperthermophilic genera that grow at ≥90 °C (italics; Chapter 14), the figure (modified from Deming and Baross, 2002) now highlights branches containing cold-adapted species, whether psychrophilic (thick black lines) or psychrotolerant (thick gray lines). Obligate psychrophiles are known to have evolved only among the Bacteria, although the degree of cold-adaptation for marine Crenarchaeota remains unknown.

as must all of the enzymes that enable life (Madigan *et al.*, 1997). More complex organisms (Eukarya), whether single-celled or multicellular (*metazoa*), also live in some types of ice – most notably the photo-synthesizing algae that flourish along with their metazoan grazers during sunlit seasons in sufficiently clear sea ice (and sometimes also on the surface of snow formations and in rock crevices). Indeed, the first organisms living in a subzero environment to be recorded were later identified as photosynthetic Eukarya or "red algae" – Aristotle noted that "living animals are produced … in long-lying snow" that "gets reddish in color" (Barnes, 1984). Here we pay more attention to *heterotrophic* bacteria – those capable of deriving nutrition from pre-formed organic compounds and thus living independently of solar radiation.

Heterotrophy is an obvious focus, given that the dissolved impurities in the liquid contained in ice formations on Earth invariably include not only min-eral salts but also organic compounds. Metabolisms independent of light are also highly relevant to an astrobiological consideration of life in ice: on Earth, many of the liquid niches in ice are not sunlit; on Mars and Europa, a persistent liquid phase of water most likely exists only at subsurface depths in the dark. We also focus on heterotrophic bacteria because of their long history on Earth, their dominance globally in terms of abundance (Deming and Baross, 1993), their profound effects on geochemistry, and their occurrence in all natural ice formations examined to date.[2]

Finally, in the course of considering the physics, chemistry, and microbiology of Earth ice and its fluid inclusions, we have built a rationale for the search for microbial life in ice formations elsewhere in the Solar System. Of interest is the detection of accidental *tourists* in extraterrestrial ice – those microorganisms not native to their icy environments but preserved upon arrival by the extreme cold and possibly recoverable upon return to conditions permissive to their growth (see also Chapter 25). Of even greater interest is the detection of cold-adapted microbial life actively metabolizing or possibly growing within an ice forma-tion. On Earth, the bacteria known to be cold-adapted, whether requiring the cold for growth (*psychrophilic*) or simply tolerating it as they grow (*psychrotolerant*), were late arrivals in the broad evolutionary scheme of things, according to molecular-phylogenetic analyses (Fig. 15.2). Psychrophilic bacteria are generally under-stood to have evolved from *mesophilic* bacteria, organ-isms that grow at room temperature and warmer (to about 40–50 °C) but are inactive below about 10 °C. Permanently cold habitats that would favor the evolution of *obligate* psychrophiles (bacteria that *must* have cold to reproduce, growing optimally below 10 °C) were not available on Earth until about half way through its history (~2.3 Ga), when the first glaciation events appear in the geological record (Section 4.2.4). Had persistent ice formations appeared sooner, cold-adapted bacteria may well have evolved sooner: the genetic steps from mesophily to psychrophily do not appear to be particularly daunting. Detection of meta-bolically active microorganisms (or their traces) in the ice formations of Mars or Europa could thus imply either a *pioneer* community struggling to colonize the ice or the presence of a well-established, *resident*

---

[2] Global occurrence in ice is only known now with respect to the domain of Bacteria, since study of the distribution of Archaea in ice has only just begun.

microbial world in milder regions (or eras) of the planet or satellite from which evolved successful ice inhabitants. Regardless of the possible types of organisms envisioned – tourists, pioneers, or residents – and regardless of the thermal history of a given planet or moon, the extraterrestrial surfaces that we can expect to examine in the foreseeable future (on Mars and possibly Europa) are frozen environments. Their successful exploration will benefit from what we can learn on Earth about life in ice.

15.2 Physics and chemistry of ice

15.2.1 Liquid water at subzero temperatures

To approach the question of how microbial life may persist and even be active in frozen environments, the basic physics and phase chemistry of liquid water and ice must first be appreciated. Understanding how impurities (salt and other freezing point depressants) in natural ice formations, as well as kinetic and other constraints, enhance the occurrence of the liquid phase even at very low temperatures is also essential. Together, the specific physical and chemical aspects of a frozen system yield a three-dimensional microstructure that defines the inhabitable liquid space for microbial life. The amount of liquid and degree of connectivity of those spaces determine in large part whether bacteria simply persist in ice or actually benefit from diffusive or advective exchange of chemical compounds (of nutritional or other value) and thus have the opportunity to evolve through collaboration or competition at very cold temperatures.

15.2.1.1 Phase diagram of liquid and solid water

The structure of the water molecule and the nature of its hydrogen bonds play key roles in the physicochemical and biochemical processes that underlie the origin and evolution of life on Earth. They also account for a remarkable range of different ice phases, each of which are stable under particular conditions that can be encountered at the surface or in the interior of the various moons and planets in the Solar System. To date, thirteen different stable crystalline phases of ice have been described (Petrenko and Whitworth, 1999; Fig. 15.3). In the terrestrial and marine environments of Earth, the only important form of ice is the hexagonal modification of ice I, termed ice Ih (throughout this chapter, the term "ice," if used without a qualifier,

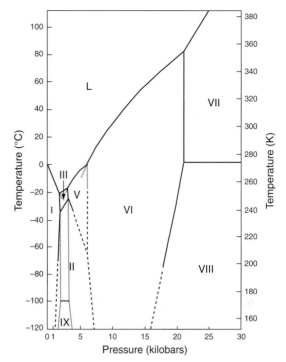

FIGURE 15.3 Phase diagram of pure water at high pressures, showing the fields of stability of liquid water (L) and of ices I through IX. Ice I is the dominant form of ice on Earth. Atmospheric pressure on Earth corresponds to 10^{-3} kbar and the bottom of a 1-km thick glacier to ~ 0.1 kbar. (From Hobbs, 1974.)

always refers to ice Ih). This form is *stable*[3] under conditions encountered throughout the Earth's atmosphere and hydrosphere.

A key property of ice Ih is apparent from the negative slope of the phase boundary (called the *liquidus*) that distinguishes the field of stability of ice Ih from that of liquid water (Fig. 15.3). Owing to the lower density of this ice ($0.917 \, \text{g cm}^{-3}$ at standard conditions of $0 \, °C$ and $1 \, \text{bar} = 0.1 \, \text{MPa}$) compared to water, the freezing/melting point of water/ice Ih decreases with increasing pressure. At the base of the Antarctic ice sheet, where sub-glacial Lake Vostok is located, the pressure of more than 3 km of ice pushes the freezing/melting point (and hence water temperature in the lake) to just below $-3 \, °C$. The other phases of ice in equilibrium with liquid water do not show this anomalous behavior; instead, the freezing/melting point increases with increasing pressure. Consequently, the lowest temperature at which liquid

---
[3] A phase of H_2O is said to be *stable* under a given set of conditions if over time it does not spontaneously change to another phase.

water can be present (in the absence of chemical freezing-point depressants such as salts or various organic compounds) is the point where the phase boundaries between ice Ih, liquid water, and ice III meet. This triple point lies at a temperature of $-22.3\,°C$ and a pressure of 2.1 kbar (Petrenko and Whitworth, 1999), a combination of conditions not likely to occur on Earth.[4]

At considerably higher pressures (100 kbar), ice can be stable at temperatures approaching $400\,°C$. While high-pressure phases of ice may be stable in the interior of some icy planets or moons, the ice crust overlying the putative ocean on Europa (Pappalardo et al., 1999) only barely, if at all, experiences pressures high enough (a few kbar) for ice II or III to appear (Lupo and Lewis, 1979). Nevertheless, the possible occurrence of modified forms of ice under high pressure may be important to consider, since the presence of any form of ice in hydrothermal environments would have substantial consequences for processes related to the possible origin and evolution of life in such settings. Furthermore, in the surface environments of some planets and moons, very low temperatures (and the resulting slowness of chemical changes) may create conditions such that a normally unstable state of ice is instead metastable, meaning that in effect it does not undergo transformations over time periods of interest.

15.2.1.2 Freezing point depression in brines

The importance of water as a prerequisite for life on Earth derives in large part from the polar nature of the water molecule and its role as a solvent for ionic and other compounds (see also Chapter 14). The same molecular-level properties and electrostatic forces governing the interaction between water, ionic compounds, and many types of organic compounds – hydrogen bonding and van der Waals interaction – are also critical for freezing-point depression that allows the survival and activity of microorganisms at subzero temperatures. The distribution of electrons within the water molecule, with a net negative charge associated with the oxygen atom and a net positive charge associated with the two hydrogen atoms, imparts a permanent dipole moment to the molecule as a whole (Petrenko and Whitworth, 1999). This inherent feature results in the hydration of ions (or other polar molecules), meaning that water dipoles orient themselves such that a layer (called a hydrate shell) surrounds individual anions or cations. The change in potential energy (Gibbs free energy) associated with such hydration, which depends on the radius and electronegativity of a specific ion, governs the solubility of any compound in water.

Hydration, however, also strongly affects phase changes, including the freezing of water, since the electrostatic interaction between ions and water molecules can substantially lower the temperature at which water undergoes a phase change from liquid to solid. In addition, the crystal lattice of ice does not allow for incorporation into the solid phase of most salt ions (F^- and NH_4^+ being notable and biologically relevant exceptions; Hobbs, 1974), causing ions to be concentrated in any liquid that remains upon ice formation. Thus, for a saturated solution of NaCl in water (23% salt), ice only forms when the temperature is dropped to $-21.2\,°C$ (Fig. 15.4); until that point water molecules in the liquid phase survive in hydrate shells.[5] Ionic salts, as well as polar molecules such as methanol or ethylene glycol (anti-freeze), hence depress the freezing point of water well below that of pure water.

The impurity effect on the availability of liquid water in frozen systems can be illustrated by a phase diagram for the two-component system $NaCl$-H_2O (Fig. 15.4). Although brines usually include several salts other than NaCl, this diagram serves as a reasonable proxy for natural systems such as permafrost soils or seawater, where the composition of the liquid brine fraction is typically more than 80% NaCl (Weeks and Ackley, 1986). The liquidus curves in the phase diagram indicate the first appearance of solid ice in a solution of given NaCl composition as the solution is cooled. Upon further cooling, more and more of the water turns to ice, with the salt becoming increasingly enriched in the remaining liquid. The amount of ice freezing out of solution and the concentration of salt in the remaining liquid is dictated by the freezing-point depression specific to each ionic compound. At the so-called eutectic point, the remaining liquid water simultaneously freezes with the salt (forming hydrohalite, $NaCl \cdot 2H_2O$), leaving a mixture of pure ice and hydrohalite (bottom of Fig. 15.4).

Although such liquid-free mixtures of ice and salt precipitates are believed to occur only rarely in natural ice formations on Earth, the eutectic point is

---

[4] Pressure and temperature at the bottom of the deepest ocean (~11 km) are ~1.1 kbar and ~2 °C.

[5] For standard seawater (3.5% NaCl), the freezing temperature is depressed to $-1.9\,°C$.

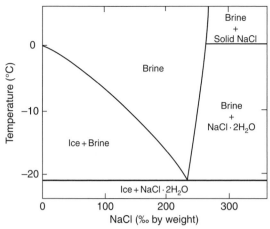

FIGURE 15.4 Phase diagram for the system NaCl-H$_2$O. A saturated solution of NaCl ($23\% = 230‰$) depresses the freezing point to $-21.2°C$. *Brine* refers to a liquid state.

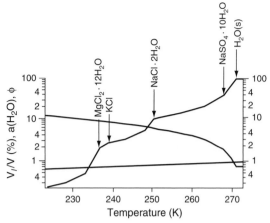

FIGURE 15.5 Modeling of the freezing of standard seawater, based on the Ringer–Nelson–Thompson pathway with eutectic point of $-54°C = 219\,K$ (the right edge of the plot is at $273\,K = 0°C$). Shown are the volume fraction of liquid within the ice, V_l/V (%), the activity of water within the brine, a(H$_2$O), and the ionic strength of the remaining liquid brine (Φ) as a function of temperature. Arrows indicate the points at which different salts precipitate. (Data from G. Marion.)

approached when severe conditions sufficiently chill an ice formation (e.g., in South Pole snow, Arctic winter sea ice, or Siberian permafrost). Presumably, as the fraction of liquid water becomes vanishingly low, bacteria also present as "impurities" in the ice would cease to function (though might well be preserved in an inactive but recoverable state) or become damaged (perhaps irreversibly). Such effects could happen as a result of (a) low water activity (see below) that prevents the maintenance of essential cellular components, or (b) physical damage from ice crystals forming either within or adjacent to the cells. The precise physical and chemical conditions leading to ice formation are of critical importance to bacterial survival. As we shall see, however, bacteria themselves, as living organic entities, have their own strategies for surviving extreme conditions and even altering to their advantage both the freezing and eutectic points of the fluid niches they inhabit.

In a simple (life-free) water system, the magnitude of the freezing-point depression can be found by solving for T_f in a derivation of the Clausius–Clapeyron equation (e.g., Fletcher, 1993):

$$\ln(a_1) = \frac{\Delta H_f}{R}\left(\frac{1}{T_f^{\cdot}} - \frac{1}{T_f}\right),$$

with ΔH_f the enthalpy of fusion (latent heat of fusion) of water, R the universal gas constant, T_f the freezing temperatures of the solution and the pure compound (superscript ·), and a_1 the activity of the water in the liquid phase fraction. The *activity* corresponds to the product of an activity coefficient and the molar fraction of the relevant component; it represents a measure

of the extent to which water molecules are available to participate in chemical reactions or phase changes. The fact that the magnitude of freezing-point depression depends on the number rather than the mass of ions of the component per unit volume is an important aspect of a system (often referred to as a *colligative* property). The activity of water decreases substantially with increasing salt concentration or ionic strength of the remaining solution (Fig. 15.5). Although theoretical modeling of low-temperature and sub-freezing processes is progressing (e.g., Marion and Farren, 1999), semi-empirical approaches are still common (Eicken, 2003).

What and where are the lowest temperatures at which we can expect to find liquid water on Earth? They will not be in ice formations derived from source waters (or vapors) with limited impurities, but rather in those derived from salty solutions. Liquid water is found routinely at fractions of several parts per thousand even in the coldest types of saline ice; for example, in Arctic winter sea ice at temperatures from $-20°C$ to below $-30°C$ (Eicken, 2003). Consider, for example, a volume of seawater with "standard" composition (six major salt components) and a total salt concentration of $35\,g$ solute per kg of solution (3.5%). Modeling by Marion and Farren (1999) shows that as the fraction of liquid water decreases with dropping temperature, the concentration (expressed as the ionic strength) of inorganic ions in solution increases (Fig. 15.5). The liquid fraction becomes supersaturated with respect to several salts that at specific temperatures begin to precipitate. The first of these precipitates (forming at about $-6.3\,°C$) is mirabilite ($NaSO_4 \cdot 10H_2O$), accounting for roughly 15% of the total salt fraction in standard seawater. At $-22.9\,°C$, with the volumetric fraction of liquid water still above 10%, hydrohalite ($NaCl \cdot 2H_2O$) begins to precipitate. Only at a temperature of $-54\,°C$, the eutectic point for seawater, does the model indicate that the liquid fraction drops to zero (upon precipitation of Antarcticite, $CaCl \cdot 6H_2O$). Another pathway, involving precipitation of gypsum ($CaSO_4 \cdot 2H_2O$), results in the liquid phase vanishing at a higher temperature of $-36\,°C$. The model assumes thermodynamic equilibrium and the absence of any other compounds in the system that could affect ice formation or salt precipitation. That the latter assumption will not always pertain to highly concentrated solutions at very low temperatures in natural systems (see below) points the way for future research.

While freezing point depressants can significantly extend the presence of liquid water into the subzero realm – and by implication the presence of living microorganisms – the benefits of acquiring a fluid niche in ice come at a cost to the organism in terms of maintaining its metabolism. The high ionic strengths and low activities of the liquid water may represent more severe challenges to microbial life than the usually cited reduced reaction rates and sluggish kinetics associated with a drop in temperature (Chapter 14). A bacterium living successfully in a salty solution (whether cold or not) is generally understood to keep its interior water activity at a sufficient level by raising interior concentrations of certain components high enough to cause H_2O (alone) to diffuse into the cell. Accumulating these so-called "compatible solutes" (examples are the sugars – sorbitol, trehalose, and glycerol) thus constitutes a primary strategy to avoid *osmotic shock*[6] and cell lysis (rupture and loss of contents) (Madigan *et al.*, 1997). Expending energy to pump salt ions out of the cell is another approach. Even with these identified strategies, however, no cultured bacterium is known to grow at water activities below 0.75 and even the hardiest of all *xerophilic* (dry-loving) organisms – some

yeast and fungi – fail to grow at water activities below 0.61 (Madigan *et al.*, 1997).

These perceived limits, however, are not based on studies of bacteria uniquely adapted to life in saline ice (e.g., the best-known users of compatible solutes are found in canned food, while the xerophiles mentioned above are mesophilic Eukarya). Appreciating the relative or combined impacts of temperature and ionic strength or water activity on life in ice is hampered by an absence of suitable model organisms. Although extreme *halophiles* (salt-lovers that grow best in solutions of 15–30% salt) and obligate psychrophiles have been known for decades, the former come from warm habitats and cannot grow in the cold, while the latter cannot grow in salt concentrations above 10%. If organisms that combine the growth traits[7] of extreme halophily and obligate psychrophily exist today, they continue to evade cultivation. Instead, progress in understanding microbial life in ice is coming from (a) the study of organisms that experience some degree of thermal and/or saline stress in ice (e.g., psychrophiles that are moderately halophilic, requiring seawater concentrations of salt [3.5%] for best growth), and (b) the direct study of natural ice formations as non-invasively as possible (Section 15.3.3).

Complicating the problem, however, is the close coupling between the temperature of any ice formation and the composition and ionic strength of its liquid fraction: each of these factors becomes more extreme simultaneously (Eicken, 2003). Effectively separating one factor from others for an experimental evaluation (as often attempted in microbiological studies) may not be practical or desirable. In the coldest ice formations, organisms must overcome the stress factors *in combination* in order to survive and grow. Recent studies of Arctic-winter sea ice suggest that bacteria can indeed succeed in this struggle (even if we do not yet have them in culture or understand their strategies for success). Significant numbers of active bacteria (Junge *et al.*, 2004b) and high concentrations of previously unrecognized organic *cryoprotectants* (compounds that prevent freeze damage to cells; Krembs *et al.*, 2002) have been detected, even at the lowest temperature examined, $-20\,°C$, and with a concomitant salinity of $\sim21\%$ (see also Section 15.3.3).

---

[6] *Osmotic shock* is a sudden change in the osmotic state of a cell, which can lead to its rupture and loss of contents (*lysis*). The osmotic state refers to the balance in solutes (largely controlled by diffusion) between the interior and exterior of a cell.

[7] Standard categorizations are based on parameters of *growth*, since reproductive success is the traditional (Eukaryotic) measure of "survival of the fittest"; but *tolerance* and *survival* traits in the microbial world (Bacterial and Archaeal), even in the absence of growth, deserve more consideration and study than they receive.

15.2.1.3 *Other mechanisms enabling liquid water in ice*

In contrast to freezing-point depression proportional to the concentration of inorganic salts, as discussed above, other processes may help to maintain liquid water in ice at temperatures extending even below the eutectic point of seawater. They involve properties of organic compounds produced by living organisms, as well as the molecular interactions that occur at interfaces between ice and liquids, gels (organic polymers), and solids such as entrained mineral grains or ice-crystal boundaries. These processes may be particularly important to evaluate when considering the potential for active or dormant (recoverable) microorganisms in environments on Mars and Europa that are colder than anywhere on Earth.

Compared to the role of inorganic solutes in lowering the freezing point, organic-solute *colligative properties* (having to do with levels of concentration and how solute molecules bind) have received much less attention. Only recently has the widespread occurrence of high levels of dissolved organic matter in such frozen environments as sea ice been well established (Thomas and Dieckmann, 2002). From the biochemical literature (often in medical and food sciences), however, certain classes of organic compounds, particularly glycoproteins, are well known to repress ice formation at temperatures well below those predicted on the basis of equilibrium processes. This phenomenon is particularly well documented for Antarctic bottom-dwelling fish that combat ice formation in their blood stream (Deluca *et al.*, 1998). Rather than increase the concentration of freezing-point depressants, an energetically costly strategy that would affect other metabolic processes, some fish produce compounds that even at low concentrations kinetically inhibit ice formation and growth. These "surface-active" freeze protectants readily adsorb in a mono-molecular layer onto free ice surfaces in a solution. This organic film then prevents or greatly reduces the rate of further ice accretion onto those surfaces at temperatures well below the equilibrium freezing point. For example, a solution containing $50\,mg\,ml^{-1}$ of a fish-derived anti-freeze glycoprotein can be supercooled by more than $3\,°C$ below the freezing point without any ice formation (Woehrmann, 1993).

Such "ice-active" compounds are also known in the microbial world, especially among bacteria studied as contaminants of frozen foods. Among natural ice formations, Antarctic summer sea ice has yielded at least one type of micro-algal community that releases protein-rich substances that are "ice-active." In addition to inhibiting ice formation (Raymond and Fritsen, 2001), some of these compounds roughen the surface of existing ice crystals on a micro-scale, which alters fluid flow and primes the surfaces for more effective bacterial colonization. In general, more attention should be paid to the ecological and geophysical consequences of bacterial processes that suppress ice formation and its ensuing damage to cells, or that physically alter the ice itself for greater habitability.

Recent studies of Arctic sea ice in winter suggest that gelatinous organic (sugar-based) polymers, known to be released by several types of microorganisms in sea ice, may prevent or reduce ice formation and hence act as cryoprotectants (Krembs *et al.*, 2002). The precise mechanism of action of these large molecular weight *exopolymers* has not yet been determined, but possibilities include the gamut of known mechanisms for other organic compounds and some new angles. In gel form, the exopolymers can fill space within ice and help to retain dissolved impurities during ice growth, thus altering the properties of the ice. These changed properties include its habitability, since gels (as "impurities" in the liquid fraction) represent occupiable space that the solid phase of ice has not encroached upon. A cell fully encased by ice is subject to direct physical damage from ice crystals; a cell first encased by gelatinous material, however, should have a physically and chemically protective advantage. Exopolymeric gels can also scavenge water directly within their molecular structures. As thick coatings around cells living in very cold brines, such gels have the potential to act as a buffer between organism and brine (or organism and ice), protecting against osmotic shock or physical ice damage by elevating the effective water activity in the immediate vicinity of the cell. Given that a variety of organic polymers can occur in significant concentrations in space environments such as comets and interstellar gas, studies of the habitability of natural ice formations on Earth need to consider in more detail these and other organic chemical processes.[8]

The abiotic formation of organic polymers in deeply frozen ice has several interesting implications for astrobiology. Just as the freezing process can concentrate dissolved salts to the point of precipitation, organic compounds can be brought to concentrations higher

---

[8] For example, how does the freezing rate of an organic-rich fluid influence its final properties and the survival of entrapped bacteria (Dumont *et al.*, 2004)?

than usually seen elsewhere in nature. Recent work capitalizing on this freeze-concentration effect has included detection of polymerization reactions without the enzymes usually required under milder conditions (Kanavarioti *et al.*, 2001), as well as enhanced rates or yields of other biomolecular reactions and improved amplification of L-enantiomers (stereoselection; see Vajda and Hollosi, 2001). Such organic reactions and products may enhance, now and in the past, the availability of liquid water at very low temperatures, and thus the persistence and evolution of life in frozen environments. Further, we perhaps should add a selection of "sorbets" to the warm or hot "primordial soup" on the menu of options for providing the biochemical precursors to life.

Also independent of biology but of potential importance to life in ice is another phenomenon known as *premelting*. Recent studies have shown that liquid or quasi-liquid water films persist at sub-freezing temperatures on interfaces such as those between an ice matrix and enclosed mineral particles, or even along ice grain boundaries. This premelting phenomenon is due to thermomolecular forces, which are interactions between molecules in the liquid and the solid that affect the free energy of the system and hence the thermodynamics of ice formation.[9] Wettlaufer (1999) has shown that the presence of ionic impurities can increase this film thickness substantially, for example to more than $0.1\,\mu m$ at a supercooling of $0.1\,°C$. (The effects of organic impurities on film thickness await study.) Such liquid films may have important implications for the survival of individual bacteria embedded within an ice matrix and for the advection of nutrients and other materials driven by thermomolecular forces. Advection of critical compounds to and from a bacterium can enable metabolic activities beyond those required simply to persist. Over 99% of all actively respiring bacteria in Arctic sea ice, when examined at a winter temperature of $-20\,°C$ (and corresponding brine salinity of 21%), have been found to be associated with surfaces such as mineral grains and organic particles encased in the ice, as well as the walls of ice crystals framing a brine pore (Junge *et al.*, 2004b) (Fig. 15.6). Further investigations of the life-facilitating aspects of these microscopic surfaces and interfaces promise to be important in considering life on Earth or elsewhere.

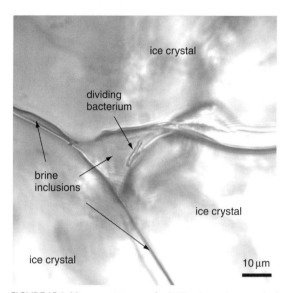

FIGURE 15.6 Microscopic image of a dividing bacterium attached to the ice wall of a brine-filled pore within an Arctic winter sea-ice sample at its *in situ* temperature of $-15\,°C$. The inhabited brine inclusion occurs at a triple juncture, formed by the conjoining of three ice crystals. (Adapted from Junge *et al.*, 2001.)

A final aspect of the liquid phase of frozen environments to consider is viscosity. Given the impurity-concentration effect inherent to the freezing process, the resulting salty and organic-rich liquids are more viscous than the source waters. Relationships between viscosity and the freezing characteristics of ice are complex and poorly known. Highly viscous fluids (imagine molasses) might be expected to pose problems to organisms that need to move or, if attached to a surface, that rely on diffusion to deliver nutrients or remove wastes. Recent work, however, shows that an obligately psychrophilic bacterium can swim easily and rapidly through a highly viscous sugar (glycerol) solution at temperatures as low as $-10\,°C$, though not in an equally cold brine solution (Junge *et al.*, 2003). Viscosity itself (just as temperature alone) may therefore not present too serious a limit on microbial activity in ice, returning us to high salt and low water activity as the primary challenge to life within the inhabitable niches of ice formations.

15.2.2 Structural constraints on life in ice

Summarizing the discussion of the previous sections, Fig. 15.7 provides a qualitative overview of the different types of frozen environments on Earth, all harboring microbial life in either active or dormant states (next section). The environments are located on the

---

[9] Premelting appears to be the main reason why ice is so slippery when one is standing on it (in shoes) (Rosenberg, 2005).

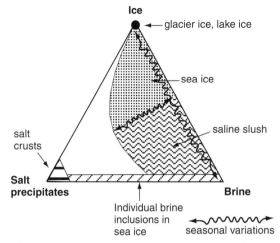

FIGURE 15.7 Diagram qualitatively depicting fractions of solid ice, liquid brine, and salt precipitates in various low-temperature environments. Typical seasonal variations are also indicated. All of the labeled environments are known to support either active or dormant microbial life.

diagram according to the fraction of their volume occupied by solid ice, liquid brine, and salt precipitates, the primary determinants for the presence of liquid water in frozen environments and hence for their habitability. Although other variables can contribute to available water in important ways (for example, the previously discussed "ice-active" organic compounds and particles known to promote premelting), they are not depicted because they may be present in any frozen environment. In this scheme, the cleanest of freshwater ice (glacier and lake) occupies one apex, while purely liquid brines (salt lakes) and salt crusts at similarly extreme temperatures occupy the others. Terrestrial permafrost, the frozen soil environment, may locate in various positions on the diagram, depending on the salinity of the initial liquid water and subsequent complex processes that can occur during seasonal temperature variations. The bulk of the diagram is occupied by the various seasonal and perennial forms of sea ice, as well as the saline slush that leads to its formation at the start of winter.

Until recently, the frozen Earth environments receiving the most widespread attention from an astrobiological perspective have been freshwater ice and permafrost; for example, deep glacial ice overlying Lake Vostok in Antarctica and permafrost in Siberia (Priscu and Christner, 2004; Rivkina et al., 2000). As we have seen, glacial and lake-ice formations cluster in a small section of the entire suite of habitats (near the ice apex in Fig. 15.7), while permafrost can occur anywhere within this spectrum of possibilities for liquid

water. Other potentially habitable worlds in the Solar System, such as Mars and Europa, are known to have salts on their surfaces, and may now or in the past have had briny environments (Chapters 18 and 19). Sea ice can thus serve as an instructive and contrasting habitat to freshwater ice and permafrost. Remember that it contains the largest volumetric fraction of liquid water of the various ice formations on Earth – and at the coldest temperatures (Table 15.1). What remains to be considered from a physical perspective is the comparative microstructure of these ice formations, in particular the three-dimensional shape of the interior (liquid-filled) spaces within the ice matrix and their degree of connectivity.

In freshwater ice with no thermal or chemical gradients, the distribution of the liquid phase can be limited to isolated single inclusions (closed pores) that remain disconnected from each other on the scale of thousands of years or more (in some glacial ices). In contrast, the microstructure of sea ice, when examined at even much lower temperatures than glacial ice, includes a connected network of liquid water that enables diffusion between pores and favors advective processes conducive to microbial activity. The distribution and activity of microorganisms in the various frozen environments of Earth (next section) supports the idea that frozen low-salt environments are more relevant to preserved (dormant) life and opportunities to detect and possibly recover ancient organisms from extraterrestrial environments. On the other hand, knowledge of frozen habitats of higher salt content may better instruct on the possibility of finding active life elsewhere.

The microstructure of natural sea ice depends on the growth environment, on boundary conditions at the advancing ice–water interface, and on the temperature and chemical composition of the ice. Among the effects of growth and boundary conditions, many of which are peculiar to Earth ice and the three-dimensional forms it takes (see Eicken, 2003), the rate of ice formation may be of particular astrobiological relevance. In general, the more rapid the freezing rate of sea ice on Earth, the greater the retention of salts and other impurities, with consequent effects on the distribution of phases within the ice (solid ice, liquid brine, salt precipitates, gas inclusions). Indeed, some of the highest concentrations of bacteria ("impurities" in this context!) in sea ice have been observed just after the period of rapid freezing in late autumn (Grossmann and Dieckmann, 1994; Delille et al., 1995). If salty ocean water on Europa were to rise through tidal cracks towards a deeply

5 mm   −21 °C   −10 °C   −6 °C

FIGURE 15.8 Magnetic resonance images showing how fluid inclusions in Arctic winter sea ice change as the ice is warmed. Brine-filled pores appear dark. Note that as the temperature rises from its *in situ* value of −21°C, brine pores become larger, more numerous, and more interconnected. (Adapted from Eicken *et al.*, 2000.)

frozen surface (mean temperature of −160 °C), as predicted by analyses of cycloidal features (Hoppa *et al.*, 1999; Section 19.7), the formation rate of new ice from that liquid would be very rapid. Any liquid water flowing to the very cold surface of Mars (−55 °C) would also experience rapid freezing. By analogy with Earth ice, salt – and any microscopic life forms also present as "impurities" – would tend to be retained in this new ice, making such formations ideal targets for exploration.

The temperature and chemical composition of the ice is also of prime importance to the morphology and connectivity of inhabitable pore spaces within an ice formation. Recent advances in appreciating these small-scale microstructural features have been made by the direct application of magnetic resonance imaging (MRI) to sea-ice samples. The specific example in Fig. 15.8 tracks the pore morphology in Arctic winter sea ice as the sample was warmed from its *in situ* temperature of −21 °C to −6 °C (Eicken *et al.*, 2000). As predicted by the phase relations in thermodynamic equilibrium, the brine fraction in the ice was observed to increase with warming. At the same time, however, the size, morphology, and connectivity of the pores (whose original spacing and orientation was determined by ice-growth conditions) clearly evolved, with pores observed to link up at warmer temperatures. The MRI images suggest that these connections exist at −10 °C, but recent microscopy with an effective magnification of more than 3200 has confirmed the interconnectedness of brine-filled pores on the μm scale at even colder temperatures of −15 °C (Fig. 15.6; Stierle and Eicken, 2002) and −20 °C (Junge *et al.*, 2001, 2004b). With an ability to exchange fluids and chemical compounds between even the smallest of brine pores in salty ice formations, bacteria are not confined to single closed pores (as they are in cold glacial ice), and thus have a better chance to metabolize and evolve.

15.3 Microbiology of ice

15.3.1 Bacterial abundance

Bacteria are known to be present in significant numbers in all types of natural ice formations on Earth; a "sterile" or bacteria-free sample of any type of natural ice formation remains to be documented. Rare reports of sterile ice in the past (reviewed by Baross and Morita, 1978) were based on highly selective culturing techniques that would have missed the vast majority of bacteria likely to be present. The known densities of bacteria in ice today, based on microscopic counts of all DNA-staining forms present, range over six orders of magnitude, from $200 \, \text{ml}^{-1}$ in South Pole snow to $>6 \times 10^7 \, \text{ml}^{-1}$ in sediment-laden glacial ice to $>10^8 \, \text{g}^{-1}$ of soil in permafrost (Table 15.2). Note that the highest concentrations of bacteria, across all types of natural ice formations, have been observed in ice that was also enriched with other organic particles or with *lithogenic* (originating in rocks) sediments (Table 15.2).

Also note that for all ice formations but permafrost, bacterial densities are reported routinely for a given volume of *melted* ice. This approach reflects the fact that, until very recently, all ice samples (including permafrost) were first allowed to melt before microscopically examining their contents to count the number of bacteria present. Melting natural ice samples provides for ready examination of their contents by established methods for water samples, but presents an osmotic shock to organisms that may have acclimated to the higher ionic strength (and organic content) of their former liquid water home within the ice. This problem pertains especially to sea ice, but also to other frozen systems, since the "impurity" concentration mechanism inherent to the freezing process leads to liquid inclusions of higher ionic strength even in freshwater ice. The loss of bacteria during melting has been investigated by comparing ice samples melted

TABLE 15.2 Bacterial abundance in ice[a]

Type of ice formation	Sampling location	Sample T (°C)	Particle-poor ice	Particle-rich ice
Snow	South Pole	−15	0.2–5×10^3	
Ice sheet	Over Lake Vostok (2–4 km)	−3	0.2–8×10^3	
	Greenland (bottom of sheet)	−9		$>6 \times 10^7$
Lake ice	Lake Bonney, Antarctica	$<$−5?	5×10^3	0.1–4×10^5
	Imikpuk Lake, Alaska	−5	7×10^4	7×10^5
Sea ice	Southern Ocean, summer	−2	0.01–3×10^6	0.02–2×10^6
	Southern Ocean, winter	−2	0.2–2×10^6	1×10^7
	Arctic Ocean, summer	−2	0.4–2×10^6	0.05–1×10^7
	Arctic Ocean, winter	−2 to −20	0.2–1×10^5	0.5–3×10^6
Permafrost	Northeast Siberia	−10		$>1 \times 10^8$

Note:
[a] Number ml^{-1} melted ice or g^{-1} soil for permafrost; data compiled from Carpenter *et al.* (2000), Delille *et al.* (1995), Gradinger and Zhang (1997); Grossmann and Dieckmann (1994); Helmke and Weyland (1995); Junge *et al.* (2001, 2003, 2004a), Karl *et al.* (1999), Priscu and Christner (2004), Rivkina *et al.* (2000), and Sheridan *et al.* (2003).

directly, melted into source waters (first made bacteria-free by filtration), and melted into a brine solution that provides for a final melt salinity equivalent to the liquid phase in the ice before melting. The latter approaches yield the higher counts of intact bacteria, up to twice as many.

Scaling the number of bacteria to the total volume of the original ice sample, even when melted in a way to minimize osmotic shock, masks another reality. As revealed by new microscopic techniques for examining intact (unmelted) ice, bacteria do not inhabit the ice uniformly in space (Junge *et al.*, 2001, 2004a). Instead, most of the organisms reside within the liquid phase of the ice (Fig. 15.6), where they become increasingly concentrated during the freezing process along with other "impurities." If the number of bacteria is scaled to the inhabitable space within an ice formation – the volume of water in the liquid phase (the brine), then *in situ* bacterial densities are of course much greater. For example, in Arctic winter sea ice, the range for bacterial abundance in "standard" terms is 0.5–3×10^6 ml^{-1} melted ice (Table 15.2); but when scaled to brine volume, the densities are $\sim$100 times higher. When considered in terms of the volume of melted ice, relationships between heterotrophic bacteria and their food (dissolved organic matter) in the ice have been ambiguous or elusive; when scaled to brine volume, the puzzle resolves, with bacterial density correlating strongly with dissolved organic matter (Junge *et al.*, 2004b).

Few researchers have considered this fluid-phase scaling approach for ice samples, but it is important

to conceptualizing and understanding how bacteria colonize, survive, and metabolize within ice, especially the colder forms of liquid-filled ice on Earth – winter sea ice and permafrost. Note that in general bacteria more homogeneously distributed in a given volume (e.g., microbial life in water) interact minimally compared to those in closer contact with each other within a biofilm or connected network. The latter scenario best describes life in the liquid phase of natural ices. The extreme conditions of Earth's coldest ice formations would seem to severely limit if not preclude microbial activity, but on the other hand the freeze-concentration aspects of ice formation should make ice a compelling setting for intensive cell-to-cell communications, including genetic exchange.

Which factors account for the universal presence of bacteria in natural ice formations on Earth? At the most basic level, the answer lies in the fact that all natural source waters (and sediments) for ice on the planet contain bacteria. Knowledge of the bacterial contents of the source water and of the concentration process inherent to freezing might thus seem sufficient to account for bacterial numbers in any final ice formation. However, the bacterial content of an ice sample (even when melted to minimize osmotic shock) rarely matches the bacterial density of the source water: bacteria are not passive particles, nor are the physics and chemistry of natural ice formation simple linear processes. Even if bacteria were to behave as relatively passive particles (expending minimal energy), their lipid- and polysaccharide-rich outer coats can result in attachment to other particles and surfaces.

The presence of particulate matter in the source water can thus influence in non-linear ways the number of bacteria ending up in the fluid phase of the ice. As discussed in the previous section, the rate of freezing also influences this process, with more rapid freezing leading to higher concentrations of impurities. Airborne and snow delivery of particles to ice forming at an atmosphere–water interface also contributes to the overall bacterial count.

The sum of these physical inputs of bacteria can be nullified, however, by physical losses from the system during ice formation. Even as impurities (bacteria) become concentrated in the liquid fraction of ice forming over a body of water, a significant portion of those fluids (and bacteria) will be expelled from the ice back into the source water. The classic case of fluid loss from an ice formation is brine rejection from young sea ice (the expulsion of liquid from within the ice pore network into the underlying ocean), creating dense water that sinks below the ocean surface and helps to drive global thermohaline circulation.

The only physical mechanism to account for *very* high bacterial numbers in an ice formation is the entrainment of high concentrations of lithogenic particles (and their attached bacteria). Such entrainment occurs in sea ice as a result of benthic storms[10] (Stierle and Eicken, 2002) and in terrestrial ice sheets at the ice–soil interface (Table 15.2). Note that soils and sediments contain many billions of bacteria per gram of particulate matter, compared to source waters for ice formation that typically contain only millions per milliliter (Deming and Baross, 1993).

An accounting of bacterial numbers in ice via physical inputs and losses will inevitably fall short, since bacteria rarely behave as passive particles. The full range of metabolic states, from dormancy to ongoing cell division, can be expected among the bacteria present in source waters (and sediments) prior to ice formation. Many bacteria will move to position themselves advantageously within the physico-chemical gradients of their aquatic environments, a process known as *chemotaxis*. If chemotaxis is possible within an ice–brine network (Junge *et al.*, 2003, and citations therein), then altered densities at favorable locations may result. Aquatic bacteria in general are known to attach to surfaces and particulate matter, either to remain in position against fluid flow (receiving new

nutrients, while waste moves downstream) or to travel more rapidly as a passenger to a new position. Attachment is a common mode of existence for bacteria within permafrost (Rivkina *et al.*, 2000) and sea ice (Junge *et al.*, 2004b), where the number of attachment sites can be very high due to the presence of sediments (Stierle and Eicken, 2002). Even when sediment particles are rarities in ice, they can serve as "hot spots" of bacterial abundance and activity (Priscu and Christner, 2004). When conditions are particularly favorable in the ice for a given bacterium, it will also grow (in size) and divide (as in Fig. 15.6), increasing population densities even under what we perceive as extreme conditions.

Once a distinct ice habitat has formed from source waters, some of the bacteria that were best or uniquely adapted to the prior aquatic conditions can be expected to succumb to their new circumstances. Complicating analysis of this loss term is that both the dissolved and particulate debris from such bacteria represent fuel for the growth of neighbors better adapted to life in ice. In a relatively closed ice system, losses might be expected to continue as the ice ages and bacteria spend more time under unfavorable conditions without replenishment of nutrients; the ice sheet overlying Lake Vostok, for example, contains only very low concentrations of bacteria (Table 15.2). Arguing against significant losses due only to the age of the ice, however, are the well-known preserving effects of freezing temperatures (and of brines) on a wide variety of bacteria from various milder habitats (Madigan *et al.*, 1997; Dumont *et al.*, 2004). Low bacterial abundance in ancient glacial ice is more likely due to low bacterial content of its source waters and the ice formation process than to massive cell lysis over time. If the latter were a common phenomenon, the base of the Greenland ice sheet (where sediments were entrained 0.1 to 2 Ma) would not still contain the observed high concentrations of bacteria (Table 15.2).

The permafrost environment, which includes some of the oldest (millions of years) and coldest frozen habitats on Earth, contains the highest numbers of intact bacteria among natural ice formations. These high densities are readily explained by the equally high abundances that characterize source (unfrozen) soils and the long-term preserving effects of subzero conditions once the soils are frozen. The extent to which high numbers may also indicate a dynamic population of bacteria, metabolizing and growing *in situ*, depends, however, upon the availability of a liquid phase. The search for dynamic bacterial populations

---

[10]   The energy of such storms reaches the seafloor and suspends benthic (bottom) sediment into the water, including surface waters that subsequently freeze.

in frozen habitats has thus been most rewarding in the most liquid-rich of ice formations – sea ice (Section 15.3.3).

In very cold ice formations and ice formed from source waters of limited impurities, the inhabitable (liquid-filled) space is so restrictive in its morphology that the most common form of bacterial loss – grazing by larger and more complex organisms (eukaryotic protists) – ceases to be a factor. Such grazers are excluded from these ice habitats by their size (Krembs et al., 2000), if not their greater fragility under extreme conditions. Bacteria cannot entirely escape predation through these spatial refugia, however, since they have other attackers in the form of nm-scale bacteriophage (bacteria-specific viruses). In frozen environments, such viral particles appear to become concentrated in the liquid phase along with salts, organic polymers, and colloids (in the same nm-size class as viruses), and their μm-sized bacterial targets. Some of the highest viral concentrations in aquatic environments have been found in sea ice ($>1 \times 10^8$ viral particles ml^{-1} melted volume of sea ice; Maranger et al., 1994). Yet, the well-known role that viruses play in bacterial mortality in milder aquatic habitats has not been studied in frozen systems. The profound role that viruses can play in the evolutionary process, by mediating gene exchange between organisms for improved survival, constitutes an essential astrobiological question, but one that has not yet been addressed for life in ice.

15.3.2 Bacterial diversity

The advent of modern molecular techniques for assessing genetic diversity within and between natural microbial assemblages, without having to bring the organisms into culture, has greatly enhanced our general understanding of bacterial diversity. Some extreme environments, especially hydrothermal vents (Chapter 14), have received considerable attention from this perspective, but by comparison the diversity of bacteria in frozen environments is poorly known.

The handful of studies available on direct phylogenetic analyses of ice well fit the history and physical-chemical nature of the various types of ice formations that have been examined. For example, in glacial ice sheets and Antarctic lake ice, where bacterial abundance is usually very low (Table 15.2), bacterial diversity reflects a cosmopolitan input of Bacterial tourists. Notably absent from these DNA libraries to date (likely reflecting low numbers not detectable by the methods in use) are any Archaea, even though

cold-adapted methanogens (methane-producing members of the Archaea) have been cultured from the source waters (Fig. 15.2). Many of the Bacterial organisms detected, typically dominated by Gram-positive bacteria (known for their thick cell walls) and several subdivisions of the common Proteobacteria, are recognizable as delivered aerially or via entrained soils. They are either associated with atmospheric indicators, adapted to dessication (thick-walled spore formers), or of known soil origin (Priscu and Christner, 2004; Sheridan et al., 2003). The detection of thermophilic bacteria in glacial ice (!) is unambiguous evidence of the preserved state of many of these airborne tourists (Priscu and Christner, 2004; Sheridan et al., 2003), since thermophiles cannot function at temperatures below about $+30\,^{\circ}$C, even if all other requirements for activity are met in glacial ice. The latter is unlikely for any type of microorganism, given the very limited fraction of liquid and rare chemical impurities that could serve as a source of energy or nutrition. The presence of photosynthetic bacteria, however, in ice overlying some permanent, surface Antarctic lakes (Priscu and Christner, 2004) does point to potential activity in situ when sunlight becomes available as a source of energy.

The detection in glacial ice of other bacteria (and viruses) normally associated with the human body has fueled speculation that ancient ice may provide a repository of preserved organisms that reflect human health conditions of ages past. Such speculation is tempered by the issue of contamination (and methodological artifacts) when applying DNA amplification techniques to natural samples. Considerable effort has thus been expended on contaminant-free collection of ancient ice samples, so that confidence is increasing in techniques that distinguish sample-derived organisms from contaminants (Sheridan et al., 2003).

Another form of freshwater ice – Antarctic snow, with very low bacterial abundance (Table 15.2) and a vanishingly low content of liquid water (Table 15.1; Warren and Hudson, 2003) – has also been examined by direct amplification of DNA. Carpenter et al. (2000), who targeted only Bacteria, recovered DNA sequences indicative of organisms known for their ability to survive stresses related to dryness or low water activity, in particular various members of the genus Deinococcus. Other bacteria related to known (cultured) marine psychrophiles also appeared in this DNA library of snow from the South Pole, suggesting that aerially delivered organisms came from other cold environments.

The ice formation expected to support the highest bacterial diversity, for both Bacteria and Archaea, is permafrost. Temperate soils (that may experience only brief periods of freezing in winter) are well known for their high microbial diversity, often one or two orders of magnitude higher than in associated bodies of water (or ice). This difference is due to the wide range of niches, including anoxic ones, present in soils and sediments. Such structural complexity and niche diversity also pertains to permafrost, even if deeply frozen by atmospheric conditions: upon warming, permafrost samples have yielded a wide variety of culturable Bacteria, as well as methanogens. The full extent of microbial diversity in deeply frozen permafrost, however, has yet to be examined using modern phylogenetic or DNA amplifying techniques; the habitat is wide open for discovery. At issue is the degree to which the severity of conditions in deeply frozen ice may alter the initially high microbial diversity prior to freezing. A further unanswered question (which can also pertain to sea ice) is the degree to which freeze–thaw cycles influence microbial diversity.

The ice formation perhaps most intriguing with regards to bacterial diversity is sea ice. Spring and summer sea ice – sunlit, near the melting point of water, and with nutrient-rich seawater infiltrating from below – is the best studied and most biologically active of all ice formations, even trumping the underlying ocean in terms of biological productivity on a per-volume (of bulk ice) basis (Arrigo et al., 1997). The total biomass of microscopic organisms, especially the pigmented photosynthetic algae that bloom in the ice during the sunlit seasons, often becomes visible to the naked eye (Fig. 15.9). Given such high levels of biological growth and production, warm-season sea ice (called "summer" sea ice, as opposed to the much colder winter sea ice) should be populated primarily by residents (as opposed to tourists) that are well adapted to conditions in the ice; and indeed it is, for every class of organism and by every type of analysis (Krembs et al., 2000; Thomas and Dieckmann, 2002). From the bacterial perspective, sea ice is a global seed bed for psychrophilic marine bacteria, providing a highly selective habitat for their competitive survival through the winter season, then releasing them to the ocean (and atmosphere) during melt seasons (Helmke and Weyland, 1995). Unlike glacial ice formations, where thermophilic tourists can be preserved in a dormant state, no thermophiles and rarely even mesophiles have been cultured from sea ice. Their absence further supports the cold-adapted and dynamic state of

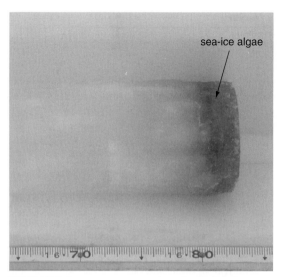

FIGURE 15.9 Summer sea-ice core from the Chukchi Sea, Western Arctic Ocean, with a visible band of algae and other organisms, including bacteria, and organic particles and gels. Ruler scale in cm; right end of the core was in contact with the seawater.

sea-ice bacteria: tourists and even pioneers from other climes are poor competitors with residents for local resources.

Although resident sea-ice bacterial communities may be predominantly psychrophilic, how diverse are they phylogenetically? The structural complexity of the ice, as well as seasonal opportunities to associate with a wide range of higher organisms (the algae and their metazoan grazers) not found in other ice formations, should favor an abundance of niches, including anoxic ones, and thus relatively high diversity. The stress factors of low temperature and high salt, however, might be expected to place limits on diversity, especially during winter. Recent applications of DNA amplification techniques to summer sea ice do suggest a gradient in Bacterial diversity, with higher diversity in algal bands near the seawater (Fig. 15.9) and lower diversity in other portions of the ice. A study of winter sea ice suggests a further reduction in bacterial diversity as the severity of conditions increases (Junge et al., 2004b). An important exception to this trend is that winter sea ice has also yielded the first evidence for Archaea in any ice formation other than frozen soil (which is a repository of Archaeal methanogens even before freezing). If these results hold, then one of the coldest of ice formations on Earth – Arctic winter sea ice – supports diversity at the Domain level unseen in other (particle-poor) frozen environments. Because the detection method for these Archaea also implied an actively metabolizing state (Junge et al., 2004b), the

implications for adaptation of life to subzero temperatures are potentially important. Future work on
winter sea ice (and possibly permafrost) may yield
organisms that define a deeper root for cold-adapted
Archaea on the Tree of Life (Fig. 15.2), changing the
way we think about temperature as a driving force in
the origin and evolution of life on Earth.

15.3.3 Bacterial activity

The heterotrophic bacteria emphasized in this chapter,
being free of any direct requirement for light, need only
an adequate supply of dissolved organic and inorganic
nutrients for their life activities. Many if not most are
facultative anaerobes, able to live independently of free
oxygen as well, using other oxidized compounds
(e.g., dissolved nitrate, sulfate, metal oxides, or organic
compounds) as electron acceptors during the metabolism that generates energy for the cell. The requirement
for "dissolved" nutrients implies the presence of a
liquid phase within the ice matrix. Liquid allows
for diffusion of compounds to and from the organism,
i.e., delivery of nutrients and removal of waste products (required for continued activity and growth). The
minimal liquid fractions and spatially restricted and
disconnected inhabitable spaces within freshwater ice
formations do not favor bacterial growth; the "dryness" of some may preclude even cellular maintenance
activities, leaving saline ice formations as the better
candidates to house dynamic populations of bacteria.
In fact, the natural ice formation on Earth most
supportive of a flourishing and diverse microbial (and
Eukaryotic) community (and most flushed with liquid
water) is spring–summer sea ice.

In sea ice (or any ice formation on Earth), the
organic nutrient supply for heterotrophic bacteria
traces back either to the byproducts of algae that also
inhabit the ice or to the dissolved organic materials
(also biologically produced) in the source seawater.
Although abiotic sources of organic nutrients may
well pertain to more deeply frozen systems on Earth
and elsewhere (Section 15.2.1), the dominant role of
photosynthesizing organisms in Earth's contemporary,
sunlit global ecosystem makes their influence on other
life forms virtually inescapable.[11] Levels of both bacterial abundance and activity in sample cores of sea ice
invariably reach their maxima in association with a
band of springtime ice algae (Fig. 15.9) and are best

understood for that reason (Thomas and Dieckmann,
2002). The levels achieved rival those known in milder
environments on Earth.

In the dark polar winter, however, Earth's photosynthetic organisms are the least influential in a given
habitat; heterotrophic organisms reign. As we have
seen, bacteria are present in significant numbers
in every cubic centimeter of ice that covers the polar
seas (or land; Table 15.2), regardless of the season, age,
or other conditions of the ice, but during the winter
season they account overwhelmingly for the total
biomass of *all* organisms present in sea ice (Delille
et al., 1995; Krembs *et al.*, 2002). A relevant astrobiological question is whether or not these bacteria
continue to be metabolically active or even reproductive under the most extreme of winter conditions
(below $-20\,°C$ and in brine solutions of $>21\%$ salt).
If the most liquid-filled niches of Earth's coldest ice
formations – winter sea ice in the high Arctic – do allow
for continued activity, what factors may ultimately
impose a limit on that activity? Could other types of
extremely cold ice formations akin to sea ice, whether
on Earth or elsewhere, also support dynamic bacterial
populations? To what extent is microbial life itself in
control of its own environment and threshold settings
for continued activity?

Empirical research to address some of these astrobiological questions has only recently begun. There
have been few direct attempts to measure bacterial
activity in natural (or simulated) ice formations under
extreme conditions, e.g., at temperatures near or
below $-15\,°C$. This temperature has sometimes been
considered a lower limit for growth of bacteria in
culture, based on extrapolations from measured
growth (production of new cells) at warmer temperatures where it can be readily detected by conventional
methods. Still speculative are much lower temperature
limits for metabolic activities that, while not supporting growth, would ensure cell maintenance and genetic
survival (undamaged DNA). Table 15.3 summarizes
recent results, but differences among the available studies in both methods and motivations make comparisons difficult. Furthermore, deeply frozen samples
must be manipulated or altered (e.g., warmed, melted,
amended, incubated), usually to a greater degree than
other types of environmental samples, in order to make
an activity measurement, sometimes calling into question the relevance of results to the *in situ* frozen environment. A microscopic approach for working with
natural ice samples that are neither warmed nor melted
does exist (Junge *et al.*, 2001) and cross-disciplinary

---

[11] Some inhabitants of the deep subsurface biosphere are important
and exciting exceptions (Chapter 14).

TABLE 15.3 Lowest known temperatures for bacterial activity in ice[a]

Type of ice formation	Samples examined	Lowest T (°C) examined	Type of activity detected	Sample manipulations
Artificial ice VI	Liquid cultures (grown at +25°C)	+25 (at 1.25 GPa)	Anaerobic formate oxidation (slow rate)	Amended with formate (no warming), pressurized, incubated 300 h at test T
Artificial glacial ice	Liquid cultures (grown at +22°C)	−15	DNA and protein production (slow rate)	Pre-frozen to −70°C, amended with liquid tracer (warmed), incubated 280 d at test T
Antarctic snow	Snow samples (at ≤−28°C *in situ*)	−12 to −17	DNA and protein production (slow rate)	Amended with liquid tracer (warmed), incubated 24 h at test T
Siberian permafrost	Soil samples (at −10°C *in situ*)	−20	Lipid production (near detection limit)	Amended with liquid tracer (warmed and mixed), incubated 550 d at test T
Arctic winter sea ice	Sea-ice sections (at −20°C *in situ*)	−20	Oxygen respiration (1–4% of cells)	Melted in −20°C brine, amended with respiratory stain (no warming), incubated 24 h at test T
Arctic winter sea ice	Sea-ice sections (at −20°C *in situ*)	−20	Possible protein production (70% of cells)	Fixed (no warming or incubation), examined microscopically using fluorescent 16 S rRNA probes

Note:
[a] data compiled from Carpenter *et al.* (2000), Rivkina *et al.* (2000), Christner (2002), Sharma *et al.* (2002), and Junge *et al.* (2004b).

interaction, one of the hallmarks of astrobiology (Chapter 28), is further advancing the issue of how to measure or deduce metabolic activity in unaltered ice (Carpenter *et al.*, 2000; Warren and Hudson, 2003; Junge *et al.*, 2004a, b; Price and Sowers, 2004).

In spite of the limited experimental work to date and the various caveats inherent to it, activity measurements made on the coldest of ice formations (Arctic winter sea ice and Siberian permafrost) indicate that the lowest temperature yet tested, −20 °C, still permits significant bacterial activity. The measured activities include those required for cellular maintenance, not necessarily growth: oxygen respiration to generate chemical energy for the cell; lipid production to keep membranes well structured and functional; and protein production to ensure critical biochemical functions. Where rates have been determined for a range of temperatures (in permafrost samples), activities are lower than under warmer conditions, as thermodynamics would predict. One theoretical treatment of data available in the literature, including a deduction of

microbial activity at −40 °C from chemical anomalies in glacial ice, suggests that an absolute temperature minimum for cellular activity may not exist (Price and Sowers, 2004). If one does exist, it may simply define the point at which critical (DNA) repair mechanisms in the cell fail to keep pace with any degradative processes, however slow they both may be in deeply frozen systems. In any case, a temperature minimum would not be the same for all frozen systems, but instead would vary according to the specific physical-chemical features of a given ice formation. Since attempts to measure DNA replication or bacterial growth in natural ice formations have not yet been reported at −20 °C (only to −15 °C; Table 15.3 and Fig. 15.6), nor have empirical tests for any type of activity below −20 °C been reported, surprises may lie in store (Junge *et al.*, 2006). Determining the physical-chemical ice features and/or biological mechanisms that enable continued bacterial activity in frozen environments at much lower temperatures represents a field of inquiry of critical importance to astrobiology.

Pursuing even colder temperatures will not be the only direction to take. Recall that under extremely high pressures, ice VI can form over the temperature range of 0 to +80 °C. Novel experiments on bacterial activity under such unusual conditions (not found on Earth) have yielded intriguing results (Sharma *et al.*, 2002; Table 15.3) that include apparent bacterial motility, pointing to the importance of an interconnected network of liquid water within an ice formation. We again emphasize that the basic biology of the resident organisms themselves is only one part of the story. The other part comprises the combined physical and chemical features of the frozen environment – its microstructure, the availability of interconnected fluids as well as attachment surfaces, the hydrated state of organic gels in which cells may be imbedded, the initial freezing rate of the source fluids, etc.

15.4 Summary and prospectus

In this chapter, we have emphasized the range of frozen environments on Earth and thus the known range of specific physical and chemical conditions encountered by microorganisms either thriving or simply managing to survive and persist within those ice formations. Perfect analogues for potential habitats elsewhere in the frozen realm of the Solar System may not exist on Earth, but examining and comparing the possibilities on this planet provides an excellent basis for developing predictions – and often novel experimental approaches to test them – of where explorers might find microbial life, and in what stage of development or survival. A synthesis approach, integrating the available information across many different types of ice, reveals new avenues for research and discovery in well-studied ice formations, as well as gaps in the astrobiological exploration of ice on Earth. In the latter case, for example, the surfaces of remote high-salt lakes in Antarctica often experience atmospheric conditions colder than those of high Arctic sea ice or permafrost, yet the microbiology of saline ice that forms on some of them during the coldest winter months awaits study. Although terrestrial[12] permafrost formations are well studied from some perspectives, modern molecular techniques for evaluating basic microbial diversity have not yet been applied to these environments that are often presented as analogues for frozen martian soils.

Regardless of where on Earth we search for life in ice, the issue of paramount importance to microbial and enzymatic processes in an ice formation is the availability of liquid water within the ice matrix. Although a threshold fraction or absolute amount of liquid water required for microbial activity in ice is not known, connectivity of the liquid on a scale relevant to bacteria, the presence of nutrients, and materials they exude all favor activity. We know that the most liquid-rich ice formations on Earth (sea ice) support very dynamic microbial populations dominated by heterotrophic bacteria adapted to the cold. In contrast, liquid-poor ice formations (e.g., glacial ice) appear to support only the limited maintenance activities that may be required for long-term persistence and survival of those microorganisms, usually in very low numbers and sometimes of diverse origins (aerially delivered thermophiles as well as psychrophiles). As we have seen from physical-chemical observations and theory, however, the amount of liquid water in ice is not a simple function of temperature considered in isolation from other factors, just as a lower temperature threshold for bacterial activity or survival may not exist. Salt and other impurities are essential to the existence of the liquid phase at temperatures below the freezing point of water. On other bodies such as Europa, high pressure may be another key factor.

Let us push the physical-chemical boundaries of ice that can support microbial life to limits that are relevant off Earth, namely −55 °C in the soils of Mars, and even colder in the saline ice covering Europa. The frozen systems of choice for the question of bacterial survival or preservation under severe extraterrestrial conditions may not be natural ice formations on Earth but rather artificial constructs in the laboratory. Under controlled conditions, study of the deep-freezing process (e.g., to −80 °C or lower) known as vitrification, whereby water freezes to a glassy state without the formation of cell-damaging ice crystals or the retention of detectable liquid water (not discussed in this chapter; see Dumont *et al.*, 2004), may be particularly illuminating. To move beyond microbial survival, however, and consider dynamic life processes, ice formations that retain some liquid should be best. Among the relatively unexplored ice formations on Earth that still retain an unfrozen fraction under the most extreme of natural conditions – lowest temperatures and highest concentrations of salt impurities (with the accompanying benefits of organic compounds and lithogenic particles, as we have discussed) – the various sea-ice formations over the Arctic Ocean

---

[12] Extensive permafrost formations that lie frozen beneath the Arctic Ocean have not at all been explored for their microbial content.

during wintertime offer great promise, with Arctic winter permafrost a close second.

In addition to being the ice formation on Earth that contains the highest fraction of liquid water, retaining liquid even at its coldest *in situ* temperatures (due to salt and other impurities), natural sea ice, especially during winter, presents within a single study area the advantage of access to a wide range of ice features and conditions (Fig. 15.10). The winter ice is wedged between severe atmospheric conditions above, which can approach the mean martian surface temperature of $-55\,°C$, and the warmer ocean below, at $-1.9\,°C$. Level pans of undeformed ice grow through the winter to a thickness of 2–3 meters as seawater continues to freeze to the bottom of an ice floe. Analysis of winter ice cores from these pans, with temperature gradients from about $-20\,°C$ (warmer than the atmosphere due to an insulating snow cover) to $-2\,°C$, has elucidated the roles of concentrated organic polymers and provided insight into bacterial abundance, diversity, and activity under extreme conditions, as discussed throughout this chapter. Cracks or leads that open and then freeze over again in winter can provide samples of rapidly frozen ice to test hypotheses on the effects of freezing rate on bacteria. The very coldest ice (well below $-20\,°C$ in winter) occurs at the surface of windblown ridges formed by compressive forces, windblown ice in rubble fields formed by shear forces, and windblown (thus nearly free of insulating snow) level ice (Fig. 15.10). The microbiology of these rapidly and/or deeply frozen environments awaits interdisciplinary study.

Although formed by large-scale planetary processes, the ability of these Arctic winter sea-ice formations – or any frozen environment – to support an active microbial population depends upon features evaluated on the scale of micrometers and less. As reinforced in this chapter, evaluating information on the dimensional scale directly relevant to an individual microorganism can reveal much about the inherent habitability of a given environment and thus whether to expect long-term residents, pioneers, or accidental tourists. Without such information from Earth, assessments of extraterrestrial habitability will be limited if not inaccurate. Nevertheless, the larger scale features of an extreme environment on Earth can be instructive and compelling; those observed in aerial images of the ice-covered Arctic Ocean (Fig. 15.10) in particular have captured the attention of planetary scientists and astrobiologists focused on the icy moons of Jupiter. Even though occurring on still larger scales and resulting

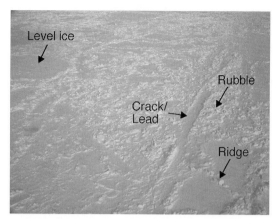

FIGURE 15.10 Aerial image of sea ice in the Chukchi Sea, Western Arctic Ocean (May 2002). The crack at right is 10–20 m wide, the pan of undeformed level ice at left is ~100 m wide, with an ice thickness of ~1.5 m. In ridged and rubbled areas ice thickness can exceed 10 m.

from deformation mechanisms that differ in some ways from those on Earth, the frozen and salty surface of Europa also clearly exhibits ridged ice, refrozen cracks, and level rafts of ice of variable thickness (Pappalardo *et al.*, 1999; Chapter 19). The possible relevance to Europa of microbial life in saline ice formations on Earth provides a powerful motivation for future study.

Studies of frozen environments on Earth can significantly help in achieving key objectives of the emerging science of astrobiology. Specifically, they further our understanding of fundamental constraints on the evolution of life at low temperatures and aid in testing conceptual models of the potential presence and detectability of life in extraterrestrial environments such as Mars or Europa. They also help to constrain scenarios of planetary development and composition that may include microbial colonization and evolution. Moreover, experience has shown that closer looks at Earth's frozen environments have almost invariably yielded remarkable and sometimes unexpected evidence for the sustainability of microbial life under ever more extreme conditions than previously imagined.

Ultimately, the interdisciplinary study of life within the full range of frozen environments on Earth may lead us towards more meaningful predictions of life – or its absence – in ice elsewhere. The liquid water boundaries on Earth have not yet been adequately explored from the microbial, enzymatic, or physical-chemical perspectives. The phase fractions of solid ice, liquid brine, and salt precipitates need to be considered in parallel with microbiological parameters.

In this chapter we have only just begun such an analysis from the existing literature – the field is wide open for discovery of microbial strategies for surviving and even thriving in ice, and the critical physics and chemistry underlying them. And beyond water ice, the natural state of H_2O in our Solar System, the possibilities for life in more exotic frozen environments, such as methane or hydrocarbon ice on Titan (Chapters 20 and 27), further beckon.

REFERENCES

Arrigo, K. R., Worthen, D. L., Lizotte, M. P., Dixon, P., and Dieckmann, G. (1997). Primary production in Antarctic sea ice source. *Science*, **276**, 394–397.

Barnes, J. (ed.) (1984). *The Complete Works of Aristotle*. The Revised Oxford Translation, Vol. 1, *History of Animals*, Book V, Chapter 19, Section 552b6–552b17, p. 871. Princeton, New Jersey: Princeton University Press.

Baross, J. A. and Morita, R. Y. (1978). Microbial life at low temperatures: ecological aspects. In *Microbial Life in Extreme Environments*, ed. D. J. Kushner, pp. 9–71. New York: Academic Press.

Carpenter, E. J., Lin, S., and Capone, D. G. (2000). Bacterial activity in South Pole snow. *Applied and Environmental Microbiology*, **66**, 4514–4517.

Christner, B. C. (2002). Incorporation of DNA and protein precursors into macromolecules by bacteria at −15 °C. *Applied and Environmental Microbiology*, **68**, 6435–6438.

Delille, D., Fiala, M., and Rosiers, C. (1995). Seasonal changes in phytoplankton and bacterioplankton distribution at the ice–water interface in the Antarctic neritic area. *Marine Ecology Progress Series*, **123**, 225–233.

Deluca, C. I., Davies, P. L., Ye, Q., and Jia, Z. (1998). The effects of steric mutations on the structure of type III antifreeze protein and its interaction with ice. *J. Molecular Biology*, **275**, 515–525.

Deming, J. W. and Baross, J. A. (1993). The early diagenesis of organic matter: microbial activity. In *Organic Geochemistry*, Vol. 6, Topics in Geobiology, ed. M. H. Engel and S. A. Macko, pp. 119–144. New York: Plenum Press.

Deming, J. W. and Baross, J. A. (2002). Search for and discovery of microbial enzymes from thermally extreme environments in the Ocean. In *Enzymes in the Environment: Activity, Ecology and Applications*, eds. R. G. Burns and R. P. Dick, pp. 327–362. New York: Marcel Dekker, Inc.

Dumont, F., Marechal, P.-A., and Gervais, P. (2004). Cell size and water permeability as determining factors for cell viability after freezing at different cooling rates. *Appl. Environmental Microbiology*, **70**, 268–272.

Eicken, H. (2003). From the microscopic to the macroscopic to the regional scale: growth, microstructure and properties of sea ice. In *Sea Ice: an Introduction to Its Physics, Biology, Chemistry and Geology*, ed. D. N. Thomas and G. S. Dieckmann, pp. 22–81. London: Blackwell Scientific, Ltd.

Eicken, H., Bock, C., Wittig, R., Miller, H., and Poertner, H.-O. (2000). Nuclear magnetic resonance imaging of sea ice pore fluids: methods and thermal evolution of pore microstructure. *Cold Regions Science and Technology*, **31**, 207–225.

Fletcher, P. (1993). *Chemical Thermodynamics for Earth Scientists*. Harlow, Essex: Longman Scientific and Technical.

Gradinger, R. and Zhang, Q. (1997). Vertical distribution of bacteria in Arctic sea ice from the Barents and Laptev Seas. *Polar Biology*, **17**, 448–454.

Grossmann, S. and Dieckmann, G. S. (1994). Bacterial standing stock, activity, and carbon production during formation and growth of sea ice in the Weddell Sea, Antarctica. *Appl. Environmental Microbiology*, **60**, 2746–2753.

Helmke, E. and Weyland, H. (1995). Bacteria in sea ice and underlying water of the eastern Weddell Sea in midwinter. *Marine Ecology Progress Series*, **117**, 269–287.

Hobbs, P. V. (1974). *Ice Physics*. Oxford: Clarendon Press.

Hoppa, G. V., Tufts, B. R., Greenberg, R., and Geissler, P. E. (1999). Formation of cycloidal features on Europa. *Science*, **285**, 1899–1902.

Junge, K., Eicken, H., and Deming, J. W. (2003). Motility of *Colwellia psychrerythraea* strain 34 H at subzero temperatures. *Appl. Environmental Microbiology*, **69**, 4282–4284.

Junge, K., Deming, J. W., and Eicken, H. (2004a). A microscopic approach to investigate bacteria under *in situ* conditions in Arctic lake ice: initial comparisons to sea ice. In *Bioastronomy 2002: Life Amongst the Stars*, eds. R. Norris and F. Stottman. San Francisco: Astronomical Society of the Pacific, pp. 381–388.

Junge, K, Eicken, H., and Deming, J. W. (2004b). Bacterial activity at −2 to −20 °C in Arctic wintertime sea ice. *Appl. Environmental Microbiology*, **70**, 550–557.

Junge, K., Eicken, H., Swanson, B. D., and Deming, J. W. (2006). Bacterial incorporation of leucine into protein down to −20 °C with evidence for potential activity in subeutectic saline ice formations. *Cryobiology*, **52**, 417–429.

Junge, K., Krembs, C., Deming, J., Stierle, A., and Eicken, H. (2001). A microscopic approach to investigate bacteria under in-situ conditions in sea-ice samples. *Annals of Glaciology*, **33**, 304–310.

Kanavarioti, A., Monnard, P.-A., and Deamer, D. W. (2001). Eutectic phases in ice facilitate nonenzymatic nucleic acid synthesis. *Astrobiology*, **1**, 271–281.

Karl, D. M., Bird, D. F., Bjorkman, K., Houlihan, T., Shackelford, R., and Tupas, L. (1999). Microorganisms in the accreted ice of Lake Vostok. *Science*, **286**, 2144–2147.

Kobalenko, J. (2002). *The Horizontal Everest: Extreme Journeys on Ellesmere Island*. Hammondsworth, England: Penguin.

Krembs, C., Gradinger, R., and Spindler, M. (2000). Implications of brine channel geometry and surface area for the interaction of sympagic organisms in Arctic sea ice. *J. Experimental Marine Biology and Ecology*, **243**. 55–80.

Krembs, C., Eicken, H., Junge, K., and Deming, J. W. (2002). High concentrations of exopolymeric substances in wintertime sea ice: implications for the polar ocean carbon cycle and cryoprotection of diatoms. *Deep-Sea Research I*, **9**, 2163–2181.

Lupo, M. J. and Lewis, J. S. (1979). Mass-radius relationships in icy satellites. *Icarus*, **40**, 157–170.

Madigan, M. T., Martinko, J. M., and Parker, J. (1997). *Brock Biology of Microorganisms*. Upper Saddle River, New Jersey: Prentice Hall.

Maranger, R., Bird, D. F., and Juniper, S. K. (1994). Viral and bacterial dynamics in Arctic sea ice during the spring algal bloom near Resolute, NWT, Canada. *Marine Ecology Progress Series*, **111**, 121–127.

Marion, G. M. and Farren, R. E. (1999). Mineral solubilities in the Na-K-Mg-Ca-Cl-SO4-H2O system: a re-evaluation of the sulfate chemistry in the Spencer–Møller–Weare model. *Geochimica et Cosmochimica Acta*, **63**, 1305–1318.

Pappalardo, R. T., Belten, M. J. S., Breneman, H. H., *et al.* (1999). Does Europa have a subsurface ocean? Evaluation of the geological evidence. *J. Geophys. Res.*, **104**, 24015–55.

Petrenko, V. F. and Whitworth, R. W. (1999). *Physics of Ice*. Oxford: Oxford University Press.

Price, P. B. and Sowers, T. (2004). Temperature dependence of metabolic rates for microbial growth, maintenance, and survival. *Proc. Natl. Acad. Sci. USA*, **101**, 4631–4636.

Priscu, J. C. and Christner, B. C. (2004). Earth's icy biosphere. In *Microbial Diversity and Bioprospecting*, ed. A. T. Bull, pp. 130–145. Washington, DC: American Society for Microbiology.

Raymond, J. A. and Fritsen, C. H. (2001). Semipurification and ice recrystallization inhibition activity of ice-active substances associated with Antarctic photosynthetic organisms. *Cryobiology*, **43**, 63–70.

Rivkina, E. M., Friedmann, E. I., McKay, C. P., and Gilichinsky, D. A. (2000). Metabolic activity of permafrost bacteria below the freezing point. *Appl. Environmental Microbiology*, **66**, 3230–3233.

Rosenberg, R. (2005). Why is ice slippery? *Physics Today*, **58**, 50–55.

Sharma, A., Scott, J. H., Cody, G. D., Fogel, M. L., Hazen, R. M., Hemley, R. J., and Huntress, W. T. (2002). Microbial activity at Gigapascal pressures. *Science*, **295**, 1514–1516.

Sheridan, P. P., Miteva, V. I., and Brenchley, J. E. (2003). Phylogenetic analysis of anaerobic psychrophilic enrichment cultures obtained from a Greenland Glacier ice core. *Appl. Environmental Microbiology*, **69**, 2153–2160.

Stierle, A. P. and Eicken, H. (2002). Sediment inclusions in Alaskan coastal sea ice: spatial distribution, interannual variability, and entrainment requirements. *Arctic, Antarctic, and Alpine Research*, **34**, 465–476.

Thomas, D. N. and Dieckmann, G. S. (2002). Antarctic sea ice – a habitat for extremophiles. *Science*, **295**, 641–644.

Vajda, T. and Hollosi, M. (2001). Cryo-bioorganic chemistry: freezing effect on stereoselection of L- and DL-leucine co-oligomerization in aqueous solution. *Cellular and Molecular Life Sciences*, **58**, 343–346.

Warren, S. G. and Hudson, S. R. (2003). Bacterial activity in South Pole snow is questionable. *Appl. Environmental Microbiology*, **69**, 6340–6341.

Weeks, W. F. and Ackley, S. F. (1986). The growth, structure and properties of sea ice. In *The Geophysics of Sea Ice*, ed. N. Untersteiner, pp. 9–164. New York: Plenum.

Wettlaufer, J. S. (1999). Impurity effects in the premelting of ice. *Phys. Rev. Lett.*, **82**, 2516–9.

Woehrmann, A. P. A. (1993). Freezing resistance in Antarctic and Arctic fishes (in German). *Ber. Polarforschung*, **119**, 1–99.

Zhang, T., Heginbottom, J. A., Barry, R. G., and Brown, J. (2000). Further statistics on the distribution of permafrost and ground ice in the Northern Hemisphere. *Polar Geography*, **24**, 126–131.

FURTHER READING AND SURFING

Lin, J.-F., Gregoryanz, E., Struzhkin, V. V., Somayazulu, M., Mao, H., and Hemley, R. J. (2005). Melting behavior of H2O at high pressures and temperatures. *Geophys. Res. Lett.*, **32**, L11306, doi:10.1029/2005GL022499. New data on the stability of high temperature/high pressure phases of ice suggest that several planetary interiors may harbor stable ice phases. Might there be deep interior icy biospheres?

Christner, B. C., Mikucki, J. A., Foreman, C. M., Denson, J., and Priscu, J. C. (2005). A model system for developing extraterrestrial decontamination protocols. *Icarus*, **174**, 572–584. Determining how to sample microbial life in Earth's glacial ice without introducing contaminants also pertains to sampling extraterrestrial ice formations with the principles of planetary protection in mind.

Gilichinsky, D., Rivkina, E., Bakermans, C., *et al.* (2005). Biodiversity in cryopegs in permafrost. *FEMS Microbiology Ecology*, **53**, 117–128. These marine-derived lenses of briny liquid at −10 °C, buried within permafrost for more than 100,000 years, provide unique and intriguing analogues for possible subsurface environments elsewhere.

www.gi.alaska.edu/snowice/sea-lake-ice/ – information on sea-ice physics and biology, including live-data feeds.

16 The evolution and diversification of life

Stanley M. Awramik, *University of California, Santa Barbara*
Kenneth J. McNamara, *Western Australian Museum*

16.1 Introduction

The fossil record conjures up images of dinosaurs, trilobites, and extinct humans. This reflects not only our anthropocentric (or better, *metazoan*[1]-centric) bias, but also a visual bias. Starting about 542 Ma, macroscopic remains of animals, traces, shells, and later bones and plants, become quite evident in the fossil record, hence the geological term *Phanerozoic*, which literally means the eon of "visible life." These 542 Myr of the fossil record demonstrate that changes have occurred in organisms and have provided compelling evidence for the theory of evolution. This chapter addresses the history of life in the two youngest eons of geological time, the Proterozoic and the Phanerozoic, which span the last 2500 Myr.[2] It is a record of contrasts. Whereas the Proterozoic was mainly a microbial world, the Phanerozoic is a world of macroscopic animals and plants. The transition from the Proterozoic to the Phanerozoic marks what may be one of the most significant evolutionary events in the history of life, when many of the modern animal phyla evolved. Highlights of this 2500 Myr record include the rise to dominance of cyanobacteria in shallow marine environments, the early evolution and diversification of eukaryotes, the first appearance of multicellular algae, the appearance of animals and their subsequent rapid diversification (the Cambrian explosion), the first land plants and their subsequent diversification, and the appearance and dominance of intelligent life (humans). In addition to these biospheric changes, the Earth's crust continued to be dynamic with active, seafloor spreading and plate tectonics; the atmosphere changed from an essentially anoxic atmosphere to an oxygenic atmosphere early in the Proterozoic; oxygen levels rose profoundly late in the Proterozoic; and the oceans responded to these and other biospheric, lithospheric, and atmospheric changes.

As far as astrobiology is concerned, an understanding of the evolution and diversification of life on Earth is critical. The Earth is the most active of all the planets in the Solar System and continues to evolve, 4,550 Myr after its formation. Moreover, it is, as far as we know, unique in possessing life, and this life has evolved in conjunction with planetary evolution. Reading the fossil and rock records provides valuable insight into life's evolution and diversification. Although the search for extant (existing) life elsewhere in the Universe is of tremendous importance, such a search would focus only on the very narrow window of the present time; i.e., what exists now. Searches should also be made at extraterrestrial sites for fossil life, since such a search takes full advantage of the immense spans of time involved with a planet's history. We can look back into the past not only by looking at the light from distant stars, but also by looking at the fossil record.

16.2 The Proterozoic

The Proterozoic spans almost 2,000 Myr. It records the further evolution and diversification of already established bacteria and eukaryotes. This long record, dominated by microbial remains and their sedimentary constructions (stromatolites), documents important evolutionary changes that shaped the biosphere on its way to a planet dominated by plants and animals. Figure 16.1 summarizes the oldest fossil evidence for various types of organisms during the past 2,500 Myr.

The term *Proterozoic* was coined from combining the Greek words *protero*, meaning "fore, anterior," and

---

[1] *Metazoans* are multi-celled animals with different functions for different cells.

[2] See Appendix C for a discussion of the names and dates associated with the Geological TimeScale. See Chapter 12 for the record of life prior to 2500 Ma (the Archean Eon).

Planets and Life: The Emerging Science of Astrobiology, eds. Woodruff T. Sullivan, III and John A. Baross. Published by Cambridge University Press. © Cambridge University Press 2007.

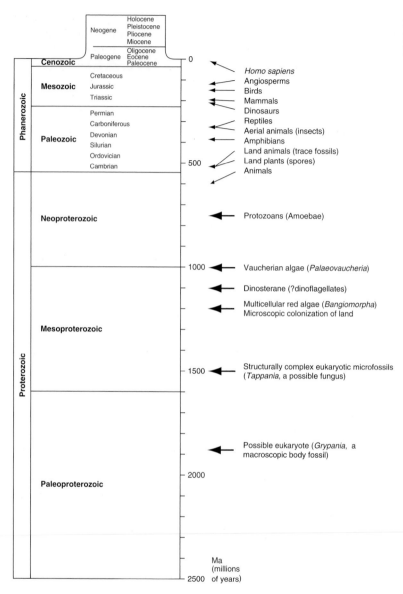

FIGURE 16.1 Some major biological landmarks in the Proterozoic and Phanerozoic history of life.

zoic, which refers to life.[3] Historically, rocks older than the Cambrian were thought (for example, by Darwin) to lack fossils. However, as the sedimentary rocks found below the Cambrian received more attention, indications of fossils were found. This suggested the possibility of a longer history of life than previously thought. Based on a variety of studies, most notably by Walcott (1914), it became established that the Proterozoic contained significant remains of life, but it was dominated by microorganisms.

All three of the domains of the tree of life – Archaea, Bacteria, and Eucarya – were present by the beginning of the Proterozoic. During the Archean (*Archean*, a geological time term for ≤4030 to 2500 Ma, and *Archaea*, a domain of life, share the same Greek root meaning "ancient") the full spectrum of metabolic diversity evolved; see Chapter 12 for a discussion. This included the establishment of anaerobic heterotrophs, lithotrophs, and photoautotrophs as well as variations on these metabolic pathways. Furthermore, during the early Proterozoic cyanobacteria had already evolved oxygenic photosynthesis (see Chapter 12), one of the most complex metabolic pathways. With the rise of an oxygenic atmosphere in the early Proterozoic,

---

[3] The Proterozoic Eon is further divided into three Eras: Paleoproterozoic (2,500–1,600 Ma), Mesoproterozoic (1,600–1,000 Ma), and Neoproterozoic (1,000–542 Ma).

microbes using oxygen greatly diversified; this aerobic metabolism was essential for the evolution of eukaryotes within the already established Eucarya domain. It was during the Proterozoic that three of the four traditional major eukaryotic groups evolved (algae, fungi, then animals, with plants not appearing until the early Paleozoic). However, these groups are probably only minor components of a much larger eukaryoic diversity that is largely microbial in nature and arose in the Proterozoic.

16.2.1 Types of Proterozoic fossils

Proterozoic fossils consist of the following major types: (a) microfossils, (b) stromatolites, (c) macroscopic carbonized compressions, (d) soft-bodied impressions, (e) trace fossils, and (f) mineralized hard parts.

Microfossils are the preserved remains of microbes. There are two primary methods of study: acid residues and petrographic thin sections. Both rely on light microscopy for examination. To prepare a petrographic thin section, a thin sliver of rock is cut from the sample and epoxied to a glass microscope slide. The sliver is further thinned by grinding to a thickness (usually 30 to 50 µm) that allows transmission of light. Chert (microcrystalline quartz; SiO_2) is the sedimentary rock of choice as it can permineralize (mineral matter precipitated in the original organic structure) the microbes, preserving them *in situ* within the rock. The other main technique involves dissolving rock samples in acid and examining the insoluble residues under a microscope. Organic-walled remains in the residues are commonly called acritarchs (see 16.2.4.1). Rocks like siltstone are most commonly used, which requires the use of hydrofluoric acid. Although light microscopy is the most common means to search for and study microfossils, various electron microscopic techniques have also been used.

Stromatolites are laminated structures produced by the sediment trapping, binding, and/or mineral precipitation of microbes, principally cyanobacteria. They are abundant and diverse in Proterozoic carbonate rocks (Awramik and Sprinkle, 1999).

Carbonaceous compressions or carbonaceous films are compressed, chemically resistant, organic material visible to the naked eye (*macroscopic*), and are known from the Proterozoic (Hofmann, 1994; as well as from the Phanerozoic, see Fig. 16.7c). Many, because of their macroscopic size and regular shape, are considered to be eukaryotic fossils.

Soft-bodied organisms can be preserved as *impressions* on bedding planes. For the Proterozoic, the most famous of these fossils is the Ediacaran biota of late Neoproterozoic age that is known from many localities.

Trace fossils are the tangible evidence of the activity of an organism (such as burrowing). There are numerous reports of trace fossils older than 600 Ma; however, most such reports have not withstood critical review.

The acquisition of durable, *mineralized skeletons* (hard parts) by eukaryotes very late in the Proterozoic gave rise to the rich, macroscopic fossil record that characterizes the Phanerozoic.

Unlike the Phanerozoic record of plants and animals, which has been studied for almost 200 years and is rather robust, the pre-Phanerozoic fossil record has received significant attention only for the past 50 years. Many new fossils are yet to be found.

16.2.2 Prokaryotes

Within the past decade or two, it has become apparent that two distinctly different groups (*domains*) comprise the prokaryote microorganisms:[4] Archaea and Bacteria. *Archaea* differ from *Bacteria* in cell wall composition (bacteria have peptidoglycan), the structure and composition of membranes, as well as other differences. There is no direct, body fossil evidence for Archaea in the Proterozoic or earlier. As discussed in Chapter 12, however, there is indirect evidence for the existence of Archaea in the Late Archean based on sediments with highly depleted $\delta^{13}C_{org}$ values, indicative of menthanogens (Archaea).

The record for Bacteria in the Proterozoic is good and is actually better than in the Phanerozoic. Although the Proterozoic microfossil record is dominated by cyanobacteria, this does not necessarily mean that they dominated the biosphere; rather, it could simply be that cyanobacteria were more likely to be preserved than other bacteria. Many of the samples are from silicified stromatolites, which were primarily built by cyanobacteria (see 16.2.3). In the Proterozoic, microbial mats were likely geochemical sites for early silicification (chert formation), which favored the preservation of cyanobacteria. In addition, the chemistry and structure of cyanobacterial sheaths and cell walls probably favored their silica permineralization over other types of prokaryotes.

---

[4] *Prokaryotes* are the simplest single-celled organisms, whereas the cells of *eukaryotes* are more complex, with a nucleus containing the genetic material and other distinctive, functional parts (organelles).

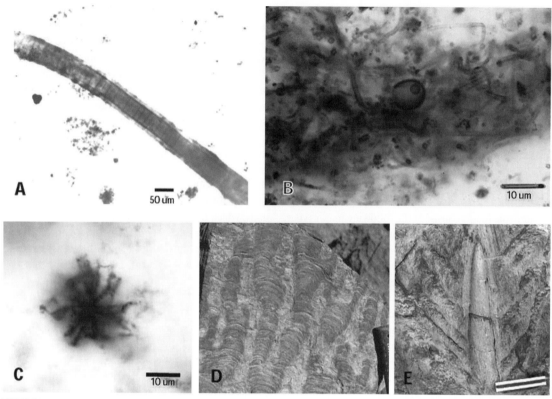

FIGURE 16.2 Proterozoic prokaryotic microfossils and stromatolites.

a. Filamentous cyanobacterium from the ∼600 to 590 Ma (Neoproterozoic) Doushantuo Formation, China, showing internal stack of cells (trichome) surrounded by layered sheath.

b. Probable cyanobacterium (center) from Order Pluricapsales from the ∼1875 Ma Gunflint Formation, Canada. The small spheroidal structure is a spore (endospore) within an open vesicle (sporangium).

c. *Eoastrion* from the ∼1875 Ma Gunflint Formation, Canada. This is a problematic microfossil that resembles the modern manganese-oxidizing bacterium *Metallogenium*.

d. Branched columnar stromatolite from the ∼1,200 Ma (Mesoproterozoic) Atar Group of Mauritania. Such branched columnar stromatolites are quite common in the Mesoproterozoic and Neoproterozoic, less common in the Paleoproterozoic, but uncommon in the Phanerozoic.

e. Branching conical stromatolite from the ∼1,200 Ma (Mesoproterozoic) Atar Group of Mauritania. Such elaborate stromatolites are known only from Meso- and Neoproterozoic strata.

The identification of cyanobacteria in many Proterozoic prokaryotic microfossil assemblages is reasonably well established based on size, morphology, comparisons to extant examples, and depositional settings. For smaller (<2 µm) spheroidal and filamentous microbial fossils, however, bacterial taxonomic assignments other than cyanobacteria are possible; the small size and simple morphology (threads and balls) are shared by many groups of bacteria. This makes assignment to any particular group of bacteria very difficult for such small microfossils. Sometimes the microbial fossil has such a distinctive morphology that modern analogues other than cyanobacteria are indicated (Fig. 16.2c). Sometimes, biomarker studies

can be used to aid in the determination of bacterial taxonomic assignments (e.g., sulphur-oxidizing bacteria; Logan *et al.*, 2001), but unfortunately such examples are few.

Most of our understanding of the development of diversity over time of Proterozoic prokaryotes is dominated by the fossil record of cyanobacteria. It is likely that by about 1875 Ma (age of the Gunflint Formation, Canada) cyanobacteria had already diversified into all but one of their orders (Fig. 16.2b). The overall record of the number of species of Proterozoic fossil prokaryotes indicates a relatively slow rate of increasing diversity that reached a peak early in the Neoproterozoic (Fig. 16.3b). This peak was followed

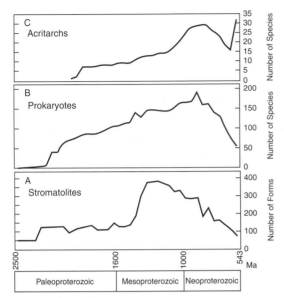

FIGURE 16.3 Three graphs showing the number of taxa (diversity) over time for three major types of Proterozoic fossils.

a. Plot of diversity versus time for stromatolite forms. Forms are roughly equivalent to species. Based on data from Awramik and Sprinkle (1999).
b. Plot of diversity versus time for prokaryote species. These are primarily from microfossils found in thin sections of chert. Most of the taxa are cyanobacteria. Based on data from Schopf (1992).
c. Plot of diversity versus time for acritarchs. Acritarchs are organic-walled microfossils of uncertain affinity that are studied from acid-residues of shale and siltstone. Most acritarchs are considered to be eukaryotic phytoplankters. Based on data from Schopf (1992) and Knoll (1994).

by a comparatively rapid decrease in diversity for the remainder of the Neoproterozoic (Schopf, 1992). The fossil record for Phanerozoic prokaryotes is too meagre for meaningful diversity analysis.

16.2.3 Stromatolites

The most obvious fossils from the Proterozoic are cm- to m-sized stromatolites.[5] Many Proterozoic shallow, marine carbonate units contain stromatolites of a variety of shapes. Shapes include wavy-laminated stratiform structures, domes, columns, branched columns (Fig. 16.2d), cylindrical cones (sometimes with branches; Fig. 16.2e), and oncoids (stromatolites with laminae that almost or entirely encapsulate a central core). The sizes of Proterozoic stromatolites span six orders of magnitude, ranging from columns several

[5] See Section 12.2.2 for a more detailed discussion of stromatolites.

mm in diameter to domes over 100 m across. As highly variable as this might seem, one of the hallmarks of Proterozoic stromatolites and of a few Archean and Phanerozoic examples, is the fact that within a given bed, along the same horizon, the shape and size of the stromatolites are often narrowly restricted. This gives the sense that there is a morphological theme to the stromatolites in that bed. Changes vertically, however, are common. For example, often the base of a bed will have laterally continuous stratiform stromatolites that, as one moves upward, become columns, followed by branched columns.

The morphological theme of relatively narrow size range, the same or nearly the same shape, and a distinctive microstructure (microscopic fabric and texture) has resulted in the taxonomic treatment of a number of stromatolites. These distinctive stromatolites are even treated like fossil species, given species-like binomial names, provided with taxonomic descriptions, and have their stratigraphic ranges determined (e.g., Grey, 1994). Although this practice is controversial (Grotzinger and Knoll, 1999) and is not always practiced in a uniformly rigorous manner, a number of stromatolite taxa can be easily recognized and used as index fossils for determining the age of the strata. About 1200 species-level taxa (forms) have been treated taxonomically, of which 90% of these taxa are from the Proterozoic (Awramik and Sprinkle, 1999). When these data are used to construct diversity plots through time (Fig. 16.3a), the pattern is strikingly similar to the diversity pattern of Proterozoic prokaryotes (Fig. 16.3b).

16.2.4 Eukaryotes

The antiquity of eukaryotes is surprising. Biomarker evidence for their existence is found in 2700 Ma sedimentary rocks in Western Australia (Chapter 12). This was well before the atmosphere was fully oxygenated (~2300 Ma). Also surprising, the first "body" fossil or physical remains is not a microfossil, but macroscopic carbonaceous compressions, millimeters in diameter, from the Negaunee Iron Formation of Michigan, USA (Han and Runnegar, 1992), which is now considered to be 1875 Ma. These compressions have been identified as the fossil alga *Grypania spiralis* (Fig. 16.4a). The subsequent Proterozoic eukaryotic fossil record consists of additional macroscopic carbonaceous compressions (Fig. 16.4b), other macroscopic fossils, and microfossils. Microfossils dominate the eukaryotic record.

In addition to the 2700 Ma sterane biomarker evidence for Eucarya mentioned above, two other

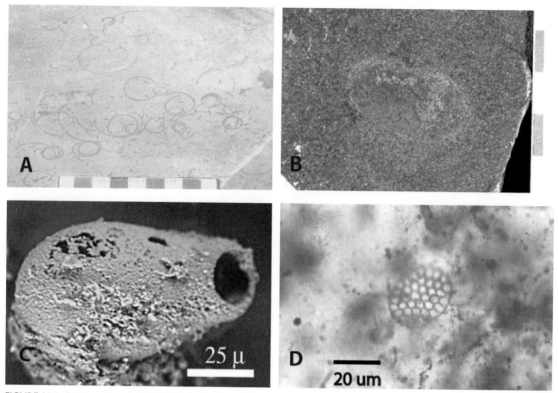

FIGURE 16.4. Proterozoic eukaryotic microfossils, macrofossils, and problematica.

a. *Grypania spiralis*, a carbonaceous compression from the ~1875 Ma Negaunee Iron Formation, Michigan, USA. Coiled to spiralled compressions are seen. This fossil is generally considered to be the oldest remains of a eukaryote. Scale is in centimeters.

b. Carbonaceous compressions from the ~1800 Ma Changzhougou Formation, Tanshan Range, north China. These compressions may be multicellular algae (Zhu *et al.*, 2000) or redeposited, thin clasts of carbon-rich material. Scale is in centimeters.

c. A scanning electron microscope image of a vase-shaped microfossil from the Neoproterozoic Chuar Group, Grand Canyon, USA. These have been interpreted as the remains of testate amoebae. Image courtesy of S. M. Porter.

d. Scale-like, siliceous microfossil (probably chrysophyte) from the Neoproterozoic Tindir Group, Canada.

biomarkers provide significant information on appearances of eukaryotes (Porter and Knoll, 2000): (1) dinosterane, which is a distinctive sterol synthesized by dinoflagellates, is known from the ~1100 Ma Nonesuch Formation, Michigan, USA, and (2) gammacerane suggestive of ciliates occurs in the ~750 Ma Chuar Group, Grand Canyon, USA.

The fossil record, biomarkers, and molecular phylogeny of the Eucarya suggest that the appearances of the major clades were spread over many hundreds of millions of years, perhaps as much as 2,000 Myr (Fig. 16.1). This appears to be different from the apparent pattern (a) in Bacteria, in which major clade diversification may have occurred within a few hundred million years and (b) within the Eucarya, for example, in Animalia, where phylum-level diversification occurred within a few tens of millions of years.

Our knowledge of the fossil record of eukaryotes is continually being revised as new and often significant fossils are discovered. The overall understanding on the origination and diversification of major eukaryote groups is ongoing and subject to change. Hypothesized patterns of origination and diversification often need to be modified or replaced with new hypotheses as these discoveries are reported. Twenty years ago a report of Paleoproterozoic eukaryotes was greeted with disbelief. Today, the "too-great-an-age" shock is reduced, a diverse record is being established, and yet healthy skepticism prevails. We now briefly discuss the major types of eukaryote fossils found in the Proterozoic.

16.2.4.1 Acritarchs

The term *acritarch* was created to accommodate organic-walled microfossils of uncertain taxonomic affinity that were found in acid-resistant residues of shale and siltstone. Acritarchs come in a variety of shapes (from spheroidal to polyhedral), with or without ornamentation (some have spines or elaborate processes), and sizes (a few micrometers to several millimeters). They are usually considered to be cysts, spores, or vegetative unicells of algal phytoplankters (microscopic algae that live in the upper part of water columns), although other origins are possible. Size is important when considering the affinity of these microfossils. Unornamented spheroidal acritarchs (sphaeromorphs) >25 μm in diameter are usually considered to be eukaryotes, but this rule of thumb must be used cautiously because prokaryotes larger than 25 μm are known (Schopf, 1992). Possibly the oldest acritarchs (sphaeromorphs around 48 μm in diameter) come from ~2100 Ma shale in Siberia, suggesting the establishment of eukaryotic phytoplankton by this time. Acanthomorphs, complex acritarchs with spines or other projections that are more confidently interpreted as planktonic eukaryotes, are known from strata ~1500 Ma in Australia. Acritarchs are the most abundant eukaryotic fossils found in the Meso- and Neoproterozoic.

Acritarch diversity patterns (e.g., Knoll, 1994) are roughly similar to prokaryote diversity patterns except for a second marked peak in diversification in the latest Neoproterozoic (Fig. 16.3c). This second peak corresponds to an interval of time, 600 to 542 Ma, when animals first appeared and a global, soft-bodied fauna, the Ediacaran Fauna, existed. Both the acritarchs and the fauna of this interval underwent significant extinctions by the close of the Neoproterozoic. This corresponds to an interval of geological time that is characterized by one or two Snowball Earth events (glaciation so extreme that ice covered the seas and parts of the continents in equatorial regions as well as at higher latitudes).

16.2.4.2 Other algae, amoebae, and fungi

Some microfossils that are morphologically distinctive can be attributed to specific eukaryotic clades (and hence are not called acritarchs). Microscopic red algae (*Bangiomorpha*) are found in 1200 Ma rocks of Arctic Canada. These are the oldest, taxonomically resolved eukaryotes yet known (Butterfield, 2000). Plate-like, siliceous structures (Fig. 16.4d) a few micrometers in diameter found in chert of the 750–600 Ma upper Tindir Group in Canada are probably scales of chrysophytes (Allison and Hilgert, 1986). Chrysophytes, also known as golden algae, are heterokonts (a group

of eukaryotes that includes diatoms, brown algae, and water molds that underwent a secondary symbiosis involving a red alga and possesses two different flagella). Calcification in algae appears to have occurred earlier than in animals. Thin, sheet-like calcareous structures, interpreted as calcareous algae, have been found in the Kingston Peak Formation (750–700 Ma), California, USA (Horodyski and Mankiewicz, 1990). It is not until the latest Neoproterozoic, at about the same time that calcareous shelled animals appeared, that calcareous algae as well as calcified cyanobacteria became more abundant and diverse. Vase-shaped microfossils (Fig. 16.4c), common from several late Neoproterozoic (~750 Ma) localities, are considered to be testate amoebae (protozoans) based on their strong similarity with living examples (Porter and Knoll, 2000). The fossil record of fungi is known almost exclusively from microscopic remains. Based on molecular clock analyses, fungi are generally considered to have originated in the Proterozoic. The aforementioned Tindir Group contains morphologically complex and distinctive microscopic remains of likely fungal origin (Allison and Awramik, 1989). One microfossil from 1,430 million-year-old rocks (Roper Group, Australia), *Tappania*, might also be a fungus (Butterfield, 2005).

16.2.4.3 Carbonaceous compressions

Millimeter and larger-sized carbonaceous compressions from the Proterozoic constitute an interesting type of fossil (Fig. 16.4a,b). Based on their macroscopic size and distinctive shapes they are considered to be multicellular eukaryotes (Hofmann, 1994). Disks, ellipses, ribbons, sausage-like shapes, leaf-like shapes, and irregular shapes are some of the morphologies found. Besides the 1875 Ma example from Michigan, which is a coiled ribbon (Fig. 16.4a), there are other Paleoproterozoic occurrences that include irregular angulate forms, irregular rounded forms, disks, and straight to curved ribbons. These shapes continue into the Mesoproterozoic; however, diversity and complexity increases in the Neoproterozoic. In all, about 100 genus-level taxa have been established for the Proterozoic. Most are considered to be formed by algae, although some annulated forms from 850–800 Ma rocks have been interpreted as worms or worm-like metazoans.

16.2.4.4 Animals

The fossil record of Proterozoic animals is controversial but the most confident remains consist of an

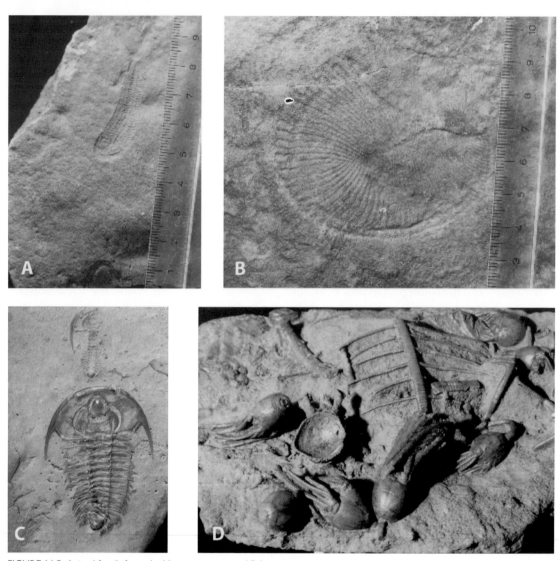

FIGURE 16.5 Animal fossils from the Neoproterozoic and Paleozoic.

a. *Spriggina*, a member of the late Neoproterozoic Ediacaran fauna. Interpretations of this fossil range from a polychaete worm to an extinct phylum. This is from the Pound Quartzite, South Australia. Centimeter-scale ruler.

b. *Dickinsonia*, a member of the late Neoproterozoic Ediacaran fauna. Interpretations of this fossil range from a polychaete worm to an extinct phylum. This is from the Pound Quartzite, South Australia. Centimeter-scale ruler.

c. The Trilobite *Redlichia* from the Early Cambrian Linnekar Limestone, Western Australia. Trilobites dominate the Cambrian fauna and are restricted to the Paleozoic. Specimen is 22 mm long.

d. The stalked crinoid *Neocamptocrinus* from the Early Permian Cundlego Formation, Western Australia. Crinoids are echinoderms (spiny-skinned animals) and these were attached to the substrate, had a stalk (a), calyx (b), and long arms (c) for gathering food. The calyx in these is 20 mm across.

assemblage of soft-bodied animals, the Ediacaran biota, from late Neoproterozoic strata worldwide (except Antarctica). The ages of these fossils range from ~565–540 Ma (a few taxa are known from the Cambrian). The Ediacarans comprise a wide variety of soft-bodied marine animals with some forms suggestive of, for example, jellyfish, worms, and sea pens (Fig. 16.5a,b), while others do not resemble any living counterparts. Because of the unique morphology of many, the animal affinities have been questioned and it has been proposed that these represent an extinct Kingdom.

In addition to soft-bodied fossils, animal trace fossils are also known from Ediacaran-aged rocks. Most traces are mm-sized horizontal trails, but some show shallow sediment penetration (burrowing). Pre-Ediacaran trace fossils are more controversial. The controversy often centers on the unexpected great age. Curiously, some of these structures would likely be accepted as trace fossils if they were found in younger (Phanerozoic) rocks. Two reports on pre-Ediacaran trace fossils are worthy of mention. One consists of mm-diameter, sinuous traces from strata in India that was thought to be 1100 Ma. This report received considerable attention, in large part because the structures looked like animal traces, but were almost 500 Myr older than the previously accepted oldest animal traces. Recently, new radiometric age dates on the rocks yielded an even older age: 1600 Ma. It is likely that these are not animal trace fossils. The second report describes possible "trace-like fossils" described from >1200 Ma rocks in Western Australia. It is not clear if these structures (a) were produced by animals, (b) represent some extinct lineage independent of the metazoans, or (c) had an origin independent of organisms (i.e., abiogenic) (Rasmussen *et al.*, 2002). The pre-Ediacarian age suggested by reports like these is not inconsistent with available molecular phylogenetic data that postulate 1500 to 700 Ma for the divergence of crown-group metazoans.

The evolution of the Eucarya, as seen from the fossil record, was one of increased morphological complexity. This complexity included an increase in intricacy of cells and their coverings (acanthomorphs), an increase in the number of cell types, the evolution of multicellularity, the development of a tissue grade of organization, and the formation of organs. Much of this was accompanied by an increase in size to macroscopic proportions. The development of eukaryotes may have been influenced by oxygen in the atmosphere. Oxygen levels rose significantly at least twice in the Mesoproterozoic and Neoproterozoic, once at about 1300 Ma and again in the late Neoproterozoic.

The close of the Proterozoic saw an end to the heyday of prokaryotes dominating the fossil record. The biosphere was beginning to be taken over by macroscopic eukaryotes, and the tempo and mode of the evolution would profoundly change as animals and plants expanded in the biosphere.

16.3 The Phanerozoic

The Phanerozoic, an eon characterized by macroscopic animals and plants, has lasted only 542 Myr. However,

the microbial world of the Proterozoic has not really disappeared; it has just been overshadowed by these large forms of multicellular life. Other chapters in this book testify to how microorganisms even today are in some sense still dominant in the planet's biosphere. The fossil record of the Phanerozoic is much better understood than the Proterozoic, not only because of its richness, but also from the fact that many more paleontologists study its plants and animals than study the fossil record of the Proterozoic. The biostratigraphic and absolute time control on the fossil record is also excellent. Hence, patterns of radiations (increased diversification of a group of organisms) and extinctions of fossil groups are quite well defined and contribute significantly to our understanding of evolution on planet Earth.

16.3.1 The Cambrian "explosion"

The appearance of organisms in the fossil record with hard parts such as shells signals the transition from the Proterozoic to the Phanerozoic. The first 54 million-year period of the Phanerozoic (the Cambrian Period) marks perhaps the most important interval in the evolution of life on Earth. It is during this time that all the major modern animal phyla either appear or greatly diversify. Known as the Cambrian "explosion," the nature and cause of this seemingly abrupt appearance of animals has been, and continues to be, the subject of much debate (e.g., Conway Morris, 2000). The question is, was this a real event – a sudden explosion in life forms – or was it due to the nature of the fossil record, preserving much more readily those organisms that had evolved hard parts? Either way, this most significant of evolutionary radiations invokes a range of causes, both intrinsic – the role of developmental genetics in the evolution of novel body plans, and extrinsic – changes in ocean and atmospheric chemistry, as well as ecological changes.

The Neoproterozoic–Cambrian boundary (dated at ~542 Ma) is available for study in many areas around the world. Studies have resulted in a general pattern with regard to the order in which the animal fossils appear in the strata. Below the boundary, low diversity faunas[6] of both body and trace fossils are abundant, while above one finds diverse metazoan assemblages, typified initially by so-called "small shelly faunas," then by more typical Cambrian faunas, dominated

---

[6] A *fauna* is a collection of associated animals, and a *flora* is similarly for plants.

by arthropods such as trilobites (Fig. 16.5c). Either fortunately, or due to the nature of the Cambrian environment, there is a disproportionately large number of exceptionally preserved fossil deposits in which organisms both with and without hard parts have been preserved. The two most important are the Early Cambrian Chengjiang biota in China and the Middle Cambrian Burgess Shale biota in Canada.

The fossil record indicates a short interval of only ~30 Myr, from the time of the Ediacaran fauna (the end of the Proterozoic) to the early Cambrian, in which most modern phyla evolved! However, studies of molecular clock data from living organisms propose a very much earlier appearance for metazoans, as much as 1500 Ma. The fossil record suggests that only five of the modern animal phyla had appeared before the Cambrian – sponges, stem-group cnidarians, anthozoans, stem-group lophotrochozoans (the group containing molluscs and annelids) and possibly platyhelminthes (Conway Morris, 2000). However, during a period in the Early Cambrian that may be as short as 10 Myr, the other major phyla appeared, including groups such as ctenophores, priapulids, nematodes, lobopodians, arthropods, molluscs, halkieriids, annelids, brachiopods, hemichordates, echinoderms, cephalochordates, and chordates (which include all vertebrates).

One of the challenges in interpreting the extent of the Cambrian evolutionary radiation (explosion) is the degree to which the early Phanerozoic faunas were related to the Ediacaran faunas. While arguments for the affinities of these organisms are legion, at least some of these earlier forms would appear to have affinities with modern phyla. The most significant of these are the sponges, anthozoan cnidarians (corals), and stem-group triploblasts. Although one of the features that characterizes the Cambrian explosion is the evolution of biomineralization (the ability to produce minerals, usually based on Ca, but also Si and P, through metabolic processes that results in shells or other hard parts), forms that frequently occur in younger rocks and are heavily mineralized, notably the cnidarians, are not very common in Cambrian strata. For example, the Chengjiang and Burgess Shale faunas indicate that arthropods and sponges were the dominant groups in the Early to Middle Cambrian seas. Following their appearance in the latest Proterozoic, sponges diversified greatly in the Cambrian. For instance, in the Chengjiang fauna more than 20 species occur. They also play a major role in the formation of the first Cambrian reefs, particularly archaeocyathid sponge reefs that are particularly common in shallow marine environments.

Arthropods comprise ~45% of the Chengjiang and Burgess Shale faunas. Functional analyses suggest that arthropods occupied a range of ecological niches, having evolved morphological adaptations linked to defence and feeding strategies. The geologically simultaneous appearance in a number of phyla of biomineralization argues for strong survival advantages for these hard, outer body coverings. This may have been for protection in response to the evolution of predator/prey relationships. The top-line predator was *Anomalocaris*, a form that links the more primitive lobopods with the more advanced arthropods (such as crustaceans and trilobites).

Studies of predation on Cambrian trilobites (Babcock, 1993) have shown that many of the healed scars on the trilobites' bodies match the size and shape of the mouth parts of *Anomalocaris*. Babcock's study revealed that the trilobite predators show evidence of having exhibited strong right–left behavioral asymmetry. Predation scars demonstrate that 70% were preferentially attacked on the right side. In modern animals this is a function of lateralized nervous systems, indicating that such systems were present in arthropods as far back as Early Cambrian times. This, combined with complex compound eyes in even the earliest trilobites, supports the contention of an evolutionary history for the arthropods that extends back into the Proterozoic.

The not infrequent occurrence of spectacularly preserved Early and Middle Cambrian faunas, such as those at Chengjiang and the Burgess Shale, raises the question of why similar soft-bodied preservation of marine ecosystems is rare in post-Cambrian strata. One reason may be the lack of tiering in Cambrian marine faunas. In other words, most organisms lived close to the sediment–water interface – few organisms had evolved the ability to burrow. With few burrowing scavengers, chances of dead organisms being fossilized would have been increased. The preservation of soft tissue could also have been the result of less decay due to somewhat lower oxygen levels in the Cambrian compared to today.

As to what triggered the Cambrian explosion, the mechanisms are still obscure. Studies of isotopic and chemical indicators, principally $\delta^{13}C$, $^{87}Sr/^{86}Sr$, and phosphogenesis, indicate that significant changes were occurring in ocean chemistry and circulation. However, the extent to which these affected or stimulated the evolutionary radiations, or rather are a reflection of it, is unclear. The Proterozoic–Phanerozoic transition is characterized by extensive phosphogenesis, deposition of black shales, and extreme negative shifts in $\delta^{13}C$. Generally considered to have resulted as a consequence of changes in global tectonics, ocean circulation, and/or nutrient supply, it is also possible that these chemical changes are a result of the Cambrian explosion arising from the development of more complex food webs (Butterfield, 1997). Analysis of the Burgess Shale community has indicated the presence of feeding habits analogous to those existing in present-day marine ecosystems (Conway Morris, 1986). Marine primary productivity is likely to have increased significantly with a major diversification of phytoplankton in the Cambrian (Vidal and Moczydlowska-Vidal, 1997). Accompanying this, and perhaps even driving it, was the evolution of zooplankton such as filter-feeding branchiopod arthropods. This may have been a key innovation that led to the evolution of larger animals (Butterfield, 1997). In addition to providing a food source for larger predators, the zooplankton may have significantly affected the ecology of the ocean floor. Zooplankton fecal pellets may have played a role in oxygenation of the seafloor, as well as increasing the availability of nutrient-rich organic compounds to the seafloor biota.

What drove the evolutionary radiations of metazoans and phytoplankton were likely to have been ecological factors, in particular the appearance of herbivores and carnivorous predators, engendering evolutionary "arms races" between predators and prey, as well as the evolution of a range of new feeding strategies. These include carnivorous behavior, scavenging, and filter feeding in the ocean. Intrinsic factors related to developmental genetics were also likely to have been of significance. Poor regulation of genes controlling development would have resulted in significant morphological diversification over very short time periods. For instance it is suggested that poor regulation of homeotic genes in trilobites during the Early Cambrian may have been a factor in driving evolutionary radiations in trilobites (McNamara, Yu, and Zhou, 2003).

16.3.2 The biosphere colonizes the land

16.3.2.1 *Bacteria and plants*

The earliest evidence for colonization of the land comes from 1200 Ma rocks in central Arizona (USA), where Horodyski and Knauth (1994) described microfossils that represent the remains of microbial communities on land. These have morphologies and sizes suggestive of

bacteria (including cyanobacteria), and are thought to have been living at or close to the land surface. Evidence for biological activity in these rocks is also supported by ^{13}C depletion, caused by CO_2 respiration into the groundwater by terrestrial[7] organisms. These microbial communities may also have contributed to observed weathering of the rocks.

The presence of spore-like structures in Middle Cambrian shales in Tennessee and the Grand Canyon (USA) may be the first indication of the presence of land plants (Strother, 2000). These simple spheres are thought to have derived from very primitive land plants, and were probably closest to simple bryophytes (mosslike plants). The large morphological diversity of these structures indicates a relatively diverse flora on land at this time. This would have had significant consequences for atmospheric evolution. The presence of a Cambrian terrestrial flora living by photosynthesis may have played a significant role in drawdown of atmospheric CO_2 (however, see Kasting and Siefert, 2002). Indeed, it has been suggested (Strother, 2000) that this might even have played a part in the Cambrian "explosion," carbon runoff affecting the trophic structure of shallow water ecosystems.

16.3.2.2 *Invertebrates*

The fossil record indicates that the first metazoans to colonize the land were arthropods. Arthropod tracks from ~500 Ma sandstones in Ontario, Canada, have been interpreted as having been made out of the water on land (MacNaughton et al., 2002). However, these arthropods are likely to have been amphibious, and would have faced immense physiological problems in making the transition from an aqueous to a dry environment. The much greater diurnal changes in the atmosphere than in the aquatic environment presented problems of potential desiccation. Moreover, different methods of gas exchange were needed. A major limiting factor in the transition to air-breathing was the problem of CO_2 excretion in air. The success of arthropods in early land colonization lay in their tough exoskeleton, pre-adapted to a life on land. As well as providing defense against predators, it strengthened the body for locomotion and feeding, and helped overcome the effectively increased gravity (due to loss of buoyancy in the water).

---

[7] The term *terrestrial* in this chapter refers to "on the land," as opposed to "*marine*."

Other evidence for animal activity on land is Late Ordovician (~450 Ma) burrows in fossil soils in Pennsylvania (Retallack and Feakes, 1987) that may have been made by millipedes. These small arthropods may have overcome high temperatures and potential desiccation by burrowing in the "soil." In addition, Sherwood-Pike and Gray (1985) have described fossilized fecal pellets from Late Silurian deposits (~410 Ma) in Sweden that contain fungal hyphae indicating the existence of fungivorous microarthropods, such as mites or millipedes. This implies the presence of a decomposer niche, with the arthropods playing an important role in the establishment of soils through reworking and increasing nitrate and phosphate levels. This would have provided a necessary habitat to allow subsequent colonization by vascular plants,[8] which had occurred by that time.

Support for millipede colonization of the land comes from subaerial (out of the water) fossil tracks attributed to these arthropods from Middle Ordovician (~470 Ma) rocks in the Lake District, England (Johnson et al., 1994). A rich variety of arthropod trace fossils from Late Ordovician rocks (~440 Ma) in Western Australia provides evidence for a diverse fauna of large terrestrial arthropods on land by this time. Tracks ranging in width from 5 to 300 mm indicate that eurypterids (possible giant scorpions, centipedes, and euthycarcinoids), as well as a range of other unknown arthropods, were walking on land. A single body fossil of the euthycarcinoid arthropod *Kalbarria brimmellae* represents the oldest known fossil of a land animal (Trewin and McNamara, 1995).

Muddy siltstones just above the Late Silurian Ludlow Bone Bed (~410 Ma) in Shropshire, England, have yielded a large quantity of fossil arthropods that were permanently terrestrial. This includes at least two types of centipedes, the trigonotarbid arachnid *Eotarbus jerami*, an arthropleurid, and a probable terrestrial scorpion, all predatory. The Early Devonian Rhynie Chert (~400 Ma) in Aberdeenshire, Scotland, a terrestrial hot-spring deposit, contains fossils of some of the earliest, simple vascular plants, in addition to algae, fungi, lichens, and a fauna dominated by arthropods. The arthropods include the fairy shrimp *Lepidocaris*; three species of spider-like trigonotarbids; the mite *Protocarus*; and the earliest insect, *Rhyniella*. The presence of book lungs similar to those in modern spiders in one trigonotarbid, *Palaeocharinus*, provides

firm evidence that the animal had achieved a major step in living on land – respiring in a gaseous atmosphere. Slightly younger Middle Devonian remains from Gilboa in New York State (USA) have yielded remains of the earliest known spider, *Attercopus fimbriunguis*. Like modern spiders it is equipped with a spinneret, providing evidence that even the earliest spiders were able to spin webs. It was also endowed with fangs and a poison gland. Like the arthropods from Rhynie and Ludlow, those from Gilboa were mainly predators, probably feeding on abundant mites and millipedes. There is little evidence for herbivorous arthropods in the fossil record until well into the Carboniferous Period, ~50 Myr later. Shear (1991) has argued that by-products from the synthesis of lignin, which was present in early vascular plants, may have been toxic to early terrestrial animals. Eating plants may only have become established when animals had evolved appropriate enzymes and digesting symbiotic bacteria in their gut.

16.3.2.3 Vertebrates

Vertebrate colonization of the land occurred some time after the arthropod invasion. The oldest tetrapod (four-legged animal) evidence is incomplete remains of *Elginerpeton* from the Late Devonian Scat Craig deposit (~370 Ma), near Elgin, Scotland (Ahlberg, 1991). This amphibian was very fish-like in its anatomy, and presumably in its behavior. The most complete of the early tetrapods are *Ichthyostega* and *Acanthostega* from the Late Devonian (~360 Ma) of East Greenland. Recent detailed work on their limbs has revealed that *Ichthyostega* had seven digits on the foot, whereas *Acanthostega* had eight on the hands and feet. Coates and Clack (1990) suggested that this was probably an adaptation for swimming, rather than for walking on land, casting doubt on the terrestrial ability of these tetrapods. Studies of the braincase, limb skeletons, and gill arches indicate that these early amphibians were little more than fishes with slightly modified skull patterns, and digits rather than fins on the ends of their limbs. They were still largely aquatic, possibly venturing onto land for short periods.

The fish–tetrapod transition was complex, involving large morphological and physiological changes. These include the development of air breathing, the ability to walk on land, the ability to hear in air, the ability to retain body moisture, and the ability to reproduce out of water. It is likely that these steps were taken one at a time. The first tetrapods did not rush out of the water, but emerged fully equipped and pre-adapted for a life on land. The earliest tetrapod

---

[8] *Vascular plants* are those with a system of tissues for circulating water and nutrients, as in all modern flowering plants and ferns.

likely to have been adapted for walking on land was *Pederpes* from the Early Carboniferous (~300 Ma) of Scotland. While the presence of a lateral line indicates *Pederpes* was partly aquatic, the structure of its foot suggests that it was also adapted for walking on land.

The selection pressure that "drove" vertebrates onto land may well have been to find new prey. Panderichthyid fishes, from which tetrapods probably evolved, are thought to have been predators, possessing very large heads relative to their body size, large fangs, and a wide gape. Their well-developed hands and feet, while not initially evolved for locomotion on land, were pre-adapted for supporting their body weight out of water. The positioning of their eyes on top of the skull meant that they were perfectly suited to viewing a new, terrestrial world.

The first tetrapods possessed a long fish-like tail, little changed from the panderichthyid fish tail. Even after the more advanced amphibians developed improved limb girdles that allowed them to walk freely on land, the tail remained as the important propulsive device for aquatic forays. Finally, the first amphibians retained a body covered by fish-like scales. Coates (1996) has suggested that as they began to invade the terrestrial habitat, on limbs of limited capability for locomotion on land, a scale-cover over the belly would have provided important protection when dragging themselves across the ground. It would also have assisted in preventing desiccation.

16.3.2.4 *The aerial niche*

It was not until middle Carboniferous time (~325 Ma) that the first organisms occupied the aerial niche. Arthropods again led the way, this time in the form of flying insects. The first insects were flightless and evolved in the Early Devonian, but then insects are absent from the fossil record for 55 Myr. During this time the first flying insects must have evolved, as ten orders of flying insects suddenly appear in the fossil record at the Middle/Early Carboniferous boundary (~335 Ma) (Labandeira, 2001). The evolution of the largest known flying insects early in the group's evolutionary history, dragonflies with wingspans up to 70 cm (!), may have been in part a consequence of the high atmospheric levels of both CO_2 and O_2 at this time, resulting in higher atmospheric pressures capable of supporting such large bodies.

16.3.3 Evolution during the Paleozoic Era

16.3.3.1 *Marine faunas*

The fossil record of the Phanerozoic is dominated by marine organisms, mainly invertebrates. This is due to

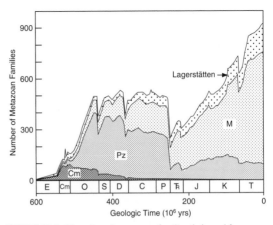

FIGURE 16.6 Diversity of metazoan families (adapted from Sepkoski, 1981). Cm = Cambrian fauna, Pz = Paleozoic fauna, and M = Modern fauna. E = Ediacaran, T = Tertiary. For the other time symbols, see Fig. C.1. Lagerstätten are unusual fossil deposits that contain a wide variety of well-preserved fossils, often soft-bodied, that provide snapshots of diverse past life.

the enhanced chances in marine settings of burial in sediments after death. Moreover, for the Paleozoic, particularly the early part, the fossil record indicates that most organisms were aquatic. Studies of changing diversity levels during the Phanerozoic have shown that the Paleozoic comprises two distinct evolutionary faunas: the Cambrian fauna, dominated by trilobites (Fig. 16.5c), and a post-Cambrian fauna, known as the Paleozoic fauna (Sepkoski, 1981). These are two of the "three great marine faunas of the Phanerozoic" of Sepkoski, the other being the post-Paleozoic Modern fauna (Fig. 16.6). The Cambrian fauna declined appreciably in diversity at the onset of the Ordovician (~500 Ma), being replaced by the evolutionary radiation of the Paleozoic fauna, which persisted for 200 Myr before declining abruptly at the Permo–Triassic mass extinction event (~250 Ma; Chapter 18), and then itself was replaced by the Modern fauna.

Whereas the Cambrian evolutionary radiation was characterized by the origin and diversification of basic body plans at the phylum and class levels, the Paleozoic fauna arose from a burst of diversification at lower taxonomic levels. There was a three- to four-fold increase of global richness of marine families during the Ordovician compared with the Cambrian. The dominant classes to radiate were primarily immobile, suspension-feeding species, such as articulate brachiopods, crinoids (Fig. 16.5d), cephalopods, gastropods, and bivalves. Many of these groups were associated with reef systems that became dominant in low latitude regions from mid-Ordovician times onward. By the

FIGURE 16.7 Phanerozoic fossils.

a. The placoderm fish *Mcnamaraspis* from the Late Devonian Gogo Formation, Western Australia. This extinct group of fishes was dominant during the Devonian. Their mineralized bone was on the outside as armored plates covering the head and front of the body. Total length 10 cm.
b. Carbonaceous compressions of the seed fern *Alethopteris* from the Late Carboniferous of Pennsylvania, USA. Centimeter scale.
c. Trackway (trace fossil) in the Late Ordovician Tumblagooda Sandstone, Western Australia, made by an unknown arthropod walking on wet sand exposed to the air. Lens cap 5 cm across.
d. The ammonite *Stephanoceras* from the early Jurassic of eastern France; 13 cm in diameter.
e. The echinoid *Linthia* from the Early Tertiary Cardabia Formation, Western Australia. Such spatangoid echinoids (heart urchins) are common elements of late Mesozoic and Cenozoic shallow marine sediments. Scale bar 10 mm.

mid-Devonian metazoan reef builders had evolved a variety of shapes of large structures, with individuals or colonies up to 5 m in diameter and reef systems many hundreds of kilometers long. The reef landscape varied appreciably during the Paleozoic. Most reef-building organisms were unable to gain secure and permanent attachment to hard substrates, as in modern reefs, implying they might not have been able to colonize environments with strong waves and currents. Tropical shallow waters supported very large populations of small erect bryozoans, corals, and sponges.

The increase in diversity during the early part of the Paleozoic evolutionary radiation (Fig. 16.6) was not the same at all places, with marked differences by latitude and environment. Diversification differed appreciably among the different Ordovician continents, with low-latitude carbonate faunas dominated by brachiopods and sponges, but high-latitude siliciclastic faunas were trilobite- and mollusc-rich. On a global scale, Sepkoski's Paleozoic fauna went into decline during the Devonian. This may have been due to increasing interactions between the predominantly immobile suspension-feeders with diversifying deep-burrowing deposit feeders. This group also benefited from that period's increased tectonic activity that led to increased supplies of siliciclastic sediments. Fishes also

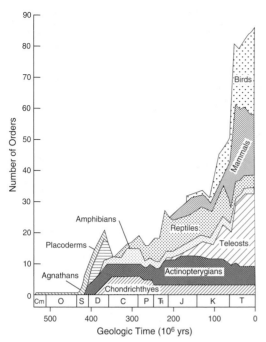

FIGURE 16.8 Changes in diversity for marine and terrestrial vertebrates (modified from Miller (2002) and Padian and Clemens (1985)).

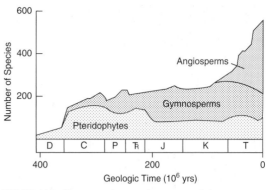

FIGURE 16.9 Changes in diversity for terrestrial plant species (modified from Miller (2002) and Niklas (1997)).

diversified greatly during mid-Paleozoic times, particularly sharks, osteichthyans (Fig. 16.7a), acanthodians, and placoderms (Fig. 16.8).

16.3.3.2 Terrestrial plants

During Cambrian–Ordovician times ($\sim$542–450 Ma), atmospheric CO_2 levels may have been up to 18 times higher than present-day levels. This played an important role in soil formation since more acidic rain increased chemical weathering of rocks, so helping soil formation and providing an environment in which vascular plants could grow and diversify. Moreover, CO_2-enrichment favored photosynthetically active organisms that promoted the decomposition and release of inorganic nutrients. However, the very high CO_2 levels were not beneficial for terrestrial plants as they contributed to very high global temperatures, with summer temperatures averaging over 40 °C (Berner, 1997). Only when global temperatures began to decline did vascular plants evolve and diversify.

Another reason why terrestrial plants did not diversify until the Devonian (Fig. 16.9) may relate to the evolution of lignin. Lignin is an important strengthening polysaccharide for structural support. In aquatic environments plants do not require a vascular system, i.e., specialized cells to distribute water and nutrients.

However, terrestrial plants need specialized conducting tissue to transport food, even to get 2 cm off the ground. The first vascular plants, such as *Cooksonia*, developed tracheids – tubes with lignified-type thickening. The ability of plants to synthesize lignin is dependent on levels of atmospheric oxygen. Moreover, to grow upright the plants required an anchoring mechanism, in the form of roots. These were also necessary to obtain nutrients and water from soils. The first evidence for roots in the fossil record is not until the Early Devonian ($\sim$400 Ma).

Over the next 50 Myr the terrestrial flora underwent its greatest diversification (Fig. 16.9), evolving from small vascular plants to dense vegetation with trees over 35 m tall (Fig. 16.7b). During the early period of vascular plant diversification in Early to Middle Devonian times, CO_2 levels were possibly 8–9 times higher than now. This was one of the most active periods in the Phanerozoic of plate movement and dramatic global climate change, and probably contributed significantly to floral diversification. During Carboniferous time global climate changed from warm, humid, and ice-free to cooler and drier with a glaciation at high latitudes in the Southern Hemisphere. This caused a drop in sea level of 100–200 m and aridity in high latitudes. But in a narrow equatorial belt, rain all year meant that vegetation accumulated in lowland swamps and led to the formation of the great coal deposits of North America and Europe. The continental blocks of Gondwana and Laurasia joined to form the supercontinent of Pangaea late in the Early Carboniferous.

From Late Devonian to Late Carboniferous times ($\sim$370–300 Ma) CO_2 levels are thought to have plummeted from 3600 to 300 ppm, contributing significantly to global cooling. CO_2 levels are estimated by using the density of stomata on fossil leaves – the greater the density the lower the CO_2 levels. This reduction in

CO_2 levels was partly due to the rapid diversification of vascular plants (Berner, 1997). As they diversified, their roots contributed to increasing weathering and consequent drawdown of atmospheric CO_2. Some of the CO_2 became dissolved bicarbonate and was transported to the oceans by rivers, while some was locked away as coal in the large accumulations of organic plant material, particularly as lignin, which is very resistant to decay.

During Devonian and Carboniferous times (~400–300 Ma) plants not only grew larger, but developed more sophisticated reproductive structures, water transport systems, and root and leaf architecture. The first land plants had stomata on their stems and were photosynthesizing from their stems. Leaves first appeared as very small structures during the Early Devonian, then became larger and broader by the end of the Devonian. This presumably was in response to the reduction in atmospheric CO_2, as more leaf area was required to accommodate more stomata. The evolution of seeds during the Devonian revolutionized the plant cycle by freeing plants from the need for external water for sexual reproduction, as well as protecting the developing embryo and providing it with nutrients. This allowed ecological expansion of plants from restricted stream habitats to inland and higher terrains. As plant size increased, so too did depth of roots. This had a further impact, producing weathering deeper into the soil profile.

The first trees appeared during mid-Devonian times (~380 Ma). For the next 80 Myr forests of spore-producing trees, plus two groups of early seed-producing trees, dominated the landscape. The spore-producing trees were lycopsids (club mosses) such as *Lepidodendron*, which grew up to 35 m and represented two-thirds of early forests, and sphenopsids (horsetails) such as *Calamites*, which flourished in swampy conditions. Other significant elements of the flora were ferns, up to 10 m tall; progymnosperms, an extinct group from which all seed plants (including conifers) evolved; pteridosperms (seed ferns; Fig. 16.7b); and *Cordaites*, a common Carboniferous plant.

By the Early Permian (~290 Ma) the development of the single supercontinent Pangaea had great impact on climate. Global temperatures increased due to the movement of Pangaea to tropical latitudes; the establishment of a larger continental landmass; and the formation of extensive uplands, which affected air flow between continents and oceans and resulted in continental heating. Against this background of warmer and drier conditions, and enhanced by increased CO_2 levels due to increased

tectonic activity, conifers diversified and seed ferns evolved into cycads, bennettitaleans, and ginkgos. These forms were to dominate the Mesozoic landscape.

16.3.3.3 Terrestrial faunas

Amphibians, particularly temnospondyls and batrachosaurs, dominated Carboniferous coal forests and continued as important aquatic animals during the Permian. During the Carboniferous (~325 Ma) the first truly terrestrial vertebrates, the reptiles, evolved. Several major reptilian lineages radiated in the Late Carboniferous, but at relatively low diversity and small size. Two lineages emerged: the diapsids (the group containing living reptiles, dinosaurs, and birds) and the synapsids. The latter, known also as the "mammal-like reptiles," were the dominant terrestrial vertebrates during the late Paleozoic.

Fossil remains indicate that synapsids had spread across Pangaea by the Permian. The first mammal-like reptiles that evolved in the Late Carboniferous (~300 Ma) were the pelycosaurs. Unlike the earliest reptiles, pelycosaurs possessed a much larger head and body. Among them were the first terrestrial vertebrate carnivores. One form, the edaphosaurs, was herbivorous and sported huge, fin-like sails on their backs. Extremely long neural spines projecting from their backbones supported a thin membrane. A rich supply of blood vessels at the base of the spines suggests that the sail functioned as a thermoregulator, allowing rapid absorption or radiation of heat. While some features of the skulls were becoming mammal-like, this inbuilt temperature regulator indicates that these animals were cold-blooded, like other reptiles.

By the late Permian (~260 Ma) pelycosaur diversity had declined and they were replaced by a new group of mammal-like reptiles – the therapsids. Looking like a cross between a hippopotamus and a crocodile, the therapsids were more mammal-like than the pelycosaurs and probably lived in different environments as they diversified in a warmer world than the pelycosaurs. Therapsids were small, lightly built carnivores. Unlike pelycosaurs, therapsids had limbs set beneath their bodies, giving them greater mobility and therefore increasing their foraging ranges.

16.3.4 Evolution during the Mesozoic Era

16.3.4.1 Marine faunas

Shallow seafloor communities experienced what has been termed the "Mesozoic marine revolution" (Vermeij, 1977) during the period 251 to 65 Ma. This was characterized by a diversification in durophagous predators (shell crushers) and by an intensification of grazing by both vertebrates and invertebrates. Durophagous habits commonly evolved in mollusc-feeders, for example, in some cephalopods and marine reptiles, such as ichthyosaurs, in the Triassic. During the Jurassic, some sharks, rays, and crustaceans evolved the habit, while during the Cretaceous it appeared in a number of groups of gastropods.

Predation and grazing increased on the seafloor during the Mesozoic, and one response was the displacement of some groups, such as stalked crinoids and some brachiopods, into deeper water habits where predation levels were lower. Another response was the adoption of a burrowing habit. Such strategies evolved in echinoids (Fig. 16.7e), gastropods, and bivalves. Increased predation pressure in seafloor species is also shown by trends to various "fortress strategies," such as increased shell thickness, "thornier" shells, and increases in body size. Others became much smaller, to avoid detection by predators.

What drove this Mesozoic marine revolution? While "evolutionary arms races," involving interactions between predators and prey contributed, nonbiological factors may also have been significant. Similar, though more restrained, correlations between predation levels and abiotic factors occur in the Paleozoic, and there, as in the Mesozoic, they appear to correlate to intervals of climatic warming, marine transgression (sea level rising and covering land), and high productivity. During the late Mesozoic, periods of massive submarine volcanism contributed enhanced nutrients to the marine environment. These, plus the elevated temperatures, would have increased levels of productivity, thus enabling the evolution of lifestyles that utilized higher energy levels, such as more calcification and higher metabolic rates supporting more active organisms. The two principal phytoplanktonic algae in the oceans today, the diatoms and coccolithophorids, also appear in the Mesozoic.

16.3.4.2 Terrestrial floras

During the Late Permian to Early Triassic ($\sim$251 Ma) terrestrial floras underwent a major transition from Paleozoic groups to the emergence and radiation of a characteristic Early Mesozoic flora (Fig. 16.9) dominated by cycads, ginkgos, bennettitaleans, and conifers (Willis and McElwain, 2002). There is some evidence for ecological trauma at the Permian–Triassic boundary, with family diversity dropping by about 50%, but over a 25 Myr period. Following a 5 Myr gap in the Early Triassic, when floras were mainly bryophytes and lycopsids, coniferous forests became dominant. Global warming, arising from a four-fold increase in CO_2 levels related to increased volcanism arising from the break-up of Pangaea, saw distinct floral changes across the Triassic–Jurassic boundary ($\sim$200 Ma). This is reflected in leaf shape, with a trend for reduction in leaf width in some groups, such as the ginkgos. It has been argued that this warming in high latitudes contributed to a 95% turnover of plant species, due to high-temperature injury to large-leafed plants.

By the Early Jurassic global floras for the first time contained a significant proportion of modern day forms. Moreover, five distinct biomes could be identified: cool temperate, warm temperate, winterwet, subtropical desert, and tropical summerwet (Willis and McElwain, 2002). Conifers and ferns were widespread through these biomes (apart from the desert), while cycads and bennettitaleans were also absent from the cool temperate biome, which was dominated by ginkgos.

The oldest evidence for angiosperms is from the Early Cretaceous ($\sim$120 Ma) while fossil evidence for stem group angiosperms may extend back to the Late Triassic ($\sim$220 Ma). The first fossil flowers and leaves occur in Early Cretaceous strata. The group underwent a rapid radiation, such that by the Late Cretaceous they exceeded all other land plants in diversity (Fig. 16.9). Their initial radiation in the Early Cretaceous again may reflect a response to increasing global temperatures. These had fallen by the end of the Jurassic, but increased rapidly during the Early Cretaceous. This period saw major changes in continent configurations with the beginning of the break-up of Pangaea. An increased plate spreading rate saw a 50–100% increase in ocean crust production. An extraordinary upwelling of heat and material from the core–mantle boundary, known as the "superplume episode," had a significant effect on the environment. Increased volcanism pumped greenhouse gases, mainly CO_2, into the atmosphere, increasing concentrations probably to four to five times higher than today. This would have raised average global temperatures by nearly 8 °C and may have promoted angiosperm evolution and diversification. It has been suggested that angiosperms were more drought resistant and able to cope with higher temperatures than gymnosperms (naked-seed plants such as pines and hemlocks). They possessed tougher, leathery leaves, often reduced in size; tougher, more resistant seed coats protecting the

embryos from drying out; more efficient water conducting cells than other groups; a deciduous habit,
which would have been critical during periods of
drought; and more rapid reproductions and faster
growth, thus better suited to unstable environments.

16.3.4.3 Terrestrial faunas

There was every bit as much an evolutionary revolution
in terrestrial vertebrates during the Mesozoic as there
was in the marine realm (Fig. 16.8). Many new groups
evolved and radiated during the Triassic: archosaurs
and cynodonts, initially, with turtles, crocodilians,
dinosaurs, and mammals later. Moreover, the first flying vertebrates, the pterosaurs, also evolved at this
time. Birds evolved during the Late Jurassic, probably
from theropod dinosaurs, and diversified greatly during the Cretaceous. During the early Mesozoic many
families of terrestrial tetrapods were global in their
distribution, despite the break-up of Pangaea. An
increase in global diversity is a prominent trend in
tetrapod evolution during the Mesozoic as well as during the Cenozoic. Another feature of reptilian diversity
in particular is increase in body size. Like with marine
invertebrates, this trend has been correlated with
predator–prey relationships, as well as greater reproductive success, increased intelligence, expanded ranges,
greater longevity, and decreased annual mortality.
Downsides, though, were the need for large quantities
of food, smaller population size, and greater likelihood
of being adversely affected during periods of environmental deterioration; these led to a greater chance of
extinction.

More specific changes are evident in one of the
more dominant reptilian groups, the dinosaurs.
Ornithopods (bipedal herbivores) evolved near the
Late Triassic–Early Jurassic boundary (~200 Ma).
Diversity increase was limited until the Late
Cretaceous when it markedly increased. This is particularly evident in the hadrosaurs and is attributed to the
evolution of more efficient jaw mechanics and denti-
tions. These later forms evolved high, horse-like snouts
and huge batteries of teeth (up to 2,000), as well as
specialized jaw muscles for chewing. These changes
occurred in stages during the Mesozoic. This particular
trend was probably in response to environmental stimuli, especially the changing nature of the plant food.
It is possible that changes in dinosaur feeding habits
may have preferentially selected for angiosperms.
The dominance of high browsers, such as sauropods,
in the Jurassic, gave way to lower browsers, the

ornithischians, during the Cretaceous (Bakker, 1978).
Periods dominated by high browsers enabled gymnosperm saplings to grow. However, low browsing, it is
argued, preferentially weeded them out, allowing
angiosperms to diversify. Early angiosperm traits of
small size, rapid reproduction, and high colonizing
ability would have given them a competitive edge over
the struggling gymnosperm saplings. The synchronicity
in angiosperm and bird evolution and diversification
also suggests a coevolutionary link between pollinator
and pollinatee.

The Late Cretaceous diversification of angiosperms
also corresponds strongly with insect diversification.
Early Cretaceous flowers have features such as stamens
with small anthers and pollen grains too big for effective
wind dispersal, indicating that many were pollinated by
insects. There is much evidence from the fossil record of
insects eating angiosperm leaves (Labandeira, 2001).
For example, Cretaceous evolutionary radiations of
leaf beetles occurred during the periods of angiosperm
diversification, suggesting coevolution. It has been
argued that the radiations of beetles that fed on angiosperms account for the evolution of the present day
100,000+ species of beetles (Labandeira, 2002). Moths
and butterflies also seem to have radiated about the time
of the angiosperms, as well as bees, whose fossil record
dates back to mid-Cretaceous times.

16.3.5 Evolution during the Cenozoic Era

16.3.5.1 Marine faunas

Biotic   recovery   was   variable   after   the
Cretaceous–Tertiary (K-T) mass extinction event
eliminated 60 to 75% of all species 65 Ma
(Chapter 17). The extinction removed a sizeable part
of the food chain, resulting in profound and extended
reorganization of nutrient and biogeochemical
cycling in the oceans (Norris, 2001). Re-establishing
pre-extinction levels of morphological, taxonomic,
and ecological diversity occurred by a series of discrete
radiations. The range of ecosystems present in the
Late Cretaceous took up to 10 Myr to become re-
established, in the case of reef ecosystems, and even
longer in others. While some groups, such as ammonites (Fig. 16.7d) and marine reptiles became extinct,
others, such as echinoids (Fig. 16.7c), saw little change
in diversity across the boundary, but later underwent
major changes. These include the evolution of the
clypeasteroids (sand dollars), which have diversified
throughout the Cenozoic. The numbers of micro-
plankton never returned to those of the Mesozoic.

Higher diversity occurred during warmer phases, notably the Paleocene–Eocene and the Miocene. Bivalve and gastropods families did not return to pre-K-T diversity levels until the Middle Eocene, 20 Myr after the event.

When compared with pre-Cenozoic times, the marine biota, once it had recovered from the K-T extinction event, experienced great increases in diversity levels through the Cenozoic (Fig. 16.6), most especially at lower taxonomic levels (species to families). However, at the level of orders, diversity remained reasonably constant (about 100 ± 15 orders) from the Ordovician onward (Benton, 1997). The Cenozoic marine biota can therefore be regarded as essentially a continuation of the Modern fauna that first appeared during the early Mesozoic. Apart from the casualties that were experienced at the K-T boundary, the most significant change in composition of the marine fauna, apart from the increased diversity levels, was the evolution of marine mammals during the Eocene and their subsequent diversification.

16.3.5.2 Terrestrial floras

During the last 65 Myr continental plates have moved into their present configurations, with extensive tectonic activity producing mountain ranges such as the Alps, Himalayas, and Rockies. In addition, global climate has changed from a very warm "greenhouse" world in the early Cenozoic to an "icehouse" world in recent times. These changes, which have occurred over what is a geologically short interval of time, have had a dramatic effect on the terrestrial flora and fauna.

Early Paleocene to Middle Eocene times (65–45 Ma) were one of the warmest periods of Earth history. For example, deep oceans reached 9–12 °C higher than present-day levels and annual sea surface temperatures around Antarctica were 15–17 °C, while mean diurnal temperatures at 45° latitude reached about 30 °C. This was due to a number of factors: changing continental configurations affecting ocean currents and therefore global climate; the Pacific Ocean enlarging and taking warmer water to high latitudes; increased mantle degassing; increased volcanic activity; methane-induced polar warming; and a period of global carbon imbalance, with the output of CO_2 to the atmosphere far exceeding burial. Thus, during this time the poles were ice-free and the continents were of lower relief than later; so precipitation occurred more inland and temperature gradients from the equator to the poles were much smaller than now.

However, a number of factors combined to cause rapid global cooling. During the Oligocene, Antarctica separated from South America, creating Drake's Passage, and Australia separated from Antarctica, initiating a circumpolar cold current. This had the effect of cooling Antarctica and creating an ice cover on the continent by ~35 Ma. Furthermore, major mountain building during this period increased aridity by restricting the flow of moist winds. Finally, temperatures fell because of a CO_2 drawdown effect due to intense weathering, and more water was trapped in ice at the poles.

During these profound environmental changes angiosperms diversified, forests shrank, and grasslands expanded. Forests and woodlands of angiosperm trees extended from pole to pole. Continents were dominated by an extensive tropical everwet biome (large-scale, major plant community) that extended to relatively high latitudes. Floras at the poles were cool to warm temperate, consisting of polar broad-leaved deciduous forests. Large leaf-size was an adaptation to low light levels in the winter at high latitudes.

The early Cenozoic saw the evolution of grasses, within the angiosperm family Poaceae. The subsequent diversification of the family, to the present-day number of more than 10,000 species, was initially relatively slow. It was not until 10–20 Ma that widespread grass-dominated ecosystems existed, even though they had started to form a significant part of the global vegetation by ~25 Ma (Willis and McElwain, 2002). The spread of grasslands was promoted in part by the cooling and increasing aridity. Many grasses have drought resistant adaptations, such as increased root growth, decreased shoot growth, reduced physiological activity during droughts, sunken stomata, and thick, dense cuticles (Archibold, 1995). Such adaptations also coped with stress arising from the increased frequency of fires associated with a more arid climate. Changing patterns of herbivorous mammals may also have promoted the diversification of grasses (see below). Other more drought-resistant angiosperm families also diversified during the Cenozoic.

16.3.5.3 Terrestrial faunas

Like reptiles that diversified into both marine and aerial niches during the Mesozoic, mammals similarly diversified during the Cenozoic, with the evolution of both cetaceans and bats. Terrestrial faunas are particularly characterized by a great increase in diversity of mammals, in terms of both morphological diversity and size. Like the flora, the evolution of the terrestrial mammal fauna was strongly influenced by the profound climatic changes of the Cenozoic. Many mammalian lineages increased in body size, in part in response to ecological changes brought on by the transition ~35 Ma from a "greenhouse" world to an increasingly "icehouse" world (Janis and Damuth, 1990). The classic example for this is the evolution of horses, from the small tropical-forest inhabiting forms to larger forms adapted for wide open grass plains and with teeth modified to eat this coarser vegetation. Many of the changes in diversity also relate to the shifting continents, and the evolution of increasingly more endemic groups, such as the many marsupial families that evolved on the Australian continent once it had separated from Antarctica.

Early Cenozoic mammals are essentially relict Mesozoic groups. Most modern orders, such as the primates, artiodactyls (pigs, antelopes), and perissodactyls (horses, rhinos) arose during the Eocene (~56–34 Ma). These were mainly herbivores. Larger carnivorous predators evolved in the Late Eocene. The more archaic groups persisted until the onset of global cooling during the Late Eocene. The very cool world of the Oligocene is reflected in a low diversity of mammal faunas, but by warmer times during the Miocene (~20 Ma) most modern families had evolved. At this time grasslands were spreading and adaptations to grazing evolved in many families of mammals. One of these adaptations was longer legs, allowing broader foraging ranges in less productive environments. The onset of global cooling in the late Miocene and the spread of the grasslands saw the diversification of both browsers and grazers in this ecosystem, as well as in the temperate woodlands at higher latitudes (such as deer and bear). It is only in the more restricted equatorial forests that Eocene-like mammals (such as lemurs) still persist. And it is out of that group that larger bipedal primates, known as the hominins, evolved 6–7 Ma. It took another 4 Myr, though, before these primates underwent an evolutionary change that had occurred in other mammal groups, but to a lesser extent – the evolution of a large brain.

Mammals in general have large brains for their body size, compared with other vertebrates (with the exception of birds). The evolution of the largest relative brain size occurred in the hominin genus *Homo*. Because the brain is a metabolically "hungry" organ, for hominids to evolve large brain-size, developmental tradeoffs were necessary to allocate energy that would enable larger brain growth (Aiello and Wheeler, 1995). This is thought to have occurred by a reduction in gut size, which necessitated a change in diet from predominantly herbivorous to omnivorous, with a large intake of meat. Increase in brain size in the *Homo* lineage correlates with a reduction not only in gut size, but also in jaw and tooth size.

The final species in this lineage, *Homo sapiens*, evolved such a well-developed neocortex and increased so much in cognitive development that it has not only manipulated the environment to a greater extent than any other single species of metazoan, but has single-handedly, directly or indirectly, instigated a level of extinction of fauna and flora greater than that achieved by any other organism in the last half-a-billion years. The overall trend of evolution – colonization first of the oceans, then the terrestrial environment, and then the atmosphere – has been taken one step further by this species: the ability to spread beyond the atmosphere into outer space. This single lineage may therefore be the first form of life in this Solar System to subsequently colonize beyond the confines of the Earth. The evolution of life on planet Earth has not been a series of continuously random episodes that have fortuitously produced the pattern we see in the fossil record. For the last 3.5 billion years or so, life has expanded along environmental gradients – such as deep to shallow water; aquatic to terrestrial; terrestrial to aerial. Moreover, the genetic changes that have occurred have been heavily constrained along pre-existing developmental pathways, resulting in many examples of convergent evolution. If we are to seek life on other planets then we need to find another active, evolving planet. The dynamic physical changes on Earth, in addition to helping drive evolution, have also been heavily affected by the life that has evolved here. Evolving life on Earth has strongly influenced the nature of the atmosphere, which in turn has affected the climate, weathering regimes, and rates of sedimentation. Life could not have evolved on a dead planet.

How inevitable is life on a planet that has evolved in the way that the Earth has? If we were to re-run the clock on planet Earth would we get a similarly evolutionary scenario? Conway Morris (2003) has argued that we would, and that ultimately what we perceive as intelligent life would also be inevitable. But if another planet is evolving under very different constraints, would life evolve and would it bear any remote resemblance to life on Earth? There are those that have argued that the formation of our Solar System occurred by a series of unusual chance events, that their repetition elsewhere is so unlikely that "we may have to face the possibility that, in the universe, we are utterly alone" (Bevan and de Laeter, 2002). So, are we utterly alone, or just one of a myriad places in the Universe where life, in whatever form, is thriving? There is no room for compromise.

REFERENCES

Ahlberg, P. E. (1991). Tetrapod or near tetrapod fossils from the Upper Devonian of Scotland. *Nature*, **354**, 298–301.

Aiello, L. C. and Wheeler, P. (1995) The expensive-tissue hypothesis. *Current Anthropology*, **36**, 199–221.

Allison, C. W. and Awramik, S. M. (1989) Organic-walled microfossils from earliest Cambrian or latest Proterozoic Tindir Group rocks, northwest Canada. *Precambrian Research*, **43**, 253–294.

Allison, C. W. and Hilgert, J. W. (1986) Scale microfossils from the Early Cambrian of northwest Canada. *J. Paleontology*, **60**, 973–1015.

Archibold, O. W. (1995) *Ecology of World Vegetation*. London: Chapman and Hall.

Awramik, S. M. and Sprinkle, J. (1999) Proterozoic stromatolites: the first marine evolutionary biota. *Historical Biology*, **13**, 241–253.

Babcock, L. E. (1993) Trilobite malformations and the fossil record of behavioral asymmetry. *J. Paleontology*, **67**, 217–229.

Bakker, R. T. (1978) Dinosaur feeding behavior and the origin of flowering plants. *Nature*, **274**, 661–663.

Benton, M. J. (1997) Progress and competition in macroevolution. *Biol. Rev.*, **62**, 305–338.

Berner, R. A. (1997) The rise of plants and their effect on weathering and atmospheric CO_2. *Science*, **276**, 544–546.

Bevan, A. and de Laeter, J. (2002) *Meteorites: a Journey Through Space and Time*. Sydney: University of New South Wales Press.

Butterfield, N. J. (1997) Plankton ecology and the Proterozoic–Phanerozoic transition. *Paleobiology*, **23**, 247–262.

Butterfield, N. J. (2000) *Bangiomorpha pubescens* n. gen., n. sp: implications for the evolution of sex, multicellularity, and the Mesoproterozoic/Neoproterozoic radiation of eukaryotes. *Paleobiology*, **26**, 386–404.

Butterfield, N. J. (2005) Probable Proterozoic fungi. *Paleobiology*, **31**, 165–182.

Coates, M. I. (1996) The Devonian tetrapod *Acanthostega gunnarit* Jarik: postcranial anatomy, basal tetrapod interrelationships and patterns of skeletal evolution. *Trans. Roy. Soc. Edinburgh: Earth Sciences*, **87**, 363–421.

Coates, M. I. and Clack, J. A. (1990) Polydactyly in the earliest known tetrapod limbs. *Nature*, **347**, 66–69.

Conway Morris, S. (1986) The community structure of the Middle Cambrian Phyllopod Bed (Burgess Shale). *Palaeontology*, **36**, 423–467.

Conway Morris, S. (2000) The Cambrian "explosion": slow-fuse or megatonnage? *Proc. Nat. Acad. Sci.*, **97**, 4426–4429.

Conway Morris, S. (2003) *Life's Solutions: Inevitable Humans in a Lonely Universe*. Cambridge: Cambridge University Press.

Grey, K. (1994) Stromatolites from the Palaeoproterozoic (Orosirian) Glengarry Group, Glengarry Basin, Western Australia. *Alcheringa*, **18**, 275–300.

Grotzinger, J. P. and Knoll, A. H. (1999) Stromatolites in Precambrian carbonates: evolutionary mileposts or environmental dipsticks? *Ann. Rev. Earth Planet. Sci.*, **27**, 313–358.

Han, T.-M. and Runnegar, B. (1992) Megascopic eukaryotic algae from the 2.1-billion-year-old Negaunee Iron-Formation, Michigan. *Science*, **257**, 232–235.

Hofmann, H. J. (1994) Proterozoic carbonaceous compressions ("metaphytes" and "worms"). In *Early Life on Earth*, ed. S. Bengtson, pp. 342–357. New York: Columbia University Press.

Horodyski, R. J. and Knauth, L. P. (1994) Life on land in the Precambrian. *Science*, **263**, 494–498.

Horodyski, R. J. and Mankiewicz, C. (1990) Possible late Proterozoic skeletal algae from the Pahrump Group, Kingston Range, southeastern California. *Amer. J. Sci.*, **290-A**, 149–169.

Janis, C. M. and Damuth, J. (1990) Mammals. In *Evolutionary Trends*, ed. K. J. McNamara, pp. 301–345. London: Belhaven.

Johnson, E. W., Briggs, D. E. G., Suthern, R. J., Wright, J. L., and Tunnicliff, S. P. (1994) Non-marine arthropod traces from the subaerial Ordovican Borrowdale Volcanic Group, English Lake District. *Geological Magazine*, **131**, 395–406.

Kasting, J. F. and Siefert, J. L. (2002) Life and the evolution of Earth's atmosphere. *Science*, **296**, 1066–1068.

Knoll, A. H. (1994) Proterozoic and Early Cambrian protists: evidence for accelerating evolutionary tempo. *Proc. Natl. Acad. Sci. USA*, **91**, 6743–6750.

Labandeira, C. C. (2001) Rise and diversification of insects. In *Palaeobiology II*, eds. D. E. G. Briggs and P. R. Crowther, pp. 82–88. Oxford: Blackwell Science.

Labandeira, C. C. (2002) Timing the radiations of leaf beetles: hispines on ginger from latest Cretaceous to Recent. *Science*, **289**, 291–294.

Logan, G. A., Hinman, M. C., Walter, M. R., and Summons, R. E. (2001) Biogeochemistry of the 1640 Ma McArthur River (HYC) lead-zinc ore and host sediments, Northern Territory, Australia. *Geochimica et Cosmochimica Acta*, **65**, 2317–2336.

MacNaughton, R. B., Cole, J. M., Dalrymple, R. W., Braddy, S. J., Briggs, D. E. G., and Lukie, T. D. (2002) First steps on land: arthropod trackways in Cambrian–Ordovician eolian sandstone, southeastern Ontario, Canada. *Geology*, **30**, 391–394.

McNamara, K. J., Yu, F., and Zhou, Z. (2003) Ontogeny and heterochrony in the oryctocephalid triobite *Arthricocephalus* from the Early Cambrian of China. *Special Papers in Palaeontology*, **70**, 103–106.

Miller, A. I. (2002) Diversity of life through time. *Encyclopedia of Life Sciences*. London: Nature Publishing Group.

Niklas, K. J. (1997) *The Evolutionary Biology of Plants*. Chicago: University of Chicago Press.

Norris, P. D. (2001) Impact of K-T boundary events on marine life. In *Palaeobiology II*, eds. D. E. G. Briggs and P. R. Crowther, pp. 229–231. Oxford: Blackwell Science.

Padian, K. and Clemens, W. A. (1985) Terrestrial vertebrate diversity: episodes and insights. In *Phanerozoic Diversity Patterns: Profiles in Macroevolution*, ed. J. W. Valentine, pp. 41–96. Princeton: Princeton University Press.

Porter, S. M. and Knoll, A. H. (2000) Testate amoebae in the Neoproterozoic Era: evidence from vase-shaped microfossils in the Chuar Group, Grand Canyon. *Paleobiology*, **26**, 360–385.

Rasmussen, B., Bengtson, S., Fletcher, I., and McNaughton, N. (2002) Discoidal impressions and trace-like fossils more than 1200 million years old. *Science*, **296**, 1112–1115.

Retallack, G. J. and Feakes, C. R. (1987) Trace fossil evidence for Late Ordovician animals on land. *Science*, **235**, 61–63.

Schopf, J. W. (1992) Patterns of Proterozoic microfossil diversity: an initial, tentative, analysis. In *The Proterozoic Biosphere*, eds. J. W. Schopf and C. Klein, pp. 529–552. Cambridge: Cambridge University Press.

Sepkoski, J. J. (1981) A factor analytic description of the Phanerozoic marine fossil record. *Paleobiology*, **7**, 36–53.

Shear, W. A. (1991) The early development of terrestrial ecosystems. *Nature*, **351**, 283–289.

Sherwood-Pyke, M. A. and Gray, J. (1985) Silurian fungal remains: oldest records of the class Ascomycetes. *Lethaia*, **18**, 1–20.

Strother, P. K. (2000) Crypotospores: the origin and early evolution of the terrestrial flora. In *Phanerozoic Terrestrial Ecosystems*, eds. R. A. Gastaldo and W. A. DiMichele, pp. 3–20. Pittsburgh: Paleontological Society.

Trewin, N. H. and McNamara, K. J. (1995) Arthropods invade the land: trace fossils and palaeoenvironments of the Tumblagooda Sandstone (?late Silurian) of Kalbarri, Western Australia. *Trans. Roy. Soc. Edinburgh: Earth Sciences*, **85**, 177–210.

Vermeij, G. J. (1977) The Mesozoic marine revolution: evidence from snails, predators and grazers. *Paleobiology*, **3**, 245–258.

Vidal, G. and Moczydlowska-Vidal, M. (1997) Biodiversity, speciation, and extinction trends of Proterozoic and Cambrian phytoplankton. *Paleobiology*, **23**, 230–246.

Walcott, C. D. (1914) Cambrian geology and paleontology III; No. 2, Pre-Cambrian Algonkian algal flora. *Smithsonian Miscellaneous Collections*, **67**, 77–156.

Willis, K. J. and McElwain, J. C. (2002) *The Evolution of Plants*. Oxford: Oxford University Press.

Zhu, S., Sun, F., Huang, X., He, Y., Zhu, G., Sun, L. and Zhang, K. (2000) Discovery of carbonaceous compressions and their multicellular tissues from Changzhougou Formation (1800 Ma) in the Yanshan Range, north China. *Chinese Science Bulletin*, **45**, 841–847.

FURTHER READING AND SURFING

Briggs, D. E. G. and Crowther, P. R. (eds.) (2001). *Paleobiology II*. Oxford: Blackwell Science. 137 short papers covering the gamut of paleobiological topics. Although technical, they provide excellent synopses of current thoughts in paleobiology.

Conway Morris, S. (2003). *Life's Solution: Inevitable Humans in a Lonely Universe*. Cambridge: Cambridge University Press. A paleobiologist, expert on the Burgess Shale fauna, details how attributes inherent in evolutionary processes would produce intelligent life if the history of life "tapes" were to be rewound and played again.

Cowan, R. (2005). *History of Life*. Fourth edn., Malden: Blackwell Science. A textbook covering the history of life on planet Earth. Aimed at the "intelligent nonspecialist."

Gould, S. J. (gen. ed.) (2001). *The Book of Life*. New York: W. W. Norton. A beautifully illustrated volume that covers some of the major evolutionary stories of life's history. Not as comprehensive as Cowan, it focuses primarily on vertebrates.

Knoll, A. H. (2003). *Life on a Young Planet: the First Three Billion Years of Evolution on Earth*. Princeton: Princeton University Press. Knoll deftly combines the fossil record with the latest knowledge from several fields – molecular biology, isotope geochemistry, paleobiology, and evolutionary theory – to produce a very readable account of the early history of life on Earth.

McNamara, K. and Long, J. (2007). *The Evolution Revolution*. Second edn., Melbourne: Melbourne University Press. Very readable and often witty account on contributions that the fossil record makes to the understanding of evolutionary patterns and processes.

www.ucmp.berkeley.edu/historyoflife/histoflife.html. University of California's Museum of Paleontology. Provides several avenues to information on the history of life on planet Earth.

www.els.net. *Encyclopedia of Life Sciences*, (Wiley) contains numerous articles on the history of life. The website has good cross-referencing, and articles are continually added and updated.

17 Mass extinctions

Peter D. Ward
University of Washington

17.1 Introduction

The frequency of animal life[1] in the Universe must be some function of how often it arises, and then how long it survives after evolving. Both of these factors may be significantly influenced by the frequency and intensity of *mass extinctions*, brief intervals of time when significant proportions of a planet's biota[2] are killed off. They are killed by one or some combination of too much heat or cold, not enough food or nutrients, too little (or too much) water, oxygen, or carbon dioxide, excess radiation, incorrect acidity in the environment, or environmental toxins. Based on the history of Earth's life, mass extinctions seemingly have the potential to end life on any planet where it has arisen.

Mass extinctions do more than threaten biota. They may also play a large part in evolutionary novelty. On Earth there have been about 15 such episodes during the last 500 Myr, five of which eliminated more than half of all species then inhabiting our planet. These events significantly affected the evolutionary history of Earth's biota: for example, if the dinosaurs had not been suddenly killed off following a comet collision with the Earth 65 Ma, there probably would not have been an Age of Mammals, since mammals were held in evolutionary check so long as dinosaurs existed. The wholesale evolution of mammalian diversity[3] took place only after the dinosaurs were swept from the scene. Mass extinctions are thus both instigators as well as foils to evolution and innovation.

On every planet, sooner or later, a global catastrophe can be expected that either threatens the existence of any animal life, or altogether eliminates it. Earth is constantly threatened by such planetary catastrophes – mainly by impact from comets and asteroids, but potentially from other hazards of space. It is also true that the diversity of life on this planet and surely other planets has been affected by "local" causes as well as extraplanetary causes, but this chapter focuses on the latter.

Astrobiology combines the larger cosmos with the origin and evolution of life on Earth. Mass extinctions are an important part of astrobiology in that they potentially provide an important link to influences from space. This chapter describes the major mass extinctions known from the fossil record, their effects on the course of life, and current ideas about their causes.

17.2 Brief history

Prior to about 1800 there was no concept that past mass extinction events occurred at all. French Baron Georges Cuvier, often considered the originator of the discipline known as comparative anatomy, was the first to draw attention to the concept of extinction by demonstrating that bones of large elephant-like animals found in Ice Age sedimentary deposits could not be assigned to any living elephant. He deduced that these bones came from an entirely extinct species.

The birth of the geological timescale in the subsequent decades of the early nineteenth century quickly demonstrated that not only did species undergo extinction, but that many had done so in short intervals of time. In order to devise some way of determining the age of rocks, European and American geologists had already begun to systematically collect fossils as a means to subdivide the Earth's sedimentary strata into large-scale units of time. In so doing they made the discovery that intervals of rock were characterized by sweeping changes in fossil content. Setting out to discover a means of calibrating the age of these rocks,

---

[1] This chapter is primarily concerned with animals (also called *fauna*), which are amongst the most complex of all organisms (see the Tree of Life in Fig. 10.1).

[2] *Biota* refers to all living organisms.

[3] The term *diversity* refers to the number of species (or genera, families, etc.), not the number of individual organisms.

Planets and Life: The Emerging Science of Astrobiology, eds. Woodruff T. Sullivan, III and John A. Baross. Published by Cambridge University Press. © Cambridge University Press 2007.

they also discovered a means of calibrating the diversity of life on Earth. They also found intervals of biotic catastrophe, which were named *mass extinctions*. In a doctrine that came to be known as Catastrophism, these were thought to be caused by a succession of worldwide floods or other disasters that killed off most or all species, followed by a reintroduction (or re-creation) of new species.

The two largest mass extinctions, recognized even as early as the mid-nineteenth century, were so profound that they were used in the 1840s by John Phillips, an English naturalist, to subdivide the stratigraphic record – and the history of life it contains – into three large blocks of time: the *Paleozoic Era*, or time of "old life," extending from the first appearance of skeletonized life 530 Ma until it was ended by the mass extinction of 250 Ma; the *Mesozoic Era*, or time of "middle life," beginning immediately after the Paleozoic extinction and ending 65 Ma; and the *Cenozoic Era*, or time of "new life," extending from the last great mass extinction to the present day.[4] Phillips also made the first serious attempt at estimating the diversity of species present on the Earth during the past. He showed that the diversity of life on Earth has been increasing, in spite of the mass extinctions, which were only short-term setbacks in diversity (Fig. 17.1, taken from his book *Life on the Earth: Its Origin and Succession*, Phillips, 1860). Somehow, after each extinction there seemed to be room for larger numbers of species than were formerly present. Far more creatures were present in the Mesozoic than the Paleozoic, and then far more again in the Cenozoic. But the mass extinctions did more than just change the number of species on Earth. They also changed the *makeup* of the Earth. A more recent view of diversity through time and the placement of mass extinctions is shown in Fig. 17.2.

Most scientists of the time rejected these Catastrophist precepts. While mass extinctions were accepted as having taken place, they were viewed as gradual, long-term events, a Uniformitarian view that extended well into the twentieth century. In 1980, however, the field was revolutionized when a multidisciplinary group of scientists (Alvarez *et al.*, 1980) published what are now known as the Alvarez hypotheses, namely that (1) the Earth was struck 65 Ma by an Earth-crossing object of some sort (asteroid or comet), and (2) the environmental

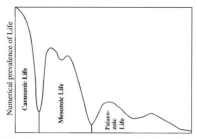

FIGURE 17.1 Diagram from John Phillips (1860: p. 66) illustrating his estimates of diversity of species through time (the present is on the left). Phillips's ordinate corresponded to the number of marine species per 1,000 feet (305 m) of strata. The two mass extinctions (Phillips called them "zones of least life") became the means of differentiating the Paleozoic, Mesozoic, and Cenozoic Eras.

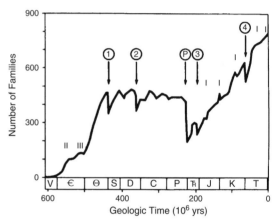

FIGURE 17.2 Diversity through time, as indicated by the number of families found in the fossil record over the past 600 Myr (the present is at zero). The so called "Big Five" mass extinctions of the Ordovician, Devonian, end-Permian, end-Triassic, and end-Cretaceous are indicated by arrows. (Sepkoski and Raup, 1982.)

effects stemming from that impact brought about a mass extinction among the late Cretaceous fauna (animals) sufficient to end the Mesozoic Era.

17.3 Studying mass extinctions

Mass extinctions are recognized by large changes in the number and types of species over short intervals of time. In actual rocks this occurs as abrupt changes in thin vertical *stratigraphic successions* (sequences of adjacent sedimentary layers). Demonstrating mass extinction thus requires close sampling of measured stratigraphic sections such as diagrammed in Fig. 17.3 and photographed in Figs. 17.4 and 17.5. When enough such sections of equivalent age have been examined, one assembles so-called *synoptic data sets* that yield worldwide extinction metrics that can be tabulated as

---

[4] Appendix C details the names and ages of the various geological periods. The cited ages are modern values; in the nineteenth century, the timing of events in the fossil record was highly uncertain and debated.

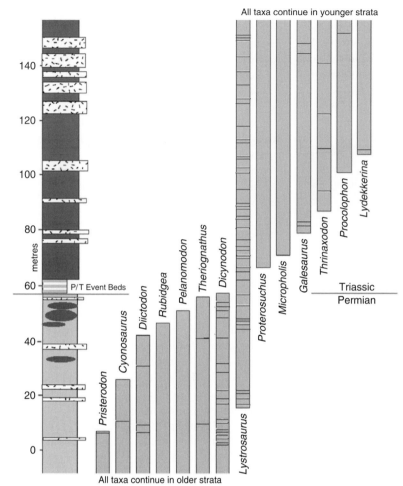

All taxa continue in younger strata

FIGURE 17.3 A measured section from Permian and Triassic strata in the Karoo of South Africa, showing the ranges of selected vertebrates over stratigraphic distance (thickness of layers). The extinction among vertebrates shows a step-wise pattern, which, on face value, would be indicative of a gradual mass extinction. Because of sparse sampling, however, such patterns can be observed even in a sudden mass extinction. It is thus possible that every Permian taxon on this range chart went extinct exactly at the P/T boundary (250 Ma). (Smith and Ward, 2001.)

All taxa continue in older strata

either the number of species disappearing, or as the percentage of the total biota disappearing.

Stratigraphic and synoptic data sets are studied in different ways to answer different questions about mass extinction. Questions about the mode and tempo of extinction can only be answered by studying the ranges of taxa[5] from actual rock outcrops. However, no single outcrop (or even group of outcrops) can answer questions concerning the percentage of taxa going extinct on a global basis. These latter questions are only answerable using synoptic data sets.

17.3.1 Types of mass extinction

Two major categories of mass extinction, perhaps having different causes, have been identified. A *gradual* *mass extinction* is characterized by a reduction of species over some period of time, i.e., smeared out over some extensive stratigraphic interval. Long-term climate change has been cited as a cause of this type of mass extinction. The second type, or *sudden mass extinction*, shows a sudden disappearance of species over a short period of time. Asteroid impact, as proposed by the Alvarez group for the end of the Cretaceous, would create such a catastrophic mass extinction. Paleontological evidence plays an important part in choosing between these rival hypotheses, for only the fossil record yields information about times and rates of extinction, which in turn is essential for establishing causation.

17.3.2 Sources of error

Studying mass extinctions in the fossil record is subject to errors due to the nature of both fossilization and sedimentation. In an ideal world, an outcrop would

---

[5] A *taxon* (plural *taxa*) is any classification category of organisms. Its *range* is the interval of strata (and thus time) over which its fossils are found.

contain the fossilized remains of every creature that once lived there – and each of these fossils would be collectable and actually collected. The reality of course falls far short of this. First of all, in all known environments the number of organisms with hard parts (and thus the potential of readily being fossilized) is always less than about a third of the total number of species present. Since the majority of organisms leave no fossils, any record is immediately biased, and we must assume that the rate of extinction of the entire biota resembles that of those that left a fossil record.

Secondly, in older strata the quality of the fossil record deteriorates. This is because over time rocks become ever more affected by heating, metamorphism, and structural change. Additionally, the amount of rocks that can be searched and studied for mass extinction patterns diminishes with age.

Thirdly, much work on extinction boundaries has now made clear that vagaries of sedimentation and the removal of sediments by erosion can effectively obscure the tempo of an extinction. Sometimes an actual gradual extinction will look sudden due to the effects of erosion on the rock record.

Fourthly, a sudden extinction will usually look gradual, due to inadequate sampling and the rarity of many species. The use of range charts to investigate tempo and mode of extinction (gradual vs. catastrophic) illustrates this error. All ranges are truncated to some extent, for it is unlikely that one can ever discover the true last occurrence of a fossil in a stratigraphic section. The amount of range truncation is a function of the rarity of the fossil in the sampled strata, and of the intensity of sampling. This truncation makes extinctions of taxa appear more gradual and appear to occur at slightly older time intervals than actually happened. The net effect of this is always to make any catastrophic, suddenly occurring extinction appear gradual. This has been named the Signor–Lipps effect, after its discoverers, Phil Signor and Jere Lipps (Signor

FIGURE 17.4 The Cretaceous/Tertiary (K/P) boundary beds at Gubbio, Italy, where in the late 1970s the Alvarez group first recognized impact debris in the uppermost Cretaceous strata.

FIGURE 17.5 A K/P boundary site at Stevns Klint, Denmark. The change of lithology from chalk (below, Paleogene) to shale above (Tertiary) is separated by a 1-mm-thick oxidized layer that contains signs of an impact: shocked quartz, tektite spherules, and large iridium enhancement. The lower sediment is white due to an abundance of planktonic coccolithophorids in the late Cretaceous sea, whereas the dark color of the shale above the impact layer is due to the absence of plankton, which all went extinct.

and Lipps, 1982). An example of correcting for this effect is given in Section 17.11.1.

17.3.3 Anatomy of a mass extinction

The typical sequence of events in a mass extinction begins with the *extinction phase*, when biotic diversity falls rapidly. During this time the extinction rate (the number or percentage of fauna going extinct in any time interval) far exceeds the "origination rate," which is the number of newly appearing taxa through evolution. After some period of time the extinction phase ends, and is succeeded by what is often called the *survival phase*. This is a time of minimal diversity but little or no further extinction. During this interval the number of species levels out, neither increasing nor decreasing. During the third, *rebound phase*, diversity slowly begins to increase. Finally, the *expansion phase* is characterized by a rapid increase in diversity due to the evolution of new species. The last three phases are grouped together in what is known as a *recovery interval*, which is followed by a long period of environmental stability (until the next mass extinction). The rate of the recovery interval is usually proportional to the intensity of extinction that triggers it: the more intense the mass extinction, the more rapid the rate of new species formation.

Three types of taxa are generally found immediately after a mass extinction: *holdover taxa*, or survivors; *progenitor taxa*, the evolutionary seeds of the ensuing recovery phase; and *disaster taxa*, species that proliferate immediately after the end of the mass extinction. Disaster taxa not only tolerate, but thrive in the stringent ecological conditions following the mass extinction event. They are generally small, simple forms capable of living and surviving in a wide variety of environments. (Today we have another term for such organisms: *weeds*.)

The recovery interval is marked by a rise in diversity. This sudden surge in evolution is generally due to the many new opportunities found within the various ecosystems following the extinction. Because so many species are lost in a mass extinction, opportunities for novel species to "make a living" are created. Darwin once likened this process to a wedge: the modern world has so many species in it that in order for a new species to survive and compete it must act like a wedge, pushing out other already entrenched species. But after a mass extinction no wedging is necessary, and virtually *any* new design will do. Many new species appear with morphologies or designs seemingly rather poorly adapted and inferior to those of species that existed prior to the extinction. Rather quickly, however, a winnowing process takes place through natural selection, and new, increasingly efficient suites of species rapidly evolve.

17.4 Earth's mass extinctions: ten real or potential events

Paleontologists have discovered many mass extinction events occurring since the "Cambrian Explosion" of 540 Ma (Section 16.3). Yet other mass extinction events of earlier time are largely unknown to us because they occurred when organisms rarely made skeletal hard parts, and thus rarely became fossils. Perhaps the long period of Earth history prior to the advent of skeletons was also punctuated by enormous global catastrophes decimating the biota of our planet, mass extinctions without record, or at least without a record that has yet been deciphered.

Of the fifteen mass extinctions officially classified as such from the last 500 Myr, six have been especially catastrophic as measured either by the number of families, genera, or species going extinct, or by their effect on subsequent biotic evolution. To this list I propose three more that probably occurred prior to 500 Ma, as well as the *current* biodiversity crisis caused by the effects of a runaway human population. Although the final extinction total for the current mass extinction cannot yet be tallied, it may be representative of what happens whenever an intelligent species arises on a planet. These ten events are discussed below. Section 17.5 more generally discusses possible causes for these mass extinctions, as well as similar events that might occur on other planets with complex life.

Our understanding of the various events is in inverse proportion to their age: the older they are, the less we know because of the increasingly corrupted fossil and geological record. The most recent of the larger, ancient events, the Cretaceous/Tertiary mass extinction 65 Ma, is by far the most studied and best understood.

17.4.1 Early bombardment (4.0–3.8 Ga)

The period of early heavy bombardment is thought to have sterilized the surface of the Earth at least several times – assuming that life had evolved at all by this time. This is at best a potential event – if life was indeed present, it may have suffered one or more mass extinctions.

17.4.2 Advent of oxygen and the first Snowball Earth (2.5–2.2 Ga)

The rise of oxygen certainly caused the extinction of many, perhaps most, anaerobic bacterial species then on Earth. (But not complete extinction: there were certainly many anaerobic environments, then as now, in the sea and in fresh water.) There is little or no fossil record of this event, which may have coincided with the first "Snowball Earth" event when geological evidence indicates that even the tropics were frozen (Section 4.2.4). Although either the rise of oxygen or the Snowball Earth could potentially have caused a mass extinction, at the present time there is no evidence for this from the fossil record.

17.4.3 More Snowball Earths (750–600 Ma)

There is almost no information known about these events, which may have included three or four separate extinctions coinciding with repeated global glaciations. There do appear to have been wholesale extinction among stromatolites and planktonic organisms called acritarchs (Section 16.2.4.1). The lack of fossilized animals from this time period obscures these events.

17.4.4 Cambrian (560–500 Ma)

Extinctions taking place immediately before and then during the Cambrian Period remain the most enigmatic of all the extinction episodes. The so-called "Cambrian Explosion" was the relatively short interval of time when all of the animal phyla (high-level taxonomic groupings) appeared. Some experts argue that the Cambrian Explosion was preceded by the first of all mass extinctions, which caused the disappearance of the so-called Ediacaran fauna. These were odd assemblages of jellyfish-like and sea-anemone-like animals that can be found in strata immediately below the base of the Cambrian from many parts of the world (Section 16.2.4.4 and Fig. 16.5). They appear to be the first diverse assemblage of animals, perhaps early forerunners of today's animals such as the Cnidaria (jellyfishes, corals, hydras) and various worms, or perhaps an assemblage of phyla now completely extinct. They disappeared in sudden and dramatic fashion immediately prior to the Cambrian period, but the cause of their disappearance remains a mystery. It may have been caused by competition with the newly evolving groups of animals typifying the Cambrian, or through sudden environmental change.

A second wave of extinctions occurred about 20 Myr after the Ediacaran crisis. This second crisis was protracted over several millions of years, and gravely affected the first reef-forming organisms (archeocyathids), as well as many groups of trilobites and early molluscs. Again, there is little direct evidence of what brought about these extinctions, other than they seem to be linked to changes in worldwide sea level and to the formation of deep anoxic bottom water.

The Cambrian extinctions remain a huge enigma. Here we have consequential events without overt cause.

17.4.5 Ordovician and Devonian (400 Ma and 370 Ma)

During the Paleozoic Era there were two major mass extinctions. In the Ordovician Period (> 400 Ma) and in the Devonian Period ($\sim$370 Ma), major events decimated the marine faunas of the time. Because our record of land life is poor for both of these intervals, there is still much to learn about the severity of these events on land. In the sea, however, it is clear that the majority of species went extinct, and that more than 20% of marine families were eliminated.

The causes of these extinctions remain obscure. While impact has been proposed for the Devonian mass extinction, little evidence for such a cause has emerged in spite of intensive searching. Nor does evidence for impact exist for the Ordovician extinction. Anoxia, temperature change, and sea-level change are the favored causes, but it seems difficult to account for the extensive extinction based on these causes. These are two major extinction events in search of causes.

17.4.6 Permian/Triassic (P/T) (250 Ma)

Based on standard extinction metrics (percentage of species, genera, or families eliminated worldwide during the event) the Permian/Triassic (P/T) mass extinction of 250 Ma appears to have been the most catastrophic. This particular event stands alone in its severity – more than 50% of marine families died out, twice as high as in any other extinction. Estimates of the number of *species* (belonging to various families) that went extinct vary from nearly 80% to more than 90%. It is clear that the vast majority of animal and plant life on the Earth was extinguished.

Although it has long been recognized as being the most catastrophic mass extinction, its cause (or causes) have been problematic. One possible major cause was an increase in atmospheric CO_2 arising from

(1) a short-term release of carbon dioxide from sediments formerly sequestered on the ocean floor, and (2) gas emitted during unusually severe volcanic eruptions (Hallam and Wignall, 1997). Such a sudden release of huge volumes of carbon dioxide would cause a direct killing of marine organisms by carbon dioxide poisoning, as well as a direct killing of terrestrial life due to a sudden and intense global heating from a greatly increased greenhouse effect. A heat spike of perhaps 5–10 °C, 10,000 to 100,000 years in duration, was the probable cause of the terrestrial extinctions (Kiehl and Shields, 2005). Another possible cause, a release of poisonous hydrogen sulfide into the atmosphere, is discussed in Section 17.5.8.

Based on the presence of ^{3}He trapped in fullerenes ("bucky balls"), the impact of a comet has also been proposed, but not confirmed, with new evidence still appearing (Becker et al., 2001, 2004; Li, 2005). The major evidence against an impact as the major environmental perturbation causing extinction comes from findings that the extinction among some groups (such as land animals) was protracted and that the isotopic record across this boundary shows multiple perturbations, not the single event one would expect with an impact (see Section 17.11.2 for details) (Payne et al., 2004; Ward et al., 2005). Also, there is no clear signal from either iridium or a layer of shocked quartz grains at the boundary, such as is found at K/P (Cretaceous/Paleogene) boundary sections (Section 17.6).

17.4.7 Triassic/Jurassic (T/J) (210 Ma)

The end of the Triassic Period witnessed a significant mass extinction, with about 50% of genera being eliminated. We have only a poor record for the fate of land life during this extinction, but it is clear that marine life was extensively and catastrophically affected. This mass extinction, like the K/P event, appears to have been brought about by the impact of a large extraterrestrial body, either a comet or an asteroid. One candidate for the impact crater is Manicouagan Crater in Quebec, Canada, which is 100 km in diameter and has been variously dated at about 210 Ma.

Environmental changes other than impact have been associated with this extinction event as well, most notably oceanic changes creating anoxia in many shallow water environments at the end of the Triassic. These changes in turn may have been caused either by extrusion of gas from flood basalts or by an impact. Information from stable isotope work (Ward et al.,

2001, 2004) indicates that the Triassic event appears to resemble more closely the P/T than the K/P event.

17.4.8 Cretaceous/Tertiary (K/P)[6] (65 Ma)

This mass extinction killed off the dinosaurs as well as 50% or more of the other species then on Earth – it has been recognized for more than a century and a half as one of the most devastating periods of mass death in Earth's history (Figs. 17.4 and 17.5). Section 17.11.1 and Figs. 17.6 and 17.7 present sample data for the K/P event. Although numerous explanations for this event have been proposed, an asteroid impact is now largely accepted as the cause; see Section 17.6 for details.

17.4.9 The modern extinction

Scientific consensus is that the number of extinctions occurring since the end of the last glacial period, some 12,000 years ago, places our own era as one of pronounced and elevated extinction rate (Ward, 1994; Hallam and Wignall, 1997). There are many estimates about how many species are currently going extinct each year, but few hard data for many regions. What is clear is that the world's forests are being felled to make way for agriculture and colonization at a momentous rate, and that the removal of forests leads to extinction. The seas are more insulated and there is little evidence of major marine extinction occurring at the present time, although this could change quickly as pressure on the world's fisheries increases. Furthermore, if the rate of chemical pollution increases over the next several centuries, the extinction rate in the sea may also substantially rise. Tallies of lost animal species vary, but all share the grave message that the Earth is rather quickly losing a great number of species (Ward, 1994). The ultimate cause of this current potential mass extinction is the runaway population of *Homo sapiens*.

17.5 Causes of mass extinctions

In one way or another, all mass extinctions appear to have been immediately caused by changes in the "global atmosphere inventory." The specific killing agents arise

---

[6] The traditional name for this mass extinction has been K/T, standing for Cretacious/Tertiary, where Tertiary was a formerly used name. K/P is the preferred term today, standing for either Cretacious/Paleogene or Cretaceous/Paleocene (see Appendix C).

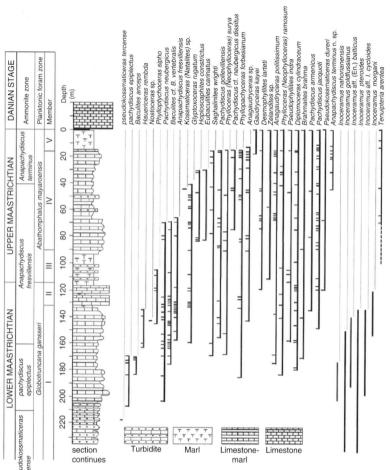

FIGURE 17.6 Ranges of ammonites from a K/P boundary site at Zumaya, Spain. The tick marks on each range refer to individual fossil finds. The K/P boundary is at zero meters, with ~ 3–4 Myr of record extending downwards. (Marshall and Ward, 1996.)

through changes in the makeup and behavior of the atmosphere, or through factors such as temperature and circulation patterns that are dictated by properties of the atmosphere. Those in turn can be caused by many things: asteroid or comet impact, loss of carbon dioxide or other gases into the oceans and atmosphere during flood basalt extrusion (when great volumes of lava flow out onto the Earth's surface), loss of gases caused by liberation of organic-rich oceanic sediments during sea-level change, or changes in ocean circulation patterns.

The geological record on this planet suggests, however, that more than a single cause is usually associated with any given mass extinction. Sometimes these multiple events occur at the same time; sometimes they are separated by hundreds of thousand of years. Perhaps one perturbation stresses the planet, making it more susceptible to the next perturbation. For example, both

the P/T and K/P calamities appear to have more than a single cause.

In the following sections I discuss possible causes for mass extinctions, whether on this planet or other Earth-like planets with complex animal life.

17.5.1 Leaving the animal "habitable zone"

Animal life requires liquid water on or near the surface, which in turn requires global temperatures allowing liquid water to exist. Any movement of a planet out of an orbit allowing such temperatures will create a planetary disaster. Though such changes of orbit are unlikely, they are still possible, and could be caused by gravitational perturbations from another planet in one's own system, or, more likely, from a star passing nearby. Such perturbations would be common, for instance, if the system were part of a star cluster.

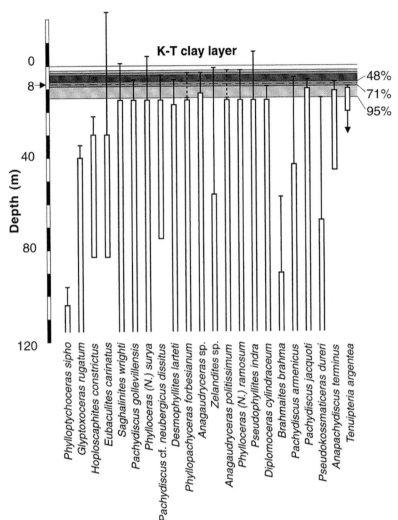

FIGURE 17.7 Confidence intervals on each ammonite species from Fig. 17.6. An open bar indicates the observed range, and the upward line extensions show the 90% statistical confidence interval for the true range of a species. (Marshall and Ward, 1996.)

17.5.2 Change in the star's energy output

Complex animal life on any planet is dependent on stellar energy. If a star's output either increases or decreases so that liquid water can no longer exist, the results will be disastrous. Short-term or long-term stellar energy output changes might be one of the most common forms of planetary extinction, and even sterilization. Some researchers are convinced that the end of life on Earth will be caused by the expected increase in the Sun's energy output. As discussed in Chapters 3 and 4, the amount of energy being produced by the Sun – and indeed by all main sequence stars – increases through time. On Earth, a relatively constant temperature has been maintained in the past

through a gradual reduction in greenhouse gases as the amount of energy from the Sun has increased, thus keeping temperatures in check. It appears, however, that we are nearing the end of having such a planetary thermostat. Very small volumes of carbon dioxide remain in the atmosphere compared to earlier periods of geological time, yet the Sun's energy output continues to increase. Some studies have predicted that temperatures on Earth could become too high for animal life within several hundred million years from now (Ward and Brownlee, 2002). Further deterioration of conditions on Earth will eventually lead to the loss of *all* life. That event, when it comes to pass, will produce the greatest – and last – mass extinction on Earth, its sterilization.

17.5.3 Catastrophic climate change: icehouse and runaway greenhouse

Under certain circumstances great changes in climate can cause mass extinction. Major glaciations or greenhouse heating are examples, and both are a function of the amount of carbon dioxide or other greenhouse gases in the atmosphere. These are the actual killing mechanisms brought about by changes in stellar output, or by changes in a planet's orbit. Climate changes of magnitude sufficient to threaten the biosphere with major mass extinction would involve great changes in mean planetary temperature, as well as relocation of oceanic current systems and global rainfall patterns.

The two most catastrophic climate changes can be called *Icehouse* (the Snowball Earth events were examples of this) and *Runaway Greenhouse*. In both cases global temperatures depart from the 0–100 °C range necessary to allow the presence of liquid water on the planet. Chapter 4 discusses these scenarios in detail.

17.5.4 Short-term heating from flood basalt events

There is increasing evidence that the occurrences of *flood basalts* in the Earth's past were coincident with mass extinctions (Hallam and Wignall, 1997). At the end of the Permian, the so-called Siberian Traps were extruded, the largest known flood basalt on Earth, while at the end of the Triassic, the Central Atlantic Magmatic Province was extruded. These events involved the short-term extrusion of large volumes of basalt onto the surface of the Earth, as well as outgassing of large volumes of carbon dioxide, a greenhouse gas, and sulfur dioxide, a poisonous gas. At the present time, however, it is not known how extrusion of large amounts of lava onto the surface of the Earth translates to a killing mechanism on land or in the sea. One idea is that short-term global heating is caused by the liberation of huge amounts of greenhouse gases, including methane (previously bound in clathrates). As mentioned in Section 17.4.6, modeling of temperatures at the end of the Permian confirm a short-term heat spike (Kiehl and Shields, 2005).

17.5.5 Impact of a comet or asteroid

Any system of planets orbiting a star is rife with cosmic debris such as asteroids and comets, the residue left over from planetary formation (Chapter 3). Great quantities of this material will eventually strike all members of a planetary system, and the energy released by such an impact creates a planetary disaster. Such disasters are now known to have caused at least one mass extinction on Earth – the 65 Ma disappearance of the dinosaurs and many other species living near the end of the Mesozoic Era was caused by the impact of a large meteor or comet on the Earth, as briefly described earlier (Alvarez *et al.*, 1980). This is discussed in detail in Section 17.6.

On a larger scale, it may be that planetary systems associated with stars farther from a galaxy's center contain smaller numbers of asteroids and comets than those more centrally located. Because of this, impact rates would be lessened, and hence there would be fewer mass extinctions. This is one of the factors contributing to the concept of a "Galactic Habitable Zone" (Section 3.4; Ward and Brownlee, 2000).

17.5.6 Nearby supernova

Another potential mechanism producing a mass extinction is the occurrence of a nearby supernova. Ellis and Schramm (1995) calculated that a star going supernova within 30 light years of our Sun would release fluxes of energetic electromagnetic radiation and charged particles (cosmic rays) sufficient to destroy the Earth's ozone layer in 300 yr or less. Removal of the ozone layer would then prove calamitous to the biosphere as it would expose both marine and land organisms to potentially lethal solar ultraviolet radiation. Photosynthesizing organisms such as phytoplankton and reef communities would be particularly affected.

Based on the average number of stars near the Sun and the rates of supernova explosions (only very rare stars, those with masses at least eight times that of the Sun, go supernova), Ellis and Schramm concluded that supernova explosions within 30 lt-yr of the Earth (and therefore presumably dangerous) occur on average every 200–300 Myr.

Results such as this, however, are uncertain because present models for how much damage a given flux of radiation would cause to our atmosphere are very uncertain. We do not yet know the answer to a basic astrobiological question: how close must a supernova be to have significant effects on Earth's biota (or on any another planet's ecosystem)? A combination of biology and astrophysics is needed to answer this, but geology can also play an important role by providing evidence in the Earth's crust of past irradiation by a supernova. Knie *et al.* (2004) have found a strong spike

2.8 Ma in the ^{60}Fe abundance of deep-ocean ferro-manganese crust. ^{60}Fe has a half-life of only 1.5 Myr and no other plausible sources exist for it than from the material generated in a supernova, so we may have the beginning of a new way to observe the cosmos. Fields *et al.* (2005) calculate that a supernova at a distance somewhere in the (admittedly large) range of 50–400 lt-yr would have produced enough ^{60}Fe to account for what has been found in Earth's crust. They calculate the likely abundances of other short-lived isotopes (such as ^{10}Be, ^{26}Al, ^{53}Mn, ^{182}Hf, and ^{244}Pu) and suggest that we may entering an era of "deep ocean crust astronomy"!

Being placed in the outer parts of a galaxy lessens the risk of supernova extinctions because of the decreased number of stars, adding further weight to the notion of a galactic habitable zone. Lineweaver *et al.* (2004) calculate that a Galactic Habitable Zone exists today for Earth-like life in the range of distances of 23,000–29,000 light years from the galactic center (the Sun is in the center of this zone). This is based on a model of our Galaxy, and habitability requirements including a star of sufficient longevity for biological evolution, enough heavy elements to form Earth-like planets, and not too frequent nearby supernovae.

17.5.7 Gamma ray burst

A new entry into the mass death "rogue's gallery" is *gamma ray bursts*, one type of which lasts a second or less, peaks at ∼200 keV energy, and is thought to arise from the merging of two neutron stars into a black hole (most observed bursts originate in very distant galaxies). Such a burst of gamma rays hitting the Earth could have many deleterious effects, such as (a) delivering enough ionizing energy to Earth's surface (after the original gamma rays are transformed into ultraviolet spectral lines) to cause significant genetic damage to organisms, despite the relatively thick atmosphere of Earth, (b) stripping the ozone layer and thus allowing solar ultraviolet radiation to damage life, (c) changing atmospheric chemistry that could lead to, for example, great changes in surface temperature or biogeochemical cycles, or (d) producing harmful radioactive species on Earth. Melott *et al.* (2004) have proposed that certain aspects of the Ordovician mass extinction can be explained by the incidence of a gamma ray burst.

A critical issue is the frequency of these postulated events in any galaxy. A detailed analysis by Scalo and Wheeler (2002) indicates that the Milky Way should have one gamma ray burst every ∼2 Myr, of which one

every ∼2–10 Myr might be of significance to complex life on Earth – thus a gamma ray burst happening *anywhere* in our Galaxy may well seriously affect Earth's life with a "photon jolt" (as well as any other life in the Galaxy, no matter where located). Long-term biological effects, however, are very uncertain regarding what levels of radiation are necessary to cause significant mutations and other evolutionary effects, and how these vary between different species. The above numbers, for example, assume that a significant biological effect results when 500 J m^{-2} of ionizing radiation is delivered to the Earth's surface, a number derived from experiments on mammals – but prokaryotes are much hardier (Table 14.2) and thus may suffer effects from gamma ray bursts a hundred times less frequently, i.e., every ∼200–1,000 Myr. But this still could be of importance, given life's 3.5–3.8 Gyr history on Earth.

17.5.8 Hydrogen sulfide poisoning

Another possible killing mechanism has been proposed by Kump and Arthur (2005), specifically dealing with the Permian Extinction. They suggest that a long period of low atmospheric oxygen observed at the end of the Permian would have created conditions in the sea favoring enhanced growth of bacteria that release hydrogen sulfide. Such a pulse of poisonous H_2S entering the oceans and atmosphere would have provided a kill mechanism for both terrestrial and marine life at the P/T boundary itself. Better than other hypotheses, this mechanism explains the relatively sudden death at the P/T boundary, which is seen in the fossil record as a spike of extinction overlaying a much longer-term interval of heightened species extinction.

17.6 Impacts and mass extinctions

The presence of numerous impact craters on every planet or satellite of the Solar System with a solid surface gives stark and impressive evidence of the frequency and importance of these events, at least early in the history of our Solar System. It is probable that impact is a hazard as well in most, or perhaps all, extra-solar planetary systems. Impacts are probably the most frequent and important of all planetary catastrophes. They can completely alter the biological history of a planet by removing previously dominant groups of organism and thus opening the way for either entirely new groups or previously minor groups. The best evidence we have for this comes from the K/P event.

17.6.1 The K/P Event

Although originally controversial, mineralogical, chemical, and paleontological data gathered during the 1980s persuaded most experts that a large comet or asteroid ($\sim$10–15 km diameter) impacted the Earth $\sim$65 Ma, and that, at the same time, more than half of the species then on Earth rather suddenly became extinct in the K/P event (Alvarez et al., 1980). The discovery of a large impact crater of precisely the right age in the Yucatan region of Mexico (the Chicxulub Crater; Sharpton et al., 1992), now suspected to be a multi-ring impact basin 300 km in diameter (Sharpton et al., 1993; Schuraytz et al., 1994), largely swept away remaining opposition to the impact hypothesis. The energy of a $\sim$10-km-diameter impactor would have been the equivalent of $\sim$10^8 Mton of TNT, $\sim$10^7 times that of the Mt. St. Helens volcanic eruption in 1980.

Two of the most important lines of evidence used to convince most workers in the field that the K/P extinction was indeed caused by large body impact were (a) the discovery of elevated iridium values within the boundary clays (Figs. 17.4 and 17.5), and (b) abundant "shocked quartz" intermingled with the iridium.

High concentrations of iridium (and other platinum-group elements) have been detected at over 50 K/P boundary sites worldwide (Hallam and Wignall, 1997). Iridium is seen as an indicator of impact because it is quite rare on Earth's surface, but occurs at much higher than Earth concentrations in most meteorites, asteroids, and comets.

Shocked quartz is seen as an indicator of impact because multiple lamellae (thin layers) on sand-sized quartz grains found at most K/P boundary sites could only have been produced by a high-pressure event, such as a large asteroid hitting rock at high velocity. No "Earthly" conditions naturally create such quartz grains with multiple shock lamellae.

In addition to iridium and shocked quartz grains, K/P boundary strata have also yielded evidence of fiery conflagration that must have occurred soon after the impact: fine particles of soot found in the same K/P boundary clays from many parts of the globe. This type of soot comes only from burning vegetation, and its quantity suggests that much of Earth's surface was consumed by forest and brush fires. The ultimate killer, according to Alvarez et al. (1980), was a several month period of darkness, or blackout as they called it, following the impact. The blackout was due to great quantities of meteoric and Earth material thrown into the atmosphere after the blast, and lasted long enough to kill off much of the plant life then living on Earth, including the plankton. With the death of the plants, disaster and starvation rippled upward through the food chains.

Models of lethality caused by such atmospheric change have been calculated by several groups. Pope et al. (1994) used a 2-D hydrodynamical model to estimate that between 0.4 and 7.0×10^{17} g of sulfur was released into the atmosphere after the K/P impact. A small portion of this was reconverted into H_2SO_4, which fell back to Earth as acid rain; this may have been a killing mechanism, but was probably more important as an agent of cooling than direct killing through acidification (Sigurdsson et al., 1992; D'Hondt et al., 1994). More deleterious to the biosphere, however, may have been the reduction (by as much as 20% for 8–13 years) of solar energy transmission to the Earth's surface through absorption by atmospheric dust particles (*aerosols*). This would have been sufficient, according to Pope et al., to produce a decade of freezing or near freezing temperatures on a world that, at the time of impact, had been largely tropical. The prolonged winter is thus the most important killing mechanism – and it was brought about by vastly increasing aerosol content in the atmosphere over a short period of time.

Covey et al. (1994) looked at the global climatic effects of atmospheric dust produced by the impact of a large (10 km diameter) asteroid or comet. They found that fine dust would be generated by impact into either an oceanic or continental target area; the after-effects would include long-term blackout (on the order of months) producing light levels below that necessary for photosynthesis, accompanied by rapid cooling of land areas. But perhaps the most ominous prediction in this model is the formerly unappreciated effect that the giant volume of atmospheric dust generated by the impact would have had on the hydrological cycle. Covey et al. (1994: p. 271) described this effect as follows: "The dust cloud not only perturbs surface temperatures but also causes a collapse of the hydrological cycle in our simulations. Globally averaged precipitation decreases by more than 90% for several months and is still only about half normal by the end of the year." In other words, the entire planet becomes cold, dark, and dry. This is an excellent recipe for mass extinction, especially for plants – and the creatures feeding on plants.

Following the success of explaining the K/P event, some investigators thought that a general model linking most or all mass extinctions to impacts would emerge. This was the thinking behind Raup and Sepkoski's

(1984) suggestion that mass extinctions show a 26 Myr periodicity. By 1990, however, it became clear that the K/P catastrophe, rather than being a typical mass extinction, appears to have been a unique event. Although a minority (most notably M. Rampino and his co-workers) still argue that much evidence for impact-related extinction exists throughout the geological column, the consensus continues to be that mass extinction boundaries recorded in sedimentary rocks of ages other than end-Cretaceous only rarely have yielded *unambiguous* evidence of impact-generated extinction. Indeed, it has been argued that the K/P was perhaps the *only* one of the major mass extinctions to have been caused by impact. It is clear that the K/P impact is in a class by itself with its widespread, extremely elevated iridium values, shocked quartz, fine soot, and spherules (droplets of molten rock that are ejected during an impact and freeze during flight to distant locations). It may have been caused by the most energetic impact event certainly of the past 500 Myr, and perhaps for much of Earth's history (Schuraytz *et al.*, 1994).

17.6.2 Variables in the efficiency of impacts

Although the evidence that mass extinctions other than the K/P have been caused by comet or asteroid impact remains equivocal, the evidence that the Earth has been repeatedly hit by objects from space is not. This leads us to the following questions. What effect do *smaller* impacts on the Earth have on its biota at the time of impact? Do impacts by bodies smaller than that causing the Chicxulub crater release sufficient energy to cause a mass extinction? And if so, what is the minimum diameter of an object that can create such a mass extinction? Conversely, if smaller objects *can* cause local (as contrasted to worldwide) mass extinction, what is the relationship between impactor size and resulting kill?

This line of reasoning led to the formulation of "kill curves" relating impactor size to percentage of species that are killed (Raup, 1990, 1991). Unfortunately, the curve itself is entirely theoretical, as very few large craters have been sufficiently well dated or studied to allow confident ties to mass extinction events. Chicxulub remains the only crater known with any degree of confidence to be associated with a mass extinction.

Raup's concept was that impact events would fall on a logistic curve of simple shape – above some minimum threshold size below which no mass extinction would be detectable, the bigger the impact event, the greater the percentage of global fauna going extinct. This idea is both simple and powerful, but there are clearly different, critical variables in the equation such as composition of the target rocks, as well as location and season of the impact. This leads to the following questions:

1. Are any other mass or pulsed extinctions (in addition to K/P) caused by impact?
2. What determines the killing potential of an impact event?
3. Are kill curves monotonic even for small impacts, or is there some threshold (in size of crater or impacting body) below which minimal or no extinction takes place?
4. Is crater size the primary or sole variable describing killing potential, or are other factors such as target materials or impact location significant?
5. Do impacting bodies kill via the same mechanisms regardless of size, or is killing mechanism related to impactor size?

Many variables must come into play, including factors associated with the incoming body (its size, composition, angle of impact, and velocity), as well as the nature of the impact area (the target area). In the case of the K/P impact at Chicxulub, for instance, the target rock was rich in sulfur, which exacerbated resultant environmental effects, for the sulfur reacted with air and water to produce a highly toxic acid rain lasting many months after the impact event itself. Moreover, not only the geology of the impact site, but its geography as well may play an important part: a low-latitude impact will produce entirely different consequences from those of a high-latitude site hit by a similar body with similar angle and speed, since the distribution of lethality across the globe may be strongly influenced by atmospheric circulation patterns. Finally, the nature of both the biota and the atmosphere at the time of impact are surely important. An impact in a highly diverse world of ecological "specialists" – animals and plants with little tolerance for environmental change – might produce more extinction than the same event on a low diversity world of "generalists," just as impact into a greenhouse world may have different effects than one where greenhouse gas or oxygen content is low. Kring (2003) reviews the many factors influencing the environmental consequences that may result from impacts.

The way in which impacts have been linked to mass extinctions has proceeded in a less than ideal fashion. As happened with the history of K/P investigations, typically a mass extinction has been identified in the stratigraphic record using paleontological data, and then evidence for impact was sought *at that same location*. If physical evidence for impact (iridium, spherules, shocked quartz, etc.) was found, the last step in the process was identifying a suitable crater. I believe, however, that the process should *start* with the crater. This, of course, requires that the crater be precisely dated, a major gap in the past. But if large impact craters can be dated with high precision, we can then go out to the stratigraphic record and search for extinctions (and physical evidence of impact) in strata with those ages, a methodology exactly opposite of that employed to date.

17.7 Frequency of mass extinctions

How often do mass extinctions take place? Perhaps the best way to analyze this question is to use the same methodology that meteorologists and hydrologists use in assessing the risk from weather and floods. Many natural phenomena – such as floods, earthquakes, and droughts – are distributed through time in a similar way. Small events are common and large ones rare. The best way to estimate frequency of the rarest events is to assemble all the available data and arrange them by mean intervals between events of a given magnitude. For instance we might ask how often in a century, or a thousand years, a given flood of some intensity occurs? We can then define "ten-year floods," events of such magnitude that we can expect one every ten years on average, and compare this to much larger 100-year events, and the even more catastrophic 1000-year events. This does not mean that we cannot get two 100-year events in successive years – just that the probability is very small. Hydrologists use a technique called extreme-value statistics to extrapolate waiting times beyond the length of historical records. These estimates are, of course, imperfect, but they permit estimates about, say, 1000-year events despite only 100 years of historical records.

This same technique has been adapted to the questions of mass extinction by Raup (1991), who noted that we have a good record for the past 600 Myr of Earth history, allowing us to define the 10-Myr and 30-Myr events with confidence. Using these statistics, Raup calculated a second kind of kill curve showing the expected time intervals for mass extinctions of varying magnitude. Raup first amassed the extinction records of more than 20,000 genera[7] of organisms based on

their geological longevities. Using the Zoological Record, a compendium of life now on Earth, Raup found the first and last occurrences of all genera of organisms, and then tabulated the results in a colossal data base.

These results give us a statistical sense of how many species go extinct over a given period of time. Raup found that there is negligible extinction during a typical 0.1 Myr interval, but rarer 1 Myr events are more consequential, with perhaps 5–10% of all species on Earth going extinct. That figure rises to 30% of all species for 10 Myr events, and to nearly 70% of all species for 100 Myr events. These are sobering numbers. If nearly three quarters of all species go extinct in a short-term planetary catastrophe of some sort every 100 Myr, it suggests that we are living on a planet subject to many contingencies. The astrobiological question of whether similar principles apply to any other inhabited planet remains open.

17.8 Longer-term effects of mass extinctions

Short of complete sterilization, are mass extinctions necessarily detrimental to planetary diversity? Perhaps it can be argued that instead of being deleterious to diversity, they are actually forces that *increase* diversity, in some fashion analogous to culling weeds from a garden. For example, it can be argued that the various Paleozoic Era extinctions caused archaic reef communities to be re-assembled into more diverse types of corals such as we have today. Mass extinctions also paved the way for a takeover of bottom communities previously dominated by brachiopods (archaic shellfish) by the more modern (and diverse) molluscs. In another case, the extinction of the dinosaurs paved the way for the evolution of many new types of mammals, and it appears that there are more types of mammal species today than there were ever dinosaur species. Perhaps mass extinctions are positive things, creating new opportunities and evolutionary innovation by weeding out decadent or poorly fit, but resource clogging and entrenched, species. If these mass extinctions had not occurred, would planetary diversity be higher or lower than it is today? Countering one's usual picture of impacts as wholly negative to life, there are definite evolutionary and ecological *benefits* of impacts (Cockell and Bland, 2005).

Interesting as these questions are, they have not yet been tested in any way. The fossil record, however, does yield some clues that mass extinctions are on the

---

[7] *Genus* (plural *genera*) is the taxonomic classification level above species.

deleterious rather than positive side of the biodiversity ledger. Perhaps the best clue comes from the comparative history of reef ecosystems following mass extinction. Reefs are the most diverse of all marine habitats; they are the rainforests of the ocean. Because they contain so many organisms with hard skeletons (in absolute contrast to a rainforest, which bears very few creatures with any fossilization potential), we have an excellent record of reefs through time. Reef environments have been severely and adversely affected by all past mass extinctions. They suffered a higher proportion of extinction than any other marine ecosystem during each of the six major extinctions of the last 500 Myr. After *each* mass extinction reefs disappeared from the planet, and usually took tens of millions of years to re-establish. It appears that complex ecosystems such as reefs take a long time to build (or rebuild). When reef systems eventually re-appeared, they were composed of entirely new suites of creatures having less diversity than before. The implication is that mass extinctions, at least for reefs, greatly reduce biodiversity.

In summary, the longer-term effects of mass extinctions are complex and require more study. They kill large numbers of organisms, but in their aftermath, there are faster than normal rates of new species formation.

17.9 Severity of mass extinctions

The conventional means of comparing severity among the various mass extinction events has been to compute the percentage of taxonomic categories going extinct. This monumental work has been carried out largely by Raup and Sepkoski (1984). Through such statistics the "Big Five" mass extinctions (Ordovician, Devonian, Permian, Triassic, and Cretaceous) were recognized. If number of families going extinct is used for comparison, the P/T mass extinction leads with a rate of of 54%, followed by 25% for the Ordovician, 23% for the Triassic, 19% for the Devonian, and 17% for the K/P event. The Cambrian extinctions do not appear as "major," although in other analyses made by Tappan and Newell, the Cambrian extinctions of *marine* families are even higher (~60%) than those of the P/T (~55%).

We have not yet found any causes for the Cambrian extinctions, but our best guess is that the "garden analogy" holds true: the garden of animals during Cambrian time had only just emerged, and while there were many different types of body plans (in fact more so than now), there were very few species in each category. Even slight changes in environmental conditions were

apparently sufficient to wipe out entire categories, entire phyla. The Cambrian was the riskiest period for animal life of all time. Its extinctions, in my view, were the most important in the history of life on Earth, including the P/T event, simply because most higher taxa (and therefore body plans) existed at low species numbers, and were therefore quite susceptible to total extinction. There is more to comparing extinctions than simply adding up the number of species killed off.

17.10 Extinction risk through time

Do the risks of extinction vary through time? This question involves two variables: (1) do environmental conditions affecting life on a given planet change through time?; and (2) does the evolutionary progression of life through time change its susceptibility to extinction? For the same reasons that astrobiologists have had to modify the concept of "habitable zone," this question cannot be answered without some qualification relating to degree of complexity, for the extinction rates of microbes are far different from those of more complex species. If, as we surmise, extinction rates vary according to an organism's complexity, we might expect rather low extinction rates during the long period prior to the evolution of animal life, followed by increasing extinction rates. But extinction rates also must depend on the frequency and severity of the physical events that drastically change conditions on a planet.

The history of major mass extinctions on Earth suggests that only two causes have been operative in Earth history: impact and global climate change. There are other possibilities that may have caused extinctions as well, such as a nearby supernova, but to date there is limited evidence that this latter mechanism has indeed occurred. With regard to impact, there is good evidence that the frequency of impacts has changed through time (Chapter 3). The most obvious of these changes was the cessation of major impacts at the end of the period of heavy bombardment ~3.8 Ga. But even after this rain of major impactors ceased, there is evidence of a long slow decline in impact rates (Grieve and Pesonen, 1992). This decrease would have tended to reduce the overall extinction rate during the same period that vulnerability of emerging animal groups would have tended to increase it. Even in the last 500 Myr, the time of complex animals, one can argue that there should have been a sufficient number of comet or asteroid strikes to totally exterminate animal life on this planet (Raup, 1990). This has obviously not happened.

17.11 Two case histories

In this section I describe two research projects that deal with the paleontological record of two major mass extinctions. These case histories illustrate how actual collected data can be used to test hypotheses of mass extinction. The first deals with the extinction of mollusks at and below the 65 Ma K/P boundary, while the second relates to the 250 Ma P/T mass extinction. These are examples of a catastrophic extinction and of a protracted, more gradual mass extinction.

17.11.1 Data for the K/P event

The K/P event of 65 Ma is now thought to have produced a worldwide species-level extinction of 50–70%, and in so doing radically changed the faunal and ecological makeup of the biosphere. Most experts accept that the Chicxulub impact was the dominant factor in causing the K/P mass extinction, but there remains considerable skepticism (especially among vertebrate paleontologists) that this event was the dominant (or even a contributing) factor. At the present time, the debate is centered on four questions. (1) Was there one extinction event, or several? (2) If composed of more than one distinct event, was the extinction gradual or sudden? (3) If there were multiple events, were the causes the same for each? (4) Were the causes terrestrial or extraterrestrial?

The mode and tempo of this extinction can be tested by sampling fossils from complete stratigraphic sections. The goal is to determine the stratigraphic *range* of each type of fossil, i.e., the number of sedimentary strata (and their thicknesses) over which the fossil can be found. Suitable sections containing both microfossils and macrofossils spanning the K/P boundary have been discovered in France and Spain, and have been analyzed using a recently developed statistical protocol called stratigraphic confidence interval methodology (Marshall, 1990).

Our database consisted of 321 occurrences from 31 species of ammonite (large-shelled cephalopod molluscs; Fig. 16.7d), as well as even richer fossil records of inoceramid bivalves (echinoids) (Marshall and Ward, 1996). These were collected from nine measured stratigraphic sections (Fig. 17.6) in France and Spain over a twelve-year interval. The sampled sections are appropriate for this type of analysis, being paleontologically complete (no missing microfossil or macrofossil zones), relatively unfaulted, and expanded (deposited in a basin characterized by high sedimentation rates). Observed patterns of extinction can be markedly compromised by stratigraphic biases such as erosion (creating artificial range truncation), bioturbation and sedimentological time averaging (which can artificially extend fossil ranges), as well as by collecting or preservation biases (the Signor–Lipps effect described in Section 17.3.2, which makes abrupt extinctions look gradual). We therefore employed statistical methods to test for artificial extinction patterns and determine confidence intervals on measured stratigraphic ranges.

The sampled lithologies were offshore limestone and marls dominated by Milankovitch-forced rhythmites, and are divided into five distinctive lithological members of Maastrichtian Age, which can be readily correlated within the basin (with a distance of 200 km between the westernmost and easternmost exposures), as well as to regions as separated as Gubbio, Italy; Caravaca and Agost, Spain; and the Republic of Georgia. Water depths were probably between 100 and 500 m. The stratigraphic levels sampled for this study range from within the Lower Maastrichtian *Gansserina gansseri* zone, to the top of the Cretaceous (top of the *Abathomphalus mayaroensis* zone), encompassing a time interval of approximately 3–4 Myr.

The stratigraphic occurrences of each taxon were tabulated for each section at meter-scale precision. For the last 150 cm of Maastrichtian strata for each section, stratigraphic ranges of the collected fossils were obtained at a centimeter scale. The boundary itself is well marked in each section by a 0.1–0.3 cm thick reducing layer containing abundant micro-spherules and enhanced iridium concentrations, overlain by a 10–20 cm thick boundary clay. Quantitative analysis was based on all occurrences projected onto a single standard composite section.

Of the 41 macro-fossil species recovered, none appear to have survived into the Tertiary in this region (Fig. 17.6). We analyzed the fossil records of 28 species of ammonites (known from 271 fossils), seven species of inoceramid bivalves (known from over 70 fossil horizons from the Zumaya section alone), the nautiloid *Eutrephoceras* sp., and an un-named spatangoid echinoid species.

We determined which species most likely became extinct well before the K/P boundary using the following equation:

$$C = 1 - (G/R + 1)^{-(H-1)}, \tag{17.1}$$

where C is our confidence that the species became extinct somewhere within some interval G above its

FIGURE 17.8 Permian/Triassic (P/T) boundary from the Karoo of South Africa. The laminated beds shown here mark the base of the Triassic, and are abiotic pond deposits. The lamination comes from the fact that bioturbating organisms (whose activities normally mix up such layers) were all killed off in the extinction event.

last appearance, R is the observed stratigraphic range, and H the number of fossiliferous strata that make up that range (Marshall, 1990; Wang and Marshall, 2004). A value of C of 90% is usually taken as a standard for the location of a species' extinction.

Application of Eq. (17.1) to the ammonite fossil record identified six species that were almost certainly extinct by the time of the emplacement of the K/P boundary clay. These species were most likely victims of "background" extinction processes, which continuously occur and are caused by non-catastrophic events. As illustrated in Fig. 17.7, all other species (22 ammonites, one echinoid, and the inoceramid *Tenuipteria argentea*) are candidate victims of the K/P mass extinctions.

The marine stratigraphic record thus shows a sudden extinction. Less clear is the pattern of dinosaur extinction on land, where the rarity of dinosaur fossils exacerbates the effects of sampling and the Signor–Lipps effect. However, the coincidence of the iridum rich layers, a single-event carbon isotope anomaly found in both marine and terrestrial sections, and the simultaneous last occurrence of key fossils both on

land and in the sea has led to the conclusion that the K/P was simultaneous on land and in the sea to within less that 1 Myr.

17.11.2 Data for the P/T event

The mass extinction ending the Permian Period (the P/T event of 251.4 ± 0.3 Ma) is universally acknowledged as the most consequential mass extinction of the past 500 Myr (Stanley, 1987; Teichert, 1990; Erwin, 1993, 1994; Hallam and Wignall, 1997). The paleontological record of this event in the best-studied sections of marine strata, located in China, indicates a major pulse of extinction, constrained to an interval of rock deposited over only 10^4 to 10^5 yr, but far less is known about the pattern and duration of the extinction among non-marine (called *terrestrial*) organisms. Until recently, the P/T boundary has been difficult to pinpoint in terrestrial rocks because of incomplete and/ or low-resolution sampling around the boundary. This situation has been improved due to (1) recognition of a "fungal spike" (a layer with mainly fungal spores that

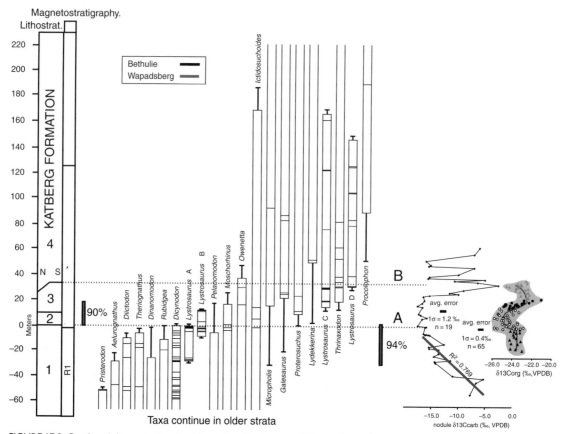

FIGURE 17.9 Combined chemostratigraphy, biostratigraphy of mammal-like reptiles, and magnetostratigraphy from the P/T boundary in the Karoo of South Africa. The boundary itself is nominally at 0 meters. The bar labeled "Magnetostratigraphy" indicates two reversals of the Earth's magnetic field over the studied interval of 10 Myr. The fungal spike (see text) occurs at 20 meters and roughly 0.5 Myr removed from the extinction of the mammal-like reptiles at ~0 meters.

has been interpreted as a disaster layer) at many terrestrial and shallow marine P/T sections around the world (Looy et al., 1999); (2) implementation of carbon isotope stratigraphy (Morante, 1996; Retallack and Krull, 1999; MacLeod et al., 2000); and (3) use of new terrestrial ecosystem event stratigraphy markers. However, high-resolution biostratigraphy of terrestrial vertebrates across this boundary interval on a regional scale had not been undertaken prior to the work of Smith and Ward (2001; see Fig. 17.3 for an example), and most recently by Ward et al. (2005). This work shows that the extinction of mammal-like reptiles and the position of the fungal spike occurred at different stratigraphic levels. Thus, unlike the K/P extinction, the P/T extinction appears to have taken place at different times in the marine and terrestrial realms.

During six field seasons, 151 identifiable, in situ vertebrate fossils (mostly mammal-like reptiles) were collected from 52 discrete stratigraphic levels spanning the uppermost Permian and lowermost Triassic from

seven stratigraphic sections in the Karoo desert of South Africa. Our collecting yielded a minimum of 15 species; eight species in Permian strata, seven in Triassic strata, and one (Lystrosaurus spp.) spanning the boundary. Not only is the pattern of species extinction and recovery similar in all studied sections, it also occurs in a similar sequence of sedimentary facies.

The P/T boundary is immediately overlain by 3–5 m of distinctively laminated maroon mud-rock made up of thinly bedded dark reddish-brown and olive-gray siltstone–mudstone couplets (Fig. 17.8). This laminated facies is a stratigraphically unique unit in all our measured sections. Intensive prospecting has failed to locate any vertebrate fossils in this interval, although subhorizontal siltstone-filled cylinders, resembling callianassid shrimp burrow casts, make their first appearance in this facies.

The disappearance of the mammal-like reptiles is shown in Fig. 17.9, which covers an interval of 10 Myr. If this extinction on land happened simultaneously with the marine extinction, this overall pattern would mimic the K/P event, and suggest an impact-caused extinction. However, the fungal spike, which in marine stratigraphic sections is found at the level where more than 80% of marine species disappear in Chinese and Pakistani sections, is found in the Karoo stratigraphically well above ($\sim$0.5 Myr) the mammal-like reptile extinction level. This suggests that the vertebrate extinction on land preceded the marine invertebrate extinction in the sea. The record of isotope anomalies from P/T sections seems to corroborate this finding, for a series of perturbations has been found, in contrast to the K/P case where but a single change in δ^{13}C is recorded, coinciding exactly with the level of mass extinction. Huey and Ward (2005) have described evolutionary consequences of this event, concluding that a combination of lowered oxygen and elevated temperatures was involved with the mass extinction (as mentioned in Section 17.4.6).

17.12 Closing

Mass extinctions are important to astrobiology if the example of our planet can be generalized to other planets where complex life may have occurred. They are one of the key manifestations of the intimate connection between the cosmos as a whole and the history of life on Earth, where they seem to have been both curse and blessing. It seems that, as for any individual, life on a planet can end through old age, or through sudden accident.

References

Alvarez, L., Alvarez, W., Asaro, F., and Michel, H. (1980). Extra-terrestrial cause for the Cretaceous–Tertiary extinction. *Science*, **208**, 1094–1108.

Becker, L., Poreda, R. J., Hunt, A. G., Bunch, T. E., and Rampino, M. (2001). Impact event at the Permian–Triassic boundary: evidence from extraterrestrial noble gases in fullerenes. *Science*, **291**, 1530–1533.

Becker, L., Poreda, R. J., Basu, A. R., Pope, K. P., Harrison, M., Nicolson, C., and Iasky, R. (2004) Bedout: an end-Permian impact crater offshore northwestern Australia. *Science*, **304**, 1469–1476.

Cockell, C. S. and Bland, P. A. (2005). The evolutionary and ecological benefits of asteroid and comet impacts. *Trends in Ecology & Evolution*, **20**, 175–179.

Covey, C., Thompson, S., Weissman, P., and Maccracken, M. (1994). Global climatic effects of atmospheric dust from and asteroid or comet impact on earth. *Global and Planetary Change*, **9**, 263–273.

D'Hondt, S., Pilson, M. E. Q., Sigurdsson, H., *et al.* (1994). Surface-water acidification and extinction at the Cretaceous–Tertiary boundary. *Geology*, **22**, 983–986.

Ellis, J. and Schramm, D. (1995). Could a nearby supernova explosion have caused a mass extinction? *Proc. Natl. Acad. Sci.*, **92**, 235–238.

Erwin. D. (1993). *The Great Paleozoic Crisis: Life and Death in the Permian*. New York: Columbia University Press.

Erwin, D. (1994). The Permo-Triassic extinction. *Nature*, **367**, 231–236.

Fields, B. D., Hochmuth, K. A., and Ellis, J. (2005). Deep-ocean crusts as telescopes: using live radioisotopes to probe supernova nucleosynthesis. *Astrophys. J.*, **621**, 902–907.

Grieve, R. A. F. and Pesonen, L. J. (1992). The terrestrial impact cratering record. *Tectonophysics*, **216**, 1–30.

Hallam, A. and Wignall, P. (1997). *Mass Extinctions and Their Aftermath*. Oxford: Oxford University Press.

Huey, R. B. and Ward, P. D. (2005). Hypoxia, global warming, and terrestrial late Permian extinctions. *Science*, **308**, 398–401.

Kiehl, J. T. and Shields, C. A. (2005). Climate simulation of the latest Permian: implications for mass extinction. *Geology*, **33**, 757–760.

Knie, K., Korschinek, G., Faestermann, T., Dorfi, E. A., Rugel, G., and Wallner, A. (2004). ^{60}Fe anomaly in a deep-sea manganese crust and implications for a nearby supernova source. *Phys. Rev. Lett.*, **93**, 171103.

Kring, D. A. (2003). Environmental consequences of impact cratering events as a function of ambient conditions on Earth. *Astrobiology*, **3**, 133–152.

Kump, L. and Arthur, M. (2005). Massive release of hydrogen sulfide to the surface ocean and atmosphere during intervals of oceanic anoxia. *Geology*, **33**, 397–400.

Li, Y., Liang, H., Yin, H., Sun, J., Cai, H., Rao, Z., and Ran, F. (2005). Determination of fullerenes (C_{60}/C_{70}) from the Permian–Triassic Boundary in the Meishan Section of South China. *Acta Geol. Sin.*, **79**, 11–15.

Lineweaver, C. H., Fenner, Y., and Gibson, B. K. (2004). The galactic habitable zone and the age distribution of complex life in the Milky Way. *Science*, **303**, 59–62.

Looy, C., Brugman, W. A., Dilcher, D. L., and Visscher, H. (1999). The delayed resurgence of equatorial forests after the Permian–Triassic ecologic crisis. *Proc. Natl. Acad. Sci.*, **96**, 13857–13862.

MacLeod, K. G., Smith, R. M. H., Koch, P. L., and Ward, P. D. (2000). Timing of mammal-like reptile extinctions across the Permian–Triassic boundary in South Africa. *Geology*, **28**, 227–30.

Marshall, C. (1990). Confidence intervals on stratigraphic ranges. *Paleobiology*, **16**, 1–10.

Marshall, C. and Ward, P. (1996). Sudden and gradual molluscan extinctions in the latest Cretaceous of Western European Tethys. *Science*, **274**, 1360–1363.

Melott, A., Lieberman, B., Laird, C., *et al.* (2004). Did a gamma-ray burst initiate the late Ordovician mass extinction? *Int. J. Astrobiology*, **3**, 55–61.

Morante, R. (1996). Permian and early Triassic isotopic records of carbon and strontium events in Australia and a scenario of events about the Permian–Triassic boundary. *Historical Geology*, **11**, 289–310.

Pavlov, A., Toon, O. B., Pavlov, A. K., and Bally, J. (2005). Passing through a giant molecular cloud: "Snowball" glaciations produced by interstellar dust. *Geophys. Res. L.*, **32**, L03705.

Payne, J. L., Lehrmann, D. J., Wei, J., Orchard, M. J., Schrag, D. P., and Knoll, A. H. (2004). Large perturbations of the carbon cycle during recovery from the end-Permian extinction. *Science*, **305**, 506–509.

Phillips, J. (1860). *Life on the Earth: Its Origin and Succession.* Cambridge: Macmillan (1980 reprint, New York: Arno Press).

Pope, K., Baines, A., Ocampo, A., and Ivanov, B. (1994). Impact winter and the Cretaceous Tertiary extinctions: results of a Chicxulub asteroid impact model. *Earth and Planetary Science Express*, **128**, 719–725.

Raup, D. (1990). Impact as a general cause of extinction: a feasibility test. In *Global Catastrophes in Earth History*, eds. V. Sharpton and P. Ward, *Geol. Soc. Am. Special Paper*, **247**, 27–32.

Raup, D. (1991). A kill curve for Phanerozoic marine species. *Paleobiology*, **17**, 37–48.

Raup, D. and Sepkoski, J. (1984). Periodicity of extinction in the geologic past. *Proc. Natl. Acad. Sci.*, **A81**, 801–805.

Retallack, G. and Krull, E. (1999). Landscape ecological shift at the Permian/Triassic boundary in Antarctica. *Aust. J. Earth Sciences*, **46**, 785–812.

Scalo, J. and Wheeler, J. C. (2002). Astrophysical and astrobiological implications of gamma-ray burst properties. *Astrophys. J.*, **566**, 723–737.

Schuraytz, B. C., Sharpton, V. L., and Marin, L. E. (1994). Petrology of impact-melt rocks at the Chicxulub multiring basin, Yucatan, Mexico. *Geology*, **22**, 983–986.

Sepkosky, J. J. and Raup, D. M. (1982). Mass extinctions in the marine fossil record. *Science*, **215**, 1501–1503.

Sharpton, V. L., Dalrymple, G. B., Marin, L. E., *et al.* (1992). New links between the Chicxulub impact structure and the Cretaceous Tertiary boundary. *Nature*, **354**, 819–821.

Sharpton, V. L., Burke, K., Camargozanoguera, A., *et al.* (1993). Chicxulub multiring impact basin – size and other characteristics derived from gravity analysis. *Science*, **261**, 1564–1567.

Signor, P. and Lipps, J. (1982). Sampling bias, gradual extinction patterns and catastrophes in the fossil record. *Geol Soc. Am. Special Paper*, **190**, 291–6.

Sigurdsson, H., D'Hondt, S., and Carey, S. (1992). The impact of the Cretaceous–Tertiary bolide on evaporite terrain and generation of major sulfuric acid aerosol. *Earth Plan. Sci. Lett.*, **109**, 543–559.

Smith, R. M. H. and Ward, P. D. (2001). Pattern of vertebrate extinctions across an event bed at the Permian/Triassic boundary in the main Karoo basin of South Africa. *Geology*, **29**, 1147–1150.

Stanley, S. (1987). *Extinctions.* San Francisco: W. H. Freeman.

Teichert, C. (1990). The Permian-Triassic boundary revisited. In *Events in Earth History*, Kauffman, E. and Walliser, O., eds. Berlin: Springer-Verlag, 199–238.

Wang, S. and Marshall, C. (2004). Improved confidence intervals for estimating the position of a mass extinction boundary. *Paleobiology*, **30**, 5–18.

Ward, P. (1994). *The End of Evolution.* New York: Bantam Doubleday Dell.

Ward, P. and Brownlee, D. (2000). *Rare Earth.* New York: Copernicus (Springer-Verlag).

Ward, P. and Brownlee, D. (2002). *The Life and Death of Planet Earth.* New York: Henry Holt.

Ward, P., Haggart, J., Carter, E., Wilbur, D., Tipper, H., and Evans, T. (2001). Sudden productivity collapse associated with the Triassic/Jurassic boundary mass extinction. *Science*, **292**, 1148–1151.

Ward, P. D., Garrison, G. H., Haggart, J. W., Kring, D. A., and Beattie, M. J. (2004). Isotopic evidence bearing on Late Triassic extinction events, Queen Charlotte Islands, British Columbia, and implications for the duration and cause of the Triassic/Jurassic mass extinction. *Earth Plan. Sci. Lett.*, **224**, 589–600.

Ward, P. D., Botha, J., Buick, R., *et al.* (2005). Abrupt and gradual extinction among late Permian land vertebrates in the Karoo Basin, South Africa. *Science*, **307**, 709–714.

Part V
Potentially habitable worlds

18 Mars

Bruce M. Jakosky
University of Colorado
Frances Westall and André Brack
CNRS Centre de Biophysique Moléculaire

18.1 Mars and astrobiology

Mars is at the center of the field of astrobiology in many ways. Today's vigorous program of ongoing exploration is largely motivated by Mars's potential to harbor life at present or at some time in the past. But astrobiology as a discipline is about more than just finding out whether there is or ever was life on Mars or on other planets and satellites. It is about determining the governing principles behind whether life will originate on a given planet or whether it can survive if transplanted there; about what the mechanisms are by which planets and biota can and do interact with each other; and about determining which fundamental factors distinguish a planet that is habitable from one that is not.

We know that the Earth meets the environmental requirements for being habitable. In Mars, we have an example of a planet that evolved under different physical and chemical conditions, allowing us to see what effects these differences can have. Today, Earth has global-scale plate tectonics, and Mars does not. Earth has global oceans at its surface, and Mars does not. Earth has a climate conducive to the coexistence of the solid, liquid, and vapor phases of water, each of which affects the geology and geochemistry of the planet, and Mars does not. In finding out whether Mars has or ever had life, we obtain a second example of whether and how life can occur on a planetary body. This will help us to understand whether the conditions that allowed life to form and evolve on the Earth are general to life or merely specific to our one example of life. If we can learn more about the general conditions that allow life to exist on a planet, we can extrapolate to other planets and satellites in our own Solar System and to Earth-like planets that might exist around other stars. Thus, in many ways the exploration of Mars is an important

step in inferring the abundance and distribution of life in the Universe.

In addition, Mars has the advantage that its relative proximity has allowed us to explore it in more detail than any other planet. Since 1965 more than fifteen spacecraft have operated successfully on or near Mars, including flyby spacecraft, orbiters, landers, and rovers, and several more spacecraft are planned for the next decade (Table 18.1). These missions have allowed us to make significant progress in understanding the present-day nature of Mars and its history, from the deep interior to the upper atmosphere. Mars is also the only planet on which we have attempted life-detection experiments, and the results of those experiments (and of analysis of rocks from Mars that have found their way to the Earth as meteorites) provide important insights into how one can search for life in general.

In this chapter we will address a number of these issues. These include the evidence that Mars meets or in the past has met environmental conditions suitable for supporting life, as well as the different martian environments in which life might occur. We will discuss how the tremendous diversity and robustness of terrestrial life point the way to the existence of potential ecological niches on Mars. We will describe the searches that have been carried out for martian life to date, including the Viking spacecraft experiments of the 1970s and the analysis of martian meteorites (ALH84001 in particular) in the 1990s. We conclude with descriptions of various approaches and missions for identifying evidence for life on Mars that do not rely on properties that might be specific to life on Earth.

We begin with brief introductions to Mars and to the requirements for terrestrial life, in order to learn the context in which we should look to Mars as a potential abode for life.

Planets and Life: The Emerging Science of Astrobiology, eds. Woodruff T. Sullivan, III and John A. Baross. Published by Cambridge University Press. © Cambridge University Press 2007.

TABLE 18.1 Successful and planned spacecraft missions to Mars[a]

Year launched	Name	Type of mission
1964	Mariner 4	Flyby
1969	Mariner 6	Flyby
1969	Mariner 7	Flyby
1971	Mariner 9	Orbiter
1973	Mars 5	Orbiter
1975	Viking 1	Orbiter + lander
1975	Viking 2	Orbiter + lander
1988	Phobos 2	Orbiter
1996	Mars Pathfinder	Lander
1997	Mars Global Surveyor	Orbiter
2001	Mars Odyssey	Orbiter
2003	Mars Exploration Rover (2)	Rover
2003	Mars Express	Orbiter
2005	Mars Reconnaissance Orbiter	Orbiter
Upcoming missions:		
2007	Phoenix	Lander
2009	Mars Science Laboratory	Lander/rover

[a] After Snyder and Moroz (1992).

TABLE 18.2 Physical characteristics of Mars[a]

Semimajor axis of orbit	1.524 AU
Eccentricity of orbit	0.0934
Orbital period	686.98 Earth day = 669.60 Mars solar days
Tilt of rotation axis to orbit	25.2°
Mean solar day (*sol*)	24 hr 39 m 35.2 s
Mass	6.42×10^{23} kg = $0.11 \times$ Earth
Mean radius	3390 km = $0.53 \times$ Earth
Mean density	3.93 g/cm^3 = $0.71 \times$ Earth
Surface gravity at equator	3.71 m/s^2 = $0.38 \times$ Earth
Total surface area	1.44×10^{14} m^2
Mean atmospheric pressure	6 mbar = $0.006 \times$ Earth
Atmospheric composition	CO_2 (95.3%), N_2 (2.7%), Ar (1.6%), other (0.4%)

[a] After Kieffer *et al.* (1992).

18.2 Mars as a planet

Before discussing specific aspects of Mars that make it potentially suitable for sustaining life, we provide a brief overview of Mars as a planet. Detailed discussion of the various components of the Mars environment can be found in Kieffer *et al.* (1992) and Kallenbach *et al.* (2001).

Mars is a rocky terrestrial planet in the inner Solar System; Table 18.2 summarizes its physical properties. It is about half the diameter of the Earth and a tenth of its mass. Its average density is 3.9 g/cm^3, denser than it would be if it consisted only of rocky material, yet less dense than if it had an interior structure similar to the Earth. From the mean density and from gravitational measurements, we know that Mars has a dense core; we cannot distinguish, however, between a small, denser Fe core or, for example, a larger, less dense FeS core. The mantle is slightly denser than the Earth's mantle, possibly due to a greater amount of Fe substituting for Mg in the minerals.

The martian surface shows evidence for a wide variety of processes having acted throughout its geological history. Volcanic structures occur in a number of places, ranging from those like cinder cones to large Hawaiian-style volcanoes. In addition, the northern plains of Mars appear to have been formed from relatively fluid lava that erupted from vents that were subsequently covered by the lava and hidden from view.

Mars does not have global-scale plate tectonics like the Earth, but does have notable features caused by tectonics (deformation or movement of the planet's crust). There are grabens[1] and faults that are related to a large bulge called Tharsis; they appear to result from the stress that resulted from placing onto the surface a large amount of volcanic material. The largest of the grabens is Valles Marineris, a 4000-km long, 500-km wide valley that runs radially outward from the center of Tharsis. The western end of Valles Marineris, closest to Tharsis, appears tectonic in origin, with indications that it formed from stretching apart of the valley walls in one or, in some places, two directions. Other parts of it are more complex, however. Within the individual canyons comprising Valles Marineris, there are layers that suggest either that it was filled at one time with liquid water that had deposited sediments, or that water-related dissolution of subsurface minerals played

---

[1] A *graben* is a depressed region with steep walls, often caused by faults.

a role in forming the canyons. The eastern end of Valles Marineris grades smoothly into catastrophic flood channels where tremendous quantities of water appear to have disgorged; this and other evidence for liquid water in Mars's past will be discussed in detail in Section 18.4.

The plains in the southern hemisphere have the greatest number of impact craters, and are therefore older than the northern lowlands. Based on the number of impact craters and on estimates of the cratering rate through time, these surfaces are thought to have been present since 3.7–4.2 Ga. Although erosion by wind and probably water has removed most morphological features that might identify the mechanism by which the plains formed, they are thought to have been volcanic as well. The plains also show evidence for the past presence of liquid water, including degradation of the impact craters and what appear to be networks of valleys formed by runoff of liquid water. This evidence suggests that the climate on ancient Mars allowed water to exist as a liquid.

The present-day climate, though, is not conducive to liquid water, being too cold (average temperature of 220 K, or −53 °C) and having too thin an atmosphere. The atmospheric pressure at the surface is about 6 mbar, less than 1% of the Earth's surface pressure, and the dominant gas is CO_2. Mars does have water, but the low temperatures dictate that it be present as vapor and clouds in the atmosphere or as ice deposits in the high-latitude subsurface (similar to permafrost on Earth) and in the polar regions. The climate today is dominated by cold temperatures controlled by heating from the Sun, winds driven by the geographical variations in temperature, and the raising and transport of dust. Local dust storms occur regularly, sometimes expanding to global proportions. Atmospheric water acts as a tracer of wind activity, rather than being the driving force seen on Earth, and its seasonal behavior indicates significant exchange with the subsurface and with the polar ice caps. The polar caps themselves contain layers that suggest that the annual exchange with the atmosphere can move large quantities of water over long periods; this transport may be driven by long-term oscillations in the amount of the planet's obliquity, or tilt of its polar axis. There are also geologically recent gullies that suggest the presence of seeps of water from the subsurface and, therefore, the recent occurrence of liquid water beneath the surface (Section 18.4.1).

Overall, Mars is somewhat of a contradiction. Despite a present-day climate that is not conducive to liquid water or to life at the surface, it appears to have had more clement conditions at the surface in its earliest history and beneath the surface during more recent epochs. This occurrence of liquid water suggests that Mars might have had life at some point in its history.

In the next section we will briefly discuss the characteristics of terrestrial life and its history, and then apply them to the search for life on Mars.

18.3 Requirements for life to originate on a planet

Before we can identify characteristics of Mars that might be conducive to supporting life, we need to discuss the general requirements for life. While these issues are described in detail in other chapters, we discuss them briefly here in order to address specific connections to Mars. Based on the characteristics of life on Earth, there are five important lines of evidence that Mars could harbor life.

The elemental building blocks of life are common. The basic elements comprising terrestrial life (*biogenic elements*) are C, H, O, and N, along with nearly two dozen other elements that play a role. The first four are very abundant in the Galaxy and in the Universe; the others, while of lesser abundance, are still widespread. The planet-forming processes are such that rocky planets that occupy the inner portions of planetary systems would be expected to have an appropriate complement of all of the necessary elements (see Chapter 3).

The molecular building blocks of life are common. Life on Earth is based on carbon. The complex organic molecules that are necessary for life include, for example, amino acids (that make up proteins) and sugars (that make up the backbone of DNA and RNA molecules). These molecules must be created out of elemental C, H, O, N, S, and P before they can be combined in the complex ways that make up life. This formation of complex organic molecules could have occurred naturally on the early Earth, for example in the energy-rich environment of hydrothermal systems. However, the molecules also could have arrived on Earth (and Mars) intact, contained in the debris out of which the planets formed. We see organic molecules today in meteorites that fall to Earth, and we have evidence that asteroids and comets also contain them. Using radio telescopes we also have identified organic molecules in interstellar gas clouds in our Galaxy (Chapter 3).

Life on Earth arose quickly after it became possible for it to exist continuously. The Earth's geological record goes back about 4 Gyr, starting about 0.5 Gyr after the Earth itself formed. Some rocks that are older than $\sim$3.4 Gyr contain evidence that life existed at that time. Although we do not know when life originated, it is likely that it already existed at 3.5 Ga, and there is moderately convincing evidence that life existed nearly to 3.9 Ga (see Chapter 12 for a full discussion). However, life could not begin and then exist continuously on Earth until the rain of planetesimals out of which the Earth formed diminished (Chapter 3). When the largest objects impacted onto the forming Earth, they would have heated the Earth up enough to vaporize any oceans, raise surface temperatures to very high values, and kill any life that did exist. If the drop in the impact rate was continuous, then the last Earth-sterilizing impact probably occurred 4.2 to 3.9 Ga (Chyba et al., 1994). However, there is some evidence that the objects comprising the late bombardment of material at about 4.0–3.85 Ga were not large enough to vaporize more than the upper few hundred meters of the oceans; if this is correct, the last Earth-sterilizing event probably occurred about 4.3 Ga. Either way, there was a relatively short time period between the time of the last killer impact and the time that we find the earliest evidence of life – perhaps only 100 Myr, and certainly less than 1 Gyr. This window during which life originated is short compared to the lifetime of a planet; if typical, an origin of life would seem to be a relatively likely event given the proper environmental conditions.

The environmental conditions that appear to be required for an origin of life or for its continued existence are relatively straightforward and are likely to be widespread. Based on our example of terrestrial life, we think that life's origin required the presence of liquid water, access to the various elements out of which life is constructed, and a source of energy to drive chemical disequilibrium (with reactions back toward equilibrium then providing the energy to support metabolism). These conditions are all likely to be met on Mars (and elsewhere): hydrogen and oxygen are sufficiently abundant that we expect water to be very widespread, and we see abundant evidence for it in the inner Solar System as well as on Mars. The elements that make up life should be incorporated into any planet, such as Mars, that formed as the Earth did. Finally, energy to support metabolism is likely to be available on any geologically active planet. While the energy could come from sunlight driving photosynthesis, chemical energy, for instance released from chemical reactions between water and rock, was more likely to have been the driving force for the first life.

The origin of life on Earth may have taken place in an environment that could occur on almost any geologically active planet, including Mars. While we are not certain where the origin of life on Earth took place, we do have some clues (see Chapters 6 and 9). In particular, the nature of the chemical environment and the availability of abundant energy from chemical reactions points toward the possibility that the origin took place at hydrothermal vents analogous to what we find at volcanic spreading centers on the seafloor today. Chemical disequilibrium occurs at these vents because of mixing between high-temperature vent water that has equilibrated with the crust and upper mantle and low-temperature ocean water that has not. Even at relatively low temperatures, chemical reactions between water and rock can occur. These reactions can be seen today in the weathering of fresh volcanic rock into clays and carbonates under water. These reactions are thermodynamically favorable (they release more energy than they use) and this chemical energy could be used to create complex molecules, stringing them together into complicated structures and supporting the metabolism of primitive organisms. This type of environment, whether hydrothermal or aqueous, should be widespread on *any* geologically active planet that has water.

Together, these characteristics suggest that an origin of life may be a relatively likely occurrence if there are habitats that can support the necessary chemical reactions. We now ask whether these environmental conditions have ever occurred on Mars and whether Mars might have had life at some time in its history.

18.4 The potential for life on Mars

In this section, we will address Mars as a planet and whether it meets the environmental requirements necessary to support an origin of life or its continued existence. In particular, we will discuss the evidence for liquid water at the surface or in the subsurface, the history of martian climate and its ability to support liquid water, access to the elements necessary to support life, and sources of energy to support metabolism. We will also address the possibility for interplanetary exchange of organisms within the inner Solar System, in particular between Earth and Mars.

18.4.1 Liquid water on Mars

The presence of liquid water is likely to be the most limiting factor in a planet's ability to support life. Liquid water is not stable at the surface of Mars today because global average temperatures are $\sim$220 K, some

FIGURE 18.1 Valley networks on the ancient terrain of the martian surface. Notice that valleys converge and coalesce downstream. Individual valleys are about a kilometer across.

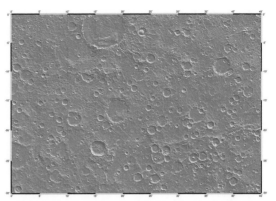

FIGURE 18.2 Heavily cratered surface on the martian highlands. Image shows a region 540 km across (at the equator). The number of impact craters indicates that the surface dates to about 3.7–4.0 Ga. Note the paucity of craters smaller than about 15 km diameter and the heavy degradation of the larger craters (see text). (Map courtesy of M. H. Carr.)

53 K below the freezing point of pure water. Even though surface temperatures regularly rise up to the melting point at midday in equatorial latitudes, the low vapor pressure of water in the atmosphere would cause any ice that was present at the surface to sublimate into the atmosphere rather than melt. Thus, water is either gaseous or solid, and liquid water can be present only as a transient, non-equilibrium phase.

There is evidence, however, that liquid water has existed at and near the martian surface at various times in its history. The evidence comes from geological observations of the morphology of the surface, from analysis of the chemistry and mineralogy of the surface and subsurface (using spacecraft observations of Mars and analysis in terrestrial laboratories of meteorites that came from Mars), and from theoretical models based on our understanding of various components of the martian system.

The geomorphological indicators provide strong evidence for the presence of liquid water at the surface and in the crust at various times in Mars's history. We look for features that are similar in appearance to terrestrial water-formed features, and ask whether they might have been formed by similar processes that involve liquid water. For example, the oldest surfaces on Mars (3.7 to 4.2 Ga) show features that suggest that liquid water played an important role in shaping the surface.

Many of these oldest surfaces have networks of valleys that are very similar to terrestrial river drainage valleys (Carr, 1996). Smaller channels coalesce downhill into larger valleys and, together, provide drainage to large, closed basins (Fig. 18.1). It is widely accepted that liquid water was involved in their formation, but there is substantial debate as to whether these channels

were (a) carved by water that drained from the surface following precipitation from the atmosphere, or (b) formed by the release of water from within the crust in a process known as "sapping" (Carr and Malin, 2000). The dendritic (branching) character of the valleys indicates that it took substantial periods of time to form them, rather than, for example, quick formation in a catastrophic flood. In turn, this suggests that during the epochs in which the valleys formed liquid water was more stable at the surface than it is today.

There is a second morphological indicator of water on these ancient surfaces. Ancient impact craters larger than about 15 km diameter have a substantially degraded appearance, as if they had been physically weathered and the debris carried away. Crater rims, ejecta blankets, and central peaks have, in many cases, been completely removed, and the crater interiors have been filled in with debris; craters smaller than about 15 km have been erased entirely (Fig. 18.2; Carr, 1996). While this erosion could have many possible causes, some craters that are only partially eroded show a "spur and valley" topography on their interiors that is very similar to features carved by liquid water in the walls of terrestrial canyons and valleys. Again, these features point to liquid water being more abundant or more stable in the past than it is today.

Given that Mars is now too cold for liquid water to be abundant enough to carve these features, it is likely that early Mars was warmer than it is today; the additional heating provided by an atmospheric greenhouse gas is a plausible way to produce the warm temperatures (Pollack et al., 1987; Chapter 4).

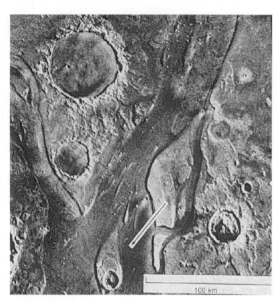

FIGURE 18.3 Flood channel occurring on relatively young surfaces. Note the well-defined margins of the channel indicating confined flow and the streamlined, tear-drop-shaped islands where erosional remnants have been left behind obstacles. Scale bar indicates 100 km.

FIGURE 18.4 Ravi Vallis, one of the source regions for the catastrophic flood channels. Note the jumbled and chaotic blocks at the head of the channel on the right, indicating that the water flowed from beneath the surface, and the linear striations that show flow toward the left. The head scarp (cliff on the right) is approximately 20 km across and 2 km deep.

CO_2 is thought to have been present on Mars in amounts that, if present in the atmosphere, could cause substantial warming; it would have been released into the atmosphere by the substantial volcanism that occurred on early Mars, associated with the formation of the volcanic plains and the Tharsis bulge (Phillips et al., 2001). Another possible greenhouse gas is methane (CH_4), perhaps formed as a waste product if there had been early life on Mars (Kasting, 1997; Kasting and Catling, 2003). In addition, the radiative effects of clouds might have affected atmospheric temperatures (Forget and Pierrehumbert, 1997), and condensation of CO_2 might have limited the amount of atmospheric CO_2 (Kasting, 1991). In the end, though, it is not clear how much greenhouse warming would have been required. It is not possible from the geological evidence to determine the surface temperature required for water to have been able to carve the visible features. Water certainly can flow long distances either in disequilibrium with its environment or protected from evaporation by a covering of ice (Carr, 1996). In Antarctica, for example, streams and lakes exist in regions that are typically at temperatures of about 245 K, nearly 30 K below the melting temperature of ice (McKay et al., 1992).

Some younger surfaces also have features that appear to have been carved by liquid water. In particular, there are large channels that are up to 150 km across from one

bank to the other and that are continuous for thousands of kilometers (Fig. 18.3; Baker et al., 1992). The channels are confined, with well-defined boundaries and edges that rule out formation by the wind. The channel interiors have striation marks that are parallel to the banks and are similar to striations formed in water-carved channels on the Earth. There are tear-drop-shaped "streamlined" islands on the channel floors, apparently formed as the water flowed around less-readily eroded obstacles such as impact craters (Fig. 18.3). Moreover, there are giant ripples that are unlike those formed by wind and very similar to those formed on Earth by catastrophic flooding.

When traced upstream, many of these large channels appear to have a source region beneath the surface. The channels originate in chaotic and jumbled debris, suggesting that the water was present within the crust and was then rapidly released to the surface, breaking up the surface rocks in the process (Fig. 18.4). The size of the blocks of this chaotic debris suggests that the water originated from a depth of ~2 km, consistent with the depth at which the geothermal gradient could raise temperatures high enough that ice would melt. Other channels originate at the eastern edge of the large Valles Marineris canyon system, and may have been formed by the sudden drainage of water filling the canyons (Baker et al., 1992). Either way, the large sizes of the channels and of the erosional features suggest that the channels were carved by the sudden and catastrophic release of large amounts of liquid water, and that most of the water must have originated from subsurface aquifer systems. The ages of the channels, based on the number of impact craters that are present, suggest that the channels formed sporadically throughout much of martian history (Baker *et al.*, 1992).

Unlike the valley networks, the catastrophic outflow channels did not have to form in a climate that is different from today's. When released to the surface in large amounts, water can flow a substantial distance as a liquid prior to freezing (Carr, 1996). Recall that it was the dendritic nature of the early valley networks and the relatively low volumes of liquid implied by their smaller size that indicated that the climate on early Mars must have been different.

How do we know that these features were carved by liquid water instead of by some other liquid? Given the similar formation processes for planets in the inner Solar System, water is expected to have been abundant on Mars, and would have been released to the surface by planetary differentiation and volcanism, for which we see substantial evidence. In addition, we see evidence that water is present at the surface today in the form of ice at the polar caps and vapor in the atmosphere (Jakosky and Phillips, 2001). Although CO_2 is present on Mars and can be a liquid under certain circumstances, those conditions would require unreasonably large amounts of CO_2 that are difficult to emplace into the crust in a stable manner. Molten rock in the form of lava flows can carve channels under some conditions, but there is no evidence in the vicinity of the channels for contemporaneous active volcanism or for volcanic features. The features that we see look so remarkably like water-carved features on Earth, and so unlike features that would be carved in other ways, that erosion by water seems to be the most likely explanation.

Even some of the youngest features on Mars appear to show evidence for liquid water. Gullies have been identified on the walls of canyons, channels, and impact craters (Malin and Edgett, 2000). These gullies (Fig. 18.5) are 10–30 m wide and appear to emanate from a region typically 100–300 m below the rims of the walls. Debris eroded from the gullies has accumulated as fans at the bottom of the slopes. The sharp, unweathered appearance of such small features and the fact that debris from them overlies features that are themselves thought to be very young (such as sand dunes on the floors of valleys) suggest that the gullies were formed very recently and possibly within only the last 1–2 Myr. That they often emanate from below the rims of the scarps suggests that they were formed by the seepage of some fluid from within the crust rather than as, say, dry avalanches. Liquid water again is the most likely fluid to have carved these gullies, as it can occur at shallow depths in places

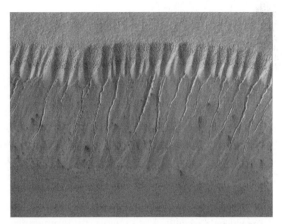

FIGURE 18.5 An example of martian gullies in a scarp (cliff) face at latitude 71°S. From top to bottom, the image (looking northeast) shows the flat surrounding surface, regions of collapse where erosion from midway down the scarp undercuts the overlying material, gullies carved directly by the water, debris aprons of material that were deposited by the flowing water (some with valleys incised into them), and the valley floor. (NASA/Malin Space Science Systems.)

where surface materials have low thermal conductivity, which means that temperatures rise very quickly with depth (Mellon and Phillips, 2001). It also can occur where either the ambient geothermal gradient or intrusive volcanism drives the convection of water within the crust and brings liquid water up very close to the surface (Travis *et al.*, 2003). It has also been proposed that it formed beneath ice deposits that might form when Mars is at high obliquity and water is driven from the polar caps and condenses onto the surface (snow) at lower latitudes; penetration of sunlight beneath the ice might then cause subsurface heating and melting at the bottom of the ice layer, and could leave behind debris layers like those that are seen in some locations (Christensen, 2003).

Liquid CO_2 has also been suggested as an agent for carving the gullies, based on the fact that the pressure within the crust at the depth from which many of the seeps occur is very close to the pressure at which CO_2 liquefies (Musselwhite *et al.*, 2001). This explanation, however, appears to be ruled out because any pressurized liquid CO_2 in the crust released to the surface would catastrophically vaporize on expansion, rather than flow as a liquid and produce gullies (Stewart and Nimmo, 2002). Furthermore, no plausible physical mechanism has been suggested by which the crust could be filled with liquid CO_2.

Other morphological features are suggestive of the presence of liquid water, but with less certainty. Features within impact craters and within Valles Marineris are suggestive of the presence of ancient lakes. Potential crater lakes are identified by the presence of channels flowing into and/or out of the craters, providing a source or sink for water, respectively (Cabrol *et al.*, 1998; Cabrol and Grin, 1999). Some deposits within craters have an appearance similar to deltas, sedimentary terraces, and shorelines, as might be formed by flowing water or within standing bodies of water. Although most of the impact craters that have these features are themselves relatively old, the age of the lakebed deposits is uncertain. There are few impact craters superposed on the lakebed deposits, which indicates either that they are very young or that they were buried for long periods and exhumed only recently. For example, the stratigraphic relationships of layered sedimentary deposits within Valles Marineris suggest an old age despite the absence of craters (Malin and Edgett, 2001). Well-defined layering has been identified within these deposits, supporting the idea of standing water, although layers formed from windblown sediments can have a similar appearance. Even if windblown, however, these sediments would have required at least small amounts of either liquid water or water ice in order to cement the grains together to form coherent layers.

There has been much debate about the possible existence of large standing bodies of water in the northern lowlands. While catastrophic flood channels did drain into this region in amounts large enough that it was likely that standing water existed, the size and longevity of these bodies is uncertain (Carr, 1996). Some morphological evidence has been interpreted as being shorelines carved by waves, which would suggest the presence of relatively long-lived seas or oceans. However, there is uncertainty over whether these features really represent shorelines or were formed by other processes (such as lava flows or wind), or even whether they exist at all. The shoreline-like features do not have a sufficiently constant or uniform altitude, there is an absence of other geological or geochemical evidence (such as carbonate minerals) for long-lived liquid water, and there are no plausible physical mechanisms for providing that much water at one time and then removing it. Overall, the possibility of ancient oceans on Mars must be considered as speculative (Carr and Head, 2003).

Geochemical and mineralogical evidence indicates that liquid water has been present on Mars. On the geochemical side, we can examine the martian meteorites for evidence of liquid water. There are more than two dozen meteorites that have been collected on Earth and that are thought to have come from

FIGURE 18.6 The martian meteorite ALH84001, in which a claim has been made for the presence of fossil evidence for martian life. The cube has dimensions of 1 cm.

Mars[2] (Fig. 18.6). A martian origin is indicated based on three lines of evidence (see McSween, 1994). First, all but one of the meteorites are relatively young volcanic rocks (180 to 1300 Myr) with compositions similar to terrestrial basalts; this requires that they formed on a planet large enough to undergo melting and fractionation relatively late in Solar System history. Second, oxygen isotope ratios in the silicate minerals group all of these meteorites together, and indicate that they did not form on the Earth or Moon. Third, elemental and isotopic compositions of gases trapped in inclusions in two of the meteorites are almost identical to those of the martian atmosphere as measured by the Viking Landers in 1976; moreover, there are also similarities between the elemental compositions and oxidation states of the meteorites and those in martian soils and rocks as measured *in situ* by spacecraft. All of these data indicate that these meteorites must have come from a Mars-like body in our Solar System, and Mars is the only known object that satisfies these conditions (McSween, 1994).

Some of the martian meteorites contain carbonate minerals that fill in voids within the rocks (e.g., Romanek *et al.*, 1994). Carbonate deposits like this are common in terrestrial rocks, and are formed when

---

[2] This collection of meteorites was referred to initially as the *SNC meteorites*, using the initials of the type examples of the three original groupings of these rocks, Shergotty, Nakhla, and Chassigny. With the discovery of another meteorite that did not fit into this classification scheme, ALH84001 (discussed below), there has been a transition to the term *martian meteorites*, reflecting both the increased diversity of the samples and the now widely accepted view that they originated on Mars.

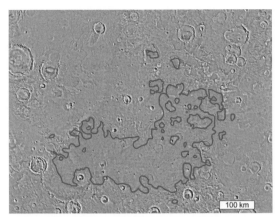

FIGURE 18.7 Map of Sinus Meridiani, showing the location of coarse crystalline hematite, as inferred from orbital spectroscopy of surface thermal emission by Mars Global Surveyor. The region outlined identifies where hematite has been identified, in abundances ranging up to about 20%. This is the landing site for the *Opportunity* Rover, whose investigations in 2004 strongly indicated that the hematite was produced through processes involving liquid water (Section 18.8).

liquid water flows through fractures and fissures in rocks and dissolved CO_2 can be precipitated. In the martian meteorite ALH84001, carbonates that have been deposited in fractures contain layers of slightly different composition that suggest multiple events of water pulsing through the rock; similar features are formed by this process on Earth, and laboratory experiments have been able to reproduce similar compositional layering (Golden *et al.*, 2001, 2002).

Hydrated minerals are present in several of the martian meteorites as well. While some of these minerals could have formed after the rocks fell to the Earth's surface, they contain an enriched ratio of deuterium to hydrogen relative to terrestrial values, similar to the enrichment seen in the martian atmosphere (Watson *et al.*, 1994). The mineral hematite has also been identified on Mars's surface using thermal emission spectroscopic measurements (Fig. 18.7; Christensen *et al.*, 2000). Hematite is an iron oxide weathering product, and is inferred to have been precipitated from water flowing through the crust (see Section 18.8 for further discussion).

It appears then that there is abundant evidence for the occurrence of liquid water on Mars. It was most likely present at the surface up until about 3.5–3.7 Ga, and has probably been present within the crust for all of martian history, including up to very recent times (Jakosky and Phillips, 2001). The transition from surface water to crustal water was probably driven by changes in the climate from an early Mars that was generally warmer and wetter to a colder, drier Mars.

18.4.2 Climate change on Mars

In order to understand what drove the availability of liquid water, we wish to understand the changing environmental conditions on Mars, the transition from surface to subsurface water, and the implications for the presence of liquid water throughout time. To do this, we need to understand the processes that control the martian climate.

The geological evidence suggests that the early surface environment was warmer and wetter than the present one, although it is not certain exactly how much warmer. Furthermore, the features indicative of surface water are, for the most part, not present on younger surfaces, suggesting that the transition in climate occurred ~3.7 Ga. The simplest explanation for both of these features is that the early, warmer environment resulted from greenhouse warming by atmospheric gases and that the transition to a colder climate occurred when the gas was removed from the atmosphere (Jakosky and Phillips, 2001).

Atmospheric CO_2 from an early, thicker atmosphere could have been removed by three main processes: (1) ejection to space by the impact of large asteroids or planetesimals during the end stages of the accretion of Mars; (2) stripping by the solar wind during earlier phases of the Sun's history when the solar wind was more intense; or (3) loss to the regolith and crust by the formation of carbonate minerals. There is evidence that each of these processes has occurred in varying amounts.

Loss through impacts

There are abundant large impact craters on the ancient surfaces, with the largest basins being up to 1,000 km across. The debris ejected into space by these impacts certainly would have dragged gas with it (Melosh and Vickery, 1989). Numerical models of this process suggest that substantial amounts of gas might have been removed. However, it is not known how much was actually removed, since, for example, water or CO_2 that was not in the atmosphere at the time of the impacts would not have been subject to ejection. The impacts whose resultant craters we see might have removed as much as 90% of the atmospheric gas that was present at the time of the onset of the surface geologic record (Brain and Jakosky, 1998).

Stripping by the solar wind

Loss to space could have occurred through two distinct processes. In the first process, ions created in the upper atmosphere by the absorption of solar ultraviolet light could have become entrained in the solar wind's magnetic field and carried away (Luhmann *et al.*, 1992). We observe upper-atmospheric ions being carried away from Mars today, and similar processes seen in more detail at Venus suggest that this effect might have operated on Mars as well. Alternatively, hydrodynamic pressure from the impinging solar wind could have removed gases from the top of the atmosphere. *In situ* spacecraft measurements show the occurrence of discrete blobs of gas whose morphology is suggestive of this type of hydrodynamic stripping (Mitchell *et al.*, 2000).

Measurements of the ratios of stable isotopes in the present-day martian atmosphere suggest that loss of martian atmosphere to space has indeed occurred in significant amounts. Gaseous diffusion enriches the relative abundance of atoms or molecules that contain lighter isotopes in the upper atmosphere; thus, the removal to space of gas from the top of the atmosphere (by any process) preferentially removes atoms or molecules having lighter isotopes. We see this enrichment very consistently in the ratios of $^{15}N/^{14}N$, D/H, $^{13}C/^{12}C$, and $^{38}Ar/^{36}Ar$, which are all greater than would be expected unless removal to space has occurred. The enrichment factors suggest that as much as 90% of each of these gases (N, H, C, and Ar) has been lost to space (Owen *et al.*, 1988; Jakosky, 1991; Brain and Jakosky, 1998).

Loss to the surface

Finally, atmospheric CO_2 can be lost by dissolving in water and then reacting with surface minerals to form carbonate minerals such as calcite ($CaCO_3$). On the Earth, this process rapidly produces limestone deposits on the continental shelves and ocean floor. On Mars, there is spectroscopic evidence for the presence of carbonate minerals at the surface in low abundances, although the crust could contain substantial quantities that are hidden from view (Bandfield *et al.*, 2003). The martian meteorites have carbonate minerals in them that are enriched in ^{13}C, suggesting that they formed from dissolved atmospheric CO_2 and indicating with some confidence that atmospheric CO_2 has indeed been sequestered in the crust (Romanek *et al.*, 1994; Jakosky and Phillips, 2001). It is not possible, however, to determine what the global inventory of CO_2 in this form might be.

Clearly, each of these three processes is capable of removing substantial quantities of gas from the atmosphere, and there is compelling evidence that each has occurred. Together, they are capable of removing the large amounts of CO_2 that could have produced early greenhouse warming. However, the information available today does not allow us to determine the extent to which each process actually operated on Mars or where the CO_2 is today.

There are potentially significant problems with the CO_2 scenario outlined above, however. It is not clear whether a thick CO_2 greenhouse is able to raise temperatures enough to allow liquid water to have existed on early Mars, especially in light of the 30% lower solar flux 4 Ga (Chapter 4). Two alternative scenarios have been discussed. One involves the potential greenhouse warming from other gases, such as methane and ammonia. These gases would have to have been produced continually, as their lifetime against breakdown by sunlight is relatively short (Kasting, 1997). These gases might have been produced in significant quantities if there had been a widespread surface biosphere.

In the second alternative, globally warmer temperatures are not necessary to produce the water-related features. Instead, it is possible that large impacts at the tail end of accretion mobilized water from the crust into the atmosphere, and that the rainout of this water was responsible for producing the valley networks and degradation on the ancient surfaces (Segura *et al.*, 2002). In this scenario, the water would have been present at the surface only for very short periods of time (years to thousands of years). In addition, it would have been so hot that conditions would not have been conducive to the existence of life. However, there are concerns with this model, involving the relative timing of events and the ability to mobilize enough water to have the necessary geological effects.

The changing early martian climate points out issues in understanding planetary habitability. For example, whether the surface of Mars had a climate conducive to the existence of life critically depends on the composition of its atmosphere. With a moderate greenhouse atmosphere (Chapter 4), Mars would be warm enough even today to sustain liquid water and life at its surface. Without greenhouse warming, Mars is too cold to have either at the surface, but it is still warm enough to have liquid water and life hidden below the surface within its crust. The importance of the atmosphere in determining planetary habitability is underscored by the changing solar temperature and luminosity output through time (see Chapter 4). Thus, the martian climate appears to have been warmer when the Sun was colder, and Mars has been cooling down even while the Sun has been heating up over the last four billion years.

There is an additional component of climate change on Mars that would have acted on much shorter time-scales. The obliquity, or tilt of the polar axis, changes with time as a result of the gravitational pull from the other planets (Jupiter in particular). It probably has varied from its current value of 25.2° to values as low as nearly 0° and as high as perhaps 60° within the last few tens of millions of years (Laskar and Robutel, 1993; Armstrong *et al.*, 2004). The changes in tilt cause corresponding changes in the summertime heating of the polar water-ice deposits. At higher obliquity, the poles get warmer and will drive water into the atmosphere; this water might condense as surface or subsurface ice at lower latitudes. Significantly, even at moderate obliquities, polar and high-latitude ice temperatures might rise enough that thin films of liquid water could form even at sub-freezing temperatures; temperatures might even rise to the melting point (Jakosky *et al.*, 2003). And, as mentioned earlier, heating of any ice deposited at lower latitudes might melt it, providing localized and short-term liquid water (Christensen, 2003).

18.4.3 Biogenic elements and energy to support life

Besides liquid water, the other two factors in Mars's potential to support life involve availability of the biogenic elements and a source of energy to support metabolism.

Any rocky planet such as those in our inner Solar System should contain a sufficient complement of elements out of which living organisms can be constructed. The lighter elements such as C, H, O, and N are present in atmospheric gases. H and O are present in water, for which there is abundant evidence (as discussed above). C and O are present as atmospheric CO_2; the ability of CO_2 to dissolve in water or be stored temporarily as minerals suggests that it is readily available everywhere on the surface and within the crust. N is present in the atmosphere as N_2. Although its present abundance is low (2.5%), the enhanced ratio of $^{15}N/^{14}N$ in the atmosphere (McElroy *et al.*, 1976) suggests that there must have been more nitrogen present earlier on. In addition, it is possible that nitrogen is present within the crust in the form of nitrates or nitrogen-bearing minerals.

Other elements that are important in terrestrial life are also present on Mars, including S, P, Ca, Fe, etc. We expect them to be present based on our understanding of how elements formed within earlier generations

TABLE 18.3 Composition of the martian crust[a]

Species	Concentration	Species	Concentration
SiO_2	43.4%	P_2O_5	0.68
Fe_2O_3	18.2	TiO_2	0.6
Al_2O_3	7.2	MnO	0.45
SO_3	7.2	Cr_2O	0.29
MgO	6.0	K_2O	0.10
CaO	5.8	CO_3	<2
Na_2O	1.34	H_2O	0–1
Cl	0.8		

[a] Based on Viking analysis of soil (expressed as oxides; from Kieffer *et al.*, 1992).

of stars and created the interstellar dust and debris out of which planets formed. More specifically, we have detected them directly on Mars with *in situ* observations from landed spacecraft and analysis in terrestrial laboratories of martian meteorites (McSween, 1994). Table 18.3 lists the composition of the martian crust.

Energy to support the construction of complex organic molecules and the metabolism of extant organisms could have come from a number of sources. Sunlight is very abundant everywhere at the martian surface, albeit about half as strong as at Earth, and can support photosynthesis through a series of chemical reactions (McKay *et al.*, 1992). However, photosynthesis involves a complex suite of chemical processes and may not have been the earliest metabolic pathway. On Earth, for example, the earliest organisms may have obtained their energy ultimately from disequilibrium chemical reactions that occur between water and minerals (Woese, 1987). Energy would have been available either when water that had equilibrated with rock at high temperatures mixed with water at lower temperatures, thereby producing a chemical system that contained species that were out of equilibrium with each other, or from chemical reactions between water and minerals at lower (ambient) temperatures that occur because the rocks are out of equilibrium with their environment (Shock, 1997; Chapter 8).

We can try to estimate the amounts of energy that might have been available on Mars from these processes, based on the inventory of volcanism that might have driven hydrothermal systems and on the amount of weathering that has occurred on Mars. Figure 18.8 shows the history of volcanism on the terrestrial planets, and indicates that Mars has had about 100 times less volcanism than the Earth throughout its history. As a

TABLE 18.4 Examples of net chemical reactions that can support metabolism in organisms

Photosynthesis	$CO_2 + H_2O + \text{sunlight} \rightarrow CH_2O + O_2$
Oxidation	$CH_2O + O_2 \rightarrow CO_2 + H_2O$
Fermentation	$C_6H_{12}O_6 \rightarrow 2C_2H_6O + 2CO_2$
Oxidation of iron	$2Fe^{+2} + 1/2\,O_2 + 2H^+ \rightarrow 2Fe^{+3} + H_2O$
Oxidation of hydrogen	$6H_2 + 2O_2 + CO_2 \rightarrow CH_2O + 5H_2O$

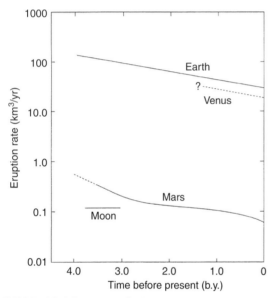

FIGURE 18.8 4-Gyr history of volcanic eruption rate (in units of km³ of magma erupted per year) on the terrestrial planets. The present-day Earth value is observed and older values are extrapolated based on the presumed radioactive heating rate. The present-day Venus value is inferred and also extrapolated backwards in time based on the expected radioactive heating rate; the dashed line reflects uncertainty over the timing and nature of episodic eruptions. The Mars rate is inferred for different epochs based on morphology of its surface; the dashed earliest epoch reflects uncertainty over whether surfaces of that age are volcanic. The lunar value is based on total volcanic volume and estimated eruption duration. From Jakosky and Shock (1998).

result, it probably had about 100 times less hydrothermal activity (assuming that the amount of crustal water would not have been a limiting factor) (Jakosky and Shock, 1998).

Table 18.4 includes examples of the types of metabolic reactions that support organisms on Earth (e.g., Nealson, 1997). On Earth these organisms are involved in a large portion of the *weathering* of the surface, i.e., alteration of surface minerals. If organisms were involved in weathering the entire martian crust in a

similar manner, a tremendous amount of energy could be released. However, it appears as if only a small portion of the crust has been weathered since altered debris at the planet's surface has been estimated to be equivalent to a global layer perhaps only a few meters thick (Christensen *et al.*, 2003). In addition, the most heavily altered of the martian meteorites has only had a couple of per cent of its mass weathered (Mittlefehldt, 1994).

The energy available from weathering reactions can be estimated from thermodynamic properties of the minerals involved, combined with global estimates of the amount of alteration (Jakosky and Shock, 1998; Varnes *et al.*, 2003). The combination of energy from hydrothermal systems and from chemical weathering reactions could have supported the formation of a layer of organisms of order 10 cm thick, integrated over the entire 4 Gyr of martian history. This is very similar in column thickness to the amount of biomass (mass of all organisms) on the present-day Earth, although on our planet this biomass decays and is renewed every 1,000 years. This means that Mars over its entire history might have had as much energy available to support the origin and maintenance of life as did the early Earth in its first 50–100 Myr. Thus, it is at least plausible that aqueous and hydrothermal systems on Mars might have been capable of supporting an origin of life (Varnes *et al.*, 2003).

18.4.4 Potential for interplanetary exchange of biota

In addition to the possibility that life may have had an independent origin on Mars, it is possible that living microorganisms could have been exchanged between the Earth and Mars. This would require that rocks be ejected from the surface of one planet into space by an impact, yet without being subjected to temperatures or pressures high enough to kill the microorganisms living inside; that they travel through space without being killed by the extreme temperatures, vacuum, or high cosmic ray flux; that they survive atmospheric entry and landing on the other planet; and finally, that they find an environment suitable for survival (Melosh, 1988). Remarkably, all of this appears to be possible.

We know from the existence of the martian meteorites that rocks can be ejected into space from one planet and fall onto the surface of another. Theoretical models of the ejection process suggest that at least some of these rocks would be lifted into space relatively gently, such that microorganisms would not be killed by heat or pressure (see Melosh, 1989). Once a rock is ejected from Earth, the travel time from Earth to Mars typically would be tens of millions of years (Gladman *et al.*, 1996). After its initial ejection, its orbit around the Sun would be continually modified due to close approaches to both Earth and Mars, and a given rock may be ejected from the Solar System, land on Mars, or land elsewhere in the inner Solar System (including back on Earth). Calculations suggest that most of the rocks that are ejected from Earth or Mars end up being ejected from the Solar System, but that close to 10% land on the other planet (Gladman *et al.*, 1996). While in space, many terrestrial organisms are very hardy and can survive the harsh environment without much difficulty (Mileikowsky *et al.*, 2000; Horneck *et al.*, 2002). When the rock (including organisms) finally impacts onto Mars, the smallest particles will burn up in the atmosphere, the largest will impact the surface at high velocity and create impact craters, and those in a wide range of intermediate sizes will land relatively gently and potentially deliver organisms to the martian surface (Pierazzo and Chyba, 1999). While only a very small fraction of the rocks exchanged would have been ejected with sufficient speed yet gentleness that they could transfer living organisms, the large number of rocks ejected, especially during the early history of the Solar System, would allow the potential for significant exchange either from Earth to Mars or from Mars to Earth (Gladman *et al.*, 1996; Gladman and Burns, 1996). One factor strongly favoring the latter route is the smaller gravity of Mars compared to Earth, making it both easier for a rock to leave Mars and more likely for it to be swept up by Earth.

If we do find evidence for life on Mars, we may discover that it shares a common origin with life on Earth and, in fact, had been transferred from Earth. It is interesting that, in the earliest epochs of the Solar System, it is possible that Venus, Earth, and Mars each may have had an environment conducive to an origin of life. Given the amounts of material that would have been exchanged between all of these planets, it is possible that life could have originated on any of these planets and subsequently been transferred to the other two. In such a situation it is not at all clear how one could determine on which planet the origin of life actually took place.

18.4.5 Summary of the potential for life on Mars

Mars appears to have met all of the environmental conditions necessary for either an independent origin or the continued existence of life – liquid water, availability of the biogenic elements, and a source of energy

that could drive metabolism. It certainly appears plausible that Mars could have life now or could have had it in the past. In a sense, then, searching for life on Mars becomes a test of our understanding of the origin and evolution of life on Earth. If Mars is found to have life that represents an independent origin, then it would seem that life can really appear as readily as many think. This, in turn, would suggest that life could be widespread throughout the Galaxy. On the other hand, if Mars is found to have no life and never to have had any, despite having all of the ingredients and environmental conditions that we think were necessary, then there would have to be something missing in our understanding of the sequence of events that led to life on Earth. At a minimum, these issues suggest that a search for life on Mars is far from being a fool's errand. Instead, it becomes a vital integrating theme in our exploration of the red planet.

18.5 Past searches for martian life

The primary searches for evidence of life on Mars took place in three different eras – Percival Lowell about 100 years ago, the Viking spacecraft in the mid-1970s, and the analysis within the last decade of the martian meteorite ALH84001 to find microscopic geochemical evidence. We now discuss each of these, with an emphasis on the last two; the first era is also covered in detail in Chapter 1 and the historical significance of the latter two in Chapter 2.

18.5.1 Percival Lowell

There was tremendous interest in the possibility of life on Mars in the late 1800s and early 1900s. This stemmed from the first systematic telescopic observations that allowed us to recognize the similarity in the length of the Mars and Earth days, the similar tilts of the polar axes with respect to the orbital plane (and the resulting similar nature of seasons), and the occurrence of polar caps on Mars that grew larger in winter and smaller in summer in a manner similar to Earth's wintertime polar caps of snow. Mars was thought at that time to have a thick atmosphere, and it was expected that surface temperatures would not be too different from those on Earth.

The Italian astronomer Giovanni Schiaparelli in the 1870s made drawings of lineations connecting discrete features on the surface. He called them *canali*, which means "channels" in Italian. The American Percival Lowell interpreted these features literally as canals and during the 1890s mapped out hundreds of them (still using drawings and the eye at the telescope) extending from the polar regions to the equator (Martin *et al.*, 1992; Section 1.7.1). Given the summertime contraction of the polar caps, he built an entire civilization around these canals, concluding that the martians were a dying civilization on a planet that was becoming increasingly desert-like, and that the canals were created to bring polar meltwater to the equatorial inhabitants.

Lowell was not the only one to believe in a martian civilization. There were articles in such notable places as the *New York Times* describing the forests and climate on Mars. Furthermore, Guillermo Marconi and Nikola Tesla, who played major roles in the early history of radio, thought at times that they were detecting radio signals broadcast to us by martians.

The martian civilization, however, was to be short-lived. The first spacecraft to visit Mars in the 1960s showed us that the atmospheric pressure was less than 1% of the Earth's pressure, that temperatures were well below the freezing point of water, and that there were no canal-like features like those mapped by Lowell.

18.5.2 The Viking Mission

The low atmospheric pressure, the temperatures too low to support liquid water, and the heavily cratered nature of the surface as seen from flyby spacecraft could not dissuade people from thinking of Mars as a possible abode for life. The discovery by the Mariner 9 spacecraft in the early 1970s of water-carved features kept the possibility open, and four Viking spacecraft in the mid-1970s were designed to search for evidence for life. Much of our present understanding of Mars comes from these missions, and they have been our only attempt so far to search for life on another planet, so it is appropriate to describe these investigations in some detail.

The Viking mission consisted of two orbiters and two landers. The orbiters had remote-sensing instruments used to identify safe and interesting landing sites (with a strong emphasis on safe). They eventually provided global imaging and thermal and water vapor measurements for more than a martian year.

The two landers landed on opposite sides of the planet in 1976 and obtained measurements from imaging, inorganic chemistry, meteorology, seismology, physical and magnetic properties, mass spectroscopy, and biology experiments (Fig. 2.3 shows a photograph of one of the landers). The mass spectrometer and the three biology experiments are of the most relevance here (Klein *et al.*, 1992). The biology instrument package consisted of a single mechanism that obtained samples and distributed them to three separate experiments (Fig. 18.9). Each of these experiments was designed to test for a different metabolic pathway in potential martian organisms.

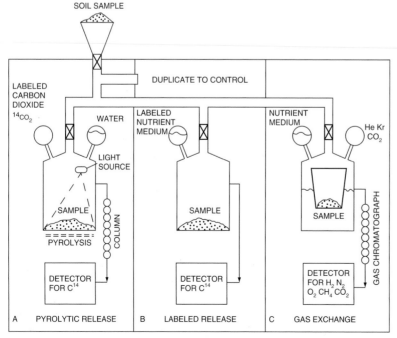

FIGURE 18.9 Schematic diagram of the three biology experiments on the 1976 Viking landers.

The carbon assimilation experiment (also referred to as the pyrolitic release experiment; labeled A in Fig. 18.9) was designed to test for organisms that could take up CO_2 or CO from their environment and incorporate them into organic material (Horowitz *et al.*, 1977). This experiment was done under conditions most like those of the ambient martian environment, and involved exposing a sample of soil to CO_2 and CO gas, either with or without water being present. The gas had been brought from Earth, and was "labeled" with the radioactive isotope ^{14}C. After a period of time during which metabolism might occur, the soil sample was heated; this would break apart any organic compounds that might have formed and release them as CO_2 (a process known as *pyrolysis*). Thus, if any CO_2 or CO had been incorporated into material in the soil, it would be released and would show itself by counters that could detect the ^{14}C isotope. This experiment was designed to test for processes such as photosynthesis, chemosynthesis, or other metabolisms that would involve the uptake of CO or CO_2.

The experiment showed that labeled carbon was, in fact, incorporated into the soil, exactly as might be expected if the soil contained active biota. The level of activity, however, was substantially lower than was expected based on experience with terrestrial soils. Moreover, heating the sample to as high as 175 °C for three hours prior to exposing it to the gases reduced but did not eliminate the positive response. Any carbon-based biological organisms within the sample would have been killed by temperatures that high; even terrestrial biota from hydrothermal systems would not survive such high temperatures. The conclusion reached was that some nonbiological, geochemical process was taking CO or CO_2 from the atmosphere and producing small amounts of carbonate mineral or carbon polymer within the soil. There was no convincing indication from this experiment that biological activity was occurring.

The second experiment was the gas exchange experiment (Oyama and Berdahl, 1977; labeled C in Fig. 18.9). Again, soil was exposed to a mixture of nutrients. The gas surrounding the soil was sampled subsequently in order to look for specific gases that might be given off by martian biota; the gases tested for were H_2, N_2, O_2, CH_4, CO_2, Ar, and Kr, the last two as calibration test gases. The nutrients included a wide range of amino acids, salts, vitamins, CHO compounds, and the bases from DNA molecules; because of the unknown "handedness" of any martian biota (see Section 18.6.1), both left- and right-handed molecules

were included. This experiment could be run either in "wet" mode, with the nutrient solution added directly into the soil, or in "humid" mode, with the nutrient allowed to evaporate along with the water vapor into the space above the soil.

Significantly, oxygen was given off by the soil immediately after being exposed to the nutrient. However, this was not thought to be a biological response for several reasons. First, the oxygen was given off in the dark, which is different from how terrestrial organisms behave. Second, the reaction was not stopped by heating the sample to 145 °C. Third, the same gases were given off (albeit at a slower rate) even when the soil was exposed only to water vapor, with no nutrients. The results could best be explained by the presence of oxidants such as hydrogen peroxide (H_2O_2) in the soil; these would be able to break the organic nutrients into smaller molecules, and these and the water molecules would displace other gas molecules that had attached themselves to the soil grains. Oxidants could be produced via photochemical processes in the atmosphere, and would have diffused into the soil (Hunten, 1979).

The third experiment was the labeled release experiment (Levin and Straat, 1977; labeled B in Fig. 18.9). This experiment tested for the presence of martian organisms able to assimilate organic compounds from their environment and, in doing so, release gas back into the atmosphere. Here, organic nutrients labeled with radioactive isotopes were put into contact with the soil. If the nutrients were utilized by organisms within the soil, then gases given off by the soil might contain some of the labeled atoms. These gases were then tested for radioactivity from the nutrients. The nutrients, consisting of molecules of formate, lactate, glucose, glycine, and sulfate in water, were labeled with radioactive ^{14}C and ^{35}S; again, both left- and right-handed molecules were utilized.

A positive result from this experiment would be detection of radioactive counts in the gas surrounding the soil upon the addition of the nutrients; this signal would increase with time and then would gradually level off as the nutrients were exhausted. In fact, that was exactly the signal that was observed (Fig. 18.10). Furthermore, heating the sample to 50 °C substantially reduced the signal, and heating it to 160 °C destroyed the signal entirely. This is the type of signature that would be indicative of the presence of martian biota. However, it is also possible that it represents the response to chemical reactions taking place within the soil. Active oxidants in the soil could react with the nutrients to produce the observed signal; interaction

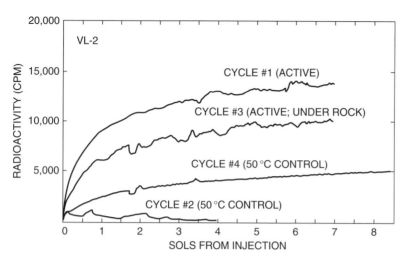

FIGURE 18.10 Results from the Viking Labeled Release experiment. The plot shows the time sequence of measured counts per minute of ^{14}C-labeled gas released from the martian soil following addition of nutrient-containing water to the soil, as described in the text. After Levin and Straat (1977).

with clay minerals in the soil or with the naturally occurring mineral limonite (an oxidized iron-bearing mineral) could also catalyze the destruction of the nutrient. Although this experiment yielded a positive signal for the detection of life based on the pre-flight evaluation criteria, we now know that this signal was not necessarily a unique indication of the presence of life (Klein *et al.*, 1992; Klein, 1999).

The final relevant experiment was not a formal part of the biology package, but was nonetheless critical in properly understanding the implications of those experiments. This was the gas chromatograph/mass spectrometer experiment. It was designed to measure the abundance of organic molecules in the soil by heating the soil until any complex organics would be broken apart into small organic molecules, evaporated, and then detected (Biemann *et al.*, 1977). The detection mechanism consisted of first passing the molecules through a gas chromatograph, which is a porous column through which different gases pass at different rates. The gas exiting the column was passed into a mass spectrometer, which could determine the abundance of different organic molecules according to their masses. The only organic molecules that this experiment detected were contamination brought along from Earth – solvents left over from cleaning the instrument before launch. The upper limit determined for the abundance of organics in the soil varied by molecule but was typically at the parts per billion level; this is well below the level expected if there were any active or even dead biota present. Further, a much higher level of organic molecules should have been measured even if the only source of organics at the martian surface was incoming meteorites and asteroids. This experiment

placed a very stringent limit on the occurrence of biological activity and again suggested the presence of some type of oxidizing molecule in the soil that could destroy organic molecules (Biemann *et al.*, 1977; Klein *et al.*, 1992).

Interestingly, the results of all three biology experiments initially were consistent with the soil samples containing active biota. This points to a very important aspect of the Viking biology experiments (and indeed of any tests for biological activity): it is important that control experiments be performed on the same soil samples that are being tested. With the Viking instruments, the control experiments consisted of doing the same analyses on samples that had first been heated to sufficiently high temperatures that a positive biological response would not be expected to occur. Without the controls, the Viking experiments could have led to the conclusion that there was life on Mars. Yet, having done the controls and obtained the entire suite of experimental results, the consensus of almost all experts is that life is *not* the best explanation for the observations as a whole (Klein, 1999; Klein *et al.*, 1992).

Although the labeled release experiment was to some extent consistent with the presence of life, the mass spectrometer indicated that organic molecules were not present at the levels required for a biological explanation. Thus, despite the apparent positive results from the labeled release experiment, the best explanation for the results is a geochemical rather than biological one. This points out a second important aspect of the Viking experiments: it is important that multiple experiments be performed to test multiple hypotheses. Although one experiment did produce an apparently positive result, the results from the ensemble of experiments and their controls provided a much clearer picture than did any single experiment. Again, the entire set of results suggests that it is extremely unlikely that biological activity is the best explanation (Klein, 1999).

In the end, the Viking results were limited by the requirement that they be designed years prior to flight and then packaged on the spacecraft with stringent volume, mass, and power requirements. This limited the range of experiments that could be performed and the hypotheses that could be tested, and in particular precluded the ability to do different follow-up experiments in response to the initial results. For example, it turns out that certain refractory salts can be formed from the oxidation of organic molecules, and that these are not volatile and would not have been detected using the Viking gas chromatograph/mass spectrometer (Benner *et al.*, 2000). Thus, although it is likely that no living and metabolizing organisms were present at the two Viking sites on Mars, there still could have been organic molecules in the martian soil.

A second severe limitation in the Viking results is that they were based entirely on the ability to ingest soil that might contain organisms, add nutrients, and observe the results. Although the three different experiments looked at three different forms of metabolism, they all were based on, in essence, trying to *culture* potentially present martian organisms.[3] On Earth, where we know a lot about the biochemistry and ecology of different organisms, only perhaps 1 in 100 to 1 in 1,000 organisms can be cultured in the laboratory. It was tremendously arrogant, at least in hindsight, to imagine that we could go to a planet with wildly different environmental conditions, use a nutrient broth made of a mix of many different organic compounds, and hope to culture organisms that may have fundamentally different properties from those on Earth.

While the Viking results were a watershed in our understanding of Mars and of how to search for life, it seems much more likely that the question of life on Mars will be answered only by bringing samples back to Earth. In a laboratory on Earth experiments can be designed in response to results from earlier ones, and newly developed hypotheses about metabolism and the chemistry of life can be tested without the decade-long development required to go back to Mars.

18.5.3 Claims for life in the ALH84001 meteorite

We turn now to the recent debate about possible fossil evidence for life in one of the martian meteorites. The issue was raised at the same time that new discoveries about extremophiles and the origin of life on Earth were just beginning to filter into the planetary science community and reopen the "life on Mars" issue. The debate about the potential for life to exist on Mars was very public, and has played an important role in the life-on-Mars discussion for the past decade. It has had a strong impact on our confidence in our ability to unambiguously identify *biomarkers* (indicators of the presence of life now or in the past) in extraterrestrial samples.

David McKay of NASA's Johnson Space Center and his colleagues (McKay *et al.*, 1996) garnered a tremendous amount of attention, both in the scientific community and with the public, with their results about possible life in the martian meteorite ALH84001 (Fig. 18.6). They based their conclusions primarily on: (a) the presence of a particular class of organic molecules known as polycyclic aromatic hydrocarbons (PAHs) that could be decay products of biological entities (Chapter 3); (b) the presence of small magnetite crystals whose chemical and crystallographic characteristics mimic those produced by certain bacteria on Earth; (c) the occurrence of minerals in close proximity to each other that were not in chemical equilibrium with each other, with biological activity being a nonequilibrium process that could have produced them; and (d) the presence of structures in carbonate globules in cracks in the rock that had an appearance similar to terrestrial microbes. They argued that, while each of the observed features could be produced either by biological or geochemical processes, the combination of features was best explained by a single process, biology, rather than by a combination of multiple geochemical processes. Their results led to a myriad of investigations into subjects as diverse as the origin and temperature of formation of the carbonate globules, the origin of the magnetite crystals, and the origin of the microbe-shaped structures, as well as analogue investigations of the formation of carbonates on Earth, fossilization of terrestrial bacteria, and non-biological formation of magnetites. To date, there is still no unanimous agreement as to whether there is or is not evidence of martian life in ALH84001. We probably will not be able to answer this question definitively until a suitable sample has been returned directly from a known location on Mars, without the possibility of

---

[3] *Culturing* refers to creating favorable conditions in which sampled microorganisms will multiply.

contamination from Earth. Here, we discuss the current status of the debate over the data and its analysis.

18.5.3.1 *The ALH84001 meteorite*

The ALH84001 meteorite has had a checkered history (Treiman, 1998). Radioactive dating indicates that the rock crystallized ~4.5 Ga from volcanic magma. It is the oldest martian meteorite known and must have formed soon after the planet solidified. The area in which the rock formed on Mars's surface would have been bombarded subsequently by impacts, and the rock still contains the traces of those shocks. It is in the fractures produced by these impacts that the carbonate globules that supposedly hold evidence for martian life were deposited from water; Borg *et al.* (1999) have dated the carbonate formation to 3.9 Ga. The rock remained on the surface of Mars until about 16 My ago when another impact ejected it into space; it orbited the Sun until about 13,000 years ago, when it was captured by the Earth's gravitational field and fell onto the ice on Antarctica. Covered by successive snow falls, the meteorite was buried in the ice until about 700 years ago when ice flow and sublimation of the ice exhumed it again. Since these volcanic meteorites are characterized by black fusion crusts, they are readily recognized against the white ice or snow, and ALH84001 was collected from the ice in 1984 by an expedition supported by the US National Science Foundation. Despite being by far the oldest of the several dozen martian meteorites (Section 18.4.1), its oxygen isotope composition clearly groups it with the other martian meteorites (Mittlefehldt, 1994).

There is ample evidence that ALH84001 underwent alteration from both shock metamorphism and aqueous weathering before it was removed from Mars. Evidence of aqueous deposition of salts is present as zoned carbonates in the form of "flattened pancakes" or flattened fissure fillings occurring in cracks in the rock (Fig. 18.11; Mittlefehldt, 1994). The zonation is chemical, the deposits being Ca- and Mn-rich in their cores and progressing to Mg-rich siderites and magnesites, often with an outer Fe- and S-rich rim. Two observations demonstrate that these carbonate deposits must be martian in origin and cannot be a terrestrial contaminant. First, their age of formation is >3.9 Gyr according to Pb-Pb and Rb-Sr isotopic dating (Borg *et al.*, 1999). Second, the carbonate deposits have been physically disrupted and penetrated by shock melts, a phenomenon that could not have taken place in the last 13,000 years while the rock was on

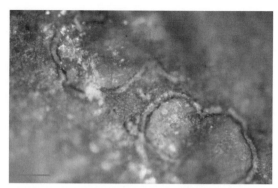

FIGURE 18.11 These carbonate globules that occur in cracks in the martian meteorite ALH84001 were probably deposited by water flowing through the rock. They consist of an orange-colored, Fe-rich carbonate center (siderite, $FeCO_3$) and a thin, black, magnetite-rich rim surrounded by a white Mg-rich carbonate halo. The carbonate globules range from 50 to 250 μm in size. The double globule in this image is 250 μm long.

Earth, but probably occurred during the impact ejection from the surface of Mars (McKay *et al.*, 2002). Early oxygen isotope measurements indicated that the carbonates formed at low temperatures (consistent with the presence of life). Although some later discussion suggested that they formed at high temperatures, a low temperature of formation from evaporation from the bed of a water body or brine infiltration is now generally accepted (McSween and Harvey, 1998).

The ancient age for ALH84001 and its apparent alteration in the presence of liquid water make it an important candidate rock for discussions pertaining to martian life. It was present on Mars during early epochs when the surface was warmer and wetter, as discussed above. There is also evidence that water percolated through the rock, and the rock has been almost completely unaltered since those early epochs. Most of the search for evidence for life has been done on the carbonate deposits within the cracks and voids in the meteorite, as it is these carbonates that were deposited by water and this is where any evidence pertaining to life would be found.

18.5.3.2 *Organics*

Numerous investigations of ALH84001 have shown that it contains components of organic molecules that are both *endogenous* (of *interior* origin, i.e., very likely from Mars) and *exogenous* (of exterior origin, i.e., from Earth after it landed). A large portion of the organic content (80%) is of terrestrial origin, as documented by ^{14}C measurements and by the release of organics at low temperature upon heating the carbonaceous

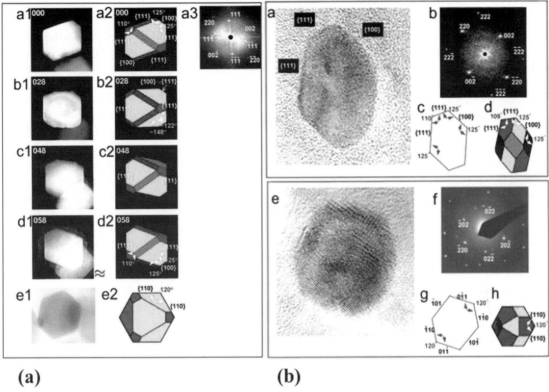

(a)   **(b)**

FIGURE 18.12 (a) Examples of a single truncated hexa-octahedral magnetite produced by magnetotactic bacteria strain MV-1 imaged with incremental stage rotation. (b) High-resolution transmission electron microscope images of hexa-octahedral magnetite particles (~35 × 25 nm) embedded in carbonate in ALH84001. (Thomas-Keprta *et al.*, 2001.)

components in the rock. The source of the terrestrial contamination appears to be predominantly bacterial infiltration; the mycelia of the filamentous bacterium *Actinomycetes*, a common endolithic (living within rocks) bacterium in Antarctica, have been observed in the meteorite (Steele *et al.*, 2000). Amino acids detected in the meteorite are also likely to be due to terrestrial contamination (i.e., breakdown products from terrestrial bacteria; Bada *et al.*, 1998). However, there also appears to be an endogenous carbonaceous component contained in the carbonate globules, reflecting both martian magmatic carbon and martian organic carbon (Grady *et al.*, 1998). The carbonaceous molecules that are most likely to be of martian origin are the insoluble, heavier molecules such as polycyclic aromatic hydrocarbons (PAHs) and aliphatic hydrocarbons (e.g., Clemett *et al.*, 1998). What is the origin of these hydrocarbons, and are they of a biological origin? They can be formed by both processes, e.g., heavy organic molecules could be produced by disequilibrium reactions when magmatic and impact-generated gases such as CO, CO_2, and H_2 cool (Zolotov and Shock, 2000).

18.5.3.3 *Magnetite crystals*

Small magnetite crystals are concentrated in the rims of the carbonate globules, and are also found in their cores. About 25% of these consist of nanometer-sized crystallites (4–100 nm in size) that are single-domain, chemically pure, defect free, elongated along the [111] crystallographic axis, and of unusual truncated hexa-octahedral crystal shape (Thomas-Keprta *et al.*, 2000; Fig. 18.12). The other 75% of the magnetites contain crystals with a poorly crystalline, irregular shape, as well as regular crystals with a whisker-like, platelet or equi-dimensional shape. Electron microscope studies have clearly demonstrated that the magnetite crystallites in the rims of the globules are not terrestrial contaminants (Friedmann *et al.*, 2001). There are a number of possible origins for the magnetites, including an intracellular or extracellular biogenic origin (Thomas-Keprta *et al.*, 2002), hydrothermal precipitation, thermal oxidation of the sulfides, and *in situ* formation by shock-induced thermal decomposition of carbonate (e.g., Golden *et al.*, 2001, 2002; Barber and Scott, 2002).

The biogenic hypothesis for their formation is based on the inferred similarities between the single-domain magnetites and those produced within the cells of one particular strain of terrestrial magnetotactic bacteria known as MV1 (Fig. 18.12) (Thomas-Keprta *et al.*, 2000). *Magnetotactic bacteria* are aquatic microorganisms that produce chains of tiny magnetic crystals (usually magnetite (Fe_3O_4) but sometimes greigite (Fe_3S_4) in the nanometer size range) within their cells. These organisms can sense the Earth's magnetic field and use it to orient themselves in the aquatic environment (a process called *magnetotaxis*), possibly in a strategy designed to keep the organisms in zones of low oxygen content near the oxic/anoxic interface. MV1 is a marine magnetotactic bacterium that uses dissolved iron to produce magnetite crystals with a specific size, crystal shape and crystallographic structure, and chemical purity. The crystals are also multiple, occurring in chains. However, the exact crystal shape of the proposed intracellular biogenic magnetites from the carbonate globules in ALH84001 has recently been questioned and, furthermore, other magnetotactic bacteria produce magnetite crystallites with very different characteristics. It has been claimed that some of the magnetites embedded within the carbonates occur in chains, as would be expected if they represented intracellular magnetite chains (Friedmann *et al.*, 2001). Such chains could also be explained by nonbiological nucleation along microfractures in the carbonate crystals, however, or by nucleation along crystal lattice planes (Barber and Scott, 2002). In any case, if some or a majority of the magnetites were of biogenic origin, how can their occurrence within fractures in the basaltic host rock be explained? One suggestion is that the crystals were washed into the cracks along with the fluid from which the carbonates precipitated (Friedmann *et al.*, 2001).

Recent experimental work has supported the possibility of an inorganic origin for these magnetite crystals through thermal dissociation of the carbonates (Golden *et al.*, 2001, 2002), although these experimentally produced crystals do not share all the characteristics of those produced by MV1 or found in ALH84001. Detection in the carbonate globules from ALH84001 of crystallographic continuity between magnetites and their host carbonate demonstrates that the crystals formed from the carbonates by exsolution in the solid state, presumably during shock impact heating. This hypothesis provides a plausible explanation for the totality of the characteristics of the magnetites in their host carbonates since the evidence is strong that the rock was indeed subjected to multiple shock events during its history, at least one event of which occurred after the deposition of the carbonate globules.

18.5.3.4 *Ovoid structures in carbonate globules*

The carbonate globules in ALH84001 contain small ovoid structures, 20–100 nm in size (Fig. 18.13). McKay

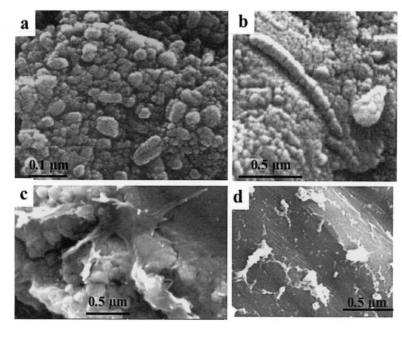

FIGURE 18.13 (a, b) So-called "nanobacteria" fossils from carbonate globules in ALH84001. (c) Contaminating microorganism *Actinomycetes* of terrestrial origin. (d) Contaminating polymer that condensed onto a pyroxene crystal surface from humid air.

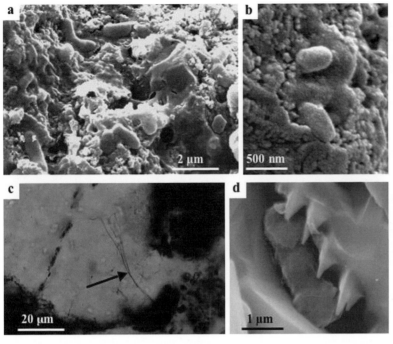

FIGURE 18.14 Examples of 3.5–3.3 Ga fossil bacteria from the Barberton greenstone belt, South Africa (Westall *et al.*, 2001). They formed biofilms at the surface of sediments deposited in shallow water environments and were later mineralized by impregnation with silica-rich hydrothermal fluids. Similar microfossils might be found in rocks on Mars. (a) A scanning electron microscope (SEM) image of a fossilized (silicified) biofilm consisting of vibroid-shaped microfossils embedded in a polymer film; (b) SEM image of small, silicified, coccoidal microfossils; (c) Thin-section image of black, filamentous microfossils still containing a substantial amount of carbon; (d) SEM image of a chain of silicified, dividing, coccoidal microfossils.

et al. (1996) stated that they resembled very small fossil bacteria on Earth (so-called *nanobacteria*;[4] Folk, 1993), and interpreted these structures as fossilized microbes. Although it was demonstrated that they were not artifacts of sample preparation (McKay *et al.*, 1997), as was first suggested, there is still debate regarding their origin. One issue involves the size of the features. A recent discussion concerning the size limits of bacteria came to the conclusion that 200 nm was the lower limit in size for modern microbes based on the minimum amount of components needed for metabolism and cell reproduction (Knoll and Osbourne, 1999), although 150-nm-sized bacteria recently have been discovered in a deep-sea hydrothermal vent (Huber *et al.*, 2002). Of course, it is possible that the martian features could represent dehydrated life forms or the equivalent of an earlier stage in microbial evolution.

[4] *Nanobacteria* is a term used for possible very small bacteria, having sizes less than 100 nm. Folk (1993) was the first to use this term for small, mineralogical structures having a morphology similar to that of bacteria. He claims to have found nanobacteria as small as tens of nanometers in size. Uwins *et al.* (1998) also claim to have found living nanobacteria in Australian sandstones. These claims have been vigorously contested and it appears that the "nanobacteria" could be either simple polymers (in a mineralized form) or other mineralogical phenomena. Note that viruses also range over 20–400 nm in size, but are not independent organisms because they need a host organism in order to be able to reproduce.

A second issue involves the distinction between fossil bacteria and similar-looking structures (bacteriomorphs) of nonbiological origin. For example, Fig. 18.14 shows examples of 3.5–3.3 Ga fossil bacteria, similar to what could be found on Mars, from some of the oldest sedimentary rocks preserved on Earth, occurring in South Africa. There are many other phenomena, both crystalline and organic, that can form small structures similar to those seen by McKay and his colleagues. Convincing identification of bacterial fossils requires the fulfillment of a series of criteria including morphology, colony-related aspects relating to biofilm formation, and biochemical indicators such as a carbon composition, and evidence for fractionation of C, S, N, and possibly Fe (Westall, 1999; Westall *et al.*, 2000; Westall and Southam, 2006). The small ovoids originally described by McKay and his colleagues do not fulfill these conditions. The simplest explanation for the "jelly-bean" bacteriomorphs is that they represent the products of aqueous corrosion of the carbonates (occurring either on Mars or on Earth). There are other morphological features within the carbonate globules in the meteorite that resemble polymer filaments. However, given the known contamination of this meteorite, it is difficult to reach a conclusion as to the likelihood of a martian or a terrestrial origin. For instance, some of the simple polymers clearly have been precipitated from humid air onto the surface of the rock in the terrestrial environment (Fig. 18.14; Westall *et al.*, 1998). The main problem in determining a martian versus a terrestrial origin for these structures lies in the fact that organic matter can mineralize very rapidly, within a day to a week (Toporski *et al.*, 2001), so that contaminating organics (whether of biogenic or abiogenic origin) can become fossilized even though they may be very young.

18.5.3.5 *Summary*

Most of the original observations that were argued to point toward fossil life in ALH84001 have now been discounted. Most of the apparent morphological fossils are thought to be too small to represent living organisms. The disequilibrium amongst adjacent minerals represented in the zonation in carbonates could be produced by multiple events of hydrothermal water coursing through the rock while it was still on Mars. Most of the organics are terrestrial in origin, and the organics of martian origin could have been produced by nonbiological processes. The magnetite grains are thought to represent the strongest evidence for life, due to their similarity to magnetite grains produced by terrestrial organisms; even these, however, might have been produced by nonbiological processes involving crystallization in hydrothermal systems.

McKay and his colleagues have continued their research into bacteriomorph structures in cracks in other martian meteorites (Nakhla, with an age of 1.3 Gyr, and Shergotty at 165 Myr), and have documented globular bacteriomorphs with characteristics (especially size) more consistent with those of Earth bacteria. There will be important consequences in terms of life on Mars if their interpretation of these other meteorites is correct, since it will indicate that martian life would have survived to relatively recent times (165 Ma). However, there are two main problems which affect analysis of all the martian meteorites – demonstration of a martian origin for the features, and demonstration that the structures are biogenic. In the end, this can only be accomplished by the return of suitable samples from the martian surface or by the development of *in situ* techniques to test for the potential biogenicity of microscopic structures.

Perhaps the most significant aspect of the debate over the biogenicity of the features in ALH84001, in the end, is the demonstration that we do not as yet understand what characteristics of an extraterrestrial sample would be indicative of the presence of life and how to uniquely distinguish biological from geochemical

phenomena. If the structures described by McKay and his colleagues are geochemical in origin or the products of terrestrial organic contamination, then what characteristics *could* we use to identify life in ancient rocks on Mars? In the following section we examine potential martian biomarkers.

18.6 Searching for life on Mars

What would it take to obtain convincing evidence that life exists on Mars? Alternatively, what body of evidence would lead scientists to conclude that there probably was *not* life on Mars? To address these questions, we will discuss the nature of biomarkers and their interpretation, the role of *in situ* analysis versus sample return, the Mars exploration program for the next two decades, and the possible outcomes of the analyses.

18.6.1 Martian biomarkers

The debate over the ALH84001 meteorite has, if nothing else, convinced researchers that it is extremely difficult to find absolute and unique evidence pertaining to the existence of life. This is especially so in an extraterrestrial environment such as Mars where life may not share a common origin, evolution, or biochemistry with terrestrial life. For example, in determining whether a given terrestrial entity is or is not biological, it is sufficient to determine that it has RNA and/or DNA molecules, uses certain amino acids, and has the proper "handedness" of those molecules that can have different shapes. But such tests are not adequate for non-terrestrial life, which could employ very different biochemistry.

Let us examine briefly some of the biomarkers used for Earth-life and see how they might apply to Mars. In particular, we will look at ratios of stable isotopes, fossilized cells, and preserved organic molecules.

Isotopic ratios
On Earth, the variable ratio of $^{13}C/^{12}C$ is a common indicator of the existence of biological activity (e.g., Schidlowski *et al.*, 1983). Because of the differing zero-point (lowest) energy states of molecules containing the two isotopes, they participate in chemical reactions with different efficiencies (Urey, 1947). In particular, it is energetically favorable for organisms using CO_2 as a source of carbon to preferentially take up $^{12}CO_2$ relative to $^{13}CO_2$. This process of *isotopic fractionation* means that organic detritus has a ratio of $^{13}C/^{12}C$ that is up to several percent lower than either the C in the

atmosphere from which it was derived (as CO_2) or carbonate minerals such as limestone that formed from atmospheric CO_2 by nonbiological processes.

With terrestrial samples, an enrichment in ^{12}C is usually strongly indicative of biological activity. However, even this can be questioned, as there are nonbiological processes that, under certain circumstances, can also produce such an enrichment. On Mars, the situation is complicated by the fact that known nonbiological processes involving the escape of atoms to space can change isotope ratios by amounts larger than those produced by biological processes (Section 18.4.2). The preferential escape of lighter isotopes to space leaves the atmosphere enriched in heavier isotopes, opposite to the fractionation produced by biology, and this enrichment can hide or confuse any biological enrichment. Thus, while isotopes might be a useful indicator of biological activity, the uniqueness of any interpretations will depend strongly on knowing the magnitude and timing of the nonbiological fractionations as well as the detailed history of individual samples.

Fossil cells

Similarly, the presence of fossil cells in terrestrial samples is used as a strong indicator of biogenicity, notwithstanding the problems of nonbiological bacteriomorphs mentioned above. Cells can be preserved for billions of years in sediments that have been consolidated into rock without being heated or squeezed to such an extent that the fossil cells are destroyed. Well-preserved samples can show morphological forms that are essentially identical in size, shape, and structure to modern, living single-celled organisms (Fig. 18.14). In addition, fossils have been found that show clear evidence for cellular features such as the cell wall, cell division, and colony and biofilm formation (Knoll, 1985; Westall *et al.*, 2006a, b). The combination of all of these features can provide convincing evidence for the biogenicity of the samples. However, even in the case of terrestrial samples there can still be uncertainty and debate.

Consider the current debate concerning some of the evidence for the presence of the oldest fossil cells in Earth rocks (see also Section 12.3). Originally described as possibly representing fossil cyanobacteria (Schopf, 1993; Schopf *et al.*, 2002), certain 3.45 Gyr-old, carbon-containing filamentous structures have recently been reinterpreted as abiogenic carbon deposits produced in a hydrothermal vent (Brasier *et al.*, 2002), where photosynthetic cyanobacteria would not be found. Much of this debate hinges on detailed geological field mapping of sample sites, an important point to take into account in discussions of any *in situ* or return-sample search for martian life. But carbon isotope studies suggest that the original material nevertheless may have been biological in origin (Brasier *et al.*, 2002; Ueno *et al.*, 2001). A reasonable solution to this conundrum could be that the carbon in the hydrothermal vent originated from the degraded remains of fauna fossilized in sedimentary strata through which the vein passed, as nearby locations contain abundant evidence for fossilized microbial mats of the same age.

For Mars, we should anticipate that samples may not be well preserved or that microbes might not have the same structural features as terrestrial organisms. In fact, this situation is similar to the debate over features in ALH84001 in which, as we have seen above, structural features still cannot be uniquely ascribed to biological activity.

Organic molecules

Organic molecular signs of life in the near-surface environment might include volatile low-molecular-weight biological organics, such as some specific alkanes. The survival of more-complex organic molecules will depend upon their chemical stability under martian conditions. Some important biomolecules (e.g., amino acids and purines) are stable in an oxidizing environment (Benner *et al.*, 2000), whereas others (e.g., nucleotides) are not. Based on an understanding of terrestrial biogenic organic matter, potential organic biomarkers of martian origin should contain compounds with structures characteristic of a biosynthetic origin (e.g., acyclic isoprenoids, terpenoids, hopanoids, and steroids), and a structurally related series of components with a non-random carbon number distribution.

Many organic molecules occur in two mirror-image forms called the L- and the D-forms, a phenomenon called *handedness* or *chirality*. Life on Earth, however, utilizes only one of these two configurations for each chiral molecule. A racemic life (one that simultaneously used the right- and left-handed forms of the same biological molecules) appears very unlikely for geometrical reasons. Thus, homochirality (exhibiting only one handedness) could be a crucial signature for life. Although homochiral molecules gradually revert to racemic mixtures over time in a dry environment such as that on Mars, the homochirality of molecules 3–4 Gyr old may still be detectable. The ratio of enantiomeric molecules (molecules with preferences for either the L- or D-form) in a martian sample could thus be used as a test for extant or extinct life: Pure enantiomeric molecules would indicate extant life, whereas partial enantiomeric mixtures would indicate extinct life (samples of different amino acids, for instance, lose their handedness at different rates). However, it will be necessary to test for handedness in more complex molecules, since there is evidence in a few meteorites for the *non*-biological production of preferential chirality (at the several percent level) in simple molecules such as amino acids (Cronin and Pizzarello, 1997).

Biological activity might also be indicated by the presence of complex molecules in the martian atmosphere that are not in chemical equilibrium with their environment. For example, the recent putative detection of methane is extremely interesting as it could be a sign of the existence of subsurface, hydrogen-fed microbes, although it could also be produced by volcanic outgassing of reduced gases from the interior (see Section 18.8).

The ability of martian soil or rock samples to confound us with unexpected responses to chemical or biological experiments has already been demonstrated by the Viking and ALH84001 experiences. It is therefore generally accepted that the most likely way in which the presence of present-day or past life on Mars might be demonstrated is by visiting the places on Mars that are most likely to have or to have had life, selecting samples of rocks that are most likely to preserve evidence of that life, and returning them to Earth for detailed analysis. It seems likely that at best one could then build a plausible case based on a number of lines of evidence and a lack of alternative, nonbiological processes that could account for them.

In the end, it would seem that there are two possible scenarios. One is that life could be so obvious that the scientific community would immediately accept the evidence and interpretations. For example, there could be obvious living or fossil cells, showing all of the characteristics of terrestrial biological samples. The second possibility is that the evidence could be ambiguous, and that further analyses still did not provide a compelling case. In such a case, we might never have a definitive answer, but only intriguing evidence.

18.6.2 The Mars exploration program

A principal science goal of the Mars exploration program is to address in a substantive manner the issues of the potential for martian life. As discussed earlier, this involves not only finding out whether there ever has been life on Mars, but also learning about the planetary context so that the results can be properly understood and integrated with our understanding of the rest of the Solar System and beyond. While Mars exploration is in a seemingly constant state of flux in terms of which specific missions will fly and when, and when and if we can return scientifically interesting samples to the Earth for analysis, we will briefly describe the current state of the program (also see Table 18.1).

As of this writing, NASA's Mars Global Surveyor (MGS) and Mars Odyssey spacecraft both continue to operate in orbit around Mars. By gathering detailed information on the present state and history of the martian surface and interior, these provide the global geological context for framing and addressing questions about martian biological potential. In addition, both spacecraft are obtaining data that will help to select the most interesting landing sites for future missions, with an emphasis on sites for which convincing evidence can be obtained regarding aqueous or hydrothermal activity.

On board MGS, the Mars Orbiter Camera is obtaining high-resolution images that show features indicative of processes that have involved the presence of liquid water, and is also mapping the geology of possible future landing sites. The Thermal Emission Spectrometer (TES) is identifying and mapping minerals that tell us the history of the crust and of possible water-related minerals; nighttime thermal-emission maps also provide information on physical properties to better judge the probable safety and utility of future landing sites. The Mars Orbiter Laser Altimeter has mapped the entire planet's topography. Along with the gravity field as determined by tracking the spacecraft's orbit, the topography provides clues to the planet's geophysical history as it pertains to water (Phillips *et al.*, 2001).

Mars Odyssey contains two instruments that are relevant here. First, the Thermal Emission Imaging System is obtaining very-high-spatial-resolution spectral maps that provide detailed mineralogical information that is complementary to that obtained by MGS TES (Christensen *et al.*, 2003). Second, the Gamma-Ray Spectrometer and Neutron Spectrometer are mapping out the abundance of near-surface hydrogen, present primarily in water molecules, and thus providing important clues on the distribution and accessibility of near-surface water ice (Boynton *et al.*, 2002).

NASA's Mars Exploration Rovers (MER) arrived at Mars in early 2004 (see Fig. 23.3). These are very capable rovers designed to conduct surface operations at sites of particular relevance to water: Sinus

Meridiani, which has considerable quantities of hematite (see Fig. 18.7), and Gusev Crater, which shows evidence of an ancient lake within the crater interior. The goal of these missions is to determine the geological and geochemical role that water has played at each site, rather than to search for life. Of central interest is to understand the geological processes that shaped the surface and to identify minerals within the rocks and soils at the surface to determine whether they were formed by aqueous or hydrothermal processes. Imaging of the surface is being carried out in the visible and near-infrared, and also at mid-infrared wavelengths using an instrument similar to the MGS TES. Individual samples are being examined using a Mossbauer spectrometer to identify the iron composition and an X-ray spectrometer to identify which elements are present. Microscopic imaging also allows determination of the texture of rocks and soil at sub-millimeter scales. Section 18.8 briefly describes key results from the MER Mission.

The European Space Agency's Mars Express mission entered orbit around Mars in December 2003. Its main objectives involve the search for water and life. Instruments on board the orbiter are observing from the main spacecraft in polar orbit, and will obtain global coverage over the mission's expected lifetime of nearly one martian year. An imager is taking detailed images of the surface and atmosphere of Mars, as well as of the entire planet in full color, in stereo, and at high resolution. A ground penetrating radar will search for water at depths up to 5 km below ground. Three complementary spectrometers are analyzing the atmosphere and the surface – the Infrared Mapping Spectrometer for the mineralogical composition of the surface, the Planetary Fourier Spectrometer for the composition of the atmosphere in the ultraviolet and infrared (see Section 18.8 for one result), and the Atmospheric Spectrometer to address the question of why the martian atmosphere is so oxidizing. Another instrument is detecting energetic neutral atoms to address the question of how the solar wind erodes the martian atmosphere.

Mars Express also jettisoned a lander (named Beagle-2 after HMS *Beagle*, on which Charles Darwin sailed), but was never heard from after its supposed landing. It contained an integrated suite of instruments to search for evidence of life on Mars in subsurface and rock interior samples.

The Mars Reconnaissance Orbiter was launched in 2005 and is now successfully in orbit. Current observations include atmospheric sounding to understand seasonal cycles, high-resolution imaging to identify features smaller than a meter across, high-resolution reflectance spectroscopy to determine mineralogy and surficial geological history, and a radar imager to search for near-surface liquid water and map subsurface structure.

The Phoenix mission is planned for launch in 2007 and landing in 2008 at a high latitude site, where near-surface ice is accessible, in order to explore the chemical and physical nature of the ice. Its goal is to determine the chemical effects of the presence of ice, whether there has ever been liquid water at the landing site, and whether there is evidence of organic chemical reactions or possibly life at the site.

Current plans also include a much more sophisticated mobile laboratory to be launched in 2009. This Mars Science Laboratory will emphasize understanding the context in which possible life might exist – the history of water and of volatiles, the mineralogical evolution of the surface, and the possible identification of organic molecules. While sample return from a promising site is anticipated to take place sometime following 2009, the difficulty of developing the necessary technology and of keeping the costs manageable makes its timing uncertain.

18.6.3 What would be needed to conclude that life is *absent* on Mars?

By the year 2016, we hope to have developed a much better understanding of the global geological context for Mars and to have martian samples in our laboratories here on Earth. We do not know what these samples will contain, of course, but it is possible that we will have an answer about martian life. However, most scientists think that it is extremely unlikely that the first samples that are returned from Mars will contain conclusive evidence for life. This skepticism is based on the relatively small fraction of terrestrial rocks that contain evidence for life, the anticipated difficulty of martian rocks being adequately preserved, and the demonstrated difficulty of biomarker analyses of non-terrestrial samples.

If we do not find life in these first samples, we should not then conclude that martian life is absent. But what kinds of analysis might it take to reach such a conclusion? This is a difficult question to answer, as it is not possible to prove that something is absent from a planet without searching the planet in its entirety at a microscopic level. However, one can make some plausibility arguments.

Suppose that we could identify a number of types of locations that we feel have the best chances of having had life. Such sites, as noted above, would probably include exposed locations that had had aqueous activity, hydro-thermal sites associated with volcanic activity, ancient crater lakes, potential aquifers that had been buried at several kilometers depth and that had been subsequently exposed (as on the walls of Valles Marineris), and polar and near-polar ice deposits that might have had transient epochs of melting or thin-film liquid water. If a number of sites of each type were explored and showed no evidence that suggested that there was a possibility of life, then it would be hard to argue that there were types of locations or places that had still gone unexplored and might be an oasis for martian life. While one cannot choose a specific number of sites to explore that would lead to such a negative conclusion, intuition suggests that this number might not be much greater than two dozen (that is, perhaps a few of each type of site).

Although we have not yet done even a single sample return, and a program that can return samples from two dozen sites will be difficult and not in the near future, this is nevertheless a tractable problem that is technologically feasible.

18.7 Concluding comments

The exploration of Mars, and the search for life as a part of that exploration, is a highly uncertain venture. Young scientists who participated in the Mars exploration program in the mid-1960s are now at the twilight of their careers. Although we now know much more about planets and planetary systems, about Mars in particular, and about the Earth and its origin and evolution of life, we are in some sense no closer to answering the question of whether there is life on Mars, or elsewhere in our Solar System, or beyond. But this does not represent a failure in our program or in our intellectual directions. Rather, it is a measure of the subtle complexity of life's signatures and our increased understanding of this. The question of what is the distribution of life in our Solar System and throughout the Galaxy is of tremendous interest and importance both to the scientist and to the general public. The large investment made over the past forty years reflects the centrality of the question. As John F. Kennedy said at the beginning of the Space Age, "We do these things not because they are easy, but because they are hard." Learning about the Universe around us and about the distribution of life profoundly affects our view of who we are here on Earth, as a species and as a civilization.

The questions about life elsewhere are fundamental, and Mars may provide the best place to get a first answer because of its proximity and its similarity to

Earth. Whatever the answer turns out to be, the next generation of scientists may be the one to find it.

18.8 Addendum: recent discoveries

Important developments since the original writing of this chapter include discoveries by the Mars Exploration Rovers on Mars, and the apparent detection of methane in the martian atmosphere. Both of these have significant astrobiological implications, and are summarized very briefly here.

The Rovers (Fig. 23.3) landed on Mars in early 2004 and, as this book goes to press, are both amazingly *still* operating over three years later. They were designed to determine whether each of two different sites had had liquid water at or near the surface, and to understand the geological and geochemical implications; Section 23.3 briefly describes the Rovers' instrumentation. The first rover, *Spirit*, landed in Gusev Crater, a site chosen because morphological evidence suggested that there had been a standing lake within the crater at one time. While evidence for liquid water was indeed identified, there was no evidence that liquid water had been present in significant quantities or as a surface lake. For example, no major deposits of materials suggestive of chemical weathering in the presence of water were seen. However, veins were identified running through some of the rocks, suggesting that liquid water had been present within the subsurface and that some minor chemical alteration had occurred. Results for *Spirit* were presented by Squyres *et al.* (2004a) and in accompanying papers.

The second rover, *Opportunity*, landed in the Meridiani region, near the equator. That site had been chosen based on remote-sensing observations that indicated the presence of coarse-grained specular hematite (Fig. 18.7). This mineral could have formed by a number of mechanisms, but the most plausible appeared to be by low-temperature alteration of basalt, suggesting the presence of liquid water. On the surface, *Opportunity* was able to examine deposits of bedrock in their original form (Fig. 18.15). There, several features were identified that compellingly pointed to liquid water. Cross-bedding of layers indicated deposition in a surface water environment such as a lake or stream (Fig. 18.16). Additionally, the deposits had a significant abundance of magnesium-rich sulfates (up to ~40%) of the type that are deposited at the bottom of a standing body of water; furthermore, the mineral jarosite, which forms as a weathering product in water, also was identified. The deposits also contained

FIGURE 18.15 Panoramic image of Opportunity Ledge, a small outcrop (height only ∼10 cm) of light-colored bedrock exposed in the interior of Eagle Crater in which Opportunity landed.

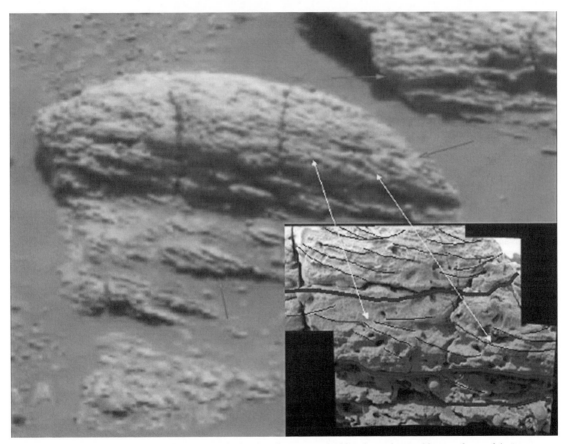

FIGURE 18.16 Image of an outcrop in Opportunity Ledge that shows cross-bedding. Inset shows a blowup of part of the image with superimposed marks tracing non-parallel individual sedimentary layers. Cross-bedding of this nature results from the back-and-forth transport of sediment by the movement of water in a surface environment such as a lake or stream.

several-mm-sized round nodules that are thought to be concretions, which form by dissolution and reprecipitation of minerals within the subsurface as liquid water flows through a rock (Fig. 18.17); these concretions are the site of the hematite that originally drew investigators to this location. This evidence points in a convincing manner to the occurrence of liquid water at some time in the past, both as a surface deposit and flowing through the rock in the subsurface. Results are reported by Squyres *et al.* (2004b) and Squyres and Knoll (2005) and in accompanying papers.

While liquid water was previously known to occur on Mars – as inferred, for example, from the geochemical evidence within the martian meteorites and from the morphological evidence seen in many places – these results allow us to point for the first time to a particular location that clearly had liquid water. In addition, they allow us to infer the chemistry of the water at the time of deposition and the potential implications for astrobiology (e.g., Sumner, 2004; Fairen *et al.*, 2004).

Another significant new finding is that three groups have independently announced the detection of methane

FIGURE 18.17 Microscopic image of exposed materials within the Guadalupe portion of Opportunity Ledge (width of image is a few cm). Tablet-shaped holes are locations where minerals such as gypsum are thought to have been dissolved by water flowing through the rock. The round concretions are several millimeters wide and were deposited by the re-precipitation of minerals dissolved by water flowing through the rock.

in the martian atmosphere, two from ground-based telescopes (Krasnopolsky *et al.*, 2004; Mumma *et al.*, 2004) and one from the Mars Express spacecraft (Foramisimo *et al.*, 2004). Average abundances are roughly 10 ppb, although some investigators have questioned the detection. Methane can be produced by biological processes, for example as a byproduct of metabolism by certain types of organisms (Chapter 11), and this is one reason for the intense interest in its detection. However, it also can be produced by nonbiological processes involving outgassing of gases associated with volcanism or the simultaneous oxidation of minerals and reduction of carbon; as such, it is not, by itself, a unique biosignature. Further analysis will be required in order to confirm the detection and to understand its implications.

REFERENCES

Armstrong, J. C., Leovy, C. B., and Quinn, T. (2004). A 1 Gyr climate model for Mars: new orbital statistics and the importance of seasonally resolved polar processes. *Icarus*, **171**, 255–271.

Bada, J. L., Galvin, D. P., McDonald, G. D., and Becker, L. (1998). A search for endogenous amino acids in martian meteorite ALH84001. *Science*, **279**, 362–365.

Baker, V. R., Carr, M. H., Gulick, V. C., Williams, C. R., and Marley, M. S. (1992). Channels and valley networks. In *Mars*, H. H. Kieffer, B. M. Jakosky, C. W. Snyder, and M. S. Matthews, eds. Tucson: University of Arizona Press, 493–522.

Bandfield, J. L., Glotch, T. D., and Christensen, P. R. (2003). Spectroscopic identification of carbonate minerals in the martian dust. *Science*, **301**, 1084–1087.

Barber, D. J. and Scott, E. R. D. (2002). Origin of supposedly biogenic magnetite in the Martian meteorite Allan Hills 84001. *Proc. Natl. Acad. Sci.*, **99**, 6556–6561.

Benner, S. A., Devine, K. G., Matveeva, L. N., and Powell, D. H. (2000). The missing organic molecules on Mars. *Proc. Natl. Acad. Sci.*, **97**, 2425–2430.

Biemann, K., Oro, J., Toalmin, P., *et al.* (1977). The search for organic substances and inorganic volatile compounds in the surface of Mars. *J. Geophys. Res.*, **82**, 4641–4658.

Borg, L. E., Connelly, J. N., Nyquist, L. E., Shih, C.-Y., Wisemann, H., and Reese, Y. (1999). The age of the carbonates in Martian meteorite ALH84001. *Science*, **286**, 90–94.

Boynton, W. V., Feldman, W. C., Squyres, S. W., *et al.* (2002). Distribution of hydrogen in the near surface of Mars: evidence for subsurface ice deposits. *Science*, **297**, 81–84.

Brain, D. A. and Jakosky, B. M. (1998). Atmospheric loss since the onset of the martian geologic record: combined role of impact erosion and sputtering. *J. Geophys. Res.*, **103**, 22689–22694.

Brasier, M. D., Green, O. R., Jephcat, A. P., *et al.* (2002). Questioning the evidence for Earth's oldest fossils. *Nature*, **416**, 76–81.

Cabrol, N. A. and Grin, E. A. (1999). Distribution, classification, and ages of martian impact crater lakes. *Icarus*, **142**, 160–172.

Cabrol, N. A., Grin, E. A., Landheim, R., Kuzmin, R. O., and Greeley, R. (1998). Duration of the Ma'Adim Vallis/Gusev Crater hydrogeologic system, Mars. *Icarus*, **133**, 98–108.

Carr, M. H. (1996). *Water on Mars.* New York: Oxford University Press.

Carr M. H. and Head, J. W. III. (2003). Oceans on Mars: an assessment of the observational evidence and possible fate. *J. Geophys. Res.*, **108**, 5042, doi:10.1029/2002JE001963.

Carr, M. H. and Malin, M. C. (2000). Meter scale characteristics of martian channels and valleys. *Icarus*, **146**, 366–386.

Christensen, P. R., Banfield, J. L., Clark, R. N., *et al.* (2000). Detection of crystalline hematite mineralization on Mars by the Thermal Emission Spectrometer: evidence for near-surface water. *J. Geophys. Res.*, **105**, 9623–9642.

Christensen, P. R., Banfield, J. L., Bell, J. F., III, *et al.* (2003). Morphology and composition of the surface of Mars: Mars Odyssey THEMIS results. *Science*, **300**, 2056–2061.

Chyba, C. F., Owen, T. C., and Ip, W. H. (1994). Impact delivery of volatiles and organic molecules to the Earth. In *Hazards Due to Comets and Asteroids*, T. Gehrels, ed. Tucson: University of Arizona Press, 9–58.

Clemett, S. J., Dulay, M. T., Gillette, J. S., Chillier, S. D. F., Mahajan, T. B., and Zare, R. N. (1998). Evidence for the extraterrestrial origin of polycyclic aromatic hydrocarbons in the Martian meteorite ALH84001. *Faraday Discuss.*, **109**, 417–436.

Cronin, J. R. and Pizzarello, S. (1997). Enantiomeric excesses in meteoritic amino acids. *Science*, **275**, 951–955.

Fairen, A. G., Fernandez-Remolar, D., Dohm, J. M., Baker, V. R., and Amils, R. (2004). Inhibition of carbonate synthesis in acidic oceans on early Mars. *Nature*, **431**, 423–426.

Folk, R. L. (1993). SEM imaging of bacteria and nanobacteria in carbonate sediments and rocks. *J. Sediment. Petr.*, **63**, 990–999.

Forget, F. and Pierrehumbert, R. T. (1997). Warming early Mars with carbon dioxide clouds that scatter infrared radiation. *Science*, **278**, 1273–1276.

Formisano, V., Atreya, S., Encrenez, T., Ignatiev, N., and Giuranna, M. (2004). Detection of methane in the atmosphere of Mars. *Science*, **306**, 1758–1761.

Friedmann, E. I., Wierzchos, J., and Winkelhofer, M. (2001). Chains of magnetite crystals in the meteorite ALH84001: evidence of biological origin. *Proc. Natl. Acad. Sci.* **98**, 2176–2181.

Gladman, B. J. and Burns, J. A. (1996). Mars meteorite transfer: simulation. *Science*, **274**, 161–165.

Gladman, B. J., Burns, J. A., Duncan, M., Lee, P., and Levison, H. F. (1996). The exchange of impact ejecta between terrestrial planets. *Science*, **271**, 1387–1392.

Golden, D. C., Ming, D. W., Schwandt, C. S., *et al.* (2001). A simple inorganic process for formation of carbonates, magnetite, and sulfides in martian meteorite ALH84001. *Amer. Mineral.*, **86**, 370–375.

Golden, D., Ming, D. C., Morris, R., Lofgren, G., and McKay, G. (2002). Inorganic formation of "truncated hexaoctahedral" magnetite: implications for inorganic processes in Martian meteorite ALH84001. *Lunar Planet. Sci. Conf.*, **33**, abstract # 1835.

Grady, M. M., Wright, I. P., and Pillinger, C. T. (1998). A nitrogen and argon stable isotope study of Allan Hills 84001: implications for the evolution of the martian atmosphere. *Meteor. Planet. Sci.*, **33**, 795–802.

Horneck, G., Mileikowsky, C., Melosh, H. J., *et al.* (2002). Viable transfer of microorganisms in the solar system and beyond. In *Astrobiology: the Quest for the Conditions of Life*, eds. G. Horneck and C. Baumstark-Khan. Berlin: Springer-Verlag, 57–76.

Horowitz, N. H., Hobby, G. L., and Hubbard, J. S. (1977). Viking on Mars; the carbon assimilation experiments. *J. Geophys. Res.*, **82**, 4659–4662.

Huber, H., Hohn, M. J., Rachel, R., Fuchs, T., Wimmer, V. C., and Stetter, K. O. (2002). A new phylum of Archaea represented by a nano-sized hyperthermophilic symbiont. *Nature*, **417**, 63–67.

Hunten, D. M. (1979). Possible oxidant sources in the atmosphere and surface of Mars. *J. Molec. Evol.*, **14**, 71–78.

Jakosky, B. M. (1991). Mars volatile evolution: evidence from stable isotopes. *Icarus*, **94**, 14–31.

Jakosky, B. M. and Phillips, R. J. (2001). Mars' volatile and climate history. *Nature*, **412**, 237–244.

Jakosky, B. M. and Shock, E. L. (1998). The biological potential of Mars, the early Earth, and Europa. *J. Geophys. Res.*, **103**, 19359–19364.

Jakosky, B. M., Nealson, K. H., Bakermans, C., Ley, R. E., and Mellon M. T. (2003). Sub-freezing activity of microorganisms and the potential habitability of Mars' polar regions. *Astrobiology*, **3**, 343–350.

Kallenbach, R., Geiss, J., and Hartmann, W. K., eds. (2001). Special issue on chronology and evolution of Mars. *Space Sci. Rev.*, **96** (Nos. 1–4).

Kasting, J. F. (1991). CO_2 condensation and the climate of early Mars. *Icarus*, **94**, 1–13.

Kasting, J. F. (1997). Warming early Earth and Mars. *Science*, **276**, 1213–1215.

Kasting, J. F. and Catling, D. (2003). Evolution of a habitable planet. *Ann. Rev. Astron. Astrophys.*, **41**, 429–463.

Kieffer, H. H., Jakosky, B. M., and Snyder, S. W., eds. (1992). *Mars.* Tucson: University of Arizona Press.

Klein, H. P. (1999). Did Viking discover life on Mars? *Origins Life Evol. Biosphere*, **29**, 625–631.

Klein, H. P., Horowitz, N. H., and Biemann, K. (1992). The search for extant life on Mars. In *Mars*, H. H. Kieffer, B. M. Jakosky, C. W. Snyder, and M. S. Matthews, eds. Tucson: University of Arizona Press, 1221–1233.

Knoll, A. H. (1985). Exceptional preservation of photosynthetic organisms in silicified carbonates and silicified peats. *Phil. Trans. Roy. Soc. Lond.*, **B311**, 111–122.

Knoll, A. H. and Osborne, M. J., eds. (1999). *Size Limits of Very Small Organisms: Proceedings of a Workshop.* Washington, DC: National Academic Press.

Krasnopolsky, V. A., Maillard, J. P., and Owen, T. C. (2004). Detection of methane in the martian atmosphere: evidence for life? *Icarus*, **172**, 537–547.

Laskar, J. and Robutel, P. (1993). The chaotic obliquity of the planets. *Nature*, **361**, 608–612.

Levin, G. V., and Straat, P. A. (1977). Recent results from the Viking labeled release experiment on Mars. *J. Geophys. Res.*, **82**, 4663–4668.

Luhmann, J. G., Johnson, R. E., and Zhang, M. H. G. (1992). Evolutionary impact of sputtering of the martian atmosphere by O+ pickup ions. *Geophys. Res. Lett.*, **19**, 2151–2154.

Malin, M. C. and Edgett, K. S. (2000). Evidence for recent ground water seepage and surface runoff on Mars. *Science*, **288**, 2330–2335.

Malin, M. C. and Edgett, K. S. (2001). Sedimentary rocks of early Mars. *Science*, **290**, 1927–1937.

Martin, L. J., James, P. B., Dollfus, A., Iwasaki, K., and Beish, J. D. (1992). Telescopic observations: visual, photographic, polarimetric. In *Mars*, H. H. Kieffer, B. M. Jakosky, C. W. Snyder, and M. S. Matthews, eds. Tucson: University of Arizona Press, 34–70.

McElroy, M. B., Yung, Y. L., and Nier, A. O. (1976). Isotopic composition of nitrogen: implications for the past history of Mars' atmosphere. *Science*, **194**, 70–72.

McKay, C. P., Mancinelli, R. L., Stoker, C. R., and Wharton, R. A., Jr. (1992). The possibility of life on Mars during a water-rich past. In *Mars*, eds. H. H. Kieffer, B. M. Jakosky, C. W. Snyder, and M. S. Matthews. Tucson: University of Arizona Press, pp. 1234–1245.

McKay, D. S., Gibson, E. K., Jr., Thomas-Keprta, K. L., *et al.* (1996). Search for past life on Mars: possible relic biogenic activity in martian meteorite ALH84001. *Science*, **273**, 924–930.

McKay, D. S., Gibson, E. K., Thomas-Keprta, K., Romanek, C. S., and Allen, C. C. (1997). Possible biofilms in ALH84001. *Lunar Planet. Sci. Conf.*, **28**, 919–920.

McKay, D. S., Clemett, S. J., Gibson, E. K., Thomas-Keprta, K., and Wentworth, S. J. (2002). Are carbonate globules, magnetites, and PAHs in ALH84001 really terrestrial contaminants? *Lunar Planet. Sci. Conf.*, **33**, abstract # 1943.

McSween, H. Y., Jr. (1994). What we have learned about Mars from SNC meteorites. *Meteoritics*, **29**, 757–779.

McSween, H. Y. and Harvey, R. P. (1998). Brine evaporation: an alternative model for the formation of carbonates in Allan Hills 84001. *Meteoritics Planet. Sci.*, **33**, A103.

Mellon, M. T. and Phillips, R. J. (2001). Recent gullies on Mars and the source of liquid water. *J. Geophys. Res.*, **106**, 23165–23179.

Melosh, H. J. (1988). The rocky road to Panspermia. *Nature*. **332**, 687–688.

Melosh, H. J. (1989). *Impact Cratering – a Geologic Process.* New York: Oxford University Press.

Melosh, H. J. and Vickery, A. M. (1989). Impact erosion of the primordial atmosphere of Mars. *Nature*, **338**, 487–489.

Mileikowsky, C., Cucinotta, F. A., Wilson, J. W., *et al.* (2000). Natural transfer of viable microbes in space. 1. From Mars to Earth and Earth to Mars. *Icarus*, **145**, 391–427.

Mitchell, D. L., Lin, R. P., Reme, H., Cloutier, P., Connerney, J., Acuna, M., and Ness, N. (2000). Crustal magnetocylinders at Mars (abstract), Spring American Geophysical Union meeting, 2000.

Mittlefehldt, D. W. (1994). ALH84001, a cumulate orthopyroxenite member of the martian meteorite clan. *Meteoritics*, **29**, 214–221.

Mumma, M. J., Novak, R. E., DiSanti, M. A., Bonev, B. P., and Russo, N. D. (2004). Detection and mapping of methane and water on Mars (abstract), Division Planetary Science annual meeting, Louisville, 2004.

Musselwhite, D. S., Swindle, T. D., Lunine, J. L. (2001). Liquid CO_2 breakout and the formation of recent gullies on Mars. *Geophys. Res. Lett.*, **28**, 1283–1285.

Nealson, K. H. (1997). The limits of life on Earth and searching for life on Mars. *J. Geophys. Res.*, **102**, 23675–23686.

Owen, T., Maillard, J. P., deBergh, C., and Lutz, B. L. (1988). Deuterium on Mars: the abundance of HDO and the value of D/H. *Science*, **240**, 1767–1770.

Oyama, V. I. and Berdahl, B. (1977). The Viking gas exchange experiment results from Chryse and Utopia surface samples. *J. Geophys. Res.*, **82**, 4669–4676.

Phillips, R. J., Zuber, M. T., Solomon, S. C., *et al.* (2001). Ancient geodynamics and global-scale hydrology on Mars. *Science*, **291**, 2587–2591.

Pierazzo, E. and Chyba, C. (1999). Amino acid survival in large cometary impacts. *Meteoritics Planet. Sci.*, **32**, 909–918.

Pollack, J. B., Kasting, J. F., Richardson, S. M., and Poliakoff, K. (1987). The case for a warm, wet climate on early Mars. *Icarus*, **71**, 203–224.

Romanek, C. S., Grady, M. M., Wright, I. P., Mittlefehldt, D. W., Socki, R. A., Pillinger, C. T., and Gibson, E. K., Jr. (1994). Record of fluid-rock interactions on Mars from the meteorite ALH84001. *Nature*, **372**, 655–657.

Schidlowski, M., Hayes, J. M., and Kaplan, I. R. (1983). Isotopic inferences of ancient biochemistries: carbon, sulfur, hydrogen, and nitrogen. In *Earth's Earliest Biosphere*, J. W. Schopf, ed. Princeton, Princeton University Press, 149–186.

Schopf, J. W. (1993). Apex chert and the antiquity of life. *Science*, **260**, 640–646.

Schopf, J. W., Kudryavtsev, A. B., Agresti, D. G., Wdowiak, T. J., and Czaja, A. D. (2002). Laser-Raman imagery of Earth's earliest fossils. *Nature*, **416**, 73–76.

Segura, T. L., Toon, O. B., Colaprete, A., and Zahnle, K. (2002). Environmental effects of large impacts on Mars. *Science*, **298**, 1977–1980.

Shock, E. L. (1997). High-temperature life without photosynthesis as a model for Mars. *J. Geophys. Res.*, **102**, 23687–23694.

Snyder, C. W. and Moroz, V. I. (1992). Spacecraft exploration of Mars. In *Mars*, H. H. Kieffer, B. M. Jakosky, C. W. Snyder, and M. S. Matthews, eds. Tucson: University of Arizona Press, 71–119.

Squyres, S. W. and Knoll, A. H. (2005). Sedimentary rocks at Meridiani Planum: origin, diagenesis, and implications for life on Mars. *Earth & Planetary Sci. Lett.*, **240**, 1–10.

Squyres, S. W., Arvidson, R. E., Bell, J. F., III, *et al.* (2004a). The Spirit Rover's Athena science investigation at Gusev Crater, Mars. *Science*, **305**, 794–799.

Squyres, S. W., Grotzinger, J. P., Arvidson, R. E., *et al.* (2004b). *In situ* evidence for an ancient aqueous environment at Meridiani Planum, Mars. *Science*, **306**, 1709–1714.

Steele, A., Goddard, D. T., Stapleton, D. V., Bassinger, V., Sharples, G., Wynn-Williams, D. D., and McKay, D. S. (2000). Investigations into an unknown organism on the martian meteorite Allan Hills 84001. *Meteoritics Planet. Sci.*, **35**, 237–241.

Stewart, S. T. and Nimmo, F. (2002). Surface runoff features on Mars: testing the carbon dioxide formation hypothesis. *J. Geophys. Res.*, **107**, 5069, doi:10.1029/2000JE001465.

Sumner, D. Y. (2004). Poor preservation potential of organics in Meridiani Planum hematite-bearing sedimentary rocks. *J. Geophys. Res. – Planets.* **109** (E12), Article E12007.10.1029/2004JE002321.

Thomas-Keprta, K. L., Bazilinsky, D. A., Kirschvink, J., *et al.* (2000). Elongated prismatic magnetite crystals in ALH84001 carbonate globules: potential Martian magnetofossils. *Geochim. Cosmochim. Acta.*, **64**, 4049–4081.

Thomas-Keprta, K. L., Clemett, S. J., Bazilinski, D. A., *et al.* (2001). *Proc. Natl. Acad. Sci.*, **98**, 2164–2169.

Thomas-Keprta, K. L., Clemett, S. J., Romanek, C. S., *et al.* (2002). Multiple origins of magnetite crystals in ALH84001 carbonates. *Lunar Planet. Sci. Conf.*, **33**, abstract # 1911.

Toporski, J. K. W., Westall, F., Thomas-Keprta, K. A., Steele, A., and McKay, D. S. (2001). The simulated silicification of bacteria – new clues to the modes and timing of bacterial preservation and implications for the search for extraterrestrial microfossils. *Astrobiology*, **1**, 1–26.

Travis, B. J., Rosenberg, N. D., and Cuzzi, J. N. (2003). On the role of widespread subsurface convection in bringing liquid water close to Mars' surface. *J. Geophys. Res.*, **108**, 8040, doi:10.1029/2002JE001877.

Treiman, A. H. (1998). The history of Allan Hills 84001 revised: multiple shock events. *Meteoritics Planet. Sci.*, **33**, 753–764.

Ueno, Y., Maruyama, S., Isozaki, Y., and Yurimoto, H. (2001). Early Archean (ca. 3.5 Ga) microfossils and ^{13}C depleted carbonaceous matter in the North Pole area, western Australia: field occurrence and geochemistry. In *Geochemistry and the Origin of Life*, S. Nakashima *et al.*, eds. Tokyo: Universal Academic Press, 203–236.

Urey, H. C. (1947). The thermodynamic properties of isotopic substances. *J. Chem. Soc.*, 562–581.

Uwins, P. J. R., Webb, R. I., and Taylor, A. P. (1998). Novel nano-organisms from Australian sandstones. *American Mineralogist*, **83**, 1541–1550.

Varnes, E. S., Jakosky, B. M., and McCollom, T. M. (2003). Biological potential of martian hydrothermal systems. *Astrobiology*, **3**, 407–414.

Watson, L. L., Hutcheon, I. D., Epstein, S., and Stolper, E. M. (1994). Water on Mars: clues from deuterium/hydrogen and water contents of hydrous phases in SNC meteorites. *Science*, **265**, 86–90.

Westall, F. (1999). The nature of fossil bacteria: a guide to the search for extraterrestrial life. *J. Geophys. Res.*, **104**, 16,437–16,451.

Westall, F. and Southam, G. (2006). The early record of life. In *Archaean Geodynamics and Environments*. K. Benn *et al.*, eds. AGU Monograph, **164**, 283–304.

Westall, F., Gobbi, P., Mazzotti, G., *et. al.* (1998). Combined SEM (secondary electrons, backscatter, cathodoluminescence) and atomic force microscope investigation of the carbonate globules in Martian meteorite ALH84001: preliminary results. *SPIE, Instruments, Methods and Missions for Astrobiology*, **3114**, 225–233.

Westall, F., Steele, A., Toporski, J., *et al.* (2000). Polymeric substances and biofilms as biomarkers in terrestrial materials: implications for extraterrestrial materials. *J. Geophys. Res.*, **105**, 24,511–24,527.

Westall, F., de Wit, M., Dann, J., van der Gaast, S., and Gerneke, D. (2001). Early Archean fossil bacteria and biofilms in hydrothermally-influenced sediments from the Barberton greenstone belt, South Africa. *Precamb. Res.*, **106**, 93–116.

Westall, F., de Vries, S. T., Nijman, W., *et al.* (2006a). The 3.466 Ga Kitty's Gap Chert, an Early Archaean microbial ecosystem. In *Processes on the Early Earth*. W. U. Reimold and R. Gibson, eds. *Geol. Soc. Amer. Special Pub.*, **405**, 105–131.

Westall, F. C., de Ronde, E. J., Southam, G., *et al.* (2006b). Implications of a 3.472-3.333 Ga-old subaerial microbial mat from the Barberton greenstone belt, South Africa for the UV environmental conditions on the early Earth. *Phil. Trans. Roy. Soc.*, **B361**, 1857–1875.

Woese, C. R. (1987). Bacterial evolution. *Microbiol. Rev.*, **51**, 221–271.

Zolotov, M. Y. and Shock, E. L. (2000). An abiotic origin for hydrocarbons in the Allan Hills 84001 through cooling of magmatic and impact generated gases. *Meteoritics Planet. Sci.*, **35**, 629–638.

19 Europa

Christopher F. Chyba, *Princeton University*
Cynthia B. Phillips, *SETI Institute*

19.1 Introduction

Europa, one of the four large satellites of Jupiter, is nearly the size of Earth's Moon. Tidal flexing driven by Jupiter's gravity and sustained by an orbital resonance with two other jovian satellites, Io and Ganymede, results in significant heat dissipation within Europa. Calculations indicate that this tidal heating is sufficient to maintain liquid water beneath Europa's ice crust. Moreover, observational evidence suggests that it is indeed probable, but not yet completely certain, that Europa harbors a subsurface ocean of liquid water whose volume is about twice that of Earth's oceans. The likely presence of abundant liquid water places Europa among the highest priority targets for astrobiology.

To support life, Europa would also require an inventory of biogenic elements and a source of sufficient free energy. The ability to support life does not of course guarantee that the origin of life took place on Europa, or that life is present today; answering these questions will require further exploration. This chapter considers Europa in an astrobiological context, distinguishing among what is known, what is supported by evidence but still uncertain, and what remains more speculative. We conclude with a discussion of future missions that will be needed to address current geological and astrobiological questions.

19.2 Jupiter and its satellites

The planet Jupiter, orbiting the Sun at 5.2 AU, is more massive than the other planets in the Solar System combined; it is 3.3 times more massive than Saturn and 318 times more than the Earth. Jupiter is a gas giant (Chapter 3), with most of its mass in the form of hydrogen (by number about 86%) and helium (over 13%). Some models suggest that it contains a solid core of rock and ice of about 10 Earth masses (Guillot,

1999). Its outer portion (20% by radius) consists of gaseous molecular hydrogen and helium, which we would traditionally call an atmosphere. Beneath this outer gaseous layer, hydrogen is transformed into a liquid metallic state. This phase can only form under extremely high pressures and temperatures, and in our Solar System is thought to exist only in the interiors of Jupiter and Saturn (Stevenson, 1982). Liquid metallic hydrogen is an excellent conductor, and the strong magnetic field of Jupiter is thought to be generated by electrical currents in this layer. The jovian magnetosphere extends out to about 10 Jupiter radii (R_J), between the orbits of Europa (9.4 R_J) and Ganymede (15 R_J). Large numbers of charged particles accelerated in Jupiter's magnetosphere cause problems for probes such as the Galileo spacecraft and would provide a severe challenge to human exploration. This radiation environment also results in interesting chemistry as these particles impact and interact with Europa's surface (Section 19.4). The stunning cloud layers in Jupiter's upper atmosphere are thought to be combinations of ammonia ice, water ice, and small amounts of sulfur, phosphorous, and other compounds that produce bands ranging in color from reds to yellows to whites. The energy that drives Jupiter's strong storm systems and other weather has a substantial internal component; in fact, Jupiter emits ~1.7 times as much energy as it absorbs from the Sun. This excess flux of energy is due to the ongoing loss of heat derived from slow gravitational contraction, a process underway ever since Jupiter formed 4.6 Ga.

Jupiter's four large satellites were discovered by Galileo in 1610 using a small telescope (Galileo, 1610). This discovery of objects orbiting a celestial body other than Earth helped replace the geocentric view of the universe with the heliocentric model, with important religious, philosophical, and scientific consequences. These Galilean satellites (Fig. 19.1) are (in

Planets and Life: The Emerging Science of Astrobiology, eds. Woodruff T. Sullivan, III and John A. Baross. Published by Cambridge University Press. © Cambridge University Press 2007.

FIGURE 19.1 Four Galilean satellites. The four large satellites of Jupiter to scale, from left to right: Io, Europa, Ganymede, and Callisto. Io is a volcanic world covered with bright volcanic deposits ranging in color from red to yellow to black to white. Europa's icy surface is covered with linear features and disrupted regions, which appear dark in this global view. The white splotch is an impact crater. Ganymede, the largest satellite in the Solar System, is larger than both Mercury and Pluto, with a surface covered with old tectonic features and bright impact craters. Callisto's surface, the oldest of the Galilean satellites, consists of dark dusty ice covered with bright impact craters of all sizes.

order from Jupiter): Io, a brightly colored, actively volcanic body; Europa, an ice-covered world with a young surface; Ganymede, a rock and ice body with some old cratered areas and some younger terrains; and Callisto, a body with an old surface superficially resembling Earth's heavily cratered Moon. Io and Europa are about the size of Earth's Moon, while Ganymede and Callisto are the size of the planet Mercury. Table 19.1 lists values for various properties of the Galilean satellites.

The Galilean satellites resemble a miniature solar system, and are thought to have condensed out of material orbiting Jupiter in much the same way as the planets condensed 4.6 Ga out of material in the dust and gas disk surrounding the young Sun (Chapter 3). There is a gradation in composition and density from rocky, dense, almost water-free Io (mean density of $3.5 \, g \, cm^{-3}$); to Europa, with an ice/liquid water layer over rock ($3.0 \, g \, cm^{-3}$); to Ganymede, with a thick water ice mantle over a large rock/metal core ($1.9 \, g \, cm^{-3}$); to lowest density Callisto ($1.8 \, g \, cm^{-3}$), which may not be completely differentiated and could have a more uniform mixture of rock and ice. This gradation is reminiscent of the change with distance from the Sun in the volatile content and density of the planets, from metallic and rocky high-density Mercury out to the icy outer Solar System (where, for example, comets are 40–50% water ice by mass). As discussed in Chapter 3, the Solar System's variation in volatile content and density is thought to have come from a temperature gradient in the primordial disk from which the planets condensed; the "snow line," where water could condense,

is thought to have been near the orbit of Jupiter, which is why the worlds of the outer Solar System have a much higher volatile content than those closer to the Sun. Similar primordial gradients could explain the Jupiter system as well, but this interpretation is complicated by the possibility that initially present volatiles may have been driven off Io, and perhaps Europa to a lesser extent, by intense geologic activity caused by tidal heating.

19.3 Tidal evolution, resonances, and heating in the Galilean moons

Each of the Galilean moons is *tidally locked* (or "spin-locked") to Jupiter, meaning that it always keeps the same face to Jupiter, that is, it rotates about its pole with a period identical to its orbital period. For Europa this period is 3.6 Earth days (Table 19.1). Such a spin-locked situation (also called *synchronous rotation*) is common among the satellites in our Solar System that orbit close to their primaries[1] (e.g., Earth's Moon always keeps the same face to its primary, Earth), and is due to torques[2] generated by the gravitational force of the primary acting on the tidal bulges raised by the

[1] A *primary* refers to the more massive of two bodies in a mutual orbit about their common center of mass. If the primary body is much more massive than the other body (the satellite), then the center of mass is in fact close to the center of the primary and one often loosely says the satellite is orbiting about the primary.

[2] A *torque* is a change in angular momentum per unit time, and is typically caused by a force acting on an off-axis portion of a body. *Angular momentum* is a measure of a body's rotational (spin) or

TABLE 19.1 The Galilean satellites and Earth's Moon[*][+]

	Io	Europa	Ganymede	Callisto	Earth's Moon
Radius (km)	1822	1561	2631	2410	1738
Mass (10^{21} kg)	89.3	48.0	148.2	107.6	73.5
Mean density (g cm^{-3})	3.53	3.01	1.94	1.83	3.34
Surface gravity (cm s^{-2})	180	131	142	124	162
Escape velocity (km s^{-1})	2.6	2.0	2.7	2.4	2.4
Visual geometric albedo	0.62	0.68	0.44	0.19	0.12
Average surface temperature (K)	118	103	113	118	253
Orbits					
Semimajor axis (R_J)	5.91	9.40	15.0	26.4	5.38
					($= 60.1\ R_{\oplus}$)
Period (days)	1.769	3.551	7.155	16.689	27.322
Eccentricity	0.004	0.0101	0.0015	0.007	0.055
Inclination (deg)	0.04	0.47	0.21	0.51	5.15

Notes:
*Data from National Space Science Data Center (2004).
[+]Jupiter also has dozens of smaller moons, whose combined mass is about one-thousandth that of Europa.

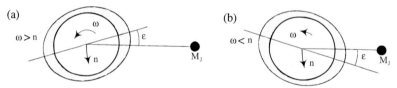

FIGURE 19.2 Tidal bulge and tidal spin-locking. Europa orbits Jupiter (mass M_J) with mean motion n while rotating at angular frequency ω. There is a lag angle ε between the radial line from Europa to Jupiter and the line of symmetry of Europa's tidal bulge. (a) If $\omega > n$, $\varepsilon > 0$ and jovian torques act to decrease ω; or (b) If $\omega < n$, $\varepsilon < 0$ and ω is increased. Drawing is not to scale: Europa's diameter is only 2.2% that of Jupiter, and its orbit is at 9.4 jovian radii. Europa's permanent tidal bulge (the outer oval) is also greatly exaggerated.

primary on the satellite.[3] These torques act either (a) to slow the satellite's rotation[4] if its spin angular velocity ω is faster than its mean orbital angular velocity n, or (b) to increase it if $\omega < n$ (Fig. 19.2). When $\omega = n$, torques go to zero for a circular orbit, and a stable state is reached (see Appendix 19.1[5]). Because tidal

    orbital momentum, and, like linear momentum, is a quantity that in general is conserved.

[3] The gravitational force between two bodies is proportional to the product of their masses divided by the distance squared between the bodies. A *tidal force* refers to the *difference* between the gravitational force of an external body acting on one side of a body (say the near side) and that acting on another side (say the far side). Tides due to an external body create symmetric distortions (bulges) in a second body of an amplitude proportional to the inverse *cube* of their separation.

[4] This process is called *despinning* and is the more common situation.

[5] The appendices at the end of this chapter present greater detail about the physics of the Jupiter–Europa system; they may be skipped without loss of continuity.

effects decline rapidly with distance from the primary and also depend on the masses of the primary and the satellite, timescales to reach spin (or orbital) end-states for various objects in the Solar System can vary from geologically short (perhaps as quick as $\sim 10^3$ yr for Europa – see Appendix 19.1) to longer than 10 Gyr.

    Tidal evolution also acts to make orbits circular, i.e., to drive satellites' eccentricities e to zero for satellites orbiting close to massive primaries (Appendix 19.2 gives the details). Figure 19.3 explains orbital elements such as *eccentricity* and *semimajor axis*. The timescale for orbit circularization is longer than the despinning timescale because satellite spin angular momenta are typically very small in comparison to their orbital angular momenta.

    Jupiter's strong gravity raises tidal bulges along the line towards Jupiter (Appendix 19.1 and Fig. 19.2), but

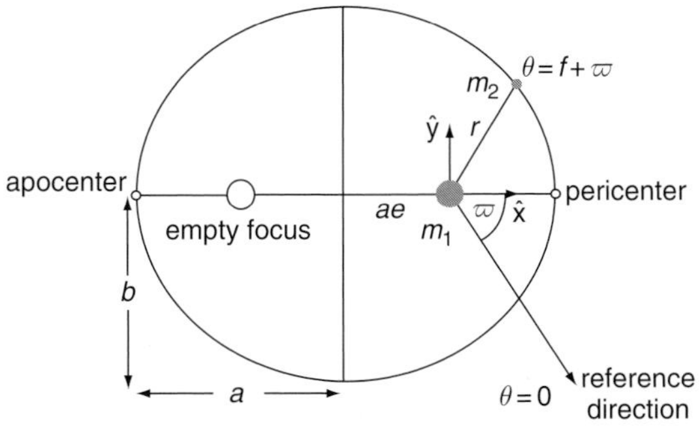

FIGURE 19.3 Orbital elements. Standard notation used to describe a Keplerian orbit of semimajor axis a, semiminor axis b, and eccentricity e. *Pericenter* and *apocenter* are the minimum and maximum distance of the satellite m_2 from the primary body m_1. The eccentricity e of the orbit, a measure of its deviation from a circle ($e = 0$) is defined as $e = \sqrt{(1 - b^2/a^2)}$. The longitude of pericenter ϖ measures the angular position of the pericenter from a chosen reference direction. (After Murray and Dermott, 2001.)

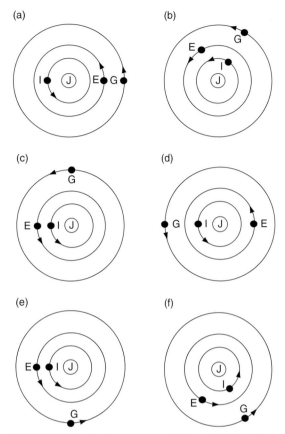

FIGURE 19.4 The Laplace resonance. (a) through (f) show the sequence of conjunctions for the three Galilean satellites (I = Io, E = Europa, G = Ganymede) that participate in the resonance as they orbit about Jupiter (J). The seventh configuration in the sequence would complete one period of Ganymede (which is two periods of Europa and four periods of Io) and return to (a). (Murray and Dermott, 2001.)

since the distance to Jupiter varies for any satellite with a non-circular orbit, Jupiter's gravitational force and thus the tidal bulge's height also vary around the satellite's orbit. The satellite's figure is flexed as the bulge moves up and down, producing an internal dissipation of energy called *tidal heating* (Appendix 19.3 contains details). Since the satellite's orbit is eccentric, its orbital velocity also varies around its orbit (faster when closer), so that its spin-locked *constant* rate of rotation does not keep perfect pace with its *variable* orbital velocity. The result is a slight "sloshing" or *libration* of the satellite's tidal bulge back and forth around the radial line to Jupiter. This libration tide also contributes substantially to the tidal heating of the satellite (Appendix 19.3). Energy dissipation (tidal heating) from both these tidal effects acts to circularize the satellite's orbit.

Once any satellite is both spin-locked and driven into a circular orbit, its tidal bulge will no longer vary around the orbit and tidal heating will cease. However, despite tidal dissipation and the resulting tendency toward orbital circularization, the orbits of the inner Galilean satellites are all slightly noncircular (Table 19.1). These "forced" eccentricities are maintained by the satellites' mutual periodic perturbations in a so-called *Laplace resonance* (Fig. 19.4), first shown by and then named after the French mathematician and physicist (1749–1827).

Io, Europa, and Ganymede are in a 1:2:4 *orbital resonance*, meaning that Ganymede's orbital period is very nearly twice that of Europa's, which is very nearly twice that of Io's. This resonance is *stable*, meaning

that the effects of any perturbations to the situation (say, caused by the gravity of a passing planet or of the moons themselves) will eventually die out and the system will remain unchanged. The configuration of the Galilean system is such that whenever two of the three inner satellites are in conjunction, the third is at least 60° away as seen from Jupiter (Fig. 19.4). Because Io and Europa, or Europa and Ganymede, line up periodically at the same points in their orbits, each pair experiences regular gravitational perturbations that do not average out over many orbits. These perturbations maintain the forced eccentricities of the satellites' orbits and thus sustain tidal heating over time. Appendix 19.4 gives the details of how the Laplace resonance works.

Since tidal heating is driven by Jupiter's gravity, which falls off with distance, the tidal heating is most intense at the innermost moon Io. Io is in fact the most volcanically active body in the Solar System, surpassing even the Earth. Io's volcanic activity was brilliantly predicted by Peale, Cassen, and Reynolds (1979) in a paper published three weeks prior to the Voyager 1 spacecraft's arrival at the jovian system. If the current rate of volcanic activity on Io is typical over its history, then there has been sufficient volcanic processing to recycle the entire crust of Io out onto the surface many times. This activity may well have driven off most volatiles over time – indeed any water present on Io at its formation would now be long gone.

Europa is farther away from Jupiter than Io (9.4 vs. 5.9 R_J), so despite its orbit's larger eccentricity (0.010 vs. 0.004) it experiences less tidal heating, which depends on eccentricity e and semimajor axis a like $e^2 a^{-15/2}$ (Appendix 19.3, Eq. (19.8)). Nevertheless, a substantial amount of heat is dissipated within Europa's core, mantle, and ice shell, although there is an ongoing debate over exactly how much is dissipated overall and within each layer (Appendix 19.3).

More distant Ganymede at 15 R_J is also subject to tidal heating, though less so. Still, Ganymede may have experienced sufficient tidal heating to drive some geologic activity and internally differentiate into an icy mantle and a rock/metal core. Ganymede's surface includes old, cratered terrain, but also younger grooved and ridged areas which are the result of more recent tectonic activity. An interesting comparison can be made between Ganymede and its neighboring satellite Callisto, which is nearly the same size as Ganymede. Callisto does not participate in the resonance with the other three Galilean satellites, and is an old, cratered body with few, if any, signs of endogenic geologic activity. Callisto also is thought not to be fully differentiated, although recent measurements suggest that it could have a small core.

19.4 Europa's space environment

Europa orbits Jupiter at a distance of 9.4 Jupiter radii (671,000 km) with an orbital period of 3.55 days, and has a diameter of 1560 km (just under the size of Earth's Moon). Its *albedo* is ~0.68, meaning that it reflects 68% of the visual sunlight incident on it, making it one of the brighter objects in the Solar System (for comparison, Earth's Moon's albedo is 0.12). Note that most of the publicly released images of Europa are considerably contrast-enhanced, and that the difference in brightness between the bright and dark regions of Europa's surface is actually small. Europa's surface temperature has been measured to be as high as 130 K

at the equator in daytime (Spencer *et al.*, 1999) and could drop as low as 50 K at the poles.

19.4.1 UV and charged particle bombardment

Europa's charged particle and ultraviolet (UV) radiation environment has been measured by the Galileo spacecraft and is summarized in detail by Cooper *et al.* (2001). The solar irradiation of its surface is smaller than that at Earth's distance from the Sun by a factor $(5.2)^2 \approx 27$ and totals $8 \times 10^{12}\,\mathrm{keV\,cm^{-2}\,s^{-1}}$; the incident ultraviolet-C (UV-C) flux[6] ($\lambda < 280$ nm) is $4 \times 10^{10}\,\mathrm{keV\,cm^{-2}\,s^{-1}}$.

Europa's harsh charged-particle radiation environmet is due primarily to the acceleration of these particles in Jupiter's magnetosphere. The incident energy flux of charged particles is $\sim 8 \times 10^{10}\,\mathrm{keV\,cm^{-2}\,s^{-1}}$, about twice that due to UV-C. It is dominated by energetic electrons in the keV to MeV range, but includes H^+, O^{n+}, and S^{n+} ions as well. The depths affected by this radiation depend on the density of the surface, which could range from that for solid ice ($0.92\,\mathrm{g\,cm^{-3}}$) to that for water frost ($\sim 0.1\,\mathrm{g\,cm^{-3}}$). Since the density of Europa's uppermost surface is unknown, penetration depths are given in terms of mass cross sections; the average stopping depth for the most penetrating charged particles, the electrons, is $0.62\,\mathrm{g\,cm^{-2}}$, corresponding to a depth of ~ 0.7 mm in $0.92\,\mathrm{g\,cm^{-3}}$ ice. These depths are much greater than those penetrated by UV photons, which affect only the top submicrometer layer; for example, Lyman-α photons ($\lambda = 121.6$ nm) have a mean penetration depth of only $0.04\,\mu\mathrm{m}$ in ice.

A crude understanding of the implications of the incident radiation fluxes cited above for materials or biology can be obtained by calculating net volume radiation dose rates as a function of depth into Europa's ice (Fig. 19.5). Depths of 1 cm in Europa's ice (assuming an ice density of $1\,\mathrm{g\,cm^{-3}}$) experience a volume dosage rate of ~ 0.3 Mrad per month (Cooper *et al.*, 2001). This dose is hundreds of times higher than the lethal dose for human beings (a lethal dose for 50% of humans is ~ 0.0006 Mrad). The most radiation-resistant terrestrial organism known, *Deinococcus radiodurans*, has 90% survival after 6 Mrad, dropping to 10^{-6} survival after 12 Mrad of ionizing radiation

---

[6] Energies greater than ~ 4.4 eV, corresponding to photon wavelengths below 280 nm, are required to dissociate H_2O, so UV-C fluxes are those of relevance to radiation chemistry.

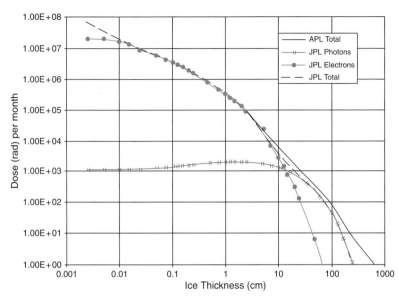

FIGURE 19.5 Radiation dose in ice. The dose in units of rad (water) per month (30.4 days) of exposure beneath a given thickness of ice for both UV photons and electrons. Information provided by researchers at the Jet Propulsion Laboratory (JPL) and the Applied Physics Laboratory (APL). (After Space Studies Board, 2000.)

dose (Auda and Emborg, 1973). Radiation doses fall rapidly with depth in Europa's ice, so that at meter depths monthly doses are ~100 rad; doses drop to values similar to those in Earth's biosphere at depths of 20 to 40 m. An ocean beneath kilometers of ice would be entirely protected; microorganisms living in putative near-surface niches of liquid water, especially if these were close enough to the surface to permit photosynthesis (Greenberg et al., 2000), would have to cope with the resulting radiation stress. This alone does not seem insurmountable for highly resistant strains; D. radiodurans could survive at the 90% level for nearly two years at 1 cm depth in Europa's ice. Using terrestrial algal mats growing on the bottoms of permanently ice-covered Antarctic lakes as an analogue, it appears that photosynthesis should be possible beneath ice thicknesses of ~15 m on Europa (Reynolds et al., 1983), where the radiation dose would be tolerable by many organisms. More recent work by Warren et al. (2002) suggests that photosynthesis is possible at light levels lower than those considered by Reynolds et al. (1983), implying that photosynthesis could be possible under a few tens of meters of europan ice. If near-surface liquid water niches do in fact exist, microbial photosynthesis on Europa would seem possible.

19.4.2 Sputtering

Sputtering is erosion of a surface by charged particle impacts. Sputtering at these high energies is explosive – many thousands of water molecules are removed from

the surface for each single incident high-energy ion. The best estimate for the sputtering erosion rate on Europa due to energetic H^+, O^{n+}, and S^{n+} ions is about $0.02\,\mu m\,yr^{-1}$ for erosion of surface molecules of H_2O (Cooper et al., 2001), although some previous estimates have been as much as 100 times higher. Over Europa's ~50 Myr surface age (Section 19.4.3), sputtering thus should have removed ~1 m of material.

Europa's tenuous atmosphere, due mostly to oxygen sputtered from its surface ice, is in fact a near-vacuum with a surface number density of O_2 gas of $\sim 10^8$–$10^9\,cm^{-3}$ (Ip, 1996), equivalent to that at ~400 km altitude in Earth's atmosphere. Atomic sodium is also present in Europa's atmosphere and could come either from a source such as material ejected volcanically from Io, or perhaps from an europan source.

19.4.3 Impacts and gardening

Europa, like all other Solar System bodies, is subject to impacts by comets, asteroids, and smaller objects. Gravitational focusing (Section 3.9.2) by Jupiter's large mass enhances the impact flux experienced by Europa. The main objects impacting Europa and the other bodies in the jovian system are thought to be Jupiter family comets (Zahnle et al., 1998), of which a spectacular example was the impact of comet Shoemaker–Levy 9 with Jupiter in 1994. Shoemaker–Levy 9 was gravitationally disrupted into multiple pieces during an earlier pass by Jupiter, and then impacted Jupiter in a series of well-documented collisions.

The number of craters on planets and satellites can be used to gauge both relative and, with additional information, absolute surface ages. Europa's surface has only about 15 impact craters with diameters greater than 10 km (Moore *et al.*, 2001), and the majority of small craters seem to be secondary craters formed by debris thrown out from these large events. This lack of craters implies a young surface, because recent (or perhaps current) geologic activity is required to erase them. Bodies with current geologic activity such as Io and Earth have very few recognizable impact craters – Io, in fact, is so volcanically active that not a single impact crater, of any size, has been found on its surface to date! The Earth has about 150 recognized craters, but many have been geologically modified and would be difficult to recognize from orbit. On the other hand, geologically inactive bodies with old surfaces such as Earth's Moon or Callisto are covered with impact craters of all sizes. The most densely cratered surfaces are said to be "saturated", as each new crater destroys old craters beneath it.

Relative ages can be deduced by comparing the number density of craters on different parts of a planet or satellite's surface. Europa's crater density is so low, however, that no part of the surface has been reliably inferred to be younger or older than any other. With respect to absolute age, various efforts have been made to establish impact fluxes throughout the Solar System's history in order to establish the actual age of a body's surface based on its crater density. From calculations of the comet and asteroid fluxes at Jupiter, the age of Europa's surface has been estimated to be only ~50 Myr. Despite Europa's geologically young surface (by comparison, the majority of Earth's surface, the ocean floors, are typically 100–200 Myr in age), no current geologic activity was found to a resolution of ~2 km over 20% of Europa's surface in a comparison between Voyager and Galileo spacecraft images taken 20 years apart (Phillips *et al.*, 2000).

Europa's surface is subject to a range of impacts extending in size from those responsible for the largest observed craters down to micrometeorites. These impacts together result in the formation of a *regolith*, a layer of broken impact debris at the surface of an airless world that is continually mixed through a process called *impact gardening*. The depth to which surface material has been mixed after a given length of time is called the gardening depth. Impact gardening can potentially preserve surface material by moving it down below the sputtering depth and even the effective radiation penetration depth. If sputtering dominates, compounds produced by radiation processing at the

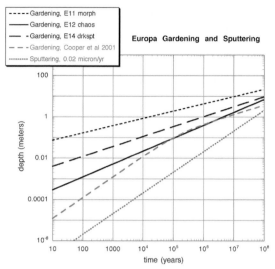

FIGURE 19.6 Impact gardening on Europa. Three estimates of gardening rates, labeled E11, E12, and E14 for different regions of the surface imaged on separate Galileo flybys of Europa, are compared with a previous gardening rate (Cooper *et al.*, 2001) and the sputtering rate. Note that all these gardening rates dominate sputtering over the presumed age of Europa's surface (~50 Myr).

surface are lost before they have a chance to be buried and possibly eventually transported to a subsurface ocean layer (Section 19.10.4). However, if gardening dominates, then material may be buried faster than most of it can be removed through sputtering.

Models for gardening on Europa remain poorly developed. Cooper *et al.* (2001) scaled from Ganymede regolith depth estimates, as well as a mass flux from studies of rings, to estimate a gardening depth of 2.6 m over a surface age of 50 Myr. The present authors have found a similar result using a model of impact gardening based on studies of the lunar regolith coupled with estimates of the impactor population in the outer Solar System (Zahnle *et al.*, 1998). Further refinement of the models, taking into account studies of the numbers of very small craters on Europa, indicates that the gardening depth could extend deeper than earlier estimates, and that depths of 1–10 m per 10 Myr could be achieved (Fig. 19.6). For a sputtering erosion rate of $0.02\,\mu m\,yr^{-1}$, any of the gardening estimates so far discussed suggests that gardening dominates sputtering by an order of magnitude over Europa's surface age.

19.5 Surface composition

We have no samples of Europa's surface nor any results from *in situ* measurements, so spectroscopy, which samples only the top few microns of the surface, is currently our only source of precise information regarding Europa's surface composition. Even before the Voyager spacecraft arrived at Jupiter in 1979, ground-based spectroscopy had shown that Europa's surface was predominantly water ice. The Galileo spacecraft's near-infrared spectrometer detected absorption features on Europa due to sulfur dioxide (SO_2) and hydrogen peroxide (H_2O_2; present at 0.13% by number relative to H_2O). Oxidants such as H_2O_2 and O_2 at Europa's surface are an expected product of charged-particle bombardment of water ice (Delitsky and Lane, 1998) and have important astrobiological implications (Section 19.10).

Observed near-infrared ice absorption bands (from 1.0 to 2.0 μm) show distortions from those for pure water; these distortions have been modeled as due to the presence of hydrated salts or other compounds, for example, magnesium sulfate hydrates (e.g., $MgSO_4 \cdot nH_2O$) or bloedite ($Na_2Mg(SO_4)_2 \cdot 4H_2O$) (McCord et al., 1999), or sulfuric acid ($H_2SO_4 \cdot 8H_2O$) (Carlson et al., 1999). Hydrated salts in which Na and Mg are the dominant cations and sulfate (SO_4) the dominant anion are consistent with experimental and theoretical studies (Kargel et al., 2000; Zolotov and Shock, 2001; but see McKinnon and Zolensky, 2003) that assume evolution from an initial carbonaceous chondrite composition (Chapter 3). The locations of these potential salts are mostly along ridges, in craters, or in the matrix of chaotic terrain (see Section 19.7 for descriptions of Europa's surface features). The composition of the inferred salts is similar from place to place on Europa, consistent with an origin from a globally mixed reservoir. One intriguing explanation for such a reservoir is a global salty water ocean, which reaches the surface in places of recent geologic activity and leaves salts behind. Even the presence of such salts, however, does not unambiguously require a global ocean.

Carbonaceous chondrite meteorites are typically several percent organic by mass, so it is reasonable to expect such compounds on Europa – though an analogy should not be mistaken for evidence. A variety of organic functional groups have been detected on Ganymede and Callisto (such as C≡N and C-H) and there are very low signal-to-noise observations of such compounds on Europa (McCord et al., 1998). Carbon dioxide has been observed on Ganymede, Callisto, and Europa; the CO_2 abundance of Europa's ice shell appears to be about 0.2% by weight (McCord et al., 1998). Charged particle irradiation of CO_2-containing H_2O ice should lead to limited production of simple organics such as formaldehyde (H_2CO) (Section 19.10).

Dalton et al. (2003) have shown that the distortion in the near-infrared ice absorption bands can be fit at least as well by the spectra of 100 K ice and bacteria mixtures as by bloedite, magnesium hexahydrites, or sulfuric acid. That is, to the extent that the spectral features provide evidence for the presence of certain salts present in Europa's ice, these features also provide evidence for europan microorganisms. Dalton's intriguing work provides an important caution about the interpretation of features at low spectral resolution such as that of the Galileo instrument ($\lambda/\Delta\lambda \approx 100$ from 0.7 μm to 5.2 μm). At these low resolutions, features are ambiguous and at risk of misinterpretation.

This is not a new lesson. A similar challenge was experienced with low-resolution near-infrared observations of comet Halley around the time of the Giotto and Vega spacecraft flybys in 1986. A broad spectral feature around 3.4 μm observed with resolution $\lambda/\Delta\lambda \approx 400$ was consistent with emission from an organic grain model based on tholins synthesized in laboratory ice irradiation experiments (Chyba and Sagan, 1987). However, these same features could also be fit by the spectral characteristics of grains rich in bacteria and viruses (Hoyle and Wickramasinghe, 1987). In the end it took 25 times higher resolution to largely resolve these ambiguities. It was ultimately demonstrated that the majority (though not all) of the 3.4 μm feature was due not to organic grains, but to molecular lines of individual organic molecules (such as methanol, CH_3OH) that could not be individually resolved at the lower resolution (Mumma et al., 1993). Higher resolution spectra of Europa's surface may likewise be essential to finding the right interpretation of the near-infrared features.

19.6 Interior models

Europa's bulk density of 3.0 g cm^{-3} is substantially lower than that of ice-free Io (Table 19.1). As noted above, spectroscopy by itself only establishes the presence of a thin layer of water ice at Europa's surface, so such observations were consistent with models of a Europa with substantial water ice below its surface, but did not require it.

Voyager images showed an absence of both very dark terrain and substantial topographic relief, suggesting that Europa was covered with at least several km of H_2O, concealing any underlying silicate topography. (Ice that is sufficiently thick and warm beneath the surface cannot support substantial topography because it will deform and flatten over geological timescales; this is called viscous relaxation (Thomas and Schubert, 1986).) Early thermal models using conductive cooling and heating from radioactive elements predicted the possibility of a differentiated Europa with water ice overlying a silicate/metal core; in this case a Europa whose silicate/metal interior had the same average density as Io ($3.5\,\mathrm{g\,cm^{-3}}$) required a surface H_2O layer 120 km thick (Cassen *et al.*, 1982), a model that would later be shown by Galileo spacecraft measurements to be broadly correct (see below).

These early models also predicted the possibility of a *liquid* water layer beneath Europa's frozen ice crust. Tidal and radioactive heating could be balanced by conductive or convective cooling, which would allow a portion of the subsurface water layer to remain liquid over geologic time. Current models have failed to resolve this issue completely (Ojakangas and Stevenson, 1989; McKinnon, 1999), mostly because the properties of ice are poorly known at the temperature of Europa and on the days-long timescales associated with tidal flexing. Also poorly known is the composition of Europa's ice; the addition of small amounts of other volatiles such as ammonia or salts could dramatically change its viscosity (McCord *et al.*, 1999). Also important is the physical state of the ice, including its grain size and degree of fracturing, which could affect its strength.

In conclusion, *theoretical* models of tidal heating and deformation suggest that a liquid water layer of thickness up to ~100 km could have been or may currently be present beneath Europa's surface, but cannot definitively say so. In the following sections, however, we present Galileo Mission measurements that further strongly support this view.

19.6.1 Gravity measurements

Radio data from the Galileo spacecraft's close flybys of Europa suggest that Europa is completely *different-iated* (separated into a dense core with less dense overlying layers). This technique uses measurements of the spacecraft's trajectory (via observing the Doppler shifting of its radio signals) to probe the moon's gravity field. Newton (1686) showed in his *Principia*[7] that a perfectly spherical distribution of mass has a gravitational field identical to that of a point of the same mass, i.e., one that falls off like the square of the distance. However, deviations of a body from sphericity (e.g., the oblateness that arises on any spinning world) leads

to higher-order terms (terms falling off at powers higher than the square of the distance) in the expression describing its gravitational field. The size of these terms depends on the internal distribution of matter within the body (Appendix 19.5 gives further details). Sufficiently accurate tracking of a spacecraft's trajectory can determine its deviation from that expected for a perfect sphere, and thereby determine the coefficients of some of the higher-order terms in the gravitational field. These in turn give information about the interior mass distribution of the object.

Such measurements from the Galileo spacecraft, combined with our knowledge of Europa's bulk density and the cosmic abundances in the Solar System (Chapter 3) indicate that Europa's most likely internal configuration includes an Fe or Fe-S central core, an anhydrous (lacking water) rocky mantle, and a surface layer of material with a density of around $1\,\mathrm{g\,cm^{-3}}$ that is ~100 km thick (with uncertainties ranging from 80 to 170 km) (Anderson *et al.*, 1998). The only cosmochemically plausible material with this abundance and density is H_2O. Many objects in the outer Solar System are extremely water-rich; comets, for example, are as much as 40–50% water ice by mass. On this interpretation Europa is ~6% H_2O by mass, consistent with the water inventory of a carbonaceous chondrite meteorite. But the gravity data cannot distinguish between liquid water and solid ice due to their nearly identical densities. If this water were predominantly liquid, its volume would be more than double that of all of Earth's oceans. The pressure at the bottom of this ocean would be around twice that in the deepest parts of Earth's oceans.

If the H_2O layer were instead ice all the way down, the most likely interior temperature profiles (Section 19.8) suggest that the ice would be ice I to the bottom.[8] Pressures at the bottom of the thickest H_2O layer (170 km) consistent with the gravity data are ~200 MPa, making ice III a possibility for warm ice at this depth. At lower pressures, unlikely but not impossible conditions of temperature and pressure would permit ice II to exist.

19.6.2 Magnetic field measurements

Perhaps the most convincing evidence for an ocean of liquid water beneath Europa's surface comes from magnetic field measurements. Data from Galileo's magnetometer (Kivelson *et al.*, 2000) show that Europa has an

---

[7] Book I, Sec. XII, Prop. LXXIV, Th. XXXIV.

[8] Section 15.2 gives a phase diagram for H_2O and includes a discussion of the various types of ice (I, II, III, etc.).

induced magnetic field that varies in direction and strength in response to Jupiter's rotating magnetic field. This results from Faraday's law of induction: time variations in a magnetic field cause (induce) an electric field, which (if a conducting material is present) in turn causes (so-called eddy) currents, which in turn create a secondary magnetic field that tends to oppose the original magnetic field (further details are in Appendix 19.6). The periodic variation in direction shows that the field is not due to a permanent internal dipole, i.e., it is not analogous to the Earth's internal field. The strength and response of the induced field at Europa require a near-surface, global, electrically conducting layer.

Similar magnetometer data suggest that Callisto and Ganymede (Kivelson *et al.*, 2002) may also harbor subsurface oceans. These latter oceans may exist in a layer sandwiched between two phases of water ice; if so they would not provide the astrobiologically more interesting rock/water interface (with possibilities for hydrothermal vents, see Section 19.10.3) that may be present at the bottom of Europa's ocean (McCollom, 1999).

The magnetic data for Europa cannot be explained by localized pockets of salty water, and require a nearly complete spherical shell. The induced field cannot be the result of a frozen ice layer, even if it had pockets of briny water, since ions in solid ice would be insufficiently mobile (Stevenson, 2000a). It is possible that a type of conducting layer other than a global salty ocean could account for the induced magnetic field, but the salty ocean explanation is the most plausible. Postulating currents induced in a metallic core does not work because the induced dipole field strength falls off with the cube of the distance and the core is simply too far away to provide the observed field. Nor are the data consistent with a field induced in Europa's ionosphere, which is too tenuous to support the electrical currents needed to explain the field strength (Zimmer *et al.*, 2000).

In addition to the (variable) eddy currents induced in Europa's ocean due to Jupiter's time-varying magnetic field, a *constant* electrical current might also flow because of Europa's orbital motion through the near-constant vertical component of Jupiter's magnetic field. This is due to the Lorentz force $F = qv \times B$ experienced by a particle of charge q moving with a velocity v relative to a magnetic field B. The resulting motion of charge should lead to currents in Europa's ocean that run from Europa's subjovian hemisphere to the anti-jovian hemisphere. However, these currents are strongly limited by the electrically insulating water ice overlying the ocean (Reynolds *et al.*, 1983; Colburn and

Reynolds, 1985). Such electrical currents in Europa's ocean could be considerably greater than those suggested by Colburn and Reynolds (1985) because Europa's ice is expected to be salty, and therefore much more conducting than pure water ice.

19.7 Europan geology

Images of Europa returned by the Voyager spacecraft in 1979 revealed a bright, icy world covered with long linear features, some of which seemed to circle nearly the entire globe. There was also a striking absence of impact craters, suggesting a young surface age. The surface was remarkably smooth, with little topography more than a few hundred meters high. Also visible in the Voyager images (limited by a maximum resolution of ~2 km/pixel) was so-called "mottled terrain," fuzzy-looking dark patches that interrupted the dark linear features and bright background regions. Galileo spacecraft imagery over 1996–2002 revealed many more details of Europa's geology (e.g., Fig. 19.1). About 10% of Europa's surface was imaged at resolutions of 200 m/pixel or higher with very limited coverage at resolutions as high as ~6–10 m/pixel (Greeley *et al.*, 2000). The mottled terrain turns out to consist of areas of disrupted surface, in some cases broken up into coherent iceberg-like blocks that seem to have "rafted" into new positions (Greenberg *et al.*, 1999). Such areas can be reconstructed by fitting the pre-existing features on the blocks back together (Spaun *et al.*, 1998). It is still the case that few large impact craters have been found and the largest impact craters are anomalously shallow (Moore *et al.*, 1998). The topography remains quite low, with just a handful of instances of features over a few hundred meters in height (Greeley *et al.*, 2000). Other features of interest on Europa's surface include regions that could possibly be due to surface flows of low-viscosity material such as liquid water.

Because our interest in this chapter is focused on astrobiology, we survey Europa's geology from the point of view of its implications for the existence of a subsurface ocean (Pappalardo *et al.*, 1999), other pockets of liquid water, and possible communication between Europa's surface and the ocean. Although many models have been proposed for formation of the surface features (see Greeley *et al.* (2004) for a review), virtually all of them take the existence of a subsurface liquid water layer as a starting assumption. The main debate now is over the *thickness* of the overlying ice layer. The thickness of this crust is of interest for astrobiology, but not crucial to prospects for the existence of subsurface life. Thick- and thin-ice models do affect, however, proposed mechanisms for providing speculative europan biospheres with the free energy they would need to persist (Section 19.10). A thin crust would also make direct exploration of the liquid layer easier for future spacecraft (although still extremely challenging). Lastly, if life does exist in the liquid layer, a thin-ice cover would seem to allow remnants to be more readily found at the surface.

19.7.1 Lineaments

The dominant surface features on Europa are linear ridges and bands called *lineaments*. Of these, *ridges* are long linear features usually a kilometer or so wide, a few hundred meters tall, and hundreds or thousands of kilometers long. The common "double ridge" (Fig. 19.7) has two symmetrical ridges separated by a central trough. *Bands*, by contrast, consist of multiple parallel linear features with little or no topographic expression. Their formation may or may not be similar to the formation of ridges. There is also a background terrain of many small ridges and cracks, with a variety of orientations.

Most models of ridge formation begin with a crack generated through tidal flexing (Hoppa *et al.*,1999a), and various mechanisms then exploit this zone of weakness to create a ridge. Models vary in requirements for ice thickness, ranging from a very thin crust overlying liquid water to completely solid-state models with a thin, brittle crust on top of a lower-viscosity, warm ice layer.

The tidal squeezing model proposed by Greenberg *et al.* (1998) assumes a thin ice layer overlying liquid water, and relies on diurnal tides to open and close a crack. Whenever the crack is open, water or slush enters the crack, and this material is then squeezed up to the surface when the crack closes again half a period later. Over time, this material is built up on both sides of the crack, producing the signature double ridge frequently seen on Europa.

The thick ice models of ridge formation again start with a tidal crack at the surface. An upwelling linear *diapir* (a buoyant plume of warmer, soft ice) then rises to exploit this zone of weakness, and warps the overlying crust when it nears the surface (Head *et al.*, 1999). This mechanism explains the observation that pre-existing features are sometimes visible on the upwarped ridge flanks, suggesting that the entire surface has been lifted rather than covered with material in a

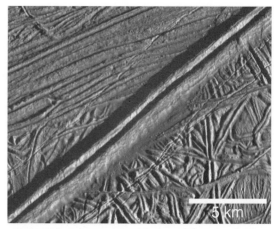

FIGURE 19.7 A ridge. The predominant ridge type on Europa is double, consisting of parallel ridges symmetrical about a central trough. Double ridges are usually ∼1 km wide, a few hundred meters tall, and can stretch for hundreds or thousands of km in length. The pictured ridge is superimposed on a background plain of older, smaller ridges.

depositional mechanism such as in the tidal squeezing model. (See the upper right-hand corner of Fig. 19.7 for an example.) The linear diapir model requires that solid state convection is taking place in a thick warm ice crust which may or may not overlie a liquid ocean layer. It also requires, for solid state convection to begin, that the ice layer has a thin brittle shell at the surface, and a warmer, lower-viscosity layer of at least 10 km thickness (McKinnon, 1999). There are other suggested ridge-formation mechanisms, including a *cryovolcanism*[9] model in which material is explosively vented onto the surface, building up ridge margins surrounding a crack (Fagents *et al.*, 2000).

A different and striking ridge type on Europa is the *cycloidal ridge* (Fig. 19.8). These appear as a series of connected arcs that march across the surface like a child's drawing of waves on an ocean. In an elegant model of global tidal stress orientations and magnitudes, cycloidal ridges have been shown to correspond in orientation and location to cracking of the surface in response to changing diurnal tidal stresses (Hoppa *et al.*, 1999b). In some cases this model predicts that cracks on the surface propagate at a speed of ∼3 km hr^{-1}, a good walking pace! This model requires the existence of a global ocean to obtain a large enough, time-varying tidal bulge to give sufficient tidal stresses

---

[9] *Cryovolcanism* refers to ice volcanism, specifically the eruption of liquid or vapour phases of water or other volatiles that would be frozen at the normal temperature of the icy satellite's surface (Geissler, 2000).

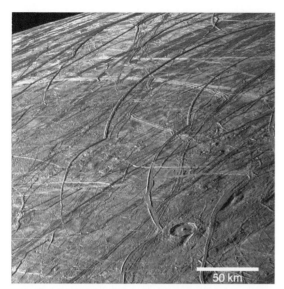

FIGURE 19.8 Cycloids and a large crater. Cycloidal ridges are thought to be formed by the changing orientation of tidal stresses during Europa's 3.6-day orbit around Jupiter. Such ridges are more often found at high northern and southern latitudes, and contain many arcs and cusps of varying sizes and orientations. The 20-km diameter crater in this image is Maeve (58° N, 75° W), and illustrates the shallow appearance of impact craters on Europa's surface.

to crack the ice. The cycloidal ridge model explains the shapes of the cracks, but does not determine how the ridge is built up once the crack exists.

It is unclear whether any of the proposed ridge formation models can explain how some ridges maintain such a uniform, linear appearance over thousands of kilometers. Clearly, the various proposed models are contradictory, and have very different implications for Europa's subsurface structure and the location or existence of a subsurface water layer. These questions and contradictions emphasize the need for a future mission to Europa.

19.7.2 Bands

Bands on Europa are similar to ridges, but clearly represent places where the surface has opened up and been pulled apart, resulting in the extrusion of new surface material in between the older pieces of crust (Fig. 19.9). Bands can be wedge-shaped or long, parallel sets of small ridges and grooves. Reconstructions of pre-existing features cross-cut by the bands show that the bands open along a fracture and then offset, rather than cover over, nearby features (Sullivan et al., 1998). Such pull-apart bands demonstrate that the surface layer has broken and moved; strike-slip motion has

also been observed along bands (Tufts et al., 1999). Such motion requires a brittle surface layer on top of a more ductile, lower-viscosity layer. Bands seem to be similar to mid-ocean ridge morphology on Earth (Prockter et al., 2002), and in some cases seem to originate along a pre-existing double ridge that was then pulled apart laterally rather than continuing to grow vertically like a typical ridge. Models exist for their formation in either a thin-ice or thick-ice regime.

An open question has been that if bands and perhaps also ridges are formed by extension, where is the accompanying compression that must be accommodated by Europa's surface? An answer may come from observations of folds detected at Astypalaea Linea (Prockter and Pappalardo, 2000), which suggest that bands could account for both extension and compression in a crustal recycling process on Europa. In this model, bands are formed through spreading of the crust as a large crack opens and pulls apart and new material fills in the gap. However, once a large band forms, it becomes an area of relatively thin, ductile ice which is weaker than the surrounding regions of colder, thicker, firmer ice. It therefore becomes a "target" for global compressional stresses that serve to fold the entire region. Over time, the large-scale topography created by this folding relaxes away, leaving only smaller-scale remnants.

19.7.3 Chaotic terrain and lenticulae

Chaotic terrain, with its resemblance to terrestrial iceberg-like blocks floating in water (Fig. 19.10), would seem to provide good evidence for the presence of liquid water beneath Europa's surface (Greenberg et al., 1999), although other models have been proposed which could form disrupted areas through solid-state ice diapirism (Pappalardo et al., 1998; Sotin et al., 2002). Regions of chaotic terrain are seen as areas of localized heat flow in a thin-crust scenario where the ice layer melted all the way to the surface (crustal thickness temporarily goes to zero) (Greenberg et al., 1999; Thomson and Delaney, 2001). In this model the blocks, buoyant remnants of a pre-existing icy crust, move about in a slushy matrix, both translating and tilting. Eventually the matrix freezes, ending the blocks' motions and preserving their final positions. Heights of the blocks (~200–300 m) above the lower-lying matrix material have been calculated through shadow measurements; these plus buoyancy physics then imply that the pre-existing icy crust was no more than several kilometers thick (Williams and

Greeley, 1998). To cause the melt-through, this model requires localized heating of the crust to be maintained for some period of time. Thomson and Delaney (2001) model Europa's ocean as weakly stratified[10] (compared with Earth's oceans), and argue that warm buoyant plumes can rise from volcanic regions on the ocean floor and reach the underside of the ice crust. Such plumes would be narrowly confined by the Coriolis force arising from Europa's rotation. Plumes with heat fluxes of $\sim10^{11}$ W, perhaps $\sim1\%$ of Europa's global heat flux, could melt through the crust in $\sim10^3$ yr.

An objection to localized melt-through events is Stevenson's (2000b) argument that a melt-through event would be impossible; the viscosity of warm ice is so low that adjacent ice would flow quickly enough to fill in any growing gap. This argument, while sometimes stated categorically, does not apply to thin conducting shell models, for which only a very thin layer of ice at the base of the ice shell has a low enough viscosity to flow appreciably (O'Brien et al., 2002).

The solid-state formation model for chaos regions suggests that ice rises to the surface in a diapir, eventually disrupting the brittle surface, perhaps even causing surface melting (Head and Pappalardo, 1999). Partial melting might occur due to positive tidal heating feedback during the ascent (Sotin et al., 2002). Chaotic terrain may form over these areas of partial melting within a solid ice shell. The pits, spots, and domes on Europa, collectively sometimes called *lenticulae*, have been suggested as the surface expression of diapirs. Others have argued that these features are rather the early stages of chaotic terrain development (Greenberg et al., 1999). Convection models have focused on the size and spacing of these features as a way of determining the thickness of the (putatively) convecting layer (see Section 19.8). Such features could form either by the intrusion of liquid material near the surface (a laccolith) or by warm, solid-state material nearing the surface (a diapir). Pappalardo et al. (1998) suggest a diapir origin, based on their nearly uniform size (~10 km) and spacing distribution of the lenticulae. Greenberg et al. (1999), however, have suggested that the uniform size and spacing is an artifact of the classification process; they find no peak in sizes at ~10 km but rather a continuous distribution. This topic remains a lively area of disagreement,

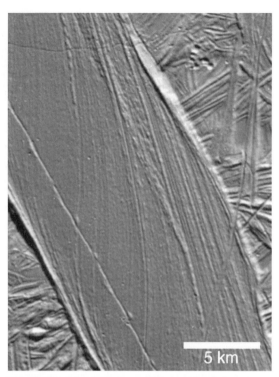

FIGURE 19.9 A band. Bands are similar to ridges, but appear to have been pulled apart laterally rather than built up vertically like double ridges. Bands can be up to 10 km wide, and hundreds of km long. This band shows small ridges inside of the main feature, perhaps formed as it spread apart.

hampered by the fact that Galileo imaged only $\sim9\%$ of Europa's surface at a resolution of 200 m or better. Since models for the formation of chaos regions have been proposed for both thin- and thick-ice regimes, and there is ongoing disagreement about the nature of the data for the lenticulae, these features do not resolve the thick ice/thin ice controversy.

19.7.4 Cryovolcanic flows and the "Puddle"

Cryovolcanic surface flows would be intriguing evidence for the presence of liquid water somewhere beneath Europa's surface, but there are few known candidate regions (Fagents et al., 2000; see Fig. 19.11 for one example). There is a buoyancy problem in their formation since it is difficult for liquid water, which is 10% denser than ice, to reach the surface. Diapirs have been suggested as formation mechanisms for various surface features on Europa, and one region, Murias Chaos (nicknamed the "Mitten"), could represent a location where warm, buoyant material was extruded onto the surface and flowed for a few kilometers (Figueredo et al., 2002).

---

[10] Weak stratification means that water density, which varies with temperature and salinity, changes only slightly with depth; the greater the density gradient, the more work is required to move a parcel up or down the water column.

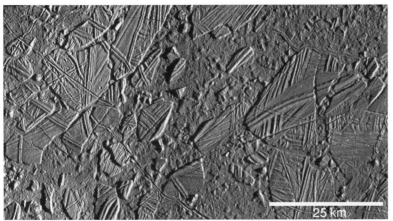

FIGURE 19.10 Conamara chaos. Disrupted, iceberg-like features are found in chaotic terrain. This image shows part of Conamara chaos, the defining feature of this type. Blocks of surface material seem to have broken apart, translated and rotated, and then refrozen in their new surface locations. The blocks can be reassembled like a jigsaw puzzle based on their shapes and features on their surfaces.

The "Puddle" is an intriguing, 3-km diameter feature (Fig. 19.11) where fluid-like material appears to have embayed (covered) pre-existing ridges and other terrain features. Underlying surface features are also visible inside the Puddle, suggesting that it consists of a thin surface veneer of fluid material that filled in low-lying regions, leaving ridges partially visible. The Puddle is one of only a few surface features exhibiting signs of cryovolcanism or other flow-like mechanisms. Small impact craters on the Puddle are secondary craters made by material thrown out by a large primary impact crater. No other geologic activity seems to post-date the formation of the Puddle.

19.7.5 Craters

Europa's surface has very few large impact craters, as mentioned before, and those that are present are different from similarly sized craters elsewhere in the Solar System. The morphology of Europa's impact craters (see Fig. 19.8 for an example) suggests that they formed within a solid target, but their shallow depths imply that they have relaxed since their formation (Moore et al., 1998). Such models suggest that most craters on Europa formed within a 5–15 km thick brittle surface layer, overlying a lower-viscosity subsurface layer. This subsurface material, however, could either be liquid water or warm, low-viscosity ice. More recent models attempt to use more subtle morphological details visible in Europa's craters to obtain better estimates of ice layer thickness. Turtle and Pierazzo (2001) modeled melt production and suggest that the ice crust must be at least 3–4 km thick to support the central peaks usually present. Schenk (2002) suggests that the ice shell could be 20 km or thicker, based on an interpretation of large, multi-ringed craters.

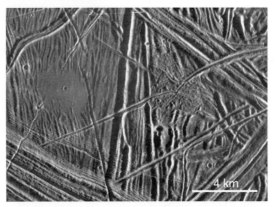

FIGURE 19.11 The "Puddle". The circular feature on the left side of this image, dubbed the "Puddle," is one of the few examples of a feature on Europa that could be the result of a cryovolcanic flow or eruption. The feature seems to embay (cover) the pre-existing terrain. Several small impact craters are visible on its smooth surface.

19.8 Heat transport and physics of the ice shell

Interpreting Europa's geology requires an understanding of the physics of Europa's ice shell. The deformation and flow of solids (*rheology*) is strongly dependent on temperature. Therefore, understanding Europa's surface features requires some knowledge of the ice shell's thermal structure, which in turn depends on how heat is applied to, generated within, and lost from the ice shell.

There are a variety of mechanisms for energy transport relevant to planetary bodies. *Conduction* is the transfer of energy by molecular collisions, and often dominates heat transfer in solids. It is a diffusive process in which molecules transfer their kinetic energy via collisions with other molecules. Heat will be conducted wherever temperature gradients are present. *Convection*, by contrast, is transfer of energy via mass motion within the medium. Both conduction and convection seem plausible as means for heat transfer through Europa's ice shell. The temperature gradient within a shell whose surface temperature is ∼100 K and whose base temperature must be that of Europa's ocean, ∼270 K, will lead to heat conduction upwards from the warmer to colder ice. But it will also encourage convection, since the warm deeper ice will, because of thermal expansion, be more buoyant than the overlying colder ice. This creates a gravitationally unstable configuration that could lead, over geological timescales, to the rising of the warm ice and sinking of the cold in adjacent convection cells. Many geological materials are *viscoelastic*, meaning that they behave as an elastic solid on short timescales (e.g., Earth's mantle rock behaves elastically on timescales shorter than a few hours) but as a viscous fluid on long timescales (for Earth's mantle, timescales longer than 10^4 yr).

The physics of conduction and convection are discussed in Appendix 19.7. Proposed subsurface temperature profiles for Europa vary greatly depending on which meachanism dominates heat loss. Candidate profiles in Fig. 19.12 have implications for europan geology (since they affect the depth at which ice becomes sufficiently warm to be ductile) and for prospects for future remote subsurface probing within the ice shell (Section 19.9).

Other mechanisms for energy transfer are also important. *Radiation equilibrium* is the balance between energy absorption (from the Sun) and emission (roughly like a black body) that ultimately determines the temperature of Europa's surface. In addition, *tidal heating* acts to heat Europa's interior through energy transport from Jupiter's rotational kinetic energy and Europa's orbital energy (Appendix 19.3).

Models of the thermal state of Europa's icy crust show that solid-state convection will begin if there exists a thin, cold, brittle, stagnant lid over a sufficiently thick, warmer, ductile ice sublayer (McKinnon, 1999). The Rayleigh number Ra provides a criterion for the onset of convection – when conditions are such that Ra is greater than some value Ra_{crit}, convection occurs. The derivation of Ra and Ra_{crit}, and often overlooked assumptions in this derivation, are discussed in Appendix 19.7.

Early models suggested that if the ice layer were thick, solid-state convection would ensue, and the resulting heat transfer would rapidly freeze any ocean solid. Squyres *et al.* (1983) found the ice thickness

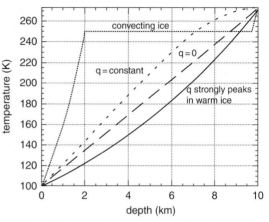

FIGURE 19.12 Theoretical temperature profiles in Europa's ice. The total ice layer thickness is assumed to be 10 km. The temperature profile labeled "convecting ice" is appropriate for an upwelling plume; descending plumes would be far colder. The volumetric heat dissipation rate *q* varies for different tidal heating models. (From Chyba *et al.*, 1998.)

needed for the onset of convection to be about 30 km. Their models combined three sources of heating to maintain Europa's liquid water ocean. Both tidal heating and radiogenic heating due to the decay of radioactive elements were taken to produce heat in Europa's core, combining to provide a heat flux at the base of the ocean of 16 erg cm^{-2} s^{-1} from the core. Tidal heating was taken to be uniform throughout the ice shell, and was the dominant heating term at 28 erg cm^{-2} s^{-1}. Thus the ice shell was heated from below by the tidal and radiogenic heating in the core (conveyed through the ocean) and from within by tidal heating. These effects combined to keep the ice layer thin, maintaining a liquid water ocean that lost its heat through conduction. Although available heating was insufficient to melt an ocean of water if ever it froze, if Europa started with liquid water, the ocean could be maintained as a liquid through geological time.

Ojakangas and Stevenson (1989) extended this work to the case of more realistic ice rheologies in which tidal heating *q* within the ice shell depended on the product of the stress and strain rate; in this case *q* depended exponentially on temperature (see Appendix 19.7). More recent work has argued that convection is in fact the dominant mechanism for heat loss, and that conductive models alone cannot explain some of the surface features (Pappalardo *et al.*, 1998). The thickness of an ice layer needed for the onset of convection has also been reconsidered. The viscosity of ice varies not only as a function of temperature, but also depends on strain rate, grain size, and other parameters.

Pappalardo *et al.* (1998) calculate that convection will occur if the subsurface layer is at least 2–8 km thick. They took the convecting layer to reside beneath a (geologically inferred) brittle *lithosphere*[11] less than 2 km thick. The observed 5–20 km spacing of supposed diapiric upwellings also supports a 2–8 km thick convection zone, since standard theory indicates that surface features should be about twice as far apart as the thickness of the convection zone.

It has been suggested that a change in ice thickness with time can be inferred from Europa's surface (Pappalardo *et al.*, 1998; McKinnon, 1999), since features such as the lenticulae, perhaps associated with a thicker crust, often cross-cut (and are therefore younger than) ridges and tectonic features that are perhaps associated with a thinner crust. Changes in ice thickness through time are in fact predicted by coupled thermal–orbital evolution models of Europa (Hussmann and Spohn, 2004).

Virtually all geophysical models for Europa now include a subsurface liquid ocean. This is because (a) tidal heating theory and magnetometer data provide strong constraints, and (b) maintaining the warm ice needed for convection *without* ever creating a liquid water layer seems like a more extraordinary, fine-tuned set of conditions than the presence of a liquid ocean.

19.9 Future means for detecting an ocean

Future missions, especially with a Europa orbiter, could determine with certainty whether a europan ocean exists (Cooper *et al.*, 2002). An orbiting laser altimeter should be able to track details of Europa's deformation over each orbit. If a global subsurface ocean exists, the amplitude of Europa's time-varying tidal bulge as its distance from Jupiter slightly varies over an orbit will be as great as ~30 m, as opposed to only ~1 m if instead the water is all frozen down to the mantle (Moore and Schubert, 2000). Accurate spacecraft tracking would also allow measurement of higher-order terms of Europa's gravity field (Section 19.6.1; Appendix 19.1) by measuring the orbiter's response to the varying gravity field caused by the tidally deforming bulge. These measurements should also provide additional information about the satellite's internal structure.

A more direct approach would be to orbit an ice penetrating radar. Detailed modeling indicates that a 20 W radar operating at 50 MHz could sound to depths of ~10 km in cold clean ice, but perhaps only to depths of several km if the ice is warm (~250 K) or if impurity concentrations are high (Chyba *et al.*, 1998; Moore, 2000). Fig. 19.12 shows that conductive ice temperature profiles are more favorable (because of colder temperatures to greater depths) than convective ice temperature profiles at the sites of upwelling plumes. However, very deep sounding might be possible at the locations of cold, descending plumes.

A Europa *lander* would present opportunities for different measurements. A single seismometer on the surface could passively monitor to determine the presence or absence of an ocean and estimate the thickness of the ice sheet (Kovach and Chyba, 2001). If two surface stations were available, coordinated seismic and magnetic measurements could not only detect the ocean, but also provide constraints on the ocean's thickness and on the deeper interior.

19.10 Habitability of Europa

Life as we know it on Earth depends on liquid water, a suite of "biogenic" elements (most famously C, but also H, N, O, P, S, Fe, and others) and a source of *free energy*, i.e., energy that is available and suitable to drive biological processes. As we have seen, it is likely, but not yet completely certain, that Europa harbors a subsurface ocean of liquid water whose volume is about twice that of Earth's oceans. Little is known about the inventory of carbon, nitrogen, and other biogenic elements, but lower bounds on their abundance can be placed by considering the role of cometary delivery over Europa's history (Pierazzo and Chyba, 2002). Sources of free energy may be limited for an ocean world covered with an ice layer kilometers thick (Reynolds *et al.*, 1983; Gaidos *et al.*, 1999), but it is possible that hydrothermal activity (McCollom, 1999) and/or organics and oxidants provided by the action of radiation chemistry at Europa's surface (Gaidos *et al.*, 1999; Chyba and Phillips, 2001; Cooper *et al.*, 2001; Chyba and Hand, 2001) and subsequent mixing into Europa's ocean could power a europan ecosystem. In this section, we assume the presence of liquid water in Europa's subsurface, and examine Europa's suitability for life from the perspective of biogenic elements and free energy. Figure 19.13 illustrates many of the processes discussed below.

---

[11] *Lithosphere* here refers not to rock, as its etymology would suggest, but to the brittle upper layer of the ice shell.

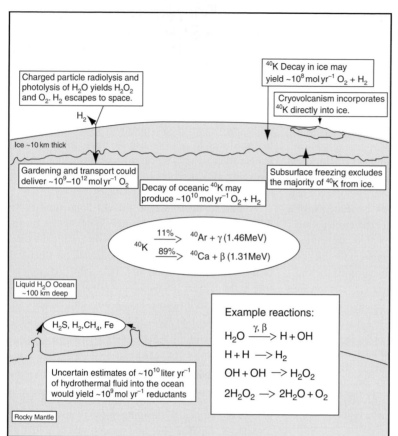

FIGURE 19.13 Free energy sources for life on Europa. Radiation effects on Europa's ice and liquid water could provide the chemical disequilibrium to fuel a biosphere. (After Chyba and Hand, 2001.)

We will argue in this section that on the basis of what we currently know, Europa's putative subsurface ocean could be habitable by microorganisms with attributes similar to those known to exist on Earth. Habitability for extant life is not the same as suitability for the origin of life, as we have emphasized elsewhere (Chyba and Phillips, 2001), and will examine in the following section. However, Europa's apparent habitability presents an example of a world that is very different from one satisfying the traditional definition of habitability that relies on stable surface liquid water (Chapter 4). It appears likely that the biological arena may be much broader than that suggested by the traditional *habitable zone* definition, namely those circumstellar distances over which liquid water is possible on a planetary surface (Chapter 3; Sagan, 1996).

19.10.1 Biogenic compounds

In Section 19.5 we summarized the spectroscopic evidence for simple organic functional groups at Europa's surface, as well as the expectation that radiation acting on CO_2 in Europa's ice should produce small amounts of simple organics such as H_2CO. A common default model for elemental composition of outer Solar System objects is that of a carbonaceous chondrite meteorite (perhaps with additional volatiles added); these meteorites are typically several percent organic by mass (Section 7.4.1). Such models for Europa's initial complement of organics (e.g., Kargel *et al.*, 2000) must be viewed with caution, however, because the circumjovian nebula in which Europa formed 4.6 Ga has a poorly known composition, and it may well have contained far less carbon, nitrogen, and other volatiles than carbonaceous chondrites.

Nevertheless, even if Europa was initially depleted in these biogenic elements, over Solar System history it would have acquired a considerable inventory of them via cometary impacts. While much material from such impacts would be lost due to Europa's low escape velocity, a significant fraction of lower velocity cometary material would be retained. Integrated over the history of the Solar System, Europa should have accumulated $\sim 10^{15}$ g of carbon in this way (Pierazzo and

Chyba, 2002) – ~0.1% of the carbon present in Earth's total mass of living organisms (*biomass*). If the CO_2 that is spectroscopically observed at a level of 0.2% by weight (Section 19.5) were present throughout Europa's ice shell, ~10^{20} g of C would be present in the ice alone, ~10^2 times more C than in Earth's biomass.

Because Europa lacks an atmosphere,[12] "soft landings" of interplanetary dust particles or small meteorites, sources that may have been important for the prebiotic organic inventory of early Earth (Section 3.8) or for the successful transfer of microorganisms between Earth and Mars (Section 18.4.4), will not occur on Europa. Delivery of intact organic molecules or viable microorganisms will therefore be far more difficult. Moreover, the probability that material launched from Earth or Mars by impacts would reach Europa is far smaller than the probability of that material exchanging between Earth and Mars. If life exists on Mars, it possibly shares a common ancestor with life on Earth, due to microbial transfer in meteorites; such common ancestry with terrestrial life is much less likely for life on Europa.

19.10.2 Photosynthesis

Photosynthesis on Europa is constrained by Europa's thick ice cover. It has been suggested that niches might exist within Europa's ice shell where transient near-surface liquid water environments could permit photosynthesis or other metabolic processes. Some models have as their context cracks or melt-throughs in a relatively thin, conducting ice shell (Reynolds *et al.*, 1983; Greenberg *et al.*, 2000), whereas another relies on dissipative heating in Europa's cracks (Gaidos and Nimmo, 2000) and works with either a conductive or convective model.

Were photosynthesis possible at Europa, for example if Europa's cracks permitted liquid water environments to reach sufficiently close to the surface, the energy available for biology from this source would likely swamp all others. Bacterial photosynthesis on Earth ranges over the wavelength range 360 nm to 1020 nm (Section 13.3); over this range nearly two orders of magnitude more energy is incident on Europa (globally averaged) in the form of photosynthetically useful photons than is incident in charged particles (Cooper *et al.*, 2001). This suggests that were life to exist on Europa,

and were it to have access to near-surface liquid-water environments, there would be a strong selection pressure for the evolution of photosynthesis at any near-surface niches (such as cracks). The possibility that the reddish brown coloration of Europa's cracks or other disrupted terrain might be due to such microbial colonization has been a topic of speculation in the literature (Dalton *et al.*, 2003). However, without excluding this possibility, we caution that it is easy to attribute unexplained features to biology, since life's potential characteristics are poorly constrained.

This is not a new caution. Pollack and Sagan (1967) warned against the temptation to attribute enigmatic martian albedo features to biology: "The varieties of life on Earth and of possible organisms that we can imagine on Mars are so large that virtually any change in surface features can be 'explained' by a necessarily vague attribution to biological processes." In our view this does not mean that biological explanations are out of bounds, but rather that they require strong evidence before being accepted.

19.10.3 Hydrothermal vents and chemical disequilibria

Microbial life on Earth often does not directly harvest solar energy, but obtains its energy from other chemical disequilibria in the environment, bringing together an electron donor (a "fuel") with an electron acceptor (an "oxidant") to liberate energy (Chapters 6, 8, and 11). On Earth, photosynthesis, coupled with organic carbon burial, produces oxidizing surface conditions that provide chemical disequilibria for biology to exploit. Gaidos *et al.* (1999) emphasize the difficulty of identifying sources of chemical disequilibrium on an ice-covered world lacking photosynthesis, and the corresponding difficulty of sustaining a large biomass on Europa. Yet the Earth itself must have hosted a biosphere prior to the evolution of photosynthesis, so the absence of photosynthesis cannot be an insurmountable obstacle to maintaining chemical disequilibria that can provide the electron donors and acceptors to power life. Indeed, there may be elements of Earth's deep subsurface biosphere, for example hydrogen-consuming methanogens, that are entirely independent of surface photosynthesis (Chapelle *et al.*, 2002).

Early models for heating in Europa's interior suggested that the heat flow at the base of the ocean might be 24 mW m^{-2} (Squyres *et al.*, 1983). For comparison, the average heat flow at the surface of Earth's Moon is 29 mW m^{-2}, one-third of Earth's. Since the Moon

---

[12] As discussed in Section 19.12.1, Europa may have had an atmosphere for its first 10^5 yr, but this would seem to be too short an interval for life to both originate somewhere else and undergo transport to Europa.

is geologically dead, this simple comparison seems discouraging for hydrothermal activity at the base of Europa's ocean. However, more recent tidal models for Europa suggest the possibility of heat flows as high as $\sim$200–300 mW m^{-2}, in which case hydrothermal activity would be expected. In these models a liquid inner core for Europa allows more tidal flexing and so more heat production in the mantle, and the resulting large heat flows produce partial melting, resulting in volcanism at the interface between the ocean and mantle (McKinnon and Shock, 2001). If, however, the convective heat transfer in Europa's mantle is efficient enough to prevent partial melting of silicates, volcanic activity might be precluded.

If hydrothermal vent activity does exist on Europa, venting of CO_2 to the ocean seems plausible (McKinnon and Shock, 2001). This CO_2 could then serve as an electron acceptor, e.g., for molecular hydrogen produced by water acting on basalt. The methane-producing biomass that could be supported by such a system (Chapelle *et al.*, 2002) is difficult to calculate, but is likely small. McCollom (1999) estimates a potential annual biomass production of $\sim$10^8–10^9 g yr^{-1} (using a dry biomass production energy efficiency φ $\sim$0.1 g kcal^{-1}), vastly less than Earth's primary production based on photosynthesis of $\sim$10^{17} g yr^{-1}. The extent to which such an ecosystem, lying at the base of a 100-km deep ocean, would be detectable by measurements at Europa's surface would depend on the amount of vertical mixing in Europa's ocean. The model of Thomson and Delaney (2001) suggests that a great deal of mixing between ocean bottom and the ocean/ice interface might occur.

19.10.4 Chemical disequilibria from particle radiation and radioactive decay

A number of authors have discussed the production of oxidants (such as molecular oxygen and hydrogen peroxide) in Europa's uppermost ice layers due to charged-particle bombardment of water ice (Cooper *et al.*, 2001). Since CO_2 is also present in the ice (McCord *et al.*, 1998), there should be a small simultaneous production of simple organics such as formaldehyde (Chyba and Phillips, 2001). The mixing of oxidants and organics into the europan ocean over the $\sim$50 Myr lifetime of Europa's ice shell would then make chemical disequilibrium energy available to the ocean.

The extent to which organics and oxidants produced at Europa's surface are available to power an oceanic ecosystem depends on the competition among

charged particle and ultraviolet processing of the surface, sputtering and gardening (Section 19.3), and the communication of the surface layers with the oceans via geological processes. There are many uncertainties in such a calculation; estimating a biomass production rate further requires that the calculated energy available from fuel-oxidant reactions be combined with estimates of the efficiency φ of converting energy into microbial biomass (dry weight). Biomass production rates given below allow for an order of magnitude uncertainty in φ, ranging $\sim$0.1–1 g kcal^{-1} (McCollom, 1999; Chyba and Phillips, 2001).

The more conservative estimates suggest that Europa's oceanic biomass could be limited to $\sim$10^{22}–10^{24} prokaryote-like cells. In these conservative scenarios, it would be difficult to supply the ocean with enough radiation-produced oxygen to reach levels capable of sustaining analogues to terrestrial macrofauna (Chyba and Phillips, 2001), which require O_2 concentrations above $\sim$20 μM. In this case, any europan ecosystem would be limited, assuming the validity of this analogy to terrestrial biology, to microscopic life. Estimates that assume that the available ice substrate keeps pace with the intense radiation bombardment – so that the uppermost ice layers are being replenished and substrate limitation is not a factor – can increase the O_2 concentration by factors of 10^3 or more, comparable to that on Earth (Cooper *et al.*, 2001), and seemingly permit macrofauna.

Conductive models involving melt-through (Greenberg *et al.*, 2000) and convective models in which such nutrient-rich ice is circulated in the solid-state down to the ocean (Chyba and Phillips, 2001) may permit the upper meters of Europa's surface to mix with Europa's ocean. Barr and Pappalardo (2003) suggest that downward convective circulation in the ice could permit material from Europa's stagnant lid to be transported to the base of the ice shell in 10^4–10^7 yr, perhaps including the upper meter of ice. If Europa has hydrothermal vents that introduce reduced fluids into the ocean, then the production of organics or hydrogen in Europa's ice is not needed, and the oxidants would be provided by radiation chemistry at the surface. This difference in oxidation state could provide the chemical energy needed by chemolithoautotrophic microbes, as it does in vent systems on Earth (Chapters 8 and 14). The availability of O_2 could also allow a metabolism based on iron reduction and oxidation (Schulze-Makuch and Irwin, 2004).

Chyba and Hand (2001) have considered additional sources of oxidants and hydrogen on Europa. In particular, they consider simultaneous production of oxygen and hydrogen in Europa's bulk ice and in the ocean itself due to the decay of radioactive ^{40}K, using the Kargel *et al.* (2000) model to estimate the concentration of potassium in Europa's ocean. This model assumes total extraction of salts from a carbonaceous chondrite precursor. Zolotov and Shock (2001) view this as unlikely, and present a range of other models; their favored model gives K concentrations $\sim$30 times lower than that used by Chyba and Hand (2001). Taking this uncertainty into account, these processes could allow a biomass production of $\sim$10^8–10^{12} g yr^{-1}. Sources of oxidants to Europa's ocean from the variety of radiation mechanisms considered here are summarized in Fig. 19.13.

Biomass production rates can be converted into estimates of steady-state biomass by using estimates of the biomass turnover time. For a turnover time of $\sim$10^3 yr, perhaps appropriate for Earth's deep biosphere (Whitman *et al.*, 1998), the above calculation permits a steady-state biomass of $\sim$10^{11}–10^{15} g, compared with the terrestrial biomass of $\sim$10^{18} g.

19.10.5 Other sources of free energy

Other sources of free energy for life on Europa have been proposed in addition to photosynthesis and chemical disequilibrium. Scenarios for life based on electrical energy and thermal energy (Reynolds *et al.*, 1983), as well as osmotic energy, kinetic energy, magnetic energy, gravitational energy, and piezoelectric energy have been considered (Schulze-Makuch and Irwin, 2004). Schulze-Makuch and Irwin conclude that osmotic and thermal gradients, as well as the kinetic energy of convection currents, represent plausible alternative sources of energy for life on Europa, although there are no known organisms on Earth that obtain their energy in these ways.

19.11 Europa as a possible analogue for early earth

Europa is of astrobiological interest primarily because of its potential as a habitat for extraterrestrial life. However, it may also provide a model for the early Earth at the time of the origin of terrestrial life. Stars like the Sun grow more luminous through time as the hydrogen in the star's core is converted to helium. Our Sun was $\sim$25% less luminous 4 Gyr ago at the time of the origin of life, with the result that Earth's surface would likely have been frozen over unless there

was a much stronger greenhouse effect at that time (Section 4.2).

One popular, but uncertain, solution to this "early faint Sun paradox" is to invoke much higher levels of the greenhouse gas CO_2 in early Earth's atmosphere. However, although this remains the favored resolution, there are reasons to remain skeptical about this solution (e.g. Sagan and Chyba (1997); Section 4.2 argues for CH_4 as the early greenhouse gas). A different possible resolution to the problem is instead to envision the early Earth as a world whose mean average temperature was only $\sim$237 K. The resulting "Snowball Earth" may also have occasionally reappeared throughout terrestrial history (Section 4.2.4).

If this model of the early Earth is correct, somehow life must have originated and then flourished, even under these conditions (Bada *et al.*, 1994). Europa provides a Solar System analogue for this model. Earth's early oceans, like those of Europa today, would not have frozen solid since the geothermal heat flux would have limited the thickness of the oceans' ice cover to an average 300 $\pm$ 100 m (via Eq. (19.25) in Appendix 19.7). Tidal heating of the ice itself would have been negligible since ice this thin is brittle, and only ductile ice can deform and be tidally heated. The challenge is to understand how the origin of life could have occurred either on such a Snowball Earth or perhaps on Europa.

19.12 An origin of life on Europa?

In Section 19.10 we argued that Europa's ocean appears habitable by microorganisms, but of course habitability by analogies to extant Earth organisms is not the same thing as life being present. Since it is unlikely that organisms have been successfully transferred between Earth and Europa (Section 19.10.1), life on Europa likely exists only if a separate origin of life took place there. The prospects for this are difficult to evaluate, especially given our limited understanding of the origin of life on Earth.

19.12.1 An early surface ocean?

One model for the formation of Ganymede, Callisto, and Titan suggests that these moons began with an atmosphere rich in water vapor and a correspondingly powerful greenhouse effect; however, in this model the resulting surface liquid water ocean rapidly cooled below the freezing point in <10^5 yr (Kuramoto and Matsui, 1994). Unless further modeling indicates that the timescale for the existence of a surface ocean on Europa could be substantially longer, or that the origin of life on Europa could happen within this extremely short period, any life that originated on Europa must have done so beneath Europa's ice cover.

Conditions on Europa shortly after its accretion may have been different in other ways important for its habitability or for the origin of life, but these effects are difficult to evaluate. Radioactive heating in Europa's core should have been several times higher than today, which (everything else being equal) should have favored a thinner ice layer. Similarly, more left-over accretional heat should have been present, as well as heat from Europa's initial rapid despinning (Appendix 19.1). The cometary impact flux may also have been substantially higher, though models for the decay of the impact flux in the early outer Solar System are uncertain. Perhaps most important, the timing of the commencement of Europa's orbital resonances with Io and Ganymede, and the implications for tidal heating, is controversial (Appendix 19.4).

19.12.2 An origin of life in the subsurface?

If life in general cannot originate at depth, then only worlds that have clement surfaces sufficiently long for life to originate (Earth, perhaps Mars) can host endemic biologies, although interplanetary transfer of microorganisms remains possible. If life can originate at depth, however, then Europa might harbor its own endemic biology. Certain prebiotic processes under hydrothermal conditions may have been important in the origin of Earth's life (Chapters 8 and 14), but it remains unclear whether the entire origin of life could have proceeded in the absence of sunlight.

Just as Titan and Europa may provide partial analogues to early Earth environments of relevance to the origin of life, so carbonaceous chondrite meteorites may provide clues into the origin of life at depth. Most researchers conclude that these meteorites experienced liquid water, based on their mineralogy and the standard interpretation of their amino acid chemistry (Section 3.9.2). Asteroid thermal history models suggest that large asteroids ($\sim$100 km diameter) could maintain liquid water interiors for $\sim$10^8 yr, but the best-studied carbonaceous chondrite, the Murchison meteorite, probably experienced water for much less time than this, perhaps only $\sim$10^4 yr (Chyba and McDonald, 1995). Such meteorites thus present actual examples of prebiotic organic synthesis in subsurface hydrothermal environments – they experienced liquid

water, potentially catalytic mineral surfaces, and abundant organic molecules. Yet while over 75 types of amino acids are present in Murchison (Tables 7.1 and 7.2), only very low concentrations of even two-amino-acid chains have been found.

19.12.3 Advantages of ice

A number of authors have remarked that low temperatures (provided liquid water exists) may favor the origin of life. Bada et al. (1994) and Bada and Lazcano (2002) emphasize the enhanced stability of prebiotic organic molecules at low temperatures. For example, hydrogen cyanide (HCN) is often viewed as a critical precursor to the synthesis of both amino acids and nucleic acid bases, yet its half-life against hydrolysis is only a few years at 50 °C but 10^4–10^5 yr at -2 °C. Steady-state concentrations of key molecules would thus have been highest in a cold ocean or lake. Moreover, freeze–thaw cycles could have concentrated prebiotic reactants in spaces between ice crystals, offsetting the slower reaction rates at lower temperatures. There is now direct experimental evidence for this in the context of nucleic acid synthesis (Kanavarioti et al., 2001). Similarly, Levy et al. (2000) find that some nucleic acid bases and amino acids are synthesized in substantial yields in dilute solutions of NH$_4$CN held at both -20 °C and -78 °C for 25 years. These authors suggest that this demonstrates the potential for prebiotic chemistry on Europa or an ice-covered early Earth.

19.12.4 Effects of a salty ocean

Some models for Europa's ocean (Kargel et al., 2000; Zolotov and Shock, 2001) imply that its contemporary salt concentration may be high enough to impede the origin of life (Hand and Chyba, 2007). Monnard et al. (2002) have shown that two processes relevant to the origin of life, membrane self-assembly and RNA polymerization, are both impeded by ionic solute concentrations much lower than even those of Earth's contemporary ocean (Section 19.6). They conclude that the origin of cellular life on Earth more likely occurred in a freshwater rather than a marine environment. For example, NaCl concentrations in the range of 80 mM (14% of present-day seawater) are sufficient to halve the longest RNA *oligomers*[13] formed in their

---

[13] An *oligomer* is a relatively short polymer, in this case containing a small number of nucleotides.

experiments; concentrations around 40 mM reduce the total yield of oligomers synthesized by a factor of two. (They do not, however, present data for $MgSO_4$.) They point out that freezing mechanisms that act to concentrate organics may also concentrate salts due to the exclusion of salts from the freezing ice matrix (Chapter 15), suggesting that freezing in a salty ocean, as discussed above, may cause as many problems as it solves. However, melts that occurred within previously existing ice, perhaps due to tidal mechanisms (e.g., Gaidos and Nimmo, 2000; Sotin *et al.*, 2002), may have low salinity because the ice already excluded most salts during its initial freezing.

Hand and Chyba (2007) derive constraints on the salinity of Europa's ocean, and the thickness of its ice cover, based on Galileo magnetometer measurements combined with geophysical models (Section 19.6). They find that these constraints are best met by (a) a global layer of water with near-saturation $MgSO_4$ or sea-salt (mostly NaCl) concentrations ($\sim$300 g salt per kg H_2O), and (b) a europan ice shell less than 10 km thick. For comparison, Earth's oceans have a salinity of $\sim$35 g sea salt (mostly NaCl) per kg H_2O. Whether Europa's ocean salinity would prevent the origin of life in that ocean depends on the identity of its conductors, any changes in the ocean over time, and the poorly known details of the prebiotic reactions actually required for the origin of life.

19.13 Searching for life

Any search for life on a distant world must establish criteria for what would qualify as evidence of success. But no general definition of life has been accepted by the scientific community (Chyba and McDonald, 1995). There appear to be fundamental philosophical reasons having to do with the nature of definition and our current understanding of life for why no compelling definition is currently possible (Chapter 5). This has implications for how we should search for life at present, as does the one historical example we have of the search for extraterrestrial life from a spacecraft, that conducted by the Viking project in 1976 (Sections 18.5.2 and 23.2).

19.13.1 Possible missions

We regard europan exploration as comparable in importance to that of Mars, demanding a systematic program of exploration. Mars is a target of such importance that its exploration requires multiple missions over many decades; these missions are interwoven in a way that incorporates new knowledge into an existing program. Missions to Europa are necessarily more challenging and expensive than missions to Mars, but a program of europan exploration need not launch to Europa every two years, which is the current pace of Mars exploration. A wide variety of missions to Europa has been envisioned.

The first lander on Europa – which may well be preceded by an orbiter or flyby mission (Cooper *et al.*, 2002) – should touch down at a site where liquid water from Europa's subsurface may have recently reached the surface. The locations, or even existence, of such sites are difficult to determine with confidence on the basis of current knowledge, but if we had to choose a landing site now, we might well choose a feature like the Puddle described in Section 19.7.5 (Fig. 19.11). Information from an orbiter mission would obviously be of great help in answering these questions.

A search for subsurface europan life via measurements performed on Europa's surface should examine ice from the youngest possible surface. Prior to or simultaneous with any experiments to search for life, chemical context should be established. This is a clear lesson from the Viking missions to Mars (Section 18.5.2; Chyba and Phillips, 2002). Measurements would include determining the abundance of cations and anions, salinity, pH, volatiles (O_2, CO_2, CH_4, etc.), and organic molecules. The last is probably the highest priority "life-detection" experiment to be performed. Additional experiments could include high-sensitivity searches for specific indicative organic molecules (such as amino acid enantiomers), determination of key stable isotope ratios (e.g., $^{12}C/^{13}C$), or fluorescent microscopy. Searches for organic molecules should be made at the greatest depth into the ice as possible, because laboratory experiments suggest that radiation near the surface will destroy the less stable biomolecules on short timescales (Bada and Lazcano, 2002).

The biological models described in Section 19.10 suggest that europan biomass may be limited. This suggests that any search for life on Europa from a lander should either survey a large amount of material so as to choose particular locations for subsequent high-sensitivity investigations, and/or take advantage of the opportunity to concentrate samples by melting and filtering ice. Chyba and Phillips (2002) have discussed the energy requirements for the latter strategy on Europa, and consequent limits on how much material might be processed.

Finally, even without landing it would be possible to sample the surface with a mass spectrometer to analyze either material sputtered off the surface (Johnson, 1998) – though this material will be the most heavily radiation processed – or blown off the surface by an artificial impact (Cooper *et al.*, 2002). Such a flyby mission would have the advantage of being less technically demanding and much less costly than an orbiter or lander.

19.13.2 Preventing contamination of Europa

There is both a legal requirement and a practical scientific objective that any environment to be searched for life not be contaminated with "false positives" inadvertently brought along from Earth. We also believe that "planetary protection" requirements flow from an ethical obligation to avoid significantly contaminating possible alien biospheres (see Chapter 25 for a full discussion).

It is credible that microbes within a terrestrial spacecraft could reach Europa and remain viable. Experiments conducted aboard a number of Earth-orbiting spacecraft indicate that a variety of common terrestrial bacteria can withstand the space environment for as long as six years provided they are shielded from ultraviolet light (Horneck, 1993). After travel through interplanetary space, the microbes would then have to (1) survive the jovian radiation environment (Section 19.4.1) during flyby or while in orbit around Europa, (2) be delivered to Europa's surface, (3) be buried in Europa's ice quickly enough so that at least some would survive the surface radiation environment, and (4) be transported into Europa's ocean (either by cracking, melting, descending plumes, or some other mechanism; see Section 19.7). Finally, and perhaps most challenging of all, (5) at least one of these organisms would then have to be able to survive and reproduce in the ocean.

In the 1970s values for such probabilities were adopted for Mars, but then later criticized as subjective and uncertain by the US National Research Council's Task Group on Planetary Protection (Space Studies Board, 1992). For Mars missions the 1992 study adopted mission categories that in general required Viking lander presterilization (clean room) requirements and, for missions that would search for life, Viking lander sterilization requirements (which added day-long baking at 110 °C) (Section 25.6). A later Task Group on the Forward Contamination of Europa attempted to quantify the probabilities for four classes of microorganisms. Majority opinion was that a more

quantitative and potentially more stringent requirement was needed: that the probability be 1 in 10,000 for any mission to contaminate a europan ocean with a viable terrestrial microbe (Space Studies Board, 2000). They argued that this criterion can be met even with minimal ground-based sterilization of the spacecraft, largely because of the post-launch sterilization effects of Europa's radiation environment.

Critics of this report suggested that the requirement should be based on a more transparent criterion, such as requiring the probability of contamination to be "substantially smaller than the probability that such contamination happens naturally." This could prove an extremely stringent requirement in the case of Europa, a world to which viable microbial transfer in terrestrial meteorites is likely extremely difficult (Section 19.10.1). However, the probability of natural contamination still needs to be carefully modeled. We are unlikely to launch more than ten missions (and probably fewer) to Europa before we either determine that a biosphere exists or lose enthusiasm for further europan exploration, so that the 10^{-4} contamination probability requirement (per mission) would mean that the chances of contaminating Europa for all missions would be $\sim$0.1%.

19.14 Europa and astrobiology

In this chapter, we have tried to distinguish between what is known with confidence, what is likely to be true, and what is speculative. It seems virtually certain that Europa has a $\sim$100-km thick layer of H_2O on its outer surface. It is likely that most of this water is liquid, comprising an ocean whose volume is roughly twice that of Earth's oceans. The evidence for this picture is strong but still indirect. Europa seems likely to be habitable in the sense that it probably has liquid water, biogenic elements, and sources of free energy capable of sustaining life. Europa also provides an analogue to at least one class of models for the Earth at the time of the origin of life. Europa thus joins Mars as the two most likely environments for extraterrestrial life in our Solar System. In one sense, the possibility of life on Europa seems the more intriguing, since were it present it would likely represent an origin separate from life on Earth.

None of this means, of course, that there is in fact a europan biosphere. It is extremely difficult, given our poor understanding of the origin of life, or the prospects for the origin of life in subsurface environments, to estimate the likelihood of the origin of life on

Europa. The only way we will discover whether there is life on Europa is to go there and find out.

Appendix 19.1 Satellite tides and synchronous rotation

This appendix explains the physics of Europa's tidal torques and how they lead to synchronous rotation.

A point P on Europa experiences a force $\mathbf{F}_J$ due to Jupiter's gravity corresponding to an acceleration $\mathbf{a}_J = GM_J d^{-3} \mathbf{d}$, where G is the gravitational constant, M_J is Jupiter's mass, and $\mathbf{d}$ is the vector distance from the point P to the center of Jupiter. (Boldface type denotes vector quantities.) P also experiences a centripetal acceleration $\mathbf{a}_c$ as it, along with Europa, orbits the system's center of mass. The difference between the two accelerations, $\mathbf{a}_T \equiv \mathbf{a}_c - \mathbf{a}_J$, is called the *tidal acceleration*, and it is not difficult, though trigonometrically tedious, to calculate $\mathbf{a}_T$ explicitly (e.g., Murray and Dermott, 2001; Section 4.2). Because $\mathbf{F}_J$ falls off with the square of the distance, the side of Europa nearer Jupiter experiences $|\mathbf{a}_J| > |\mathbf{a}_c|$, with the situation symmetrically reversed on the far side where $|\mathbf{a}_J| < |\mathbf{a}_c|$. The magnitude of the difference is greatest at the points nearest and farthest from Jupiter. The result is that Europa's figure experiences a position-dependent tidal force that distorts it into a prolate spheroid with its major axis oriented toward Jupiter. (Put another way, Europa is stretched along the radial line to Jupiter because the *difference* in $\mathbf{F}_J$ experienced by one point at a distance x from Jupiter's center and another at a distance $x + \delta x$ is proportional to $x^{-2} - (x + \delta x)^{-2} \approx 2(\delta x) x^{-3}$.) That is, the tidal bulge is symmetric on both the subjovian and antijovian sides of Europa, and is proportional to the inverse cube of the distance to Jupiter. The magnitude of the distortion depends not only upon $\mathbf{F}_J$, but also on Europa's gravity and rigidity, which resist the distortion.

Gravity is a conservative force, so the vector $\mathbf{a}_J$ can be written as the gradient of a scalar potential, as can the centripetal acceleration. Tidal calculations are usually framed in terms of gravitational potential functions; forces and accelerations can be determined by taking the gradient of their potential. Ignoring a constant, we can then express the tidal acceleration $\mathbf{a}_T$ as $\mathbf{a}_T = \nabla V_T$, where V_T is Jupiter's tide-raising potential. For spherical coordinates r and θ (measured from the Europa–Jupiter radial line), and europan radius r_e, to second order,

$$V_T(r_e, \theta) = -(GM_J r_e^2 / r_{eJ}^3) P_2(\cos \theta), \quad (19.1)$$

where $P_2(\cos \theta) = (1/2)(3\cos^2\theta - 1)$ is the second degree Legendre polynomial and r_{eJ} is the distance between the centers of Europa and Jupiter (Stacey, 1977).

The distortion of Europa's surface caused by the potential V_T is $\delta(\theta) = h_2 V_T(r_e, \theta)/g$ (giving a prolate spheroid), where g is Europa's gravitational surface acceleration and h_2 is the tidal Love number, which corrects for the rigidity of Europa and a second degree disturbance of the tidal potential due to the deformation itself. For a homogeneous sphere of radius r_e, density ρ, and rigidity μ, a measure of the force needed to deform an elastic body is:

$$h_2 = (5/2)[1 + 19\mu/2\rho g r_e]^{-1}.$$

Now consider Europa early in its history, when its rotational period was likely shorter than its orbital period about Jupiter. (If its rotational period were in fact longer, an analogous argument to the one given below would apply and give the same conclusion.) As each longitudinal section of Europa rotates past the Europa–Jupiter radial line, the tidal bulge of this section rises to maximum height and then falls. If Europa could deform instantaneously in response to the tidal potential, the tidal bulge would always be perfectly aligned with the radial line to Jupiter, and there would be no net torque acting on Europa's tidal bulge. Since no real material can deform instantaneously, however, Europa's elastic response lags behind the periodic tidal potential, resulting in an angular offset of the maximum bulge a bit past the subjovian point (Fig. 19.2). This offset tidal bulge produces a torque. Since gravity falls off with distance, the torque is slightly greater on the bulge nearer Jupiter (tending to slow down the rotation) than on the symmetric bulge on Europa's far side (tending to speed up the rotation). The sum of the torques therefore does not quite cancel, and acts to "despin" the satellite, i.e., slow down its rotation.

The distorted shape of Europa leads in turn to a distortion in its own gravitational potential field, introducing a quadrupole term

$$V_D(r, \theta) = k_2 V_T(r_e, \theta)(r_e/r)^3, \quad (19.2)$$

where k_2 is called the second-order gravitational Love number; for a homogeneous sphere, $k_2 = (3/5)h_2$. Europa's total potential at point (r, θ) then becomes $V_e = Gm_e/r + V_D(r, \theta)$. The magnitude of the torque Γ acting to despin Europa becomes

$$\Gamma = |\mathbf{r} \times \mathbf{F}_J| = rM_J(1/r)(\partial V_D/\partial \theta), \quad (19.3)$$

evaluated at $r = r_{eJ}$ and $\theta = \varepsilon$, where ε is the tidal lag angle (Fig. 19.2). One finds from Eqs. (19.1) to (19.3),

$$\Gamma = 3/2 \; k_2 G M_J^2 r_e^5 r_{eJ}^{-6} \sin \; 2\varepsilon. \qquad (19.4)$$

This torque acts to despin the satellite until its rotation period becomes identical to its orbital period ($\omega = n$), after which $\varepsilon = 0$ and the system is in equilibrium. The satellite is then in synchronous rotation with its primary (as is the Moon with the Earth). The despinning timescale τ for Europa to become spin-locked with Jupiter is given by dividing its primordial spin angular momentum $C\omega_0$ by the despinning torque: $\tau = C\omega_0/\Gamma \propto r_{eJ}{}^6 (\sin \; 2\varepsilon)^{-1}$, where $C = \alpha m_e r_e^2$ is the moment of inertia, ω_0 is the intial spin angular velocity, and $\alpha \approx 0.35$ for Europa (see Appendix 19.5).

The energy dissipation rate in Europa from its periodic tidal forcing is related to the lag angle ε. The effective tidal dissipation function Q^{-1} is defined as

$$Q^{-1} = \frac{1}{2\pi E_0} \oint (-\dot{E}\,) \mathrm{d}t \equiv \frac{\Delta E}{E_0}, \qquad (19.5)$$

where $\dot{E}$ is the rate of energy dissipation, E_0 is the peak potential energy stored in the tidal bulge, and the integral extends over one tidal period (Goldreich and Soter, 1966). Q^{-1} is the part, ΔE, of the total energy stored, E_0, dissipated in a complete tidal period.[14] Integrating $\dot{E}$ for a sinusoidally varying external potential with a phase lag ε in the amplitude of the tidal motion, one finds $Q^{-1} = \tan(2\varepsilon)$ (MacDonald, 1964; Section 6). Since Q is generally large, $Q^{-1} \approx \sin(2\varepsilon) \approx 2\varepsilon$. As we will see below, Q for Europa is a matter of some controversy, but $Q \approx 100$ may be a reasonable guess for icy satellites in general (Squyres *et al.*, 1985). Europa's despinning must then have been extremely rapid, causing substantial but short-lived early tidal heating due to dissipation, perhaps as short as $\tau \sim 10^3 \; (Q/100)$ yr.[15]

[14] Some authors define Q to be one-half the value used here.

[15] There are two different, and not entirely mutually consistent, formulations for Q in the literature. The approach used here dates to MacDonald (1964), and assumes that the satellite's tidal bulge is always offset from the line to the primary by a constant angle of magnitude ε whose sign depends only on the sign of the difference between ω and n. In a different formalism, called Kaula perturbation theory and dating back to George Darwin (son of Charles), coordinates of the perturbing potential are complicatedly but spectacularly transformed from spherical polar coordinates into polynomial functions of the orbital elements (Kaula, 1966).

Appendix 19.2 Tidal circularization of orbits and nonsynchronous rotation

Consider the effects of tidal heating (dissipation of energy) in a satellite that is spin-locked but in an eccentric orbit (Burns, 1976). The *specific* (per unit mass) *angular momentum H* for an orbit of semimajor axis a and eccentricity e is

$$H = [GM_J a(1 - e^2)]^{1/2}, \qquad (19.6)$$

and the *specific energy* of the orbit is

$$E = -GM_J/2a. \qquad (19.7)$$

The radial component of the tide (along the line connecting the centers of Europa and Jupiter) causes energy loss without causing a torque, that is, while conserving H. As E decreases due to energy dissipation, a must also decrease (the energy of the bound orbit becomes more negative), so that for H to remain constant, e must decrease as well (by Eq. (19.6)), leading to orbit circularization. For a given H, a circular orbit ($e = 0$) has the smallest a, so the minimum (largest negative) energy. Orbit circularization is opposed by the effect of torques due to tides raised by the satellite on the planet (called "planet tides"), but in most cases dissipation due to satellite tides dominates planet tides, and drives the eccentricity to zero (Goldreich and Soter, 1966).

For Europa, however, its eccentricity (and therefore tidal heating) is nonetheless kept from reaching zero by its orbital resonance with Io and Ganymede (Appendix 19.4). This may have important implications for Europa's spin rate. In Appendix 19.1 we showed that tidal evolution should lead to despinning of Europa on extremely short timescales. However, synchronous rotation is not necessarily the final rotation state for satellites whose orbits have nonzero eccentricity. The orbital distance from its primary of a moon in an eccentric orbit varies from $a(1 - e)$ to $a(1 + e)$. Because tidal torques (Eq. (19.4)) vary so strongly with the Europa–Jupiter distance, the torque Γ around Europa's orbit does not average to zero. In fact, $\langle \Gamma \rangle = (171/8\pi) \; k_2 G M_J^2 r_e^5 r_{eJ}^{-6} Q^{-1} e$ (Goldreich, 1966; Greenberg and Weidenschilling, 1984). The average torque is positive, since the torque is positive at perijove (point closest to Jupiter) where the tidal bulge is largest. This net nonzero torque may result in nonsynchronous rotation.

Were Europa's orbit circular, $\langle \Gamma \rangle$ would average to zero. Even for $e \neq 0$, however, synchronous rotation often occurs, due to counter-torques from permanent asymmetries in the satellite. In such a case a mean deviation from spherical symmetry (e.g., due to an internal mass distribution asymmetry or a surface bulge) gives rise to a torque that balances the net nonzero tidal torque when the axis of minimum moment of inertia (e.g., the long axis of an ellipsoid) is oriented slightly off the direction to the planet. This orientation is stable and depends on the magnitude of the internal mass asymmetry and $\langle \Gamma \rangle$ (Greenberg and Weidenschilling, 1984). Most rocky bodies easily support such local mass asymmetries and so may be despun to synchronous rotation even for orbits with $e \neq 0$. Consider Earth's Moon, which maintains its synchronous rotation despite its eccentric orbit ($e = 0.055$) because substantial mass asymmetries are present. On the other hand, Europa has a thin ice shell incapable of supporting topography of much relief, and the ice shell is decoupled from its rocky interior by a liquid ocean. One could have mass asymmetries in the rocky interior that spin-lock the interior to Jupiter even while the ice shell responds to $\langle \Gamma \rangle \neq 0$ and rotates nonsynchronously.

Such non-synchronous rotation has been invoked to explain the orientation of tectonic features on Europa's surface that are thought to have formed due to tidal fracturing of the surface, but which are not in locations or orientations currently coinciding with the greatest stresses predicted by tidal modeling. Geissler et al. (1998) unravel the ages and orientations of linear features in a northern high-latitude region of Europa, and find evidence for a systematic clockwise rotation of feature orientation with time that is consistent with Europa's surface spinning faster than synchronous and being decoupled from the interior.

A search for evidence of this non-synchronous rotation by comparing feature locations from Voyager to Galileo images provides a lower limit of $\sim 10^4$ yr for this rotation period (Hoppa et al., 1999c). Greenberg et al. (1998) found that the lineament orientations examined by Geissler et al. (1998) suggested that Europa had rotated about $60°$ during the time that these tectonic cracks formed. For an estimated surface age for Europa of ~ 50 Myr from cratering studies (Section 19.3.3), this suggests that Europa's nonsynchronous rotation period may be some tens of millions of years, although a later study (Kattenhorn, 2002) suggests a much longer period.

If cracks form only under favorable tidal flexing conditions, then at any given time there is a limited set of locations and orientations where such features will be active. In the Greenberg et al. (1998) model in which cracks provide conduits of liquid water to Europa's surface, nonsynchronous rotation has implications for possible biological niches in the cracks (Section 19.10). At most, organisms living in cracks would have millions of years before their habitat became inactive (and frozen) and they were forced to hibernate or migrate to a more favorable location (Greenberg et al., 2000).

The positions and orientations of various features have also been argued as evidence for *polar wander*, in which a decoupled ice shell slowly reorients itself in a north–south direction, producing an apparent change in position of the poles with respect to surface features (Sarid et al., 2002).

Appendix 19.3 Tidal heating

A spin-locked satellite orbiting its primary will have a fixed tidal bulge with its axis of symmetry pointing towards the primary. If the satellite is in an eccentric orbit, there will also be *radial tides* (due to the tidal bulge rising and falling with respect to the fixed bulge as the satellite's distance from the primary varies around the orbit) and *librational tides* (due to the oscillation of the tidal bulge across the surface of the satellite as its velocity varies about the eccentric orbit). There may also be *obliquity tides* if the satellite has an obliquity, that is, if the obliquity angle ψ between its axis of rotation and the orbit normal is nonzero. In this case the tidal bulge will also oscillate above and below the satellite's equator throughout an orbit.

All of these tides cause dissipation of energy within the satellite, i.e., tidal heating $\dot{E}$. If ΔE is the energy dissipated in the satellite over one tidal cycle (one rotational period T of the satellite), for a synchronously rotating satellite we may write $\dot{E} = E/T = nE/2\pi$ or, by Eq. (19.5), $\dot{E} = E_0(n/Q)$, where E_0 is the peak energy stored in the tidal bulge. For a Keplerian orbit, $n^2 = G(M_J + m_e)a^{-3}$. Calculating $\dot{E}$ exactly is complicated; it was first done for Earth's Moon in a model of a homogeneous, incompressible sphere, and extended to a two-layer lunar model in which a rigid mantle overlies a soft core (Peale and Cassen, 1978). This model turned out to be readily adaptable to a subsurface ocean model for Europa.

For an incompressible homogeneous satellite of radius r_e, tidal dissipation is given by (Murray and Dermott, 2001):

$$\dot{E} = 3/2(n/Q)k_2 G M_J^2 r_e^5 a^{-6}(7e^2 + \sin^2 \psi), \quad (19.8)$$

where the various quantities were defined in Appendix 19.1. This model was used by Peale,

Cassen, and Reynolds (1979) to predict volcanism on Io, then applied to the case of Europa (Cassen *et al.*, 1982), in which a function $f(r_c/r_e)$ was introduced as a factor on the right-hand side of Eq. (19.8). This function determines the ratio of total dissipation in a mechanically decoupled shell (such as an ice layer floating on a liquid water core of radius r_c) to dissipation in a homogeneous world (Peale and Cassen, 1978). As r_c/r_e increases from 0 to 1, for realistic parameter values $f(r_c/r_e)$ increases nonlinearly by over an order of magnitude (Cassen *et al.*, 1982). Physically, dissipation increases because, in the case of Europa, the distortion of the ice shell is much greater when it rides over a liquid ocean than it would be over ice. Indeed, estimates of the time-varying tidal bulge amplitude for Europa range from about 1 m for a solid ice shell with no liquid water layer to about 30 m for tens of km of ice over a liquid water layer (Moore and Schubert, 2000).

More recent theoretical work specific to tidal heating on Europa has focused first on modeling the tidal heating of the ice shell in a more realistic fashion, then on reconsidering the heating of Europa's core. Ojakangas and Stevenson (1989) calculated the energy dissipation in Europa's ice shell for two different candidate ice rheologies, assuming that heat loss through the ice occurs via conduction. The dissipation depends on ice viscosity, which in turn depends exponentially on temperature; the result is that tidal heating is strongly concentrated in the lowest, warmest layer of the ice shell. Indeed, the resulting temperature distribution in the ice shell is exponential with depth (Chyba *et al.*, 1998), rather than the linear temperature gradient familiar for a conductor heated simply from below. Heating of the ice varies as a function of latitude and longitude by a factor of five or so. Uncertainties in these models (Ojakangas and Stevenson, 1989; McKinnon, 1999) arise because the rheology of ice is poorly known at very low temperatures and at the low frequencies (corresponding to Europa's 3.6 day orbital period) associated with tidal flexing. Also unknown are the physical state of Europa's ice, especially its grain size and degree of fracturing, as well as its composition: the addition of small quantities of volatiles such as ammonia or salts to the ice could dramatically alter its viscosity.

As discussed in Section 19.10.3, models indicate that europan heat fluxes could be as high as $\sim$300 mW m^{-2} (McKinnon and Shock, 2001). Or, if one simply uses Eq. (19.8) to scale from the net tidal heating rate measured for Io (10^{14} W) to Europa's silicate/metal interior, one finds a value of 190 mW m^{-2} (O'Brien *et al.*,

2002).[16] Tidal heating may also vary substantially through time (Hussmann and Spohn, 2004).

Appendix 19.4 Orbital resonance and the Laplace resonance

The orbital periods of Io, Europa, and Ganymede are in a ratio of 1:2.007:4.045, very close to the integer ratios 1:2:4 (Fig. 19.4; Table 19.1). There are many examples in the Solar System where just two satellites have orbits near an analogous 1:2 commensurability.

Any small-integer commensurability leads to periodic mutual perturbations, whose effects are reinforced through repetition. In such a case, the perturbed satellite's natural frequency (given by its orbital mean motion $n = 2\pi/T$, where T is the orbital period) is nearly proportional to the forcing frequency (the frequency of strongest mutual gravitational interaction) of the system. As with many physical systems, when the forcing frequency and the natural frequency are nearly equal, a resonance results, referred to in this case as an "orbit–orbit" resonance. If the resulting enhanced mutual perturbations maintain the commensurability against other disruptive influences, the resonance is *stable*, or *locked*.

Here we examine the basic physics of orbit–orbit resonances. We restrict ourselves to a simplified description of a resonance involving only two satellites, since most of the important physics can be illustrated in this simpler system. At the end of the discussion, we will return to the three-satellite Laplace resonance of the jovian system.

Consider two satellites orbiting a primary. Subscript 1 labels variables pertaining to the inner satellite s_1, and 2 is for the outer satellite s_2. The natural frequency of s_1's orbit is just its mean motion n_1. The forcing frequency of the gravitational perturbation s_1 experiences because of the close approaches of s_2 is given by the difference $n_1 - n_2$, which measures the mean motion of s_1 as seen by an observer orbiting with s_2. Resonance will occur if the period of the natural frequency is an integer multiple of the forcing

---

[16] This scaling's validity depends, *inter alia*, upon whether heat flows resulting from Io's partially melted interior are reliably extrapolated to Europa. Since partially melted interiors result in greater tidal heating, which in turn makes a partially melted interior more credible, the extrapolation in some sense assumes its conclusion. These very different tidal heating estimates have important implications for both the thickness of Europa's ice layer (predicted to be as thin as 2 km for the most extreme value of heat flow cited here) and prospects for hydrothermal activity at the floor of its ocean.

frequency, or when $n_1 = j(n_1 - n_2)$, where j is an integer. This gives the resonance condition $(n_1/n_2) = (T_2/T_1) = j/(j-1)$ – resonance will occur if the mean motions (or equivalently, periods) of the satellites are in the whole number ratios $j/(j-1)$. In the case where $j = 2$, for example, s_1 orbits twice for every single circuit of s_2, so that a conjunction between the two satellites at a particular position (called the longitude of conjunction) in their orbits will be followed by a conjunction at the same orbital longitude a time t later, given by $n_1 t - n_2 t = 2\pi$, or $t = 2\pi/(n_1 - n_2) = 2\pi j/n_1 = jT_1$. Such an orbital commensurability exactly satisfies the condition of resonant forcing.

Now consider the net effect of this resonance (Murray and Dermott, 2001). We first try to explain the effects physically, then proceed to a more formal account. For illustration, consider the special case where s_1 has negligible mass and is moving in an eccentric orbit whose pericenter (the point in its orbit closest to the primary) has longitude ϖ_1, and s_2 is moving on a circular orbit in the same plane (Fig. 19.14). Consider a conjunction of s_1 and s_2 that occurs shortly after s_1 has passed pericenter. (Refer to Fig. 19.3 to review the definitions of orbital elements used in this example.) The orbital elements of s_1, including its semimajor axis a_1, eccentricity e_1, and longitude of pericenter ϖ_1, will change as a result of this encounter. Because a and n are related by Kepler's third law according to $n^2 = G(M_J + m_e)a^{-3}$, n_1 and T_1 must change as well.

It is not hard to see why a_1 and e_1 must change. Consider the case of a conjunction that occurs when the objects are moving away from pericenter (Fig. 19.14). They are converging in their orbits because s_1 is moving outward from the primary. s_1 experiences a force $\mathbf{F}$ due to s_2; this force has both radial component $\mathbf{F}_r$ and tangential component $\mathbf{F}_t$. Exactly at conjunction $\mathbf{F}$ is purely radial and in an outward direction. This acts to increase the energy of s_1's orbit, therefore increasing a_1 and decreasing n_1. Since n_1 has decreased with negligible impact on n_2, the subsequent conjunction of s_1 and s_2 must occur sooner in s_2's orbit: conjunction after conjunction, the longitude of conjunction is pulled slowly back toward pericenter.

A similar analysis for a conjunction that occurs immediately *prior* to pericenter shows that in this case the longitude of conjunction is pushed forward toward pericenter. That is, the longitude of conjunction always experiences a kind of restoring force that drives it toward pericenter. Conjunction at pericenter is therefore a stable equilibrium point; the effect of the resonance is to reinforce this configuration, and an

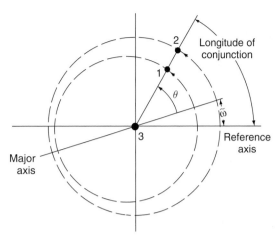

FIGURE 19.14 Satellites in conjunction. Since satellites 1 and 2 are moving away from pericenter, they will experience a kind of restoring force pushing the longitude of conjunction back towards pericenter. (From Greenberg, 1982.)

analogy to a simple spring system suggests that we might well expect bounded oscillations or librations to occur about this equilibrium point. These are in fact observed.

Immediately prior and subsequent to conjunction, in addition to $\mathbf{F}_r$ acting outward, s_1 also experiences a tangential force component $\mathbf{F}_T$. $\mathbf{F}_T$ immediately before conjunction occurs while s_1 is overtaking s_2, so acts in the direction of s_1's motion in its orbit, whereas $\mathbf{F}_T$ immediately after conjunction occurs after s_1 has passed s_2, so acts in the opposite direction. However, for a conjunction after s_1 has passed ϖ_1, $\mathbf{F}_T$ immediately before conjunction is smaller than $\mathbf{F}_T$ immediately after, because the orbits are converging and the gravitational force increases as distance decreases. The net effect of these forces therefore sums to decrease the orbital angular momentum H of s_1. By Eq. (19.6), with both a_1 and H changing, it is clear that s_1's eccentricity e_1 must in general also change.

These physical arguments can be put onto a mathematical footing by writing expressions for changes in specific energy E and angular momentum H due to a perturbing force $\mathbf{F}$ (Burns, 1976). $\dot{E}$ here is the work done per unit mass on a body per unit time by a disturbing force, so by the definition of work,

$$\dot{E} = \mathbf{F} \cdot d\mathbf{r}/dt = (dr/dt)F_r + r(d\theta/dt)\, F_T , \quad (19.9)$$

where F_r and F_T are the magnitudes of $\mathbf{F}_r$ and $\mathbf{F}_T$ and r and θ are the instantaneous radius and true longitude, respectively, of the orbit (defined in Fig. 19.3). Similarly, the time rate of change of H is just

$$\dot{H} = |\mathbf{r} \times \mathbf{F}| = rF_T. \quad (19.10)$$

The normal force N does not appear in either Eq. (19.9) or (19.10) since we assume coplanar orbits. Differentiating Eq. (19.7), then using (19.9) provides an equation for da/dt in terms of a, e, F_r, and F_T; similarly Eqs. (19.6), (19.9), and (19.10) provide an expression for de/dt. Analogous perturbation equations are derivable for the other orbital elements (Burns, 1976); in particular $d\varpi/dt = \dot\varpi_1$ also depends on a, e, F_r, and F_T. In fact, one finds $\dot\varpi_1 \propto e^{-1}$. Therefore ϖ also varies due to orbit–orbit perturbations. Because of the resonance, these effects are reinforced over successive orbits.

Because the orbital longitude λ_1 of s_1 is measured with respect to ϖ_1, which itself is changing, the resonance condition $n_1 = j(n_1 - n_2)$ must be appropriately modified. If ϖ_1 shifts, then λ_1 shifts to $\lambda_1 - \varpi_1$. But since

$$\lambda_1 = \int n_1 dt, \tag{19.11}$$

n_1 shifts to κ_1, where $\kappa_1 \equiv n_1 - \dot\varpi_1$. κ_1 is simply the natural frequency of s_1 taking into account that the orbit itself is precessing at $\dot\varpi_1$. The condition for resonance therefore becomes $\kappa_1 = j(n_1 - n_2)$, or

$$jn_2 - (j-1)n_1 - \dot\varpi_1 = 0. \tag{19.12}$$

Equation (19.12) shows that observed resonances in the Solar System will not have exact whole number ratios between satellite periods or mean motions; these ratios will deviate slightly from whole number ratios due to the presence of small terms like $\dot\varpi_1$.

Finally, as previously noted, $\dot\varpi_1 \propto e^{-1}$, so that in order for $\dot\varpi_1$ to satisfy Eq. (19.12), e_1 must have a specific value. This eccentricity is the forced eccentricity that is set and maintained by the orbit–orbit resonance. (The importance of this forced eccentricity for the Galilean satellites was overlooked for many years, with the result that tidal heating went unpredicted; see Greenberg (1982) for a discussion.)

Laplace three-body resonance

The only known *three-body* resonance in the Solar System is the Laplace resonance among Io, Europa, and Ganymede. Adopting notation analogous to that used in our discussion of the two-body case, and as shown in Fig. 19.4, Io, Europa, and Ganymede are involved in a resonance where the mean motions satisfy

$$n_1 - 3n_2 + 2n_3 = 0. \tag{19.13}$$

This relationship among the three satellite mean motions, unlike that in Eq. (19.12), is strictly equal to

zero (Murray and Dermott, 2001), with precision measured to eleven significant figures! The stability of the three-body Laplace resonance can only be sustained if the precessions of the satellites' orbits are equal. Expressing this mathematically, subtracting Eqs. (19.12) for the two satellite pairs, Eq. (19.13) follows only if $\dot\varpi_1 = \dot\varpi_2$.

Equation (19.13) corresponds by Eq. (19.11) to a relation among the satellites' orbital longitudes: $\lambda_1 - 3\lambda_2 + 2\lambda_3 = \phi \approx 180°$. The approximation sign is a reminder that ϕ librates about $180°$, with an amplitude of $0.064°$ and a period of 2071 days (Murray and Dermott, 2001). The fact that $\phi \approx 180°$ guarantees that no triple conjunction of the three satellites ever takes place, as shown in Fig. 19.4. For example, whenever Europa and Ganymede are in conjunction ($\lambda_2 = \lambda_3$), Io must be $180°$ away, i.e., $\lambda_1 = \lambda_2 + 180°$.

How did something as marvelous as the Laplace resonance arise? In one theory (Yoder, 1979), the Laplace resonance was assembled from an originally non-resonant configuration of the satellites' orbits by the differential expansion of the orbits due to tides raised on Jupiter by the satellites. To understand this effect, recall the discussion in Appendix (19.1). Equation (19.4) is the torque experienced by the tidal bulge raised on the satellite due to the primary. An analogous expression can be derived for the torque experienced by the primary, due to a tidal bulge raised on it by the gravity of the satellite. This expression is identical to Eq. (19.4), except with the substitutions $m_e \leftrightarrow M_J$, $r_e \leftrightarrow R_J$, and $k_2 \sin 2\varepsilon = k_2 Q^{-1}$ now refers to the primary rather than the satellite. This torque acts to despin the primary (without much effect in the case, say, of Io and Jupiter), but by action–reaction, the torque slowing the planet acts as well on the satellite's orbit – which expands in response. (In the Earth–Moon system this expansion is thought to have driven the Moon from its formation near Earth out to its present distance of 60 Earth radii.)

We derive an equation for this tidally driven orbital expansion for the special case where the orbit is circular and has zero inclination. (For the more general case, see Chyba *et al.*, 1989.) The torque Γ and angular momentum H are related by

$$\Gamma = \dot H, \tag{19.14}$$

where Γ is given by Eq. (19.4) with the substitutions just described and H is from Eq. (19.6) with $e = 0$. Integrating the resulting equation for $\dot a$ gives

$$a^{13/2} = a_0^{13/2} + (13/2)\beta(t - t_0) \tag{19.15}$$

for the semimajor axis of the satellite's orbit as a function of time, measured with respect to initial time t_0 and semimajor axis a_0. Here $\beta = (39/2)(k_2/Q)_J m_e R_J^5 (G/M_J)^{1/2}$, and $(k_2/Q)_J$ refers to the primary. From Eq. (19.15), the timescale for orbital expansion $\tau \propto (a^{13/2} - a_0^{13/2})$; that is, satellites evolve outwards more quickly when they are closer to the primary. In the distant past Io's orbit therefore expanded more rapidly than Europa's, caught up to Europa, and drove it into the configuration where $n_1/n_2 \approx 2:1$. The two satellites were then captured into this stable resonance so that as Io's orbit continued to expand due to angular momentum exchange with Jupiter, Io transferred to Europa part of its increased angular momentum. In a similar way, the three-body lock of Io–Europa and Ganymede was subsequently established as the orbits of Io and Europa expanded and a 2:1 commensurability between the mean motions of Europa and Ganymede was approached. There has not been sufficient time since Solar System formation for Callisto to have been brought into the resonance.

This scenario, while elegant, requires Jupiter's Q_J to be $\sim 10^5$–10^6 or smaller for sufficient expansion of the orbits to have occurred (Goldreich and Soter, 1966), and as low as $\sim 10^4$ to maintain the current configuration in equilibrium (Peale and Lee, 2002). An alternative, primordial origin for the Laplace relation has been proposed out of concern that these values for Q_J may be unreasonably small (Greenberg, 1982). This model involves differential migration of the newly formed Galilean satellites due to their interactions with the circumjovian accretion disk (Peale and Lee, 2002).

Appendix 19.5 Gravity and interior models

The gravitational potential for a non-spherical body may be expanded as an infinite series of spherical harmonics (Kaula, 1966). An approximation through second degree and order is:

$$V = \frac{Gm_e}{r}\left[1 - \frac{1}{2}J_2\left(\frac{r_e}{r}\right)^2(3\sin^2\phi - 1) + 3C_{22}\left(\frac{r_e}{r}\right)^2\cos^2\phi\cos 2\lambda\right], \quad (19.16)$$

where the coefficients J_2 and C_{22} measure the contributions to the gravitational potential of the spherical harmonics of degree l and order m for $l = 2$, $m = 0$ and $l = 2$, $m = 2$, respectively. The spherical coordinates are fixed in Europa (radius r, latitude ϕ, longitude λ), with longitude measured from the Europa–Jupiter line. The

first term is the gravitational potential for a spherical body. J_2 arises from Europa's oblateness due to its rotation about its spin axis. Longitude-dependent C_{22} is required by the asymmetry (about its rotation axis) in Europa's permanent tidal bulge.

Both J_2 and C_{22} are directly related to Europa's internal mass distribution. $J_2 \approx (C - A)/m_e r_e^2$ and $C_{22} \approx (B - A)/4m_e r_e^2$, where C is Europa's axial moment of inertia and A and B are its equatorial moments. For a body in rotational and tidal equilibrium, C_{22} is related to the "rotational response" parameter $q = \omega^2 r_e^3/Gm_e$ (the ratio of centrifugal to gravitational acceleration at Europa's equator), where ω is the satellite's rotational angular velocity. In fact,

$$C_{22} = 3\alpha q/4, \quad (19.17)$$

where α is a dimensionless coefficient that depends on the distribution of mass within the satellite. With q known and C_{22} obtained from spacecraft Doppler measurements, α may be determined by Eq. (19.17); Anderson et al. (1998) have measured $\alpha = 0.35$. The satellite's axial moment of inertia C then follows from the relationship (Anderson et al., 1996):

$$\frac{C}{m_e r_e^2} = \frac{2}{3}\left[1 - \frac{2}{5}\left(\frac{4 - 3\alpha}{1 + 3\alpha}\right)^{1/2}\right]. \quad (19.18)$$

For Europa $C/m_e r_e^2 = 0.346 \pm 0.005$, which is less than 0.4, the value for a sphere of constant density, implying that Europa must have a concentration of mass towards its center. Europa's mean density and C then in turn provide constraints on two- or three-layer models for the internal mass distribution (Anderson et al., 1998).

Appendix 19.6 Induced magnetic fields

Jupiter's magnetic field axis is anchored in the planet, tilted by 9.6° from its spin axis and offset by 0.13 R_J from its center (NSSDC, 2004). Therefore, over one jovian rotational period (synodic period of 11.23 hr, i.e., as experienced by Europa), Europa experiences a periodic time-varying magnetic field. Consider a projection of Jupiter's magnetic field on to Jupiter's equatorial plane. At Europa this can be resolved into two components, $\mathbf{B}_1$ in the x-direction (the radial direction, pointing toward Jupiter) and $\mathbf{B}_2$ in the y-direction (the tangential direction of Europa's motion in its orbit). $\mathbf{B}_1$, for example, varies sinusoidally with a synodic period of 11.23 hours. By Faraday's law of electromagnetic induction, $\mathbf{B}_1$ induces an electric field given by $\nabla \times \mathbf{E} = -\partial \mathbf{B}_1/\partial t$, which in turn causes a current I_1 to flow within Europa's conducting ocean. This eddy current I_1 (which changes direction twice each synodic period) encircles $\mathbf{B}_1$, setting up a secondary magnetic field of its own whose orientation is approximately antiparallel to the primary field, and which opposes the change in magnetic flux (Lenz's law). The Galileo spacecraft magnetometer indeed measured changes in the total magnetic field consistent with the presence of a secondary field whose direction changed in response to Jupiter's primary field (Kivelson et al., 2000).

A primary magnetic field (from Jupiter) oscillating at frequency ω along a direction with unit vector $\mathbf{e}$ may be written $\mathbf{B}_p = B_p e^{-i\omega t} \mathbf{e}$. The total time-varying field is the sum of $\mathbf{B}_p$ and the secondary (induced) field $\mathbf{B}_s$. For a simple shell model of Europa, where a spherical conducting shell of conductivity σ is enclosed by an insulating core and insulating outer shell, the solution to the induced field at a vector distance $\mathbf{r}$ outside the conducting shell is $\mathbf{B}_p + \mathbf{B}_s$, where

$$\mathbf{B}_s = -A e^{-i(\omega t + \phi)} B_p [3(\mathbf{r} \cdot \mathbf{e})\mathbf{r} - r^2 \mathbf{e}] r_e^2 / 2r^5, \quad (19.19)$$

which is a dipole field. Here A and ϕ are real numbers that depend on the thickness of the conducting layer within Europa and the conductivity of that layer. Comparing this simple shell model to Galileo observations suggests that Europa's ocean has a conductivity corresponding to near-saturation concentrations of salts with an ice layer less than 10 km thick (Hand and Chyba, 2007).

Appendix 19.7 Heat conduction and convection

We explain here how the subsurface conductive temperature gradients shown in Fig. 19.12 result from solutions of the one-dimensional time-independent thermal diffusion equation, given different assumptions about the nature of tidal heating. We then explain the origin of the Rayleigh number and show why it provides a criterion for whether heat loss will occur primarily through conduction or convection.

Conduction

Consider heat flow through Europa's ice shell in a coordinate system where z measures the depth into the ice below the surface. Fourier's empirical law says that the heat flux H (erg s^{-1} cm^{-2}) is directly proportional to the temperature gradient $\partial T / \partial z$:

$$H = k \, \partial T / \partial z .^{17} \quad (19.20)$$

The thermal conductivity k of ice is given by

$$k(T) = a/T + b, \quad (19.21)$$

where $a = 4.88 \times 10^7$ erg cm^{-1} s^{-1} and $b = 4.68 \times 10^4$ erg cm^{-1} s^{-1} K^{-1} (Hobbs, 1974). For mathematical tractability in analytical work, either $k(T)$ is often taken to be a constant and assigned its value at some median temperature (say 185 K), or b is set to zero, a good approximation for temperatures below $\sim 10^3$ K.

The rate at which a unit volume in a given subsurface layer heats up is due to the volumetric tidal heating rate within that layer, q (erg cm^{-3} s^{-1}), and the difference between heat flows H in and out. For an infinitesimal layer, this becomes

$$q + \partial H / \partial z = \rho C_P \, \partial T / \partial t, \quad (19.22)$$

where ρ is the density of the material and C_P is the specific heat capacity at constant pressure (the amount of heat needed to raise the temperature of one gram of ice by 1 K at constant pressure). Both C_P and ρ for ice are temperature-dependent (Hobbs, 1974).

Combining Eqs. (19.20) and (19.22) yields the thermal diffusion equation

$$q + \partial(k \partial T / \partial z) / \partial z = \rho C_P \, \partial T / \partial t. \quad (19.23)$$

We want the time-independent equation, for which $\partial T / \partial t = 0$:

$$\partial(k \partial T / \partial z) / \partial z = -q. \quad (19.24)$$

Solutions of Eq. (19.24) give steady-state temperature gradients in Europa's ice; Eq. (19.23) is required for cases where melting (O'Brien et al., 2002) or refreezing (Buck et al., 2002) is being modeled.

Were k constant and $q \approx 0$, Eq. (19.20) gives the thickness of the ice layer as

$$h = k\Delta T / H, \quad (19.25)$$

where $\Delta T = T_h - T_s$ and T_s and T_h are the temperatures at the surface and base of the ice layer, respectively. In this case Eq. (19.24) yields a simple linear temperature profile through the ice,

$$T(z) = T_s + z\Delta T / h. \quad (19.26)$$

---

[17] The right-hand side of Eq. (19.20) is positive, which is opposite to the usual sign convention for Fourier's law, because of our choice of z increasing downward into the ice shell.

However, q is positive in Europa's ice due to tidal heating in the shell; if its value is taken to be constant throughout the shell, Eq. (19.24) gives

$$T(z) = T_s \exp[(H + qh)z/a - qz^2/2a]. \quad (19.27)$$

But for realistic ice rheologies, q is extremely temperature dependent and the assumption of constant q is poor. For example, $q \propto \exp[l(1 - T_m/T)]$, where T_m is the melting temperature and $l \approx 24$ at T_m (Ojakangas and Stevenson, 1989). In this case, tidal heating is confined to the warmest ice, i.e., to a thin layer at the base of the ice shell. Equation (19.24) can be solved for the case where q is strongly peaked in warm ice; one finds (Chyba *et al.*, 1998)

$$T(z) = T_s \exp[(z/h)\ln(T_h/T_s)]. \quad (19.28)$$

Plotted in Fig. 19.12 are the temperature gradients, corresponding to Eqs. (19.26) to (19.28), assuming the same ice layer thickness in each case. Equation (19.28) is most likely the best approximation for Europa, assuming that the shell loses heat primarily through conduction rather than convection, and is thick enough for tidal heating to be important.

Convection

Will Europa's ice lose heat primarily through conduction or convection? The usual criterion to answer this question is the value of the *Rayleigh number Ra* (Rayleigh, 1916). The Rayleigh number that is usually cited in the Europa literature as determining the onset of convection (e.g., Pappalardo *et al.*, 1998; McKinnon, 1999) is:

$$Ra = \rho g \alpha_V \Delta T h_c^3/\eta\kappa, \quad (19.29)$$

where g is the surface gravity, α_V the volumetric coefficient of thermal expansion, η the dynamic viscosity, $\kappa = k/\rho C_P$ is the thermal diffusivity, and h_c is the thickness of the ice beneath the stagnant lid. Convection will occur when Ra exceeds some critical value Ra_{crit}, usually taken to be $\sim 10^3$; since $Ra \propto h_c^3$, there is a critical thickness h_c above which convection occurs.

This expression for Ra is derived by applying linear stability analysis to a set of four differential equations describing thermal convection for a particular system, but these assume that k is constant and $q = 0$ in Eq. (19.23). Given the importance of tidal heating within Europa's ice shell, it is not obvious why investigators have chosen this formulation for Ra. For example, a Rayleigh number appropriate for a layer heated uniformly from within is (Turcotte and Schubert, 1982):

$$Ra = \rho^2 g \alpha_V q h_c^5/k\eta\kappa. \quad (19.30)$$

Ra_{crit} for Eq. (19.30) is again $\sim 10^3$. The difference between Eqs. (19.29) and (19.30) should lead to caution in evaluating conclusions based solely on Eq. (19.29). An additional uncertainty concerns the observation that η depends not only on temperature but also on other properties such as the grain size of the ice.

References

Anderson, J. D., Sjogren, W. L., and Schubert, G. (1996). Galileo gravity results and the internal structure of Io. *Science*, **272**, 709–712.

Anderson, J. D., Lau, E. L., Sjogren, W. L., *et al.* (1998). Europa's differentiated internal structure: inferences from four Galileo encounters. *Science*, **281**, 2019–2022.

Auda, H. and Emborg, C. (1973). Studies on post-irradiation degradation in *Micrococcus radiodurans*, strain $R_{II}5$. *Radiation Research*, **53**, 273–280.

Bada, J. L., Bigham, C., and Miller, S. L. (1994). Impact melting of frozen oceans on the early Earth: implications for the origin of life. *Proc. Natl. Acad. Sci. USA*, **91**, 1248–1250.

Bada, J. L. and Lazcano, A. (2002). Some like it hot, but not the first biomolecules. *Science*, **296**, 1982–1983.

Barr, A. C. and Pappalardo, R. T. (2003). Numerical simulations of non-Newtonian convection in ice: application to Europa. *Lunar Planet. Sci.* **XXXIV**, abs. 1477. Houston, TX: Lunar and Planetary Institute [CD-ROM].

Buck, L., Chyba, C. F., Goulet, M., Smith, A., and Thomas, P. (2002). Persistence of thin ice regions in Europa's ice crust. *Geophys. Res. Lett.*, **29**, 2055, doi:10.1029/2002GL016171.

Burns, J. A. (1976). Elementary derivation of the perturbation equations of celestial mechanics. *Am. J. Phys.*, **44**, 944–949 (Erratum: **45**, 1230).

Carlson, R. W., Johnson, R. E., and Anderson, M. S. (1999). Sulfuric acid on Europa and the radiolytic sulfur cycle. *Science*, **286**, 97–99.

Cassen, P., Peale, S. J., and Reynolds, R. T. (1982). Structure and thermal evolution of the Galilean satellites. In *Satellites of Jupiter*, ed. D. Morrison, pp. 93–128. Tucson: University of Arizona Press.

Chapelle, F. H., O'Neill, K., Bradley, P. M., *et al.* (2002). A hydrogen-based subsurface microbial community dominated by methanogens. *Nature*, **415**, 312–314.

Chyba, C. F. and Hand, K. P. (2001). Life without photosynthesis. *Science*, **292**, 2026–2027.

Chyba, C. F. and McDonald, G. D. (1995). The origins of life in the solar system: current issues. *Ann. Rev. Earth Planet. Sci.*, **23**, 215–249.

Chyba, C. F. and Phillips, C. B. (2001). Possible ecosystems and the search for life on Europa. *Proc. Natl. Acad. Sci. USA*, **98**, 801–804.

Chyba, C. F. and Phillips, C. B. (2002). Europa as an abode of life. *Orig. Life Evol. Biosphere*, **32**, 47–68.

Chyba, C. F. and Sagan, C. (1987). Infrared emission by organic grains in the coma of comet Halley. *Nature*, **330**, 350–353.

Chyba, C. F., Jankowski, D. G., and Nicholson, P. D. (1989). Tidal evolution in the Neptune–Triton system. *Astron. Astrophys.*, **219**, L23–L26.

Chyba, C. F., Ostro, S. J., and Edwards, B. C. (1998). Radar detectability of a subsurface ocean on Europa. *Icarus*, **134**, 292–302.

Colburn, D. S. and Reynolds, R. T. (1985). Electrolytic currents in Europa. *Icarus*, **63**, 39–44.

Cooper, J. F., Johnson, R. E., Mauk, B. H., Garrett, H. B., and Gehrels, N. (2001). Energetic ion and electron irradiation of the icy Galilean satellites. *Icarus*, **149**, 133–159.

Cooper, J. F., Phillips, C. B., Green, J. R., *et al.* (2002). Europa exploration: science and mission priorities. In *The Future of Solar System Exploration, 2003–2013*, ed. M. Sykes, pp. 217–252. San Francisco: Astronomical Society of the Pacific.

Dalton, J. B., Mogul, R., Kagawa, H. K., *et al.* (2003). Near-infrared detection of potential evidence for microscopic organisms on Europa. *Astrobiology*, **3**, 505–529.

Delitsky, M. L. and Lane, A. L. (1998). Ice chemistry on the Galilean satellites. *J. Geophys. Res.*, **103**, 31,391–31,403.

Fagents, S. A., Greeley, R., Sullivan, R. J., *et al.* (2000). Cryomagmatic mechanisms for the formation of Rhadamanthys Linea, triple band margins, and other low-albedo features on Europa. *Icarus*, **144**, 54–88.

Figueredo, P. H., Chuang, F. C., Rathbun, J., *et al.* (2002). Geology and origin of Europa's "Mitten" feature (Murias Chaos). *J. Geophys. Res.*, **107(E5)**, 5026, doi:10.1029/2001JE001591.

Gaidos, E. J. and Nimmo, F. (2000). Tectonics and water on Europa. *Nature*, **405**, 637.

Gaidos, E. J., Nealson, K. H., and Kirschvink, J. L. (1999). Life in ice-covered oceans. *Science*, **284**, 1631–1633.

Galileo, G. (1610). *Sidereus Nuncius, or, The Sidereal Messenger* (trans. and intro. by A. van Helden). Chicago: University of Chicago Press (1989).

Geissler, P. E. (2000). Cryovolcanism in the outer solar system. In *Encyclopedia of Volcanoes*, ed. H. Sigurdsson, pp. 785–800. San Diego, CA: Academic.

Geissler, P. E., Greenberg, R., Hoppa, G., *et al.* (1998). Evidence for non-synchronous rotation of Europa. *Nature*, **391**, 368.

Goldreich, P. (1966). Final spin states of planets and satellites. *Astron. J.*, **71**, 1–7.

Goldreich, P. and Soter, S. (1966). Q in the Solar System. *Icarus*, **5**, 375–389.

Greeley, R., Figueredo, P. H., Williams, D. A., *et al.* (2000). Geologic mapping of Europa. *J. Geophys. Res.*, **105**, 22559–22578.

Greeley, R., Chyba, C. F., Head, J. W., *et al.* (2004). Geology of Europa. In *Jupiter: the Planets, Satellites and Magnetosphere*, eds. F. Bagnold, T. E. Dowling, W. B. McKinnon, *et al.* Cambridge: Cambridge University Press.

Greenberg, R. (1982). Orbital evolution of the Galilean satellites. In *Satellites of Jupiter*, ed. D. Morrison, pp. 65–92. Tucson: University of Arizona Press.

Greenberg, R. and Weidenschilling, S. J. (1984). How fast do Galilean satellites spin? *Icarus*, **58**, 186–196.

Greenberg, R., Geissler, P., Hoppa, G., *et al.* (1998). Tectonic processes on Europa: tidal stresses, mechanical response, and visible features. *Icarus*, **135**, 64–78.

Greenberg, R., Hoppa, G. V., Tufts, B. R., Geissler, P. E., and Reilly, J. (1999). Chaos on Europa. *Icarus*, **141**, 263–286.

Greenberg, R., Geissler, P., Tufts, B. R., and Hoppa, G. V. (2000). Habitability of Europa's crust: the role of tidal-tectonic processes. *J. Geophys. Res.*, **105**, 17,551–17,562.

Guillot, T. (1999). A comparison of the interiors of Jupiter and Saturn. *Planet. Space Sci.*, **47**, 1183–1200.

Hand, K. H. and Chyba, C. F. (2007). Empirical constraints on the salinity of the europan ocean and implications for a thin ice shell. *Icarus*, in press.

Head, J. W. III and Pappalardo, R. T. (1999). Brine mobilization during lithospheric heating on Europa: implications for formation of chaos terrain, lenticula texture, and color variations. *J. Geophys. Res.*, **104**, 27,143.

Head, J., Pappalardo, R. T., and Sullivan, R. J. (1999). Europa: morphological characteristics of ridges and triple bands from Galileo data (E4 and E6) and assessment of a linear diapirism model. *Geophys. Res. Lett.*, **104**, 24,223–24,236.

Hobbs, P. V. (1974). *Ice Physics*. Oxford: Oxford University Press.

Hoppa, G., Tufts, B. R., Greenberg, R., and Geissler, P. (1999a). Strike-slip faults on Europa: global shear patterns driven by tidal stress. *Icarus*, **141**, 287–298.

Hoppa, G. V., Tufts, B. R., Greenberg, R., and Geissler, P. (1999b). Formation of cycloidal features on Europa. *Science*, **285**, 1899–1902.

Hoppa, G. V., Greenberg, R., Tufts, B. R., *et al.* (1999c). Rotation of Europa: constraints from terminator and limb positions. *Icarus*, **137**, 341–347.

Horneck, G. (1993). Responses of *Bacillus subtilis* spores to space environment: results from experiments in space. *Origins Life Evol. Biosphere*, **23**, 37–52.

Hoyle, F. and Wickramasinghe, N. C. (1987). Organic dust in comet Halley. *Nature*, **328**, 117.

Hussmann, H. and Spohn, T. (2004). Thermal-orbital evolution of Io and Europa. *Icarus*, **171**, 391–410.

Ip, W.-H. (1996). Europa's oxygen exosphere and its magnetospheric interaction. *Icarus*, **120**, 317–325.

Johnson, R. E. (1998). Sputtering and desorption from icy surfaces. In *Solar System Ices*, ed. B. Schmitt, C. de Bergh, and M. Festou, pp. 303–334. Dordrecht, The Netherlands: Kluwer.

Kanavarioti, A., Monnard, P., and Deamer, D. W. (2001). Eutectic phases in ice facilitate nonenzymatic nucleic acid synthesis. *Astrobiology*, **1**, 271–281.

Kargel, J. S., Kage, J. Z., Head, J. W., *et al.* (2000). Europa's crust and ocean: origin, composition, and the prospects for life. *Icarus*, **148**, 226–265.

Kattenhorn, S. A. (2002). Nonsynchronous rotation evidence and fracture history in the bright plains region, Europa. *Icarus*, **157**, 490–506.

Kaula, W. M. (1966). *Theory of Satellite Geodesy*. New York: Dover.

Kivelson, M. G., Khurana, K. K., Russell, C. T., Volwerk, M., Walker, R. J., and Zimmer, C. (2000). Galileo magnetometer measurements: a stronger case for a subsurface ocean at Europa. *Science*, **289**, 1340–1343.

Kivelson, M. G., Khurana, K. K., and Volwerk, M. (2002). The permanent and inductive magnetic moments of Ganymede. *Icarus*, **157**, 507–522.

Kovach, R. L. and Chyba, C. F. (2001). Seismic detectability of a subsurface ocean on Europa. *Icarus*, **150**, 279–287.

Kuramoto, K. and Matsui, T. (1994). Formation of a hot proto-atmosphere on the accreting giant icy satellite: implications for the origin and evolution of Titan, Ganymede, and Callisto. *J. Geophys. Res.*, **99**, 21183–21200.

Levy, M., Miller, S. L., Brinton, K., and Bada, J. L. (2000). Prebiotic synthesis of adenine and amino acids under Europa-like conditions. *Icarus*, **145**, 609–613.

MacDonald, G. J. F. (1964). Circulation and tides in the high atmosphere. *Planet. Space Sci.*, **10**, 79–87.

McCollom, T. M. (1999). Methanogenesis as a potential source of chemical energy for primary biomass production by autotrophic organisms in hydrothermal systems on Europa. *J. Geophys. Res.*, **104**, 30,729–30,742.

McCord, T. B., Hansen, G. B., Clark, R. N., *et al.* (1998). Non-water-ice constituents in the surface material of the icy Galilean satellites from the Galileo near-infrared mapping spectrometer investigation. *J. Geophys. Res.*, **103**, 8603–8626.

McCord, T. B., Hansen, G. B., Matson, D. L., *et al.* (1999). Hydrated salt minerals on Europa's surface from the Galileo near-infrared mapping spectrometer (NIMS) investigation. *J. Geophys. Res.*, **104**, 11,827–11,851.

McKinnon, W. B. (1999). Convective instability in Europa's floating ice shell. *Geophys. Res. Lett.*, **26**, 951–954.

McKinnon, W. B. and Shock, E. L. (2001). Ocean Karma: what goes around comes around on Europa (or does it?). *Lunar Planet. Sci.*, **XXXII**, Abs. #2181. Houston, TX: Lunar and Planetary Institute.

McKinnon, W. B. and Zolensky, M. E. (2003). Sulfate content of Europa's ocean and shell: evolutionary considerations and some geological and astrobiological implications. *Astrobiology*, **3**, 879–897.

Monnard, P., Apel, C., Kanavarioti, A., and Deamer, D. (2002). Influence of ionic inorganic solutes on self-assembly and polymerization processes related to early forms of life: implications for a prebiotic aqueous medium. *Astrobiology*, **2**, 139.

Moore, J. C. (2000). Models of radar absorption in europan ice. *Icarus*, **147**, 292–300.

Moore, J. M., Asphaug, E., Sullivan, R. J., *et al.* (1998). Large impact features on Europa: results of the Galileo nominal mission. *Icarus*, **135**, 127–145.

Moore, J. M., Asphaug, E. Belton, M. J. S., *et al.* (2001). Impact features on Europa: results of the Galileo Europa mission (GEM). *Icarus*, **151**, 93–111.

Moore, W. B. and Schubert, G. (2000). The tidal response of Europa. *Icarus*, **147**, 317–319.

Mumma, M. J., Weissman, P. R., and Stern, S. A. (1993). Comets and the origin of the Solar System: reading the Rosetta Stone. In *Protostars and Planets III*, eds. E. H. Levy and J. I. Lunine, pp. 1177–1252. Tucson: University of Arizona Press.

Murray, C. D. and Dermott S. F. (2001). *Solar System Dynamics*. Cambridge: Cambridge University Press.

National Space Science Data Center (2004). Planetary Fact Sheets. nssdc.gsfc.nasa.gov/planetary/planetfact.html. Visited 11/17/2004.

Newton, I. (1686). *Philosophiae Naturalis Principia Mathematica*. In *Sir Isaac Newton's Mathematical Principles of Natural Philosophy and his System of the World*, ed. F. Cajori. Berkeley: University of California Press (1934).

O'Brien, D. P., Geissler, P., and Greenburg, R. (2002). A melt-through model for chaos formation on Europa. *Icarus*, **156**, 152–161.

Ojakangas, G. W. and Stevenson, D. J. (1989). Thermal state of an ice shell on Europa. *Icarus*, **81**, 220–241.

Pappalardo, R. T., Head, J. W., Greeley, R., *et al.* (1998). Geological evidence for solid-state convection in Europa's ice shell. *Nature*, **391**, 365–368.

Pappalardo, R. T., Belton, M. J. S., Brereman, H. H., *et al.* (1999). Does Europa have a subsurface ocean? Evaluation of the geological evidence. *J. Geophys. Res.*, **104**, 24,015–24,055.

Peale, S. J. and Cassen, P. (1978). Contribution of tidal dissipation to lunar thermal history. *Icarus*, **36**, 245–269.

Peale, S. J., Cassen, P., and Reynolds, R. T. (1979). Melting of Io by tidal dissipation. *Science*, **203**, 892–894.

Peale, S. J. and Lee, M. H. (2002). A primordial origin of the Laplace relation among the Galilean satellites. *Science*, **298**, 593–597.

Phillips, C. B., McEwen, A. S., Hoppa, G. V., *et al.* (2000). The search for current geologic activity on Europa. *J. Geophys. Res.*, **105**, 22,579–22,597.

Pierazzo, E. and Chyba, C. F. (2002). Cometary delivery of biogenic elements to Europa. *Icarus*, **157**, 120–127.

Pollack, J. B. and Sagan, C. (1967). Secular changes and dark-area regeneration on Mars. *Icarus*, **6**, 434–439.

Prockter, L. M. and Pappalardo, R. T. (2000). Folds on Europa: implications for crustal cycling and accommodation of extension. *Science*, **289**, 941–944.

Prockter, L. M., Head, J. W. III, Pappalardo, R. T., *et al.* (2002). Morphology of Europan bands at high resolution: a mid-ocean ridge-type rift mechanism. *J. Geophys. Res.*, **107**, 5028, doi:10.1029/2000JE001458.

Raleigh, L. (1916). On convection currents in a horizontal layer of fluid when the higher temperature is on the underside. *Phil. Mag. Ser.* 6, **32**, 529–546.

Reynolds, R. T., Squyres, S. W., Colburn, D. S., and McKay, C. P. (1983). On the habitability of Europa. *Icarus*, **56**, 246–254.

Sagan, C. (1996). Circumstellar habitable zones: an introduction. In *Circumstellar Habitable Zones*, ed. L. R. Doyle, pp. 3–14. Menlo Park, CA: Travis House.

Sagan, C. and Chyba, C. (1997). The early faint sun paradox: organic shielding of ultraviolet-labile greenhouse gases. *Science*, **276**, 1217–1221.

Sarid, A. R., Greenberg, R., Hoppa, G. V., Hurford, T. A., Tufts, B. R., and Geissler, P. (2002). Polar wander and surface convergence of Europa's ice shell: evidence from a survey of strike-slip displacement. *Icarus*, **158**, 24–41.

Schenk, P. M. (2002). Thickness constraints on the icy shells of the galilean satellites from a comparison of crater shapes. *Nature*, **417**, 419–421.

Schulze-Makuch, D. and Irwin, L. N. (2004). *Life in the universe. Expectations and constraints.* Berlin: Springer-Verlag.

Sotin, C., Head, J. W., and Tobie, G. (2002). Europa: tidal heating of upwelling thermal plumes and the origin of lenticulae and chaos melting. *Geophys. Res. Lett.*, **29**, 74–1, 1233, doi:10.1029/2001GL013844.

Space Studies Board, National Research Council (1992). *Biological Contamination of Mars: Issues and Recommendations.* Washington, DC: National Academy Press, www7.nas.edu/ssb/bcmarsmenu.html.

Space Studies Board, National Research Council (2000). *Preventing the Forward Contamination of Europa.* Washington DC: National Academy of Sciences Press, www7.nas.edu/ssb/europamenu.html.

Spaun, N. A., Head, J. W., Collins, G. C., Prockter, L. M., and Pappalardo, R. T. (1998). Conamara Chaos region, Europa: reconstruction of mobile polygonal ice blocks. *Geophys. Res. Lett.*, **25**, 4277–4280.

Spencer, J. R., Tamppari, L. K., Martin, T. Z., and Travis, L. D. (1999). Temperatures on Europa from Galileo Photopolarimeter-Radiometer: nighttime thermal anomalies. *Science*, **284**, 1514–1516.

Squyres, S. W., Reynolds, R. T., Cassen, P. M., and Peale, S. J. (1983). Liquid water and active resurfacing on Europa. *Nature*, **301**, 225–226.

Squyres, S. W., Reynolds, R. T., and Lissauer, J. J. (1985). The enigma of the Uranian satellites' orbital eccentricities. *Icarus*, **61**, 218–223.

Stacey, F. D. (1977). *Physics of the Earth*, second edn. New York: Wiley.

Stevenson, D. J. (1982). Interiors of the giant planets. *Ann. Rev. Earth Planet. Sci.*, **10**, 257–295.

Stevenson, D. J. (2000a). Europa's ocean – the case strengthens. *Science*, **289**, 1305–1307.

Stevenson, D. J. (2000b). Limits on the variation of thickness of Europa's ice shell. *Lunar Planet. Sci.*, **XXXI**, Abs. #1506. Houston: Lunar and Planetary Institute.

Sullivan, R., Greeley, R., Homan, K., *et al.* (1998). Episodic plate separation and fracture infill on the surface of Europa. *Nature*, **391**, 371–373.

Thomas, P. J. and Schubert, G. (1986). Crater relaxation as a probe of Europa's interior. *J. Geophys. Res.*, **91**, 453–459.

Thomson, R. E. and Delaney, J. R. (2001). Evidence for a weakly stratified Europan ocean sustained by seafloor heat flux. *J. Geophys. Res.*, **106**, 12355–12366.

Tufts, B. R., Greenberg, R., Hoppa, G., and Geissler, P. (1999). Astypalaea Linea: a large-scale strike-slip fault on Europa. *Icarus*, **141**, 53–64.

Turcotte, D. L. and Schubert, G. (1982). *Geodynamics: Application of Continuum Physics to Geological Problems.* New York: John Wiley & Sons.

Turtle, E. P. and Pierazzo, E. (2001). Thickness of a Europan ice shell from impact crater simulations. *Science*, **294**, 1326–1328.

Warren, S., Brandt, R. E., Grenfell, T. C., and McKay, C. P. (2002). Snowball Earth: ice thickness on the tropical ocean. *J. Geophys. Res.*, **107**, 31–1, 3167, doi:10.1029/2001JC001123.

Whitman, W. B., Coleman, D. C., and Wiebe, W. J. (1998). Prokaryotes: the unseen majority. *Proc. Natl. Acad. Sci. USA*, **95**, 6578–6583.

Williams, K. K. and Greeley, R. (1998). Estimates of ice thickness in the Conamara Chaos region of Europa. *Geophys. Res. Lett.*, **25**, 4273–4276.

Yoder, C. F. (1979). How tidal heating in Io drives the Galilean orbital resonance locks. *Nature*, **279**, 767–770.

Zahnle, K., Dones, L., and Levison, H. F. (1998). Cratering rates on the Galilean Satellites. *Icarus*, **136**, 202–222.

Zimmer, C., Khurana, K., and Kivelson, M. G. (2000). Subsurface oceans on Europa and Callisto: constraints from Galileo magnetometer observations. *Icarus*, **147**, 329–347.

Zolotov, M. Y. and Shock, E. L. (2001). Composition and stability of salts on the surface of Europa and their oceanic origin. *J. Geophys. Res.*, **106**, 32815–32827.

FURTHER READING AND SURFING

galileo.jpl.nasa.gov. Galileo spacecraft website; for images and general information.

Achenbach, J. (1999). *Captured by Aliens: the Search for Life and Truth in a Very Large Universe*. New York: Simon and Schuster. Terrific romp through the search for extraterrestrial life, including at Europa.

Bagnold, F., ed. (2004). *Jupiter: the Planet, Satellites and Magnetosphere*. New York: Cambridge University Press. The most up-to-date summary of the Jupiter system, including a synthesis of Galileo spacecraft data. Graduate-level text.

Beatty, J. K., ed. (1999). *The New Solar System*, fourth edn. Cambridge, MA: Sky Publishing. Fabulous introduction to the Solar System, suitable for a general audience and for scientists.

Demy, M. W. (1993). *Air and Water: the Biology and Physics of Life's Media*. Princeton: Princeton University Press. Compilation of the physics relevant to water's role in biology.

Greenberg, R. (2005). *Europa, the Ocean Moon*. New York: Spinger-Verlag. Inside information on the Galileo mission, and a summary of scientific results for Europa.

Johnson, T. V. (2004). A look at the Galilean satellites after the Galileo mission. *Physics Today*, **57**, 77–83 (April). Nice overview for nonexpert scientists.

Lewis, J. S. (1995). *Physics and Chemistry of the Solar System*. San Diego: Academic Press. Solar System chemistry and physics.

Sagan, C. (1994). *Pale Blue Dot: a Vision of the Human Future in Space*. New York: Random House. Visionary work on planetary habitability and astrobiology.

20 Titan

Jonathan I. Lunine and Bashar Rizk
University of Arizona

20.1 Introduction

Ten times farther from the Sun than the Earth is, shrouded in an orange haze, preserved at temperatures near 100 K, Saturn's moon Titan seems an unlikely astrobiological target. In fact, its extremes suggest images of death rather than life.

Yet this planet-sized moon possesses a dense atmosphere of nitrogen and methane, which, over time and with the action of ultraviolet radiation, may have generated 10^{16} tons of hydrocarbons and nitriles – the constituents of life – that were then deposited on its surface in liquid and solid form. Within these vast organic deposits, at those times when water ice might liquefy because of volcanism or of impact heating, some of the organic chemical steps leading toward the origin of life might be replicated and then preserved for study on the surface of Titan.

Titan's dense atmosphere has easily won it the high status we reserve for terrestrial planets with atmospheres (Venus, Earth, and Mars) although it is consigned to the outer Solar System. It possesses methane-driven meteorology, and its large size for an ice-rock world likely means a wealth of tectonic activity in its interior. A variety of observed surface landforms produced, apparently caused by erosional processes driven by winds and surface liquid hydrocarbons. In short, other than the likely absence of extant life because of the extreme cold, Titan exhibits a broad range of atmospheric and geologic processes that rival in complexity those of Mars and perhaps of Earth.

In late 2004, the ambitious US/European Cassini–Huygens Mission began detailed exploration of this distant world (as well as Saturn, its rings, and the many other saturnian moons). The present chapter is a status report on the properties of Titan, initial results from Cassini–Huygens (in particular the successful descent to the surface of Titan by the Huygens Probe), and prospects for future astrobiological exploration of its atmosphere and surface. We cover the earliest telescopic studies, the Voyager and Cassini–Huygens Missions (through early 2006), and the models of Titan they engendered. Where specific references are not given, the reader is referred to the excellent compendium of our knowledge of Titan by Coustenis and Taylor (1999).

20.2 Titan's place in the Solar System

Titan has been viewed as an intriguing place, albeit mysterious, throughout the age of Solar System exploration. It is the second largest moon in the Solar System and by far the largest natural satellite of the Saturn system. Its equatorial radius is 2575 ± 2 km, smaller by only 2% than that of the Solar System's largest moon, Jupiter's Ganymede, and 5% larger than Mercury. Titan moves about Saturn in a slightly elliptical orbit (eccentricity 0.03, meaning its distance from Saturn varies by $\pm 3\%$) with a semimajor axis of 1.22×10^6 km. Titan's mass, 2% of that of the Earth, makes it the giant in a system of modest-sized moons; the next largest is Rhea, with a mass and radius only 1.7% and 30%, respectively, of those of Titan. Indeed, Titan is much closer in bulk properties to the two outermost jovian satellites Ganymede and Callisto (Fig. 19.1 and Table 19.1), which closely bracket Titan in size and density, than it is to any of the other saturnian satellites.

Titan's average density of 1.87 g cm^{-3} suggests a roughly equal mixture by mass of rocky and icy components in its interior (liquid water at normal pressures is 1 g cm^{-3}, and silicates are $\sim$3 g cm^{-3}). If one considers that the most abundant ice-forming molecule in the outer Solar System is H_2O (because of the high cosmic abundance of hydrogen and oxygen), then it is a safe assumption that close to half the mass of Titan's interior is water ice. Furthermore, water ice has actually

Planets and Life: The Emerging Science of Astrobiology, eds. Woodruff T. Sullivan, III and John A. Baross. Published by Cambridge University Press. © Cambridge University Press 2007.

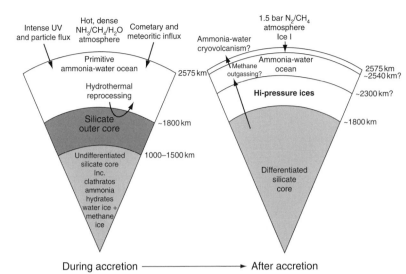

FIGURE 20.1 Possible models for Titan's interior, during and after accretion, based on Lunine and Stevensor (1987), Stevenson (1992), and Grasset *et al.* (2000). "During accretion" refers to the structure of Titan while it was growing as rock and ice planetesimals were added; "after accretion" is the structure after growth ceased and materials sorted out according to density.

been observed spectroscopically on and sampled emerging from the surface of Enceladus, one of Saturn's other moons (Porco *et al.*, 2006), and been observed spectroscopically on the surfaces of Ganymede and Callisto (McCord *et al.*, 1997). Ganymede at least seems to have a layered structure with a water ice mantle overlying a rocky core; we expect Titan to be similar. Titan might have other kinds of ices mixed with or dissolved into the water ice, including (but not limited to) methane and ammonia (Lunine and Stevenson, 1987); a possible cut through Titan is shown in Fig. 20.1.

The similarity in mass and size among the Solar System's three largest moons suggests a potential link in terms of processes leading to their origin. Titan's nearly circular orbit around Saturn, embedded within a group of other, so-called "regular" satellites, argues against its capture from another system or from solar orbit and argues instead for formation in place in a process of accretion from smaller bodies not unlike the formation of the planets themselves. In the case of all three satellites, the maximum thermal energy gained (per unit mass) from their assembly out of smaller particles (a process called accretion – see Section 3.9) turns out to be roughly equal to the latent heat of sublimation of water ice, L. That is,

$$GM/r \approx L, \qquad (20.1)$$

where G is the universal gravitational constant, M is Titan's mass and r its radius. Putting in the numbers from above one obtains 3500 J/g of accretional energy, while the latent heat of sublimation of water is 2600 J/g. The same result is obtained (within 25%) for Ganymede and Callisto.

The potential significance of this relationship for limiting the size of any icy satellite can be appreciated by considering the accretion process. Small bodies (planetesimals) of rock and ice fall into the gravitational well of the growing satellite with increasing velocity. As the satellite nears its present mass, and the relationship in Eq. (20.1) obtains, water ice is preferentially sublimated rather than accreted, while the silicates continue to accrete. Since water ice dominates over rock by a factor of a few (by mass), the net accretion rate drops by that same factor, while the forming satellite itself becomes progressively rock-rich. This appears to describe Titan well, since the satellite's density, equivalent to half-rock/half-ice, is significantly larger than that of its much smaller and more ice-rich neighbors. The same appears to have happened for Ganymede and Callisto.

Accretion for all three of these giant satellites was then finally terminated by depletion of the available rock, or perhaps simply by the general clearing of the protoplanetary disk and associated gases and solids from which the giant planets formed. Interestingly, the mass of rock is about the same not only in the three large satellites discussed so far, but also in Jupiter's other two main satellites, Io and Europa (Chapter 19), suggesting some commonality of supply or of accretion timescales. Titan (along with Ganymede and Callisto) thus appears to represent the largest rock-ice satellite that a giant planet is capable of producing.

But despite these similarities, Titan's dense atmosphere is remarkably different. This is what distinguishes its appearance and, it turns out, its evolution from that of its jovian counterparts with their extremely thin atmospheres. Indeed, were it not for this atmosphere, Titan would be neither the target of the Huygens spacecraft (see below), nor a chapter unto itself in this volume. It is to the atmosphere, therefore, that we turn next.

20.3　Titan's atmosphere

The Dutch astronomer and physicist Christiaan Huygens discovered Titan in 1655. Little else was known about the satellite until 1908, when the Catalan astronomer Jose Comas Solà measured a strong decline in brightness from center to limb (the apparent edge) on the disk of Titan. Such a limb-darkening effect is usually associated with the presence of an atmosphere, and Comas Solà suggested this possibility. Theoretical studies in the following decades showed that a large moon could retain an atmosphere, though the actual mass of Titan was highly uncertain. In 1943–44 the Dutch–American astronomer Gerard Kuiper used the 82-inch reflecting telescope at the McDonald Observatory in Texas (then among the largest telescopes in the world) to detect spectral lines of methane (CH_4) in the red part of Titan's spectrum (Kuiper, 1944). Kuiper recognized that, given plausible temperatures at the surface and the general appearance of the spectral lines, he was seeing gaseous methane, and hence an atmosphere. He also deduced, correctly, that he could not tell whether methane was the primary gas in a thin atmosphere, or a secondary gas in a thick atmosphere composed mostly of a gas that happened to have no spectral lines in the visible region. Speculations about the nature of the principal gas continued through the 1970s. Further spectroscopic observations revealed hydrocarbons additional to methane in the atmosphere, which led to the realization that Titan's atmosphere – however dense or rarefied – hosts interesting organic photochemistry, that is, chemistry initiated by the solar-ultraviolet breaking of carbon–hydrogen bonds (Strobel, 1974). Careful measurements of the limb brightness revealed that Titan's visible disk was a haze layer, lying at some unknown altitude (since the radius of the solid body was not known) above the solid surface.

The breakthrough in understanding the chemical and physical nature of Titan's atmosphere came during the Voyager 1 spacecraft's close flyby of Titan in 1980, with additional information from the more distant Voyager 2 flyby a year later. Voyager 1 passed within 4400 km of the surface of Titan and returned many, nearly featureless, images, confirming that at optical wavelengths (up through at least 700 nm) the haze is impenetrable.

Had the Voyager spacecraft been equipped only with imagers, we would have learned little about Titan from the mission. But three other experiments provided a thorough characterization of the atmosphere in a *tour de force* of synergistic science. The spacecraft was capable of sending radio waves back to Earth at two highly stable frequencies when the data transponder was turned off. As Voyager 1 passed behind Titan as seen from the Earth (an event called an occultation), these tones were beamed through Titan's atmosphere. The weakening of the signal as the beams moved deeper in Titan's atmosphere was due primarily to refraction of the signals, providing a measure of the atmospheric mass density ρ versus altitude z. It was then possible to derive the temperature T divided by the mean molecular mass m of the atmosphere, as well as the pressure gradient dP/dz, via the equation of hydrostatic equilibrium for a stable atmosphere (Chapter 4)

$$dP/dz = -\rho g, \qquad (20.2)$$

and the ideal gas law

$$P = nkT = \rho kT/m. \qquad (20.3)$$

Here n is number density, g the gravitational acceleration at Titan's surface (135 cm/s^2, or 1/6 that at Earth's surface), and k is Boltzmann's constant. Furthermore, the point at which the signal was cut off as Voyager moved deeper into its occultation by Titan, compared to the point at which the signal re-appeared on the other limb, provided the diameter of the solid body. This, plus the tracking of Doppler frequency shifts caused by the changing orbital path perturbed by Titan's mass, yielded the bulk density reported above.

The number density of the atmosphere at Titan's surface proved to be four times that at sea level on Earth, or $\sim 1 \times 10^{20}$ molecules/cm^3, even assuming a generous range of possible molecules. This is the second highest atmospheric density among the four solid Solar System bodies with atmospheres, Venus having the highest with Earth and Mars lower. To proceed further required data from other instruments. The infrared spectrometer on Voyager determined temperature as a function of altitude by measuring the emitted flux at different wavelengths in the mid-infrared. The characteristic shape of the profile of temperature divided by molecular weight was similar to Earth's, with (assuming a slowly varying molecular weight) a declining temperature from the surface upward and then a subsequent increase. The minimum was tied to the lowest observed infrared brightness

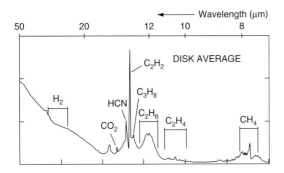

FIGURE 20.2 Infrared spectrum of Titan's atmosphere obtained by Cassini's composite infrared spectrometer (CIRS) in 2004. A rich spectrum of organic molecules is visible. (CIRS team, NASA.)

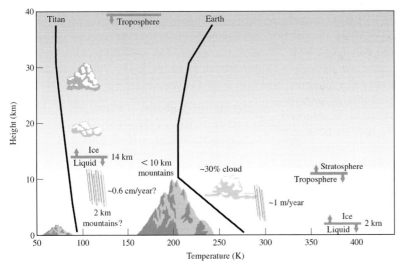

FIGURE 20.3 Schematic profile of temperature versus altitude on Titan (left) and Earth (right). "Ice/liquid" and the clouds refer to methane on Titan and water on Earth.

temperature of 70 K. Using this value led to a mean molecular mass of the atmosphere of 28 atomic mass units (amu), well above that of methane at 16. With the assumed small variation in molecular weight with altitude, the surface temperature was determined to be 95 K and the surface pressure 1.5 bars (Lindal *et al.*, 1983). This high pressure and low temperature eliminated methane, with its low vapor pressure, as the primary gas; in fact, it limited methane to being at most 10% by number near the surface and 1.5% at the temperature minimum. And the high molecular weight did so, too; considering cosmic abundances, the most plausible composition was either mostly N_2 or CO (both 28 amu), with some additional argon and methane.

The final instrument in the trio resolved the compositional ambiguity. The ultraviolet spectrometer measured atomic emission lines in the upper atmosphere, as well as ultraviolet absorption lines observed when Voyager measured the Sun's spectrum as seen "looking back" through Titan's atmosphere. The predominant ions were of nitrogen, and oxygen was limited to

10 parts per million relative to nitrogen. Hence, the dominant gas (by number) in Titan's atmosphere was established as molecular nitrogen (90 to 98%, varying with altitude), with methane next (0.5 to 4%, also varying with altitude), ~0.1% molecular hydrogen, ten parts per million carbon monoxide, a very small admixture of carbon dioxide (at the parts per billion level, discovered in the infrared spectra), and a host of hydrocarbons and nitriles (organic molecules containing CN) (Fig. 20.2). Argon was suspected to be present, but only an upper limit of 7% was obtained by Voyager. The temperature profile was then also pinned down, and found consistent with convective transport of heat in the lower 4 km of Titan's atmosphere, radiative equilibrium above that, and a temperature minimum ("tropopause") at an altitude of 40–50 km (Fig. 20.3).[1] Further, Voyager data implied that methane *clouds* with some dissolved nitrogen should condense in the lower atmosphere; and indeed, various

---

[1] The nitrogen gas is so cold at 75 K that it deviates from ideal gas by about 10%, necessitating adjustments in atmosphere models.

lines of evidence, in particular near-infrared spectra from Earth, revealed clouds (Griffith *et al.*, 2000).

The similarity in shape of Earth's and Titan's atmospheric temperature profiles, and the identity of their primary gas as N_2, is remarkable. Titan's atmosphere provides a classic greenhouse warming (Section 4.1.1) of an additional 10 K above the 85 K that it would otherwise have. Because of the low temperatures, however, water *vapor* is almost nonexistent (trace amounts, presumably from micrometeoroids, were detected by the Infrared Space Observatory (Cousteins *et al.*, 1998)) and carbon dioxide gas is scarce. In their place, the greenhouse is effected through absorption of mid- to far-infrared thermal radiation by nitrogen, methane, and molecular hydrogen. This is possible only in the high density lower atmosphere where molecular collisions distort the shapes of these simple molecules (Courtin *et al.*, 1995), shifting the wavelengths of their spectral absorption. In the region above the tropopause, absorption of solar ultraviolet radiation by hydrocarbons, as well as of solar blue light by haze particles, creates a warm stratosphere and filters out all but 10% of the sunlight before it reaches the lower atmosphere (McKay *et al.*, 1991).

20.4 Hydrocarbon chemistry and oceans

The rich suite of hydrocarbons and nitriles seen in Titan's upper atmosphere are a testament to the action of sunlight on methane and nitrogen and led in the 1980s to the prediction that the surface of Titan should be covered by a methane–ethane ocean. That this has turned out not to be the case does not detract from the chain of reasoning that led to this prediction, much of which has been supported by the latest observations. The absence of a clearly identifiable methane reservoir that operates today as it must have operated historically is still something of a mystery.

The photochemical cycle begins hundreds of kilometers above the surface where solar ultraviolet light at wavelengths below 160 nm breaks the carbon–hydrogen bonds in methane (CH_4). Many different reaction chains occur, but one typical example is

$$2\ (CH_4 + UV\ photon \rightarrow {}^*CH_2 + H_2),$$
$$2\ ({}^*CH_2 + N_2 \rightarrow CH_2 + N_2),$$
$$2\ CH_2 \rightarrow C_2H_2 + H_2; \tag{20.4}$$

net reaction:

$$2CH_4 \rightarrow C_2H_2 + 3H_2.$$

Here C_2H_2 is acetylene and ${}^*CH_2$ is the excited state of the radical CH_2. Acetylene is the second-most abundant product of Titan *photolysis* (break-up of molecules by the action of light), with the most abundant being ethane (C_2H_6), based on infrared spectra of the stratosphere from spacecraft and from the Earth-orbiting Infrared Space Observatory (see Fig. 20.2). This is somewhat surprising, because dissociation of methane would be expected to make the simpler acetylene much more readily.

The answer lies in the photodissociation of acetylene itself, which requires less energy and can occur at longer wavelengths of light (up to 200 nm), for which far more photons come from the Sun and penetrate deeper into Titan's atmosphere. The resulting radicals "sensitize" the production of ethane, C_2H_6 (Yung *et al.*, 1984) in reaction sequences such as

$$2\ (C_2H_2 + UV\ photon \rightarrow C_2H + H),$$
$$2\ (C_2H + CH_4 \rightarrow C_2H_2 + CH_3),$$
$$2CH_3 + N_2 \rightarrow C_2H_6 + N_2. \tag{20.5}$$

The net reaction is:

$$2CH_4 \rightarrow C_2H_6 + 2H,$$

where acetylene itself catalyzes the conversion of methane into ethane.

The high abundance of N_2 ensures that the final step in reaction (20.5) is not rare in Titan's atmosphere. Further, note that for both ethane and acetylene, hydrogen is liberated from carbon. This hydrogen escapes rapidly from the atmosphere, making the conversion of methane to higher hydrocarbons and nitriles irreversible. That this really happens was confirmed by a substantial circum-Titan corona of atomic hydrogen observed by Voyager's ultraviolet spectrometer.

This simple picture encountered a complication when one noted that the amount of methane that the atmosphere can hold is limited by the molecule's saturation vapor pressure. The Voyager-determined tropospheric abundance of methane is very roughly half or less of the maximum that would be attained if the entire atmosphere were saturated in methane, a result supported by the methane relative humidity of 50% detected at the Huygens landing site. This amount of methane, however, will in general be depleted within a few tens of millions of years, or only ~1% of the age of the Solar System (Yung *et al.*, 1984). Therefore either methane is continually resupplied to the atmosphere, or a relatively methane-rich atmosphere, such as today, represents a rare occurrence in Titan's history.

Resupply from impact or volcanism is possible, but it is difficult to fine-tune these processes to "refill" the atmosphere without saturating it and producing a reservoir of liquid methane at the surface. So it looked like there might well be a methane ocean on Titan. Other lines of evidence came from models of the aerosols (small suspended solid particles) in the atmosphere.

Photochemical models all predicted similar production rates for the major hydrocarbons and nitriles seen in the infrared spectra. Since essentially all of the products are much less volatile than methane, most of them quickly condense in the form of aerosols. Some of the material remaining in the gas phase, however, is processed further to very heavy hydrocarbons and nitriles. This material appears, on the basis of matching colors with laboratory simulants, to account for the orange haze obscuring Titan's surface. It has been coined *tholin* by Carl Sagan and colleagues (Khare *et al.*, 1984), and serves as nucleating seeds for aerosols, thus falling to the surface as well. One authoritative laboratory synthesis program identifies these hydrocarbons as polyacetylene $C_{2n}H_2$, vinylacetylene $C_{2n}H_{2n}$ and polyvinyl $C_{2n}H_{2n+2}$ (Dimitrov and Bar-Nun, 2003), i.e., particles with plastic cores.

Such aerosols were thought to acquire acetylene/ethane/nitrile mantles and outer crusts of liquid or solid methane from the troposphere, then drop to Titan's surface (Coll *et al.*, 1999) so gradually that the finest Earthly drizzle would seem a torrent. The acetylene would remain solid, but ethane and propane, C_3H_8, melt at 91 K and hence exist on Titan's surface as liquid. If photochemistry has occurred in steady-state over the 4.6 Gyr history of the Solar System, Titan would possess a layer of liquid hydrocarbon a few hundred meters thick, maybe double that, and a solid acetylene and nitrile layer 100 meters thick! Because the solubility of the solid material in the liquid is limited, it was expected the lower density liquid would overlay the solid organic sediments. Table 20.1 gives the equivalent layer thicknesses hypothesized over 4.6 Gyr for the major products of Titan photochemistry.

The ethane carried to the surface as part of the aerosol mantles would mix with the methane deposited during episodes of impact or volcanism to produce a euphonious ethane-methane ocean (Lunine *et al.*, 1983). The coexisting methane vapor pressure would be less than that for pure methane at the same temperature, and this would be consistent with the Voyager-derived limits on methane abundance near Titan's surface. In addition, nitrogen would dissolve in such an ocean.

TABLE 20.1 Depth of hydrocarbons and nitriles generated on Titan over the age of the Solar System

Species	Depth of layer, m
C_2H_6 ethane	600
C_2H_2 acetylene	100
C_3H_8 propane	20
HCN hydrogen cyanide	20
HC_3N cyanoacetylene	2
CN cyanogen	0.8
CO_2 carbon dioxide	0.02

For a long period such an ethane–methane–nitrogen surface ocean seemed a straightforward way of explaining steady-state methane photolysis over the age of the Solar System – the ocean would be the source (methane) and sink (ethane) of photolysis, and would be drawn down in volume as methane was converted to ethane and acetylene with loss of hydrogen.

Such an ocean, though, would have observable consequences on the shape of Titan's orbit, which is slightly eccentric. Unlike the large satellites of Jupiter, which mutually maintain their orbital eccentricities against dissipation by tidal interactions with Jupiter (Section 19.3), Titan has no such buffer. An ocean would experience strong tidal currents, which would dissipate orbital energy and make Titan's orbit circular on short timescales, in a process analogous to how Earth's tides are transferring angular momentum to the Moon and slowing down our planet's rotation rate.[2] If Titan's ocean exists, it must be either very deep (damping tidal currents and submerging all topography completely) or confined to individual crater basins (Sears, 1995). But by the mid-1990s the former possibility was ruled out by new observations from Earth, described next. And now the latter possibility has been ruled out by observations from Cassini–Huygens, described in Section 20.7.

20.5 Earth-based remote sensing

Many attempts were made to view Titan's surface from the Earth through the 1980s. Initial successes were obtained by bouncing radar signals off the surface, a technological challenge for such a small object 1.5 billion km away. The haze particles are much too small to scatter radio waves, and the atmosphere provides

[2] See Section 19.3 for details of this physics as applied to the Jupiter/Europa system.

insufficient attenuation at microwave wavelengths (this is not inconsistent with the Voyager radio science experiment, which measured *refraction* of a narrow beam, rather than direct absorption). Initial results immediately ruled out a global ocean, because the returned signal indicated a reflectivity too large for liquid hydrocarbons; furthermore, substantial variability over Titan's surface was inferred (Muhleman *et al.*, 1998). Further results with the 305-m diameter Arecibo radio telescope (Appendix D) have three intriguing characteristics (Campbell *et al.*, 2003). First, the reflected returns vary greatly, from values low enough to be consistent with patchy hydrocarbon seas, to those high enough to be water ice. Second, by comparing the strength of the returned signal in two different polarizations, it has been found that in some locations the surface is smooth (specular), like a liquid, and in others rough. The specular reflection is consistent with partial coverage of the surface by liquid hydrocarbons.

By the late 1980s near-infrared observers had also been successful in penetrating the haze of Titan. The small size of the haze particles, as determined from Voyager observations at different angles relative to the Sun, implies that beyond the near-infrared (800 nm wavelength) the scattering efficiency of the haze particles decreases. However, one must then contend with methane absorption bands that absorb most of the light from the surface. So to observe Titan's surface in the near-infrared one must look at wavelengths between the methane bands, and indeed there one finds surface variability (Griffith, 1993). Such images of Titan's surface (Smith *et al.*, 1996) by the Hubble Space Telescope (HST) revealed a large, Australia-sized bright area, called Xanadu, surrounded by mottled terrain.

With the advent of adaptive optics techniques applied to large telescopes on Earth, partially removing the blurring effects of Earth's atmosphere, images of Titan at somewhat better spatial resolution and much higher signal to noise were obtained (Combes *et al.*, 1997). They showed that some areas have reflectivities of 5% or less, consistent with liquid or solid dark hydrocarbons (Gibbard *et al.*, 1999). Recent HST images at around 2 μm wavelength (less affected by the haze than those at 1 μm) provide further evidence that the dark areas are hydrocarbons (Meier *et al.*, 2000). It has become clear that Voyager's inability to see Titan's surface was purely because the spacecraft imagers contained vidicon tubes whose sensitivity dropped off beyond 700 nm.

The presence or absence of large amounts of methane on Titan's surface has profound implications for the evolution of Titan's atmosphere and surface over time. To supply methane for billions of years of photolysis requires hundreds of meters depth of liquid methane, and this is either stored in a porous ice crust (Stevenson, 1992), concentrated in crater basins which are at the yet unobserved high latitudes, or largely absent from the surface–atmosphere system (Coustenis and Taylor, 1999). These and other questions of Titan's origin and early evolution, as well as the nature of the surface organics, required a detailed look that only *in situ* and orbiter-flyby spacecraft could provide.

20.6 The Cassini–Huygens Mission

20.6.1 Genesis of the mission

Despite the success of the Voyager flybys around 1980, budget cuts at NASA meant that a return to Saturn would take two decades. In the end, combining NASA and European Space Agency (ESA) resources led to a single mission that would include an American Orbiter of Saturn and a European Probe to study Titan, with instruments designed by consortia on both continents. Titan's low gravity, nitrogen–methane atmosphere, and solid surface, presented difficult engineering challenges. How would one design a probe to sample a large region of the atmosphere slowly enough to get meaningful data, yet not hang up in the lower atmosphere for so long that the Orbiter would be out of sight (and communications) before impact? By 1990 these technical problems had been solved and the mission was funded. The Orbiter was named Cassini, after Gian Domenico Cassini, the seventeenth-century astronomer who made key discoveries about the nature of Saturn's rings and was the first to see several moons of Saturn. The Probe was named after Christiaan Huygens, the discoverer of Titan. In 1997, after seven years of development, budgetary battles, and about US$2–3 billion spent between the United States and Europe, Cassini–Huygens was launched for its seven-year journey to Saturn and Titan.

20.6.2 The Cassini Orbiter

The Orbiter and Probe together represent one of the most heavily instrumented planetary spacecraft ever assembled (Fig. 20.4). The Orbiter payload, which arrived at Saturn in 2004 and will remain operational until at least 2008, contains 12 instruments divided into two broad categories. First, half-a-dozen experiments measure particles, fields, and waves. They assess the magnetic, charged particle, dust, and elemental

FIGURE 20.4 The Cassini Orbiter spacecraft; the Huygens Probe is the covered saucer facing the camera. The dish at the top is for radio communications with Earth. Note the "bunny-suited" workers for scale. (NASA)

environment of the Saturn system in different ways, including mapping of electromagnetic fields, direct sampling of particles, and particle imaging of the entire magnetosphere.

An Ion and Neutral Mass Spectrometer (INMS) directly samples material when the Orbiter dips within the atmosphere of Titan at an altitude of 950 km. This direct sampling of the upper atmosphere, by an instrument that is similar to one that samples the deeper atmosphere on the Huygens Probe, provides important comparisons of the major elements C, N, and H. How the elemental abundances change with altitude is an important input to models of the evolution of the atmosphere, as discussed below.

A second broad category of instruments gathers information using electromagnetic radiation. Also six in number, they sense photons from the ultraviolet to the radio, some with imaging and spectroscopic capability.

Some of these remote sensing instruments are jointly steerable so that all can simultaneously observe a target. They include an ultraviolet spectrometer, two CCD imaging cameras for optical and near-infrared wavelengths, a near-infrared spectral imager (which produced the spectrum of Fig. 20.2), a polarimeter to measure the polarization of light reflected from bodies, and a mid-infrared spectrometer that also functions as a radiometer to accurately measure temperatures. They observe Titan on a large fraction of the 40-odd close flybys that the Orbiter makes during the mission.

A further set of instruments work in the radio part of the spectrum. A radar system transmits and receives through the main dish antenna, producing images with resolution up to several hundred meters. The radio transmitters used for communication can, as on Voyager, measure atmospheric refractivity and hence generate pressure–temperature profiles. They can also be used during close flybys to measure the mass of

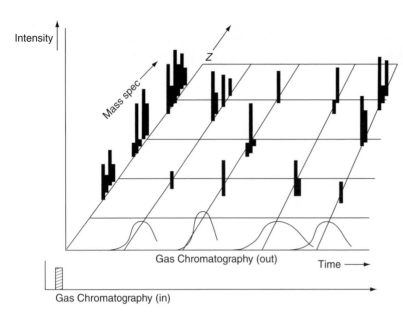

FIGURE 20.5 Schematic illustration of how gas chromatography (GC) yields a series of separate samples that, when analyzed through a mass spectrometer (MS), provide better separation than a mass spectrometric analysis of the bulk sample. The intensity–time plane defines a standard GC run, showing different peaks as a function of time that represent different types of molecules (light organic, polar, etc.) exiting the device. The z-plane illustrates the coupling of an MS to this analysis – without the GC, the MS peaks, schematically shown as black bars, overlap and are ambiguous. But if the MS sample is run on GC-separated classes of molecules, as shown in the figure, then fewer types of molecules are present in each run, reducing ambiguity. Adapted from Niemann et al. (1997).

Titan and the distribution of mass within the moon's interior, that is, to determine whether the interior rock and ice have separated from each other in the interior (or *differentiated*, as shown in Fig. 20.1).

To save on spacecraft complexity and hence cost during development, the scientific instruments are mounted on fixed pallets rather than moving scan platforms. In consequence, only one group of instruments can observe at a given time, and certain instruments within each group cannot be used simultaneously. Furthermore, since Titan is the only satellite in the system large enough to deflect the path of the spacecraft and send it on to other targets, the number of Titan flybys during the Orbiter's four-year tour (2004–08) is determined by celestial mechanics and the desire to observe closely other saturnian moons, Saturn itself, and its rings and magnetosphere.

20.6.3 The Huygens Probe

The Huygens Probe separated from the Orbiter in January 2005, entered the atmosphere of Titan at hypersonic velocity (6.2 km/s), slowed to transonic speed via a heat shield, and collected data over a 2.5 hr period, from 170 km altitude down to the surface, while descending on several parachutes (Lorenz and Mitton, 2002; Lebreton *et al.*, 2005).

The Probe was instrumented with six experiments. One used an oscillator to create a precise radio frequency whose Doppler shift was tracked by ground stations as the Probe descended; shifts from the

nominal trajectory provided a measure of wind speed. Another measured temperature, pressure, electrical potential, and other physical parameters as the Probe descended. Upon landing, accelerometers provided data on the hardness of the surface as well as measuring the surface's acoustic properties via a sounder that operated before impact. Another multipurpose instrument, the Descent Imager/Spectral Radiometer (DISR), made near-infrared panoramic images of the surface during descent, measured cloud properties through observations of optical effects such as aureoles, acquired up- and down-looking spectra in the visible and near-infrared, before impact took a near-infrared spectrum of the surface using artificial illumination, and finally took images of the surface after landing.

Finally, two instruments formed a crucial tandem to analyze Titan's atmosphere *in situ*. The Gas Chromatograph and Mass Spectrometer (GCMS) was the first such tandem device to be deployed on a planet since the 1976 Viking mission to Mars (Section 18.5.2). A gas chromatograph works by separating compounds according to how effectively they are trapped by substances within the column. The mass spectrometer used a quadrupole mass filter and could separate species ranging from 2 to 141 amu (Niemann *et al.*, 1997). Operated together, the chromatograph separated species so that they were analyzed in separate batches in the mass spectrometer, eliminating key ambiguities such as between nitrogen and carbon monoxide (Fig. 20.5). The mass spectrometer rapidly took many samples as the Probe descended; a subset of these were preceded by the gas chromatographic separation. At preselected altitudes the sample was put through a pyrolyzer that vaporized solid and liquid aerosols and cloud particles for analysis in the GCMS. Since the Probe survived its landing on the surface, and had one of its two antennas pointed upward, communication from the surface to the Orbiter was possible for an hour. During that time, the warm inlet of the GCMS was able to vaporize surface hydrocarbons and do a surface chemical analysis.

The GCMS determined abundance versus altitude for various organic species, major gases, noble gases that otherwise are difficult to detect, and isotopes of several major elements including C, N, and H. This single chemical assay from stratosphere to surface was supplemented with the Orbiter's mass spectrometer analysis at the top of the atmosphere, and the infrared studies conducted by Orbiter instruments. Together these represent a detailed assessment of the bulk and organic composition of Titan's atmosphere, which in turn is now allowing a number of key questions, elaborated below, to be addressed.

20.7 Cassini–Huygens's view of Titan

The experiments of Cassini–Huygens have tremendously expanded our knowledge of Titan. Here we focus on the issues with relevance to astrobiology: the origin of Titan's atmosphere; the history of its surface and implications for the working models of Titan accepted before the mission, especially the concept of the global ethane–methane–nitrogen ocean; Titan's liquid cycle; and any evidence for life. This summary is primarily based on the first wave of Cassini–Huygens publications, including those by Porco et al. (2005) on imaging results from the Orbiter; by Elachi et al. (2005) on radar mapping from the Orbiter; and, in a special issue of Nature, with Owen (2005) and Lebreton et al. (2005) summarizing the Huygens results, followed by many detailed papers such as Tomasko et al. (2005) and Niemann et al. (2005). The mission has at least two more years to run, and so this can only be a progress report.

20.7.1 Origin of Titan's atmosphere

Although Titan's atmosphere contains compounds that have also been found in comets, though not in similar relative abundances (Iro et al., 2002), before Cassini–Huygens the relationship between the origin of Titan and that of other outer Solar System bodies

remained unclear. It is possible that Titan accreted much like Ganymede and Callisto, and then acquired its volatile materials from solar-orbiting comets that were scattered by the giant planets (Griffith and Zahnle, 1995). Alternatively, the immediate environment around Saturn, during that planet's formation, may have altered the chemical state of carbon- and nitrogen-bearing species, yielding methane rather than carbon monoxide, and ammonia rather than nitrogen, species that could have condensed or otherwise been incorporated into water ice in large amounts. Such a Titan would have been rich in carbon- and nitrogen-bearing volatiles from the start.

Measurements of the isotopes of carbon, nitrogen, hydrogen, and oxygen provide constraints on the amount of volatile material with which Titan was initially endowed (Lunine et al., 1999). Noble gases provide yet another constraint. If Titan acquired its nitrogen largely in the form of N_2, which would suggest addition of material from solar orbit, then argon (of similar volatility to nitrogen) should have an abundance within a factor of ten of its solar abundance relative to molecular nitrogen (Owen, 1982). If, on the other hand, Titan formed in an environment rich in ammonia (NH_3), which was then converted to molecular nitrogen (as might happen in Saturn-orbiting planetesimals), then Titan could have acquired very large amounts of nitrogen, since ammonia bonds much better than argon to water ice. In this case, the measured argon-to-nitrogen abundance ratio should be less than 1% that of the Sun.

Initial results from Cassini–Huygens favor the latter model. The GCMS on the Huygens Probe found the $^{15}N/^{14}N$ ratio in Titan's troposphere and stratosphere to be enriched relative to its abundance in the Earth's atmosphere, as did the INMS for the upper atmosphere. This implies that perhaps five times the amount of nitrogen currently in Titan's atmosphere has escaped since the satellite formed. The GCMS also found the concentration of ^{36}Ar to be depleted by more than six orders of magnitude from its solar value when referenced to the concentration of N_2, and by more than two orders of magnitude from the concentration of ^{40}Ar in the atmosphere. No primordial noble gases other than ^{36}Ar were detected – no Kr, no Xe, not even ^{38}Ar. This result makes it likely that the nitrogen in Titan's atmosphere arrived not as N_2, trapped in amorphous or clathrate[3] water ice, but as NH_3

---

[3] In a clathrate one compound is trapped within the crystal structure of another.

bound to water ice within the ∼100-m-sized planetesimals that built Titan in the proto-solar, proto-saturnian nebula. Had it arrived directly as N_2, then primordial Ar and heavier noble gases would be present today since the accepted mechanisms for their incorporation into early Titan are through entrainment in these rather exotic forms of water ice.

The implications of these findings for CH_4 are even more profound. They preclude CH_4 being brought to Titan locked in clathrate form and instead point to CO_2 as the main carbon-bearing constituent at Titan's formation. CO_2 was detected in amounts comparable to ^{40}Ar by the GCMS. The current conception is that CO_2 has been and is now converted to CH_4 through a reductive process that occurs in Titan's interior – the methane is then continually supplied to the surface throughout the satellite's history, probably in episodic fashion. The $^{13}C/^{12}C$ ratio of 0.0122 is hardly different from the terrestrial ratio of 0.0111, supporting this scenario since similar processes occur on Earth.

20.7.2 History of the surface

The signature of impact cratering can be seen on the other moons of the saturnian system, just as it can be seen throughout the Solar System. Because each giant planet had its own distinctive population of planetesimals during formation, there should be variations in the numbers of craters per unit area (usually called crater frequency) versus crater size for objects in one system compared to another, and even within a given satellite system, since the outermost satellites are more affected by Sun-orbiting debris. On a body with a thick atmosphere, the crater size-frequency distribution is distorted by the screening effect of the atmosphere. For incoming projectiles that are relatively weak, and hence shatter under the effect of the dynamic pressure of the atmosphere ρ_{atm}, an impactor of size R and density ρ will break up if it is smaller than a size R_O:

$$R_O \sim (\rho_{atm}/\rho)H/\pi, \qquad (20.6)$$

where $H = kT/mg$ is the pressure scale height[4] of the atmosphere. The larger the atmospheric density, or the more extended the atmosphere, the larger the threshold impactor size that breaks up in the air. Once shattering occurs, the pieces themselves are slowed and may shatter again, eventually yielding debris that falls onto the surface at such slow speeds that it fails to make any

crater. This effect is seen clearly on Venus, where the relative abundance of large craters is much higher than for the Moon or Mars. (Most of the smaller craters that are found on Venus are secondary impacts of debris tossed up in the initial impact of a large body.) The diameter of the crater formed in a hypervelocity impact depends upon a number of factors, but is roughly 10 times the diameter of the impactor (Holsapple, 1994).

Titan's large atmospheric density and scale height should make it a candidate for a screening effect similar to that on Venus, and detailed calculations predict that craters less than a few kilometers in diameter should be rare on Titan compared to what is seen on its companion satellites without atmospheres (Ivanov et al., 1997). Cassini's instruments are easily capable of seeing any craters below the predicted cutoff size, which provides a potential means of gauging the history of the atmosphere. If Titan's atmosphere was much thinner in the past, it would have allowed survival to the surface of impactors of a size much smaller than those screened out by the present-day atmosphere. The record of this epoch might exist in places on the surface in the form of a size-frequency distribution that includes a much larger number of small craters (Engel et al., 1995).

Cassini–Huygens has so far shown us a highly resurfaced world with very few craters detected. The Orbiter's mosaic image in the near infrared (at a specific wavelength of 938 nm at which methane does *not* absorb and thus allows one to see the surface well) reveals many clouds and surface features, but very few large crater-like features (Porco et al., 2005; Fig. 20.6). The various radar swaths so far acquired also exhibit very few. The Probe's images identified only a single possible crater-like feature. The number of craters of different sizes yields a rough estimate of ∼130–300 Myr for the age of the surface on the largest scale. One large crater, imaged by the Cassini radar, dubbed Circus Maximus, has a diameter of ∼400 km and appears to be of impact origin (Fig. 20.7). It also exhibits long channels on its outer slopes, perhaps caused by flowing liquid methane.

Based on the small fraction of Titan's surface that has to date been intensively studied, four candidate processes have been identified to erase craters: (1) wind carrying solid methane particles, higher hydrocarbons, and water ice grains; (2) *cryovolcanism* (volcanism in which the working fluids are not silicates, but water, ammonia, and other low melting point fluids); (3) flowing liquid methane; and (4) perpetual deposition of plastic aerosol particles over the surface. How much each contributes is just beginning to be determined,

---

[4] *Scale height* in an atmosphere is the change in altitude necessary for the pressure to change by a factor of $e = 2.78$.

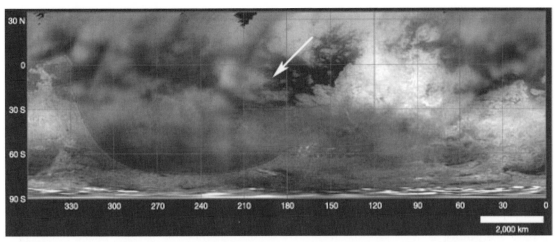

FIGURE 20.6 Composite image of Titan by the Cassini Orbiter, with resolutions ranging from ~10 to ~180 km. All images were taken at a wavelength of 938 nm (where methane does *not* absorb), allowing better visibility through the haze to the surface. The image shows a mixture of clouds and surface features; for example, the prominent white feature at 80° to 130° longitude is Xanadu Regio (already known from ground-based observations), but the white spots in the south polar region are convective clouds. The arrow shows the Huygens Probe landing site. (Porco *et al.*, 2005; NASA.)

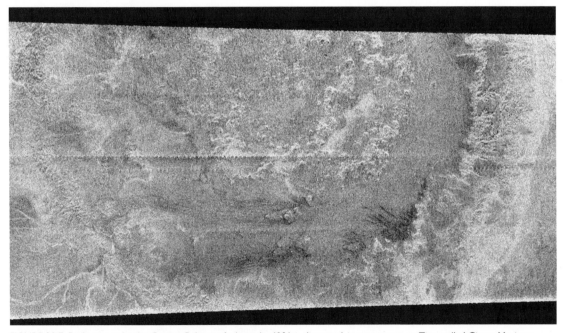

FIGURE 20.7 Radar image by the Cassini Orbiter of a large (~ 400 km diameter) impact crater on Titan called Circus Maximus. Light coloration corresponds to areas of greater roughness (more back-scattered signal). Apparent fluvial channels coursing down the outer crater slope, perhaps cut by liquid methane, can be seen in the lower left. Light features in the center are ~25 km-sized elevated regions. (NASA.)

as is whether these are the only processes that contribute and over how much of Titan's total area they contribute. Together, they have shaped a landscape that belies the cryogenic temperatures at which Titan's surface exists.

The operation of wind is readily observable from space. A series of parallel linear features hundreds of km long and 1–2 km wide are observed in the radar images (Fig. 20.8). Initially dubbed "cat-scratches," they are believed to be dunes (Lorenz *et al.*, 2006),

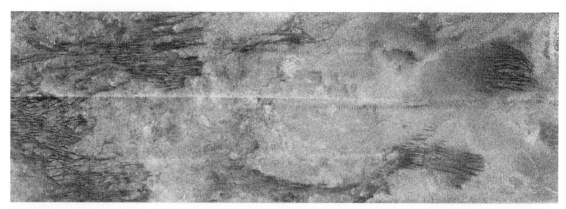

FIGURE 20.8 Radar image of Titan showing a suite of parallel, east–west dark "cat-scratch" features that are interpreted as longitudinal dunes. The image is 300 km high and almost 900 km long. The dunes themselves can extend for hundreds of km and are up to 2 km wide. (NASA.)

molded by the planet's possible methane-driven circulation (see below).

Evidence for cryovolcanism is also readily apparent from the Orbiter, where imaging reveals features that appear similar to low-viscosity flows on Earth (Mauna Loa) and on Mars (Olympus Mons) (Sotin *et al.*, 2005). The Huygens DISR image mosaics also show features indicative of cryovolcanic flows, an active geology and a possible energy source. Images such as Fig. 20.9a have a lack of impact craters, marked albedo differences between lighter highlands (upper left) and darker plains suggesting a past flow, round features that look like volcanic craters, and dark circular features that suggest subterranean upwelling.

Evidence of fluid flow at the Huygens Probe landing site is compelling. The DISR images show bright highlands terrain, at an elevation of 50–200 m, networked by dark drainage channels (Fig. 20.9a) leading to an obvious shoreline adjacent to a large flat dark plain, possibly a lakebed (Tomasko *et al.*, 2005). Although the plain appears to have been flooded in the past, it is currently dry – we know this because Huygens landed and did not splash[5] (Fig. 20.9b)! Its low contrast suggests that it is covered uniformly with a single material both darker and bluer than that covering the bright highlands.

The river channels in the highlands must have been cut (into a "bedrock" of thick ice) by some liquid, and the best candidate is liquid methane, fed either by methane rain or subsurface springs. How variable is this rain and therefore the flows in the channels? The

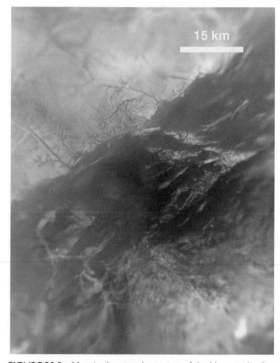

FIGURE 20.9a Mosaic showing the region of the Huygens landing site, which was at the exact center. Dark lines that look rectilinear and a network of drainage channels drain the brighter highlands area (upper left) into the darker plain, or lakebed, separated from the highlands by an evident shoreline. (ESA/NASA.)

GCMS detected an increased level of methane while on the surface, suggesting that liquid methane is mixed with the surface material and was mobilized when the Probe's heat was transferred into the surface. The measured deceleration of the impact indicated that the physical consistency of the surface was something like that of lightly packed snow or wet sand.

---

[5] However, *if* the Probe had landed in hydrocarbon liquid, it was designed to float and keep on working!

FIGURE 20.9b The surface of Titan as seen by the Huygens Probe (with the aid of a lamp). The 10–15 cm-sized "rocks" are believed to be cobbles of water-ice, covered with dark hydrocarbons. (ESA/NASA.)

Finally, the fourth process, the slow and steady deposition of the ubiquitous aerosol haze, provides the material upon which aeolian (wind) and fluid transport can operate. Initial results indicate an aerosol population that is fairly uniform in number density and size with altitude, with a scale height significantly larger than the background gas, and a fractal internal geometry that grows more compact as the altitude decreases and consists of agglomerations of several hundred 0.1 μm-diameter monomers (Tomasko *et al.*, 2005). This agrees with evidence from contemporary laboratory syntheses that the aerosols manufactured well above Titan's tropopause cannot just keep growing, but reach a size threshold past which they do not grow (Dimitrov and Bar-Nun, 2003). The Huygens observations also showed that the haze permeates the troposphere and lower stratosphere to a greater degree than expected, all the way down to the surface, as if on

a smoggy Los Angeles day. But Los Angeles's smog has its source near the ground and its concentration with altitude shows this by falling off exponentially, whereas Titan's smog has its source at high altitude (>500 km), from where it rains out, so it exhibits a much more distributed presence in Titan's atmosphere.

20.7.3 Titan's liquid cycle and the history of methane

The Cassini–Huygens observations speak of a planetary history for Titan every bit as rich as the Earth's and for similar reasons: wind, volcanism, smog, and the operation of a universally available compound very close to its triple point at the current surface temperature. This is CH_4, playing the role of H_2O on our planet, moving in and out of its solid, liquid, and gaseous phases, transporting energy and materials and participating in a complex planetwide chemical factory whose operation creates hydrocarbon compounds that on Earth would be biologically interesting. To take a midday stroll on Titan would be to explore a strangely familiar, cold, perhaps rainy, terrain, suffused by a general orange glow (from the much greater amount of atmospheric scattering of sunlight) no brighter than on Earth ten minutes after sunset.

Cassini's radar system has revealed over a hundred lakes in the high northern hemisphere, ranging in size from ∼1 km to hundreds of kilometers across. Their morphology, radar-darkness and thermal response at radio wavelengths suggest they contain methane or a mixture of ethane and methane (Stofan *et al.*, 2007). Modeling of these lakes (Mitri *et al.*, 2007) suggests that they contain enough material to control the methane humidity on a global scale with respect to the meteorology, but are not a long-term supply of methane nor a sink of ethane over geologic time. Thus, the locations of the source of methane and of the sink of ethane remain open questions, as does the extent of photochemistry over the age of the Solar System.

Perhaps chemical processing involving the aerosols is vigorous and extensive enough to as not stop at ethane, but to continue until much more polymerization occurs. The unknown products might then be as solid as the plastic cores of the aerosols themselves, and therefore unable to form liquids at Titan's surface temperature. The chemical hardening discussed in the last paragraph of the previous section might be a symptom of this effect. This notion is supported by DISR surface spectra that show a "remarkable absence" of absorption features such as ethane, acetylene, propane, ethylene, hydrogen cyanide, and their polymers (Tomasko *et al.*, 2005). The spectra exhibit a featureless blue slope between 800 and 1500 nm that matches no mixture of laboratory spectra of ices, organics, or tholins. It is also supported by the detection of benzene (C_6H_6), an organic product more complex than either ethane or propane, by the GCMS on the surface (Niemann *et al.*, 2005).

This question of a methane reservoir for Titan is important in much the same way that the abundance and history of water is for Mars or for Earth. If Titan's atmosphere has had access to a steady supply of methane for 4.6 Gyr, then the atmosphere should have remained stable, and the former models pictured that hundreds of meters of hydrocarbons would have been deposited on Titan's surface. If not, then the surface may lack significant deposits of organics, and the atmospheric history would have been more dynamic.

The Huygens Probe results strongly suggest that a liquid cycle currently operates, or has operated recently, especially when one considers: (1) the rounded, size-selected, and size-layered rocks on the surface, positioned as though located in the bed of a stream within the large dark lakebed (Fig. 20.9b); (2) the dozen or so canyon-looking features imaged from a few km above the surface (Fig. 20.9a); (3) several brighter areas in the higher altitude images that seem to be elongated along a direction parallel to the main bright/dark boundary (the "shoreline"), like islands or drift deposits; and (4) the fact that no obvious small impact features appear in any of the views in contrast with most other satellites in the Solar System (e.g., the saturnian satellite Hyperion). The images suggest that any aerosols deposited on the redder highlands have been washed into the bluer plains. However, the evident dryness of the lakebed, if lakebed it is, suggest that deluges are episodic and monsoon-like, possibly separated by time periods that might be numbered in multiples of a titanian year (29 years, the period of Saturn's orbit).

Furthermore, initial observations by the Cassini radar instrument show that the particles from atmospheric chemical processing have evidently collected into dunes and other landforms. Either we are privileged to be witnessing an active period in Titan's history, or the satellite has always possessed a robust geological and hydrological (or better, "*methano*-logical") cycle.

If Titan's atmosphere does not have access to a steady-state source of methane from the surface or interior, then there may have been epochs in the past, perhaps of long duration, when photochemistry depleted the atmosphere of methane. This would also remove the molecular hydrogen from the atmosphere on a timescale comparable to the removal of the methane. With two of the three major greenhouse gases gone, the atmosphere would cool. On the other hand, the photochemical haze, which prevents 90% of the sunlight from reaching Titan, would be removed as well, and this could encourage a warmer surface. Which effect wins?

Radiative transfer modeling of Titan's atmosphere reveals that the net effect is a cooling of the atmosphere (Lorenz *et al.*, 1997a), but the *extent* of cooling is more difficult to predict. There is the potential for a large "collapse" of the atmosphere if the nitrogen gas in the atmosphere were cooled below its saturation vapor point, which for the current 1.5 bar surface pressure is 82 K. Whether this threshold is crossed depends on the history of solar illumination of Titan. For the current solar luminosity, removal of all the methane (with the consequences described above for hydrogen and the aerosols) would only modestly lower the temperature, to somewhere between 85 and 90 K, with a resulting atmospheric density similar to that observed today. No change in the crater size-frequency distribution is predicted for this case. On the other hand, if the Sun has indeed increased in brightness roughly linearly by 30% over the past 4.6 Gyr (Section 4.2 and Fig. 4.2), then the situation up to 2 Ga would have been quite different. Loss of methane prior to that point would have lowered the surface temperature below 82 K, precipitating rainout of the atmosphere (Lorenz *et al.*, 1997a). How complete the rainout would have been depends on the relative rates of fall of the equatorial and polar temperatures, which today are nearly equal; preliminary models predict substantial thinning of the atmosphere (Lorenz *et al.*, 2001b).

Titan's atmospheric history in the absence of a steady state supply of methane for photochemistry could thus have been a roller-coaster ride of cold and warm epochs, up to approximately 2–3 Ga when the Sun brightened sufficiently to stabilize a pure nitrogen atmosphere (Fig. 20.10). Examination of cratered terrains of different ages could have determined whether this is the case, but the surface of Titan seems to be too young to have preserved the cratering record from billions of years ago. Alternatively, perhaps the Sun has brightened to a lesser extent than predicted by astrophysical models (Sagan and Chyba, 1997). While the predicted behavior of the Sun's luminosity history, buttressed by nuclear physics and observations of other stars, seems secure, it is noteworthy that evidence for the faint early Sun is also lacking in martian and terrestrial history.

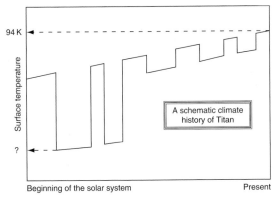

FIGURE 20.10 Schematic history of Titan's surface temperature over the past 4.6 Ga in the absence of a steady supply of methane for atmospheric photochemistry. As described in the text, the N₂ atmosphere cycled between a surface phase (probably liquid) and gaseous form early in Titan's history until the increasing solar luminosity reached a certain threshold, after which the atmosphere became stable.

Looking at the *future*, we can calculate that when the Sun becomes a red giant star some 5 Gyr hence, Titan's surface will be sufficiently warmed to support a liquid water–ammonia ocean, should ammonia be present in the crust (Lorenz *et al.*, 1997b). This is an (extreme) example of the effect discussed in Section 3.3 whereby habitable zones generally move outwards from a star as it evolves. The transformation process will be a complex one for Titan, beginning with loss of a portion of the atmosphere as the Sun blows away part of its own envelope in becoming a red giant. The increased solar luminosity (a factor of ∼100) wins out over the mass loss, however, heating the surface of Titan to approximately 200 K, allowing mixtures of water and ammonia to melt (although pure water ice will stay frozen). Moreover, the small amount of ultraviolet radiation from the red-giant Sun means that the production of haze will greatly diminish, and the atmosphere should become transparent (at least down to the troposphere, where water–ammonia clouds may form). A space-faring intelligent race (we dare not claim humankind, which would have to exceed by factors of thousands the typical mammalian species lifetime of a million years) would see a very different Titan than we view today.

20.7.4 No evidence of life

Huygens made two observations related to the question of whether life exists on Titan today. In the first, the GCMS observed a $^{12}C/^{13}C$ ratio of ∼82, compared to the typical ratio of ∼95 associated with organic molecules participating in terrestrial biological processes (Section 12.2.3). This result argues against the existence of a titanian biota, because we would expect that, as on Earth, evolution should drive biological carbon to be enriched in ^{12}C relative to ^{13}C. However, we have no knowledge of a definitely abiotic value of $^{12}C/^{13}C$ *on Titan* (a subsurface value might serve), and therefore do not have a reliable point of comparison for titanian isotopic measurements.

The second observation is more straightforward. In the 70-minute period after the Huygens Probe soft-landed on Titan's surface and could still transmit data from its location at latitude 10.2°S, longitude 167.6°E, ∼80 images showed no movement suggestive of any organism, no branches, protrusions, interesting shapes, discolorations, or patterns that might indicate macroscopic life. If life is present on Titan, by analogy with Earth's life one would expect it to be ubiquitous. Therefore sampling any particular single location should be representative of the entire surface. With this premise, it appears that Titan does not have any macroscopic life anywhere.

Definitive deduction of the absence of *microscopic* life will require far more than simple imagery. However, in the field, given the lack of anything better, evidence of microscopic life is often discovered by smelling or visually observing macroscopic morphological evidence. It is therefore noteworthy that DISR and GCMS, Huygens's eyes and nose, indicated an absence of evidence for life, including microscopic life. A more sophisticated and sensitive answer must await the results of a dedicated *in situ* sampling.

20.8 Future astrobiological exploration of Titan

Although Titan's surface is too cold to support stable liquid water, or even water/ammonia mixtures (which have a minimum (eutectic) liquid temperature of 173 K), other aspects of Titan's surface and atmosphere system make it attractive as an astrobiological target. Because of the relative scarcity of atomic and molecular hydrogen, Titan has a dense atmosphere that is not strongly reducing, unlike the giant planets. In this regard, its oxidation–reduction (redox) state may more closely resemble that of the early Earth, prior to the rise of oxygen, than does any other current planetary atmosphere. And even though the actual molecular constituents (other than nitrogen) are different on Titan compared with the prebiotic Earth, methane and nitrogen are molecules of great interest for the synthesis of important precursors to biology (Raulin and Owen, 2002). Were oxygen to be added, via reactions with liquid water, amino acids and carboxylic acids could be produced (the possibility of transient episodes of liquid water is considered below).

Although Titan receives just 1% of the sunlight that the Earth does, varied sources of energy in the atmosphere and surface may lead to continued slow processing on the surface of molecules made in the upper atmosphere. While cosmic rays and solar ultraviolet radiation dominate in the upper atmosphere, impacts and possible volcanism are principal sources of energy at the surface. As well, energy stored in the bonds of certain unsaturated hydrocarbons such as acetylene, formed in the stratosphere, may be released at the surface during polymerization and hence become available to drive other chemistry. While of smaller magnitude than other energy sources, this chemical energy may be as much as 10^{-2}–10^{-3} of the solar energy flux over Titan's history. Titan's solid surface, covered perhaps in places by hydrocarbon liquids, may also play host to templating reactions. Finally, Titan is a large body that has existed for billions of years; it provides an organic chemistry in the absence of life that operates on temporal and spatial scales unachievable in a terrestrial laboratory.

The initial energy of accretion of Titan, some of which remains as heat in the interior today, supplemented by the decay of uranium, thorium, and potassium isotopes in the silicate portion of the interior, could episodically drive cryovolcanism over geologic time. If Titan's interior contains some ammonia, as is expected, it should consist partially of a liquid ammonia–water layer (Grasset *et al.*, 2000). Although the layer overall is stably contained underneath a crust of water ice (see Fig. 20.1), it is possible – via impacts, for example – that at certain times and in specific locations liquid water/ammonia solutions could reach the surface. A stochastic but more predictable mechanism for having liquid water at the surface is impact of km-diameter or larger bodies with Titan's surface. According to computer simulations, such an impact forms a crater, ejects material, and leaves a few percent of liquid water at the base of the crater (Artemieva and Lunine, 2003). Those same simulations show that, for oblique impacts, a lightly shocked tongue of organics and water ice from the lee side of the impact drops into the liquid water (Fig. 20.11). A crust of water ice forms quickly over the liquid water and organics, but complete freezing is delayed for 10^2–10^3 years, the longer timescale obtaining for an ammonia–water mixture (Thompson and Sagan, 1992).

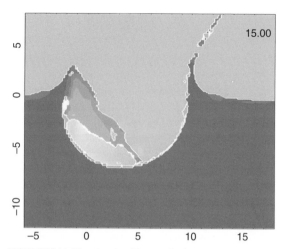

FIGURE 20.11 The situation 15 seconds after an impact of a km-sized comet into Titan with an initial velocity of 7 km/s and an angle of impact of 45° from the left side. The grey indicates water ice, the lighter shades liquid water. Organics are contained within the grey tongue of water ice at the lee side (to the left) of the impact, but are not shown in this figure. Size scales on the axes are in kilometers. From Artemieva and Lunine (2003).

The exciting prospect associated with transient pools of liquid water is that the organic sediments on Titan's surface, generated in the stratosphere in a very oxygen-poor environment, are suddenly able to react with liquid water. Oxygen-bearing compounds such as carboxylic acids and amino acids should form quickly (Lorenz *et al.*, 2001a). Whether enough time, volume, surfaces, etc. are available to go further is uncertain. Might some chiral amino acids, or chiral sugars, undergo amplification of small enantiomeric excesses? Could simple peptides form in the time available? All of these possibilities are speculation, because it is extremely difficult – in spite of sophisticated numerical models of chemical reactions – to predict a priori what the results will be when a complex mixture of organics is exposed to an aqueous medium for timescales much longer than possible in the laboratory. Whatever is produced will be well preserved, however, as the last of the aqueous medium freezes over, since the chances late in Titan's history of a second impact in the same place are slim, and the surface is well protected from ultraviolet light and most cosmic rays. Drilling several meters into a crater bottom to access the organics may be required. Should Cassini–Huygens identify places on the surface where organic deposits are associated with geological features, follow-on missions to examine these regions in detail would be of keen interest to astrobiology.

A possible strategy for identifying and sampling interesting organics on Titan's surface would be to use a blimp to examine potentially interesting sites at close range, and then land or drop a small package for analysis of the organics. Titan's high-density atmosphere, low gravity, and light surface winds are ideal for exploration with small aerial vehicles. A blimp would be deployed to one of the sites of recent geologic activity (or of a recent impact) in which organics were known to be present from the Cassini data. A remote sensing package on the blimp consisting of visible imaging and high-resolution ultraviolet and infrared spectroscopy would examine the site at close range. The goal of the search would be to identify organic samples that appear to have been altered by liquid water. Once a plausible set of deposits were identified during this phase, the blimp would set down on, or drop a package on, that site. An organic analysis package might consist of a gas or liquid chromatograph with chiral support, or other chiral detection schemes such as a quartz crystal microbalance, a mass spectrometer, or miniature Raman or NMR spectroscopy (Welch and Lunine, 2001). The total oxygen content, the pattern of abundance of organic molecules as a function of carbon number, the presence of chiral molecules, and the enantiomeric composition would be analyzed in the products obtained by breaking down polymers in the sample; in addition, the structures of the polymers themselves would be analyzed. If the sample looked promising, additional analyses and even drilling (which is difficult!) would be scheduled; otherwise, the blimp would be raised from the site to begin a new reconnaissance.

20.9 Some astrobiological speculations

Titan's surface and subsurface offer three approaches to understanding organic systems of relevance to astrobiology.

20.9.1 Chemistry in non-aqueous organic systems

Even in the absence of water, organic chemical systems may evolve over long timescales, fueled by the released enthalpy of polymerization or Titan's equivalent of geothermal energy. Polymer formation could in some cases exhibit self-organizing properties in terms of the structure of the chain. For example, how asymmetric monomers are aligned relative to each other (a property called *tacticity*) often strongly

determines the bulk material properties of their polymers. Acrylonitrile (CH_2CHCN) is an example of a possible photochemical product made in and above Titan's stratosphere (Coll *et al.*, 1999) whose polymer could exhibit preferred tacticity should self-organizing chemistry be possible in the organic sediments (Dougherty, 2001). Going one step further, one cannot rule out the possibility that a form of life, organic but biochemically different from Earth life, might be able to use, in lieu of water, the liquid hydrocarbons as its solvents for living processes. Exactly how this would work is unclear, because the nonpolar nature of liquid hydrocarbons precludes defining the inside and outside of polymers in the way that water gives rise to hydrophobic and hydrophilic sides of a molecule. Further study of the viability of hydrocarbon liquids, though, is worthwhile, particularly in regard to the role of polar molecules that might dissolve in the liquid, and the possible roles of stereoregular polymers in the organic sediments at the base of the hydrocarbon liquid. See Chapter 27 for further discussion of non-Earth-like life.

20.9.2 Chemistry in aqueous organic systems

As discussed earlier, it is possible that organic molecules on Titan have been exposed to liquid water for modest periods in the basins of newly formed craters. The cratering rate for the saturnian system, integrated over the age of the Solar System, suggests that a significant fraction of the surface of Titan could have been struck with large impactors (Thompson and Sagan, 1992). Only the freshest craters, however, are likely to yield accessible, aqueously altered organics – the older craters have been tectonically modified, mechanically stirred by secondary crater formation from nearby large impacts, and filled in by many hundreds of meters of hydrocarbon sediments over the aeons after formation. It is difficult to predict a priori what might be found on Titan should it be possible to locate a fresh crater, identify organic sediments, and drill through the upper ice layer to access the deep-frozen products. Some of the more interesting potential products of non-aqueous organic synthesis and polymerization could provide interesting biotic building blocks. Considering specifically the case of polyacrylonitrile, hydrolysis could lead to amides or carboxylic acids, or if the nitriles are reduced to amines, possibly aldehydes (Dougherty, 2001). To predict with present knowledge what might be found as the product of subsequent autocatalytic cycles would be foolhardy; it is necessary to go and sample.

20.9.3 Life in Titan's interior?

Finally, there is the tempting region below Titan's water ice crust that might be liquid ammonia–water. This region, particularly if it plays host to methane and other hydrocarbons, might contain both the raw materials and perhaps the energy necessary for life (Fortes, 2000). The kilobar pressures in the layer appear not to be a problem either, based on recent laboratory experiments on Earth organisms (Sharma *et al.*, 2002). However, even if life of some kind were to exist there, the thickness of the ice layer above it (tens of kilometers) rules out direct sampling. Possible isotopic signatures of life outgassed to the atmosphere would be subtle and difficult to distinguish from abiotic processes. For the detection of subterranean biota, Europa is a better (but still very difficult) target (Chapter 19). Titan, farther from the Sun but with a surface more benign and easier to land on than Europa's, is of most interest with regard to the organic chemical steps leading to the origin of life on Earth, and (less plausibly) as an abode for exotic life forms of its own.

REFERENCES

Artemieva, N. and Lunine, J. I. (2003). Cratering on Titan: impact melt and the fate of surface organics. *Icarus*, **164**, 471–480.

Campbell, D. B., Black, G. J., Carter, L. M., and Ostro, S. J. (2003). Radar evidence for liquid surfaces on Titan. *Science*, on-line 10.1126/science.1088969.

Coll, P., Coscia, D., Smith, N., *et al.* (1999). Experimental laboratory simulations of Titan's atmosphere aerosols and gas phase. *Planet. Space Sci.*, **47**, 1331–1340.

Combes, M., Vapillon, L., Gendron, E., Coustenis, A., Lai, O., Wittemberg, R., and Sirdey, R. (1997). Spatially-resolved images of Titan by means of adaptive optics. *Icarus*, **129**, 482–497.

Courtin, R., Gautier, D., and McKay, C. P. (1995). Titan's thermal emission spectrum: reanalysis of the Voyager infrared measurements. *Icarus*, **114**, 144–162.

Coustenis, A. and Taylor, F. (1999). *Titan: the Earth-like Moon*. Singapore: World Scientific.

Coustenis, A., Salama, A., Lellouch, E., *et al.* (1998). Evidence for water vapor in Titan's atmosphere from ISO/SWS data. *Astronomy and Astrophysics*, **336**, L85–9.

Dimitrov, V. and Bar-Nun, A. (2003). Hardening of Titan aerosols by their charging. *Icarus*, **166**, 440–443.

Dougherty, D. (2001). Polymer stereochemistry: an opportunity and a challenge on Titan. *Enantiomer*, **6**, 101–106.

Elachi, C., Wall, S., Allison, M., *et al.* (2005). Cassini radar views the surface of Titan. *Science*, **308**, 970–974.

Engel, S., Lunine, J. I., and Hartmann, W. K. (1995). Cratering on Titan and implications for Titan's atmospheric history. *Planet. Space Sci.*, **43**, 1059–1066.

Fortes, A. D. (2000). Exobiological implications of a possible ammonia–water ocean inside Titan. *Icarus*, **146**, 144–152.

Gibbard, S. G., Macintosh, B., Gavel, D., *et al.* (1999). Titan: high-resolution speckle images from the Keck Telescope. *Icarus*, **139**, 189–201.

Grasset, O., Sotin, C., and Deschamps, F. (2000). On the internal structure and dynamics of Titan. *Planet. Space Sci.*, **48**, 617–636.

Griffith, C. A. (1993). Evidence for surface heterogeneity on Titan. *Nature*, **364**, 511–514.

Griffith, C. and Zahnle, K. (1995). Influx of cometary volatiles to planetary moons: the atmospheres of 1000 possible Titans. *J. Geophys. Res.*, **100**, 16907–16922.

Griffith, C. A., Hall, J. L., and Geballe, T. R. (2000). Detection of daily clouds on Titan. *Science*, **290**, 509–513.

Holsapple, K. A. (1994). The scaling of impact processes in planetary sciences. *Ann. Rev. Earth Planet. Sci.*, **21**, 333–373.

Iro, N., Gautier, D., Hersant, F., Bockeleé-Morvan, D., and Lunine, J. I. (2002). An interpretation of the nitrogen deficiency in comets. *Icarus*, **161**, 511–532.

Ivanov, B. A., Basilevski, A. T., and Neukem, G. (1997). Atmospheric entry of large meteoroids: implication to Titan. *Planet. Space Sci.*, **45**, 993–1007.

Khare, B. N., Sagan, C., Thompson, W. R., *et al.* (1984). The organic aerosols of Titan. *Advances in Space Research*, **4**, 59–68.

Kuiper, G. P. (1944). Titan: a satellite with an atmosphere. *Astrophy. J.*, **100**, 378–383.

Lebreton, J.-P., Witasse, O., Sollazzo, C., *et al.* (2005). An overview of the descent and landing of the Huygens probe on Titan. *Nature*, **438**, 758–64.

Lindal, G. F., Wood, G. E., Hotz, H. B., Sweetnam, D. N., Eshleman, V. R., and Tyler, G. L. (1983). The atmosphere of Titan – an analysis of the Voyager 1 radio occulation measurements. *Icarus*, **53**, 348–363.

Lorenz, R. D. and Mitton, J. (2002). *Lifting Titan's Veil*. Cambridge: Cambridge University Press.

Lorenz, R. D., McKay, C. P., and Lunine, J. I. (1997a). Photochemically-induced collapse of Titan's atmosphere. *Science*, **275**, 642–644.

Lorenz, R. D., Lunine, J. I., and McKay, C. P. (1997b). Titan under a red giant sun: a new kind of "habitable" moon. *Geophys. Res. Lett.*, **24**, 2905–2908.

Lorenz, R. D., Lunine, J. I., and McKay, C. P. (2001a). Geologic settings for aqueous organic synthesis on Titan revisited. *Enantiomer*, **6**, 83–96.

Lorenz, R. D., Lunine, J. I., McKay, C. P., and Withers, P. G. (2001b). Titan, Mars, and Earth: entropy production by latitudinal heat transport. *Geophys. Res. Lett.*, **105**, 1859–1865.

Lorenz, R. D., Wall, S. D., Reffet, E., *et al.* (2006). Radar imaging of giant longitudinal dunes: Namib desert (Earth) and the Belet sand sea (Titan). *Lunar and Planetary Sci. Conf. No. 33*, 1249–1250.

Lunine, J. I. and Stevenson, D. J. (1987). Clathrate and ammonia hydrates at high pressure: application to the origin of methane on Titan. *Icarus*, **70**, 61–77.

Lunine, J. I., Stevenson, D. J., and Yung, Y. L. (1983). Ethane ocean on Titan. *Science*, **222**, 1229–1230.

Lunine, J. I., Yung, Y. L., and Lorenz, R. D. (1999). On the volatile inventory of Titan from isotopic abundances in nitrogen and methane. *Planet. Space Sci.*, **47**, 1291–1303.

McCord, T. B., Carlson, R. W., Smythe, W. D., *et al.* (1997). Organics and other molecules in the surfaces of Callisto and Ganymede. *Science*, **278**, 271–275.

Meier, R., Smith, B. A., Owen, T. C., and Terrile, R. J. (2000). The surface of Titan from NICMOS observations with the Hubble Space Telescope. *Icarus*, **145**, 462–473.

Mitri, G., Showman, A. P., Lunine, J. I., and Lorenz, R. D. (2007). Hydrocarbon lakes on Titan. *Icarus*, **186**, 385–394.

Niemann, H. B., Atreya, S. K., Bauer, S. J., *et al.* (2005). The abundances of constituents of Titan's atmosphere from the GCMS instrument on the Huygens probe. *Nature*, **438**, 779–784.

Niemann, H. B., Atreya, S. K., Bauer, S. J., *et al.* (1997). The gas chromatograph mass spectrometer aboard Huygens. In *Huygens: Science, Payload and Mission*, **SP-1177**, pp. 85–107. Noordwijk, The Netherlands: European Space Agency.

Owen, T. C. (1982). The composition and origin of Titan's atmosphere. *Planet. Space Sci.*, **30**, 833–838.

Owen, T. (2005). Huygens rediscovers Titan. *Nature*, **438**, 756–7.

Porco, C. C., Baker, E., Barbar, J., *et al.* (2005). Imaging of Titan from the Cassini spacecraft. *Nature*, **434**, 159–68.

Porco, C. C., Melfenstein, P., Thomas, P. C., *et al.* (2006). Cassini observes the active south pole of Enceladus. *Science*, **311**, 1393–1401.

Raulin, F. and Owen, T. (2002). Organic chemistry and exobiology on Titan. *Space Sci. Rev.*, **104**, 377–94.

Sagan, C. and Chyba, C. F. (1997). The faint early sun paradox: organic shielding of ultraviolet-labile greenhouse gases. *Science*, **276**, 1217.

Sears, W. D. (1995). Tidal dissipation in oceans on Titan. *Icarus*, **113**, 39–56.

Sharma, A., Scott, J. H., Cody, G. D., Fogel, M. L., Hazen, R. M., Hemley, R. J., and Huntress, W. T. (2002). Microbial activity at gigapascal pressures. *Science*, **295**, 1514–1516.

Smith, P. H., Lemmon, M. T., Lorenz, R. D., Sromovsky, L. A., Caldwell, J. J., and Allison, M. D. (1996). Titan's surface, revealed by HST imaging. *Icarus*, **119**, 336–339.

Sotin, C., Jaumann, R., Buratti, B. J., *et al.* (2005). Release of volatiles from a possible cryovolcano from near-infrared imaging of Titan. *Nature*, **435**, 786–789.

Stevenson, D. J. (1992). The interior of Titan. In *Proceedings of the Symposium on Titan*, **ESA SP-338**, pp. 29–33. Noordwijk, The Netherlands: European Space Agency.

Stofan, E., Elachi, C., Lunine, J. I., *et al.* (2007). The lakes of Titan. *Nature*, **445**, 61–64.

Strobel, D. F. (1974). The photochemistry of hydrocarbons in the atmosphere of Titan. *Icarus*, **21**, 466.

Thompson, W. R. and Sagan, C. (1992). Organic chemistry on Titan – surface interactions. In *Proceedings of the Symposium on Titan*, **ESA SP-338**, pp. 167–176. Noordwijk, The Netherlands: ESA.

Tobie, G., Grasset, O., Lunive, J. I., Mocquet, J., and Sotin, C. (2005). Titan's orbit provides evidence for a subsurface ammonia–water ocean. *Icarus*, **175**, 496–502.

Tomasko, M. G., Archinal, B., Becker, T., *et al.* (2005). Rain, winds and haze during the Huygens probe's descent to Titan's surface. *Nature*, **438**, 765–78.

Welch, C. J. and Lunine, J. I. (2001). Challenges and approaches to the robotic detection of enantioenrichment on Saturn's moon, Titan. *Enantiomer*, **6**, 69–81.

Yung, Y. L., Allen, M. A., and Pinto, J. P. (1984). Photochemistry of the atmosphere of Titan: comparison between model and observations. *Astrophys. J. Supplement Series*, **55**, 465–506.

FURTHER READING AND SURFING

Coustenis, A. and Taylor, F. (1999). *Titan: the Earth-like Moon*. Singapore: World Scientific. A graduate-level summary of the literature on Titan as of 1998; slightly dated but still useful compendium.

Harland, D. (2003). *Mission to Saturn: Cassini and the Huygens Probe*. London: Praxis Press. Popular-level description of the Cassini–Huygens mission to the Saturn system.

Lorenz, R. D. and Mitton, J. (2002). *Lifting Titan's Veil*, Cambridge: Cambridge University Press. Well-written popular account of the scientific investigation of Titan and the development of the Cassini–Huygens, from a behind-the-scenes perspective of an expert intimately involved in both.

saturn.jpl.nasa.gov. NASA website for the Cassini–Huygens mission.

www.esa.int/SPECIALS/Cassini-Huygens. European Space Agency website for the Cassini–Huygens mission.

21 Extrasolar planets

Paul Butler
Carnegie Institution of Washington

21.1 Introduction

While the notion of worlds beyond our Earth is ancient, the specific idea of planets orbiting distant stars is relatively new. Over two millennia ago Epicurus stated "there are infinite worlds both like and unlike this world of ours," but he was not speaking of Earth-like planets orbiting Sun-like stars. Indeed, planets orbiting stars would have been a meaningless issue to the Greeks, as the Sun was not recognized as a star, nor the Earth as a planet (Chapter 1).

One of the earliest and most eloquent spokespersons for what is now called astrobiology, and among the first to grasp the implications of the Sun being a star and the Earth a planet, was the mystical Roman Catholic monk Giordano Bruno. In *On the Infinite Universe and Worlds* (1584) he wrote:

> There are countless suns and countless earths all rotating around their suns in exactly the same way as the seven planets[1] of our system. We see only the suns because they are the largest bodies and are luminous, but their planets remain invisible to us because they are smaller and non-luminous. The countless worlds in the universe are no worse and no less inhabited than our Earth.

Bruno then concludes with the revolutionary slogan:

> Destroy the theories that the Earth is the center of the Universe!

Bruno's reward for this prescience and for other heresies was condemnation by the Church, followed by immolation in a public square in Rome in 1600.

Planets orbiting other stars are of extreme interest to both scientists and the public because of their central importance to the question of life in the Universe. Extrasolar planets function as Petri dishes, laboratories where liquid water, organic material, and energy sources can be mixed in various proportions, and then shaken and stirred for millions of years. For the case of the Earth, we know that the end result of this was the formation and subsequent evolution of life. Empirically we cannot know if this is a common or exceedingly rare event until we develop the means to search for life on extrasolar planets, and then systematically search the nearest few thousand planetary systems for the presence of life.

After a century of increasingly detailed and technologically complex searches for extrasolar planets, the status of the field in the mid-1990s was summarized by Dick (1996):

> The existence of planets around even the closest stars could not be confirmed directly by observation or indirectly by observation of their effects. Nor could circumstellar shells that were directly observed or indirectly inferred be conclusively linked to planetary systems.

The lack of indisputable evidence did not slow the "planet frenzy" over the last century as various groups vied for the honor of "completing the Copernican Revolution" (Black, 1995), i.e., further removing our Earth from any special status. The graveyard of claims to have found the first extrasolar planet shows how easily errors can fool even skeptical scientists when chasing after a big discovery (Section 2.2.2 briefly covers this history). And since the discovery of the first *bona fide* extrasolar planet ushered in a new era a decade ago (Mayor and Queloz, 1995; Marcy *et al.*, 1996), the rate of false planet discoveries has actually significantly increased.

The ~250 extrasolar planets known as of 2007 are enough to guide theorists to increasingly sophisticated

---

[1] These seven planets were the five naked-eye planets of today plus the Earth and Moon.

Planets and Life: The Emerging Science of Astrobiology, eds. Woodruff T. Sullivan, III and John A. Baross. Published by Cambridge University Press. © Cambridge University Press 2007.

models of planet formation and evolution. The biggest surprise to date is that most planetary systems do not resemble our Solar System, with its stately circular orbits and careful segregation between small, warm, inner rocky planets and cold distant gas giants (Chapter 3).

Planet hunting is still in its infancy. With today's techniques only gas giants like Saturn and Jupiter, 100 Earth masses and larger, are readily detectable. While these huge worlds lack solid surfaces and are presumably poor places to look for life, life has proved remarkably adaptable to the most inhospitable niches on Earth. In addition, these planets may have moons (as with Jupiter and its moon Europa – see Chapter 19) which might provide the necessary real estate for the development of life.

The US National Academy of Sciences (NAS) cited the discovery of extrasolar planets as the most significant accomplishment in astronomy of the 1990s (NAS, 2001), but the future promises to be even more exciting. The improving sensitivity of existing planet detection techniques have already begun to yield Neptune-mass (~15 Earth-mass) planets and smaller, and by the end of the decade will have revealed the first true Solar System analogues. Several space-based missions are scheduled to launch in the next ten years with goals including transit detection, astrometric detection, and infrared detection of extrasolar planets down to the terrestrial mass range. NASA and ESA both have long-range plans for space telescopes to directly image and obtain spectra of extrasolar planets.

This chapter will discuss the various means of detecting extrasolar planets, what has been learned from the extrasolar planets that have been discovered to date, and what we hope to learn over the next decade.

21.2 Stars

Planet-mass bodies presumably orbit a wide variety of stars, including massive stars, degenerate stars such as black holes and pulsars (Wolszczan and Frail, 1992), and the so-called "failed stars," or brown dwarfs.[2] These cases will not be considered here as none of these environments is likely to support life. In particular, massive stars (say, having a mass $>2\ M_\odot$, where $M_\odot$ is the mass of the Sun) have lifetimes of <1 Gyr, and their radiation is rich in ultraviolet and X-rays.

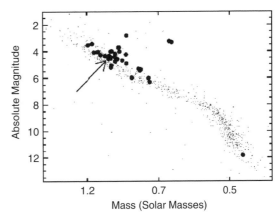

FIGURE 21.1 Stellar luminosity vs. mass. The logarithm of luminosity (total power output) is expressed as a so-called "absolute magnitude"; a difference in five magnitudes corresponds to a factor of 100 in luminosity. The band running from the upper left (highest luminosity and mass) to the lower right (lowest luminosity and mass) defines the main sequence, where stars spend 90% of their lifetimes. The small dots are all stars within 65 lt-yr, ranging in spectral type F through M (from the Hipparcos satellite catalog), while the large dots are stars that are known to have planets. Stars above the main sequence have just begun to evolve off the main sequence; those few scattered below are white dwarfs. The location of the Sun is indicated by the arrow. (California & Carnegie Planet Search.)

Degenerate stars primarily produce X-rays and gamma rays and very little thermal (blackbody) radiation. While brown dwarfs don't produce deadly high energy radiation, they also don't produce much light or heat.

Unlike more massive stars, stars of ~1 $M_\odot$ and smaller have hydrogen fusion lifetimes of 10 Gyr and longer, sufficient for life to form and evolve. The radiation of such stars is rich in optical and infrared, with relatively little high-energy ultraviolet and X-ray radiation that would tend to destroy complex organic molecules. It is thus these stars that are of the greatest interest to astrobiologists.

Figure 21.1 shows the relationship between the mass of a star and its intrinsic *luminosity* (total power output in watts). The small dots are all the stars within 65 light years (lt-yr) from the Hipparcos satellite catalog (Perryman *et al.*, 1997). The band of stars running from the upper left to the lower right marks the so-called *main sequence*. Stars spend about 90% of their lives on the main sequence, quiescently fusing hydrogen into helium in their cores. The large dots indicate planet-bearing stars, while the location of the Sun, a G star, is indicated by the arrow. As Fig. 21.1 shows, most known extrasolar planets orbit stars similar to the Sun. The low luminosity stars at the bottom right of

---

[2] A *brown dwarf* has an intermediate mass between those of the largest planets (~10–15 M_J, where M_J is the mass of Jupiter) and the smallest stars (~80 M_J, or 0.08 solar masses).

Fig. 21.1, the M dwarfs, are low-mass red stars with less than half the mass of the Sun. While these stars produce only a few percent of the Sun's luminosity, they constitute about 70% of all known stars. They have been under-represented in previous planet searches due to their intrinsic faintness.

It is the nearest stars (out to ~150 lt-yr) that are of the greatest interest to astrobiologists and planetary scientists. This volume of space includes about 2,000 Sun-like stars, a tiny fraction of the more than 2×10^{11} stars in our Galaxy. Yet these are the stars that can be studied in the greatest detail with emerging technologies such as interferometry and direct imaging, and these will be the first targets of probes if humans solve the challenges of interstellar travel.

21.3 Extrasolar planet detection techniques

21.3.1 Direct detection

The most obvious way to search for extrasolar planets is simply to point a telescope at nearby stars and look for the tiny, faint orbiting planets. The problem is that a planet is completely overwhelmed by the brilliant host star. As Fig. 21.2 shows, a Jupiter-like planet is ~10^9 times fainter than a Sun-like star at visible wavelengths. The situation is somewhat improved in the mid-infrared, with the planet only being outshone by a factor of ~10^5. The ability to image exceedingly faint companions within a fraction of an arcsecond of a star is beyond current technology, but plans are proceeding on space-based instruments that might overcome this obstacle (Section 21.7.4).

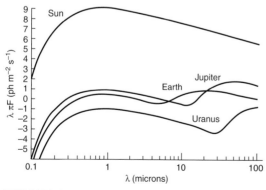

FIGURE 21.2 Spectra of the Sun and Solar System planets; ordinate is log (photons m^{-2} s^{-1}) as observed from a common distance. In the visual region (~0.6 microns) the Sun is ~10^9 brighter than Jupiter. This contrast improves to ~10^5 in the mid-infrared, while the Earth is ~10^6 times fainter than the Sun in the mid-infrared. (NASA/JPL/Caltech.)

21.3.2 Astrometry

Since the technology to directly image an extrasolar planet does not exist, a number of indirect detection techniques have been devised. These look for the subtle effects, primarily gravitational, that an orbiting planet exerts on its host star.

The common notion that planets orbit stars is correct only to first order. Newton's third law demands that for every action there is an equal and opposite reaction. Thus a planet and a star both orbit a common center of mass. In the case of Jupiter and the Sun, the common center of mass lies just outside the surface (photosphere) of the Sun. While Jupiter makes a grand 12-year orbit about this point at a distance of 5 AU (where 1 AU is the mean Sun–Earth distance), the Sun moves in a mirror image orbit that is 1050 times smaller (1050 is the mass ratio of the Sun to Jupiter). Alien astronomers could thus infer the presence of Jupiter by noting that the Sun periodically wobbles as seen against the backdrop of distant stars. Each planet produces its own Sun wobble of amplitude proportional to the planet's mass and orbit size; the Sun's total motion is then the sum of all these wobbles. Figure 21.3 shows this *astrometric*[3] wobble of the Sun, which is due primarily to Jupiter (12-year periodicity) and to Saturn (30-year periodicity). Note that *Earth's* effect on the illustrated motion over 30 years would be 30 tiny wiggles of amplitude ~5% of the width of the plotted lines in Fig. 21.3! For most of the past century this has been the only viable means of detecting extrasolar planets.

The magnitude of a stellar wobble on the sky decreases with stellar distance, so astrometry is most useful on the nearest stars with massive planets in distant orbits. As Fig. 21.3 shows, a Sun-like star at 10 parsecs[4] (33 lt-yr) with a Jupiter-like planet ($M_J \approx 0.001\ M_\odot$) wobbles with a period of 12 years by an angle of ~0.001 arcsecond relative to distant background stars. For comparison, 0.001 arcsecond (1 arcsecond is 1/3600 of a degree) is the size of a typical coin seen from 10,000 km away. Unfortunately, there are fewer than 50 single,[5] Sun-like stars suitable for monitoring within 10 parsecs.

---

[3] *Astrometry* is the branch of astronomy dealing with precise measurements of positions of objects on the sky.

[4] A parsec (pc) is an arcane, but standard, unit of distance in astronomy: 1 pc = 3.26 lt-yr = 3.1×10^{16} m.

[5] A *single star* is one that is not part of a *binary star system* where two stars orbit each other. Almost all planet searches avoid binary systems (which are very common) because calculations indicate that stable, long-term planetary orbits are less likely.

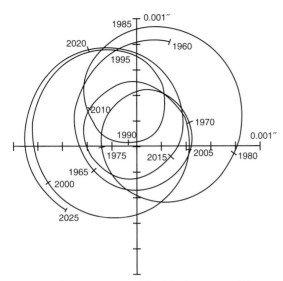

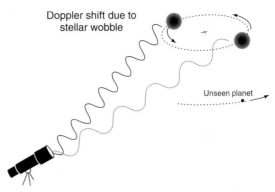

FIGURE 21.4 Doppler shift technique. An unseen orbiting planet gravitationally tugs a star into a small counter-orbit. This stellar motion can be detected by means of the Doppler effect, yielding the orbital characteristics and a minimum mass for the planet. (California & Carnegie Planet Search.)

FIGURE 21.3 Astrometric wobble of the Sun as it would be seen from a distance of 10 pc (33 lt-yr) over the period 1990 to 2020. The Sun mainly describes a periodic wobble every 12 years due to Jupiter. Its amplitude is only ~0.001 arcsecond, an angle barely able to be measured with today's techniques. This 12-year wobble is itself modulated by the other, smaller planets, primarily Saturn in its 30-year orbit. (NASA/JPL/Caltech.)

A reliable astrometric detection requires observations spanning more than a full orbit, and so programs such as Peter van de Kamp's survey lasted more than 50 years (Section 2.2.2). Current astrometric techniques are able to measure the position of a star to a precision of ~0.001 arcsecond, and although a number of claimed detections have been made over the last 50 years, none have survived scrutiny.

Nevertheless, the development of optical interferometers at the Keck Observatory in Hawaii and at the European Very Large Telescope Interferometer (VLTI) in Chile could improve astrometric precision by a factor of 10 to 100 within the next few years. These interferometers could begin yielding extrasolar planets by 2020. In addition, the first space-based interferometer (SIM) is scheduled to be launched in 2015 (Section 21.6.2). All of these programs are focusing on stars within 50 lt-yr, of which ~150 are Sun-like stars.

21.3.3 Doppler velocities

The vast majority of extrasolar planets (including all planets orbiting nearby stars) have been found by the precision Doppler technique, which like astrometry depends on the gravitational tug of a planet on the host star. Rather than searching for the side-to-side motion of a star relative to background stars, however,

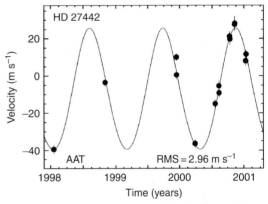

FIGURE 21.5 Doppler velocity curve of the Sun-like star HD 27442 from the Anglo-Australian 3.9-m diameter telescope. The dots are the Doppler velocity measurements, while the line is the best-fit orbital curve, implying a circular orbit with a period of 1.14 years. This planet has one of the most Earth-like orbits yet found. (California & Carnegie Planet Search.)

precision Doppler spectroscopy detects changes in the *radial velocity* (component of the velocity vector along the line of sight) of a star by means of the Doppler effect (Fig. 21.4). During every orbit the measured radial velocity cycles between maximum approach velocity, zero (when the star is moving perpendicular to the line of sight), maximum recession velocity, and again zero (Fig. 21.5).

Jupiter tugs the Sun with a velocity of $13 \, \mathrm{m \, s^{-1}}$, slightly faster than human sprinting speed, and this kind of effect can be detected as a subtle periodic shift in the wavelength of a star's spectral lines. Prior to 1980 astronomical Doppler velocity measurements typically had errors of $1{,}000 \, \mathrm{m \, s^{-1}}$ or more. Bruce Campbell and Gordon Walker pioneered a technique using an absorption cell as a wavelength reference that achieved precision of $13 \, \text{m s}^{-1}$. Currently two groups are struggling to reach long-term precision of $1 \, \text{m s}^{-1}$, which corresponds to a precision in measuring the positions of spectral lines to three parts in 10^9, or ~ 0.0003 of a pixel's width on the CCD detector at the back of the spectrometer mounted on the telescope. Among the challenges faced in making these measurements are subtle variations in keeping the star exactly on the entrance slit of the spectrometer as the telescope tracks the moving sky, and tiny changes in temperature, air pressure, and point-spread-function of the spectrometer, all of which would otherwise dwarf any velocity variations of less than $20 \, \text{m s}^{-1}$.

Figure 21.5 shows the Doppler signal of a planet orbiting the star HD 27442. Two immediate details emerge from the Doppler velocity curve: the orbital period of the planet and the amplitude of the velocity variation. The orbital period P is the time interval from velocity peak to peak (1.14 years in this case), while the semiamplitude K is $31 \, \text{m s}^{-1}$. The near-sinusoidal shape of the velocity curve indicates that the orbit is almost circular; highly elliptical orbits produce non-sinusoidal shapes (as in Fig. 21.10).

The orbit size is expressed as the semimajor axis a of the ellipse, which can also be thought of as the average star–planet distance over an orbit. The determination of a follows immediately once P and the mass M of the star (found from a detailed analysis of its spectrum) are known. The equations for a two-body orbiting system yield:

$$a^3 = G[(M+m)/4\pi^2]P^2,$$

where G is the gravitational constant. By assuming that the planet's mass m is much less than M (i.e., $M + m \approx M$), we have Kepler's famous Third Law of Planetary Motion ($a^3 \propto P^2$) and can solve for a. Note that if M is close to one solar mass, then we know immediately, e.g., that $P = 1$ year means $a = 1$ AU (an Earth analogue) and that $P = 12$ years means $a = 5$ AU (a Jupiter analogue). For other values of M, however, one must use the full equation.

Again assuming the planet mass is small compared to the host star, the orbital equations can be used to solve for the planet mass:

$$(m \sin i)^3 = M^2 K^3 (P/2\pi G)(1 - e^2)^{3/2},$$

where i is the inclination of the orbital plane to the line of sight ($90°$ for an edge-on orbit and $0°$ for face-on), and e is the *eccentricity* of the orbit ($e = 0$ is circular and higher values are more "oval" ellipses).

Doppler velocity measurements yield all parameters of the orbit except the orbital inclination i. We cannot discern whether (a) the orbit is seen fully edge-on, in which case the measured radial velocities capture all of the orbital velocity, or (b) the orbit is tilted from the line of sight and we measure only a fraction ($\sin i$) of the full orbital velocity. Without knowledge of i it is only possible to calculate the *minimum* mass of a planet, i.e., the quantity $m \sin i$. One can say no more about any *individual* planet's mass, although a statistical statement can be made about any sample of many such determined masses: assuming orbital inclinations are randomly aligned with respect to the Earth, the real mass is typically 30% larger than the minimum mass. The real mass will be within a factor of two of the minimum mass for two-thirds of all cases. Extreme "face-on" cases where the real mass is 10 times larger than the minimum mass happen by random alignment only once per thousand systems.

Unlike astrometry, the magnitude of a Doppler velocity signal is independent of the distance to the star. Since changes in stellar velocity rather than stellar position are tracked, the fastest planets, those with the shortest orbital periods, produce the largest Doppler signals. Like astrometry, the Doppler signal increases with the mass of the planet. Thus the precision Doppler technique is most sensitive to massive planets in small orbits.

The "discovery space" for Doppler planets is shown in Fig. 21.6. The semiamplitude (K) of an extrasolar planet's signal is shown as a function of planet mass and semimajor axis. A $1 \, M_\odot$ star at a distance of 50 lt-yr is assumed – there are roughly 160 Sun-like stars within this distance. The upward-sloping parallel lines show Doppler semiamplitudes of 3 and $10 \, \text{m s}^{-1}$, the detection limits for surveys that reach accuracies of 1 and $3 \, \text{m s}^{-1}$, respectively. The dots are extrasolar planets, all discovered from precision Doppler surveys, and most with semiamplitudes greater than $10 \, \text{m s}^{-1}$. The parallel downward-sloping lines are astrometric semiamplitudes of 3 and 50 micro-arcseconds, consistent with the goals for the Space Interferometry Mission (SIM) and the ground-based VLTI, respectively. Doppler precision of $1 \, \text{m s}^{-1}$ is more sensitive to planets than SIM for orbital distances out to ~ 0.5 AU, and more sensitive than VLTI out to nearly 4 AU. Demonstrated long-term Doppler precision is now approaching $1 \, \text{m s}^{-1}$, while SIM and VLTI will not begin taking data for several years. For the nearest 1,000 Sun-like stars within 100 lt-yr, the advantage is further shifted toward precision Doppler surveys as the astrometric signal decreases with distance to the star. In addition to measurement

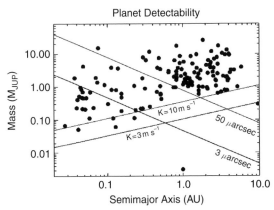

Planet Detectability

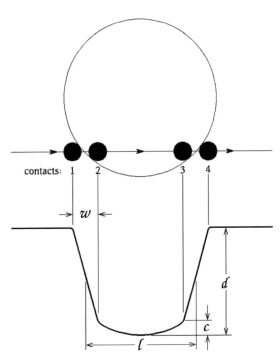

FIGURE 21.6 Planet detectability for Doppler velocities and astrometry, shown as a function of planet mass and semimajor axis. Plotted points are extrasolar planets, except that Earth is the lowest point at 1 AU, Saturn is the rightmost point at 10 AU, and Jupiter is at 5.2 AU and 1.0 M_{Jup}. The lines labeled "$K = 3\,\mathrm{m\,s^{-1}}$" and "$K = 10\,\mathrm{m\,s^{-1}}$" show the detection thresholds for Doppler surveys that reach precision of 1 and $3\,\mathrm{m\,s^{-1}}$, respectively. The lines labeled 3 and 50 micro-arcsec show the detection thresholds for, respectively, the space-based SIM mission and ground-based VLTI astrometry for stars out to 50 lt-yr, encompassing the nearest 150 Sun-like stars.

FIGURE 21.7a Transit of a planet in front of the Sun-like star HD 209458. As the planet moves from position 1 to 4, the measured intensity of starlight is shown in the lower plot. The fraction of starlight blocked by the planet is the ratio of the surface area of the planet to that of the star. Knowing the star's diameter, we can find the diameter of the planet. (Brown *et al.*, 2001.)

precision, detection of true Solar System analogues (e.g., Saturn and Jupiter-mass planets beyond 4 AU) requires more than a decade of data for both the Doppler and astrometric techniques. It is likely that precision Doppler surveys will continue to yield the vast majority of planets orbiting nearby stars into the 2020s.

21.3.4 Transits

For orbits that are very nearly edge-on ($i \sim 90°$), a planet periodically passes directly in front of the host star as seen from Earth, and thus blocks some of the star's light. The fraction of starlight blocked is simply the ratio of the surface area of the planet to that of the star. A Jupiter-sized planet transiting a Sun-like star blocks about 1% of the starlight. Even standard photometric techniques on small- and medium-size ground-based telescopes can detect brightness changes at this level (Fig. 21.7a).

The probability that a planet will transit its star falls off inversely with its orbital distance. Thus for planets in very close (0.05 AU) orbits, there is about a 10% chance of transit. For planets orbiting at 1 AU the chance of transit is reduced to 0.5%, and at Jupiter's orbital distance of 5 AU it is only 0.1%. Transit detection techniques are thus best suited to planets that are physically large and in short period orbits; the short period also allows for many transit observations in a relatively short time.

There are two strategies to search for transit planets. For the case where a planet's orbit is already known from Doppler velocities, the time of transit can be precisely calculated, assuming the orbit is edge-on. Observation of a single transit event is then sufficient to prove that the orbit is indeed edge-on as seen from Earth, and to reveal the physical size of the planet. The advantage of this technique is that it strongly favors finding transits around nearby stars (<100 pc). All three of the brightest known transit systems were discovered this way.

Alternatively, it is possible to monitor fields with thousands of stars using dedicated small-to-medium-sized telescopes, looking for a star that shows a repeated drop in brightness at regular time intervals, and then follow up with precision Doppler velocity measurements. About 20 transit planets have been discovered this way, e.g., by the OGLE survey (Optical Gravitational Lensing Experiment;[6]

---

[6] Although OGLE is primarily designed to look for gravitational lens events, the data can also be used to search for transiting planets that have nothing to do with gravitational lenses. On the other hand, gravitational lens events have (rarely) indicated the presence not only

observing from Chile). These stars tend to be more than 1,000 pc distant, making follow-up study difficult.

Transit detections are complementary to precision Doppler velocities in that they provide information about the orbital inclination angle and the physical size of the planet, while Doppler velocities yield all other orbital parameters. Since a transiting planet must have $i \sim 90°$, sin i is ~ 1.00 and hence the determined value m sin i from Doppler velocities is in fact the true mass. Since both the physical size and true mass are known for a planet detected by both transit photometry and Doppler velocities, the mean density of the planet can be immediately calculated and an educated guess made regarding its bulk composition. Rocky planets like the Earth are roughly 5 times the density of water, while gas giants such as Jupiter and Saturn have about the same density as water.

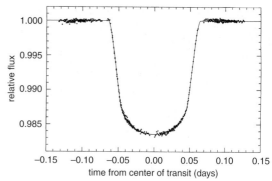

FIGURE 21.7b Extremely high-accuracy measurements made by the Hubble Space Telescope of the HD 209458 transit light curve. Observations of portions of four different transit events (each lasting for $\sim$3 hours of the planet's 3.5-day orbit) have been combined into one curve; the gap and irregularities at the center of the transit arise from incomplete data. The quality of these data allowed the determination of the planet's size to 4% accuracy and the inclination of the orbit to within 0.15°. (Brown *et al.*, 2001.)

21.4 Classes of planets

Prior to the discovery of extrasolar planets, it was assumed that planetary systems would share the same general architecture as the Solar System: planets orbiting in nested concentric circles, all in a single orbital plane co-aligned with the equatorial plane of the central star, huge gas giants in stately distant orbits gravitationally protecting the small, warm, rocky worlds of the inner planetary system. None of the planets found to date resemble this. Discovery of *bona fide* Solar System analogues, i.e., Saturn and Jupiter-mass companions in circular orbits beyond 4 AU, remains a central goal of precision Doppler surveys.

The known extrasolar planets fall into three broad classes, two of which were completely unexpected prior to their discovery. These classes are now discussed.

21.4.1 51 Peg-like planets ("hot Jupiters")

The first Sun-like star[7] to yield a planet was 51 Pegasi, announced in October 1995 by Michel Mayor and Didier Queloz (1995) (Section 2.2.2 and Fig. 2.6). The orbital characteristics of the 51 Peg planet were completely unexpected. The 4.23-day orbital period implied

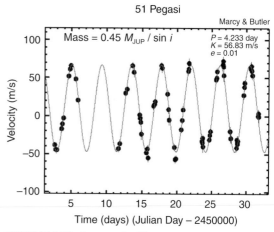

FIGURE 21.8 The first month of Doppler velocities of 51 Peg measured at Lick Observatory in 1995. With an orbital period of 4.2 days, many orbits can be observed in just a few weeks. Like all the other planets with such short orbital periods, the Doppler velocity curve is exquisitely sinusoidal, indicating a circular orbit. (California & Carnegie Planet Search.)

that the planet's orbit was 0.05 AU in size, but the mass of the planet was roughly the same as Jupiter. As this "hot Jupiter" was theoretically unexpected, and because of the abysmal history of previous claims of extrasolar planets, it was not until other groups confirmed this discovery that it became widely accepted. Figure 21.8 shows the first 30 days of confirmatory data from Lick Observatory (Marcy *et al.*, 1996).

To date, about 50 such planets have been found with orbital periods of a day to a week, much shorter even than blistering hot Mercury, the planet with the

---

of an (unseen) star, but of an accompanying planet. These events are thus also a way of detecting extrasolar planets, but this technique is not covered in this chapter because there is no possibility of any follow-up work on any given case.

[7] Earth-sized planets orbiting a pulsar had been discovered three years earlier by Wolszczan and Frail (1992) using a radio timing technique, but it is not clear if and how these relate to the concerns of astrobiology.

FIGURE 21.9 Disk–planet interaction. Artist's impression of a giant planet gravitationally carving a gap in a protoplanetary disk, as indicated by some theories. Tidal forces between the planet and the inner and outer disk drive the planet in toward the protostar at the center. (Geoff Bryden and Doug Lin, UC Santa Cruz.)

smallest orbit in the Solar System. In contrast to Mercury, which has ∼0.0002 of the mass of Jupiter, the "51 Peg-like" planets have masses ranging from ∼0.03 to 4 M_J. They are also all characterized by circular orbits, arising from strong tidal coupling to the nearby host star.[8] The implied temperatures of these planets are ∼1,000–1,500 K (compared to ∼110 K for Jupiter).

Current theory strongly argues that large planets cannot form so near to their host stars. Theorists have offered two scenarios to explain the situation, both of which require that the planet form far out in the protostellar disk (Section 3.9.1). One model suggests that gravitational interactions between a planet and a thick protostellar disk can "viscously" pull the planet in toward the star, as illustrated in Fig. 21.9. Once the planet gets very close to the star, the migration can be stopped and the orbit stabilized by tidal or magnetic forces from the rapidly rotating protostar. Another possibility is gravitational interactions between planets in systems with multiple Jupiter-like planets; these could result in one planet being thrown into a close orbit with other planets being thrown to more distant orbits. In both of these scenarios some planets could crash onto the star itself.

The hot Jupiters orbit so close to their stars that there is about a 10% chance, as discussed above, that any one of these planets will transit its host star as seen

from Earth. The first "transit planet" turned out to be the sixth hot Jupiter discovered from precision velocity surveys (Henry *et al.*, 2000; Charbonneau *et al.*, 2000). As shown in Fig. 21.7b, 1.4% of the light from the Sun-like star HD 209458 is periodically blocked as the planet passes in front of the star (Brown *et al.*, 2001). In addition to independently confirming the Doppler detection of this planet, the brightness measurement immediately solved the orbital inclination and provided a direct determination of the planet's diameter and mean density. The planet's diameter is 40% larger than Jupiter's and its mean density of 0.35 g cm^{-3} is about half that of Saturn. The mean density of all the transiting hot Jupiters but one is consistent with compositions similar to Saturn, Jupiter, and the Sun, namely dominated by hydrogen and helium. The one exception is the planet transiting HD 149026, whose mean density and size imply a much higher abundance of heavy elements, presumably in a large core. The accuracy with which these mean densites can be calculated has begun to seriously challenge models of planetary interiors, with some hot Jupiters having densities greater than expected from models, some less; such deviations may yield clues about the evolutionary history of any given planet.

To first order, a transiting planet simply acts like an opaque disk, blocking light from its host star. But, more accurately, the encircling atmosphere around the outer edge of the disk is somewhat transparent and the tiny fraction of starlight travelling through this atmosphere will be imprinted with absorption lines characteristic of the atmosphere's composition. The difference between extremely high signal-to-noise spectra of the star taken (a) during transit, and (b) outside of transit, will thus yield the spectrum of the planet's atmosphere. In the case of HD 209458, spectra from the Hubble Space Telescope (Charbonneau *et al.*, 2002) have detected the sodium "D" lines, but no other unambiguous signal.

In addition to the transit, there is a so-called secondary eclipse, when the planet disappears behind the host star. While the contribution of the planet to the total flux of the star–planet system is negligible at optical wavelengths, in the infrared it can amount to a part in one thousand or more. This drop in flux has already been detected at secondary eclipse for several of the nearest transit planet systems with the Spitzer Infrared Space Telescope. These measurements helped to pin down the atmospheric temperature of the planet to ∼1100 K.

Although theoretically troublesome, hot Jupiters are the easiest type of planet to detect because of their

[8] Chapter 19 (Appendix 19.2) explains the physics of this tidal circularization.

strong gravitational tugs on host stars, resulting in large Doppler shifts. Moreover, their short orbital periods allow many orbits to be observed in just a few months. About 0.75% of solar-type stars have these odd companions.

From the perspective of astrobiology, the 51 Peg-like planets are not very interesting. With temperatures of 1,500 to 2,000 K, these planets are completely sterile and incapable of supporting life.

21.4.2 Eccentric planets

The largest group of planets found to date are the "eccentric planets." These have noticeably elliptical-shaped orbits ($e > 0.1$), typically with periods longer than 20 days. Figure 21.10 shows a typical Doppler velocity curve for an eccentric planet, with its characteristic non-sinusodial velocity curve. Figure 21.11 schematically shows the orbit of this planet in comparison with our own Solar System.

Eccentric planets have motivated theorists to construct increasingly complex models of planet formation and evolution that take account of planet–planet and planet–disk interactions. Planets gravitationally interact with each other as well as with the central star and (early in their history) the protoplanetary disk. Models indicate that a system of multiple Jupiter-mass planets might survive in stable orbits for millions or even billions of years, but then chaotically interact, throwing some planets into smaller orbits and others out to interstellar space (Levison et al., 1998). The orbits after such an interaction are of the eccentric variety. Something like this would have happened in our Solar System if Saturn had ended up twice as massive as it did. A Jupiter–Saturn interaction, like billiard balls on a gravitationally warped pool table, would be devastating to the orbital stability of the small rocky planets of the inner Solar System, and particularly devastating to the possibility of advanced life on Earth.

Because the planet–star distance varies so greatly in an eccentric system (Fig. 21.11), the planet oscillates over a large temperature range. While this does not rule out the possibility of life (e.g., subsurface microbial life), it perhaps makes it more difficult for life to get started and to develop complexity.

Figure 21.12 shows the eccentricity of the known extrasolar planets as a function of orbital size. About 10% of Sun-like stars surveyed for at least 10 years by state-of-the-art precision Doppler techniques yield eccentric planets. The eccentric planets are by far the dominant type of planet found to date, accounting for

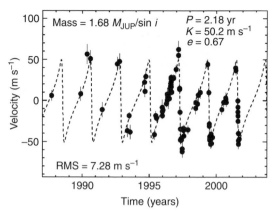

FIGURE 21.10 Doppler velocity curve of the Sun-like star 16 Cyg B from the Lick Observatory 3-m diameter telescope. The dots are the Doppler velocity measurements, while the dashed line is the best-fit orbital curve. The orbital period and velocity semiamplitude of the planet are 2.18 years and 50 m s^{-1} respectively, yielding a minimum planet mass of 1.68 M_J. The sawtooth shape of the velocity curve indicates that this orbit is extremely eccentric ($e = 0.67$), as shown in Fig. 21.11. (California & Carnegie Planet Search.)

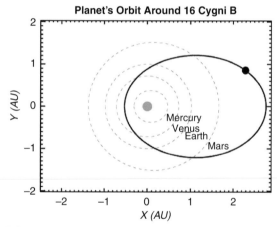

FIGURE 21.11 The orbit of 16 Cyg B's Jupiter-like planet is shown as the black line, superimposed for comparison on the inner planets of our Solar System. The extreme orbital eccentricity carries the planet from less than 0.5 AU out to nearly 3 AU, thus exposing the planet to large temperature variations. The existence of a planet with such an orbit also rules out the possibility of any planets in stable Earth-like orbits, since they would be quickly disrupted. (California & Carnegie Planet Search.)

roughly 90% of the known planets with orbital periods longer than 20 days.

21.4.3 Planets in circular orbits

Solar System planets are characterized by nested coplanar circular orbits, with the small rocky planets orbiting close to the Sun and the gas giants farther out. The

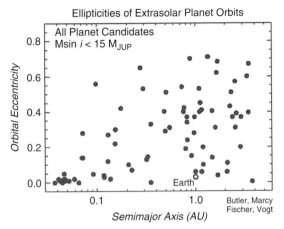

Ellipticities of Extrasolar Planet Orbits

FIGURE 21.12 Eccentricity vs. semimajor axis for a sample of extrasolar planets. 90% of the extrasolar planets found orbiting beyond 0.2 AU are in eccentric (e > 0.1) orbits. The Earth is shown for comparison. The circularity of planets in the smallest orbits (~0.05 AU) is expected due to strong tidal effects from the host stars. For planets orbiting between 0.2 and 4 AU, eccentric orbits are the rule, not the exception. (California & Carnegie Planet Search.)

detection of Solar System analogues is anthropocentrically exciting because these planets remind us of our Solar System, and are of the greatest interest to the astrobiology community.

In addition to providing a reminder of Solar System architecture, giant planets in distant circular orbits may play a crucial role in the development and evolution of life. While our current middle-aged Solar System is filled with planets that orbit in neat circles, we know that the situation was different in the early Solar System. The heavily cratered faces of bodies such as the Moon and Mercury, the Moon's very formation via collision, and the extreme tilt of the axis of Uranus tell of a violent, impact-racked early Solar System. During the first 100 Myr it is estimated that there was a new bright comet in the sky of Earth every week. Many of these comets and other large bodies would have impacted Earth. Life would have found it difficult to establish a toe-hold on the Earth during this time. As George Wetherill (1990) has pointed out, we owe the relative quiescence of the current Solar System primarily to Jupiter, which acts like a gravitational vacuum cleaner in clearing out the debris left over from the era of planet formation (see Section 3.9 for further discussion). The earliest evidence for life on Earth appears shortly after the end of the period of heavy bombardment.

With the discovery that most extrasolar planets are in eccentric orbits, it appears that Jupiter also protects Earth from rogue eccentric planets that might otherwise alter our orbit. As Jupiter dominates the planets of the Solar System, it would gravitationally eject any planet that began to significantly deviate from a circular orbit. By analogy, the existence of a Jupiter-mass planet orbiting at a distance from 0.3 to 3 AU from its star precludes the possibility of an Earth-like planet (except perhaps as a moon).

For planets orbiting within ~0.2 AU, any initial orbital eccentricity is erased by tidal interactions with the star. Planets orbiting beyond this distance preserve their orbital eccentricity. Roughly 10% of the planets orbiting beyond 0.2 AU are in circular ($e < 0.1$) orbits. These few circular planets, however, still do not qualify as Solar System analogues. A Solar System analogue must have a Jupiter-mass planet in a circular orbit beyond 4 AU, thus protecting Earth-like orbits. Doppler detection of true Jupiter-analogues orbiting beyond 4 AU requires maintaining Doppler precision of $3 \, \mathrm{m \, s^{-1}}$ for more than a decade. Existing Doppler programs should soon find many Jupiter-mass planets orbiting beyond 4 AU.

21.4.4 Multiple planets

Existing precision Doppler surveys are typically less than 15 years old, and therefore are primarily sensitive to planets of periods no more than ten years, i.e., having orbital sizes within 4–5 AU. Given this current technological hurdle, multiple-planet systems can only be found if systems of giant planets all orbit within a few AU of a Sun-like star. Remarkably, about 25 such systems have already been discovered.

Figure 21.13 shows the orbits of the first confirmed multiple-planet system, Upsilon Andromedae, relative to the planets in the inner Solar System (Butler *et al.*, 1999). The first Upsilon Andromedae planet is a hot Jupiter with an orbital period of 4.6 days, a circular orbit, and a mass of roughly 1 M_J. Long-term monitoring of this system has subsequently found two more planets with periods of 8 months and 3.5 years, and masses of approximately 2 and 4 M_J, respectively. Figure 21.14 shows the Lick Doppler velocity data for the outer two planets alone, i.e., with the effect of the inner planet removed.

The dynamic stability of the Upsilon Andromedae system has been studied by several groups. Unless the orbit is nearly face-on ($i < 30°$) and thus the planets are of very high mass, this system is stable in that the orbits never cross. As the planets interact with each other, however, the orbital parameters will evolve in a measurable way within a decade. Dynamic simulations

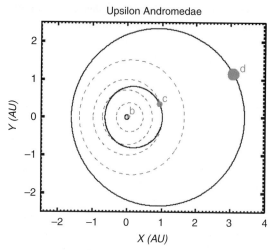

Upsilon Andromedae

FIGURE 21.13 Orbits of the Upsilon Andromedae system. The inner planet "b" is a "hot Jupiter" in a circular 4.6-day orbit. The outer planets "c" and "d" are in eccentric orbits ("a", not shown, refers to the star itself). The orbits of Mercury, Venus, Earth, and Mars (dashed lines) are shown for comparison. The masses of "b", "c", and "d" are approximately 1, 2, and 4 M_J. (California & Carnegie Planet Search.)

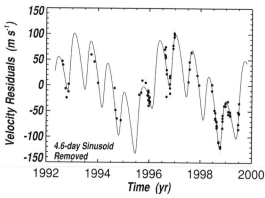

FIGURE 21.14 Doppler velocities of the outer planets in the Upsilon Andromedae system. The dots are the measured Doppler velocities, taken with the Lick Observatory 3-m diameter telescope over a 10-year period. The effect of the inner planet has been removed; the line is the best-fit orbital model for the effects of two additional planets. The 8-month and 3.5-year periodicities can clearly be seen. The subtle effects of gravitational interactions between the two outer planets are expected to become apparent within a decade. (California & Carnegie Planet Search.)

of the orbital evolution, when compared with observations, will provide further constraints on the orbital inclination of this system.

Orbital resonances are common in the Solar System, including Earth–Venus, Neptune–Pluto, and the Galilean satellites of Jupiter (Chapter 19). Resonances also appear to be common in other multiple planet systems. These resonances occur when the values of orbital period of two planets are in a small-integer

ratio such as 2:1 or 3:2. Such a situation typically leads to very strong gravitational interactions between the planets. Perhaps the extreme example is Gliese 876, a low-mass ($\sim$0.2 $M_\odot$) main sequence star (spectral type M4, temperature $\sim$3,000 K) with multiple planets. With periods of 30.1 and 61.0 days, the orbits of two large planets (of masses 0.8 and 2.5 M_J, respectively) are separated by less than 0.1 AU at closest approach. As the orbital periods are not precisely 2:1, Laughlin and Chambers (2001) showed that this system would experience "severe dynamical instabilities within five years," but that long-term stability is maintained by libration of the planets' orbital elements about the resonance. With an additional four years of precision Doppler data, this libration has indeed been observationally confirmed. Furthermore, residuals from the two-planet model indicate that a third planet, of only 7.5 Earth-masses ($\pm$10%), resides in a 1.9-day orbit at a distance of only 0.021 AU (Rivera *et al.*, 2005). In this case, the quoted planet mass is *not* a lower limit, as is usual for the Doppler technique, because the gravitational interactions of the two large planets with each other allow one to determine the inclination of the orbital system as $i \sim 50°$. So here we have a "super-Earth" extremely close to and zipping around a small, cool star; its temperature is probably in the range of 400–650 K.

Recently another "super-Earth" of minimum mass 5.0 Earth mass was discovered in a system that also has two other low-mass planets (minima of 8 and 15 Earth masses) (Udry *et al.*, 2007). These three planets all circle the M dwarf star Gliese 581, whose luminosity (power output) is only 0.01 of that of the Sun. Depending on assumptions made regarding the planets' sizes, compositions, histories and atmospheres (Sec. 3.3 and Chap. 4), it is possible that one of these could be in this star's habitable zone, defined as where water can be stable in the liquid state for a long enough period for life to originate and evolve (von Bloh *et al.*, 2007).

At least half of all stars with planets have either a known second planet or long-term velocity residuals suggesting multiple planets (Fischer *et al.*, 2001). As precision Doppler surveys reach higher precision and span more years, many more multiple planet systems will emerge, providing natural laboratories for the studies of planet formation, evolution, and dynamical interaction.

21.5 Masses of extrasolar planets

Of the $\sim$250 known substellar companions to normal Sun-like stars, only 4% have been found with minimum mass greater than 10 M_J. This scarcity is surprising as it is these very massive companions that are the easiest to

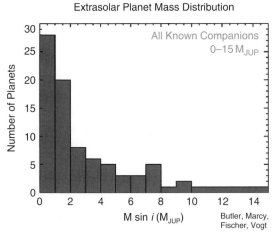

FIGURE 21.15 Substellar mass distribution from precision Doppler surveys. Out to 2 AU these surveys are complete for companions of more than 3 M_J. Incompleteness is greatest for the smallest mass bin. The discontinuous and abrupt rise in the distribution below 10 M_J motivates setting the upper mass limit of planets near 10 M_J. (California & Carnegie Planet Search.)

detect. As Figure 21.15 shows, most of the known planets have minimum masses of 2 M_J or less, although these are the hardest companions to detect. It is at the lowest masses, less than 1 M_J, that current precision Doppler programs are least sensitive to planets, yet it is this mass interval that has the largest number of planets. Given the incompleteness of the surveys in detecting the smallest mass companions of $M \sin i < 1$ M_J, the true mass distribution must rise even more steeply toward lower masses than the observed distribution.

As Fig. 21.15 shows, the number of planets abruptly increases below 10 M_J, and continues to rise down to the detection limit. This empirical result motivates setting the upper mass limit of planets near 10 M_J. Coincidentally, the theoretical boundary for the lowest-mass object that can sustain deuterium nuclear reactions (deuterium "burning") is 13 M_J. The core of objects less massive than 13 M_J never generate enough heat and pressure to ignite deuterium fusion, nor do they generate significant energy from other nuclear reactions. Just as the analogous hydrogen-fusion boundary at 80 M_J (0.08 $M_\odot$) divides stars from brown dwarfs, by analogy it has been suggested that the deuterium boundary should serve as the dividing line between planets and brown dwarfs. Planets would thus differ from stars and brown dwarfs in that they have never undergone any nuclear processing.

While brown dwarfs can exist both as orbital companions to stars and as lone objects, planets must orbit stars. Without this additional constraint, extrasolar planets would have no connection to Solar System planets, or to the 2500-year history of the word *planet*. From the perspective of astrobiology, hypothetical "free-floating planets," between the stars without a stellar energy source, are not of much interest.

21.6 Stellar metallicity and extrasolar planets

Figure 21.16 shows the fraction of stars surveyed by precision velocities that have yielded planets as a function of their "metallicity" (relative abundance of elements other than hydrogen and helium). Among the most metal-rich stars, 20% have already yielded planets. Stars that are more metal-rich than the Sun are three times more likely to have yielded planets than stars less metal-rich.

Typical planet-bearing stars are metal-rich relative both to general field stars and to the Sun. In particular, planets have been detected around three stars that are more metal-rich than any single G star within 65 lt-yr, suggesting that super metal-rich stars may commonly have detectable planets orbiting within 3 AU. At the other extreme, stars with metallicity lower than the Sun have yielded only about 10% of the known planets, yet such stars constitute the majority of targets.

There are several competing hypotheses to explain the metal-rich nature of the observed planet-bearing stars. Most obviously, planets may form more readily in metal-rich environments. Alternatively, planets orbiting metal-poor stars may typically be less massive than Saturn or in distant orbits, beyond 4 AU, a regime just now being probed by precision Doppler surveys. Both of these possibilities will be explored over the next decade.

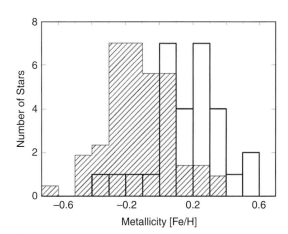

FIGURE 21.16 Histogram of "metallicities" of planet-bearing stars, compared to the nearest field stars (cross-hatched region). Metallicity is here defined as the logarithmic ratio of the iron/hydrogen abundance ratio in a star relative to that in the Sun (iron is used as a representative heavy element). The majority of the planet-bearing stars are metal-rich relative to the Sun, which is itself metal-rich relative to nearby field stars. (California & Carnegie Planet Search.)

21.7 Future of extrasolar planet searches

After 400 years of speculation, the discovery of planets orbiting nearby Sun-like stars has been a great triumph for the fledging field of astrobiology, and has been widely cited as a primary motivation for the creation of the NASA Astrobiology initiative.

Even with these recent successes, the study of planetary systems is still in its infancy. Since 1995 when there was essentially no field of "extrasolar planets," the subject has been driven exclusively by empirical data, primarily from precision Doppler surveys, but increasingly by high-resolution infrared imaging of nearby young stars and their disks. Photometric and spectroscopic follow-up of the few nearby transit planets have opened up the direct physical study of the planets, including their diameter, density, and composition.

For the foreseeable future, the field of extrasolar planets will continue to be "discovery driven." The extraordinary complexities involved in the formation and subsequent evolution of protostars, disks, and protoplanets are only beginning to be probed by computer models. Ultimately, questions such as which stars have planets, which planetary systems are similar to the Solar System, and which planets harbor life can only be addressed with empirical data.

21.7.1 Precision Doppler surveys

The great bulk of the known planets have been found by the precision Doppler surveys undertaken by the California and Carnegie Planet Search and by the Geneva Observatory, working in the past with precisions mostly in the range of 3 to 10 m s^{-1}. The two groups are now working toward the goal of surveying 2000 of the nearest Sun-like stars (and some of lower mass, too), out to 50 pc (160 lt-yr), with a precision of 1 to 3 m s^{-1}. These surveys, with existing ground-based telescopes and spectrometers, are the least expensive and the only proven technique for finding planets around nearby stars. They represent the first reconnaissance of planets orbiting nearby stars, for which we have heretofore been completely ignorant, and provide justification for the much more expensive, space-based missions to follow. Within 20 years these Doppler surveys will provide the first solid data on two unknown Drake equation parameters:

(1) What fraction of stars have planets?
(2) What fraction of planetary systems are similar to the Solar System?

21.7.2 Astrometric interferometry

Astrometry is complementary to precision Doppler velocities in that it is most sensitive to long-period planets. Like precision Doppler velocities, astrometric planet detections require observation of an entire orbital period. Groups at VLT and Keck have already achieved interferometric fringes, and VLT appears ready to begin a major astrometric survey for planets within the next few years. The first batch of astrometric planets could emerge from this in less than ten years. Since the amplitude of the astrometric signal decreases with stellar distance, the nearest 150 Sun-like stars within 50 lt-yr will be the primary targets.

NASA has funded the Space Interferometry Mission (SIM), scheduled to launch in 2015 or later. SIM consists of two 0.3-m telescopes, separated by 10 m, whose goal is to provide narrow angle astrometry with a precision of 1 to 3×10^{-6} arcseconds, 1,000 times better than the long-term precision of present astrometric surveys. As with other astrometric techniques, SIM will be primarily sensitive to planets around the nearest stars in relatively long-period orbits (up to the mission lifetime). If SIM achieves long-term precision of 1 $\times 10^{-6}$ arcseconds it will be able to detect planets in Earth-like orbits with masses as small as three times that of the Earth. For planet hunting, SIM will concentrate on 200 of the nearest Sun-like stars, all of which are currently being monitored by precision Doppler surveys. These surveys are already more sensitive than SIM to planets orbiting within 0.5 AU, and their much longer time baseline will improve SIM sensitivity to longer period planets.

21.7.3 Photometric transit surveys

A number of groups are pursuing ground-based transit searches. The goal of these programs is to find hot Jupiters with orbital periods of one to five days. Since tens of thousands of stars must be searched in order to have a realistic chance of finding a transit planet, two broad strategies have been devised. The first is the "deep and narrow" search, which makes use of 1- to 4-m diameter telescopes to monitor all stars within narrow ($\sim 1°$-wide) fields. The stars in these surveys are typically fainter than fifteenth magnitude and at a distance of many hundreds or thousands of parsecs, making follow-up by other techniques difficult.

The second ground-based transit technique makes use of small wide-field telescopic cameras that monitor 6° patches of the sky down to about thirteenth magnitude. Transit candidates found by this method can be confirmed with precision Doppler velocity measurements on 8-m diameter telescopes and with space-based telescopes.

These surveys have shown that nature has many ways to generate "transit-like" signals, including, among others, grazing eclipsing binaries, triple systems that include an eclipsing M dwarf pair and a brighter F star, and distant eclipsing binaries that lie along the line of sight of a brighter star. Typically 20 to 150 candidates for planet transit signals are investigated for every one that is eventually verified as a planet. About 20 transit planets have been discovered in all.

A much more ambitious transit experiment is the NASA Kepler mission scheduled to launch in 2008. While Jupiter-sized companions cause a roughly 1% photometric dip during transit, the photometric signature of an Earth-sized transit planet is only ∼0.01%. This level of precision cannot be reached by ground-based telescopes due to fluctuations in measured intensities caused by the Earth's atmosphere. The Kepler mission will get around this limitation by placing a 1-m diameter telescope in an Earth-trailing heliocentric orbit.

The Kepler mission will stare at a single star field in Cygnus for three to four years, continuously monitoring the brightness of 100,000 main sequence stars brighter than fourteenth magnitude. Without any Doppler velocities, three successive transit events separated by identical time intervals must be recorded to provide minimal evidence of a transit planet. The mission is thus limited to detecting planets with orbital periods of about one year or less.

The ability of Kepler to make the first unambiguous detection of an Earth-like planet depends on a number of unknowns, including the frequency of false signals due either to astrophysical or instrumental effects. The experience of ground-based transit surveys in searching for much easier 1%-level photometric events suggests that false signals will be very common. Any Kepler candidates for a hot Jupiter transiting a bright star can be confirmed by precision Doppler velocity measurements on ground-based 8-m diameter telescopes. Confirmation of giant-planet candidates would add credibility to any Kepler

detections of smaller planets, which cannot be similarly checked.[9]

21.7.4 Imaging and spectroscopy

By far the most ambitious missions currently under consideration are the space-based programs designed to directly detect extrasolar planets and obtain low resolution spectra of their atmospheres. Of particular interest would be the spectroscopic discovery of potential atmospheric biomarkers such as oxygen, methane, and water. The technical challenges are formidable. Even giant planets such as Jupiter are 10^5 to 10^9 times fainter than the central star (depending on the operational wavelength), and smaller Earth-like planets are much fainter still. As the difficulty and cost of these missions have become more defined, the projected launch dates for NASA's Terrestrial Planet Finder and the European Space Agency's Darwin missions have unfortunately been slipping about two years per year since 1997. Direct imaging and spectroscopy of Earth-like planets is likely at least 20 years off – if and when it *is* achieved, it will herald a new era in this field no less than the discovery of 51 Pegasi's planet in 1995.

REFERENCES

Black, D. C. (1995). Completing the Copernican revolution: the search for other planetary systems. *Ann. Rev. Astron. Astrophys.*, **33**, 359.

Brown, T. M., Charbonneau, D., Gilliland, R. L., Noyes, R. W., and Burrows, A. (2001). Hubble Space Telescope time-series photometry of the transiting planet of HD 209458. *Astrophys. J.*, **552**, 699–709.

Butler, R. P., Marcy, G. W., Fischer, D. A., *et al.* (1999). Evidence for multiple companions to υ Andromedae. *Astrophys. J.*, **526**, 916–927.

Charbonneau, D., Brown, T. M., Latham, D. W., and Mayor, M. (2000). Detection of planetary transits across a Sun-like star. *Astrophys. J. Lett.*, **529**, L45–48.

Charbonneau, D., Brown, T. M., Noyes, R. W., and Gilliland, R. L. (2002). Detection of an extrasolar planet atmosphere. *Astrophys. J.*, **568**, 377–384.

Dick, S. J. (1996). *The Biological Universe*. Cambridge: Cambridge University Press.

Fischer, D. A., Marcy, G. W., Butler, R. P., Vogt, S. S., Frink, S., and Apps, K. (2001). Planetary companions to HD

[9] The COROT mission, primarily sponsored by the French space agency CNES and in orbit since 2006, also is sensitive to transiting planets, and reported its first detection of a transiting Jupiter in 2007.

12661, HD 92788, and HD 38529 and variations in Keplerian residuals of extrasolar planets. *Astrophys. J.*, **551**, 1107–1118.

Henry, G. W., Marcy, G. W., Butler, R. P., and Vogt, S. S. (2000). A transiting "51 Peg-like" planet. *Astrophys. J. Lett.*, **529**, L41–44.

Laughlin, G. and Chambers, J. E. (2001). Short-term dynamical interactions among extrasolar planets. *Astrophys. J. Lett.*, **551**, L109–113.

Levison, H. F., Lissauer, J. J., and Duncan, M. J. (1998). Modeling the diversity of outer planetary systems. *Astron. J.*, **116**, 1998–2014.

Marcy, G. W., Butler, R. P., Williams, I. E., Bildsten, L., Graham, J. R., Ghez, A. M., and Jernigan, J. G. (1996). The planet around 51 Pegasi. *Astrophys. J.*, **481**, 926–35.

Mayor, M. and Queloz, D. (1995). A Jupiter-mass companion to a solar-type star. *Nature*, **378**, 355–9.

NAS (Astronomy & Astrophysics Survey Committee) (2001). *Astronomy and Astrophysics in the New Millennium.* Washington, DC: National Academic Press.

Perryman, M. A. C., Lindegren, L., Kovalevsky, J., *et al.* (1997). The Hipparcos Catalog. *Astron. Astrophys.*, **323**, L49–L52.

Rivera, E. J., Lissauer, J. L., Butler, R. P., *et al.* (2005). A ∼7.5 Earth-mass planet orbiting the nearby star, GJ 876. *Astrophys. J.*, **634**, 625–40.

Udry, S. Bonfils, X., Delfosse, X., *et al.* (2007). The HARPS search for southern extra-solar planets. XI. Super-Earths (5 & 8 Earth masses) in a 3-planet system. *Astron. and Astrophys.*, **469**, L43–47.

von Bloh, W., Bounama, C., Cuntz, M., and Franck S. (2007). The habitability of Super-Earths in Gliese 581. *Astron. and Astrophys.*, submitted.

Wetherill, G. W. (1990). Formation of the Earth. *Ann. Rev. Earth Planet. Sci.*, **18**, 205–256.

Wolszczan, A. and Frail, D. (1992). A planetary system around the millisecond pulsar PSR 1257 + 12. *Nature*, **355**, 145–7.

FURTHER SURFING

www.dtm.ciw.edu/boss/iauindex.html. International Astronomical Union Working Group on Extrasolar Planets.

www.exoplanets.org. California & Carnegie Planet Search. The group led by Marcy and Butler that has discovered about two-thirds of all extrasolar planets.

obswww.unige.ch/~udry/planet/planet.html. Geneva Planet Search. The group led by Mayor that discovered the first extrasolar planet and many others since.

planetquest.jpl.nasa.gov/index.html. "PlanetQuest: the Search for Another Earth." Information on current and planned NASA missions dealing with extrasolar planets; catalog of known planets.

sci.esa.int/home/darwin/index.cfm. DARWIN. Planned mission of the European Space Agency for imaging and spectroscopy of Earth-like planets.

Part VI
Searching for extraterrestrial life

22 How to search for life on other worlds

Christopher P. McKay
NASA Ames Research Center

22.1 Introduction and summary

In the original *Star Trek* series, Episode 26 ("The Devil in the Dark"), Mr. Spock uses a hand-held *tricoder* to remotely detect a silicon-based lifeform known as a "Horta." Unfortunately, NASA engineers have not yet invented the tricoder to aid in our own search for life on Mars or Europa. In fact they are not even close to understanding the principle by which a tricoder is able to distinguish lifeforms, either carbon or silicon, from non-living matter. Even *The Physics of Star Trek* (Krauss, 1995) is notably silent on the operating physics behind the tricoder. How then do we achieve the Prime Mission of Astrobiology: to boldly go and seek out new lifeforms on distant worlds? The answer, not surprisingly, is that we base our search for life elsewhere on what we know about life on Earth. The basic elements of this approach can be summarized as follows.

1. There is no specific definition of life that *usefully* contributes to the search for life (see Chapter 5). Don't wait for one.
2. In searching for either extant or extinct life the most useful guide is the short list of the ecological requirements for life. These are: (a) energy, (b) carbon, (c) liquid water, and (d) other elements such as N, P, and S. On the planets of our Solar System liquid water is the limiting factor.
3. Life is composed of, and produces, organic (carbon-based) matter. Forget silicon-based life for now.
4. Organic matter of biological origin can be differentiated from other organic matter because life preferentially selects and uses a few specific organic molecules. This selectivity is probably a general feature of life.
5. Small microbial life is more probable and widespread than large multicellular life. Bring a microscope.
6. Metabolism and movement are the only two ways to determine if something is alive. The Viking missions to Mars searched for microbial metabolism and macroscopic motion, as well as for organic matter.
7. Fossils recognized on the basis of morphological or organic evidence provide strong evidence of past life. Studies of the martian meteorite ALH84001 and the oldest fossils on Earth, however, highlight the difficulties with identifying microfossils.
8. Fossils are not enough because they do not inform us as to the biochemical nature of the life they record. The search for life must ultimately be directed toward determination of the biochemical, genetic, and possibly ecological, details of any alien lifeforms.
9. Photosynthetic life can create global scale changes in a planetary environment that can be observed across space.
10. Intelligent life can generate signals and artifacts detectable across space.

The following ten sections of this chapter briefly discuss each of these points, with particular emphasis on the search for microbial life in our Solar System.

22.2 Definitions for life

There have been many attempts to provide a succinct definition of life (see Chapter 5). None are generally accepted and none provide guidance in how to search for life elsewhere. From a biological point of view, probably the best definition of Earth life is the Darwinian definition: life is a material system that is subject to reproduction, mutation, and natural selection. This captures probably the most important distinction between life and non-living matter. However,

Planets and Life: The Emerging Science of Astrobiology, eds. Woodruff T. Sullivan, III and John A. Baross. Published by Cambridge University Press. © Cambridge University Press 2007.

it is not a very useful definition in terms of searching for other lifeforms. Fossils, biomarkers, dormant cells, and even single individual organisms may fail to show Darwinian evolution and yet all would be of keen interest to astrobiology as evidence of life. Similar issues limit the scope of definitions of life based on reduction of entropy, metabolism, compartmentation, etc. One might wonder how biology has progressed so far without a definition of its subject matter. Most biologists tend to take an approach similar to that famously expressed by Justice Potter Stewart when he said "I shall not today attempt to further define the kinds of material [pornography] but I know it when I see it" (Jacobellis vs. Ohio, 22 June 1964).

22.3 Requirements for life

In searching for life the most useful guide is the short list of ecological requirements for life. These are listed in Table 22.1 along with their prevalence in the Solar System (e.g., McKay, 1991).

Energy[1]

All self-organizing systems need an external source of energy. This is true for a hurricane, for a forest fire, and for active lifeforms. On Earth life uses only two types of energy: sunlight and chemical energy. There are no known lifeforms that use gravitational energy, low frequency electric or magnetic fields, or thermal gradients. Therefore, on other worlds we should search for life deriving its energy from sunlight or from chemical reactions produced by geothermal heating.

On Earth, sunlight provides the power source, directly or indirectly, for the vast majority of life. Indeed, we know of only two small isolated ecosystems that operate independently of sunlight. These are the two methanogenic microbial ecosystems within which *primary production* (of organic matter) is due to methane generation from hydrogen and carbon dioxide (Stevens and McKinley, 1995; Chapelle *et al.*, 2002). The hydrogen is produced by the reaction of water with the surrounding volcanic basalt, as deep as 5 km below the surface. But this life undoubtedly migrated from the surface to its present location sometime in the past, and survives today without any necessary connection to surface organic material or oxygen. There are complex ecosystems found at the ocean bottom surrounding deep-sea vents that do not depend on photosynthesis.

TABLE 22.1 Ecological requirements for life (McKay, 1991)

Requirement	Occurrence in the Solar System
Energy	Common, mostly sunlight – photosynthesis at light levels out to 100 AU – chemical energy, e.g., $H_2 + CO_2 \rightarrow CH_4 + H_2O$
Carbon	Common as CO_2 and CH_4
Liquid water	Rare, only on Earth for certain
Other elements: N, P, S	Common

However, these are not independent of sunlight since the basis of production there is dissolved oxygen in the water oxidizing hydrogen sulfide. The oxygen in turn comes from photosynthesis at or near the ocean surface.

The average flux of sunlight reaching the surface of the Earth is $\sim$240 W m^{-2}, about 4,000 times larger than the total outward geothermal flux at the surface, 0.06 W m^{-2}. This geothermal flux arises from (a) the planet cooling off from its gravitational heat of formation, and (b) decay of long-lived radioactive elements. Note that only a small fraction of the geothermal heat is converted into chemical energy, whereas about half the solar flux occurs at wavelengths that are usable for photosynthesis. Arguments from thermodynamics indicate that this must be the case.[2] The case of Earth tells us that a biosphere can have effects on a global scale only when it becomes photosynthetic. Chemosynthetic life would, by dint of energy restrictions, always remain small and globally insignificant. Life is able to use light at very low levels. The limit of photosynthesis is about 4–10 nmoles of photons m^{-2} s^{-1} (Raven *et al.*, 2000) or $\sim$10^{-5} of the solar flux at Earth. Even at the orbit of Pluto light levels exceed this value by a factor of $\sim$60.

This view of the energetics of life has implications for the search for life on Mars and Europa. If Mars had a surface biosphere that was global, then it was also photosynthetic. If there is life under the deep ice of Europa that is chemosynthetic, then the productivity is likely to be very small, and detectable levels of biomass could accumulate only after long times.

---

[1] Also see Section 9.5 for a full discussion of energy sources available on Earth.

[2] The entropy of any heat flow is proportional to Q/T where Q is the heat flow and T is the temperature. Thus sunlight (with an effective temperature of 5800 K) has much lower entropy per unit energy than geothermal heat (whose mean temperature is 288 K = 15 °C).

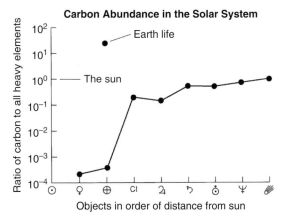

Carbon Abundance in the Solar System

FIGURE 22.1 Planet Earth is low in carbon, but Earth-life is extremely high in carbon abundance. Shown is the ratio of carbon atoms to total atoms heavier than He (relative to the solar value) for various Solar System objects. Note the depletion of carbon in the inner Solar System. The x-axis is not a true distance scale but objects are ordered by increasing distance from the Sun: symbols represent Venus, Earth, Type I carbonaceous chondrites (CI), Jupiter, Saturn, Uranus, Neptune, and comets. Mars is not shown since the size of its carbon reservoir is unknown. (Adapted from McKay, 1991.)

Carbon

Life must be composed of elements, and on Earth carbon has the dominant role as the backbone molecule of biochemistry.[3] In our Solar System carbon is generally abundant (Fig. 22.1), although surprisingly Earth is depleted in comparison to other objects.

Liquid water

As indicated in Table 22.1, sunlight and the elements required for life are common on the other worlds in the Solar System. What appears to be the ecologically limiting factor is the availability of liquid water. Liquid water is key to biochemistry because it acts as the solvent in which biochemical reactions take place. Water also forms hydrogen bonds with parts of large molecules called hydrophilic groups and repels other parts called hydrophobic groups. This forces these molecules to fold up with their hydrophobic groups in the interior and their hydrophilic groups on the exterior in contact with the water (Section 9.6).

Life needs *liquid* water, not vapor or ice. Certain organisms, notably lichen and some algae, are able to obtain water from the vapor phase if the relative humidity is high enough and maintain liquid water in their cells. Many organisms can continue to metabolize

at temperatures well below the freezing point of pure water because their intracellular material contains salts and other solutes that lower the freezing point (Chapter 15). Other organisms, such as the snow algae *Chlamydomonas nivalis*, thrive in liquid water associated with ice, but in these circumstances the organisms are the beneficiaries of external processes that melt the ice. There is no known occurrence of an organism using metabolic energy to overcome the latent heat of fusion of ice and thereby liquefy it.

Because liquid water is universally required for life on Earth and because it appears to be rare in our Solar System, designing missions to search for life must start with a search for liquid water. Its presence at a given locale would not prove that life is present, but it makes the case worth pursuing.

Other elements

In addition to carbon and the hydrogen and oxygen in water, life of any kind almost certainly requires other elements for its metabolic reactions. Although Earth-life utilizes many of the elements at some level (usually trace), this does not mean that all these elements are absolute requirements for life. Indeed, the elements N, S, and P (plus C, H, O) are probably the only required elements for Earth-like life.

Biochemicals of life

Life on Earth is morphologically diverse, ranging from the smallest bacterium to the largest whale. However, this impression of vast diversity masks a deeper similarity at the level of fundamental biochemistry. All life on Earth is constructed from a small, distinct set of biochemicals. The fundamental building blocks are the 20 left-handed amino acids and five nucleotide bases. Added to these are a few right-handed sugars, from which are made the polysaccharides, and the simple alcohols and fatty acids that are the building blocks of lipids. Life is not only composed of these organic molecules but it spreads them about in its environment. Thus organic matter provides an important clue in the search for life, past or present.

22.4 Weird life

There have been suggestions of lifeforms based on different combinations of the elements; Chapter 27 has a full discussion. Common examples of this type of alternative life are the substitution of ammonia (NH_3) for water, or of silicon for carbon. Certainly

---

[3] All molecules with carbon are called *organic*, whether or not their origin is biological (*biogenic*) or nonbiological (*abiotic*).

ammonia is an excellent solvent – in some respects better than water. The temperatures over which it is liquid are common on planets (standard melting point is $-78\,°C$ (195 K), boiling point $-33\,°C$ (240 K), liquid at room temperature when mixed with water), and its constituent elements are abundant in the cosmos. Silicon has been suggested as a substitute for carbon in alien lifeforms. However, silicon does not form long polymeric chains as readily as carbon does, and its bonds with oxygen (SiO_2) are much stronger than those of carbon with oxygen (CO_2), rendering its oxide essentially inert.

Although speculations of alien life capable of using silicon and ammonia are intriguing, no experiments elucidating such alternative biochemistries have been performed. There are no strategies for where or how to search for such alternative life or its fossil remnants. Perhaps more significant is that these speculations have not contributed to our general understanding of life. One can only conclude that speculation on alternative chemistries for life is premature. In the future we may develop general theories for life or, more likely, discover other lifeforms on other worlds, and thus generalize from the comparison. Until then, basing our search on the properties of Earth-like life should be considered a practical consequence of our state of knowledge and not a fundamental limitation on the phenomenon of life.

22.5 Establishing biological origin

For organic materials to be a useful indicator of life we must have a way to differentiate between biological and abiological origin. But although biological systems and their environs are rich in organic matter, organic carbon is also produced abiologically. Organics are found in the meteorites, on the moons, and in the atmospheres of the outer Solar System, in comets, and in the interstellar medium (Chapter 3).

Finding a large, complex, and recognizable molecule like chlorophyll would be good evidence for life. For smaller organic molecules, however, the distinction between biochemicals and abiotic organic matter is not found in a particular molecule, but in considering the *pattern* of the many types of molecules in a sample (McKay, 2004). For example, abiotic processes do not in general distinguish between similar isotopes or between left- and right-handed varieties of a molecule. If we consider a generalized multi-dimensional phase space that describes all possible organic molecules, then for abiotic production mechanisms the relative

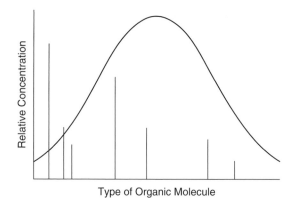

FIGURE 22.2 A highly schematic view of the distributions of the concentrations of organic material in samples of biological and nonbiological origin. Nonbiological processes produce smooth distributions of organic material, illustrated here by the curve. Biology, in contrast, selects and uses only a few distinct molecules, shown here as spikes (e.g., the 20 left-handed amino acids on Earth). Such selectivity in a sample of organic material from Mars or Europa would be a strong indication of biological origin. (From McKay, 2004.)

concentration of different types will be a smooth function. This is illustrated in Fig. 22.2 (McKay, 2004), which schematically shows one dimension of this multi-dimensional phase space of possible organics. In contrast to abiotic mechanisms, biological production *does* discriminate, selecting a few types and building up its biochemistry from that restricted set. Thus organic molecules that are chemically very similar often have widely different concentrations in a sample of biological organics. An example of this on Earth is the selection by life for only the left-handed version of the 20 amino acids used in proteins.[4] A sample of organic material from Earth thus shows high concentrations of these particular 20 left-handed amino acids, compared to both the right-handed versions and to similar amino acids not within the set of 20.[5] The selectivity of biological processes is shown schematically in Fig. 22.2 by the distribution of spikes in contrast to the smooth nonbiological distribution. General arguments regarding thermodynamic efficiency and specificity of enzymatic reactions suggest that this molecular selectivity is required for biological function and is a general result of natural selection.

---

[4] Each amino acid can be found in two mirror-image structures: D (dextro) and L (levo), respectively, right- and left-handed versions. This property of molecules is called *chirality*.

[5] There are ~150 other amino acids that occur in various minor roles in biology and that are no more complex than the 20 actually used in proteins; in principle, these could serve just as well. The total number of theoretically possible amino acids is very large.

This approach has immediate practical benefit in the search for biochemistry in the Solar System (Lovelock, 1965; McKay, 2004). Samples of organic material collected from Mars and Europa, for example, could be easily tested for the prevalence of one chirality of amino acid over the other. More generally, a complete analysis of the relative concentration of different types of organic molecules might reveal a pattern that is biological even if that pattern does not involve any of the biomolecules familiar from Earth-life. Interestingly, if a sample of organics from Mars or Europa showed a preponderance of right-handed amino acids, this would confirm the presence or past presence of life and at the same time show that this life is distinct from Earth-life. This same conclusion would apply to any clearly biological pattern that is distinct from the pattern of Earth-life.

Organic material of biological origin eventually loses its distinctive patterns (such as chirality) as a result of thermal and radiation effects. Examples of this include the thermal racemization of amino acids, whereby a beaker of initially left-handed amino acids at room temperature become equally left- and right-handed after ~10,000 years (Bada and McDonald, 1995). But at the low temperatures (of order −70 °C) in the martian permafrost, calculations suggest that there has been little thermal alteration for any organics present (Kanavarioti and Mancinelli, 1990; Bada and McDonald, 1995) over the past 4 Gyr. And on Europa, an interesting unanswered question is how long organic material frozen into the surface ice would retain a biological signature in its strong radiation environment (Chapter 19).

22.6 Life is small

Our everyday experience is with lifeforms that are large multicellular creatures. For most of Earth's history, however, life was microscopic, and even today microorganisms dominate the planet in the most diverse and extreme environments. For these reasons it is thought that if life exists on another planet, it would have started as microbial life and might still be in that stage. General arguments based on physical diffusion, bimolecular size, and information content have been used to argue that the size of microorganisms will be larger than ~0.25 μm and range up to a few μm (Koch, 1996; National Research Council, 1999). This conclusion regarding size is tentative, but provides a starting point for searching for life elsewhere.

On Earth all large multicellular lifeforms require oxygen, and the history of the development of multicellularity appears to follow the rise of oxygen (Knoll, 1992; Chapters 4 and 12). Following this logic we might expect to find multicellular life, or fossils thereof, on a planet only if that planet had geological conditions that favored the rise of oxygen. It has been suggested that early Mars may have been such a planet (McKay, 1996a).

22.7 Movement and metabolism

Finding organic material with a biological signature would indicate that life was present in the past, but not necessarily indicate that it was extant. The distinction is crucial. Fossils, skeletal remains, organic biomarkers, or even a dead rabbit, on the surface of Mars would all be compelling evidence of life on Mars, but none would be living organisms. There is a key distinction then between evidence for life and evidence for something being alive. There are only two tests to determine if an organism is alive: movement and metabolism.

Movement, microscopic or macroscopic, in response to stimuli or the presence of nutritional gradients, is a convincing indicator of being alive, but not a required one. Bacteria swimming under a microscope, the breath or heartbeat of an animal, or the nodding head of a sleeping student in a classroom are all examples of movement that would be convincing evidence that the creature is alive. However, movement is not an attribute of all living things: lichen and moss are examples of life that do not move.

In general terms, metabolism is the ability of an organism to incorporate nutrients in order to maintain its life functions or to accumulate biomass that either enlarges its own mass or is used to produce offspring. General considerations of the second law of thermodynamics and the prevalence of background radiation imply that metabolism at some minimal level is required if an organism is to remain alive. These minimal levels are not well understood. Key questions include: How long can an organism remain dormant between bouts of metabolism? What is the minimal rate of metabolism as a function of temperature? Dormant lifeforms may be all we can detect on Mars and we may never be able to figure out how to revive them. Note that dormant lifeforms such as seeds or spores are alive, but this can only be shown by reactivating them into the form of vegetative cells that exhibit metabolism or movement. Even on Earth, we are confronted all the time with the issue of how to determine whether or not a biochemically intact cell is dead or alive.

The Viking missions to Mars focused on the search for life and in particular the detection of living organisms. The instruments involved in the search for life included: a camera capable of detecting macroscopic motion, a combination gas chromatograph and mass spectrometer (GCMS) capable of detecting and characterizing organic molecules, and a life detection package comprising three incubation experiments to determine the presence of metabolism in the martian soil. These experiments are discussed in detail in Section 18.5.2. Here we note that two of the life-detection experiments gave positive results for metabolism taking place in the martian soil, and remain controversial decades later (Klein, 1999).

The GCMS, however, did not detect the presence of any organic molecules in the soil at the level of one part per billion (Biemann, 1979). This instrument received martian soil samples from the same sampling arm that provided soil to the biology experiments. The sample was then heated to release any organics that might be present. Any decomposed organics were then carried through the gas chromatograph and identified by the mass spectrometer, but the only signal was due to cleaning agents used on the spacecraft before launch. While it is true that a part per billion of organic material in such a soil sample is the equivalent of over a million bacteria each the size of a typical *E. coli* (Klein 1978, 1979), on Earth not only is all life composed of organic material, but its environments are also rich in organic material. It is difficult to imagine life without associated organic material and this is the main argument against a biological interpretation of the other ostensibly positive results. It is generally thought that the reactivity observed by the other instruments was due to one or more oxidants that had been produced by ultraviolet sunlight in the atmosphere and deposited onto the soil surface.

Consistent with the apparently negative results of the Viking biology experiments, the environment of Mars appears to be too dry for life. Indeed conditions are such that liquid water is rare and transient, if it occurs at all (Chapter 18).

The Viking results are an important cautionary tale in the search for living organisms – metabolism and movement, the only two indicators of the living state, can be nonbiological as well. For example, fire arguably has metabolism and movement (see Section 5.2). Thus these tests must be used only with additional tests for the biological origin of a sample. For microorganisms, a straightforward complementary test is the determination of the presence of biogenic organics. In the end the Viking biology results failed to demonstrate the presence of living organisms to the scientific community because of the negative results of this complementary test.

22.8 Fossils: two case studies

Other than humans, most species on Earth today represent a small fraction of their total ancestral populations. And more species have gone extinct than are present on Earth today. The fossil record preserves evidence of this earlier life. On other planets it might be possible that there is no extant life and only the remnants of past life can be found. Fossils can be recognized and proven to be evidence of past life based on morphological or organic evidence. For multicellular organisms this has proven to be relatively straightforward, but for microorganisms the issues are more complex. Two case studies under current debate illustrate the issues we would likely confront in any discussion over putative microfossils returned from Mars.

McKay[6] *et al.* (1996) reported four lines of fossil evidence in the martian meteorite ALH84001 consistent with past life on Mars. These are: (a) complex organic material (polycyclic aromatic hydrocarbons, or PAHs) is present inside ALH84001; (b) carbonate globules indigenous to the meteorite are enriched in ^{13}C to an extent that, if found on Earth, can indicate organic matter from biogenic activity (Fig. 18.11); (c) there are microscopic forms that could be actual fossils of microbial life (Fig. 18.13); and (d) magnetite and iron-sulfide particles are present with a distribution and shapes identical to those produced by magnetotactic bacteria on Earth (Fig. 18.12).

These suggestions have each been further studied, and all but the last, the magnetite, are considered unlikely to be indicators of life. It is widely agreed that the PAHs in ALH84001 are of martian origin. However, there is no evidence that they are of biological origin; rather, there is in fact considerable evidence for abiotic formation of PAHs in the early Solar System. Similarly, the isotopic signature in the carbonates cannot be tied to a biological source – on Earth such isotopic shifts are found in organic matter, not in carbonates. Thirdly, the observed fossil shapes do not convincingly tie to biology. They have no internal structures that would be indicative of microorganisms, and their size may be too small to be consistent with life (Koch, 1996).

---

[6] This is *David* McKay, no relation to the author.

The only possible evidence for life in the ALH84001 meteorite that is still seriously considered by some researchers is the presence of magnetite grains. There have been several important developments since the original work of McKay *et al.* (1996). Thomas-Keprta *et al.* (2002) have refined the arguments for the biogenic origin of some of the magnetite grains in ALH84001 based on size and shape, and Friedmann *et al.* (2001) have reported chains of magnetite similar to that found in many magnetotactic bacteria on Earth (Blakemore, 1975). On the other hand, several groups (Golden *et al.*, 2001; Barber and Scott, 2002) have developed a coherent nonbiological scenario for the formation of the magnetite grains due to shock alteration and differentiation of the carbonate. This debate continues and its resolution may ultimately require samples from Mars of the highly magnetic surface material. For a longer discussion of the ALH84001 evidence, see Section 18.5.3.

Even on Earth fossil identification of ancient microorganisms can be contentious. For fifteen years the oldest fossil evidence for life on Earth was taken by the scientific community to be microfossils and associated stromatolites from 3.5 Ga (Schopf and Parker, 1987). These have been suggested as the standard against which fossils from Mars should be compared (McKay and Davis, 1999). However, a debate has emerged as to whether these are fossils at all. Brasier *et al.* (2002) argue that the structures are produced by hydrothermal processes and are not biological, while further study by Schopf *et al.* (2002) defends their validity. Section 12.2.1 discusses this controversy in detail. Figure 22.3 shows a stromatolite from an ancient (1 Gyr old) carbonate formation. Microbial mat structures still form the basis for the oldest firm evidence of life on Earth at 3.4 Gyr ago (Tice and Lowe, 2004).

The Viking results, the studies of the ALH84001 meteorite, and the debate over the earliest fossils on Earth all show that *several, independent* indications of biological origin are needed to make the case for life.

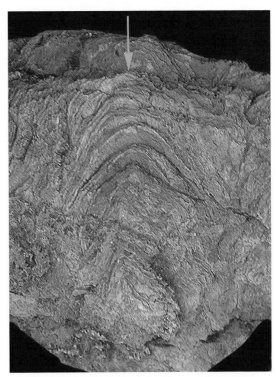

FIGURE 22.3 Stromatolite formed by phototrophic organisms over 1 billion years ago in the Crystal Springs Formation, Inyo County, California. Stromatolites (also see Figs. 12.6 and 12.7) are an important form of fossil evidence of life because they are large structures formed by microorganisms of a type that could perhaps be found on Mars.

22.9 Fossils are not enough

As discussed above, life on Earth is formed from a small distinct set of biomolecules. Pace (2001) argues that this is likely an optimal set and we can expect to find the same biomolecules composing alien life elsewhere. In his view, everything from the amino acids used to the triplex coding that specifies each amino acid has been optimized on Earth by an evolutionary selection that would act in an identical manner for life

anywhere. Variation would be expected only at higher levels of biochemistry and genetic coding. However, this suggestion that the details of biochemistry are optimal and not contingent on historical details is not compelling. Indeed, even if it happens that the optimization of biochemistry inexorably leads to the same amino acids and sugars, there is no reason why left-handed amino acids and right-handed sugars would always be selected for independent forms of life. Just for this one aspect there are four permutations of biochemical systems that are functionally and energetically identical to Earth-life. More broadly, we do not know enough about possible biochemistries to argue that there is but one global optimum in the possible "phase space" of biochemical function. For example, for a long time it was thought that only 20 amino acids were coded for in the genetic code for Earth-life, but now two additonal ones have been identified, e.g., the 22nd amino acid, pyrrolysine, is used in methane-producing proteins in the microorganism *Methanosarcina* (Srinivasan *et al.*, 2002; Hao *et al.*,

2002). An even more compelling argument against the inevitability of the coupled genetic and protein system seen in Earth-life is the demonstration of synthetic production of unnatural base pairs for incorporating amino acid analogue into proteins (Hirao *et al.*, 2002). In a system as complex as biochemistry it is likely that there are numerous "local" optima in the complex phase space – the details of the particular optimum found by evolutionary selection on any world would thus likely depend on the initial conditions and random developments in the early biological history of that world.

A goal of astrobiology is to explore these alien biochemistries. Fossil evidence for life on another world, however, would not adequately address the question of biochemistry and separate origin. Fossils would tell us that life was present but would give little insight into the biochemical or genetic nature of that life. To address these deeper questions we must find living organisms or their remains.

Ideally, any second example of life would comprise a functioning, reproducing ecosystem. In this case the structure and function of the biomolecules could be observed. The role of the genetic material in reproduction and development could be studied. In addition community interactions among different species could be monitored. The comparisons and contrasts to life on Earth would found a new scientific discipline: comparative biology.

Failing to find extant life, the next most desirable possibility would be discovery of dead, yet still intact, organisms. Perhaps they died from radiation or thermal shock. From such micro-corpses it would be possible to determine the structure of the alien biochemistry and the range of biomolecules used. It might also be possible to develop sufficient knowledge of the genetics of the alien lifeform to be able to revive the organisms using sophisticated genetic repair techniques customized to the alien biochemistry. This would likely be a large and complex undertaking requiring a level of study of the alien biochemistry comparable to that of Earth-based biochemistry. However, the task would be worth the reward if a true second genesis could be achieved.

Fortunately, there exist highly specific and sensitive methods for detecting and recognizing Earth-life (e.g., Chapter 12 and Pace, 2001). Thus it is unlikely that we will mistakenly identify Earth contamination as alien life of another biochemistry. By the same token, however, if life beyond the Earth shares a common ancestor with Earth-life, then it might well be mistaken for contamination (Chapter 25).

22.10 Global biology

Life on Earth has caused significant global changes. The most obvious is the presence of oxygen in the atmosphere. Models of oxygen enrichment on the Earth all require the presence of oxygenic photosynthesis (Kasting, 1993; Catling *et al.*, 2001; Chapters 4 and 12). It has long been suggested that even over interstellar distances ozone produced photochemically from biologically produced oxygen could be a useful biomarker (Owen, 1980). Simultaneous detection of reduced gases such as methane would further indicate the presence of life (DesMarais *et al.*, 2002). However, nonbiological mechanisms, such as photochemistry and escape of hydrogen to space, may combine to also produce the simultaneous presence of high oxygen and reduced gases in a planetary atmosphere, so there may be ambiguity as to the biological origin of such an atmosphere.

No markers have been identified that would indicate the presence of life in an *anaerobic* microbial world.

22.11 Intelligent life

The search for signals from extraterrestrial civilizations has been an active field since 1960; both radio and optical searches are currently underway (Chapter 26). In addition there have been some searches for artifacts of galactic-scale civilizations (Weinberger and Hartl, 2002). It is disappointing but not entirely unexpected that these searches have found nothing (Zuckerman and Hart, 1995; McKay, 1996b; Ward and Brownlee, 2000). Nonetheless, the search for extraterrestrial intelligence is an essential part of the search for life and only by searching can we possibly succeed. No amount of theory can substitute for looking.

22.12 Searching Mars and Europa

Given the above ten principles, what should be the approach for the search for life in our Solar System? Cursory examination shows that there are no global biospheres or intelligent civilizations in our Solar System and the search therefore focuses on microbial ecosystems.

Of the ecological requirements for life listed in Table 22.1, liquid water is the one that is most restrictive. There is direct evidence that Mars had liquid water early in its history (Fig. 22.4) and it is plausible that liquid water exists even today within the subsurface of Mars (Chapter 18). There is also evidence that Europa, a large

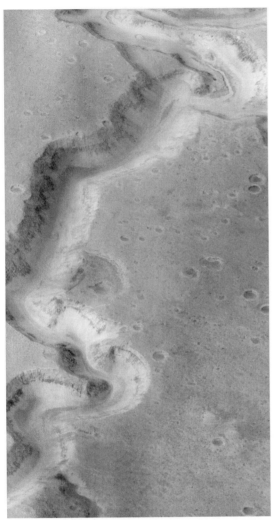

FIGURE 22.4 Water on another world. Mars Global Surveyor image showing Nanedi Vallis in the Xanthe Terra region of Mars. Image covers an area 9.8 km by 18.5 km; the canyon is about 2.5 km wide. This image is the best evidence we have of liquid water anywhere outside the Earth. (NASA/Malin Space Sciences.)

FIGURE 22.5 The search for life on Mars, Step 1: search sites on Mars where layered sedimentary rocks are found. This image shows a small sedimentary outcrop obtained by the Mars Exploration Rover *Opportunity* (note the wheel tracks) at Meridiani Planum. The image is one meter across. (NASA.)

satellite of Jupiter, has an ocean beneath its surface layer of ice (Pappalardo *et al.*, 1999; Greenberg *et al.*, 2000; Chapter 19). Arguments for liquid water habitats on the parent asteroids of the meteorites and on comets are less compelling than for Mars and Europa. Titan, although it has an organic rich atmosphere, is too cold to support liquid water (Chapter 20). The thick atmospheres on Venus, Jupiter, and possibly Saturn may have liquid water present, but only as clouds. Thus Mars and Europa are the prime targets.

For Mars, the first step is to determine if life was ever present. The easiest way to do this might be to search for fossils in sedimentary deposits (McKay,

1997). The layered rocks discovered at the Meridiani Planum site by the Mars Exploration Rover *Opportunity* show geochemical and morphological evidence of liquid water. These include: (a) layering of the rocks (Fig. 22.5); (b) variable concentrations of sulfur as would be expected if salts were transported; and (c) the presence of small spherical concretions in the layered rocks (Squyres *et al.*, 2004; see Section 18.8 for more discussion). However, given the oxidizing nature of the martian soil, it is probable that any organic remains of past life would be destroyed and only fossils would remain. Nonetheless, if we found fossils similar in quality to those from the early Earth, this would be convincing evidence that life had been present on Mars.

If we found fossil evidence for past life on Mars, the next step would be to try to find preserved biochemical remains of that life. This would allow for the direct study of the biochemistry and genetics of martian life, and would be necessary to determine if that life (a) represented a second genesis of life, or (b) shared a common origin with Earth-life. The question is raised because the exchange of material between Mars and Earth (via rocks such as ALH84001) might have allowed for life from one planet to have spread to the other (Section 18.4.4; Sleep and Zahnle, 1998; Weiss *et al.*, 2000).

Recent spacecraft data suggest locations for obtaining ancient frozen material on Mars. Magnetic crustal features detected in the southern regions of Mars

FIGURE 22.6 The search for life on Mars, Step 2: search for remnant biochemistry in the ancient permafrost on Mars. This polar terrain (near 70°S, 180°W) shows ancient craters (small dots scattered all over the southern hemisphere) with strong remnant magnetism (indicated by the banded strips). This site may harbor the oldest, coldest, undisturbed permafrost on Mars. In sedimentary deposits here we could drill to retrieve long-frozen martian organisms. If such preserved biological material were found, we could characterize the biochemistry of that life and determine whether or not it represented a second genesis of life, distinct from Earth. (After Smith and McKay, 2005.)

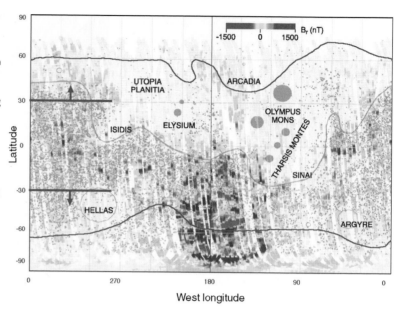

(Fig. 22.6) are amongst the oldest known features on Mars (Acuña *et al.*, 1999; Connerney *et al.*, 1999). The overprint of the large craters Hellas and Argyre indicate that the magnetic features predate the heavy bombardment, predate the crustal dichotomy between the northern and southern hemisphere, and predate the formation of large geological features such as Tharsis and Valles Marineris. The persistence of the magnetic features indicates that the crust at these locations has not warmed above the Curie point ($\sim$850 K), although it does appear to have undergone some spreading and infilling by later material.

Thus the magnetic crustal features may be sites of the oldest undisturbed terrains on Mars. They may provide a window into the earliest environments and may contain material that dates back to these early times. In particular, sites near the South Pole (70°S, 180°W) that contain magnetic crustal features may be the oldest permafrost on Mars. Searching in these materials for frozen, probably dead, remains of martian organisms is of prime interest to astrobiology (Smith and McKay, 2005). Like the mammoths extracted from the ice in Siberia, such martian microbes, if found, would be dead but organically preserved. If the martian life proved to have identical biochemistry to Earth's, it would then be possible, in principle, to determine the trans-planetary phylogenetic relationship. This would be essential to determine if life started independently on Mars or shared a common origin with Earth-life. This information cannot be obtained from fossils.

An alternative approach to the search for life on Mars is to drill deeply into any subsurface aquifers that might be discovered on future missions that carry electromagnetic or seismic sounding instruments. If there are living organisms on Mars, they would likely be present in such subsurface ecosystems (Boston *et al.*, 1992; McKay, 2001).

On Europa the strategy for the search for life also begins with the confirmation of the presence of liquid water under the ice cover. If water is indeed present, then the next question is that of a chemical energy source, since the ice is presumably too thick for photosynthesis. If such sources prove plausible, it would motivate attempts to reach the ocean and obtain samples of any organic material and/or living organisms present.

An alternative strategy for Europa, and one that may be achieved sooner, is to search for organic material on the surface, possibly brought up through cracks in the ice cover (Chapter 19). If such organic material were to have signatures of biological origin, then this would indirectly show that water and suitable energy were present and might even prove that the europan life had a separate origin from Earth-life. See Sections 19.10 through 19.13 for further discussion.

22.13 Conclusion

Everything we know about life we have learned from the one example we find here on Earth. Despite the wonderful morphological diversity of life on this planet there is a basic underlying biochemical sameness and genetic relatedness. With only this one type of life to study we are not in a position to formulate general theories about life and our best approach is an empirical one. Initially we will search for lifeforms that are like us. Eventually we might find a more diverse menagerie of life spanning the periodic table. In any case, we must go forth and search.

For now, we will conduct the search on planets. We already know of many more planets orbiting other stars than in our own Solar System. On these distant planets around other stars we may find evidence for life in the form of spectral signatures of oxygen, ozone, or methane, but it is unlikely we will be able to directly access samples of any life we thus detect. But, fortunately, our own planetary system has at least two worlds, Mars and Europa, that are promising targets for detailed searches for another, second type of life. Life on these worlds might be alien but still carbon- and water-based and thus similar enough to be recognized. Prospects are compelling enough that the search for life on Mars and Europa will proceed in the next few decades. This could be the time when humanity finally answers the question: "Are we alone?"

References

Acuña, M. H., Connerney, J. E. P., Ness, N. F., *et al.* (1999). Global distribution of crustal magnetism discovered by the Mars Global Surveyor MAG/ER experiment. *Science*, **284**, 790–3.

Bada, J. L. and McDonald, G. D. (1995). Amino acid racemization on Mars: implications for the preservation of biomolecules from an extinct martian biota. *Icarus*, **114**, 139–143.

Barber, D. J. and Scott, E. R. D. (2002). Origin of supposedly biogenic magnetite in the martian meteorite Allan Hills 84001. *Proc. Natl. Acad. Sci. USA*, **99**, 6556–61.

Biemann, K. (1979). The implications and limitations of the findings of the Viking Organic Analysis Experiment. *J. Molec. Evol.*, **14**, 65–70.

Blakemore, R. P. (1975). Magnetotactic bacteria. *Science*, **190**, 377–379.

Boston, P. J., Ivanov, M. V., and McKay C. P. (1992). On the possibility of chemosynthetic ecosystems in subsurface habitats on Mars. *Icarus*, **95**, 300–308.

Brasier, M. D., Green, O. R., Jephcoat, A. P., *et al.* (2002). Questioning the evidence for Earth's oldest fossils. *Nature*, **416**, 76–81.

Catling, D. C., Zahnle, K. J., and McKay, C. P. (2001). Biogenic methane, hydrogen escape, and the irreversible oxidation of early Earth. *Science*, **293**, 839–843.

Chapelle, F. H., O'Neil, K., Bradley, P. M., *et al.* (2002). A hydrogen-based subsurface microbial community dominated by methanogens. *Nature*, **415**, 312–315.

Connerney, J. E. P., Acuña, M. H., Wasilewski, P., *et al.* (1999). Magnetic lineations in the ancient crust of Mars. *Science*, **284**, 794–798.

DesMarais, D. J., Harwit, M. O., Jucks, K. W., *et al.* (2002). Remote sensing of planetary properties and biosignatures on extrasolar terrestrial planets. *Astrobiology*, **2**, 153–181.

Friedmann, E. I., Wierzchos, J., Ascaso, C., and Winklhofer, M. (2001). Chains of magnetite crystals in the meteoxite AlH84001: evidence of biogenous origin. *Proc. Natl. Acad. Sci. USA*, **98**, 2176–2181.

Golden, D. C., Ming, D. W., Schwandt, C. S., Lauer, H. V., and Socki, R. A. (2001). A simple inorganic process for formation of carbonates, magnetite, and sulfides in Martian meteorite ALH84001. *Amer. Mineralogist*, **8**, 370–375.

Greenberg, R., Geissler, P., Tufts, B., and Hoppa, G. V. (2000). Habitability of Europa's crust: the role of tidal-tectonic processes. *J. Geophys. Res.*, **105**, 17551–17562.

Hao, B., Weimin Gong, W., Ferguson, T. K., James, C. M., Krzycki, J. A., and Chan, M. K. (2002). A new UAG-encoded residue in the structure of a methanogen methyltransferase. *Science*, **296**, 1462–1466.

Hirao, I., Ohtsuki, T., Fujiwara, T., *et al.* (2002). An unnatural base pair for incorporating amino acid analogue into proteins. *Nature Biotechnol.*, **20**, 177–182.

Kanavarioti, A. and Mancinelli, R. L. (1990). Could organic matter have been preserved on Mars for 3.5 billion years? *Icarus*, **84**, 196–202.

Kasting, J. F. (1993). Earth's earliest atmosphere. *Science*, **259**, 920–926.

Klein, H. P. (1978). The Viking biological experiments on Mars. *Icarus*, **34**, 666–674.

Klein, H. P. (1979). The Viking mission and the search for life on Mars. *Rev. Geophys. Space Phys.*, **17**, 1655–1662.

Klein, H. P. (1999). Did Viking discover life on Mars? *Orig. Life Evol. Biosph.*, **29**, 625–631.

Knoll, A. H. (1992). The early evolution of eukaryotes: a geological perspective. *Science*, **256**, 622–627.

Koch, A. L. (1996). What size should a bacterium be? A question of scale. *Ann. Rev. Microbiol.*, **50**, 317–348.

Krauss, L. M. (1995). *The Physics of Star Trek*. New York: Basic Books.

Lovelock, J. E. (1965). A physical basis for life detection experiments. *Nature*, **207**, 568–570.

McKay, C. P. (1991). Urey prize lecture: Planetary evolution and the origin of life. *Icarus*, **91**, 92–100.

McKay, C. P. (1996a). Oxygen and the rapid evolution of life on Mars. In *Chemical Evolution: Physics of the Origin and*

Evolution of Life, J. Chela-Flores and F. Raulin, eds., pp. 177–184. Dordrecht: Kluwer.

McKay, C. P. (1996b). Time for intelligence on other planets. In *Circumstellar Habitable Zones*, L. R. Doyle, ed., pp. 405–419. Menlo Park: Travis House Publications.

McKay, C. P. (1997). The search for life on Mars. *Orig. Life Evol. Biosphere*, **27**, 263–289.

McKay, C. P. (2001). The deep biosphere: lessons for planetary exploration. In *Subsurface Microbiology and Biogeochemistry*, eds. J. K. Fredrickson and M. Fletcher, pp. 315–327. New York: John Wiley.

McKay, C. P. (2004). What is life and how do we search for it on other worlds? *Public Library of Science Biol.* **2**, 1260–1263.

McKay, C. P. and Davis, W. L. (1999). Planets and the origin of life. In *Encyclopedia of the Solar System*, P. Weissman, L. McFadden, and T. Johnson, eds. San Diego: Academic Press, 899–922.

McKay, D. S., Gibson, E. K., Thomas-Keprta, K. L., *et al.* (1996). Search for past life on Mars: possible relic biogenic activity in martian meteorite ALH84001. *Science*, **273**, 924–930.

National Research Council (1999). *Size Limits of Very Small Microorganisms: proceedings of a workshop*. Washington, DC: National Academic Press.

Owen, T. (1980). The search for early forms of life in other planetary systems: future possibilities afforded by spectroscopic techniques. In *Strategies for the Search for Life in the Universe*, M. Papagiannis. ed. Dordrecht: Reidel, p. 177.

Pace N. (2001). The universal nature of biochemistry. *Proc. Natl. Acad. Sci. USA*, **98**, 805–808.

Pappalardo, R. T., Belton, M. J. S., Breneman, H. H., *et al.* (1999). Does Europa have a subsurface ocean? Evaluation of the geological evidence. *J. Geophys. Res.*, **104**, 24015–24055.

Raven, J. A., Kübler, J. E., and Beardall, J. (2000). Put out the light, and then put out the light. *J. Mar. Biol. Ass. UK*, **80**, 1–25.

Schopf, J. W. and Packer, B. M. (1987). Early Archean (3.3-billion to 3.5-billion-year-old) microfossils from Warrawoona Group, Australia. *Science*, **237**, 70–73.

Schopf, J. W., Kudryavtsev, A. B., Agresti, D. G., Wdowiak, T. J., and Czaja, A. D. (2002). Laser–Raman imagery of Earth's earliest fossils. *Nature*, **416**, 73–76.

Sleep, N. H. and Zahnle, K. (1998). Refugia from asteroid impacts on early Mars and the early Earth. *J. Geophys. Res.*, **103**, 28,529–28,544.

Smith, H. D. and McKay, C. P. (2005). Drilling in ancient permafrost on Mars for evidence of a second genesis of life. *Planet. Space Sci.*, **53**, 1302–1308.

Squyres, S. W., Grotzinger, J. P., Arvidson, R. E., *et al.* (2004). In situ evidence for an ancient aqueous environment at Meridiani Planum, Mars. *Science*, **306**, 1709–1714.

Srinivasan, G., James, C. M., and Krzycki, J. A. (2002). Pyrrolysine encoded by UAG in Archaea: charging of a UAG-decoding specialized tRNA. *Science*, **296**, 1459–1462.

Stevens, T. O. and McKinley, J. P. (1995). Lithoautotrophic microbial ecosystems in deep basalt aquifers. *Science*, **270**, 450–454.

Thomas-Keprta, K. L., Clement, S. J., Bazylinski, D. A., *et al.* (2002). Magnetofossils from ancient Mars: a robust biosignature in the martian meteorite ALH84001. *Appl. Environ. Microbiol.*, **68**, 3663–72.

Tice, M. M. and Lowe, D. R. (2004). Photosynthetic microbial mats in the 3,416-Myr-old ocean. *Nature*, **431**, 549–552.

Ward, P. D. and Brownlee, D. (2000). *Rare Earth*. New York: Copernicus Books.

Weinberger, R. and Hartl, H. (2002). A search for 'frozen optical messages' from extraterrestrial civilizations. *Int. J. Astrobiology*, **1**, 61–69.

Weiss, B. P., Kirschvink, J. L. Baudenbacher, F. J., Vali, H., Peters, N. T., Macdonald, F. A., and Wikswo, J. P. (2000). A low temperature transfer of ALH84001 from Mars to Earth. *Science*, **290**, 791–795.

Zuckerman, B. and Hart, M. H. (1995). *Extraterrestrials: Where are They?* (second edn.) Cambridge: Cambridge University Press.

23 Instruments and strategies for detecting extraterrestrial life

Pamela G. Conrad
Jet Propulsion Laboratory

23.1 Introduction

The appeal of astrobiology lies in two primary features, the first being the important questions astrobiologists ask: "Are we alone? How did life begin? How will it all end?" The second attractor is the opportunity to work across many disciplines. Astrobiology challenges us to draw from many intellectual resources in the attempt to answer these questions – biology, chemistry, physics, astronomy, geology, and engineering are all required. Life detection, in particular, requires a strong interdisciplinary approach. In this chapter we focus on life detection within the context of Solar System exploration; techniques for detecting planets and possible associated life beyond the Solar System are discussed in Chapters 21 and 26. Within our Solar System life detection efforts have been and still are primarily focused on Mars (Chapter 18), and so we will use Mars exploration as a model for discussion, though our approach is applicable to any potential habitat for life.

The success of a life detection mission to another planet should not be focused solely upon whether or not *live* organisms are detected, but rather it must be able to correctly classify observations as evidence of (a) life, (b) non-life, (c) once-alive-but-now-dead, or (d) made-by-life (*biogenic*). It is essential that the scientific community agree on which measurements should be made, and then how to interpret those measurements. Mission success is dependent upon such scientific consensus, which is required through every step of the process, from defining the measurable parameters of life, to working with technologists and engineers on how to measure these parameters, to interpreting the meaning of the results robotically obtained $\sim 10^8$ km away. It is also important to select universal and unambiguous features that do not assume that extraterrestrial life is almost identical to Earthly life with which we are so familiar. Finally, *in advance* of any extraterrestrial life detection experiment we must agree on the criteria that define a positive result.

One viable strategy for life detection is to define a set of the most basic measurable features that could be associated with life (*biosignatures*), and then go about quantitatively distinguishing expected measurements of these features from those associated with the non-biological environment (Nealson and Conrad, 1999; Conrad and Nealson, 2001). For instance, one basic measurement is simply the amounts of chemical constituents and how they are arranged in three spatial dimensions over time. This yields four data axes: *what* chemistry, *how much* of it, over what spatial volume (*where*), and with how much change through time (*when*). This way of describing a sample works equally well for a fossil, a living entity, non-living planetary materials (such as rocks, atmosphere, or ice), or for portions of the spacecraft that delivers the experiment. We need not know what extraterrestrial life looks like, nor what exotic and unfamiliar biochemistry might be associated with it if we constrain the investigation to measuring basic chemical features such as element abundances, organic inventory, isotopic fractionation, flux of volatiles, oxidation states of metals, etc. Such measurements will always yield important geochemical information about the planetary environment, and for biology they can also obviate the Earth-centric bias inherent in prospecting for DNA, ATP, or other important terrestrial biochemicals.

Space exploration can be accomplished in several different ways: remote sensing, *in situ* robotic exploration, manned exploration, and robotic return of planetary samples. Each exploratory mode has its advantages and challenges, and often compromise and combination strategies are employed. While many people would prefer to send human explorers off to investigate destinations in our Solar System, the

Planets and Life: The Emerging Science of Astrobiology, eds. Woodruff T. Sullivan, III and John A. Baross. Published by Cambridge University Press. © Cambridge University Press 2007.

lengthy commute and medical challenges associated with zero-G conditions, not to mention the huge cost, presently prohibit human exploration of even reasonably accessible Mars. In the meantime we will focus on the other three methods of exploration.

23.2 The Viking mission

The two American Viking missions in the 1970s provide a historical context (also see Chapter 2) for the notion that life detection experiments are essentially chemical and structural investigations (Klein, 1978; Klein et al., 1992). The Viking Lander camera (Levinthal et al., 1977) probed the latter aspect, looking for potentially biogenic structures and color changes on rocks that might be associated with organisms or their activities. The camera also looked for movement within its field of view.

The chemical experiments were designed to look at the reactivity of martian material in a metabolic context. The three life detection experiments were the "gas exchange experiment," the "labeled release experiment," and the "pyrolytic release experiment," and are described in detail by Brown et al. (1978). The instrument that performed the chemical analysis for all three of these experiments was a gas chromatograph/mass spectrometer (GCMS) with a resolution of 1:200 and a range of M/Z from 10 to 220. Sample ovens were designed to hold ~0.1 g of martian soil, which could be heated in successive steps to 50, 200, 350, and 500 °C. The gas chromatograph was programmed for a temperature range of 50 to 200 °C. In the following we briefly summarize the main GCMS results; further details are in Section 18.5.2.

23.2.1 Gas exchange experiment

The gas exchange experiment (Oyama et al., 1976, 1977) looked for evidence of biological activity in the martian soil by measuring the concentration of gases above a partially filled cell of martian soil. The assumption underlying the experiment was that living things produce or consume gases by metabolism and growth, and that these processes would be observable by measuring gas fluxes. After an incubation gas mixture of helium/oxygen/carbon dioxide was added to the soil-sample cell, the experiment was conducted in one of two ways: (a) the soil was humidified and "fed" aqueous nutrient solution in an attempt to stimulate microbial growth, or (b) the soil was exposed to water only indirectly through the atmosphere in the cell. The cell was periodically sampled and measured with the

GCMS to look for changes indicative of biological activity. Changes in the gas mixture were in fact observed in both the "humidity only" and "nutrient-contact-with-soil" experiments, but for several reasons it was deemed unlikely that the observed changes were of biological origin, e.g., the same type of changes were still seen even at a temperature of 145 °C.

23.2.2 Labeled release experiment

The gases that evolve from the biological metabolism of carbon-based nutrients were used as the basis for the labeled release experiment (Levin and Straat, 1976a). Simple nutrients containing radioactive ^{14}C were introduced to a martian soil sample. If there were organisms present, they would use the radioactive nutrients, and evolved gases would be detectable by a radiation detector sensitive to beta-decay radiation from the ^{14}C. Detector counts over a period of at least 14 martian days (or sols) were then interpreted as a measure of the activity of the martian soil.

The addition of labeled aqueous nutrient to the martian soil produced a rapid release of radioactive gas, which the investigators interpreted as a result consistent with biology. There are problems with that interpretation, however, because the rapid initial result would require for its production a large number of organisms, which was inconsistent with the results of the gas exchange experiment. There were several other reasons why a biological explanation was questionable: (1) the reaction slowed and stopped with more than 90% of the organic substrate still available; (2) each time more liquid nutrient was added to the soil, about 30% of the previously evolved gas went back into solution, and there is no biological explanation for this uptake; (3) the labeled release experiment gave no positive response when it was repeated on the soil sample after two to four months of storage (also inconsistent with the gas exchange experiment); and (4) there was strong evidence from the gas exchange experiment that the labeled release results were more likely to be due to the presence of a highly oxidizing material in the martian soil. Although the investigators had agreed on the criteria for evaluating a result as positive in advance of the mission (Hubbard, 1976; Levin and Straat, 1976b; Oyama et al., 1976) – criteria which when applied to these results gave a negative result – the principal investigators of this experiment to this day remain a small minority in maintaining that the results obtained could be indicative of biological activity (Levin and Straat, 1977a, b). This disagreement illustrates why it is as important to achieve consensus regarding how the results will be interpreted as it is to agree on what experiments will generate them in the first place.

23.2.3 Pyrolytic release experiment

The pyrolytic release experiment (Horowitz *et al.*, 1976, 1977; Hubbard, 1976) was essentially designed to look for evidence of photosynthetic organisms. Martian soil was incubated in the presence of a CO/CO_2 mixture that was tagged with radioactive ^{14}C to determine if the gas mixture was being "fixed," or used to synthesize organic materials in the sealed cell in the presence of a xenon arc lamp simulating sunlight. A dark experiment was also conducted as a control. After venting the tagged gas mixture, the soil was subsequently pyrolyzed at $635\,^{\circ}C$ and the volatile organic products sampled by the GCMS. Any products that had been synthesized from the tagged gas mixture could then be sensed as radioactive.

Initially, results showing some degree of carbon fixation were interpreted as promising for the possibility of life on Mars, but the amount was small in comparison with what one might expect in terrestrial photosynthesis; moreover, the reaction also occurred in dark conditions. Ultimately, the investigators felt that it was difficult to invoke a biological explanation for the small observed amounts of organosynthesis because the reactions occurred even at temperatures that would be likely to impede a biological reaction.

23.2.4 Lessons learned from Viking

While the Viking experiments did not provide conclusive evidence for life on Mars, they nevertheless give us several important lessons. First, the notion that life will somehow chemically announce itself in the environment is a good one. If one accepts that premise, then there are several scientific issues that must be considered.

- Over what spatial footprint does this occur?
- Which factors might obscure detection of the evidence?
- Is there a sharp discontinuity between the chemistry representative of organisms and that of the surrounding environment?
- After death of the suspect organisms, over what timescales does chemical equilibrium between them and the surrounding environment occur?

- What density of organisms is required in order to detect any particular chemical imprint upon the environment?
- How ubiquitous must life be on a planet in order for it to be statistically likely to be found at any given site? In other words, how much material must one sample to obtain a statistically sound sample?

Thinking about the Viking experiments within the context of these questions leads an astrobiologist to ask "What should we do differently next time to obtain more conclusive results?" The state of technology, as well as budget and mass constraints, will also force the mission team to ask "What *can* we do differently?" Technical constraints are often imposed on the desired science. For example, only a tiny amount of surface material could be sampled by the Viking missions. Given the fraction of the martian surface volume that the Viking samples represented, was it large enough and a fair enough statistical sample to come to any meaningful conclusions?[1] Thus the first lesson from Viking is that sample size and location are critical.

The conclusion that there is a strong oxidant in the topmost martian soil should cause us to consider whether traces of life or even nonbiogenic organic materials would even be observable in this soil. This leads us to seek *sub*surface samples. Chemical evidence of life, past or present, might be preserved in rocks lying beneath the soil and dust, particularly if shielded from solar ultraviolet radiation. On Earth, in the dry, cold desert of Antarctica's McMurdo dry valleys, organisms survive by inhabiting rock (Friedmann, 1982).

But looking for organisms in rock presents other scientific and technical challenges. The Antarctic rocks are sandstones – porous and translucent to sunlight. We do not know the extent of this type of rock on Mars, though the Mars Exploration Rovers have identified another type of sedimentary rock type, chemical precipitates, on a local scale at their landing sites (Sections 23.3 and 18.8) – chemical precipitates are sometimes as porous as sandstone, but not always. Sandstone is not the exclusive host for endolithic (rock-dwelling) organisms on Earth, but igneous rocks are less favorable as a habitat because the minerals are interlocked and closer together than in

---

[1] If the depth of the "surface" region is taken as 1 m, then the Viking landers sampled $\sim 10^{-20}$ of the entire planetary surface! Yet such a small random sample would still be satisfactory on Earth, due to life's ubiquity.

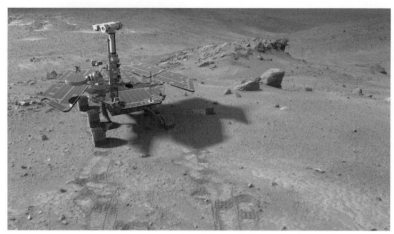

FIGURE 23.1 An artist's model of the Mars Exploration Rover has been digitally placed in a landscape actually traversed and then imaged by *Spirit* – note the wheel tracks in the soil. For scale, the stereo camera at the top of the mast (Pancam) is ∼ 1.5 m above the ground. The terrain is located above the plain of Gusev Crater, on the flank of Husband Hill, several kilometers from *Spirit*'s landing site in January 2004. The image was made on sol 438 (27 March 2005). (NASA/JPL-Caltech.)

sandstone. Although microbes do colonize Earthly igneous rocks within cracks and fractures, space is limited (also for water) and more porous rocks such as sandstone are much better habitats for microbial communities.

Organisms that live in rock tend to have a slow metabolism (Sun and Friedmann, 1999) to accommodate to the available space by not depleting limited nutrients and building up waste faster than diffusive or other processes can remove it. This slower metabolism can present a more subtle chemical signature in terms of temporal fluxes, and may present yet another detection challenge. If there is a strong oxidant in the martian surface soil, as the Viking results suggest, organisms may have moved into a subsurface realm to avoid its effects. Whether inside the rind of small rocks or in general beneath the martian surface, chemical biosignatures will be more difficult to detect if we must dig deeper, to a level where they can survive.

A third important lesson from the Viking experiments is that we should look at a lot of material in a variety of ways before making a judgment regarding the planet's capacity to have evolved life. One way to do this is to use a suite of instruments, the first of which rapidly scan large areas for potential regions of interest, then progressively constrain the search space for analytical methods that either require a long time for data collection or are destructive to their samples.

In summary, the Viking biology experiments accomplished what was possible at the time, and despite somewhat ambiguous and controversial data, provide us with a good foundation to design the next generation of life detection experiments.

23.3 The Mars Exploration Rovers

The most famous recent *in situ* Mars investigation is the Mars Exploration Rover (MER) mission, still operating after over three years on Mars. In January 2004 two NASA rovers literally bounced onto the Martian surface, safely cocooned in big airbags. The rovers *Spirit* and *Opportunity* landed in Gusev Crater and Meridiani Planum, respectively, three weeks apart (Fig. 23.1). Although the primary mission objective was not life detection, the MER investigations have enabled important progress toward the assessment of habitability on the martian surface (see Section 18.8 for further details). Two important discoveries have led the investigation team to conclude that there was once a standing water sedimentary environment on Mars: (1) the discovery of a chemical precipitate mineral called jarosite; and (2) sedimentary structures – cross bedding of the type that is formed by deposition under water rather than by winds (Fig. 18.16) (Squyres *et al.*, 2004). Hence we have confirmation that standing water has been present on the martian surface, an important first step in the assessment of potential habitability.

The rovers were equipped with identical payload instruments called the "Athena" payload, under the direction of Steven Squyres of Cornell University. This instrumentation is listed in Table 23.1. In an unprecedented manner, the chemical, mineralogical, and imaging data have been made available to the public as soon as possible, with daily mission updates posted on the internet. MER has set the stage for the next step in the characterization of potential habitability for life on Mars – the detection and characterization of organic molecules. This will be the goal of the Mars Science Laboratory, presently scheduled for launch in 2009.

TABLE 23.1 "Athena" payload on each Mars Exploration Rover

Instrument	Utility
Two panoramic imagers	Visual characterization of the environment
Microscopic imager	Detailed images of rock textures
Miniature thermal emissions spectrometer	Identification of minerals
Alpha proton X-ray spectrometer	Element identification
Mössbauer spectrometer	Identification of oxidation state of iron and abundance measurements of iron-containing minerals
Rock abrasion tool	Exposure of fresh rock surfaces

23.4 Approaches to collecting data

23.4.1 General principles

So how can we build upon the Viking heritage and construct a sound program for extraterrestrial life detection? What are the measurable parameters of life? How do we avoid biases thrust upon us by Earthly life? How should we make the measurements, and at what scales? And how should we interpret results? Resolution of these issues provides the foundation for a strategy for life detection (Table 23.2), as well as a roadmap for instrument selection (Nealson and Conrad, 1999; Conrad and Nealson, 2001).

In addition to this basic strategy, there are other important factors to be considered. One is *scale* of observation. While it is clear that there are a variety of spatial scales over which one may observe chemical and structural biosignatures, ranging in size from the atmospheric mixture and fluxes of an entire planet to the presence of small molecules in a cell or in the geologic record, there are other scales of observation as well, such as temporal and spectral scales. We must be careful not to confuse the scale of a measurement with its *magnitude*. Furthermore, it is equally important to know *when* and *how often* to search for a biosignature, and *where* to seek it. And the "where" could as easily refer to where in the electromagnetic spectrum as to where with respect to the Cartesian coordinates on a planetary surface.

We do not believe that any one measurement will definitively prove the existence of life, so a suite of instruments performing several experiments will have to be carefully ordered. The experiments that rapidly scan large areas non-destructively should be done first, with confirmatory, longer data collection experiments next. On the other hand, some tests require extensive sample preparation, which may ruin a sample for subsequent experiments. Do those experiments last!

Definitions of life abound, with no consensus as to their validity or completeness (Chapters 5 and 6), but there are some characteristics of life that typically appear: it possesses chemistry and structure distinct from its environment; it requires an energy source, which it consumes as it transforms nutrients into metabolic products (some of them waste); it reproduces with a high degree of fidelity, though it possesses the capacity to incorporate enough random change to evolve and generally improve its ability to survive; and it is able to move meaningfully in order to escape its waste products and seek new supplies of nutrients.

These measurable parameters of life can be grouped into two categories: (1) features that describe what life *is*, and (2) features that describe what life *does*, i.e., its activities. This is an important distinction because activities can be thought of as processes, which then require an experiment to be conducted at some sampling rate appropriate to their duration. Processes that occur over a very long timescale, such as evolution, would not be practical to observe within the constraints of a planetary mission. Therefore, life activities that are to be *directly* observed must have a timescale shorter than the mission lifetime. This is not to say that one might not *indirectly* observe evidence of life's longer-term activities, such as fossil sequences suggestive of an evolutionary response to changing environmental conditions. Note, however, that every time we choose to select an indirect measurement over a direct one, a further inference, subject to its own criticisms, must be made.

23.4.2 Chemical and structural biosignatures

Life is made of chemical components that are related to, but in different proportions from, the surrounding environment. One approach to life detection is thus to measure a candidate volume in terms of its chemical and structural properties, as compared to the properties of its surrounding environment. By using this "compare and contrast" strategy, one can more reliably recognize life by accurately identifying those environmental components that are definitely *not* life, e.g., minerals and other solid, liquid, or gas phases

TABLE 23.2 A strategy for life detection

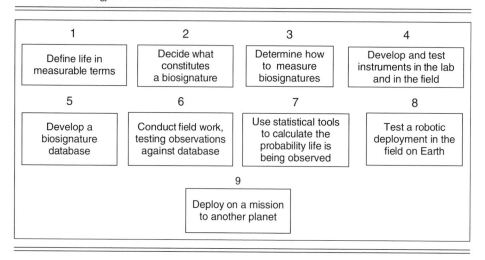

1	2	3	4
Define life in measurable terms	Decide what constitutes a biosignature	Determine how to measure biosignatures	Develop and test instruments in the lab and in the field

5	6	7	8
Develop a biosignature database	Conduct field work, testing observations against database	Use statistical tools to calculate the probability life is being observed	Test a robotic deployment in the field on Earth

9

Deploy on a mission to another planet

TABLE 23.3 Chemical and structural biosignatures

Chemistry	Structure
Element abundance ratios	Presence of cell-like and other biology-like forms
Chirality	Energy uptake/conversion (perhaps observable as sharp gradients in redox potentials)
Charge state distribution	
Stable isotope fractionation	Biogenic trace fossils
Presence and type of organic compounds	Three-dimensional distribution of chemicals over a specific volume

inconsistent with organisms and their activities. Some examples of chemical and structural aspects by which one might distinguish life are given in Table 23.3.

Just as geochemists use the ratios of chemical elements in different rocks to seek relationships that suggest common genesis, we can use element ratios to distinguish between rock and organisms living on or in it (Fig. 23.2).

Recognizing life by its element abundances may be a good way to avoid the assumption that life beyond the Earth will require any specific biochemicals. Although element abundance ratios are just one factor to be considered, they are the most general chemical indicator that could distinguish mineralogical or atmospheric environmental materials from biological materials. They are largely non-Earth-centric, and as a group have the capacity to lead one to infer that life is being observed.

Structures are harder to quantify, save the case of spectral imaging, where one uses the spatially resolved measurable parameters of some area or volume to create an image of a distribution of properties. Images must be rendered quantitative in some way, or else their interpretation will be seen as entirely subjective. Information from each pixel of a digital image can be stored, processed and assigned various parameters, enabling a sophisticated statistical comparison of images, as is commonly done in remote sensing of the Earth and interpretation of astronomical images in various wavelength bands. Again, various scales of observation are vital, from the structure of a living cell to sedimentary geomorphology (land forms), all at a variety of spectral wavelengths.

23.5 Instrumentation

The available tools for life detection are defined not only by the desired science, but also by the technological constraints of space flight. On Earth, accuracy, precision, limits of detection, etc. become the specifications that primarily define development of an analytical instrument. When an instrument is developed for space flight, however, miniaturization becomes key, as available power, mass, and volume are limited. Sometimes, miniaturization of a well-proven Earthbound instrument is not practical, hence a new or different technology must be developed. For example, the element abundances shown in Fig. 23.2 were obtained from an environmental scanning electron microscope (ESEM) equipped with an energy dispersive X-ray fluorescence

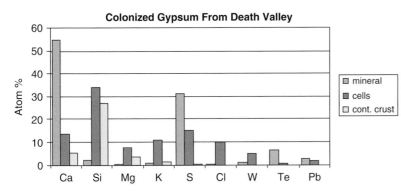

Colonized Gypsum From Death Valley

FIGURE 23.2 Abundances in a sample of gypsum (a hydrous calcium sulfate mineral) that was colonized by a layered microbial community. The distribution of element abundances from areas of the rock that are pure mineral differs distinctly from that of the microbial cells. Abundances from average continental crust are also shown.

detector. Such an instrument cannot be sufficiently miniaturized for space flight, but there are other ways of obtaining element inventory information, e.g., miniature mass spectrometers.

In any given instrument, or when comparing instruments, there are always compromises in balancing the specifications. For example, if one substitutes a miniature mass spectrometer for an ESEM, a destructive technique is substituted for a non-destructive technique. In a second example, for a portable gas chromatograph one must carefully juggle its limits of detection against its mass range – if one cools the detector to a temperature suitable for detecting low masses, large molecules may foul the instrument, rendering it inoperable. Another common problem with miniaturizing any instrument is that it may have insufficient means to dissipate heat generated in its operation.

Most astrobiologists agree that it is important to look for organic chemicals as potential evidence for life, and there are a variety of mass spectrometers and gas chromatograph mass spectrometers under development. But there are other means of detecting the presence of organic chemicals prior to their detection by these destructive means. Optical methods such as laser-induced native fluorescence detection, Fourier transform infrared spectroscopy and Raman spectroscopy are all viable methods of quickly detecting organic materials.

Table 23.4 lists some examples of *in situ* life detection instruments, with the science measures they address and their current technical readiness.

23.5.1 Technical readiness

Serious constraints are imposed by an instrument's "technical readiness level" (TRL), which determines how much time is needed before it is ready for launch. NASA's defined TRLs (Table 23.5) are judged by metrics that help instrument designers and mission planners evaluate the likelihood that a given instrument will be mature enough to make a launch date (Mankins, 1995). The TRL is a risk indicator, and instruments that have not already achieved a certain level of technical readiness cannot be accepted as candidates for a NASA mission.

23.5.2 Mission context

Let's construct a scenario based upon a hypothetical martian life detection mission in the near future. This will provide a framework from which to consider how to design and select potential life detection instruments.

The martian surface has now been extensively mapped by Mars Global Surveyor (1997–2006), and further characterization is in progress by Mars Odyssey (since 2001), by the European Space Agency's orbiter Mars Express (since 2004), and by Mars Reconnaisance Orbiter (since 2006). For example, Mars Odyssey carried a thermal emission imaging system that was instrumental in identifying minerals and helping select landing sites for the two Mars Exploration Rovers. We think it unlikely there are large organisms trooping across the martian surface, so we focus the search on microorganisms. And because we know from the Viking experiments that there is likely to be a strong oxidant on the martian surface, we surmise that any organisms on Mars are living in the subsurface. The first three tasks to accomplish in this case are to (1) collect what we know about endolithic (living within rocks) organisms on Earth, (2) define mission site-selection criteria by understanding habitats on Earth and looking for analogues in the martian environment, and (3) try to understand the specific geological history of potential sites (how was the subsurface deposited? how old is it? what is it made of?) so as to have the best chance of finding biosignatures of past or present life. After selecting the most likely landing sites, we can then propose instruments to search for structural and chemical biosignatures.

TABLE 23.4 Life detection instruments

Instrument	Science measure addressed	Development status	
		Lab/prototype	Flight configured
Scanning electron microscope	Structure and chemistry	X	-
Micro CT scanner	Structure	X	-
X-ray fluorescence spectroscopy & X-ray diffractometer	Chemistry	X	-
Reflectance spectrometer	Chemistry	-	X
Broadband infrared spectrometer	Chemistry	-	X
GCMS	Chemistry	X	X
Laser induced breakdown spectrometer	Chemistry	X	-
UV visible fluorescence imaging & spectrometer	Structure and chemistry	X	-
Raman spectrometer	Chemistry	X	-
Wet chemistry & molecular assays	Chemistry	-	X
Mass spectrometer	Chemistry	X	X

TABLE 23.5 Technology readiness levels

Level 1	Basic principles observed and reported
Level 2	Technology concept and/or application formulated
Level 3	Analytical and experimental critical function and/or characteristic proof-of-concept
Level 4	Component and/or breadboard[a] validation in laboratory environment
Level 5	Component and/or breadboard validation in relevant environment
Level 6	System/subsystem model or prototype demonstration in a relevant environment (ground or space)
Level 7	System prototype demonstration in a space environment
Level 8	Actual system completed and "flight qualified" through test and demonstration (ground or space)
Level 9	Actual system flight-proven through successful mission operations

[a] A *breadboard* is a prototype circuit designed to test feasibility and accuracy, without regard to packaging and size.

Next, for each proposed instrument, we must:

- Define the instrument delivery platform, as well as its sampling strategies for all individual and integrated instrument combinations.
- Determine whether rock samples need to be crushed, or examined intact. Also, for control experiments, how does one achieve a statistically equivalent division of samples?
- Evaluate instrument status in terms of its specifications (e.g., mass, volume/size, power, sensitivity) and stage of development (e.g., currently flight tested, working lab prototype, design stage only).
- Develop near-term and far-term life detection implementation strategies.
- Prioritize a technical development strategy based upon the science to be returned, state of instrument technical readiness, and the likelihood that the instrument will in a timely manner meet flight and performance constraints (e.g., instrument sensitivity, resolution, clarity of data).

An example of a series of complementary measurements to investigate organic materials on Mars might consist of images collected with a UV fluorescence camera/spectrometer, then spectra of atoms collected with a Raman spectrometer or Fourier transform infrared spectrometer, and finally an inventory of organic chemicals taken with a mass spectrometer or other analytical tool. This sequence would allow one to cover progressively smaller volumes with increasingly longer times of data collection. The sequence also moves from least destructive to most invasive techniques. And these data, if gathered with high spatial resolution, allow one to determine whether or not there are volumes where the results significantly vary from what one might expect for minerals (or the pore spaces between them) in sediment or rock. Note that this strategy also requires one to characterize all contextual minerals both chemically and structurally. It is critical to measure and understand the planetary materials against which one is seeking evidence of life.

23.5.3 Site selection

How does one select a site for exploration? To further the goal of life detection, NASA has recently been using a "follow-the-water" theme for its missions, so we should choose sites where liquid water has been or is currently present. Potential sedimentary environments consistent with past water might be alluvial fans and/or deep basins. Examples of sites where liquid water might have been or still be present are: deep basins, subsurface aquifers, and polar caps (where we know there is both frozen H_2O and frozen CO_2). A martian polar cap would be an interesting destination because its ice, if melted, would enable one to sample large quantities of material quickly. Additionally, it possesses the advantage of chemical simplicity relative to a multi-mineral rock-like basalt. The Phoenix mission is planned to deliver a lander to a high-latitude martian site in 2008. In the case of the Mars Exploration Rovers, selection of the landing sites was influenced by several factors including accessibility, geomorphology suggestive of the former presence of water, and mineral identification from orbit. In turn, the landing site for the 2009 Mars Science Laboratory mission (a much larger rover) will undoubtedly be influenced by the science returned from the Mars Exploration Rovers.

In addition to these factors, landing site selection may be influenced by other technical criteria such as requirements for sufficient exposure to sunlight for recharging solar batteries, necessitating latitudes and seasons that provide a reasonable length of daylight. The timing of a mission is also heavily dictated by the position of Mars relative to Earth and possible spacecraft trajectories.

23.5.4 Delivery platforms

There are a number of ways to deliver the science instruments once a spacecraft has arrived at its planetary destination. The spatial scale of the required observations helps to determine whether or not the instruments should operate from an orbital platform, a fixed landed platform, or a mobile platform such as a rover. For example, orbiters perform global scale surveys, looking with remote sensing tools for chemical signs of life such as atmospheric water vapor, surface water, sulfate and sulfide distribution, nitrogen, etc. In addition to these chemical searches, surface features can be mapped and evidence sought for weathering and/or sedimentary deposition under water. Instrument technology for remote global surveys is well established from both Earth-observing satellites and orbiters that have been deployed to study planets, such as Mars Global Surveyor and Mars Odyssey mentioned above. A third orbiter, Mars Reconnaissance Orbiter, entered martian orbit in 2006 in order to map the surface with high resolution imaging and spectroscopy (for mineralogy), as well as the shallow subsurface with a sounding instrument supplied by the Italian Space Agency.

Landed delivery platforms undergo different criteria during the selection process. What platform is compatible with the site selected? Is mobility required? If so, a rover would be the appropriate selection; if not, a fixed platform might be sufficient. There are also "passive mobility" platforms such as a cryobot, which can bore through ice by melting its way through (Zimmerman et al., 2004). Just as site and instrument selection must accommodate the sampling of relatively large amounts of material, so must the delivery platform. In addition, the delivery platform must help preserve the pristine nature of the samples, particularly if they are to be stored for later sample return. It must also accommodate both invasive and non-invasive investigations, with sufficient mass, power, and volume for the preparation of samples.

Mobility is only one feature of the delivery platform that affects the ability to accommodate the instrument payload. Other characteristics such as data handling and transmission capacity can limit the amount of science that can be accomplished. The inability of a delivery platform to provide sufficient sample processing and analysis may preclude certain instruments. In the case of a rover, size obviously affects the mass of the instrument payload. In addition, the type of power supply, e.g., solar or nuclear power, affects mission length, which facilitates certain types of studies and inhibits others.

23.6 Summary

Successful life detection missions cannot be defined by whether or not evidence of life is discovered beyond the Earth, but by whether scientists can agree on an interpretation of the results of their life detection experiments. This means building consensus throughout the process of planning and executing a mission.

Life must be defined in universally measurable parameters to avoid Earth-centrism, and a well-established library of chemical and structural bio-signatures must be developed in advance of making measurements on another planet.

Life detection instruments are selected on the basis of several criteria, including their ability to make the desired measurement reliably and precisely, the instrument's level of technical readiness at the time it is proposed, resources available on the spacecraft (e.g., power, mass, volume), and the challenges of integrating it with companion instruments.

Instrument selection is intimately related to the selection of an appropriate delivery platform for the site of interest. Both delivery platform and payload should be driven by the science requirements, even though they may have to be constrained by technical limitations.

Life detection tools and methods are an exciting manifestation of "applied astrobiology." If we plan meticulously and rigorously, we can have confidence that our robotic measurements on another planet will yield reliable evidence for whether or not we have discovered life beyond the Earth.

REFERENCES

Brown, F. S., Adelson, H. E., Chapman, M. C., *et al.* (1978). The biology instrument for the Viking Mars mission. *Rev. Sci. Instrum.*, **49**, 139–182.

Conrad, P. G. and Nealson, K. H. (2001). A non-Earth-centric approach to life detection. *Astrobiology*, **1**, 1.

Friedmann, E. I. (1982). Endolithic microorganisms in the Antarctic cold desert. *Science*, **215**, 1045–1053.

Horowitz, N. H., Hobby, G. L., and Hubbard, J. S. (1976). The Viking carbon assimilation experiments: interim report. *Science*, **194**, 1321–1322.

Horowitz, N. H., Hobby, G. L., and Hubbard, J. S. (1977). Viking on Mars: the carbon assimilation experiments. *J. Geophys. Res.*, **82**, 4659–4662.

Hubbard, J. S. (1976). The pyrolytic release experiment: measurements of carbon assimilation. *Origins Life*, **7**, 281–292.

Klein, H. P. (1978). The Viking biological experiments on Mars. *Icarus*, **34**, 666–674.

Klein, H. P., Horowitz, N. H., and Biemann, K. (1992). The search for extant life on Mars. In *Mars*, eds. H. H. Kieffer, B. M. Jakosky, C. W. Snyder, and M. S. Matthews. Tucson: The University of Arizona Press.

Levin, G. V. and Straat, P. A. (1976a). Viking labeled release biology experiment: interim results. *Science*, **194**, 1322–1329.

Levin, G. V. and Straat, P. A. (1976b). Labeled release: an experiment in radiorespirometry. *Origins Life*, **7**, 293–311.

Levin, G. V. and Straat, P. A. (1977a). Recent results from the Viking labeled release experiment on Mars. *J. Geophys. Res.*, **82**, 4663–4668.

Levin, G. V. and Straat, P. A. (1977b). Life on Mars? The Viking labeled release experiment. *BioScience*, **9**, 165–174.

Levinthal, E. C., Jones, K. L., Fox, P., and Sagan, C. (1977). Lander imaging as a detector of life on Mars. *J. Geophys. Res.*, **82**, 4468–4478.

Mankins, J. C. (1995). *Technology Readiness Levels: a white paper*. Washington, DC: NASA Office of Space Access and Technology.

Nealson, K. H. and Conrad, P. G. (1999). Life: past, present and future. *Phil. Trans. R. Soc. Lond.*, B **354**, 1923–1939.

Oyama, V. I., Berdahl, B. J., Carle, G. C., Lehwalt, M. E., and Ginoza, H. S. (1976). The search for life on Mars: Viking 1976, gas changes as indicators of biological activity. *Origins Life*, **7**, 313–333.

Oyama, V. I., Berdahl, B. J., and Carle, G. C. (1977). Preliminary findings of the Viking gas exchange experiments and a model for martian surface chemistry. *Nature*, **265**, 110–114.

Squyres, S. W. and Mars Exploration Rover team (2004). The Spirit Rover's Athena science investigation at Gusev Crater, Mars. *Science*, **305**, 794–799.

Sun, H. J. and Friedmann, E. I. (1999). Growth on geological timescales in the Antarctic cryptoendolithic microbial community. *Geomicrobiology J.*, **16**, 193–202.

Zimmerman, W., Anderson, F. S., Carsey, F., Conrad, P., Englehardt, H., French, L., and Hecht, M. (2002). The Mars 07 north polar cap deep penetration Cryo-Scout mission. *IEEE Aerospace Publication #15*.

24 Societal and ethical concerns

Margaret S. Race
SETI Institute

People have wondered for centuries whether we are alone or share this Universe with extraterrestrial beings. Questions about the origin of life and the possible existence of extraterrestrial life have deep roots in the history of both science and culture (Chapters 1 and 2; Dick 1997, 1998). In ancient times, the debate centered mainly on our uniqueness versus the plurality of worlds. Following the advances of the Copernican era, scientists gradually accepted notions about the plurality of solar systems and recognized the large-scale nature of the Universe. Today, cosmic evolution, from the Big Bang to the evolution of intelligence, has become a working hypothesis for astrobiology, one that combines characteristics of both biological and physical universes into a "biophysical cosmology." This new world view recognizes the enormity of the Universe and hypothesizes that life is one of its basic and essential properties, a cosmic imperative rather than an accidental or incidental property found only on Earth. The current key questions in astrobiology are whether *biological* laws reign throughout the Universe, whether Darwinian natural selection is a universal phenomenon rather than simply a terrestrial one, and consequently whether there may be other biologies, histories, cultures, religions, and philosophies beyond Earth. In short, is the ultimate outcome of cosmic evolution merely planets, stars, and galaxies – or life, mind, and intelligence? The answers to these questions raise a multitude of issues in the realms of both science and society.

24.1 Operating principles for astrobiology

We live in a time when answering these questions is a distinct possibility because of advanced scientific knowledge, sophisticated research methodologies, and improved technological capabilities to study Earth and beyond. In the past decade, the emerging field of astrobiology has rekindled broad interest in the topic of extraterrestrial life by superimposing an organized, systematic approach centered around three fundamental scientific questions.

- How does life begin and evolve?
- Does life exist elsewhere in the Universe?
- What is the future of life on Earth and beyond?

Behind each of these simple questions is a complex list of multidisciplinary science questions that amount to a research plan, or "roadmap," aimed at developing a comprehensive understanding of the origin, evolution, and fate of life in the Universe. NASA's Astrobiology Roadmap (2003) provides a good overview of the type of science goals and objectives that define the field. With such a program we may soon know whether we are alone in this Universe or whether we inhabit only one of many planets with the right environmental conditions for life to flourish. Scientifically, these two possibilities yield fundamentally different worldviews describing humanity's place in the Universe. If extraterrestrial life is confirmed as having arisen independently of life on Earth, this new worldview of a biological Universe and a biophysical cosmology challenges orthodoxy at every turn and raises numerous, provocative questions.

In recognition of the potentially enormous significance of future findings, NASA's Astrobiology Roadmap includes an important, but often overlooked, set of principles along with the science goals and objectives that guide the field. As presented in the roadmap, the four principles fundamental to the astrobiology program are:

- Astrobiology is multidisciplinary in its content and interdisciplinary in its execution. Its success depends critically upon the close coordination of diverse

Planets and Life: The Emerging Science of Astrobiology, eds. Woodruff T. Sullivan, III and John A. Baross. Published by Cambridge University Press. © Cambridge University Press 2007.

FIGURE 24.1 As astrobiologists explore new worlds beyond Earth, they are like Columbus and other explorers. We can only speculate how future advances may change our views of the universe and our capabilities. (Courtesy of Currier and Ives Founding Fathers Prints. www.currierandives.info/founding.)

scientific disciplines and programs, including space missions.

- Astrobiology encourages planetary stewardship through an emphasis on protection against biological contamination of planets and moons and recognition of the ethical issues associated with exploration.
- Astrobiology recognizes a broad societal interest in its endeavors, especially in areas such as achieving a deeper understanding of life, searching for extraterrestrial biospheres, assessing the societal implications of discovering other examples of life, and envisioning the future of life on earth and in space.
- The intrinsic public interest in astrobiology offers a crucial opportunity to educate and inspire the next generation of scientists, technologists, and informed citizens; thus a strong emphasis upon education and public outreach is essential.

These principles amount to a formal acknowledgement of the societal and non-scientific ramifications of astrobiological research and the importance of anticipating and addressing more than just science. We live in a democratic society that views science as an enterprise, no different from others, whose applications should be scrutinized for both their public benefits and costs. Ultimately, success in astrobiology is likely to require repeated consideration of more than just scientific facts provided by experts. Societal issues, which may at times seem to scientists secondary or minor, could prove consequential for both our scientific field and our planet.

How is it possible to address the implications of a field *before* research is completed or discoveries are made? According to historian of science Steven Dick (1997), one way to frame our thinking about the implications of extraterrestrial life is to use history as a guide. By analyzing historical responses to past discoveries and theories, we can gain a perspective on how worldviews have changed and cosmologies have developed. Enormous changes have occurred in society over the past several hundred years, many of which were built on a foundation of vastly expanded scientific knowledge, bold technological innovations, and a spirit of adventure that encouraged experimentation and exploration. In some ways, we are not that different from Christopher Columbus and others who sought to discover the New World in the fifteenth century (Fig. 24.1). We fully anticipate that our collective future will experience great changes, but we can only speculate how advances in the coming years may change our views of the Universe and our capabilities to strike out in bold new directions. Yet it is incumbent on us to anticipate these changes and to consider now what they might mean.

While historians study shifts in worldviews that develop over the scale of centuries, individuals are rarely afforded the luxury of such a broad perspective. Surrounded by day-to-day research in a fast paced scientific world, most people experience incremental change over much shorter timescales. Thus, to understand the implications of astrobiology in the lifetime of the individual scientist or citizen, it may be more productive to focus on recent history, on a scale of years or decades. In this shorter time frame, we can track how scientific information and scientists themselves are involved in sociopolitical processes in democracies. In essence, it is an attempt to look back in order to look forward; to understand lessons learned in other fields that will help us handle similar issues in the future.

This chapter begins with a comparison of the societal issues that have recently arisen in another field, namely biotechnology. By reviewing the history of biotechnology we hope to anticipate and frame the issues and concerns that have already begun to emerge in astrobiology. Using Mars exploration as a specific case example, we next highlight approaches to sociopolitical issues, regulatory complications, public concerns, and ethical questions. Because much of astrobiology's most exciting research involves planetary missions, ethical concerns raised by various space policies (e.g., use of nuclear materials on spacecraft) also become societal concerns for astrobiology, and are thus also briefly discussed.

24.2 Comparison of astrobiology with biotechnology

To better understand astrobiology in the context of science in society, it is useful to compare the emerging issues and concerns of astrobiology with experiences in other scientific fields. In general, whenever there are fast-paced shifts in our understanding and applications of complex multidisciplinary science – particularly when there are potential impacts on living organisms, the environment, or society at large – it is not surprising that science as an enterprise becomes intensely scrutinized. While some issues and concerns can only be recognized well in hindsight (e.g., those associated with the industrial revolution, nuclear sciences, and modern agriculture), others unfold more quickly, such as those raised by the recent revolution in biotechnology. Although not an exact parallel to astrobiology, many of the experiences of biotechnology during the past quarter century are relevant. Biotechnology's young history is useful for understanding how scientific information and scientists themselves can be drawn into the sociopolitical realm where they must justify their work.

Like astrobiology, biotechnology is a relatively new, multidisciplinary scientific field. It began in the early 1970s and has expanded in three decades to become a major enterprise with significant societal implications. The parallels between biotechnology and astrobiology are striking (Table 24.1). Biotechnology arose as a practical undertaking when two elements were in place: (1) sufficient scientific understanding of the principles and processes that determine the properties of living organisms from micro to macro scales, and (2) the tools to readily modify the properties of living organisms (Schacter, 1999). This combination of pure science's desire "to know" and applied science's practical enthusiasm "to do" has the potential to disturb the Universe in unnatural ways. Over the years, a great infusion of public and private funding has shifted many scientists, either directly or indirectly, from the world of small-scale, basic research experiments to a world focused on projects with multi-billion dollar price tags and great expectations (e.g., the Human Genome Project, commercialization in the pharmaceutical industry, and development of medical technologies). Even as advances and applications of this field have brought benefits to people worldwide, new proposals continue to be scrutinized, leading to debate over everything from stem cell research, genetic testing and gene therapy, to genetically modified crops, cloning, and patenting rights. Critics argue that the societal implications of such rapid progress have not been adequately considered. Even after decades of progress, there are still opponents who continue to call for a broad public debate over "what is likely to be the most radical experiment humankind has ever carried out on the natural world" (Rifkin, 1998).

Astrobiology as a practical undertaking has strikingly similar starting elements to those found in biotechnology. Viewed as a scientific enterprise, it is likewise rapidly developing a broad, new understanding of the world – in this case, an understanding of the physical and biological principles and processes involved in the formation and evolution of the Universe. Moreover, it has the requisite tools (e.g., spacecraft, rovers, telescopes, microscopes, and complex analytical equipment) and large budgets to allow humankind to study, visit, modify, and interact in significant ways with environments on global scales, both on Earth and beyond. This combination is already reflected in a growing list of proposed actions in astrobiology. For example, sample return missions, human exploration of planets, asteroid impact warning systems, and communication with extraterrestrial intelligence all have potential implications for Earth and its biota in this and future generations. In essence, as we study the Universe and determine whether life is a strictly "local" phenomenon or not, we are simultaneously evaluating the potential for expanding life's evolutionary trajectory beyond its place of origin. One can argue that we may well be on the verge of altering the course of human evolution. In such an

TABLE 24.1 Comparison of societal aspects of biotechnology and astrobiology

	Biotechnology	Astrobiology
Key elements:		
(1) Basic scientific knowledge	(1) Understanding of the principles and processes that determine the properties of living organisms	(1) Understanding of the principles and processes that determine the formation, properties and evolution of the universe and its component parts
(2) Technological capabilities	(2) Tools to modify the properties of living organisms more or less at will	(2) Tools to visit, experiment, modify and interact with environments far from Earth
Scientists involved in developing early policies and guidelines	Expert discussions on the implications of the field and early agreement on the need for voluntary guidelines for continuation of research (Asilomar Conference, 1975)	Early scientific discussions led to UN Outer Space Treaty of 1967 and to COSPAR guidelines on prevention of harmful cross-contamination
Scientists involved in expert policy making	US Recombinant DNA Advisory Committee for review of policies; development of biocontainment and laboratory guidelines; reviews of adequacy of laws and agency approval and permit processes	Development and updating of COSPAR international Planetary Protection Policies; involvement in advisory studies of US Space Studies Board; mission planning and reviews; expert advisory workshops with specialists, etc.
Scientists on both sides in public decision-making	Sociopolitical reviews and legal challenges of individual proposals or advances (field testing with genetically engineered microbes; patenting; cloning; stem cell research; human gene therapy; genetically modified foods and livestock, etc.)	Sociopolitical debates and specific legal reviews or challenges (nuclear materials on Cassini mission; Mars sample return and biocontainment; concerns about asteroid impacts; risk analyses of launch accidents and mission activities, etc.)
Ethical and philosophical questions	Ethical concerns related to activities (e.g. cloning, stem cell research, *in vitro* fertilization, therapeutic versus germ line interventions, etc.)	Ethical questions related to planetary protection and cross-contamination; rights of extraterrestrial life forms; terraforming; human colonization, commercialization, resource use, etc.

environment, it should not be surprising that astrobiology is likely to face serious, insistent questions about its environmental, health, economic, social, and ethical implications both on Earth and afar.

In biotechnology, scientists actively took part in the public debates and problem solving as soon as serious questions arose. It was scientists who called for a voluntary moratorium on recombinant DNA experiments until questions of public safety could be debated. They organized the Asilomar Conference in 1975 to debate the ethical issues raised by recombinant DNA research

and to discuss the legal liabilities that might arise from such research. They helped develop strict guidelines for safety and biocontainment that led to eventual standards for research, and they helped write and revise laws and regulations for laboratory and subsequent outdoor experiments. They also sought to share information widely with the public via the mass media (for excellent reviews of the early history of the field, see NRC, 1989a; Schacter, 1999).

Why then do so many issues and questions remain? In hindsight, although the scientists were responsible in

FIGURE 24.2 Apollo astronauts inside the mobile quarantine module before transport to the Lunar Receiving Lab at Johnson Space Center.

An elaborate quarantine (three weeks long) and testing protocol were used during the Apollo Program to ensure that astronauts had not inadvertently brought extraterrestrial life or potential biohazards to Earth when they returned from the Moon. (NASA.)

thinking ahead about the risks associated with their field, they probably did not go far enough, nor did they engage in open dialogue with those outside the field. Grobstein (1986) suggests that the early discussions among scientists made four important mistakes: (1) they ignored the possible effects resulting from widespread adoption of the technology; (2) they overlooked the possibility of efforts to *deliberately* create harmful organisms; (3) they were unable to comprehensively address the environmental and ecological impacts of an accidental or deliberate release of genetically exotic organisms; and (4) they did not anticipate how the new techniques would affect the human species in current and future generations.

Astrobiologists could easily repeat similar shortcomings by not considering the broad, long-term societal implications. Consider for a moment the implications of some of the larger projects, many of which involve broader issues of space activities, that have been discussed to date: nuclear reactors and propulsion in space; deliberate transport of organisms or materials between Earth and Mars for research and life support; terraforming (engineering an entire planet to be more like the Earth); commercial ventures in space; engineering of new lifeforms adapted to live on other worlds; deflection of Earth-crossing asteroids with missile systems; and construction of permanent bases on the Moon or Mars.

Just as in biotechnology, it is difficult to address in a comprehensive way the environmental and ecological impacts of accidentally or deliberately released exotic organisms. Because there are significant uncertainties about the existence of putative extraterrestrial life, there are likewise persistent questions about the possible effects of forward contamination[1] of extraterrestrial locations where life may be discovered (NRC 1992, 2000) and back contamination of Earth during sample return missions (NRC, 1997, 1998, 2002a). While scientists believe the risks from interactions with extraterrestrial life or samples are extremely low, they also know they are not zero. Like researchers in biotechnology, astrobiologists must proceed with exploration in the face of significant uncertainties about novel lifeforms, always using current best knowledge to make judgments, but knowing that future judgments may well vary. For example, strict *planetary protection*[2] controls and deliberately conservative approaches are routinely adopted for Solar System missions to minimize cross-contamination (Chapter 25). Until we know much more about the environments on diverse celestial bodies, and the hazards posed to humans both in space and on Earth, we must proceed with special caution (Fig. 24.2).

---

[1] *Forward contamination* refers to the transport of organisms or harmful contamination from the Earth to another planetary body; *back contamination* refers to the return of possible extraterrestrial life or harmful contamination to the Earth's biosphere.

[2] *Planetary protection* refers to the biological isolation of the (known and possible) lifeforms on various planets from each other; Chapter 25 discusses this issue in detail.

In astrobiology, there are also unanswerable questions with intergenerational implications. For example, what are the potential effects on humans of long-duration missions or colonization? Could there be any long-term impacts of martian microbes (should they exist) if they were accidentally released from containment on Earth? What are the physiological and psychological risks to children who might someday be conceived or born far from Earth? What are the implications of engineering new Earth-life forms adapted to live on other worlds? It is not far fetched to predict that successful exploration in astrobiology might someday lead to future activities or situations that could alter the human species and even entire planets in unforeseen ways.

Finally, new knowledge and capabilities in science can raise a host of challenging ethical, philosophical, and theological questions with no clear answers. For example, in biotechnology, scientific advances have forced us to reconsider the very nature of human nature, the relationship between humankind and the environment, and the theological implications of new creations that could perhaps alter the evolutionary course of humankind. No-one can predict the kinds of changes in individuals or in the human species as a whole that could be effected by applying our newfound knowledge about the human genome. Who should determine when it is morally justified to make changes in individuals or species? What kind of problems could be caused inadvertently by transferring genes across species? Will it be possible to devise ethically justifiable processes to guide major decisions with the potential to impact evolutionary trajectories? The questions are complicated and without obvious answers; the implications are staggering.

In astrobiology, it may be decades or centuries before we need to grapple with such issues, some of which may seem like topics from *Star Trek*. But one can never be sure how fast a field will progress. Here again, perhaps we can learn from biotechnology. In the early years of biotechnology (~1970–85) people did not fully understand the promises and perils ahead. At the time, molecular biologists, policy leaders, media pundits, and writers dismissed many predictions as "alarmist" or far-fetched, and argued that realization of the science applications were at least a hundred years, perhaps several hundred years, away (Rifkin, 1998). Many felt there was little need to examine the environmental, economic, social, and ethical implications of a "hypothetical" future. Today, we recognize that many of the predicted scientific and technological

breakthroughs in biotechnology have already occurred. As we witness boycotts of genetically modified foods, sabotage of research facilities, and headlines about stem cell research and cloning, it is clear that the debates are hardly resolved.

Could the same pattern be repeated in astrobiology decades hence? Can societal issues ever be resolved? Rather than viewing persistent societal questioning as a direct assault on science by people with narrow perspectives, perhaps it is better to view such inquiries in a positive light. In our democratic society, questions surrounding new science and technologies are neither abstract nor remote, regardless of whether they arise in biotechnology or astrobiology. Quite the contrary, they represent significant and legitimate concerns about how we will face a future rich with scientific potential and opportunities, yet fraught with societal costs and risks. Proactively addressing the issues will not guarantee that opposition will fade away, but it may allow scientists and citizens together to guide the progress of the field through comprehensive exploration of the issues.

24.3 Examples from Mars exploration

By examining astrobiology in the Solar System, in particular Mars exploration, it is possible to see in more detail how societal issues can impact on-going research and exploration (Race, 1996, 1998). As in Table 24.1, it is helpful to be aware of four areas where scientists and scientific information can be involved: (1) expert decision-making, (2) public decision-making and oversight, (3) education and risk communication, and (4) ethical and philosophical debates. Science plays an important part, but only one part, in all these areas. In order for scientists to play an effective role in discussion outside the strictly scientific realm, it is important to understand the strengths and limitations of scientific data, as well as to recognize where objective science must be blended with judgment and non-scientific concerns.

24.3.1 Scientists as expert decision makers

The first phase of the sociopolitical process generally involves experts making decisions about how to use scientific and technical knowledge, a process that typically takes place with little or no public input or oversight. Before science can be applied, there is usually a need for policies, treaties, and laws to guide and regulate activities. And later, with the accumulation of more knowledge, or the advent of new technologies over time, there may be a need for incremental refinement or revisions to policies or regulations. To understand the iterative nature of this aspect of sociopolitical decision-making, it is useful to focus on how planetary protection policies came about and have been applied to Mars exploration. From the earliest days of space exploration, scientists were involved in setting policy to avoid harmful contamination of planets and celestial bodies during space exploration. Their decade-long discussions eventually led to the passage of the UN Outer Space Treaty in 1967, which has been the primary source of international space law ever since (Cypser, 1993). The provisions of the treaty instruct space-faring nations to conduct exploration and activities in a manner that avoids harmful contamination of both celestial bodies and Earth.

Translating the lofty goals of the Outer Space Treaty into policies and guidelines for mission planning has occurred in a series of steps, with scientists involved through a variety of international and domestic venues (DeVincenzi et al., 1998). The work of developing, reviewing, and updating policies for Solar System exploration is a task that rests with the Committee on Space Research (COSPAR), a permanent committee of the International Council of Scientific Unions. COSPAR provides recommendations to the international space science community on how to determine the appropriate level of planetary protection controls for missions. These policies guide the regulation of spacecraft operations and the measures used to avoid forward and back contamination, and are discussed in detail in Chapter 25.

Within the United States, NASA's Planetary Protection Office maintains responsibility for ensuring that missions are planned and implemented in accord with appropriate treaties, laws, and policies. NASA often consults with both internal and external advisory groups to obtain recommendations. For example, since the time of Sputnik, reports by the Space Studies Board (SSB) of the National Academy of Sciences have been leading sources of expert information for translating international policies into recommendations for mission planning. Over the years, input from the SSB has yielded recommendations on a wide variety of topics relevant to Mars exploration, e.g., forward contamination control (NRC, 1992, 2006), Mars sample return (NRC, 1997), sample return from small bodies (NRC, 1998), quarantine and containment of Mars samples (NRC, 2002a), and robotic precursor studies prior to human missions to Mars (NRC, 2002b). In addition to

SSB advice, NASA also seeks input from scientific and technical experts through workshops and studies that are charged with specific tasks. Examples include: orbiting quarantine facilities (De Vincenzi and Bagby, 1981); quarantine and containment (DeVincenzi et al., 1999); sterilization (Bruch et al., 2001); sample-testing protocols and facility requirements (Rummel et al., 2002); and human missions (Race et al., 2003).

24.3.2 Public decision-making

Even as experts formulate polices and develop project plans, the next phase in the sociopolitical process expands the circle to involve a greater public. In contrast to the lengthy, iterative expert decision-making described above, this phase is more likely to occur in reaction to specific mission proposals (e.g., a launch with nuclear materials on board; siting or construction of a facility; or an Earth flyby or re-entry). Among the cast of characters typically involved in this phase are the lay public, government agencies, lawyers, individual citizens, environmental and other non-governmental organizations, legislative bodies, the mass media, and sometimes the courts. In the United States, it is during this phase that the public has the opportunity to ask questions about proposed actions and to review and criticize expert decisions. NASA and other governmental space agencies have an obligation to disclose and explain information in detail. If there is little or no opposition to a proposal, this phase transpires with little notice as reviews are conducted, permits are granted, and projects are implemented. However, if concern or opposition arises about unprecedented or potentially risky activities, the situation can be far from routine. Unlike expert decision-making processes, where scientists typically lead the planning and provide advice on policies, this public decision-making phase puts scientists and government officials on the defensive, trying to explain or rationalize their proposals to a non-expert citizenry. While scientific and technical details are usually at the heart of these public deliberations, non-scientific factors can also be introduced. Questions may arise that are sometimes difficult or impossible to answer with quantitative data. For example: Is it legal? Is it safe? Has it been reviewed properly? Does it have the required permits and approvals? Why will it be done here, rather than somewhere else? Why should the government even do it? Scientists who are generally comfortable with their involvement in expert decision-making may be less enthusiastic about their participation in the public phase where scientific and technical considerations are typically mixed with a diverse set of non-scientific factors. The process can be fraught with legal ambiguities or challenges that play out over a long time, sometimes causing banner headlines and surprising or undesirable outcomes (e.g., launch delays, interference with construction plans, forced recalculation of risk assessments, and questions about personal integrity or government conspiracy). Considering the potential for such adverse outcomes, many scientists feel frustrated when working in this phase of the process.

For any future US-sponsored Mars sample-return mission the most significant public hurdle will probably occur in this phase of the process. NASA must comply with the National Environmental Policy Act by writing an environmental impact statement (EIS) prior to launch. An EIS must provide details about a full range of risk assessments, potential impacts, project alternatives, worst-case scenarios, mitigating actions, and uncertainties for all phases of the mission. Considering the complexity of such a mission, it could take several years to complete the necessary documentation, public hearings, agency consultations, step-wise reviews, and public announcements required by law (Race, 1998). Unlike the Outer Space Treaty and COSPAR policies requiring avoidance of "harmful cross contamination," an EIS must assess a multitude of specific impacts in detail (e.g., effects on air quality, water quality, wetlands, fish and wildlife, farmlands, population, health and safety, employment, and historical and archaeological sites (NASA, 2001).

The task of completing an EIS for a future sample-return mission will be complicated by the lack of risk information involving putative martian life. In all likelihood, there will be no way to know definitively whether life exists on Mars, whether it will be found in returned samples, what characteristics it might have, and how it could affect Earth biota. Yet these basic, poorly understood scientific questions are central to issues of statutory authority over review and handling of returned samples and could translate into legal debates. For example, different laws and agencies could be invoked depending on whether the presumed life in a sample is characterized as an infectious agent, an exotic species outside its normal ecological range, a truly novel organism (e.g., a genetically modified organism), or a hazardous material. It is uncertain how these questions about the nature of martian life will be resolved prior to sample return and how they may contribute to public concerns or legal issues.

NASA must also satisfy the laws and regulations of many other government agencies whose mandates cover particular mission activities or actions on Earth. For example, if even small amounts of nuclear materials are used on a spacecraft (in radioisotope power sources or heater units; see the following section), mission plans must include a review of the potential for large-scale adverse environmental impacts (Race, 1996). Moreover, other government agencies may need to oversee activities such as facilities construction, transportation of returned samples, containment and quarantine of hazardous materials, or waste disposal. Based on past experiences, jurisdictional and legal ambiguities are highly probable (Race, 1996, 1998; Robinson, 1992).

The discussion above about the public phase of decision-making has been presented primarily in the context of United States/NASA policies and procedures. But there are many other nations involved in astrobiology and space exploration, and we can expect more international launches and partnerships in the future. In addition, the future may see private sector activities in space, some of which are already the subject of debate with regard to government regulation (e.g., Zubrin and Wagner, 1996; van Ballegoyen, 2000). Other nations are likely to have different laws, cultures, social concerns, ethical perspectives, or religious views that could complicate planning. It is conceivable that missions and proposals with international partners could require additional pre-launch review and approval in the contexts of other legal systems or courts, or be scrutinized by a public with different concerns or cultural values. For example, differences in French and American regulations regarding informed consent, confidentiality, human subjects, and medical monitoring have been identified during discussions to develop the handling and testing protocols for returned martian samples (Rummel et al., 2002). Such legal differences could complicate assembling an international staff for preliminary handling and testing of materials at a sample receiving facility. Even if no international partners are involved in a mission, it is possible for citizens of one nation to challenge missions of another nation because of transboundary concerns. For example, prior to the launch of NASA's Galileo mission in 1989, German environmental activists joined an American group in a legal challenge in US courts. This alliance sought to stop the mission, based on arguments about potential launch accidents that could cause widespread radioactive contamination like the then-recent Chernobyl nuclear reactor accident.

The mixture of scientific uncertainty, legal ambiguity, and complicated reviews can easily become headline news of great interest to the public, particularly when opponents repeatedly raise questions through venues such as news conferences, street protests, Internet websites, legislative hearings, or lawsuits. In order for citizens to be effectively involved in decision-making, they need accurate, timely information to help them make informed judgments about these complex topics. They need to understand the risks and benefits of the proposed actions as well as the science and technology behind them. This brings us to the next important topic, that of public education and risk communication.

24.3.3 Education, public engagement, and risk communication

NASA has an education and outreach strategy aimed at engaging the public and keeping them informed and educated about science in general and NASA's discoveries in particular (e.g., NASA, 2006). These programs are aimed at multiple audiences and include formal curricular materials for kindergarten through college, as well as informal education materials for outlets like museums, planetaria, science centers, youth groups, and the Internet. However, educating the public about astrobiology is challenging. It is a fast-paced multidisciplinary field that is only gradually being covered in textbooks and educational settings. There is an inevitable time lag between scientific research results and their assimilation by the public. In addition, the public's keen interest in extraterrestrial-life topics is a double-edged sword. The public can often mix accurate knowledge with misinformation bombarding them in the form of science fiction, Hollywood movies, and sensational headlines in marginal news sources (Shostak, 1998).

In addition to the traditional education and outreach efforts, there is need for specialized *risk communication* in situations where hazards or controversies are involved. Risk communication aims to provide citizens with understandable information about risks in order that they can make more informed choices about how risks should be managed (NRC, 1989b, 1996). Clear, accurate scientific details that are translated into lay language are a vital component of successful risk communication. A central premise of democratic government – the existence of an informed electorate – implies a free flow of information among all involved parties. Risk communication is thus more than an educational task *per se*. It is complementary

to the public decision-making process and requires a combination of science expertise, communications skills, public engagement, and detailed technical information. Risk communication is not the same as public relations or advocacy – it incorporates a dialogue about the anticipated risks so that citizens can actively participate, along with experts, in key decisions.

Risk communication plans for missions or controversial activities must include the preparation of information for varied audiences and diverse venues. It is therefore important to be aware of the public's level of scientific understanding, the nature and extent of their specific concerns, as well as their general attitudes and perceptions about space exploration and mission risks. For example, with Mars sample return missions in mind, surveys have been conducted to gauge the public's general level of knowledge about extraterrestrial life, the degree to which they believe it may pose a risk to Earth, and the levels of interest and support for space exploration (MacGregor and Slovic, 1994, 1995; Race, 1998). In general, respondents did not view sample return and biological risks from martian microbes as major risk issues; nonetheless they urged a conservative approach that considers returned materials as hazardous until proven otherwise. Subsequent surveys of microbiologists and other experts revealed similar concerns about biological risks from martian microbes, but indicated satisfaction that current methods of dealing with biological risks are satisfactory for handling martian material (Race and MacGregor, 2000). Another study analyzed claims made by mission supporters and opponents on the Internet about the plutonium energy source on the Cassini spacecraft (Rodrigue, 2001). That study focused on assertions and misinformation made online about topics including degree of risk, citizens' control over risk exposure, fairness of distribution of risks, benefits from the mission, and trust in government decision makers. Collectively, these kinds of studies help assess factual knowledge, levels of uncertainty, topics of concern, and differences in views among various audiences.

Although the lay public indicated little concern about biological contamination risks from a Mars sample-return mission, nuclear issues loomed larger than expected (Race, 1998). Nuclear systems in space exploration have had a long and successful history that stretches from the Apollo lunar missions of the 1960s through recent missions to Mars, Jupiter, and Saturn (e.g., Viking, Pathfinder, Galileo, and Cassini). Radioisotope heater units (RHUs) have been used to keep equipment warm and functioning in the extreme

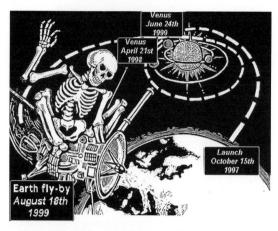

The Cassini space probe with 72.3lbs (32.8kg) of plutonium on board will come hurtling towards Earth at 42,000mph on 18th August 1999. NASA plans for it to pass 700 miles above Earth, but in the event of an accident it may come crashing into the atmosphere, burn up and disperse deadly plutonium over the world's population.

We Say: No to Nuclear Power in Space
No to Weapons in Space
Stop the Cassini Fly-by

... Cassini is part of the plan to increase the use of nuclear power in space and to fight wars from space ...

More Details from: Yorkshire CND, 22 Edmund Street, Bradford, BD5 0BH Tel: (01274) 730795	Local Contact Information

FIGURE 24.3 Some citizens groups are opposed to the launch of any rockets with nuclear materials onboard, as seen in this protest poster against the Cassini mission to Saturn. (Global Network Against Weapons and Nuclear Power in Space: www.space4peace.org.)

cold of space environments, and radioisotope thermo-electric generators (RTGs) have provided electric power to spacecraft and instruments far from the Sun. Looking ahead, the continuing or expanded use of nuclear technologies will be required for long-duration missions, both robotic and human, as well as for exploration to places where solar cells alone are insufficient. Additionally, proposals for the future use of nuclear propulsion (instead of chemical propellants) offer promise of speedier, long-distance missions to the outer planets.

While the space community may see nuclear power as a vital enabling technology, opponents view it differently. For example, protesters against the Cassini launch in 1997 and its subsequent Earth flyby in 1999 raised concerns about its payload of 33 kg of ^{238}Pt in the RTGs used to power the spacecraft during its multi-year mission to explore Saturn (Fig. 24.3). They argued that a launch or flyby accident could expose Earth's entire northern hemisphere to nuclear fallout capable of causing increased cancers and deaths, despite the fact that risk estimates by NASA and the US Department of Energy showed otherwise (NASA, 1995). Others have waged vigorous protests against the launch of nuclear materials of any type into space, whether for research missions in the Solar System or for a possible defensive strike against a menacing asteroid. These opponents argue the importance of avoiding even small steps toward the weaponization or militarization of space.

The most vigorous opponents seek to halt *all* missions with nuclear materials on board, regardless of their scientific merit. Tracking opinions about technology and space exploration, whether on the Internet, through survey research, or in face-to-face meetings, can help risk managers and decision makers become aware of the issues and opinions that may be amplified at a later time. Such awareness will be extremely important in the future as the public becomes increasingly familiar with the risks, benefits, and costs of astrobiology and space exploration.

24.3.4 Ethical, philosophical, and theological questions

In addition to the various legal and societal considerations mentioned in the foregoing sections, there are also persistent questions in the ethical, philosophical, and theological realms, all of which are beyond the professional purview of most scientists. As with biotechnology, astrobiology is not immune from having its underlying presumptions meticulously examined.

Philosophers and theologians interested in the possibility of extraterrestrial life have engaged in lively debates for millennia (Chapters 1 and 2). These discussions are in realms where competing hypotheses about origins of life and cosmology are evaluated from completely different perspectives, and where no amount of analytical expertise will necessarily resolve differences.

The various versions of the origin of life debate center around two basic hypotheses: biological determinism versus panspermia (Davies, 1998, 2002; see Chapter 6). In the former, the Universe is a bio-friendly place, where cosmic evolutionary processes lead to dynamic, chemically enriched environments that inevitably yield life and progress towards intelligence after a sufficiently long time. On the other hand, panspermia holds that life has been spread from one planet to another across interstellar distances after arising from some event of low probability. In this view intelligence likewise is a low-probability, random occurrence. During recent decades the details of the debate have undergone various elaborations, and at present biological determinism is the favored philosophy among astrobiologists.

Finding life on Mars someday may or may not help resolve the philosophical and theological debate. If we find evidence for past or present *Earth-like* life on Mars, it would be extremely interesting scientifically, but less so theologically or philosophically because it could be explained as the result of dispersal between neighboring bodies; the panspermia idea would then be a strong hypothesis. If, however, martian life were found to use a completely different biochemistry, it would be suggestive of an independent origin of life, with significant philosophical and theological implications. Similarly, the prospect of a credible, validated SETI signal would hold profound theological and philosophical meaning (Dick, 2002; Sections 26.9–26.10). While these debates are intellectually provocative, there are no direct implications for current exploration missions or astrobiological research. Even so, as Dick suggests, it may be time to seriously consider a "cosmotheology" in light of what we now know about the Universe and what we are likely to know in the future.

There are also ethical issues that must be considered, with more immediate implications for current astrobiology missions and activities both on Earth and elsewhere. As humans with an ability to rationally choose, we are responsible for the reasonably foreseen consequences of our actions. Unlike ethical dilemmas in biotechnology that mainly have ramifications at the level of applied personal ethics (e.g., medical intervention, human clinical trials, cloning, genetic counseling), the ethical dimensions of astrobiology are primarily those of normative ethics (Boylan, 2000), which concerns itself with norms or standards of human conduct in groups.

One of the earliest attempts to address ethical issues beyond Earth was published in a landmark study two decades ago that considered a host of social, environmental, human, and political dimensions of space exploration (Hargrove, 1986). Many of the moral and ethical issues raised by exploration and possible colonization within the Solar System remain as challenges today for astrobiology. For example, if we make contact with extraterrestrial life, how should our future actions be governed? What moral and ethical considerations relate to the possible discovery of extraterrestrial life (Race and Randolph, 2002)? Should we protect or exploit celestial resources and environments? What

FIGURE 24.4 Future activities such as colonization of Mars raise numerous ethical questions. (NASA.)

should be our approach to commercialization, tourism, or colonization on celestial bodies? Should extraterrestrial environments have special status for their preservation and protection (Fig. 24.4)?

It is interesting to pursue some of these questions in the context of environmental ethics, but this approach immediately runs into trouble because the entire field has been developed with *Earth* as the subject. Existing ethical systems are both anthropocentric and geocentric in nature, meaning they consider human beings as the most significant entity in the Universe, and are based on Earth as the center of perspective and valuation. There are essentially three basic approaches to current environmental ethics on Earth:

1. *preservation*, the belief that human action in nature should be minimized;
2. *stewardship*, the principle that humans must use nature wisely for their own benefit;
3. *intrinsic worth*, the supposition that there exists a class of objects that have intrinsic worth regardless of their utility or relationship to humans.

Rolston (1988) suggested that we still do not have an adequate ethic for our own planet and its biologically diverse communities of life. It is hardly surprising that we have not yet developed formal approaches applicable to extraterrestrial life and its environments. The difficulties arise because ethical systems are attached to a particular worldview, which clearly is in a state of flux when considering astrobiology. Looking ahead, it may be necessary to broaden the framing of ethical discussions and develop a "cosmocentric" ethic that incorporates considerations of more than just humans and the environments on Earth (Lupisella, 1997, 1998; Lupisella and Logsdon, 1997; Randolph *et al.*, 1997).

As a way of understanding the unusual and complicated ethical difficulties ahead, consider the situation on Mars. The current uncertainties about life on Mars raise issues about potential risks of back contamination from exploration missions. How do we determine an ethically acceptable level of risk to Earth when returning samples from Mars? Furthermore, if life exists on Mars, then we are faced with ethical questions of how to relate to it in our future activities. Does any such life belong to the martians, even if they are microbial?[3] If life on Mars is unique, does it have intrinsic value that warrants efforts to enhance its survival by deliberately induced planetary modifications? Some have suggested that near-term technologies could be used to warm Mars so that the planet would be habitable for Earth plants and microorganisms (McKay *et al.*, 1991). Would it be morally justified to implant Earth life there or attempt terraforming – planetary scale engineering to make it more Earth-like (McKay, 1990; Haynes and McKay, 1990; McKay and Marinova, 2001)? Will we someday have to decide whether to leave Mars for the martians, or introduce Earth-life to make Mars more habitable for us (McKay, 2001)? Replacing a martian life form, however simple, with life from Earth clearly represents a new extreme in biological interference and ethical dilemmas. If however, Mars proves to be lifeless, is there an imperative to preserve it unaltered, or could we utilize the planet for human purposes such as colonization? Do rocks and dirt and an atmosphere have value in and of themselves, making terraforming Mars unethical? In many ways, the situation has not changed much since Rolston (1986) suggested the need for serious "value explorations in space, for a philosophy of the solar [system]

environment to complement that of the biospheric environments." It will be critically important to consider the ethical dimensions of space activities well in advance of future, potentially irreversible, actions.

24.4 Procedures following discovery of extraterrestrial life

Current technological limitations, combined with a conservative approach to planetary protection, make it likely that, for the foreseeable future, we will do everything possible to avoid harmful cross-contamination and only minimally impact extraterrestrial environments. Even so, we must begin to prepare for the ultimate in success – what if we find what we are looking for? Considering current Mars exploration scenarios, it may be possible to discover extraterrestrial life in several ways – either robotically on the planet, in samples returned to Earth, or during future human missions. Most discussions to date have emphasized practical aspects of how to undertake robotic searches for evidence of life in a safe and appropriate manner, but provide no clear guidance on what to do if and when life is detected.

Despite a dramatic increase in Solar System missions of all types during the past decade, discussions about the societal, ethical, and practical implications of discovering *non*-intelligent, microbial life have not been formally undertaken. Since discovery of extraterrestrial life in the Solar System could occur in widely different locations or ways, it is important to anticipate what kinds of operational and long-term considerations might be appropriate for various scenarios. One way to focus the discussions about discovery of microbial life is to take a lesson from the SETI community (the Search for ExtraTerrestrial Intelligence – Chapter 26), which has developed guidelines on how to respond following a detection of extraterrestrial intelligence (ETI).

More than a decade ago, recognition of the profound significance of discovering ETI prompted leaders in the SETI community to initiate discussions about how to respond if and when a signal is detected. Rather than presuming anything about the nature of the extraterrestrial intelligence itself, they focused instead on the human response anticipated in the face of a discovery. Over a period of several years the SETI Committee of the International Academy of Astronautics developed a "Declaration of principles concerning activities following the detection of extraterrestrial intelligence" to serve as a guideline. The Declaration, which is not legally binding and has no enforcement provisions, was approved by the Academy in 1989 and subsequently endorsed by

---

[3] Carl Sagan (1980: 130) asked this question and then firmly answered yes: "The existence of an independent biology on a nearby planet is a treasure beyond assessing, and the preservation of that life must, I think, supersede any other possible use of Mars."

TABLE 24.2. Suggested guidelines related to discovery of non-intelligent extraterrestrial life

---

- When detected, do no harm; take no intrusive action without consultation
- Verify and confirm that the lifeform is truly extraterrestrial
- Prior to public announcement, confirm the discovery through additional independent research
- If discovery is credible, inform the UN and appropriate government agencies
- Make all data available to the scientific community
- Protect and preserve the extraterrestrial lifeform
- Undertake no further missions or activities prior to international consultation

---

Source: Adapted from Race and Randolph (2002.)

numerous major organizations (complete text in Billingham *et al.*, 1999). The key points of the Declaration provide step-by-step guidance as to how to proceed when an alleged signal is detected. In brief, the Principles are:

- verify the source as extraterrestrial;
- prior to public announcement, confirm the discovery by independent observations at other sites;
- if the signal is still credible, inform the United Nations and appropriate government and professional bodies;
- announce the confirmed detection promptly, openly, and widely via scientific and public media channels, with the privilege of announcement reserved for the discoverer;
- make all data necessary for confirmation of the detection available to the scientific community;
- continue to confirm and monitor the discovery, and permanently store all data;
- if the detection is in the form of electromagnetic signals, protect the appropriate frequencies through international agreement;
- make no response to the signal without detailed consultation.

As currently formulated, the SETI principles amount to a set of operational guidelines for initially disseminating information about a discovery, but they deliberately sidestep any recommendations for the longer term. Instead they prescribe a policy of "consultation" prior to sending a response to any signal, with the intent of involving a wide range of governments, agencies, and peoples.

When analyzed in detail (Race and Randolph, 2002), the SETI principles are based on a foundation of just two

ethical principles: (1) follow proper scientific procedures with honesty and integrity, and (2) communicate widely, seeking the participation of all members of the Earth community. In contrast, the underlying ethical principles applicable to the discovery of microbial, *non*-intelligent extraterrestrial life are actually more complex than for intelligent life because they must also address concerns about the risks of back contamination of Earth, and questions about the obligations humans may have to preserve extraterrestrial life and ecosystems. These ethical differences arise because most of the discovery scenarios for *non*-intelligent life involve space travel in some form or another, whereas searches for electromagnetic signals are Earth-based.[4]

Using the SETI principles as a guide, Race and Randolph (2002) have developed a set of preliminary guidelines suggesting how to respond to the discovery of *non*-intelligent extraterrestrial life (Table 24.2). The guidelines incorporate four underlying ethical principles:

- cause no harm to Earth, its life, or its diverse ecosystems;
- respect the extraterrestrial ecosystem and do not substantively or irreparably alter it (or its evolutionary trajectory);
- follow proper scientific procedures with honesty and integrity during all phases of exploration;
- ensure international participation by all interested parties.

The preliminary guidelines based on these ethical principles (Table 24.2) are intended to prompt further discussion, and to lead ultimately to development of more comprehensive guidelines.

24.5 The codependence of scientists and society

Aside from teaching about astrobiology issues in the classroom, it is likely that the majority of scientists in astrobiology will only occasionally, if ever, become directly involved in sociopolitical processes or ethical debates. Even so, they should try to develop a basic understanding and appreciation of the world beyond their labs, beyond hypothesis testing and scientific theories. It is counterproductive to think that only scientists know enough to make the right decisions. Non-experts have legitimate interests and a role to play in determining our collective societal future, even if their

---

[4] We do not here consider the possibilities of interstellar travel that might be instigated by an initial electromagnetic discovery.

opinions or scientific sophistication are dissimilar from those of the experts. Non-expert perceptions of potential risks and benefits can also have significant impacts on either public support or opposition to funding research.

Sinsheimer (1978) wrote a provocative essay entitled "The presumptions of science" that still provides sage insight to astrobiologists. In this he argued that there is an intensity of focus in the scientific perspective that is both its immediate strength and its ultimate weakness. All too often, while engrossed in the search for knowledge, scientists are blind to other concerns, adopting the position that more knowledge is the key to solving all human problems. In pondering the future directions of any scientific endeavor, we cannot afford to ignore the social and cultural contexts within which science and technology are embedded, lest we render society incapable – or unwilling – to support the scientific enterprise on which we depend.

Rather than hope that societal concerns will go away, scientists should prepare themselves to tackle the diverse and legitimate non-scientific questions that flow from their work. Although it may sometimes seem otherwise, the public does not necessarily want to stop what scientists and engineers are capable of doing, but it does want to participate in deliberations and decision-making at a substantive level. There is an implicit pact between scientists and the public, one with relentless tensions in the sociopolitical realm. As noted by Grobstein (1986):

> For scientists, who find their activities of increasing social consequence, the issue is how to manage their side of the implicit moral contract with society to do no harm while preserving sufficient independence to promote vigorous inquiry. For non-scientists, whose values and ways of life are increasingly pressured by the seemingly uncontrollable thrust of science and science-derived technology, the question is how to derive benefits but still have a voice in assessing risk and choosing among options that are often presented in incomprehensible terms and contexts.

As astrobiology explores new places far from Earth, bolstered by our ever-growing scientific understanding and technological capabilities, we must always be mindful of this codependence between science and society. Our very future depends upon it.

REFERENCES

Billingham, J. Heyns, R., Milne, D., et al. (eds.) (1999). *Social Implications of the Detection of an Extraterrestrial Civilization.* Mountain View, CA: SETI Press, SETI Institute.

Boylan, M. (2000). *Basic Ethics.* Upper Saddle River, NJ: Prentice Hall.

Bruch, C. W., Setlow, R. B., and Rummel, J. D. (eds.) (2001). *Workshop 2a: Sterilization Workshop, Proceedings and Final Report NASA/CP 2001-210924.* Washington, DC: NASA.

Cypser, D. A. (1993). International law and policy of extraterrestrial planetary protection. *Jurimetrics: Journal of Law, Science and Technology,* 33(2), 315–339.

Davies. P. (1998). *Are We Alone? Philosophical Implications of the Discovery of Extraterrestrial Life.* New York: Basic Books.

Davies, P. C. W. (2002). Biological determinism, information theory, and the origin of life. In *Many Worlds: the New Universe, Extraterrestrial Life and the Theological Implications,* S. J. Dick (ed.). Philadelphia, PA: Templeton Foundation Press.

DeVincenzi, D. L. and Bagby, J. R. (eds.) (1981). *Orbiting Quarantine Facility (OQF): the Antaeus Report, NASA SP-454.* Washington, DC: NASA.

DeVincenzi, D. L., Race, M. S., and Klein, H. P. (1998). Planetary protection, sample return missions and Mars exploration: history, status and future needs. *J. Geophys. Res.,* 103(E12) 28, 577–585.

DeVincenzi, D. L., Bagby, J., Race, M., and Rummel, J. D. (1999). *Mars Sample Quarantine Protocol Workshop, NASA CP-1999-208772.* Moffett Field, CA: NASA Ames Research Center.

Dick, S. J. (1997). Humanity and extraterrestrial life. In *Life on Mars. What Are the Implications? Proceedings of a Symposium (11/26/96),* Logsdon, J. M., Ziegelaar, B., and Burns, A. M. (eds.), Proceedings of a Symposium "Life in the Universe: what can the Martian fossil tell us?", Nov. 22, 1996. Washington, DC: Space Policy Institute, George Washington University.

Dick, S. J. (1998). *Life on Other Worlds: the Twentieth Century Extraterrestrial Life Debate.* Cambridge: Cambridge University Press.

Dick S. J. (ed.) (2002). *Many Worlds: the New Universe, Extraterrestrial Life and the Theological Implications.* Philadelphia, PA: Templeton Foundation Press.

Grobstein, C. (1986). Asilomar and the formation of public policy. In *The Gene-Splicing Wars,* ed. R. Zilinskas and B. Zimmerman. New York: Macmillan Publishing Co.

Hargrove, E. C. (ed.) (1986). *Beyond Spaceship Earth: Environmental Ethics and the Solar System.* San Francisco, CA: Sierra Club Books.

Haynes, R. H. and McKay, C. P. (1990). Should we implant life on Mars? *Scientific American,* December, p. 144.

Lupisella, M. (1997). The rights of Martians. *Space Policy,* 13(2), 89–94.

Lupisella, M. (1998). Astrobiology and cosmocentrism. *Bioastronomy News,* IAU Commission 51, 10(1), 1–2, 8. (Published by The Planetary Society, Pasadena, CA.)

Lupisella, M. and Logsdon, J. (1997). Do we need a cosmocentric ethic? International Astronautical Federation

Congress (Turin, Italy), Paper IAA-97-IAA.9.2.09. Paris: International Astronautical Federation.

MacGregor, D. G. and Slovic, P. (1994). The Planetary Exploration Survey. *Planetary Report*, **14**(4), 20a–21a.

MacGregor, D. G. and Slovic, P. (1995). The Planetary Exploration Survey: What society members think about planetary protection. *Planetary Report*, **15**(2), 4–6.

McKay, C. P. (1990). Does Mars have rights? An approach to the environmental ethics of planetary engineering. In *Moral Expertise*, pp. 184–197, ed. D. MacNiven. New York: Routledge.

McKay, C. P. (2001). Let's put Martian life first. *Planetary Report*, **21**, July/August, 4–5.

McKay, C. P. and Marinova, M. M. (2001). The physics, biology, and environmental ethics of making Mars habitable. *Astrobiology*, **1**, 89–109.

McKay, C. P., Toon, O. B., and Kasting, J. F. (1991). Making Mars habitable. *Nature*, **352**, 489–96.

NASA (1995). *Final Environmental Impact Statement for the Cassini Mission*. Washington, DC: Solar System Exploration Division, NASA.

NASA (2001). *Implementing the National Environmental Policy Act and Executive Order 12114*, Procedures and Guidelines NPG 8580.1, (Effective Dates: 11/26/01 to 11/26/06). Washington, DC: NASA.

NASA (2003). *Astrobiology Roadmap*. Moffett Field, CA: NASA Astrobiology Institute, NASA Ames Research Center. astrobiology.arc.nasa.gov/roadmap/roadmap.pdf.

NASA (2006). *NASA Education Strategic Coordination Framework: A Portfolio Approach*, Washington, DC: NASA, education1.nasa.gov/about/strategy/ (accessed April 19, 2007).

NRC (1989a). *Field Testing Genetically Modified Organisms: Framework for Decisions*. Washington, DC: National Academy Press.

NRC (1989b). *Improving Risk Communication*. National Research Council. Washington, DC: National Academy Press.

NRC (1992). *Biological Contamination of Mars: Issues and Recommendations*. Task Group on Planetary Protection, Space Studies Board, NRC. Washington, DC: National Academy Press.

NRC (1996). *Understanding Risk*. National Research Council. Washington, DC: National Academy Press.

NRC (1997). *Mars Sample Return: Issues and Recommendations*. Task Group on Issues in Sample Return. Washington, DC: National Academy Press.

NRC (1998). *Evaluating the Biological Potential in Samples Returned from Planetary Satellites and Small Solar System Bodies*, Space Studies Board, NRC. Washington, DC: National Academy Press.

NRC (2000). *Preventing the Forward Contamination of Europa*. Washington, DC: National Academy Press.

NRC (2006). *Preventing the Forward Contamination of Mars*, Washington, DC: National Academy Press (available online: www.nap.edu).

Race, M. S. (1996). Planetary protection, legal ambiguity and the decision-making process for Mars sample return. *Adv. Space Res.*, **18**, No.1/2., 345–350.

Race, M. S. (1998). Mars sample return and planetary protection in a public context. *Adv. Space Res.*, **22**(3), 391–399.

Race, M. S. and Mac Gregor, D. G. (2000). Integrating public perspectives in sample return planning. *Adv. Space Res.*, **26**(12), 1901–1909.

Race, M. S. and Randolph, R. (2002). The need for operating guidelines and a decision-making framework applicable to the discovery of non-intelligent extraterrestrial life. *Adv. Space Res.*, **30**(6), 1583–1591.

Race, M. S., Criswell, M. E., and Rummel, J. D. (2003). *Planetary Protection Issues in the Human Exploration of Mars*. Paper Number 2003–01–2523. International Conference on Environmental Systems (ICES), SAE Technical Papers, www.sae.org.

Randolph, R., Race, M. S., and McKay, C. (1997). Reconsidering the ethical and theological implications of extraterrestrial life. *CTNS Bulletin*, **17**(3), 1–8. Berkeley, CA: Center for Theology and Natural Sciences.

Rifkin, J. (1998). *The Biotech Century: Harnessing the Gene and Remaking the World*. New York: Penguin Putnam Inc.

Robinson, G. (1992). Exobiological contamination: the evolving law. *Annals of Air and Space Law*, **17**, 325–367.

Rodrigue, C. M. (2001). The Internet and plutonium on board the Cassini–Huygens spacecraft. *Risk: Health, Safety & Environment*, **12**, 221–254.

Rolston, H. (1986). The preservation of natural value in the solar system. In *Beyond Spaceship Earth: Environmental Ethics and the Solar System*, H. C. Hargrove (ed.). San Francisco, CA: Sierra Club Books.

Rolston, H. (1988). *Environmental Ethics: Duties to and Values in the Natural World*. Philadelphia, PA: Temple University Press.

Rummel, J. D., Race, M. S., DeVincenzi, D. L. (eds.) (2002). *A Draft Test Protocol for Detecting Possible Biohazards in Martian Samples Returned to Earth*. NASA/CP-2002-211842. Washington, DC: NASA.

Sagan, C. (1980). *Cosmos*. New York: Random House.

Shostak, S. (1998). *Sharing the Universe: Perspectives on Extraterrestrial Life*. Berkeley, CA: Berkeley Hills Books.

Sinsheimer, R. L. (1978). The presumptions of science. *Daedalus*, **107**, 23–35.

Schacter, B. (1999). *Issues and Dilemmas of Biotechnology: a Reference Guide*. Westport, CT: Greenwood Press.

Van Ballegoyen, A. F. (2000). Ownership of the Moon and Mars? *Ad Astra*, Jan/Feb, 35–37.

Zubrin, R. and Wagner, R. (1996). *The Case for Mars: the Plan to Settle the Red Planet and Why We Must*. New York: Simon and Schuster.

25 Planetary protection: microbial tourism and sample return

John D. Rummel
NASA Headquarters

25.1 Introduction

Picture a future triumph in robotic space exploration: in a complex mission to Mars, a sample has been collected from the martian subsurface near a newly discovered hydrothermally active site at 30°N latitude. Ten years in the detailed planning and execution, the mission's Earth return capsule with its sample canister has landed in the Utah desert. The sample is now under extensive analysis and testing in an ultra-clean containment facility – and initial observations have shown positive indications that it contains life. Only after later testing is completed, checked, rechecked, and repeated is it shown unequivocally that the lifeform contained within the sample is a soil bacterium common to the dirt of an old Soviet launch facility in Baikonur, Kazakhstan, and which has apparently been alive on Mars since a spacecraft crash-landed there in 1972 . . .

Or picture, as did novelist Michael Crichton (1969) in the very year of the first lunar sample-return mission (Apollo 11), a spacecraft returning to Earth containing a dangerous extraterrestrial organism – *The Andromeda Strain* – not related in any way to Earth-life and operating by rules scarcely understood even after hundreds of humans have met their grisly demise . . .

Once you have those events in mind, you are developing a feel for what planetary protection might be, and what it is meant to prevent. The case of Earth organisms inadvertently being taken to another Solar System body is known as *forward contamination*, while the case of alien microbes being brought to Earth is known as *backward* or *back contamination*. Avoiding forward contamination is necessary to preserve planetary conditions for future biological and organic constituent exploration, and perhaps for significant ethical reasons, while avoiding back contamination is simple prudence – a necessary step to protect Earth and its biosphere from potential extraterrestrial sources of contamination – especially important when that potential contamination could have consequences to our biological systems.

Preventing interplanetary cross-contamination may be conceptually simple, but its practical implementation is complicated by our ignorance about two main aspects that govern the success or failure of planetary protection provisions – the nature and ultimate capabilities of life, even Earth-life, and the nature and extent of suitable planetary environments for life – and the extent of their intersection. It is a bit like a biological Heisenberg Uncertainty Principle, where we can't know about life on other worlds unless we or our machines (and their associated lifeforms) are there to observe it, but we may destroy either that life or ourselves in the process of making those observations. The major problem is, as de Fontenelle said in his 1686 book, *Conversations on the Plurality of Worlds*, "the trouble is, we want to know more than we can see" (Section 1.4). Such is the essential challenge – in order to know what there is that might need protection, and whether there is anything or anybody out there to protect against, we have to go out there and "see" for ourselves. And so far, what we have seen – both on Earth and on other Solar System bodies – is of compelling interest. We are learning that life on Earth is amazingly robust under extreme conditions (Chapters 14 and 15), and we are beginning to learn about locations on or under the surfaces of other planets and moons where Earth-life, at least, might make a very reasonable living. Whether such bodies indeed contain their own life, and if so, whether it is related to ours, are matters of critical astrobiological (and planetary protection) interest.

25.2 Microbial tourists and invaders

25.2.1 Tourists?

The practical aspects of a microbe getting around the Solar System are daunting. Between the planets is an

Planets and Life: The Emerging Science of Astrobiology, eds. Woodruff T. Sullivan, III and John A. Baross. Published by Cambridge University Press. © Cambridge University Press 2007.

environment almost completely devoid of both atmospheric gases and liquid water, exposed to harsh ultraviolet, X-ray, gamma-ray, and particle radiation, and subject to temperature extremes both above and below 0 °C. Earth organisms traveling on their own are sure to be killed by one or the other of these conditions – yet natural and human-provided "tourism" opportunities still exist for microbial life.

Earth microorganisms are tough, and have been found living in extreme environments that would quickly kill most large organisms. Some of these extreme environments may have analogues on other planets or moons. As discussed in detail in Chapter 14, microorganisms found at deep-sea hydrothermal vents characteristically can be found at pressures greater than 250 atm, and have been shown in the laboratory to grow at temperatures up to 121 °C, which due to the limitations of microbial culture techniques is still not likely to be the highest growth temperature found in nature. Even in surface waters, microorganisms can be found in volcanic pools and geothermal vents in boiling water, while at the other extreme there are lichens living inside Antarctic dry-valley sandstone in areas where the yearly average daily temperature is well below freezing. Bacteria are also known to be active in brines at −15 °C within pores inside Arctic ice (Chapter 15). Microbes have been found in communities deep (>3km) below the continental surface and the seafloor, and living in highly acidic mine seepage with a pH of <1. Moreover, in space itself, microbes such as the spore-forming soil bacterium *Bacillus subtilis* have been shown to survive for extended periods under desiccation, as well as under exposure to all but the harshest UV radiation (Nicholson *et al.*, 2000). Finally, other microbes such as *Deinococcus radiodurans* have been shown to withstand Mrad doses of gamma rays (Chapter 14).

It is entirely possible that some of these tough Earth microbes have left the Earth in the past, and that some of them have ended up on other Solar System bodies. Building on earlier work dating back to Melosh (1984), Mileikowsky *et al.* (2000) examined the potential of a natural transfer of microorganisms among Solar System bodies based on high-velocity ejection of soil and rock resulting from impact bombardment by comets, asteroids, and other small bodies. The basic scenario is that a microbe ensconced inside a bit of rock is able to survive (i) getting knocked off its home planet, (ii) a many-Myr trip in orbit between Mars and Earth (although some trip times are much shorter), (iii) an eventual fiery entry into the much thicker Earth atmosphere, and (iv) a

crash on to the surface (Section 18.4.4). They paid particular attention to the potential transfer of organisms from Mars to Earth, where their main conclusion was "that if prokaryote microbes existed or exist on Mars, viable transfer to Earth must be considered not only possible but highly probable." They also estimated that while "the number of ejecta from Earth landing on Mars from 4.0 Ga to present time ... was one to two orders of magnitude lower than the number of martian meteorites landing on Earth during the same period of time," it was "nevertheless still very substantial, about a billion during the last four billion years."[1] Further studies of places like Mars may shed light on the existence and fate of any transplanted Earth microbes, as well as the question of whether any extraterrestrial microbes could have ever reached Earth. It is a major goal of planetary protection controls on forward contamination to preserve the planetary record of such natural processes and not erase it through human-caused microbial introductions.

And where are those introductions most likely to cause harm? On just those planetary bodies that are of most interest to astrobiology, because they show evidence of the materials and environments that support life on Earth – and thus might support any Earth-life transported to those bodies. As discussed below, the leading candidates at this time for extraterrestrial life are Mars and Europa. Either destination might prove irresistible to a microbial tourist which has just spent years aboard a cold, dry, aluminum robot. For example, a recent study by Cockell *et al.* (2005) found that a hardy cyanobacterium (*Chroococcidiopsis* sp. 029, found in the driest Earth deserts) could survive for many hours (testing was not done for longer periods) even under the high ultraviolet flux at the martian surface if it was shielded by only 1 mm of soil.

25.2.2 Alien invaders?

It takes little imagination to anticipate that Earth organisms might grow and thrive if presented with the appropriate environment and energy sources, so issues of forward contamination control are typically controversial only because of their operational and monetary costs (Section 25.6). Back contamination control, on

---

[1] The "conveyor belt" for interplanetary delivery runs much more strongly from Mars to Earth than vice versa because (1) Mars's gravity is much weaker than Earth's; (2) Earth's effective cross section for impacts is much larger than Mars's (Chapter 3); and (3) Earthward travel is aided by the Sun's gravity.

the other hand, is undertaken within the boundaries of the plausibility of life elsewhere, knowing that natural interchanges may have already occurred among the planets, but realizing that even on Earth there are living organisms that are disagreeable, or even dangerous, when transported to the wrong place. What are the chances that extraterrestrial life exists in this Solar System? It cannot be ruled out, and our ignorance about both life and other Solar System environments is profound. Mars, for example, is a world that seems to possess all of the materials and some of the environments that we associate with life on Earth, including accessible energy sources and occasional access to imported Earth rocks. While the planet's current surface is quite cold and extremely dry, Mars has been considered by some to be nearly Earth's twin, especially early in Solar System history. Mars shows geological evidence of ancient watercourses, and also possible evidence of present-day water on its surface (Malin and Edgett, 2000, Malin *et al.*, 2006). In early 2004, the Mars Exploration Rover *Opportunity* yielded further compelling evidence that Meridiani Planum, at least, was once the home of a salty, acidic sea, replete with evidence of water flow, sulfate minerals, and hematite concretions (Squyres *et al.*, 2004). Moreover, conditions *under* the martian surface may be quite different from those on the surface. See Chapter 18 for a full discussion. It is interesting to note that the discovery of abundant, macroscopic life surrounding deep-sea hydrothermal vents on Earth was not made until seven months *after* NASA's Viking missions had landed on Mars in 1976, and some had declared Mars a dead planet (see Horowitz, 1986). The astrobiological exploration of Mars is just beginning.

At an even earlier stage in its astrobiological exploration is Europa (Chapter 19). There is strong evidence for a water ocean below its ice-covered surface. If a means to exchange reduced material from the interior with oxidized material from the surface also exists, Europa may be a compelling site for life. And because Jupiter's deep gravity well and harsh radiation environment probably severely limit the potential for natural Earth–Europa cross-contamination, it may be an even more important site than Mars for examining the prospects for an *independent* origin of life in our Solar System. Europa is a good example of a world that even a few decades ago was not considered as a potential life-site, being too far from the Sun to have a liquid-water surface. Now, however, there is growing appreciation that Europa is part of a continuum of worlds where plentiful liquid water is hidden from view, and where sources of organic material and energy may be available

for life. Moreover, only recently have we realized that Earth may have, for short periods at least, appeared Europa-like from space – completely covered with ice ("Snowball Earth"; Section 4.2.4).

Elsewhere in the Solar System may be other worlds capable of supporting life, under conditions that are still being discovered. One example is Saturn's largest moon Titan, a Mercury-sized world with a nitrogen-rich atmosphere denser than that of Earth, with abundant atmospheric methane but at a chillingly cold 94 K (Chapter 20). The titanian stratosphere is exposed to solar ultraviolet radiation, which generates a suite of saturated and unsaturated hydrocarbons, as well as nitriles. The products of methane photolysis condense as aerosols and reach the surface, where some hydrocarbons can exist as liquids and others as solids. How these may be recycled, and whether or not subsurface liquid water might exist, is dependent on available energy sources (ultraviolet light, cosmic rays, accretional and radiogenic heating), chemical composition, and potential cryovolcanism. The Cassini mission, and in particular its Huygens probe that descended to the surface of Titan in 2005, has provided much more information (Chapter 20).

25.3 A planetary protection policy

The implications of planetary protection were appreciated long before space travel was a reality, most notably in H. G. Wells's (1898) *The War of the Worlds* with the triumph of Earth microbes over invading martians – saving the day for earthlings. The post-Sputnik timeframe saw the introduction of quarantine standards by the International Council of Scientific Unions (ICSU) in 1958 (CETEX, 1959), as well as recommendations for non-contaminating spaceflight practices by the US National Academy of Sciences (Derbyshire, 1962; Section 2.2.1). For a review of the early history, consult Phillips (1974).

By 1967 – even prior to the first robotic landing on any Solar System body other than the Moon – there was general agreement that interplanetary contamination should be regulated. Article IX of the United Nations Outer Space Treaty, which entered into force on October 10, 1967, placed obligations on spacefaring nations that:

> ... parties to the Treaty shall pursue studies of outer space including the Moon and other celestial bodies, and conduct exploration of them so as to avoid their harmful contamination and also adverse changes in the environment of the Earth resulting from the introduction of extraterrestrial matter and, where

TABLE 25.1 COSPAR Planetary Protection Mission categories

Mission Category	Mission type	Planet status
I	Any	Not of direct interest for understanding the process of chemical evolution. No protection of such planets is warranted (no requirements).
II	Any	Of significant interest relative to the process of chemical evolution, but having only a remote chance that contamination by spacecraft could jeopardize future exploration.
III	Flyby, orbiter	Of significant interest relative to the process of chemical evolution and/or the origin of life, or having a significant chance of contamination by spacecraft that could jeopardize a future biological experiment.
IV	Lander, probe	As above.
V	Earth return	Any Solar System body from which a sample is to be returned.

necessary, shall adopt appropriate measures for this purpose. (UN, 1967.)

In anticipation of a coming space age, ICSU formed an interdisciplinary committee on space research (COSPAR) that from its inception has been the focal point of much of the international discussion and consensus on planetary protection issues. Both COSPAR and the International Astronautical Federation consult with the United Nations Committee on the Peaceful Uses of Outer Space (COPUOS). These discussions have resulted in current COSPAR planetary protection policy (Rummel, 2002; www.cosparhq.org/ppp), as well as NASA policy (NASA, 1999a, b).

Planetary protection measures associated with specific missions and targets are subject to continual re-evaluation as we refine our knowledge of environments that may contain life or that could be contaminated by Earth-life. Also, the ability of mission personnel to characterize biological contamination has been revolutionized over the last several decades through the advent of molecular methods that allow markedly superior measurement of the sources of forward contamination. For NASA, policy changes are based on internal and external recommendations, most notably from the Space Studies Board (SSB) of the US National Academy of Sciences.

Policies to avoid biological and organic contamination of other worlds are based on five different categories of space missions, depending on the nature of the mission and its target body (Table 25.1). For Category I missions, where no controls are deemed necessary, none of these measures need be taken for planetary protection purposes. For Category II missions, where

the target body is of interest relative to organic chemical evolution and the origin of life, but biological contamination is not thought to be possible, a mission must provide careful documentation of spacecraft trajectories and an inventory of organic materials on board, and possibly also archival storage of certain spacecraft materials. In certain special cases, there may be a restriction on organic materials carried. For Category III missions, which fly by or orbit a planet that could be contaminated by Earth organisms, the measures include those for Category II, plus others that are more restrictive. Spacecraft operating constraints such as a biased trajectory to avoid launch vehicle impact and minimum orbital lifetime requirements may be imposed, and a strict reduction in the amount of biological contamination allowed on the spacecraft may be required if the mission cannot meet orbital lifetime requirements. For landed spacecraft in Category IV, all of the restrictions in Categories II and III (save orbital lifetime) may be imposed on the mission, and in addition the restrictions on biological contamination are generally more severe, and may require a comprehensive decontamination of the spacecraft using sterilizing procedures such as dry heat processing. Finally, Category V missions may either be judged as "unrestricted Earth return" missions, with no additional planetary protection requirements on the return portion of the mission, or as "restricted Earth return" missions. These latter missions may include all of the restrictions placed on Category II, III, and IV missions, plus severe restrictions on the handling of returned samples until their biological status is determined.

TABLE 25.2 Planetary protection studies by the US Space Studies Board

1992:   *Biological Contamination of Mars: Issues and Recommendations.* Measures to protect Mars from contamination by Earth organisms, as well as overall policy guidance (SSB, 1992).

1997:   *Mars Sample Return: Issues and Recommendations.* Specific advice on handling returned samples from Mars (SSB, 1997).

1998:   *Evaluating the Biological Potential in Returned Samples from Planetary Satellites and Small Solar System Bodies: Framework for Decision Making.* Considerations for sample return missions from small bodies, including places like Europa, asteroids, and comets (SSB, 1998).

2000:   *Preventing the Forward Contamination of Europa.* Measures to prevent the contamination of Europa by Earth organisms (SSB, 2000).

2002:   *The Quarantine and Certification of Martian Samples.* Actions to implement containment and biohazard testing measures recommended in 1997 (SSB, 2002a).

An example of a Category I mission was the NEAR-Shoemaker mission, which was launched by NASA in 1996 to rendezvous with the asteroid Eros – considered to be an S-type asteroid with no particular interest to the study of chemical evolution and the origin of life, and therefore not requiring protection under the policy.

For the Galileo mission to Jupiter and its moons, on the other hand, the situation was different. Galileo launched in 1989 with an expectation that it would deploy a probe into Jupiter itself, and also make close approaches to each of the four largest satellites: Io, Europa, Ganymede, and Callisto. At the time of launch there was no concern about contaminating Jupiter, though there were theoretical reasons to think that Europa, in particular, might have the potential for a liquid-water ocean. Accordingly, the Galileo mission was assigned to planetary protection Category II, but a specific provision was made to negotiate an end-of-mission to preclude impact with any satellite that might require more protection than Category II. During the mission, it was determined that Europa, certainly, and possibly also Ganymede and Callisto, might be affected by contamination on the spacecraft – and the decision was made to preclude a possible future impact on these bodies by ending the Galileo mission with a plunge into Jupiter's atmosphere, which occurred on 21 September 2003.

25.4 Recommendations to NASA on contamination control

NASA's planetary protection policy provides for structured advice and recommendations from two different sources – primary recommendations from the SSB, as noted above, and programmatic and operational advice from the Planetary Protection Advisory Committee of the NASA Advisory Council. The SSB has been providing recommendations on planetary protection to NASA for decades. Table 25.2 lists five reports that have been provided since 1992, covering the forward contamination of Mars and Europa and addressing back-contamination issues and requirements for Mars and the small bodies of the Solar System, including Europa. A short summary of each of the reports is provided below.

Forward Contamination of Mars (SSB, 1992). This report examined results from the Viking landers that arrived on Mars in 1976, as well as from Viking and Mariner program orbiters. Juxtaposed with the picture of Mars that emerged were the capabilities of various organisms found in extreme environments on Earth, many of which were completely unknown in 1976. On balance, however, the report's authors "unanimously agreed that it is *extremely unlikely that a terrestrial organism could grow on the surface of Mars*" [italics in the original]. They thus recommended that full "Viking-level" sterilization (a level of cleanliness equivalent to a spacecraft with fewer than ~30 aerobic spores aboard; see Section 25.6) is not required for missions to the martian surface, unless life-detection is a goal. They also recommended that new technologies to detect life (in particular, molecular methods) are important and should be adopted to measure a spacecraft's *bioload* (the kind and number of organisms accompanying a spacecraft). They further recommended that "a sequence of unpiloted missions to Mars be undertaken well in advance of a piloted mission," since "missions carrying humans to Mars will contaminate the planet."

Forward Contamination of Europa (SSB, 2000). This report (after the Galileo mission) recommended that Europa-bound spacecraft be cleaned, sterilized, and/or subject to sufficient radiation prior to contact with Europa so that the probability of contaminating a possible europan ocean with viable terrestrial organisms is less than 10^{-4}. The report concluded that NASA's then current cleaning and sterilization techniques were satisfactory to meet the needs of future space missions to Europa, but that current culture-based methods used to determine the bioload on a spacecraft should be supplemented by screening tests for specific types of extremophiles, such as radiation-resistant organisms. Echoing the 1992 Mars report, the report stated that modern molecular methods, such as those based on the polymerase chain reaction, might prove to be quicker and more sensitive to detect and identify biological contamination than NASA's established culturing protocols. Also recommended were investigations to reduce the uncertainty in calculating the probability of contaminating Europa – especially with respect to the likelihood that organisms might survive the jovian radiation environment, or might have access to the europan subsurface if deposited on Europa's surface ice.

Mars Sample Return (SSB, 1997). This report concluded that there was a very small, yet nonzero, probability that a returned martian sample would contain a hazardous lifeform, but that nonetheless a conservative approach was essential. Accordingly, the report recommended that (1) samples returned from Mars should be contained and treated as though potentially hazardous until proven otherwise; (2) if sample containment cannot be verified en route to Earth, the sample and spacecraft should either be sterilized in space or not returned to Earth; (3) the integrity of sample containment should be maintained through re-entry and transfer to a receiving facility; (4) controlled distribution of unsterilized materials should only occur if analyses determine that the sample does not contain a biological hazard; and (5) planetary protection measures adopted for the first sample return should not be relaxed for subsequent missions without thorough scientific review and concurrence by an appropriate independent body. The report stressed the importance of avoiding "false positives," i.e., positive indications regarding life on Mars. when in fact one was being misled by contamination from Earth. It also advocated that NASA thoroughly inform the public about the mission and the precautions taken with the sample.

TABLE 25.3 Space Studies Board questions on parameters for life (SSB, 1998)

Does the preponderance of scientific evidence indicate that:

1. there was never liquid water in or on the target body?
2. metabolically useful energy sources were never present?
3. there was never sufficient organic matter (or CO_2 or carbonates and an appropriate source of reducing equivalents) in or on the target body to support life?
4. subsequent to the disappearance of liquid water, the target body has been subjected to extreme temperatures (i.e., $>160\,°C$)?
5. there is or was sufficient radiation for biological sterilization of terrestrial life forms?
6. there has been a natural influx to Earth, e.g., via meteorites, of material equivalent to a sample returned from the target body?

For containment procedures to be necessary, an answer of "no" needs to be returned to all six questions.

Small Body Sample Return (SSB, 1998). This report established a framework whereby NASA could judge the likely safety of returning a sample from any particular target body. The six questions in Table 25.3 provide criteria for the possible presence of extant life on a target body and the potential danger of returning a sample to Earth. The report distinguished between (a) bodies from which samples could be returned uncontained and where no sample containment and handling is warranted beyond what is needed for scientific purposes, and (b) planetary satellites and small Solar System bodies whose samples must be contained and treated as potentially hazardous until proven otherwise. In the latter case, sample return provisions for contained samples are the same as recommended for Mars returned samples. The recommendation for containment of interplanetary dust particles was dependent on the nature of their parent body and on their time of exposure to the space environment. The report recommended, with a high degree of confidence, that no containment should be required for our Moon, Io, and new comets (e.g., the Stardust comet-sample-return mission); and, with a lesser degree of confidence, that no containment should be required for all other comets, Phobos, Deimos, Callisto, C-type asteroids, undifferentiated metamorphosed asteroids, and differentiated asteroids. In contrast, the report recommended strict containment and handling for samples from Europa, Ganymede, and P-type and D-type asteroids (these asteroids are included in this category due to our ignorance of them).

The Quarantine and Certification of Martian Samples (SSB, 2002a). This report addressed the quarantine and biosafety certification of extraterrestrial samples, including the criteria to be satisfied before samples are released from a quarantine facility, the optimal techniques for isolating and handling planetary materials, and how to determine their geochemical and biological characteristics. The report also examined to what extent lessons could be learned from the Apollo quarantine experience and from recent developments in the biotechnology and biomedical communities. A basic assumption of the study was that life would be carbon-based and microbial. A series of tests were envisioned that could provide evidence of viable or recently dead organisms, detect chemical fossils or probable biological molecules (biomarkers), and also quantify contamination by terrestrial microbes and organic compounds. It was felt that detection and identification of terrestrial contamination, both microbial and organic, would be crucially important, and that establishing unequivocal evidence of martian life would dictate an elaborate plan of handling, curation, and study. While this plan was not developed, the report recommended that if unmistakable evidence of life as we know it were to be found in the Mars samples, they should then be dedicated to biological studies. But in the more likely event that initial examination of the Mars samples neither proved nor definitively ruled out evidence that they harbored a living entity, there should be the capability to promptly sterilize samples and send them from the quarantine facility to specialized laboratories for biological and geochemical studies elsewhere. Additional recommendations concerned the need for research to resolve uncertainties in sample handling procedures, the location of the quarantine facility, and the principle that the quarantine facility should have the smallest size consistent with its role as a biological containment and clean room facility. The report stated that no scientific investigations should be carried out in the quarantine facility that could be executed on sterilized samples outside of the facility, and that planning and construction of the facility be initiated at least seven years before any anticipated return of Mars samples.

25.5 How would Earth respond to an exotic life form?

While the alien invaders envisioned by Wells quickly succumbed to Earth's less obvious dangers, there is no guarantee that organisms introduced to alien ecosystems would even notice the defense mechanisms they encounter, much less be felled by them. On Earth there has been a long, coevolutionary dance between predator and prey, parasite and host, and those who have failed in that dance are no longer here to bear witness. Faced with a truly novel introduced species, there is no way to predict with any confidence the response of Earth-life.

The experience we have with more familiar invaders on this planet is not encouraging. Often, the invaders are not problematic in their native environments because of natural predators or parasites that keep them under control. When they reach a novel environment and are released from natural controls, however, introduced species can cause extensive environmental damage – witness the prickly pear cactus (*Opuntia stricta* and *O. aurantiaca*) in Australia and the zebra and quagga mussels (*Dreissena polymorpha* and *D. bugensis*) in the North American Great Lakes. Most of the damage is done by a soaring population of the introduced species, which may crowd out, starve out, or otherwise smother native species in the invaded ecosystem. Other introduced species, such as the *Lymantria dispar* gypsy moth or the *Endothia parasitica* chestnut blight in North America directly prey upon or parasitize native species. Sometimes, though, the introduced organism is not capable of becoming a threat on its own, but only becomes a concern when paired with another species, which may be a native. For example, the fungus *Ophiostoma ulmi* responsible for Dutch elm disease is most widely spread in North American elms by the action of bark beetles. One species of beetle that spreads the fungus is itself an invading species that brought the fungus from Europe to America in 1930. Since that time, however, the native North American bark beetle *Hylurgopinus rufipes*, normally not harmful to healthy trees, has become an important vector for the fatal fungus.

An important feature about each of the invasions listed above – and many more that could be mentioned (e.g., Enserink, 1999; Ruiz *et al.*, 2000) – is that each represents an invasion of one part of Earth by another Earth organism. Despite the existence of many mechanisms that *might* have allowed them to invade (or otherwise be exchanged) at an earlier time, they had not done so. But when they did, the consequences were dire. Similarly, despite the excellent modeling and analysis done by Mileikowsky *et al.* (2000) and others, we do not know if such a Mars–Earth exchange ever has taken place. Thus, even in the face of a nonzero possibility that extraterrestrial material could already have introduced alien life to the Earth by natural means, without good evidence of such an invasion we cannot be assured that an introduction of an extraterrestrial organism would be benign. In fact, some have suggested that all life on Earth may have had an extraterrestrial origin – perhaps on Mars or beyond (Thomson, 1871; Mileikowsky *et al.*, 2000). If so, both the possible close biological relationships (with the attendant possibility of related pathogens) and the potential for additional effects on Earth's environment (if anything remains alive out there) further emphasize the need for caution.

But it is also important that that caution not be overblown. Proper precautions can prevent the release of microorganisms that might be contained in a returned sample, and spacecraft can be thoroughly cleaned to prevent round-trip contamination that might masquerade as extraterrestrial life, in the same way that they are cleaned to avoid Earth organisms invading in the other direction. And even if there were an introduction, the effects may be benign. Nonetheless, when dealing with the unknown, measured steps are prudent. Addressing the prospects of a Mars sample return mission, the SSB (1997) report stated that "in the event that living martian organisms were somehow introduced into Earth's environment, the likelihood that they could survive and grow and produce harmful effects is judged to be low," and thus the "contamination of Earth by putative martian microorganisms is unlikely to pose a risk of significant ecological impact or other significant harmful effects." But their report also stated that "the risk is not zero," that we have no experience with any extraterrestrial life forms (or even different forms of life on Earth), and that we should therefore take a conservative approach when dealing with samples returned from Mars and other Solar System bodies (also see SSB, 1998).

25.6 Spacecraft sterilization

Preventing microbial tourism depends both on operational restrictions and on our ability to reduce biological contamination that could be transferred by the spacecraft. Preventing organic chemical contamination is a more tractable problem because of strict limitations

(for other reasons) on the organic materials a spacecraft can carry.[2] Nonetheless, for mission Categories II–IV, an organic inventory and archive is required to aid in determining the correct source of any detected organic material. For missions in Category III–IV, where biological contamination is thought to be possible, measures must be taken to prevent the inadvertent impact of spacecraft with their target bodies, or insure sufficient cleanliness when a landing or impact is planned. The stringency of that cleanliness depends on the target body's potential to support Earth life, the nature of the contact planned, and the science to be performed on the mission. For missions returning a sample to Earth, spacecraft cleanliness is further required to prevent round-trip contamination, as well as to ensure that samples are not returned to Earth on the outside of the spacecraft. Even once contained, a sample from a body that may support life must be delivered to its quarantine in a very dependable manner.

The two Viking missions of the mid-1970s were a watershed in the preparation of spacecraft for planetary exploration. Each Viking mission was launched on a Titan III rocket and consisted of an orbiter and a lander (Fig. 25.1). Because one of the primary goals of the missions was an attempt to detect life on the martian surface (Chapters 2 and 18), extensive procedures were undertaken to reduce to an absolute minimum the bioload carried by the landers – and thus avoid false positive indications of life on Mars. These procedures were even more extensive than those thought necessary only to prevent the contamination of Mars. Under the allocation of allowable bioload in the Viking project, each lander was first cleaned until the surface bioload was $\leq 3 \times 10^5$ bacterial spores (as measured by the standard NASA assay technique), with an average of ≤ 300 bacterial spores per square meter, and a total bioload (including both the surface bioload and the bioload between and within spacecraft parts) of $\leq 5 \times 10^5$ spores. The entire lander spacecraft (including its two radioisotope power sources) was then packaged in a fully enclosing "bioshield" (like a large, light, casserole dish) and placed in an oven (Fig. 25.2) in a dry-heat sterilization procedure designed to achieve a temperature for all spacecraft parts of at least 111.7 °C for 30 hours. After heat treatment, the entire package

---

[2] Spacecraft designers limit organic materials because they tend to outgas (exude vapors) in a vacuum, which can create havoc with various components in the spacecraft compartment, especially optical surfaces.

Bioshield Cap
& Equipment Module ⟶

Parachute System &
Base/Aeroshell Cover ⟶

Lander System ⟶

Aeroshell &
Heat Shield ⟶

Bioshield Base ⟶

Orbiter/VLC
Interface Truss

Orbiter System ⟶

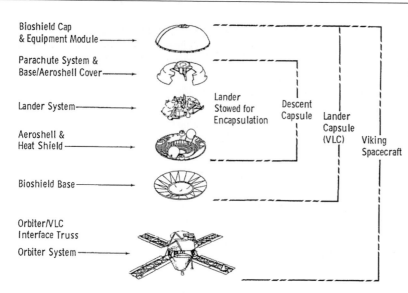

Lander
Stowed for
Encapsulation

Descent
Capsule

Lander
Capsule
(VLC)

Viking
Spacecraft

FIGURE 25.1 The Viking Mission "stack," showing the configuration of the spacecraft assemblies for launch. (Martin Marietta.)

FIGURE 25.2 A Viking "casserole." Technicians wheel the assembled Viking lander, within its bioshield, into the dry-heat oven attached to the Spacecraft Assembly and Encapsulation Facility at NASA's Kennedy Space Center. (NASA.)

was placed atop the launch vehicle, and the lander was not released from its casserole dish until it was en route to Mars.

The Mars that was explored by Viking turned out to be (on the surface) much drier and less benign than some had hoped, and although there were results from the Viking Lander that were suggestive of life (e.g., Levin and Straat, 1979), there was no consensus that the surface of Mars could support life. In fact, as noted above, the SSB (1992) recommended that forward

FIGURE 25.3 Microorganisms are sampled for a final assay of the bioload on the Pathfinder spacecraft before it left for Mars in 1996. (NASA/JPL Robert Koukol.)

contamination requirements for Mars be altered from those used for the Viking missions, since they did not believe that Earth organisms could grow there, although survival was considered possible in sheltered areas of the spacecraft. Subsequently, both COSPAR and NASA altered their requirements for Mars lander missions, stipulating that the Viking pre-heat-treatment surface bioload should be the requirement for landers that were *not* attempting life detection experiments, while the full Viking requirement was appropriate if life detection (especially cultivation experiments of the type Viking performed) were a goal of the mission. For orbiting missions, the Viking total pre-sterilization bioload level was defined as the requirement if the mission's probability of remaining in orbit around Mars for 20 years was $< 99\%$, or for 50 years $< 95\%$. This latter time period coincided with the "period of biological exploration" originally defined for Mars by COSPAR, though it was originally defined under more optimistic expectations for the frequency of successful Mars missions than has proven to be the case.

Today, spacecraft destined for Mars are kept clean and monitored by a variety of techniques. Perhaps most important are relatively clean manufacturing processes for the original spacecraft components, and the use of cleanroom techniques during spacecraft assembly, test, and launch operations. Active bioload reduction is achieved through a variety of means, such as alcohol wiping, dry heat treatment, and hydrogen peroxide sterilization systems (Debus *et al.*, 1998). Radiation treatment is an available option for some assemblies. Increasingly, attention is being paid to the use of a

combination of molecular methods to characterize microbial bioload (e.g., *Limulus* Amebocyte Lysate, polymerase chain reaction techniques, ATP detection, etc.; see SSB, 2002c), but for the time being none of these has replaced cultivation-based assays (NASA, 1999c) as a standard assay method (Fig. 25.3). In spite of their advantages, the new techniques have not been adopted because of a lack of comparability between them and the well-established cultivation methods, and because of the additional personnel, training, and equipment that would be required. Time pressure in spacecraft assembly, however, makes molecular methods very attractive because a measurement of bioload that may take three days with the cultivation assay can be approximated in less than one hour using a molecular method. In view of this advantage, in 2003 *Limulus* Amebocyte Lysate (LAL) was used for the first time to supplement the cultivation-based assays in the preparation of the Mars Exploration Rover spacecraft.

The preparation of a spacecraft to perform a dedicated mission to Europa will require a different approach from that currently applied to Mars missions. Because the martian atmosphere is a significant factor in the potential spread of Earth microbes, the greatest challenge for Mars spacecraft is *surface* cleanliness, with the spacecraft interior being of less concern because it is not intended to be exposed to the martian environment – although in the unfortunate circumstance of a crash on Mars there may be some interior exposure. For Europa, however, the exterior is not of particular concern because there is no atmosphere and because there exists an extremely high flux of energetic cosmic ray particles. Anything on the outside of the spacecraft, even *Deinococcus radiodurans*, will be killed by the brutal jovian radiation environment. On the other hand the spacecraft's extensive radiation shielding, designed to protect the electronics, will also nicely protect any microbes inside. To prevent such Earth microbes from finding their way onto the europan surface in a way that would allow them to survive until they somehow got introduced into a europan ocean, the spacecraft interior is thus of great concern. Species-specific assays may help to eliminate some of the more radiation tolerant organisms from the spacecraft. Furthermore, its design will have to prevent the spacecraft interior from being recontaminated after clean assembly is accomplished.

For sample-return missions, it is also important to attend to forward contamination concerns both to protect the other Solar System body and to prevent Earth hitchhikers from being misidentified when the sample is tested after being returned. If the only concerns about the return are with science and mission safety (sample condition and landing site targeting), the mission is classified as an "unrestricted Earth return" in Category V. For "restricted Earth return" missions, however, the chief difficulty is to devise a system that, while fitting into the challenging mass and cost considerations of spaceflight, maintains sample containment on the way back to Earth with a high reliability (perhaps as high as 99.9999%, although no firm requirement has been established). Also important is to ensure that uncontained material from the target body is not inadvertently brought back to Earth – say on the outside of the return vehicle. System designers are just beginning to seriously address these issues, and the monetary costs for implementing stringent requirements on a Mars sample return mission are significant – about 5–10% of the entire mission budget. What may be more troubling, though, are how these systems may affect overall mission success: even if one fully embraces the necessity for the precautions, very few would be pleased to see a fully contained sample fail to return to Earth because of mission complications caused by planetary protection provisions. At the present time, we are only beginning to plan missions that involve both sample return and human explorers and determine requirements for related precursor missions, both because of high cost and operational difficulties of human planetary missions, as well as the conflicts that can arise when the safety of individual humans must be balanced against the safety of the entire Earth, as happened during the Apollo Moon missions of the late 1960s and early 1970s (Allton *et al.*, 1998).

25.7 A sample handling protocol

Sample return missions may be essential to understanding other Solar System bodies and to calibrating the results of instrumentation on robotic spacecraft *in situ*. Nothing that we can send into space will have the analytical power that we can achieve in Earth's laboratories under the watchful eye of skilled technicians. Yet the benefit of this analytical power must be weighed against the potential risks. The SSB's (1997) careful consideration of a future Mars sample return provided conservative guidelines for the handling of martian samples. One of the most significant recommendations of the report with respect to risk reduction was that "controlled distribution of unsterilized materials should only occur if analyses determine the sample not to contain a biological hazard."

In response to this recommendation, and in the face of Mars mission planning that has placed a sample-return mission launch anywhere from 2003 to 2015 to indefinite, NASA, in collaboration with France's Centre National d'Etudes Spatiales (CNES), has developed criteria that should be met by Mars samples before they are released from post-flight containment (Rummel *et al.*, 2002). In addition to assessing the requirements for sample containment, hazard testing, and subsequent release, Rummel *et al.* specified the tests and procedures needed to safeguard the sample against the various threats to its scientific value caused by the Earth's environment. It was assumed that the sample collection site was at a certain martian location on the basis of data from previous missions, that the samples were not sterilized prior to return to Earth, that the sample containers were opened only in a receiving facility, and that the mass of material returned was only ~500–1,000 grams, consisting of rock cores, pebbles, soil, and atmospheric gases. Of that material, the amount of sample used in biohazard testing was designed to be the minimum necessary.

A list of the issues that drive the development and implementation of a sample handling protocol are listed in Table 25.4, while the various protocol elements that were recommended by the SSB (2002a) are listed in Table 25.5. Space limitations do not permit an extensive review of the draft protocol itself, but its overall scheme is shown in Fig. 25.4, and Table 25.6 lists the questions and protocol strategy for addressing them.

The protocol's criteria for release of any sample are (Rummel *et al.*, 2002):

TABLE 25.4 Biohazard issues in analysis and testing of a returned sample (SSB, 2002a)

- What criteria must be satisfied to show that the samples do not present a biohazard?
- What will constitute a representative sample for testing?
- What is the minimum allocation of sample material required for analyses exclusive to the protocol, and what physical/chemical analyses are required to complement biochemical or biological screening of sample material?
- Which analyses must be done within containment, and which can be accomplished using sterilized material outside of containment?
- What facility capabilities are required to complete the protocol?
- What is the minimum amount of time required to complete the protocol?
- How are these estimates likely to be affected by technologies brought to practice by two years before a sample is returned?

TABLE 25.5 Elements of a Mars sample analysis protocol

- Prior to introduction of the Mars samples, sterilize and cleanse the quarantine facility of organic contamination
- Place samples in the facility
- Inventory and carry out preliminary analyses of the samples
- Search for evidence of biological activity
- Assess whether the samples contain biohazardous material
- Sterilize aliquots of the samples in preparation for their removal from the facility
- Remove samples from the facility
- Store samples within the facility

- No solid sample shall be released from containment in the Mars receiving laboratory until it or its parent sample undergoes preliminary examination, baseline description, cataloguing, and any necessary repackaging.

- Samples containing any genuine active martian form of life, be it hazardous or not, will be kept under appropriate level of containment, or be thoroughly sterilized before release.
- Samples providing indications of life-related molecules, including proteins, nucleic acids, or molecular chirality will require more extensive testing, including additional biohazard testing, prior to their release.
- Samples may be released if they are first subjected to a sterilizing process involving heat, radiation, or a combination of these agents to ensure safe analyses outside of containment. A sample that is 'safe' is stipulated to be free of any viable self-replicating entities or entities able to be amplified.
- Samples may be released if biohazard testing does not yield evidence of live, extraterrestrial, self-replicating entities, or of harmful effects on terrestrial lifeforms, or on the environment under Earth-like conditions.

25.8 Future considerations

Solar system exploration is a compelling activity. It places within our grasp answers to basic questions of profound human interest: Are we alone? Where did we come from? What is our destiny? Solar System Exploration Survey Committee (SSB, 2002b).

Questions of this sort are now driving Solar System exploration and astrobiology, and it is likely that they will be central for decades to come. Accordingly, considerations of planetary protection and the prevention of biological contamination will continue to affect future missions and their operations. As scientific knowledge of the Solar System grows, the implementation of planetary protection provisions will also grow more precise, guided both by the potential for Earth contamination and our ability to measure and control it, and by the potential to uncover extraterrestrial life either advertently or inadvertently during planetary exploration missions. As the Solar System unfolds, it is expected that COSPAR will continue to promulgate planetary protection knowledge, policy, and plans to prevent the harmful contamination of space, and to provide an international forum for exchange of information in this area.

And those of us who look up, search the sky, and wonder about life elsewhere will have a better chance of having our questions answered.

TABLE 25.6 Strategies to answer questions about returned martian samples (Rummel *et al.*, 2002)

Question	Strategy
Is there anything that looks like a lifeform?	Microscopy; beam synchrotron or other non-destructive high-resolution analytic probe, particularly one that would allow testing un-sterilized (yet still contained) samples outside the main facility.
Is there a chemical signature of life?	Mass spectroscopy and other analytical measurement systems (to be used in containment) that would identify biomolecules, chiral asymmetry, special bonding, etc.
Is there any evidence of self-replication or replication in a terrestrial living organism?	Attempts to grow in culture, in cell culture, or in defined living organisms.
Is there any adverse effect on workers or the surrounding environment?	Microcosm tests; medical surveillance of workers and monitoring and evaluation of living systems in proximity of the receiving facility.

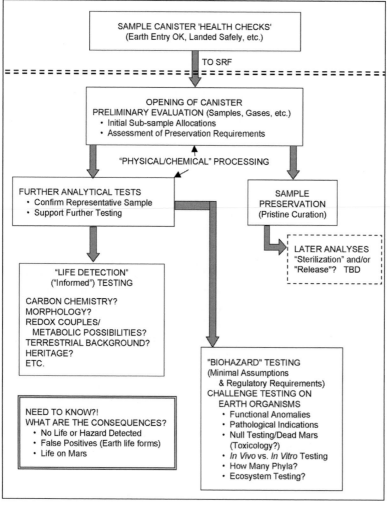

FIGURE 25.4 Outline of a Mars sample handling protocol (Rummel *et al.*, 2002).

REFERENCES

Allton, J. H., Bagby, J. R., Jr., and Stabekis, P. D. (1998). Lessons learned during Apollo lunar sample quarantine and sample curation. *Adv. Space Res.*, **22**, 373–382.

CETEX (1959). Contamination by extraterrestrial exploration. *Nature*, **183**, 925–928.

Cockell, C. S., Schuerger, A. C., Billi, D., Friedmann, E. I., and Panitz, C. (2005). Effects of a simulated martian UV flux on the cyanobacterium, *Chroococcidiopsis* sp. 029. *Astrobiology*, **5**, 127–140.

Crichton, M. (1969). *The Andromeda Strain*. New York: Alfred A. Knopf.

Debus, A., Runavot, J., Rogoski, G., *et al.* (1998). Mars 96 small station biological decontamination. *Adv. Space Res.*, **22**, 401–409.

Derbyshire, G. A. (1962). Résumé of some earlier extraterrestrial contamination activities. In *A Review of Space Research*, Chapter 10, p. 11. Washington, DC: National Academy of Sciences.

Enserink, M. (1999). Biological invaders sweep in. *Science*, **285**, 1834–1836.

Horowitz, N. (1986). *To Utopia and Back: the Search for Life in the Solar System*. New York: W. H. Freeman.

Levin, G. V. and Straat, P. A. (1979). Completion of the Viking labeled release experiment on Mars. *J. Mol. Evol.*, **14**, 167.

Malin, M. C. and Edgett, K. S. (2000). Evidence for recent groundwater seepage and surface runoff on Mars. *Science*, **288**, 2330–2335.

Malin, M. C., Edgett, K. S., Posiolova, L. V., *et al.* (2006). Present-day impact cratering rate and contemporary gully activity on Mars. *Science*, **314**, 1573–1577.

Melosh, H. J. (1984). Impact ejection, spallation and the origin of meteorites. *Icarus*, **59**, 234–260.

Mileikowsky, C., Cucinotta, F., Wilson, J. W., *et al.* (2000). Natural transfer of viable microbes in space, part 1: from Mars to Earth and Earth to Mars. *Icarus*, **145**, 391–427.

NASA (1999a). *Biological contamination control for outbound and inbound planetary spacecraft*. NPD 8020.7E. Washington, DC: NASA.

NASA (1999b). *Planetary protection provisions for robotic extraterrestrial missions*. NPG 8020.12B. Washington, DC: NASA.

NASA (1999c). *NASA Standard procedures for the microbial examination of space hardware*. NPG 5340.1C. Washington, DC: NASA.

Nicholson, W. L., Munakata, N., Horneck, G., Melosh, H. J., and Setlow P. (2000). Resistance of *Bacillus* endospores to extreme terrestrial and extraterrestrial environments. *Microb. Mol. Biol. Rev.*, **64**, 548–572.

Phillips, C. R. (1974). *The planetary quarantine program: origins and achievements, 1956–1973*. NASA SP-4902. Washington, DC: NASA.

Ruiz, G. M., Rawlings, T. K., Dobbs, F. C., Drake, L. A., Mullady, T., Huq, A., and Colwell, R. R. (2000). Global spread of microorganisms by ships. *Nature*, **408**, 49.

Rummel, J. D. (2002). *Report of the workshop on planetary protection*. Paris, France: COSPAR.

Rummel, J. D., Race, M. S., DeVincenzi, D. L., Schad, P. J., Stabekis, P. D., Viso, M., and Acevedo, S. E. (2002). *A Draft Test Protocol for Detecting Possible Biohazards in Martian Samples Returned to Earth*. NASA/CP – 2002–211842. Moffett Field, CA: NASA.

Space Studies Board, National Research Council (US) (1992). *Biological Contamination of Mars: Issues and Recommendations*. Task Group on Planetary Protection. Washington, DC: National Academy of Sciences.

Space Studies Board, National Research Council (US) (1997). *Mars Sample Return: Issues and Recommendations*. Task Group on Issues in Sample Return. Washington, DC: National Academy of Sciences.

Space Studies Board, National Research Council (US) (1998). *Evaluating the Biological Potential in Samples Returned from Planetary Satellites and Small Solar System Bodies*. Task Group on Sample Return From Small Solar System Bodies. Washington, DC: National Academy of Sciences.

Space Studies Board, National Research Council (US) (2000). *Preventing the Forward Contamination of Europa*. Task Group on the Forward Contamination of Europa. Washington, DC: National Academy of Sciences.

Space Studies Board, National Research Council (US) (2002a). *The Quarantine and Certification of Martian Samples*. Committee on Planetary and Lunar Exploration. Washington, DC: National Academy of Sciences.

Space Studies Board, National Research Council (US) (2002b). *New Frontiers in the Solar System: an Integrated Exploration Strategy*. Solar System Exploration Survey Committee. Washington, DC: National Academy of Sciences.

Space Studies Board, National Research Council (US) (2002c). *Signs of Life: a Report Based on the April 2000 Workshop on Life Detection Techniques*. Committee on the Origin and Evolution of Life. Washington, DC: National Academy of Sciences.

Squyres, S. W., Grotzinger, J. P., Arvidson, R. E., *et al.* (2004). In situ evidence for an ancient aqueous environment at Meridiani Planum, Mars. *Science*, **306**, 1709–1714.

Thomson W. (Lord Kelvin) (1871). Inaugural address of Sir William Thomson, L.LD., FRS, President (at the British Association for the Advancement of Science). *Nature*, **4**, 262–270.

United Nations (1967). *Treaty on Principles Governing the Activities of States in the Exploration and Use of Outer Space, Including the Moon and Other Celestial Bodies.* Article IX, U.N. Doc. A/RES/2222/(XXI) 25 Jan 1967; TIAS No. 6347.

Wells, H. G. (1898). *The War of the Worlds.* London: William Heinemann.

FURTHER READING AND SURFING

The NASA Planetary Protection Office planetary protection.nasa.gov. Summaries of NASA reports dealing with planetary protection planetaryprotection.nasa. gov/pp/summaries/.

26 Searching for extraterrestrial intelligence

Jill C. Tarter
SETI Institute

26.1 Technology, not intelligence

SETI (Search for Extra Terrestrial Intelligence) can be defined as the branch of astrobiology looking for inhabited worlds by taking advantage of the deliberate technological actions of extraterrestrial organisms. This definition usually draws a chuckle during public lectures, but it underscores why this chapter is somewhat different from the preceding ones. As in other parts of astrobiology, one must consider the diversity of physical environments in the cosmos, and the limitations imposed by them. But with SETI one must also consider modifications to the environment that are not just the byproduct of life, but the result of deliberate actions by intelligent organisms intended to achieve some result.

For millennia people have speculated about the existence of other habitable worlds, and their inhabitants (Chapter 1), but the rules of the game underwent a profound change in the second half of the twentieth century. The publication of the initial scientific paper on SETI (Cocconi and Morrison, 1959) and Drake's (1961) first radio search (Project Ozma, described in Section 1.9) turned speculation into an observational science. No longer were priests and philosophers the sole respondents to the "Are we alone?" question; scientists and engineers could work on finding an answer empirically. Following the first flurry of observing programs in the US and the Soviet Union (Chapter 2), the acronym SETI became the accepted name for this new exploratory activity. But, in fact, SETI is a misnomer because there is no known way to detect intelligence directly across interstellar distances. Even on Earth we argue about exactly what constitutes intelligence, and we have no reliable way of measuring it at a distance (either spatial or temporal). In the case of extraterrestrial intelligence, the best we can do is to search for some manifestation of another technology. Having detected it, we can infer the existence of intelligent technologists, who may or may not still be associated with the detected technology. This distinction is far more than semantic, it defines what we, with our early twenty-first-century technology, can and cannot attempt to do.

26.2 Which technologies?

Which technology might an extraterrestrial civilization utilize, and what are the observable consequences therefrom? As with so much of astrobiology, we are forced to extrapolate from what we know, even though we cannot be sure that it is appropriate for life-as-we-do-not-yet-know-it. From our own experience we deduce that a civilization might develop indirectly observable technologies for energy production, for waging war, for transportation (including perhaps interstellar travel), and for exchange of information. This is not an exhaustive list, but after decades of discussion, these remain the most commonly cited examples. With the exception of interstellar transportation (since this opens the possibility that "they" might come here), detecting these technologies requires remote sensing equipment. Over the past four decades, more than 100 searches have been made for specific examples of each of these potential applications of technology.[1]

26.2.1 "Local" technologies

Although it is very risky to speculate on the motivations of an unknown, extraterrestrial civilization, waging war, generating energy, and local transportation are examples of technologies likely to be employed

---

[1] See the archive of SETI searches maintained by the author at www.seti.org/searcharchive.

Planets and Life: The Emerging Science of Astrobiology, eds. Woodruff T. Sullivan, III and John A. Baross. Published by Cambridge University Press. © Cambridge University Press 2007.

for the sole use of the civilizations that have invented them. In this case, there is no reason to believe that they would make any effort to enhance the probability that another civilization would ever discover them. They would be visible only through unintentional manifestations of their technology, and perhaps it then follows that the best search strategy is to explore the Universe with all possible tools, in every possible way, and conduct a robust observational program of astronomy. If and when an anomalous phenomenon appears, one that cannot be easily explained by current astrophysics, researchers should ask whether that phenomenon might be the hallmark of some form of "astro-engineering" or other technology.

26.2.2 They send spacecraft

In contrast, when considering interstellar travel and interstellar information exchange, one can argue that these technologies might be manipulated with us (or other emerging technologies like us) in mind. "They" might actually come here or have done so in the past, or "they" might actively generate signals for the precise purpose of attracting our attention and transferring information. With respect to the first possibility, there is no proof that "they" have visited Earth (notwithstanding contrary claims that are spectacular but scientifically not established). However, to be completely honest, there is also no evidence proving that "they" have *not*. The physicist Enrico Fermi was sufficiently impressed with the apparent lack of visitation that he once asked his luncheon companions "Where is everybody?" (Jones, 1985), thus originating the so-called *Fermi paradox*. If, the argument goes, there had ever been a single other intelligent, technological civilization within our Milky Way Galaxy, then they would have developed the technology for interstellar travel quickly (relative to cosmic timescales of billions of years), and used it to colonize the Galaxy. For a wide range of scenarios, this colonization would have taken place in a time much shorter than the 10 Gyr lifetime of the Galaxy. But they are evidently not here. Therefore, such a civilization can never have existed at any prior time in the Milky Way.

Since seemingly simple paradoxes often lead to revealing conclusions, a great deal has been written about ways to explain away or answer the Fermi paradox. Webb (2002) summarizes 50 possible solutions, grouped under three headings: (1) they are here; (2) they exist but have not yet communicated; and (3) they do not exist. Webb himself subscribes to the third solution. His discussion of Group 2 contains many

relevant arguments about the enormous energy costs of interstellar travel, as well as reasons why we might not have detected deliberate signals (though this is not strictly a part of the Fermi paradox). In discussing Group 1 he dismisses all the unsubstantiated claims of visitation as well as the idea that "we are they" (via a program of directed panspermia[2]). Webb fails, however, to consider seriously what may be the fundamental answer, and why the Fermi paradox is no paradox after all. Humans have so poorly explored our own environment on Earth, and the surrounding Solar System, that we cannot in fact say "they are not here." This is particularly true if "they" are represented by some small (perhaps even nanoscale) surrogate technologies. We can only rule out the presence of large objects filled with macroscopic examples of biology (such as the crew of Starship *Enterprise*) in a few locations near Earth, but not even elsewhere in the Solar System. For example, NASA's Spaceguard Survey is attempting to locate all potential Earth-crossing asteroids greater than 1 km in diameter (Morrison, 1992). Yet even this thorough search of nearby space, looking for large objects lacking cloaking devices, is incomplete and subject to surprises. Objects can sneak up from the sunward direction and not be discovered until after they pass into the evening sky, as was the case with the 100-m-sized asteroid 2002 MN. Small, self-replicating, robotic colonizers could certainly have gone unnoticed. In case any nearby intelligent probe of non-Earth origin should exist, one group (coordinated by Canadian futurist Allen Tough) has invited it to log on to the Internet and announce itself.[3]

26.2.3 They emit electromagnetic radiation

The majority of searches for extraterrestrial intelligence in the decades since Project Ozma have instead concentrated on finding signals that are the result of exchanging information; either unintentional leakage, or deliberate beacons. The seminal Project Cyclops Report (Oliver and Billingham, 1972: 31; Section 2.2.4) specified optimal requirements for transmitting information over interstellar distances. The best information carrier should:

- require minimum energy per bit of information
- have the maximum possible velocity

---

[2] *Directed panspermia* is the idea that a civilization could purposely spread the germs of its form of life to other habitable locales (Crick and Orgel, 1973).

[3] The invitation can be found at www.ieti.org/.

FIGURE 26.1 An illustration of one way in which large amounts of data (encoded onto the disks) can be moved slowly between two destinations, and yet achieve a data transfer rate that is faster than today's so-called 'broadband' wired protocols. The question is whether data inscribed onto a dense physical memory medium and propelled between the stars at spacecraft speeds might be the modality of choice for deliberate interstellar communication. (Courtesy of Herbert Bishko, with permission from *Annals of Improbable Research*.)

- be easy to generate, launch, and capture
- not be appreciably absorbed by the interstellar medium
- go where aimed.

The last requirement rules out any charged particles, since they are deflected by the general interstellar magnetic field. Particles with mass also require a large amount of energy to accelerate close to the speed of light c (the cosmic speed limit as far as we know, 3×10^{10} cm/s). Specifically, if we denote as β the ratio of a particle's velocity v to c, the theory of special relativity indicates that a moving particle has a mass given by

$$m = \frac{m_0}{\sqrt{1 - \beta^2}}, \qquad (26.1)$$

where m_0 is the mass the particle has at rest. Unless the velocity gets close to c, the mass is little increased, but as β approaches 1 (relativistic velocity), the mass grows rapidly, as does the energy needed to accelerate it. For example, the mass of a single relativistic electron traveling at $0.5c$ is increased by a factor of 1.15 above its rest mass. Its kinetic energy is then 1.25×10^{-7} erg, fully 10^{10} times the energy of a single microwave photon.

Photons, the quanta of electromagnetic radiation, are ideal carriers of information because they are massless, travel at c, and have very small energies. The energy of a photon is proportional to the frequency of the wave associated with the electromagnetic radiation. Therefore radio and microwave photons, being of such low frequency, have very low energy. As with any wave phenomenon, the product of the frequency and the wavelength gives the speed of propagation: $c = \nu \lambda$. As

a benchmark, a frequency of 1,000 MHz = 1 GHz corresponds to a wavelength of 30 cm.

Other exotic, massless particles proposed by theoretical physicists may also travel at light speed, but we cannot now manipulate them, even if they do exist. If such exotic particles are the choice of technologies more advanced than our own, the only strategy for detecting such signals is to survive as a technological species until we learn to generate and capture them ourselves. Rose and Wright (2004) have recently suggested that if time is no concern (thus eliminating the second criterion in the list of properties for information carriers), then an extraordinary amount of information can be deliberately transferred over interstellar distances by inscribing a message into a very dense physical memory device. This missive could then travel between the sender and intended receiver at slow speeds to conserve energy. Figure 26.1 recently appeared on the cover of the *Annals of Improbable Research* (Ben-Bassat *et al.*, 2005). It humorously illustrates this concept; a giant African snail pulling two densely encoded data disks in a "feed-forward" transport mode can indeed exceed the data transfer rates achievable with many broadband systems available today. The energy costs of the redundancy required to insure successful receipt of the physical SETI messengers have not been adequately addressed, nor has the required strategy for discovery by the receiver. As we have already noted, using the Spaceguard Survey as an example, small objects in our Solar System can easily go undetected. So for the foreseeable future, photons remain the best bet for SETI.

26.3 The nine-dimensional cosmic haystack

Having settled on a search for electromagnetic signals as the methodology for SETI, we now must decide *where* to search (three spatial dimensions), *when* to search (one temporal dimension), and *what* to search for (a frequency, two possible polarizations, a modulation scheme, and a signal strength). The "cosmic haystack" to be scoured for the proverbial "needle" is thus nine-dimensional. Consider the possible scale of each of these dimensions.

A signal might be coming from any **direction** on the sky, and from a **distance** of as much as 100,000 light years (lt-yr) if it originates within our Milky Way Galaxy. The nearest neighbor galaxies are millions of light years away, so a signal coming from any of them would have to be much stronger in order to be detectable on Earth (signal strength drops off as $1/r^2$, where r is the distance to the transmitter). Section 26.5.3 discusses the issues of where to search in more detail.

The **time** at which a signal arrives could be a critical part of a search if the transmitting civilization has decided, for example, to broadcast only for one hour every year. If the signal is always present, it makes our job easier, but puts more burden on the resources of the transmitting civilization. Section 26.5.5 discusses one proposed scheme for when we should look, but in fact very few searches have carried out their observations at any such special times.

Electromagnetic radiation can have two orthogonal senses of **polarization**, and both must be examined to avoid missing a signal. Right and left circular polarizations are often used in search programs because they are unmodified by propagation through the galactic magnetic field and the interstellar medium, unlike linear polarizations.

As Harwit (1981) illustrated, the **frequency** range for acquiring information over cosmic distances via photons is vast, but finite. Frequencies lower than $\sim100\,\text{kHz}$ do not propagate through the interstellar medium because they are absorbed by its rarefied plasma (typical density of ~0.03 electrons per cm^3).[4] Conversely, if a photon has enough energy (high enough frequency), then when it interacts with one of the ubiquitous cosmic microwave background photons left over from the Big Bang, it can spontaneously transform into an electron and its antiparticle, a positron.

This high frequency cutoff for sending information by photons through the interstellar medium is $\sim10^{29}\,\text{Hz}$, which corresponds to a photon energy of $\sim4 \times 10^{14}\,\text{eV}$, well beyond the observed gamma-ray range. Within this huge allowable frequency range, Section 26.5.4 discusses reasons for choosing specific frequencies and bands.

The **modulation** parameter space is difficult to constrain. *Modulation* refers to the specific techniques for encoding information onto a signal (a familiar example is amplitude modulation, or AM). In the absence of an agreed-upon scheme between sender and receiver, efficient communication is difficult. In general, the more information that is contained within a signal, the more noise-like it appears and the more difficult it is to disentangle from the natural sky background and from receiver noise. To date SETI searches have concentrated on very simple classes of signals such as narrowband continuous tones and regular pulses. As computing capability becomes more affordable, it will be possible to search for more complex signals, although Section 26.5.2 suggests that in the case of deliberately generated signals, this may not be necessary.

Finally, **signal strength** is unknown. but it makes sense to search with the greatest possible sensitivity, so that a signal of a given strength can be detected at the farthest distance. The following section outlines some of the principles when considering signals and the inevitable competing noise.

26.3.1 Signal and noise

The strength of any arriving signal depends on the power of the transmitter and its distance, as well as the fraction of the signal that is actually collected by the receiver. Consider the factors that relate the received strength to the various properties of one antenna transmitting to another at a distance r. P_R, the amount of signal power (watts or W) collected by a receiving antenna with an *effective area*[5] A_R, is

$$P_R = \frac{P_E A_R}{4\pi r^2}. \qquad (26.2)$$

where P_E is the *effective isotropic radiated power* of the transmitter (see below). The equation states that the received fraction of transmitted power is just the fraction of the area of a sphere of radius r that is covered by the

---

[4] For ground-based observations, the plasma of the Earth's ionosphere sets a higher low-frequency limit of $\sim10\,\text{MHz}$.

[5] The *effective area* of an antenna is always less than its geometrical area, and depends on a number of efficiency factors such as the electrical properties and configuration of its materials, accuracy of its reflecting surfaces, blockage, etc.

Surface area of sphere = $4\pi r^2$

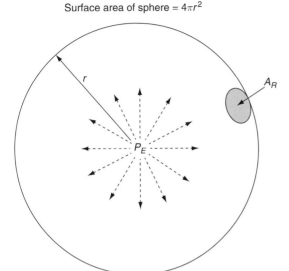

FIGURE 26.2 A transmitter of power P_E emits isotropically. A receiving antenna at a distance r will collect a fraction of the transmitted power. That fraction is simply the ratio of the effective area of the receiving antenna A_R to that of the surface of a sphere of radius r, or $4\pi r^2$.

receiving antenna's effective area (Fig. 26.2). Note that the transmitter may or may not be radiating its power in all directions (*isotropically*) – at the receiving end of a signal it is not possible to know exactly how it was transmitted, nor whether it was beamed at the receiver. As an example, terrestrial television broadcast antennas concentrate all their power into a thin "fan beam" that radiates towards the horizon, since they do not currently have any potential customers high in the atmosphere.

The effective isotropic radiated power P_E is defined as $P_T \times G_T$, where P_T is the actual transmitted power and G_T is the *antenna gain*, determined by the size and shape of the transmitting antenna. The antenna gain can be thought of as the ratio of the entire sky's solid angle[6] to that of the transmitting antenna's beam (Ω_T): $G_T = 4\pi/\Omega_T$. Unless the transmission is isotropic ($G_T = 1$), P_E is larger than P_T. Antenna theory also shows that G_T can be expressed as $4\pi\, A_T/\lambda^2$, where A_T is the effective area of the transmitting antenna, and λ is the operating wavelength. Thus the larger the antenna, the higher the gain and the more concentrated the beam; and for a given radio telescope, the gain is higher at shorter wavelengths (higher frequencies). In the case of the fan beam emitted by a TV transmitter, G_T is

about 5. Rearranging Eq. (26.2) produces the following elegant version of the free space transmission law:

$$\frac{P_R}{P_T} = \frac{A_R A_T}{\lambda^2 r^2}. \qquad (26.3)$$

We do not know what values of P_E another technological civilization might muster for transmitting, but on Earth today our cell phones typically radiate $P_E < 1\,\mathrm{W}$, commercial radio stations broadcast $\sim$10–100 kW, television stations generate up to 1 MW, and our most powerful radar transmitter, attached to the large telescope in Arecibo, Puerto Rico (Fig. 26.3), transmits a P_E value of 2×10^{13} W when it images the surfaces of distant planets and asteroids. Our SETI searches over the past four decades have not in fact been sensitive enough to detect our current level of television leakage radiation if it originated from the vicinity of nearby stars (at distances of $\sim$5–10 lt-yr), yet some searches could have detected the equivalent of the Arecibo planetary radar transmitter from as far away as 1500 lt-yr. Table 26.1 gives the number of stars that exist closer than the distance that Project Phoenix (Backus *et al.*, 2002), currently the most sensitive microwave SETI search (Section 26.6), could have detected transmitters with power analogous to those of terrestrial leakage. Some SETI programs are efficient at detecting either leakage radiation or purposeful beacons, while others are optimized for only one of the signal types. Note also that, because the Sun and its planetary retinue are all in common motion, a transmitting civilization aiming for us would need to point at the spatial location where the Earth would be when the signal arrived after many years (just like "leading" a moving duck in a shooting gallery). The transmitter would thus need to know the motion of the Sun through the Galaxy very accurately.[7]

The practical question of "Can we detect a certain signal?" depends not just on its strength, but also on the sensitivity that receiving equipment and environmental factors allow, i.e., how much random *noise* competes with the signal. For a signal to be reliably claimed (*detected*), the received power P_R must exceed by a certain factor, call it m, the always-present "competition" from fluctuations in the average noise power $\langle P_N \rangle$.[8] The brightness of the sky, the roughness of the

---

[6] *Solid angle* is a measure of the area of a patch of sky (in steradians or square degrees); the entire sky-sphere contains 4π steradians just as a circle contains 2π radians.

[7] This aiming problem becomes even more severe at the shorter optical wavelengths where gains of $\sim 10^{12}$ (beams of $\sim 1''$ diameter) might be effected (Section 26.8.1).

[8] For a radio telescope $\langle P_N \rangle = k\, T_{\mathrm{sys}}\, B/\sqrt{B\tau}$ per polarization, where B = bandwidth of the receiver in frequency (assumed to be at least as broad as the bandwidth of the signal), τ is the duration of the observation, and T_{sys} is the system temperature, which by definition

TABLE 26.1 Detectability of leakage radiation from twenty-first-century Earth with Project Phoenix sensitivity

P_E of transmitters	$P_E = P_T \times G_T$	Range r (lt-yr) of Project Phoenix	number of stars within range
Cell phones: 1 W	$1\,\text{W} \times 1$	3×10^{-4}	0
FM radio: 10–100 kW	$2\text{--}20\,\text{kW} \times 5$	0.03–0.1	0
TV: 300 kW	$60\,\text{kW} \times 5$	0.2	0
Airport radars: $\sim 10^8$ W	$35\,\text{kW} \times 2200$	3.3	~ 1 (Proxima Centauri is at 4.3 lt-yr)
Ionospheric radars: 2×10^{11} W	$150\,\text{kW} \times 1 \times 10^6$	150	$\sim 3.5 \times 10^5$
Arecibo radar: 2×10^{13} W	$1\,\text{MW} \times 2 \times 10^7$	1500	$\sim 5 \times 10^8$

FIGURE 26.3 The world's largest reflector radio telescope, at Arecibo, Puerto Rico, is 305 m in diameter and equipped with a transmitter for radar studies of Solar System objects and the Earth's ionosphere. Most of its time, however, is used for (passive) radio astronomy studies. From 1998 to 2004, ~5% of its time was used for Project Phoenix run by the SETI Institute. See Appendix D for more information about this facility. (Courtesy of Arecibo Observatory.)

surface on the receiving antenna and any deviations in its shape, as well as the random motions of electrons or photons in the electronic receiving devices, all contribute to the average noise power. Every search must set its detection threshold m high enough to reduce the statistical probability of an apparent signal actually being the result of a fluctuation in the competing noise power. The theory of the statistics of noise, the details of the signal detection hardware and software, and experience with sources of interfering signals in the vicinity guide the choice of the multiplier m (typically $\sim 2\text{--}20$). A detection is claimed when

$$P_R \geq m\langle P_N\rangle \quad \text{or} \quad \frac{P_E A_R}{4\pi r^2} \geq m\langle P_N\rangle. \quad (26.4)$$

is the physical temperature of a resistor that would produce the same equivalent blackbody noise power into bandwidth B.

This last expression allows us to calculate the range r_{max} to which a signal of a given power can be detected (as in Table 26.1), given the characteristics of a particular search project:

$$r_{\text{max}} = \sqrt{\left(\frac{A_R}{4\pi m}\right)\left(\frac{P_E}{\langle P_N\rangle}\right)}. \quad (26.5)$$

26.4 How many technical civilizations might there be?

Having investigated the size and shape of the cosmic haystack, it would be desirable to know *how many* technological civilizations (if any) produce signals that might be detectable. This would permit an estimate of how much of the haystack will need to be searched before there is a reasonable expectation that a signal will be found. It is of course impossible to know the answer in advance of success, but the *Drake*

Equation (Drake, 1962) allows us to think about the problem in an organized manner. This equation tells us that N, the number of civilizations in the Milky Way Galaxy whose electromagnetic emissions (whether intended for communication or not) are now detectable by us, can be estimated by starting with the average rate of star formation in the Galaxy, R_*, and then multiplying that by various factors representing conditions that we think necessary for technological civilizations to arise. This forms an estimate for the average number of technological civilizations arising in the Galaxy each year, which is then multiplied by the average longevity of emitting technologies to estimate the total number N that might now be detectable. The longevity may or may not be identical to the actual longevity of the intelligent species that first invented the technology. It could, for example, be longer – on Earth, civilizations have risen and fallen many times, but some of their technologies have been adopted by subsequent civilizations. The technology could also transcend its manufacturers, and continue to generate itself. The longevity of emissions could of course also be much shorter than that of the civilization, for various economic, technical, or social reasons (see below).

The Drake Equation can be written

$$N = R_* \cdot f_\odot \cdot f_p \cdot N_e \cdot f_l \cdot f_i \cdot f_c \cdot L, \qquad (26.6)$$

where R_* is the average rate of star formation in the galaxy, $f_\odot$ is the fraction of all stars that are "Sun-like," i.e., not so massive that they fuse hydrogen to helium in their cores in a time too short for intelligent life to evolve (probably billions of years), nor so low in mass that their dim glow may offer insufficient heat to sustain life in their vicinity (Chapter 21). f_p is the fraction of Sun-like stars that have planets in orbit around them, while N_e is the average number of Earth-like planets in any planetary system. Here we admit our bias for Earth-like planets as the home for any life that eventually evolves into a technical civilization. f_l is the fraction of terrestrial planets on which life actually *does* start, and f_i is the fraction of all life-starts that eventually evolve intelligence. f_c is the fraction of intelligent species that develop a civilization using a technology that generates some form of detectable emission. Finally, L is the longevity of that emission.

If the Drake Equation seems like a synopsis of the contents of this book, it is no accident. Astrobiology concerns itself with a suite of interdisciplinary programs to study life on Earth, and to search for life off Earth, and in so doing, it provides the best possible estimates of the terms in the Drake Equation. Frank Drake himself favors a value for N of $\sim 10^4$.[9] Astronomers have determined that R_* is ~ 20 per year, with reasonable accuracy. The value of $f_\odot$ is ~ 0.1, considering all stars whose mass is within a factor of two of that of the Sun, but note that this could rise significantly if ongoing deliberations conclude that small dwarf stars might, after all, host habitable planets. Our best census of giant extrasolar planets (Chapter 21) yields a value for f_p of ~ 0.1–0.2. The Kepler spacecraft that will launch in 2008 (Chapter 21) should inform us whether N_e is <1 or >1, but in any case it is unlikely to be >10. All other fraction-type terms in Eq. (26.6) are <1. Therefore, to continue discussing SETI strategies, it is sufficient to use a simple version of the Drake Equation, namely:

$$N \leq L \text{ (with } L \text{ measured in years)}, \qquad (26.7)$$

without focusing on the actual magnitude of the inequality. This simple form leads to a profound conclusion: "emitting" civilizations will not be both spatially and temporally coincident (i.e., near one another in the Milky Way at the same period during its 10 Gyr lifetime) unless their emissions typically persist for a long time. For example, if $N \sim 100$, typical separations are $\sim 10^4$ lt-yr and if $N \sim 10^4$ (Drake's preferred value), typical separations are $\sim 1,000$–$2,000$ lt-yr, which would make for far fewer candidate stars to search before likely success.

SETI is unlikely to succeed if L is short. But there are two other special conditions worth mentioning. L may be short because the inventors of technology turn it off for some good reason, and continue thriving in its absence. This is the case for the Chinese in the fifteenth century. All the great "treasure fleets" of Admiral Zheng He (Cheng-Ho), which had already navigated along the east coast of Africa and perhaps around the tip of South Africa, were called home and dismantled or left to rot on the beaches under orders from the Confucian bureaucrats who replaced the Yong-Lo Emperor. This led to China turning inward for the next 300 hundred years (Finney, 1985). Or L may only appear to be short because we are the first such technological species in the Galaxy (as asserted by the Fermi paradox), and we are still very young, with no way to know our future longevity. Gott (1993) used Bayesian statistics to estimate that there is a 95% chance that the human race will last between another 5,000 and 8 million years. If our technological longevity turns out to be at the long end of Gott's prediction,

---

[9] If you wish to calculate your own estimate for N, you can do so on the SETI Institute website at www.seti.org/drake-eq-calc.

then even if we are the first technology, SETI will perhaps eventually succeed whenever subsequent technological species emerge.

Whether or not we are the first, technological civilizations younger than us are extremely unlikely to be detected across interstellar distances. Thus, if SETI searches succeed in detecting evidence of another technology in the next few centuries, we can infer both that they are likely much older than us and that the average value of L is large.

26.5 Search strategies

26.5.1 The astrophysical background

Figure 26.4 displays the average background sky intensity over the full range of electromagnetic frequencies accessible to modern astronomy. To be detectable at a given frequency, a transmitted signal, or the portion of it that enters a particular detector, must have an intensity that can successfully compete with this natural sky background, as well as the instrumental noise in the receiver. This background radiation is due to many different classes of astrophysical sources. Stars are bright at optical frequencies, while the regions of warm gas and dust between the stars are most readily detectable in the infrared and millimeter bands. At very low radio frequencies, electrons spiralling around

galactic magnetic field lines emit synchrotron radiation, and the high-frequency, high-energy sky (X-rays, gamma-rays) exhibits emission from energetic explosions and hot gas in clusters of galaxies. In addition, the 2.73 K afterglow of the Big Bang (called the cosmic microwave background or CMB) fills the Universe in all directions and is most detectable in the microwave and infrared regions of the spectrum.

In practice, spatial, spectral, or temporal filters are used to exclude different types of background and make signals more detectable. For searches made from the ground, there are also unavoidable filters imposed on observations by the opacity of our atmosphere at some frequencies, and by human-caused interference at others. Figure 26.5 shows the height above sea level to which radiation at any given wavelength can penetrate. Although infrared and mm waves penetrate the interstellar dust that scatters and absorbs optical photons between the stars, water vapor in the Earth's atmosphere obscures almost all of these waves. Observing infrared and mm waves thus necessitates mountain-top observatories that can avail themselves of a few narrow, unabsorbed frequency bands, or orbiting telescopes operating above the atmosphere. Likewise, ultraviolet, X-ray, and gamma-ray frequencies are blocked by the ozone, oxygen, and nitrogen in the atmosphere (fortunately for our survival; Chapter 4) and observations at these frequencies also require telescopes in space.

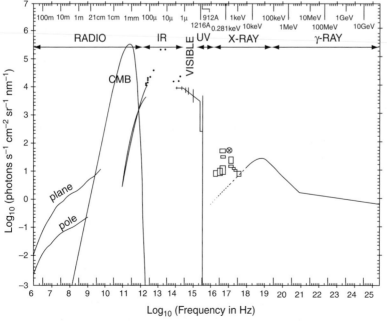

FIGURE 26.4 Spectrum of the natural sky background radiation from astrophysical sources. *Plane* and *pole* refer to values of radio background (primarily synchrotron emission), when looking in the plane or towards the pole of our Milky Way Galaxy; CMB is the 2.7 K cosmic microwave background; infrared (IR) emission comes from warm dust and gas between the stars in our Galaxy; the optical (visible) emission is a combination of light from distant galaxies and stars in the Galaxy; the ultraviolet (UV) is red-shifted Lyman-alpha emission from ionized gas within distant galaxies; and the X-ray and gamma-ray backgrounds arise from energetic processes in extragalactic and galactic sources, as well as in the intergalactic gas between galaxies. The radio portion of the spectrum comprises all frequencies $\lesssim 10^{12}$ Hz.

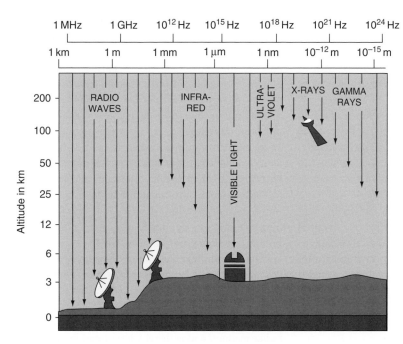

FIGURE 26.5 Atmospheric windows for electromagnetic radiation. The arrows indicate the altitude above sea level to which radiation of a given wavelength (or frequency) can penetrate the Earth's atmosphere before it is absorbed by molecules such as CO_2, O_2, O_3, and H_2O. The cartoon also illustrates the regions in which ground level, high altitude, or space-based observing platforms are appropriate. (Adapted from an image by the Dominion Radio Astronomy Observatory.)

SETI observations have traditionally concentrated on microwave radio searches (the portion of the radio spectrum from 1 to 10 GHz) where the natural background is low and where the atmospheric transparency approaches 100% (Section 26.5.4). More recently, searches have also been conducted in the optical part of the spectrum, where filters excluding all but nano-second-length pulses also nicely exclude most of the background photons from starlight. A small optical telescope with a square meter of collecting area, observing without any spectral filters, receives $\sim 10^6$ visible light photons each second from a solar-type star at a distance of 1,000 lt-yr. Therefore, the arrival of many photons (say 10–100) in only ~ 1 ns would represent a pulse signal of very high statistical significance. On the other hand, success in finding any *continuous* optical SETI signal would require extremely powerful transmitters (lasers) on "their" part to compete with their star, and long observing times on our end to average out fluctuations of the background starlight.

26.5.2 "Natural" or artificial signals?

Consider the challenge of generating some sort of a transmission that will attract the attention of an emerging technology such as ourselves. What might a deliberate beacon look like?

A case could be made that it would not be recognizable (either invisible to us, or completely inexplicable)

because any advanced technologies will only be interested in attracting the attention of other advanced technologies, and therefore their beacons would be based on science and/or technology that we currently lack. If a detected signal were completely inexplicable, one might heed Arthur C. Clarke's "third law" (Clarke, 1984): "Any sufficiently advanced technology is indistinguishable from magic." In sum, if their beacon technology is too far beyond our ken, SETI will not succeed until terrestrial technology attains the necessary competence.

Two classes of beacon signals that we could apprehend suggest themselves. The first is a signal that mimics the emission from astrophysical sources, but contains some subtle difference. The transmitting civilization would reason that when a young technology begins to explore the Universe around it, the development of certain types of astronomical detectors could be predicted by the nature of the cosmos itself. They would expect that deliberate signals, for example, resembling pulsars or quasars or gamma-ray bursters, would be registered routinely by astronomers elsewhere as they survey their environment. It might take time for these "almost natural" signals to be recognized as beacons, but the transmitting technology would have a fair degree of confidence that their efforts would eventually succeed. For instance, pulsars (rapidly rotating neutron stars) are the most precise clocks in the Universe, but physics requires that they must slow down over time. An apparent pulsar whose period did not change at all, or which oscillated between two precise values, would attract serious attention, and might finally be recognized as someone else's technology. A star whose light was 100% polarized with its sense of polarization reversing periodically would be hard to explain without technology, as would a solar-type star whose spectrum displayed an enhancement in the rare-earth elements that constitute the fissile waste products of nuclear power production (e.g., praseodymium, neodymium, zirconium). Tritium (a radioactive isotope of hydrogen containing a proton and two neutrons) has a half-life of only 12.3 years. It also has a radio frequency emission line at 1516.7 MHz (the analogue of the 1420.4 MHz spin-flip transition of neutral hydrogen atoms – Section 26.5.5). If discrete emissions were detected at the tritium frequency anywhere except in the vicinity of a recent supernova explosion, where it might have been created, technology would be a plausible explanation. These are only a few examples to illustrate serendipitous results that might attend our future exploration of the Universe.

At the other extreme, a beacon might have attributes that *cannot* be produced by astrophysics (so far as we currently understand), but can easily be generated by technology. In particular, compression in time and/or frequency could indicate a beacon. There are limits, however, to the allowed compression: basic physics indicates that the frequency range (bandwidth) $\Delta\nu$ of emissions from any observed phenomenon and the timescale Δt over which the phenomenon varies in intensity are related: the time–bandwidth product $\Delta t \Delta\nu$ must be >1. Astrophysical emissions, say, a sunspot or an interstellar cloud of gas, have very large time–bandwidth products because they result from a very large ensemble of particles (atoms, molecules, ions). The kinetic and thermal energy of these particles means that they are moving with respect to one another, and even if each particle emits radiation at precisely the same frequency, the Doppler shifts of the moving particles produce a very large bandwidth $\Delta\nu$ for the observed ensemble emission. The intensity of such emission usually also has a large value of Δt, varying only slowly if at all (see below). In contrast, our technology controls the motions of particles (e.g., within electronic devices) and produces much smaller time–bandwidth products, even approaching the minimum value. For example, $\Delta\nu$ values are very small for the carrier wave used in radio and television broadcasting or for the monochromatic beam of a laser.

It is also difficult for an astrophysical ensemble of particles to produce variable emissions with very short

time durations Δt. Because no physical effects can propagate at speeds $> c$, the size of a particle ensemble fluctuating in a coherent manner can be no bigger than $c\Delta t$, and there must be enough particles within that volume to produce a detectable emission. For example, pulsars show periodic behavior on timescales of seconds to milliseconds, but nanosecond variations have not been established.[10] A "light-nanosecond" is only 30 cm and conventional wisdom asserts that nature has no mechanism for producing detectable pulses from the particles in a volume of only ~ 0.03 m^3. In contrast, our technology can easily accomplish large compression in time. An example is the $\sim 1 \times 10^{15}$ W laser with a pulse duration of 0.4 ns developed at Lawrence Livermore Laboratory (Perry and Mourou, 1994).

Any transmitting civilization designing a deliberate beacon would also need to consider how signal propagation through the interstellar medium can modify the signal. For example, any monochromatic, continuous signal suffers from scattering off electrons in the interstellar medium and is thereby broadened in frequency. At microwave frequencies, it thus makes no sense to look for signals with $\Delta\nu \leq 0.01$ Hz – in fact, most current SETI searches employ narrowband spectrometers with $\Delta\nu \sim 1$ Hz. Furthermore, as pulsar observers are well aware, any pulsed radio signal becomes dispersed in time due to interstellar electrons, with the lower frequency components of the pulse arriving later than the high frequency ones. Since we do not know the amount of dispersion to be expected from our sought signal, we must search through a wide range of plausible values of dispersion to find any short pulse, which adds significantly to the detection problem. Neither of these effects is a problem at optical frequencies.

Although some attempts have been made to find signals mimicking astrophysics, most SETI searches have focused on signals with small time–bandwidth products. With due consideration for all the above factors, microwave SETI searches in practice have optimized their electronics to be sensitive to narrowband continuous signals and/or to long duration pulses, while at optical wavelengths broadband nanosecond pulses have been sought.

26.5.3 Targets or sky sweeps?

There are two strategies to search systematically for signals in the cosmic haystack: sweep the sky and

---

[10] Observations of giant radio pulses in the Crab nebula pulsar (Hankins *et al.*, 2003) may challenge this statement.

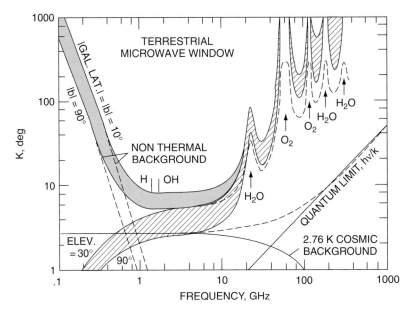

FIGURE 26.6 Observed noise temperature background for radio observations from the surface of Earth; in general the most sensitive observations are possible where this background is smallest. The range 0.1 to 1000 GHz in frequency corresponds to wavelengths of 3 m to 300 μm. Atmospheric molecules responsible for the various absorption bands are indicated. The "microwave window" is the low-noise region at ∼1–10 GHz defined by non-thermal galactic background emission at low frequencies and by atmospheric absorption at high frequencies. "Quantum limit" refers to a fundamental minimum in receiver noise proportional to photon energy.

look in all possible directions, or focus the search on directions that seem a priori more likely to contain a technological civilization. Since the only such civilization we know about has evolved on a planet in orbit about a G2 V star, solar analogues are the usual targets for the focused strategy. This so-called targeted search strategy, however, may be unnecessarily restrictive, the result of drawing conclusions from a sample of one. For example, an advanced technology may have moved away from its stellar birth place, or the correct solar analogue may be so distant that scientists compiling a list of target stars to be investigated would not know of its existence; a sky survey covers these possibilities. So ideally, every SETI search should utilize both strategies, but in practice this is seldom possible. Large telescopes with detectors that can analyze data for a long time to achieve good sensitivity on weak signals are routinely used for targeted searches. Smaller telescopes, with larger beams on the sky, and detectors that can respond well in the short time available to look at any particular direction on the sky, are better suited to sky surveys. Although the achievable sensitivity is poorer, sky surveys look in directions that would not otherwise be selected. In general, targeted searches are superior for finding weak, nearby transmitters, and sky surveys excel at finding more powerful (and presumably rarer) distant sources. If the distribution of the output powers (P_E) of all extraterrestrial transmitters were known, it would be possible to calculate statistically whether sky surveys or targeted searches had a higher probability of success

over a given time. In the absence of such knowledge, researchers have developed different figures of merit to compare the efficacy of various search strategies. These figures of merit disagree in detail, but agree that searching more stars over more bandwidth is always better.

26.5.4 Which frequency ranges?

Which frequency ranges are optimum for searches? In this section we will examine more closely the radio noise that competes with any signal that we are trying to detect. Figure 26.6 presents a more detailed look at the background radiation encountered by radio telescopes, and shows that a frequency band of ∼1–10 GHz (∼30–3 cm in wavelength) defines a low-noise "terrestrial microwave window" ideal for sensitive observations from the Earth's surface.[11] The high background noise defining the low-frequency edge of the window is due to synchrotron radiation generated by electrons spiralling around magnetic field lines that thread through the Milky Way and is stronger in the direction of the galactic plane than towards the poles.

---

[11] The level of radio emission at low frequencies rises in Fig. 26.6 (measured in terms of noise temperature) and falls in Fig. 26.4 (measured in photon flux); the reason for this seeming paradox is that the two figures plot different measures of radio emission. Both are useful, but for signal-to-noise considerations as in this section, noise temperature (as defined in footnote 8) is the quantity of interest.

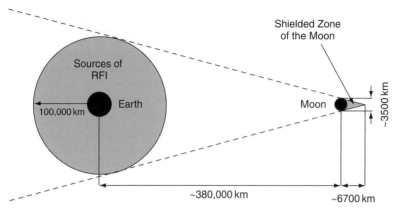

Shielded Zone
of the Moon

~3500 km

Sources of
RFI

100,000 km Earth

Moon

~380,000 km

~6700 km

FIGURE 26.7 Shielded zone of the Moon as defined by ITU Radio Regulations Article S22.22 – S22.2. SETI observations from the far side of the Moon would be free of the deleterious effects of radio frequency interference (RFI) from Earth (diagram not to scale). (Courtesy of Guillermo Lemarchand.)

(In fact, part of the static that can be heard on an FM radio or television set tuned between broadcast stations comes from this.)

The high-frequency edge of the terrestrial microwave window is caused by the noise background from absorbing atmospheric water vapor and molecular oxygen. This noise can be eliminated by going into space orbits (or the lunar far side), where only the cosmic microwave background hinders at frequencies up to ~60 GHz, at which point the "shot noise" in the receiver from the individual photons becomes increasingly troublesome ($\propto h\nu/k$; labeled "quantum limit" in Fig. 26.6). Some authors have speculated that an advanced civilization would select a frequency for its beacon requiring reception by a space-based platform, thus enforcing a minimum technological threshold for any civilization worthy of being detected.

Current microwave SETI programs confine themselves to the terrestrial microwave window; SETI from space is presently too expensive to undertake. Nevertheless, radio astronomers have become interested in one particular extraterrestrial locale. Because the Moon's synchronous rotation means that its far side is never visible from Earth, this is the one place in our Solar System whose sky never experiences human-generated interference from the Earth (a huge problem for SETI – see Section 26.6). Cognizant of this growing radio frequency interference (RFI) on Earth and in Earth orbit, researchers convinced the International Telecommunications Union in 1979 to protect the shielded zone of the Moon (as defined in Fig. 26.7) from all RFI (e.g., from future transmitters on spacecraft orbiting the Moon). If lunar bases are ever developed for other reasons, then SETI (and "traditional" radio astronomy) may some day be conducted from this shielded zone, investigating higher frequencies as well as those frequencies now contaminated by terrestrial RFI.[12]

Optical SETI for pulses is currently being carried out without imposing any spectral filters, obviating the need to search through individual optical frequencies. As previously mentioned, very short temporal filters are applied to eliminate stellar photons, and there do not appear to be any natural sources of background optical emission with nanosecond temporal variations. Instead, the challenge for optical SETI observers is that the detectors themselves produce events from corona discharge, ion feedback, and cosmic rays, thus requiring two or three such detectors working in coincidence to reduce the number of false positive events. Extending these observations into the infrared will require high-altitude sites and/or spacecraft, as well as developing suitable detectors. Observations in the infrared would avoid the problem of interstellar absorption by dust grains, which limits optical SETI to distances of ~1,000 lt-yr. Within that volume, however, there are about one million Sun-like stars.

26.5.5 Magic frequencies, places, and times

Today's microwave SETI detectors require spectrometers with $\geq 10^8$ spectral channels, in order to conduct systematic searches for signals buried somewhere in the terrestrial microwave window from 1 to 10 GHz (requiring 10^{10} channels of 1 Hz width). An alternative to covering the entire band is a search based on preferred or *magic* frequencies that can be argued as likely

---

[12] Ironically, one of the reasons to develop a lunar base might be to support a satellite launch and repair facility, now being contemplated by NASA, at the L_2 Lagrange point directly above the lunar farside, thereby disturbing the radio-quiet shielded zone.

to be mutually adopted by transmitter and receiver. Since water is so fundamental to our form of life, some researchers have suggested that a preferred portion of the terrestrial microwave window is the "cosmic water hole" marked by the natural emission lines of H and OH, the dissociation products of water. Figure 26.6 indicates the 21 cm H hyperfine-transition emission line at 1.42 GHz (from the most abundant element in the Universe) and the highest of the four maser lines associated with the OH radical at 1.72 GHz. Frequencies obtained by multiplying or dividing the 1.42 GHz H line frequency by fundamental constants such as π and e and the fine structure constant α ($\sim 1/137$) have also been promoted as potential interstellar communication channels (Blair, 1986); these have the advantage of avoiding the ubiquitous natural emission from the Galaxy's hydrogen at 1.42 GHz. But by far the greatest amount of time spent on any magic radio frequency has been on the H line itself, as first proposed by Cocconi and Morrison (1959). At optical frequencies, on the other hand, the broadband nature of the sought optical pulses means that no particular frequencies are singled out.

There are a few distinctive *locations* and directions in the Milky Way that have attracted attention from SETI researchers: (a) the galactic center is unique; (b) 90% of the stars reside within a few degrees of the galactic plane; (c) the rotational axis of the Galaxy is well defined; and (d) there is a small number of outstanding astrophysical sources that represent oft-observed directions. Transmissions by an extraterrestrial civilization aimed towards or away from these astrophysical sources could result in our detection of a beamed signal during the course of our routine astronomical studies. For those sources that are themselves masers (such as from OH, water, and methanol molecules), there is an added bonus. Molecular emission in these clouds would provide strong amplification of any signal transmitted through them at the correct frequency; a properly aligned detector on the output side of the maser would benefit from a free amplifier in space and receive a much stronger signal than was originally transmitted (Gold, 1976).

It would be extremely useful if there existed some marker in *time* that could logically be deduced by both transmitter and receiver as the moment for signal reception. Novae and supernovae are relatively rare events in the Galaxy, and might be used to synchronize the timing and aiming of deliberate transmission and reception of signals. Lemarchand (1994) has described the "SETI ellipsoid" (Fig. 26.8). In this scheme it is suggested that SETI researchers on Earth should begin observations of a given target star at the time

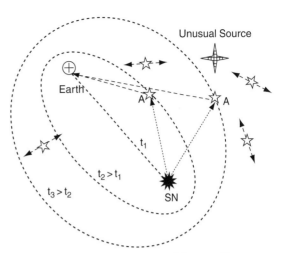

FIGURE 26.8 SETI strategies in time and direction. The star symbols represent transmitting civilizations. Two transmission strategies are illustrated: (1) continuous transmission in the direction towards and opposite (indicated by short arrows) an unusual astrophysical source that is likely to be well studied by other technological civilizations such as ours; and (2) a synchronized transmission and reception based on the rare occurrence of a supernova (SN) or other event. In the latter case, any possible transmitter (two are shown with label A) begins broadcasting towards likely candidate receivers (e.g., the Sun/ Earth) at the moment that it detects the existence of the supernova. Receivers on Earth begin observing particular target stars when they first fall on the boundary of an expanding "SETI-ellipsoid" whose foci are the Earth and the supernova. (Courtesy of Maggie Turnbull.)

when that star first appears on the surface of the expanding ellipsoid whose foci are a recent supernova and the Earth. Sullivan (1991) attempted to define magic periods or time durations for pulsed signals. There is little to constrain possible periodicity or duration, however, except in the case of the broad optical pulses, where a pulse of $\Delta t \leq 1$ nanosecond would seem more consistent with a technological source rather than astrophysics since we have not yet discovered any natural sources of radiation that vary this rapidly. Search programs that have used these magic, or hypothesis-constraining, observational strategies can be found in the archive of SETI searches given in footnote 1.

26.6 Current searches

Table 26.2 lists the parameters of current and recent SETI sky surveys and targeted searches, along with brief notes about the search strategies being employed, and a reference to the website of each project. All of these projects operate with funding from philanthropic sources. Since the termination of NASA's SETI project

TABLE 26.2. Recent and current SETI projects (2005). A complete archive of all past SETI searches is maintained at www.seti.org/searcharchive

---

META II

Start date:	1990
Observers:	Lemarchand
Site:	Institute for Argentine Radioastronomy
Instr. size (m):	30 (one of two)
Search freq. (MHz):	1420.4, 1667, 3300
Freq. res. (Hz):	0.05 and 33
Objects:	Sky survey of southern skies, 90 target stars, and OH masers
Flux limits (W/m^2):	1×10^{-23} to 7×10^{-25}
Total hours:	Ongoing
Reference:	<http://www.iar.unlp.edu.ar/ES/seti-boston.htm>
Comments:	Search for signals that have been Doppler compensated to rest frame of Solar System barycenter, Galactic center, or cosmic background radiation. A duplicate of Harvard's former META system built by Argentinian engineers and financed by the Planetary Society.
	Simultaneous observations with META over the declination range $-10°$ to $-30°$. Major upgrades in 1996 to permit long integration times, and switching between antennas. Search also of OH maser sources, looking for amplified signals with META II and digital correlator.

ARGUS

Start date:	1996
Observers:	SETI League
Site:	Multiple sites worldwide (currently ~130 backyard projects)
Instr. size (m):	~3–10 (satellite TV dishes)
Search freq. (MHz):	1420–1720
Freq. res. (Hz):	1
Objects:	Objective is to cover all the sky
Flux limits (W/m^2):	~1×10^{-21} (varies)
Total hours:	Ongoing
Reference:	<http://www.setileague.org>
Comments:	Attempt to organize radio amateurs to provide continuous sky coverage for strong, transient signals using systems that can be bought and built by individuals. SETI League currently has 1,456 members running 130 sites in 23 countries. Their website is very active, but they are financially challenged, and have not engaged in systematic observations and archiving.

SERENDIP IV

Start date:	1996
Observers:	Werthimer *et al.*
Site:	Arecibo
Instr. size (m):	305
Search freq. (MHz):	1420 ± 50
Freq. res. (Hz):	0.6
Objects:	Random survey of 30% of sky visible from Arecibo
Flux limits (W/m^2):	5×10^{-24}
Total hours:	Ongoing

TABLE 26.2. (cont.)

Reference:	\<http://seti.ssl.berkeley.edu/serendip/serendip.html\>
Comments:	Commensal search occurring at twice sidereal rate in backwards direction while radio astronomers track targets using Gregorian system. Covers sky in a random fashion every three years; candidate signals are those that recur at same frequency and location. The best candidate signals are re-observed occasionally with directed observations.

Project Phoenix

Start date:	1998–2004
Observers:	Seti Institute
Site:	Arecibo Observatory and Lovell Telescope at Jodrell Bank
Instr. size (m):	305 and 76
Search freq. (MHz):	1200 to 3000 dual polarizations
Freq. res. (Hz):	0.67
Objects:	850 nearby stars
Flux limits (W/m^2):	1×10^{-26}
Total hours:	2,300 hours
Reference:	\<http://www.seti.org\>
Comments:	Continuation of NASA targeted search survey of nearby stars, using real-time signal processing systems and a pair of widely separated observatories to help discriminate against RFI

Southern SERENDIP

Start date:	1998–2003
Observers:	Seti Australia
Site:	Parkes
Instr. size (m):	64
Search freq. (MHz):	1420.405 ± 8.82
Freq. res. (Hz):	0.07 to 1200
Objects:	Random southern sky survey
Flux limits (W/m^2):	4×10^{-24}
Total hours:	~20% duty cycle
Reference:	\<http://seti.uws.edu.au\>
Comments:	Commensal search that used two of the 13 beams of Parkes focal plane array to discriminate against RFI

SEVENDIP

Start date:	1998
Observers:	Werthimer *et al.*
Site:	Leuschner Observatory
Instr. size (m):	0.8
Search wavelength:	300–650 nm
Freq. res. (Hz):	Entire optical band
Objects:	7,225 solar-type stars, 104 galaxies to date
Flux limits (W/m^2):	1.5×10^{-9} peak during 1 ns pulse, or 1.5×10^{-20} average per 100 second observation
Total hours:	200 (ongoing)
Reference:	\<http://seti.ssl.berkeley.edu/opticalseti/\>
Comments:	First optical search to use two high time resolution photomultiplier tubes in coincidence to look for nanosecond pulses; since upgraded with three PMTs to improve false alarm rate.

TABLE 26.2. (cont.)

Harvard Optical SETI

Start date:	1998
Observers:	Horowitz *et al.*
Site:	Oak Ridge Observatory
Instr. size (m):	1.5
Search wavelength:	350–700 nm
Freq. res. (Hz):	Entire optical band
Objects:	13,000 solar-type stars of which 4000 observed to date
Flux limits (W/m^2):	4×10^{-9} peak in <5 ns pulse, or 4×10^{-20} average per 500 second observation
Total hours:	Ongoing
Reference:	<http://seti.harvard.edu/oseti/>
Comments:	Search for nanosecond laser pulses, with hybrid avalanche photodiodes in coincidence. Piggybacks on nightly searches for extrasolar planets. Now operated in coincidence with cloned detector on 0.9 m telescope at Princeton, using GPS and internet for timing.

SETI@home

Start date:	1999
Observers:	Werthimer and Anderson
Site:	Arecibo
Instr. size (m):	305
Search freq. (MHz):	1420.405 ± 1.25 MHz
Freq. res. (Hz):	0.6 Hz
Objects:	Data taken from SERENDIP IV – sky visible from Arecibo
Flux limits (W/m^2):	5×10^{-25}
Total hours:	Ongoing
Reference:	<http://setiathome.ssl.berkeley.edu>
Comments:	Hugely successful experiment in distributed computing. Has permitted more detailed processing of a fraction of SERENDIP IV data by harnessing idle CPU cycles of a total of 5 million personal and corporate computers (of which at most $\sim$500,000 at any one time).

Princeton Optical SETI

Start date:	2001
Observers:	Groth *et al.*
Site:	FitzRandolph Observatory
Instr. size (m):	0.9
Search wavelength:	350–700 nm
Freq. res. (Hz):	Entire optical band
Objects:	Solar-type stars being observed by Harvard Optical SETI project
Flux limits (W/m^2):	4×10^{-9} peak in <5 ns pulse, or 4×10^{-20} average per 500 second observation
Total hours:	Ongoing
Reference:	<http://observatory.princeton.edu/oseti/>
Comments:	Search for nanosecond laser pulses, with hybrid avalanche photodiodes in coincidence. Operates in coincidence with Harvard Optical SETI project, using GPS and internet for timing.

(officially known as the High Resolution Microwave Survey) in 1993, there has been no governmental funding available for SETI observing programs in the US. With the current focus on astrobiology and a NASA mission statement that asks "Are we alone?", it is possible that this will soon change.

The table attempts to be inclusive so as to illuminate the various approaches to signal detection, as well as the techniques used to discriminate against radio frequency interference (RFI), thus reducing the number of false positive events and improving search efficiency. RFI is the bane of SETI, for our own technology's activities often lead to signals that mimic precisely the sort of signal we seek from another technology. The problem is also steadily worsening as more and more of the Earth's atmosphere is filled with radio waves from garage-door openers, wireless connections, satellite communications, navigational radars, etc. Even for the most ingeniously designed radio telescope, these signals become mixed in with the authentic signals from the patch of sky being investigated, and time-consuming procedures must be undertaken to discriminate between the wheat and the chaff.

The large number of entries in Table 26.2 may give the misleading impression that a large community of scientists and amateurs are routinely and continuously conducting effective SETI explorations. In truth, with the exception of the *commensal*[13] SERENDIP, SETI@home, and Harvard optical SETI projects, most of the search projects are not on the air on most days (or nights). For many SETI campaigns a large radio telescope is available for only a small percentage of the time. This was true most noticeably for Project Phoenix that relied on gaining observing time on some of the largest radioastronomical instruments in the world. Phoenix started observations in 1995 renting the 64-m diameter Parkes radio telescope in Australia (*The Dish*) paired with another remote 26-m dish for exclusive use over a six-month period. From 1996 to 1998, Phoenix used about 20% of the time on the decommissioned National Radio Astronomy Observatory 43-m antenna at Green Bank, West Virginia (site of Drake's original Project Ozma in 1960), and a 30-m dish in Georgia built from a former satellite ground station by the students and faculty of Georgia Institute of Technology. The

observations of nearby stars were then finished between 1998 and 2004, simultaneously observing with the 305-m dish at Arecibo and the 76-m Lovell Telescope at Jodrell Bank, England. The end result was the observation of 850 Sun-like stars, all within 150 lt-yr of Earth, over the frequency range from 1.2 to 3.0 GHz (for a few minutes at each frequency). The resultant sensitivity was sufficient to detect any transmitter stronger than several hundred times the power of a typical airport acquisition radar ($P_E \sim 10^{11}$ W; Table 26.1).

While Project Phoenix scientists had access to an impressive number of hours of telescope time (through outright purchase or competitive proposals) – far more than the average astronomer – this still left the targeted searches off the air much of the time. In order to alleviate this problem, Section 26.8.2 describes a new telescope being built as a dedicated SETI facility and simultaneous radio astronomy facility. When Project Phoenix did have access to the sky, it used a suite of near-real-time signal detection algorithms to look for patterns in time and frequency indicative of (a) narrowband continuous signals that may or may not change their frequency over time, and (b) drifting (in frequency) narrowband pulses. Detected candidate signals were compared to a database of all signals seen within the previous week. Signals that matched against the database, despite the telescope looking in a different direction, were discarded as RFI, whereas unmatched signals became candidates to be immediately re-observed and compared with the results from the second observatory. Candidate signals seen at *both* observatories, with the expected difference in Doppler shift and frequency drift for the two sites (primarily due to the Earth's own spin relative to a distant stellar target), triggered automatic follow-up observations that continued until the signal could be demonstrated to be either of human origin or still classifiable as a potential ETI candidate. Over the decade of Project Phoenix observations, this latter event happened very infrequently, and usually when there was some problem with making simultaneous observations with widely separated telescopes. A viable candidate ETI signal would have initiated a request for an independent confirmation from another observatory and, if still passing all tests, a press conference to tell the world and to release discovery data to all astronomical facilities to encourage additional observations. During Project Phoenix, our false positive events never escalated to this last step.

In contrast to Phoenix, SERENDIP and SETI@home have maximized their time on the sky (virtually 100%) by giving up on any requirement of

---

[13] A *commensal* (or piggyback) project is one that gains telescope observing time by being willing to look at whichever part of the sky is chosen by another astronomer who has charge of the telescope. The SETI electronics used to detect and process incoming signals is entirely separate from that of the "normal" observations.

where in the sky the observations take place. Given the unique geometry of the Arecibo telescope, it is possible for SERENDIP and SETI@home to conduct a continuous, random survey of the 28% of the sky that is accessible. Over several years, much of the sky by chance gets observed multiple times, and that becomes the key to finding candidate signals. Only signals detected repeatedly whenever the same direction on the sky is accessed warrant additional attention. The SERENDIP system keeps up with the real-time signal detection required during its continuous observations by applying only a very simple algorithm that records events exceeding a high threshold within its 100 MHz bandwidth. Information about all these events is stored and post-processing filters used to recognize and discard many types of RFI. Signals that appear to reoccur from the same direction on the sky are noted and checked later during scheduled, targeted observations. No candidate signal has yet been re-acquired during these more sensitive follow-up sessions.

SETI@home combines the complex pattern recognition algorithms from Project Phoenix with the sky survey strategy of SERENDIP. The central 2.5% of the SERENDIP bandwidth are recorded directly to tape, and the data sliced up into time-bandwidth data packets and shipped off to eager volunteers around the world. As many as 500,000 have participated in any given week (with a total of 5 million over the period 1999–2005), donating a prodigious amount of off-line computer power to look for signal patterns in frequency and time. The vast quantity of CPU cycles available through this innovative distributed computing project allows SETI@home to look for continuous and pulsed signals over a much wider range of frequency drift rates, pulse widths, and pulse repetition rates than Project Phoenix could, but only over a very narrow range of frequencies surrounding the 1420.406 MHz line of hydrogen.

So which is better – SERENDIP, SETI@home, Project Phoenix, or some other search strategy like optical SETI? That's like asking whether apples are better than oranges – both belong in a healthful diet, along with grapes and nuts. Until we succeed, it isn't possible to know what is the *best* way to search, but it does seem reasonable to suggest that inclusion rather than exclusion may lead to a higher probability of success.

26.7 Should we be transmitting?

All of the discussions in this chapter have been concerned with *receiving* signals transmitted by another civilization. But since the first half of the twentieth century, our own technologies have been inadvertently broadcasting radio signals that could be detected by any other civilization with technological capabilities significantly better than current terrestrial standards. Moreover, in 1974 the Arecibo Observatory celebrated the upgrade of its antenna surface by sending a 1679-bit, pictorial message in the direction of the globular cluster M13 (Staff of the NAIC, 1975).[14] On at least one occasion a for-profit group has rented the large Evpatoriya telescope in the former Soviet Union to transmit to the stars the messages from their paying customers. These transmissions have been so short-lived that the probability of their reception by another civilization is vanishingly small. But should we regularly and deliberately broadcast signals in the direction of nearby targets, or throughout the galactic plane, or to distant galaxies, or in any of the other "magic" directions that have been previously considered for a receiving program?

As discussed in Section 24.4, most SETI research groups have signed a voluntary Post-Detection Protocol in which they agree that no reply to a detected signal should be sent until global consultation has approved the idea of transmission and the contents of a reply. This protocol, however, does not specifically refer to the *ab initio* transmission of deliberate broadcasts. In a series of workshops Ekers *et al.* (2002) considered whether deliberate transmission should be a strategy of choice in the near future, but decided against it.

For at least the next decade or so, our radio leakage will continue, although as we use the spectrum more efficiently, the signals will look more and more like noise. Therefore, we will continue to be detectable for the near term, and we can take time to consider whether we should initiate deliberate transmitting activities. Transmitting is a harder job than receiving, so humans, as an emerging technological species, should listen first. The job of transmission is harder not just because it is necessary to pay for the transmitted power, but because it is necessary to reach a global consensus on the ethical, political, and societal questions of who will speak for Earth and what they will say. Finally, it is necessary to "say" whatever we select for a *very* long time. The longevity of the transmitting program might be one practical definition of the value of L in the Drake Equation (26.5), and we have already seen

---

[14] See www.seti-inst.edu/science/a-message.html for details.

that unless L is large, success is unlikely. We are not mature enough as a species to seriously consider plans that stretch thousands or millions of years into the future. This discussion reinforces the conclusion of Section 26.4: any technology that near-term SETI efforts succeed in detecting will be older than our own.

26.8 Future searches

Even without US governmental funding for SETI, several new telescopes dedicated to searches for extraterrestrial signals at optical and radio wavelengths are under construction. A dedicated instrument for optical SETI sky surveys is being built by Harvard University. A partnership between the University of California at Berkeley and the SETI Institute is currently constructing the Allen Telescope Array to simultaneously conduct continuous targeted SETI searches and traditional radio astronomy.

26.8.1 Harvard Optical SETI Telescope

The number of optical SETI projects looking for broadband pulses from stars is rapidly growing, as are innovative schemes for lowering the false positive rates from instrumental and natural backgrounds. With funding from The Planetary Society and the Bosack-Kruger Charitable Foundation, the optical SETI group at Harvard University has constructed a 1.8-m diameter, dedicated all-sky survey optical telescope at Oak Ridge Observatory in Massachusetts. The telescope remains fixed on a given night and observes the sky as it passes by (so-called drift scans). Mirrors have been manufactured inexpensively because the system does not require image quality optics. This new telescope searches for powerful transmitters from a large collection of stars by conducting scans of the sky in $1.6° \times 0.2°$ strips (with a dwell time, defined by the Earth's rotation, of about one minute). The search is optimized for detecting broadband optical nanosecond pulses by using two arrays of 512 photodiodes. The visible sky ($\sim$60% of the entire sky) can be scanned in approximately 150 clear nights; by March 2007 half of this had been achieved. An overall improvement in sensitivity of a factor of 1.6 beyond the value listed in Table 26.2 for the Harvard optical SETI project is obtained, but the survey will cover a much larger number of stars. This instrument is unique in the world, but its cost is sufficiently low that it could be easily copied, say to survey the southern sky.

26.8.2 The Allen Telescope Array for SETI and radio astronomy

The Allen Telescope Array (ATA) is currently under construction at the Hat Creek Radio Observatory in northern California, with initial funding provided by the Paul G. Allen Family Foundation. The ATA represents a revolution in the way large cm-wavelength antennas are built. The collecting area will be 10^4 m^2, consisting of 350 arrayed antennas, each 6.1 m in diameter. The system temperature will be about 40 K in each of two linear polarizations over an extremely wide bandwidth from 0.5 to 11 GHz captured with a single low noise amplifier chip and a log-periodic feed. Because of the small size of the antenna elements, the ATA will be able to image a very large field of view in any direction on the sky, and the architecture of the telescope electronics permits it to simultaneously synthesize up to 32 narrow beams within that field of view, at up to four different frequencies. The cost of the ATA will be about 20% of the cost of traditional telescopes of comparable size because of mass production technologies and adaptation of inexpensive electronics developed primarily for the telecommunications industry. Figure 26.9 shows the first ATA antennas at Hat Creek.

Starting with an all-sky catalog of several million target stars, in any field of view chosen for astronomical study and imaged with a radio astronomy correlator, there will be on the order of 10 candidate SETI stars visible, and this is the key to simultaneous and continuous use of the ATA for both SETI and radio astronomy. Independent beams will be formed on as many of those stars as the availability of SETI signal processors permits, while other beams could also study other astronomical point sources (e.g., pulsars) within the field of view being imaged. The continuous availability of the ATA and the simultaneous observation of multiple target stars and/or frequencies will improve the speed of targeted SETI searches by at least two orders of magnitude. Within the next decade, it should be possible to observe $\sim$1 million stars from 1 to 10 GHz at sensitivities comparable to the best target searches to date. The stars will be selected on the basis of their mass, spectral type, age, metallicity, information about close companion stars, and orbiting planets. As we learn more about the exact conditions within our cosmic environment that enabled the origin and evolution of life on Earth, the criteria for inclusion in a catalog of "habstars" (Turnbull and Tarter, 2003a, b) will become more refined. Current catalogs contain only about a quarter of a million stars. The catalogs

FIGURE 26.9 Aerial view of the first thirty 6.1-m diameter Allen Telescope Array (ATA) antennas at the Hat Creek Observatory in northern California, including the construction tent. (Photo courtesy of Seth Shostak.)

will grow over the lifetime of the ATA as future space-craft missions such as SIM and GAIA[15] make fundamental measurements of more distant stars. Until then, observations will concentrate on the lower frequencies, where ATA's observing beam is larger, and current catalogs provide adequate numbers of targets. The first 42-dish array began operation in 2007, and the full 350 element array will be completed at a rate determined by fundraising success.

One of the goals for the ATA is to be able to observe at any frequency in the range 0.5 to 11 GHz, despite the problem of RFI. The choice of the remote Hat Creek site minimizes interference from ground-based populations, but satellite interference is inescapable. However, this array's 350 antennas, observing in two polarizations, yield 700 ways that signals can be combined to form the desired images or beams. The ATA will be the first array to mitigate aggressively against satellite interference in real time by continuously tracking satellites throughout observations and forming nulls ("beams" with nearly zero sensitivity) on their fast-changing sky positions.

26.8.3 The more distant future

In order to improve the sensitivity of both optical and microwave SETI observations it will be desirable to utilize even larger telescopes in the future. Today, 10-meter-class telescopes are the state of the art in optical astronomy, but initial design studies are underway for

30 to 100 meter telescopes utilizing sophisticated adaptive optics. Unlike the ATA, where small array elements provide large fields of view that enable simultaneous targeted SETI observations and traditional radio astronomy research, the fields of view of these extremely large optical instruments are tiny. Even if some sort of camera is placed at the focus to broaden the view, it is unlikely that a SETI stellar target and an astronomical object of interest will occupy the same small field. Thus it seems that substantial improvement in the sensitivity of future optical targeted searches will require a dedicated, large optical SETI telescope, but there are no plans for such an instrument at this time. However, several arrays of 10-meter-class, segmented optical "light buckets" (having limited imaging quality) are being planned or constructed by astrophysicists to study the highest energy gamma-rays (see, e.g., veritas.sao.arizona.edu). Each telescope has a dense cluster of photodiodes located at its focus. These photodiodes are intended to detect the light radiated from air showers generated by the interactions of energetic photons with the atmosphere. Having many widely spaced telescopes then allows calculation of an accurate direction of arrival for the triggering gamma-ray. Because many photodiodes on each antenna are expected to register during a spatially extended air shower, *single* photodiode events are discarded as noise. It may be possible for SETI teams to develop fast electronics that could "piggyback" on the main experiment and sense when single photodiode events occur simultaneously from a single sky direction and on multiple telescopes in the array.

At microwave frequencies, there will be several opportunities for exploring more of the cosmic

[15] See planetquest.jpl.nasa.gov/SIM/sim_index.html and astro.estec.esa.nl/GAIA/ for details of the planned missions.

haystack in coming years: (a) observing more of the sky all of the time in order to search for transient signals, (b) extending the sensitivity of targeted searches by using bigger antennas, and (c) surveying much of the galactic plane over a wide range of frequencies.

The SETI Institute sponsored a series of workshops to examine the most productive opportunities for searching in the decades to come. Their deliberations (Ekers *et al.*, 2002) led to implementation of the first searches for nanosecond optical pulses (as previously discussed), the basic design of what has become the Allen Telescope Array, and a concept for an Omnidirectional Sky Survey (OSS) instrument to stare at all (or most) of the sky continuously at microwave frequencies, looking for short-duration signals that appear only infrequently. The OSS would make use of a huge number of very small antennas, little more than dipoles or small spiral coils ($\sim$30 cm in diameter). Extraordinary amounts of computer capacity will be required to combine the output of so many small antennas together to image the entire sky and search for ETI signals from every different direction in the sky. Today we can afford to build the antenna elements, and are learning how to build very inexpensive digital receivers for each antenna that will provide wide frequency coverage, but we cannot yet afford to combine these antennas and process the received bandwidth for SETI signals. However, if present trends in increasing computer power continue, the estimated 4×10^{16} ops (computer operations per second) needed to combine an array of 4096 elements (while searching over 4 GHz of bandwidth) may be affordable by 2020. And two decades later the 2×10^{21} ops of computer power needed for an array of 10^6 elements (and a bandwidth of 1,000 GHz) might be affordable. In the meantime, Project Argus, a small array of 64 elements and 1.6 GHz of bandwidth (Ellingson, 2002), is serving as a prototype to develop algorithms to deal with the inevitable plethora of RFI with which such an ominidirectional array must contend.

Over the past decade, an international team of radio astronomers has been developing plans to build a telescope 100 times as large as the ATA. Called the Square Kilometer Array (SKA) because of its 10^6 m^2 of collecting area, this array is intended for studying a wide variety of astronomical problems. There is a number of different concepts for building this enormous telescope at an affordable cost perhaps a decade from now, and most of them would allow the same sort of simultaneous SETI observations already described for the ATA. A factor of 100 in collecting area translates into a factor

of 100 in sensitivity, so we could detect the same strength transmitter 10 times farther away. Assuming that the Square Kilometer Array gets built in an appropriate fashion, within two or three decades we should be able to survey a large fraction of the Milky Way Galaxy over much of the microwave spectrum.

It may also be possible to conduct a survey of much of the plane of the Milky Way in the nearer future. Extremely large spectrometers and SETI signal processors utilizing $\sim$10^{10} channels should shortly be affordable. If one or more of the 34-m diameter antennas that are part of NASA's Deep Space Network of satellite tracking antennas could be provided with very wideband receivers and feeds (similar to those used on the ATA), then a survey of the galactic plane covering 2–23 GHz could be accomplished within a decade.[16]

If all of these planned searches are completed and yet no signal is found, such a null result will begin to become significant for microwave and optical technologies. There is always the chance that in coming years we will discover new technologies for other parts of the electromagnetic spectrum that may be useful in searches for signals or artifacts of astro-engineering. SETI researchers reserve the right to get smarter over time, and will aggressively pursue searches with any new technologies to the extent that funding and time permit.

26.9 What if we succeed?

Because there is great public interest in the question of whether there are other sentient creatures in the Universe, it is important that any potential claim of discovery be accompanied by very credible evidence for the detection. A signal seen just once and/or at a level not much above the background noise is not convincing. Although there are reasonable classes of transmitters that might produce such transient signals (e.g., a high gain antenna beam steered rapidly around the sky to target a large number of stars sequentially), most SETI researchers today demand that candidate signals repeat and stand up to independent confirmation attempts by others at a distant observatory. The criterion of long-term repeatability is a hallmark of SETI and immediately separates it from the claims of UFOs, etc. Independent confirmation is also the best way to guard against hoaxes. In a proactive plan for success,

[16] This galactic plane survey has not yet been funded; an overview of the instrumentation required to do the job can be found at **seti**athome.ssl.berkeley.edu/~aparsons/papers/2004-01-08_URSI_Presentation.pdf.

members of the SETI Permanent Study Group of the International Academy of Astronautics have developed the "Rio Scale," somewhat analogous to the Torino Scale used to classify the potential likelihood and degree of disaster from newly detected, near-Earth asteroids. The Rio Scale rates both the significance and the credibility of announced candidate signals and evolves over time as more cases are experienced.[17] Since there have been few opportunities to exercise or publicize this metric, it has so far been calibrated roughly only against science fiction stories that deal with the discovery of extraterrestrial intelligence.

The SETI Post-Detection Protocol (Section 24.4) outlines the actions that should follow the detection of a signal or discovery of other evidence of the existence of another civilization: carefully verify the suspected discovery, attempt to get an independent confirmation, tell the world, and do not reply until there is global consensus. Although SETI researchers who have adopted this protocol can be expected to abide by it, others might not. Following a public announcement, it is possible that many people around the world with access to transmitters might decide to transmit their own replies. The resulting cacophony might be the most accurate, but least informative, representation of the multi-cultural twenty-first-century planet Earth.

How would a successful SETI program change our future? If polls taken in the US are any guide, the immediate impact might be slight. A Gallup poll conducted in 1999 found that 61% of those questioned thought there is life on other planets in the Universe and 41% thought it might be something like humans.[18] Interestingly, the poll respondents were more cynical about the possible presence of life on Mars. If so many people already believe in the existence of extraterrestrial life, even intelligent life, success in SETI will simply reaffirm their convictions. A detection announcement could cause little disruption to the activities of worldwide society if education of the public and media has been actively and effectively pursued by SETI researchers, and if the discovery team follows the Post-Detection Protocol and makes use of the Rio Scale. As Ashkenazi (1992) has shown, the world's major religions should have little difficulty incorporating this new information into their existing dogmas.

The social scientists, historians, religious leaders, and diplomats who joined SETI researchers to consider the various cultural aspects of a successful detection (Billingham et al., 1999) concluded that short-term reaction to an announcement would unfold in accordance to the personal, religious, and political belief systems in place at the time.

In the long run, however, a successful SETI detection will change humanity's view of itself and of its place in the Cosmos, just as the work of Copernicus and Darwin did in the past. Philip Morrison, co-author of the initial journal article on SETI, has called SETI the "archeology of the future." By this he meant that a signal will tell us about the transmitter's past or "archeology" because of the time that it will have taken the transmission to reach us. However, because of the large-L argument, it will also tell us that humans on Earth may in fact reasonably anticipate a long-term future too. Some scientists have postulated that a signal from ETI will be a sort of *Encyclopedia Galactica* telling us many things about the Universe and offering solutions to the myriad problems faced by our emerging technology. This is probably an overly anthropocentric speculation. However, even if the signal is only a cosmic dial-tone that proves nothing more than the fact of their existence, Morrison's conclusion about the future should provide additional incentive for humanity to solve its own problems. Knowing that another society has found a solution to long-term survival, when many current indicators would suggest otherwise for us, might be critical to our own future on Earth.

26.10 What if we don't succeed?

In this chapter, I have argued that within decades we can expect to have made microwave searches of significant portions of our Milky Way Galaxy, with sensitivities sufficient to detect the technology analogues of twenty-first-century Earth. In that same time, optical targeted searches and sky surveys can probably be extended to search the million or so solar-type stars within 1,000 lt-yr of Earth for signals as strong as the lasers on our drawing boards today. An omnidirectional microwave search for transient signals will have been conducted, and probably many other surveys, as well.

What if all this turns up nothing? At every epoch in the future, humans will need to reassess whether the resources necessary for SETI to keep searching in the same ways or in new ways are justified by the importance of the question. So far, humans have not yet lost

---

[17] The Rio Scale calculator is available at www.setileague.org/iaaseti/rioscale.htm.

[18] "Life on Mars?" news release, 27 February, 2001, Gallup Poll News Service.

interest in answering the question "Are we alone?". For our current generation, that question has been deemed sufficiently important to justify the continuing search. Humanity today seems to concur with the conclusion of the original SETI paper: "The probability of detection is difficult to estimate, but if we never search the chance for success is zero" (Cocconi and Morrison, 1959). Future generations will have to make their own judgments.

REFERENCES

Ashkenazi, M. (1992). Not the sons of Adam: religious responses to ETI. *Space Policy*, **8**, 341–349.

Backus, P. R. and the Project Phoenix Team (2002). Project Phoenix: SETI observations from 1200 to 1750 MHz with the upgraded Arecibo Telescope. In *Single-Dish Radio Astronomy: Techniques and Applications*, eds. S. Stanimirovic, D. Altschuler, P. Goldsmith, and C. Salter, pp. 525–527. San Francisco: Astronomical Society of the Pacific.

Ben-Bassat, A., Ren-David-Zaslow, R., Schocken, S., and Vardi, Y. (2005). Sluggish data transport is faster than ADSL. *Annals of Improbable Research*, **11**, 4–9. Also at www.ece.vt.edu/~swe/argus/arch2002.pdf.

Billingham, J., Heynes, R., Milne, D., and Shostak, S. (eds.) (1999). *Social Implications of the Detection of Extraterrestrial Civilization: Report on Workshops on the Cultural Aspects of SETI*. Mountain View, CA: SETI Press.

Blair, D. G. (1986). The search for extraterrestrials. *Nature*, **319**, 270.

Clarke, A. C. (1984). *Profiles of the Future: an Inquiry into the Limits of the Possible*. New York: Holt, Rinehart, and Winston.

Cocconi, G. and Morrison, P. (1959). Searching for interstellar communication. *Nature*, **184**, 844–846.

Crick, F. H. and Orgel, L. E. (1973). Directed panspermia. *Icarus*, **19**, 341–346.

Drake, F. D. (1961). Project Ozma. *Physics Today*, **14**, 40–46 (April).

Drake, F. D. (1962). *Intelligent Life in Space*. New York: Macmillan.

Ekers, R. D., Cullers, D. K., Billingham, J., and Scheffer, L. K. (eds.) (2002). *SETI 2020: a Roadmap for the Search for Extraterrestrial Intelligence*. Mountain View, CA: SETI Press.

Ellingson, S. W. (2002). Argus "2002"architecture. Unpublished report of Electroscience Lab., Ohio State University Available at www.ece.vt.edu/~swe/argus/arch2002.pdf.

Finney, B. (1985). The prince and the eunuch. In *Interstellar Migration and the Human Experience*, eds. B. Finney and E. Jones, pp. 196–208. Berkeley, CA: University of California Press.

Gold, T. (1976). "Through the looking glass," talk presented at URSI Meeting in Amherst, MA. October 1976.

Gott, J. R. (1993). Implications of the Copernican Principle for our future prospects. *Nature*, **363**, 315–319.

Hankins, T. H., Kern J. S., Weatherall, J. C., and Eilek, J. A. (2003). Nanosecond radio bursts from strong plasma turbulence in the Crab pulsar. *Nature*, **422**, 141–143.

Harwit, M. (1981). *Cosmic Discovery: the Search, Scope, and Heritage of Astronomy*. New York: Basic Books.

Jones, E. M. (1985). Where is everybody?: an account of Fermi's question. *Physics Today*, **38**, 11–13 (August).

Lemarchand, G. A. (1994). Passive and active SETI strategies using the synchronization of SN1987A. *Astrophysics and Space Research*, **214**, 209–222.

Morrison, D. (1992). The Spaceguard Survey: protecting the earth from cosmic impacts. *Mercury*, **21**, 103–106, 110.

Oliver, B. M. and Billingham, J. (eds.) (1972). *Project Cyclops: a Design Study of a System for Detecting Extraterrestrial Life*. NASA CR-114445. (Reprinted (1996) by Stanford University Press.)

Perry, M. D. and Mourou, G. (1994). Terawatt to petawatt subpicosecond lasers. *Science*, **264**, 917.

Rose, C. and Wright, G. (2004). Inscribed matter as an energy efficient means of communication with an extraterrestrial civilization. *Nature*, **431**, 48–49.

Staff of the National Astronomy and Ionosphere Center (1975). The Arecibo message of November, 1974. *Icarus*, **26**, 462–466.

Sullivan, W. T., III (1991). Pan-galactic pulse periods and the pulse window for SETI. In *Bioastronomy: the Exploration Broadens*, eds. J. Heidmann and M. J. Klein, pp. 259–68. Berlin: Springer-Verlag.

Turnbull, M. C. and Tarter, J. C. (2003a). Target selection for SETI. I. A catalog of nearby habitable stellar systems. *Astrophys. J. Suppl.*, **145**, 181–198.

Turnbull, M. C. and Tarter, J. C. (2003b). Target selection for SETI. II. Tycho-2 dwarfs, old open clusters, and the nearest 100 stars. *Astrophys. J. Suppl.*, **149**, 423–36.

Webb, S. (2002). *Where Is Everybody?: Fifty Solutions to the Fermi Paradox and the Problem of Extraterrestrial Life*. New York: Copernicus Books.

FURTHER READING AND SURFING

Drake, F. and Sobel, D. (1994). *Is Anyone Out There?* New York: Delta. An informal autobiography of Frank Drake.

Goldsmith, D. (ed.) (1980). *Quest for Extraterrestrial Life: a Book of Readings*. Mill Valley, CA: University Science Books. An excellent collection of reprints of classic articles dealing with extraterrestrial life and SETI.

Harrison, A. A. (1997). *After Contact: the Human Response to ETL*. New York: Plenum. A psychologist looks at all aspects of SETI, especially the possible effects of success.

Shklovskii, I. S. and Sagan, C. (1966). *Intelligent Life in the Universe*. New York: Dell. Reissued (1998) by Emerson-Adams Press (Boca Raton, FL) The pioneering book that first popularized SETI, still worth reading.

Tarter, J. (2001). The search for extraterrestrial intelligence (SETI). *Ann. Rev. Astron. Astrophys.*, **39**, 511–548. A review of the field more technical than this chapter.

Zuckerman, B. and Hart, M. H. (1995, second edn.). *Extraterrestrials: Where Are They?* Cambridge: Cambridge University Press. Dissenting opinions regarding the worth of SETI.

www.seti.org. The SETI Institute's website.

setiathome.ssl.berkeley.edu. SETI@home's website.

URLs for all active SETI search programs are given in Table 26.2.

setiharvard.edu/oseti/allsky. Details of Harvard's optical SETI project.

27 Alien biochemistries

Peter D. Ward, *University of Washington*
Steven A. Benner, *University of Florida*

27.1 Introduction

Common among the many definitions of life (Chapter 5) is mention of sets of chemical reactions that allow metabolism, replication, and evolution. The specifics of those reactions are generally not part of these definitions, although a century of study of the metabolisms that support life on Earth has given us a rich repertoire of illustrative biochemical examples. Unfortunately, because all known life on Earth is descended from a single common ancestor, our study of *terran*[1] biochemistry, no matter how extensive, cannot provide a comprehensive view of the full range of possible reactions that might generally support life.

This leaves open an important question. If life exists elsewhere in the Cosmos, will its chemistry be similar to the chemistry of life on Earth? Recent articles addressing this issue are by Irwin and Schulze-Makuch (2001), Crawford (2001), Bains (2004), and Benner *et al.* (2004), and several popular books are listed under "Further reading" at chapter's end. As in many areas in contemporary astrobiology, no clear methodology exists to address the question of "weird life." Chemistry, however, including the skills outlined in Chapter 7, provides one set of tools for constructing hypotheses about possible alternative biological chemistries.

The emerging field of *synthetic biology* (Benner and Sismour, 2005) provides another set of tools. Here, chemists attempt to give substance to concepts about alternative life forms by synthesizing molecules that might support alternative genetic systems or alternative metabolisms. These can be examined in the laboratory, thus learning about biochemical possibilities in ways inaccessible to those who simply analyze terran life.

Despite difficulties in generating hypotheses on which to base experimental research, these two strategies

have led to a consensus, of a sort, about a hierarchy of "weirdness": (1) alternative biochemistries that are clearly possible, (2) those that are well within the range of possibility, (3) those that are conceivable but (at least for now) confined to the *Star Trek* scripting room, and (4) those that are excluded as being inconsistent with fundamental physical law, or nearly excluded as being difficult to conceptualize within any realizable chemical model.

In some sense, consideration of how alien life might be constructed is the future of astrobiology. It certainly has important ramifications in the practical question of how one searches for extraterrestrial life, and where it might arise. This chapter briefly outlines current thoughts on some relevant biochemical issues; for the basics of terran biochemistry, see Section 6.3 and Chapter 7.

27.2 Life with different biopolymers

27.2.1 Different amino acids

One of the most compelling observations supporting the notion that all life on Earth is descended from a common ancestor is the planet-wide use of the same 20 amino acids (Fig. 7.5) as the standard components of encoded proteins. This biochemical uniformity is not obviously demanded by prebiotic chemistry. Nor is it required by terran metabolism, or constraints imposed by the machinery that terran life uses to biosynthesize encoded proteins.

For example, the famous Miller experiment generated a wide range of amino acids via nonbiological chemical processes (Section 6.2.2). A variety of amino acids is also found in meteorites, almost certainly generated without biology (Sections 3.8.2 and 7.4). No correlation exists, however, between the likelihood that an amino acid will be generated nonbiologically

---

[1] *Terran* means "associated with Earth or Earth's form of life."

Planets and Life: The Emerging Science of Astrobiology, eds. Woodruff T. Sullivan, III and John A. Baross. Published by Cambridge University Press. © Cambridge University Press 2007.

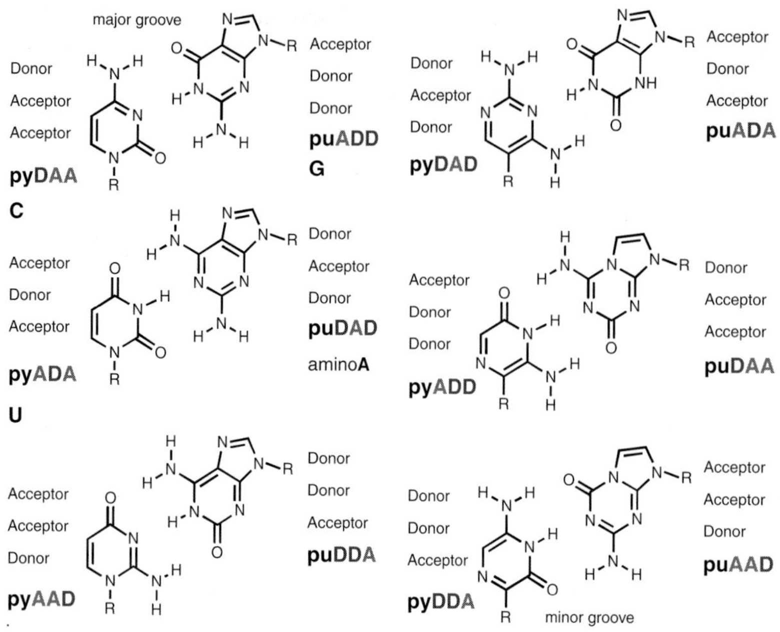

FIGURE 27.1 An expanded genetic alphabet of artificial DNA. The standard Watson–Crick nucleobases pair small nucleobases (left partner) with large nucleobases (right partner), via three hydrogen-bond donor groups pairing with three hydrogen-bond acceptor groups. The specificity of the pairing is determined by the choice of the donor (D) and acceptor (A) groups, which are listed in order from the DNA's major groove (top) to its minor groove (bottom). The standard nucleobase pairs are the top two in the left column. Note that in the standard nucleobase adenine, the third hydrogen bond donor group is missing; what is shown is therefore aminoadenine. Adapted from Benner et al. (2004).

and its presence in the standard set of amino acids in terran proteins. For example, the non-standard amino acid norvaline, not found in encoded proteins on Earth, is more abundant in meteorites than many of the amino acids found in the standard set.

Likewise, terran metabolism generates many amino acids that do not end up as encoded parts of terran proteins. Ornithine and citrulline, for example, are amino acids biosynthesized in many forms of life on Earth, where they are metabolic precursors for the encoded amino acid arginine and serve as intermediates in pathways that metabolize nitrogen. Because of their abundance and prevalence, they are quite available to be part of encoded proteins; they just are not.

Experiments have shown that many more than the standard 20 amino acids are compatible with the machinery that life on Earth uses to synthesize encoded proteins (Hecht et al., 1978; Bain et al., 1989, 1992; Noren et al., 1989; Chin et al., 2003). Given the appropriate charged tRNA and an appropriate message, a ribosome is fully capable of incorporating virtually any amino acid into a protein via encoded translation. This includes the alpha methyl amino acids that are abundant in meteorites.

This makes inescapable the possibility that the amino acid inventory in terran proteins could have been quite different. A fortiori, this implies that non-terran life, if it exists, could use quite different amino acids, even if its catalytic molecules have the same backbone as proteins on Earth (Bain et al., 1992; Hohsaka and Masahiko, 2002). It is thus possible that DNA-based life might exist in the Cosmos, similar in all other respects to terran life, except for the use of different amino acids.

27.2.2 Chemically different "DNA"

An analogous conclusion for terran genetic matter is now possible based on many recent experiments in synthetic biology. On Earth, the Watson–Crick nucleobase pairing between the two strands of DNA follows two rules of complementarity. *Size complementarity* requires that large purines pair with small pyrimidines, while *hydrogen-bonding complementarity* requires that hydrogen bond donors from one nucleobase pair with hydrogen bond acceptors on the other. Thus, cytosine is a small pyrimidine that presents a donor–acceptor–acceptor pattern for its hydrogen bonds, which is complementary to guanine, a large purine that presents an acceptor–donor–donor hydrogen bonding pattern (Fig. 27.1). Uracil is a small pyrimidine that presents an acceptor–donor–acceptor pattern (major to minor groove) that is complementary to adenine, a large purine that presents a donor–acceptor pattern (adenine is missing the third hydrogen bond donor group).

These rules can be generalized (Geyer et al., 2003; Benner, 2004), where the large and the small components are always joined by three hydrogen bonds.

In this general structure, instead of the two hydrogen bonding patterns used by terran life, $2^3 = 8$ hydrogen bonding patterns are conceivable, of which six are readily accessible (Fig. 27.1). Synthetic biologists have not only made all of these in the laboratory, but further shown that they are all competent to support genetic recognition (Geyer *et al.*, 2003; Benner, 2004; Sismour and Benner, 2005).

As with alternative amino acids, we cannot exclude these alternative nucleobases based on their incompatibility with the DNA polymerases (the enzymes that drive assembly of the DNA polymer chain). Given only modest modification of terran polymerases, DNA molecules built from expanded genetic alphabets can be copied, and then copied again (Benner, 2004; Sismour *et al.*, 2004). This artificial genetic system has sustained up to 20 generations of replication (Sismour and Benner, 2005). They can even be copied with mutations, where the mutations themselves are replicated. Thus, these synthetic genetic molecules, together with standard terran polymerases, are artificial Darwinian chemical systems.

As with alternative sets of amino acids, these experiments suggest the possibility that the components of terran DNA could have been different. The case is not as strong for DNA as for proteins, however, since the alternative nucleotides, with few exceptions, are not as prominent in terran metabolism as are alternative amino acids. Further, we have no arguments based on possible routes to prebiotically synthesize nucleic acids (few good prebiotic routes are known even for the standard A, T, G, and C). Nevertheless, experiments in the laboratory make it quite conceivable that non-terran life, if it exists, could use quite different nucleobases in its genetic molecules, even if those molecules have the same backbone as DNA on Earth.

Several groups have now connected alternative genetic alphabets with alternative protein alphabets. For example, the Benner group encoded a non-standard amino acid, iodotyrosine, using a nucleobase pair that had its hydrogen bonding groups shuffled (Bain *et al.*, 1992).

Other changes might be made in the way DNA and proteins interact. For example, the number of "letters" that encode an amino acid "word" might be increased or decreased. In making this change, the four traditional nucleobases of DNA (adenine, thymine, cytosine, and guanine) might still be used, but instead of a triplet being used to specify an amino acid, four or perhaps two nucleotides might encode an amino acid, allowing for, respectively, $4^2 = 16$ or $4^4 = 256$ possible amino acids.

What features of DNA might be universal in genetics throughout the Cosmos? Again, synthetic biology has provided some clues (Benner and Hutter, 2002). The backbone structure of DNA almost certainly reflects its need to function in water. The DNA double helix is built from two long strands of nucleosides joined by phosphates that at neutral pH each carry a negative charge. This repeating negative charge allows the DNA to be soluble in water. Further, the repeating charge on DNA dominates the physical properties of the molecule, meaning that replacement of the nucleobases in the DNA strand changes very little the overall physical properties of the DNA molecule. This allows mutation to occur without changes in biophysics, something that is very important for Darwinian evolution. Synthetic biological results (Huang *et al.*, 1993) therefore suggest that a repeating charge will be a universal structure of genetic molecules in water (Benner and Hutter, 2002).

The other component of the backbone is the sugar, ribose in RNA and 2′-deoxyribose in DNA. Many efforts have been made to change this part of the genetic molecule, in part because of perceived difficulties in forming ribose and 2′-deoxyribose under prebiotic conditions (Section 7.6). As just one recent example, the glycerol backbone has been synthesized and studied (Zhang *et al.*, 2005).

These experiments suggest that genetic matter could come in many varieties of languages. It would be interesting to know if, early in Earth's history, many separate kinds of DNA competed against each other. Is a twelve-nucleotide DNA more or less efficient than our familiar four-nucleotide DNA? Was there competition among a whole series of slightly different DNAs, with our current version proving competitively superior? Or was ours simply the first to achieve a tolerable ability to support life, suppressing a variety of equally, or even more, effective competitors through the advantage of incumbency?

27.3 Life with a different solvent

As synthetic biologists attempt to take steps away from standard terran biopolymers, they have become increasingly constrained by water, the solvent that surrounds terran life (Saenger, 1987). General experience in chemistry suggests that metabolism can efficiently operate only when metabolites are dissolved. Water is an excellent solvent by many measures, yet many compounds are not soluble in it. As one inspects terran biochemistry, one observes how terran life exploits *differential* solubility: (1) to manage compartmentalization (using insoluble membrane components), (2) to build macroscopic structure (as in cellulose, for example, where the insolubility of a very polar compound arises from the high stability of a semicrystalline phase), and (3) to achieve genetic regulation (for example, the solubility properties of steroids are key to their use in higher organisms).

Much has been made about the virtues of water as a biosolvent, at least at temperatures and pressures on the surface of Earth. It is an excellent solvent for salts. Water has a great ability to store heat. Bodies of water thus tend to stabilize their surroundings against rapid swings in temperature, in the same fashion that a coastal region is buffered against rapid temperature swings that are experienced in a desert. Water also floats when it freezes, insulating liquid water below.

Water has, however, many disadvantages as a biosolvent. Water ice has a higher albedo than water liquid, for example. Thus, when water ice floats, it reflects more light from the sun, which leads to more cooling, more ice on the surface, a higher albedo, and still more cooling. The fact that water ice floats thus causes water to amplify, not damp, perturbations in the flux of energy coming to a planet – e.g., runaway glaciation of this sort may have occurred on Earth in the past, including within the past few million years.

The chemical reactivity of water, especially when considering RNA and DNA, also creates problems with its use as a biosolvent. Cytidine reacts with water to give uridine, losing ammonia in the process, with a half-life of $\sim$70 years in water at 300 K. Adenosine likewise loses ammonia by reaction with water to give inosine, and guanosine loses ammonia to give xanthosine. As a consequence, terran DNA in water must be continually repaired.

Further, *liquid* water is not the usual form in the Cosmos. Water molecules are abundant, but generally in the form of ice on planets and moons (Chapter 15) and ice coatings on interstellar dust grains (Section 3.8), or alternatively water vapor in cool star atmospheres, in interstellar space, and in some planets' atmospheres (Chapter 4). Water can exist as a liquid on planetary surfaces only in a narrow range of distances from a particular star, and then only if the pressure/temperature of the planet's atmosphere is such that liquid water is stable. On Mars, for example, liquid water is not possible on the surface at any temperature, simply because the atmospheric pressure is too small. Note, however, that liquid water *can* more readily occur *beneath* the surface of planets and moons, as discussed in Chapter 18 for Mars and Chapter 19 for Europa.

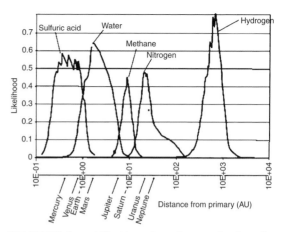

FIGURE 27.2 Potential liquids on the surface or subsurface of a planet or moon as a function of (the logarithm of) distance from a Sun-like star. Positions of the solar system's planets are indicated for comparison. Possible solvents for a form of life are discussed in the text. Adaped from Bains (2004).

Furthermore, at yet deeper levels (where pressures are very high and temperatures warmer), the crusts and mantles of rocky bodies also contain a great deal of water, some of which must be liquid, in accord with the phase diagram of water (Section 15.2).

These considerations have led many to consider alternative biosolvents that could work on a planetary body's surface with un-Earth-like conditions. Some of these, such as concentrated sulfuric acid, have very high boiling points and might support life at very high temperatures (Schulze-Mackuch *et al.*, 2004). Others, such as ammonia, dinitrogen (N_2), and supercritical dihydrogen-helium mixtures have very low freezing points, and thus might work at the other extreme (Bains, 2004; Benner *et al.*, 2004).

Bains (2004) has outlined the potential abundance of different liquids as one moves farther from a Sun-like star within a planetary system (Fig. 27.2). Less abundant solvents that have been considered include methyl alcohol, hydrogen sulfide, hydrogen fluoride, hydrogen cyanide, and hydrogen chloride.

Basic chemical principles suggest that life could easily be sustained in these solvents, given different core chemical structures and different cellular morphologies. The chemistry supporting this life would necessarily be, from a terran perspective, exotic and difficult to recognize for those familiar only with terran life. Furthermore, we lack much of the needed basic chemical data to evaluate such substances as potential biosolvents. Consider, for example, liquid dinitrogen, likely to be abundant in the outer planets and their larger moons. We do not know the solubilities of

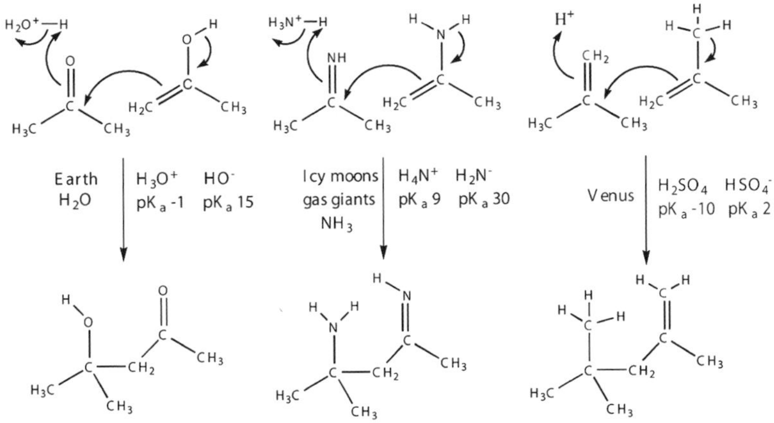

FIGURE 27.3 Different solvents would favor different, although analogous, chemical reactions to support metabolisms in life residing in different temperature regimes. Here are shown three analogous mechanisms for forming carbon–carbon bonds. The desired reactivity is conferred upon the reacting species by (left) a C=O unit (favored in water), (center) a C=N unit (favored in ammonia, such as in outer Solar System bodies), or (right) a C=C unit (favored in strongly acidic solvents such as sulfuric acid, as in Venus's atmosphere). Adapted from Benner et al. (2004).

most interesting organic species in this solvent at terran atmospheric pressure, let alone at the temperatures and pressures found on Triton, a moon of Neptune. Nor is much known about the chemical reactivity or self-assembly of key organic species in liquid dinitrogen. Thus, while it is clear that terran biochemistry could not function in liquid dinitrogen, it is not clear what chemistry might take its place.

Ammonia (NH_3) is more like water as a solvent, and is a useful solvent in the organic chemistry laboratory. The solubility of organic species and their reactivity have been better studied in ammonia than in most exotic solvents, and ammonia has long been discussed as an alternative solvent for life (see Bains, 2004). This discussion has intensified given recent evidence from the Cassini–Huygens mission on Titan (Sections 20.7 and 20.9).

Life in a water–ammonia mixture would need a different compartmentalization strategy, since liposomes, a major structural part of Earth-life cell walls (Sections 9.6 and 9.7), dissolve in ammonia. Benner et al. (2004) have suggested features of an alternative metabolism in ammonia, following the chemical principles outlined in Chapter 7. Whereas terran life in water exploits compounds containing oxygen doubly bonded to carbon (the carbonyl unit), life in ammonia would instead exploit compounds containing nitrogen doubly bonded to carbon (Fig. 27.3).

27.4 Life using different elements

Alternative solvents bring us in the hierarchy to the point where life becomes truly "weird." The next step considers alternative biochemistries that exploit different chemical elements.

Terran life is commonly called "carbon-based," or better, CHON-based (carbon, hydrogen, oxygen, nitrogen). But this ignores the vital contributions made by phosphorus and sulfur atoms to the physical properties and chemical reactivity of carbon. Terran life would be better described as CHONPS life.

There is little doubt that alternative forms of life could use these elements in different ratios, or introduce elements that are largely absent in terran biochemistry. Boron and silicon, which play only minor roles in the life that we know, could be used more, even perhaps in another form of terran life. Nevertheless, many investigators have persuasively argued that carbon is essential as a major component of life anywhere in the Universe (Pace, 2001).

This may indeed be so in water as a biosolvent. A change in solvent might, however, change the preferred elements. As mentioned above, life in ammonia would almost certainly switch the relative contributions of nitrogen and oxygen – CHNOPS life would be preferred over CHONPS life. A different environmental redox potential could also change the preferred elements.

These would be, however, only small changes. For much more profound changes, we now explore two other concepts.

27.4.1 Silicon life

As described above, cold is a challenge for carbon-based biochemistry. The bond energies intrinsic to functionalized organic species, and mechanisms for forming and breaking bonds to carbon, make common reactions slow at temperatures below 270 K. Likewise, the low temperatures that dominate the Cosmos create low solubilities, even in solvents that are liquid at those temperatures (such as dinitrogen, ethane, and methane).

Silicon-based compounds have been discussed for their potential for solving both problems (Bains, 2004). The silicon–oxygen single bond in compounds known as silanols is considerably more reactive than the analogous carbon–oxygen single bond. Further, silanes and silanols are soluble in a variety of solutes over a wide range of temperatures, including the very cold temperatures where dinitrogen is a liquid.

Furthermore, silicon–silicon bonds are well known, even in chains as long as 30 atoms. These are analogous, at least in size, to the carbon-based polymers necessary for terran life (Brook, 2000; West, 2002). The bonds are weaker and easier to transform at low temperature than the analogous bonds to carbon.

Bains (2004) has recently discussed these advantages for life in liquid dinitrogen. But again, little is known about the fundamental chemistry of silicon compounds in such a solvent. Much further work is needed.

27.4.2 Silicon/carbon clay life

A still more speculative type of life would allow minerals to play a larger role. These include silicon, but in its oxidized state as silicate (SO_3). Such ideas were suggested by Cairns-Smith (1982), who discussed a possible early form of life that exploited mineral-like crystals as structural elements. His model envisioned a form of life based on crystals of clay that would actively grow, and evolve as they did so.

27.5 Life with a different architecture

A variety of architectural features of terran biochemistry can be adjusted to create still more exotic life. For example, modern terran life uses three encoded biopolymers to manage its affairs: (1) *DNA* is used for genetics, (2) *proteins* are used to solve most of its structural and catalytic problems, and (3) *RNA* is used as an encoded biopolymer to act as an intermediate between the two, as well as to perform certain structural and catalytic roles. This combination almost certainly reflects contingencies in the history of life on Earth, as well as certain functional and vestigial features of these biopolymers.

An alternative architecture, however, might employ just one encoded biopolymer. Such a lifeform in fact is proposed in the "RNA World" models for the origin of life on Earth (Sections 6.5 and 8.2.1). Therefore, it does not stretch current theory to propose that life based on a single biopolymer might be more abundant in

the Cosmos than life based on multiple biopolymers (Benner *et al.*, 2004).

There are, however, competing demands placed on a biopolymer by catalysis and genetics. A biopolymer specifically adapted to be used as a catalyst will be optimally built of many elemental and molecular building blocks, allowing a wider spectrum of chemical reactions. On the other hand, a biopolymer specifically adapted for genetics will perform better with the smallest possible number of separate components to ensure more accurate replication. Further, a molecule adapted to be a catalyst must be able to easily fold into multiple shapes; an information-carrying molecule cannot afford to fold easily if it is to serve as template (Benner & Hutter, 2002; Baross & Deming, 1995). A molecule adapted to be a catalyst should also change its physical properties easily through a small number of building-block substitutions; this allows it to explore "behavior space" most efficiently. In contrast, a genetic molecule should not change its physical behavior at all as a consequence of building-block substitutions – this allows Darwinian evolution to proceed unconstrained by the details of chemistry.

This illustrates why a lifeform might in general create two classes of biopolymers that are radically different from one another. The building blocks of each are adapted to the specific roles that the biopolymer is asked to perform. This dual-polymer strategy affords a far broader range of possibilities in both the nature of information stored, and the diversity of chemical reactions that can be undertaken. Thus, on Earth the amino acids found in proteins may be chosen to give an optimal framework for structure and catalysis, without concern for genetic demands. Conversely, it appears that, for a genetic molecule, the choice of deoxyribose as the sugar in terran DNA is superior to the ribose of RNA. Any single-biopolymer life will be far more limited in both respects.

Yet there may be environments where a form of life exploiting a single biopolymer has advantages. Such forms of life might be smaller, offering advantages where space is an important constraint (Benner, 1999). Single-biopolymer forms of life may require fewer resources, providing advantages where certain elements are scarce.

27.6 Summary

In this brief chapter, some potential kinds of "alien" biochemistries have been proposed. These are summarized in Table 27.1. While these kinds of "life as we do

TABLE 27.1 Possible biochemistries

Name	Scaffold element	Genetic material	Solvent	Scaffold element source	Energy source	Possible Solar System habitats
Terran life	Carbon	DNA	Water	CO_2, other organisms	Many	Earth, Mars, Europa, Titan
RNA life	Carbon	RNA	Water	Organic molecules	?	Titan, Mars, Europa, early Earth?
Protein life	Carbon	Proteins	Water	Organic molecules	?	Titan, Mars, Europa, Earth?
Silicon life	Silicon	?	cryosolvents?	?	?	Titan, Triton

not know it" may be theoretically plausible, they may not be present in the Cosmos simply because no pathway existed by which they could evolve. Nevertheless, continued research into how life different from our own might be built will provide valuable insights into how our own kind of life works, and help guide the design of missions that search for life in our Solar System in environments that differ significantly from Earth.

REFERENCES

Bain, J. D., Diala, E. S., Glabe, C. G., Dix, T. A., and Chamberlin, A. R. (1989). Biosynthetic site-specific incorporation of a non-natural amino acid into a polypeptide. *J. Am. Chem. Soc.*, **111**, 8013–8014.

Bain, J. D., Chamberlin, A. R., Switzer, C. Y., and Benner, S. A. (1992). Ribosome-mediated incorporation of non-standard amino acids into a peptide through expansion of the genetic code. *Nature*, **356**, 537–539.

Bains, W. (2004). Many chemistries could be used to build living systems. *Astrobiology*, **4**, 137–167.

Baross, J. A. and Deming, J. W. (1995). Growth at high temperatures: isolation and taxonomy, physiology, and ecology. In *The Microbiology of Deep-sea Hydrothermal Vents*, ed. D. M. Karl, pp. 169–217. Boca Raton, FL: CRC Press.

Benner, S. A. (1999). How small can a microorganism be? In *Size Limits of Very Small Microorganisms: Proceedings of a Workshop*, eds. Steering Group on Astrobiology of the Space Studies Board, pp. 126–135. Washington, DC: National Research Council.

Benner, S. A. (2004). Understanding nucleic acids using synthetic chemistry. *Accounts Chem. Res.*, **37**, 784–797.

Benner, S. A. and Hutter, D. (2002). Phosphates, DNA, and the search for nonterran life. A second generation model for genetic molecules. *Bioorg. Chem.*, **30**, 62–80.

Benner, S. A. and Sismour, A. M. (2005). Synthetic biology. *Nature Rev. Genetics*, **6**, 533–543.

Benner, S. A., Ricardo, A., and Carrigan, M. A. (2004). Is there a common chemical model for life in the universe? *Curr. Opin. Chem. Biol.*, **8**, 672–689.

Brook, M. A. (2000). *Silicon in Organic, Organometallic and Polymer Chemistry*. Toronto: John Wiley.

Cairns-Smith, A. (1982). *Genetic Takeover and the Mineral Origins of Life*. Cambridge: Cambridge University Press.

Chin, J. W., Cropp, T. A., Anderson, J. C., Mukherji, M., Zhang, Z. W., and Schultz, P. G. (2003). An expanded eukaryotic genetic code. *Science*, **301**, 964–967.

Crawford, R. L. (2001). In search of the molecules of life. *Icarus*, **154**, 531–539.

Geyer, C. R., Battersby, T. R., and Benner, S. A. (2003). Nucleobase pairing in expanded Watson–Crick like genetic information systems. The nucleobases. *Structure*, **11**, 1485–1498.

Hecht, S. M., Alford, B. L., Kuroda, Y., and Kitano, S. (1978). Chemical aminoacylation of transfer-RNAs. *J. Biol. Chem.*, **253**, 4517–4520.

Hohsaka, T. and Masahiko, S. M. (2002). Incorporation of non-natural amino acids into proteins. *Curr. Opin. Chem. Biol.*, **6**, 809–815.

Huang, Z., Schneider, K. C., and Benner, S. A. (1993). Oligonucleotide analogs with dimethylene-sulfide, -sulfoxide and -sulfone groups replacing phosphodiester linkages. *Meth. Molec. Biol.*, **20**, 315–353.

Irwin, L. N. and Schulze-Makuch, D. (2001). Assessing the plausibility of life on other worlds. *Astrobiology*, **1**, 143–160.

Noren, C. J., Anthony-Cahill, S. J., Griffith, M. C., and Schultz, P. G. (1989). A general method for site-specific incorporation of unnatural amino acids into proteins. *Science*, **244**, 182–118.

Pace, N. R. (2001). The universal nature of biochemistry. *Proc. Natl. Acad. Sci. USA*, **98**, 805–808.

Saenger, W. (1987). Structure and dynamics of water surrounding biomolecules. *Ann. Rev. Biophys. Biophys. Chem.*, **16**, 93–114.

Schulze-Makuch, D., Grinspoon, D. H., Abbas, O., *et al.* (2004). A sulfur-based UV adaptation strategy for putative phototrophic life in the Venusian atmosphere. *Astrobiology*, **4**, 11–18.

Sismour, A. M., Lutz, S., Park, J.-H., Lutz, M. J., Boyer, P. L., Hughes, S. H., and Benner, S. A. (2004). PCR amplification of DNA containing non-standard base pairs by variants of reverse transcriptase from human immunodeficiency virus-1. *Nucl. Acids Res.*, **32**, 728–735.

Sismour, A. M. and Benner, S. A. (2005). The use of thymidine analogs to improve the replication of an extra DNA base pair: a synthetic biological system. *Nucl. Acids Res.*, **33**, 5640–5646.

West, R. (2002). Multiple bonds to silicon: 20 years later. *Polyhedron*, **21**, 467–472.

Zhang, L., Peritz, A., and Meggers, E. (2005). A simple glycol nucleic acid. *J. Am. Chem. Soc.*, **127**, 4174–4175.

FURTHER READING

Feinberg, G. and Shapiro, R. (1980). *Life beyond Earth*. New York: William Morrow. An interesting early reference on potential for non-terran biology.

Grinspoon, D. (2004). *Lonely Planets*. New York: Harper Collins. A philosophical look at astrobiology, including interesting recent work on potential for non-terran biology.

Koerner, D. and LeVay, S. (2000). *Here Be Dragons*. Oxford: Oxford University Press.

National Research Council (2007). *The Limits of Organic Life in Planetary Systems* (National Academy of Sciences Task Group Report). Washington, DC: National Academy Press. A task force of experts considers the possibilities for non-Earth-like life.

Ward. P. (2005). *Life as We Do Not Know It*. New York: Viking Penguin. A popular treatment of the potential for non-terran life.

Part VII
Future of astrobiology

28 Disciplinary aspirations and educational opportunities

Llyd Wells, John Armstrong, and Julie A. Huber[1]
University of Washington

28.1 Astrobiology: a new discipline?

The controversial assertion in 1996 that the martian meteorite ALH84001 contained evidence for ancient microbial life stoked public, political, and scientific interest in astrobiology. Two years later, the NASA Astrobiology Institute (NAI) was launched, its mission to "study the origin, evolution, distribution, and future of life on Earth and in the Universe." This mission, though ambitious, is not really new. Indeed, according to this definition, astrobiology has been studied for decades if not centuries (Chapter 1). Astrobiological research would include James Watson and Francis Crick's decipherment half a century ago of the double helix and with it (according to Crick) the "secret of life." It also would include Stanley Miller and Harold Urey's in vitro synthesis of amino acids by electric discharge, the Viking Mission of 1976 to search for life on Mars, the discovery of hydrothermal vents, and the recognition that a wayward impactor probably drove the dinosaurs extinct. Yet none of this research was carried out under the auspices of astrobiology or required a specific astrobiological framework. Recently, however, during the so-called "Astrobiology Revolution" of the 1990s (Section 2.3; Ward and Brownlee, 2000), increasing attention has been devoted to developing precisely such a framework, as implied by the formation of NAI. Programs oriented to astrobiology research or training have acquired various degrees of formalization throughout the world. Astrobiology journals have been created. You are reading a textbook devoted to astrobiology. These milestones both represent and fuel ambitions to transform astrobiology into a new discipline.

Although ambition may motivate the development of astrobiology as a discipline, it does not necessarily justify that development. In fact, while considerable thought and effort have been devoted to the establishment of various disciplinary trappings (including training programs, journals and textbooks), less effort has gone towards understanding the role that a discipline of astrobiology should or could fulfill. What could be accomplished with a *discipline* of astrobiology that is not already being accomplished?

To answer this question, it is instructive first to consider astrobiology's emergence in the context of that of another relatively young discipline, molecular biology. The rise of molecular biology was also fomented by the convergence of dramatic research breakthroughs, public attention, cooperation of scientists from different disciplines, and institutional dedication. In particular, dramatically contributing to molecular biology's development was the discovery of bacteriophage (bacterial viruses, or simply phage) by Frederick Twort and Félix d'Herelle (independently) in the early twentieth century (Summers, 1999). This discovery generated considerable excitement, in part due to the perceived medical application of phages: d'Herelle in particular argued vociferously that they were natural antagonists to the germs causing human infection. Popularized by famous novels like Sinclair Lewis's *Arrowsmith* and in an age before antibiotics, these claims stirred the public's imagination. Even after antibiotics outstripped viruses as the treatment of choice for bacterial infections, the phage remained important as a powerful biological tool. For example, Alfred Hershey and Martha Chase used bacteriophage in 1952 to demonstrate that the molecular agent of heredity is DNA, not protein.

Molecular biology thus gained impetus from its popular appeal, as well as its immediate practical and technological utility. At the same time, it steadily entrained and was transformed by the interests and techniques of

---

[1] The authors of this chapter were all graduate students in the University of Washington's Astrobiology Program when this chapter was written. Wells is currently at Sterling College, Vermont, Armstrong at Weber State University, Utah, and Huber at the Marine Biological Laboratory at Woods Hole.

Planets and Life: The Emerging Science of Astrobiology, eds. Woodruff T. Sullivan, III and John A. Baross. Published by Cambridge University Press. © Cambridge University Press 2007.

scientists from other disciplines. The discovery of phage re-emphasized an old but unanswered question: what, after all, is "life?" The scope of this challenge attracted not just biologists, but also physicists such as Erwin Schrödinger, Francis Crick, and Max Delbrück (student of Niels Bohr), and chemists like Linus Pauling, Max Perutz, and Erwin Chargaff. A multidisciplinary institutional framework became established, notably at the Cavendish Laboratory of Physics in Cambridge University, the Pasteur Institute in Paris, and the California Institute of Technology (Fruton, 1999). Together, the shared interests of these scientists and their recognition of the congruence of their disciplinary fields exerted major influences on the development of theoretical and experimental approaches in molecular biology. In fact, it was by using new physical–chemical approaches and by integrating data from many disciplines that Francis Crick could make his famous (but incorrect) pronouncement about discovering the secret of life. Overshadowed in the stir was the further entrenchment of molecular biology as a discipline.

Astrobiology's development shows important similarities to and differences from that of molecular biology. Like its disciplinary forebear, it has generated substantial public and political enthusiasm. The assertion that ALH84001 contained evidence for ancient martian life (Section 18.5.3), for example, was met by a special announcement from President Clinton. Similarly, astrobiology has drawn to it disparate scientists representing many disciplines, just as did molecular biology. Some of the very questions that stimulated interdisciplinary interest in molecular biology persist for astrobiology: we still do not know what we mean by "life" or, therefore, "the origin of life." The ambiguous nature of life exposed by the discovery of phage has not been resolved. Finally, responsive to both its public and its interdisciplinary appeal, institutional frameworks have developed to support astrobiology, as they did for molecular biology, including not only the NAI and its associated US institutions, but also others in Spain, Australia, and Russia. Yet, despite these similarities, astrobiology also differs significantly from molecular biology in its nascent stages. For example, it does not devote itself to a previously overlooked subject matter. It does not identify (or at least has not yet identified) a specific new scale at which to ask questions. And, in stark contrast to molecular biology, it has not yet been associated with any practical utility. Instead, the claim appears to be that many different disciplines should now be applied to a class of questions perceived as broadly unified, and that such an amalgamation justifies a new

discipline (or even metadiscipline), astrobiology. The fact that research questions in astrobiology have long been asked in a collaborative framework, however, begs the question: why is the old framework no longer sufficient?

We think an answer begins with the recognition that some of the ignorance exposed by astrobiological questions reveals not the boundaries of scientific knowledge, but instead the boundaries of individual disciplines. Furthermore, collaboration by itself does not address this ignorance, but instead compounds it by encouraging scientists to rely on each other's authority. Thus anachronistic disciplinary borders are reinforced rather than overrun. In contrast astrobiology can motivate challenges to disciplinary isolation and the appeals to authority that such isolation fosters. An immediate practical role for a discipline of astrobiology then emerges, one that reclaims the etymologic sense of the word *discipline*, from the Latin *disciplina*, "teaching." What astrobiology can offer as a discipline is, paradoxically, a critique of the disciplinary structure of scientific education and a platform to develop instead a more unifying approach.

No one knows how best to do this. As Astrobiology Certificate students at the University of Washington (UW) in one of the world's first such graduate programs, we have seriously considered this problem. Based on our experiences at the graduate level, both successes and failures, we will outline how astrobiology can productively be used to improve science education. Along the way we suggest steps that an educational program in astrobiology could follow. This is not meant to advocate a particular curriculum, but rather to share some of our solutions to the many challenges that the pursuit of an interdisciplinary education poses. It is important to emphasize that such an education will not haphazardly emerge by assembling well-intended scientists who represent different disciplines. Instead, it will require an integrative strategy, in combination with significant, long-term, personal and organizational commitments. Finally, if astrobiology is truly to emerge as a discipline, it should not only ambitiously attempt to synthesize historically segregated subjects, but also to convey the integrity and nuances of scientific approaches and discoveries to the public at large.

28.2 Astrobiology and ignorance

28.2.1 Ignorance as opportunity

To understand why we think an educational program in astrobiology is important, we begin by examining the appeal of astrobiology itself. Why did we UW

Box 28.1 Why did students choose astrobiology?

Nicolas Pinel, a graduate student in astrobiology and microbiology, interviewed his fellow graduate students to understand some of the reasons motivating their involvement.

For some of us, childhood fantasies were filled with images of rockets and magical trips to stellar bodies. For others, our innocent imagination created grandiose biological expeditions to wildly remote jungles in search of unknown organisms. Astrobiology, despite its etymological allusion, is not simply the amalgamation of the dreams of astronauts and biologists. In fact, few of us knew about astrobiology before our entry into university-level science. Our backgrounds and goals widely vary, and there was no common a priori aspiration that led each of us to this nascent field.

Some of us had long cultivated curiosity in many of the fields within astrobiology. The program has also served as a catalyst to revive and nourish dormant scientific interests that otherwise would have had to await the completion of graduate school for renewed consideration. Some of us were lured into astrobiology by the research autonomy it promised. These students wanted the power to cross barriers, to sprint to the opposite end of campus to share an idea or a question with people working under different labels but with complementary interests. As one of us expressed it while reminiscing on the days of graduate school selection, "I decided that if I was going to do science, I'd rather do it my way," not according to the reductionism of traditional research projects. Astrobiology can provide precisely that kind of intellectual freedom.

For others, pondering the decision to join the Astrobiology Program ignited serious personal debates. Skepticism was provoked by the dubious respectability of the field in some quarters, as well as the uncertain potential for professional advancement. Once these students began interacting with other astrobiologists, however, and once the community started to gain momentum, they felt their decisions were vindicated.

The specific research projects that we conduct may or may not significantly be influenced by the Astrobiology Program. In either case, however, astrobiology has enriched our ability to frame the questions we ask and created a new context in which to interpret our results. Astrobiology provides us with the opportunity to make fundamental questions part of our everyday thinking. In the words of one student: "It would seem hard for anyone, given the opportunity, not to do astrobiology, not to take the opportunity to address the most fundamental questions in science."

---

students choose to pursue astrobiology, sometimes against the advice of mentors and advisors? Why were we willing to take on more coursework and more requirements to identify ourselves with a newly emerging field of dubious respectability? Certainly, there are both general and personal answers to these questions. Some of the more general reasons are discussed below, while personal responses are elaborated in Box 28.1.

Part of a general response is revealed by the disparate and sometimes non-disciplinary backgrounds that we had as undergraduates (Fig. 28.1). Dissatisfaction with disciplinary approaches to fundamentally interdisciplinary questions also led many of us to major in more than one field. This is not to deny the importance of reductionist approaches or the advances stimulated by them. Rather, as undergraduates we wanted to integrate the results of reductionist science. Such synthesis is often poorly accommodated by disciplines that have evolved, especially in academia, to become insular, autonomous departments. Despite the importance of

synthesis for many basic scientific questions, it is rarely attempted in research and too often relegated solely to popular science books.

Some of our interest in astrobiology, then, represents our broad curiosity, the scope of the questions we are asking, and the recognition that no single discipline can answer them. Most importantly, what do we mean by "life?" How did it originate? What conditions are necessary for it to exist? How is it distributed in the Universe? But there is another, very different type of question that also stimulates us. For example, why do few departments encourage young scientists to study the origin of life? Why is scientific research funneled toward what is thought, a priori, to be knowable, rather than toward what is fundamentally unknown? How is it that most undergraduate biology classes do not begin with a confession that, despite the formulation of evolution as an organizing biological principle, despite the discoveries of genetics and the advances of molecular biology, despite the recognition of

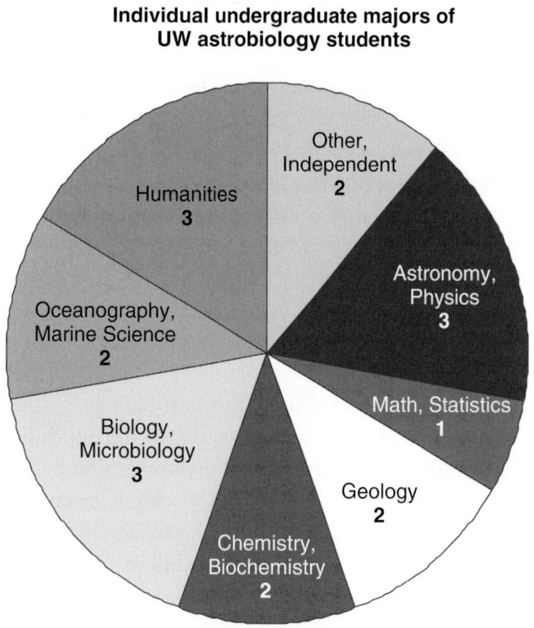

**Individual undergraduate majors of
UW astrobiology students**

FIGURE 28.1 The undergraduate majors of the first 14 UW astrobiology graduate students. For the four students with multiple majors, all are shown.

extremophiles and of the complexities of ecosystem function, still no one knows what life *is*? Ignorance should motivate scientific inquiry and education, rather than be marginalized. Yet disciplinary segregation and the tendency to hand down supposed facts to students, so characteristic of education at all levels, encourage a naive and smug sense of certainty against which ignorance can falsely be reviled.

Astrobiology can change this by challenging the ignorance fostered by disciplinary structure while pursuing the creative ignorance underlying genuine inquiry. Because of its integrative questions, interdisciplinary nature, and enormous public appeal, astrobiology emerges as an ideal vehicle for scientific education at the graduate, undergraduate, and even high school levels. Astrobiology permits treatment of traditionally disciplinary subjects as well as areas where those subjects converge (and, sometimes, fall apart!). At the same time, astrobiology is well suited to reveal the creative ignorance at scientific frontiers that drives discovery. Recognition of this potential for education is an important part of astrobiology's appeal to us.

Are we *astrobiologists*? We are often asked this question, which seems to imply that for astrobiology to emerge as a discipline, it will need astrobiologists to practice it. Yet, at the UW, astrobiology students are first accepted into a traditional department, whose

requirements and standards must be met foremost. Thus, all of us have additional disciplinary affiliations as astronomers, oceanographers, geologists, microbiologists, engineers, and historians of science. Which are we: these, or astrobiologists? The question creates a false dichotomy. As we have argued above, astrobiology is an opportunity to integrate science. Formalization as a discipline would recognize the unique advantages astrobiology offers science education. Further formalization, whether through the identification of special initiates ("astrobiologists") or the creation of astrobiology departments, would carry sociological significance, but in our view would be counterproductive to the type of education we seek. Indeed, such formalization poses a risk to astrobiology as a discipline (at least in the sense of that word proposed here): namely, that what began as a critique of the disciplinary and authoritarian structure of knowledge in science will instead become an example of it. If astrobiology is to erode barriers between departments, it should hardly begin by erecting walls. Thus, we believe it is inconsequential whether or not we *are* astrobiologists – this is just a label and, under some circumstances, a veiled appeal to authority. What matters is that astrobiology is the type of science we *do*. In Box 28.2, two examples are considered of scientists who have long been involved in astrobiology, even if under a different name.

28.2.2 Ignorance as challenge

The ignorance motivating scientific inquiry will never wholly be separated from the ignorance hindering it. The disciplinary organization of scientific education encourages scientists to be expert on specialized subjects and silent on everything else. By creating scientists dependent on each other's specializations, this approach is self-reinforcing. A discipline of astrobiology should attempt something more ambitious: it should instead encourage scientists to master for themselves what formerly they deferred to their peers. From the outset, however, it is clear that no individual can know everything. This foreseeable shortcoming caused many of us to hesitate before joining the program. Some of us were afraid that having a certificate in astrobiology would signal a lack of scientific intensity to prospective employers. We were also concerned (and remain concerned) that, due to the financial opportunism to which basic research is frequently driven, sponsored research would be only superficially astrobiological, and thus not seriously advance the field. Imagine our dilemma

---

Box 28.2 Why did established scientists choose astrobiology?

Two interviews conducted by Diane Carney, history of science and astrobiology graduate student.

As graduate students in astrobiology, we are simultaneously participating in and shaping the future of graduate education in this interdisciplinary field. We learn from one another, the faculty, and other professionals. Before the existence of an astrobiology program, even before the common use of the term *astrobiology*, today's leaders in the field were fostering interdisciplinary interactions and educating themselves across disciplinary boundaries. How did they get started? As examples, here are excerpts from interviews with two scientists who described their introduction to the field.

Don Brownlee (University of Washington)

Q: *Describe your education and training in science.*

A: I was an undergraduate in electrical engineering at Berkeley, but when I was a senior I got interested in astronomy. I went to graduate school at the UW in Astronomy ... which was an interesting switch going from engineering to science. Then I got involved in planetary science and space studies.

Q: *When did you first become interested in astrobiology?*

A: I've always been extremely interested in space and that has entanglements with astrobiology, but I never really took it terribly seriously. I guess it was when I got involved with teaching part of an Astronomy 201 course, "The Universe and the Origin of Life." We used [the book by] Sagan and Shklovskii and I had to self-teach myself a lot related to that. We covered all the basics of Miller–Urey synthesis, and so forth. There was a lot unknown at that time. There wasn't anything known about extra-solar planets, and there wasn't much known about exobiology – as it was known at that time – so I've been involved in astrobiology from that time. In fact some other people on campus were involved in that. I think Woody Sullivan was involved in that in some early stage.

One of the early beginnings of astrobiology at UW has to be when Milt Gordon from Biochemistry would come up and give some talks on microbiology to our Astronomy 201 class. [It] was an amazing thing because he got nothing back from that, but he liked astronomy and he did it for many years. As far as I know, that's the first interdisciplinary activity on campus related to astrobiology.

Q: *How long has that course been taught? Was that one of the first times?*

A: Well, yeah. Paul Boynton was the official professor on it and it started around 1971 or 1972 and has been taught over a long period of time. Although the emphasis changes depending on who is available: sometimes the life part disappears and it is mostly "the Universe," but at least to me, that played a role and formed some contact between the south campus and the upper campus [where biological and physical sciences are located].

Margaret Race (SETI Institute)

Q: *Describe your education and training in science.*

A: I went to the University of Pennsylvania – it was a biology degree but I was interested in water and you can't do a lot of marine biology in Philadelphia. So I began taking classes in other departments – geology, engineering – and I now realize that what I did was cobble together an environmental sciences major. So I had a multidisciplinary perspective because Penn had an independent major. I think that made a big difference.

... [Eventually] I went back to school at UC Berkeley ... [studying] exotic species. I looked at an example of something that was moved around – not by ballast water – but by railroad. These oysters were brought into San Francisco Bay after the gold rush and on the oyster shells were a lot of East Coast organisms. They were put into San Francisco Bay, so San Francisco is a kind of unique island experiment, if you will, it's an experiment where a lot of organisms from the East Coast are now on the West Coast in one bay. And so you had a controlled situation that was an experimental interaction between native species and introduced ones. So that was really influential. I just chose it as a dissertation topic, but [it] turned out [that] when the University of California was involved in genetically engineered organisms, the questions were [similar]: what would be the impact of having a genetically engineered organism deliberately introduced to the outdoors? Whose laws do they fit under, what agency, what kind of experiments would you do, what kind of data would you want? So that whole situation played out in real time in the public. As a result of that, NASA asked me to come down for a workshop and talk about what would happen when they brought back samples from Mars.

Q: *What was your first introduction to astrobiology? Was it the invitation to that workshop?*

A: Absolutely. It was a phone call that asked whether I would come, . . . what kinds of things might NASA bump into when they went [to Mars], and did I see any parallels with what the university went through in genetic engineering?

Q: *What was your immediate reaction to that request?*

A: Well it was a long phone call! I was delighted to go because I had thought a lot about science controversies and science decision making. I used to teach at Stanford in environmental decision making. I had already begun to think about how science fits in a process and what kind of data you'd need to make what kind of decisions and where it *didn't* matter if you had the scientific data [because] people were coming at it from a different perspective and their views or ethical questions were something you couldn't address. So I think the invitation to think about it and to do it formally in front of a workshop was the beginning of my experience with astrobiology.

as incoming students: did we want to deny ourselves the refuge of specialization and the appeal to experts, instead risking the label of "dilettante?" Did we want our research associated carelessly with the casual work of opportunists?

Comforting answers to these questions would be welcome, but are not likely forthcoming. Disparate opinions about astrobiology pervade the scientific community. This is apparent from a sample of non-astrobiology UW faculty members, each asked whether astrobiology should be considered a serious science. One respondent was very positive: "It is a serious science of considerable general interest awaiting a break-through." Another was not so encouraging: "I think it should be a part of standard biology. It is useful to consider how life forms may exist under other conditions, but calling it astrobiology seems like a gimmick to me." Finally, a third respondent acknowledged that his views were in a state of flux and that he was "growing much warmer to the field as observational techniques become more integrated into the research." Clearly, the jury is still out.

What astrobiology can mean as a science and discipline is yet to be decided, for it must face the two-fold challenge of cross-disciplinary ignorance that disciplinary education itself enforces. First, ignorance cannot be skirted by deferral to experts, or by other implicit invocations of the disciplinary mold that astrobiology should instead critique. Second, ignorance must actually be *recognized*. This is not trivial: how do you know what you do not know? Is it possible to understand a general principle without also understanding

the assumptions and caveats underlying it? Knowledge superficially "understood" is self-affirming. For example, the meaning of a molecular tree of life may appear unproblematic to an astronomer who has learned that the branch lengths represent evolutionary distance, but will that astronomer even know to consider the hidden assumptions about rate constancy by which the tree is derived? Similarly, images from the surface of Mars showing evidence of running water are prevalent in the media, yet how often will a biologist be exposed to alternative explanations for these geological forms, or to significant evidence to the contrary? What is needed is an extension of Carl Sagan's (1995) colorfully termed "baloney detection kit," a way he proposed to discriminate between science and pseudo-science. In our case the "baloney" is not necessarily pseudo-science, but uncritically accepted results of science.

Unfortunately, there exists no simple baloney detection kit. Instead, the onus is on the student of astrobiology to exceed superficial familiarity with traditionally disciplinary knowledge outside her or his own field. Some would argue that the accompanying risk of dilettantism is sufficient reason to curtail astrobiology's unifying ambitions. Yet does that risk, or the need for a baloney detection kit, ultimately reflect *astrobiology's* shortcomings? Is it not instead a reflection on the inadequacies of the disciplinary structure of scientific education? Too often, the scientific community and academia facilely redress ignorance by appealing to the testimony of experts. This does not resolve ignorance. It fosters it. The problem facing astrobiology is nothing less than the problem inciting astrobiology's development.

28.3 The UW graduate program

28.3.1 First try at a curriculum

The initial attempt at the UW to grapple with these challenges was instructive but not successful. Instead it delineated, in many cases for the first time, the issues we are describing here.

Originally (1999), the core curriculum instinctively reverted to the discipline-style training that each faculty member had. The graduate students were separated according to their undergraduate majors into two broad groups representing "physical" and "biological" scientists. Thus, in this very first step, the program segregated students trying to be interdisciplinary into disciplinary groups. Depending on its identified expertise, each group then took a two-quarter sequence of courses addressing the perceived disciplinary gaps. Those of us who were physical scientists took "The Evolution of Life: Biology for Astrobiologists" while the biological scientists took "The History of Planets: Planetology for Astrobiologists." There were two problems with this. First, with no prior student input (there had been no prior students!), and with the faculty trained in traditionally structured disciplines, there was no consistent assessment of what students did or did not know. Should the astronomy professor teaching a room full of presumed biologists begin with planets or basic physics? The second major problem was that we students did not fit into the neat little boxes implied by our undergraduate majors. For example, a student with a Geology and Geophysics degree would be placed on the "physical" side, while a student with a Marine Biology degree would be placed on the "biological" side, although both students had pursued extensive undergraduate research in geomicrobiology. While not surprising in hindsight, the people attracted to an interdisciplinary graduate program often already had interdisciplinary backgrounds. The disciplinary mold they were trying to escape still did not fit.

The culmination of the first year curriculum was an integrative class called "Astrobiology," in which the two groups of graduate students were reunited. A few weeks of unconnected, revolving-door lectures, however, disheartened us. As a group, we realized that we could not speak the language of the many disciplines in astrobiology and that we lacked the basic information to consider their claims critically. Instead, this attempt at an integrative approach provided only a superficial introduction to the major contributions of each discipline to astrobiology. How can critical science be built on a superficial foundation? Major gaps in our backgrounds still needed to be addressed. In addition, we realized it was necessary to direct ourselves toward a more specific goal. What types of scientific information did we most need? What level of mastery should we aspire to? At the same time, catalyzed by our regular interactions in the class, we students realized that we learned the most (and enjoyed ourselves the most) in each other's interdisciplinary company. While each of us had major gaps in our basic knowledge, as a group we could begin to fill many of them. These insights drove us, together with the faculty, to redesign the curriculum.

28.3.2 The present curriculum

Our experiences in the first year led to a reformulation of our educational objectives. It was realized, first, that no training program could *de novo* prepare students of such diverse backgrounds sufficiently in all subjects. It was also realized that students were each other's greatest assets and that most of what we would learn as astrobiologists would come, not from lectures, but from interactions with each other and other scientists. This led to the following restatement of program goals (also see Staley (2003) for a full description).

1. *Mastery of every subject is not feasible.* Instead, students should at least know the major terminology, tools, resources, and questions of other disciplinary approaches. We call this "basic competence." Students should have sufficient familiarity with other disciplines to recognize possible areas of ignorance and reliable reference sources to begin redressing those areas if they choose.
2. *Besides being competent, students should pursue specialization in more than one field.* Their primary specialization is their home department field of study, in which they receive their Ph.D. In addition, students – through auxiliary coursework and research experiences – should obtain a secondary specialization in at least one other field.
3. *Opportunities for students and faculty to interact, educate each other, present results to each other, and collaborate must be fostered.*

These goals led to the curriculum outlined in Fig. 28.2. To address the first goal, competence, two ten-week core courses were designed for all students to take together. "Astrobiology Disciplines" addresses the need for students to learn a great deal of basic information about other disciplines in astrobiology.

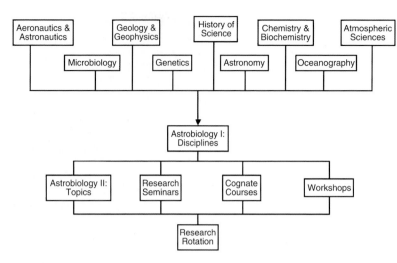

FIGURE 28.2 The UW Astrobiology Curriculum. Participating departments contribute faculty and students to the Astrobiology Program. Certificate students take the "Astrobiology Disciplines" course to identify basic areas of disciplinary knowledge about which they are ignorant. The "Astrobiology Topics" course, seminars, cognate courses, and annual workshops are opportunities to investigate specific subjects in greater depth. The program culminates in an out-of-field research rotation.

In its first redaction, the class was self-taught by the participating students: each student spent 1–6 classes introducing the others to the fundamentals of his or her science, including the basic resources available to pursue more detailed study. The current course, now mentored by faculty, still contains a strong graduate student presentation and discussion component. The second core course, "Astrobiology Topics," applies an interdisciplinary approach to expose and explore specific astrobiological subjects, such as hydrothermal vent ecology, the geochemical evidence of ancient life, or the habitability of Mars. The goals of both courses are not to produce experts, but rather to establish a foundation of basic knowledge and to help students identify subjects outside of their home departments that they might further pursue. These courses ultimately are not the means to competence but provide self-selecting students the impetus and direction to become more competent.

To meet the second goal, specialization in a separate discipline outside of the home department, students are required to take cognate courses and do a research rotation. Most of the cognate courses are pre-existing classes offered by both astrobiology and non-astrobiology faculty. These courses are the means by which students can refine specific interests identified by the two core courses. Ideally, the cognate courses are also used in part to prepare the student for his or her research rotation. This rotation is two quarters of research with an advisor outside the student's primary field of study. In many ways, the research rotation can be the pinnacle of the academic program, requiring the student to design and implement a short research project and then present the results. A productive rotation

can involve one quarter of preparation, including faculty-guided review of the relevant literature, and at least one quarter of research. Past examples of research rotations include a microbiologist studying the characteristics of stars with extrasolar planets, a biological oceanographer working with an engineer to develop a hydrothermal vent sampler, and a historian of science assessing microbial diversity in Arctic sea-ice.

Finally, the third goal, promoting interaction, is implicit to everything described so far. Students are no longer divided up into bins of biological or physical scientists. They take the core courses together. Through their cognate coursework, the students interact with other departments and other students, culminating in research rotations that expose them not only to new ways of doing things, but also to new people doing them. Students and faculty are brought together by a continuous seminar series examining astrobiological problems as well. Each year this seminar series includes one quarter of public talks, mostly by guest speakers; one quarter of in-house presentations of student and faculty research; and a final quarter, again in-house only, used to explore as a community a specific subject about which only a few participants are well informed. This latter is particularly important. During each session, a graduate student presenter is paired with a faculty mentor to cover a subtopic within the seminar theme. For example, in a seminar series devoted to planetary atmospheres, a graduate student in oceanography teamed with an atmospheric chemist to present a lecture on Earth's early atmosphere. The point is for students to learn new subjects and approaches for themselves (i.e., become competent), rather than rely on an expert's appraisal.

FIGURE 28.3 The annual astrobiology workshop provides an occasion for students and faculty members to share ideas and explore astrobiological ideas in the field. At one workshop, held at Friday Harbor Laboratories on San Juan Island, we studied morphological radiation of animals during the "Cambrian explosion" of $\sim$60 Ma (Chapter 16). A dredge in Puget Sound was sorted (panel a) to obtain samples of the vast range of body plans that developed in the early Cambrian (panel b), including the five-fold symmetry of the sea urchin and other echinoderms (panel c). This was complemented by a trip to nearby Sucia Island for *in situ* observation of fossilized marine organisms from the Cretaceous Period (panel d).

Further advancing the third goal are group activities designed to build community and share differing scientific perspectives. For example, we hold annual 3–4 day workshops, which are field trips to places of astrobiological interest. Past workshops have included visits to the Pacific Northwest National Laboratory, where deep aquifers were sampled in a search for subsurface autotrophic microorganisms; to NASA Ames to learn about astrochemistry and the Mars lander site selections; and to Friday Harbor Laboratories, where the "Cambrian explosion" was explored by comparing the body plans of extant organisms to the fossil record (Fig. 28.3). While the goal of each trip is to further develop interdisciplinary competence, participants also learn about the diverse backgrounds and approaches of fellow members of the program. More informal (but regular) meetings also cement relationships and bridge disciplines. Students and faculty meet once per month for a social hour where scientific and unscientific discourse alike are catalyzed by a friendly atmosphere and a few drinks. Similarly, we students have formed UWAB, the UW Astrobiology student

---

Box 28.3 Why did I choose astrobiology?

Diane Nielsen

Before starting graduate school in oceanography, I taught middle and high school science. One of my primary goals as a teacher was to get students excited about science by showing them how creative science can be and how much there is still to discover. Equally important to me was to have my students gain an awareness that all science is inextricably linked. These same reasons drew me to oceanography and astrobiology, particularly hydrothermal vent research, to address fundamental questions that cross the traditional fields within science. I wanted to work with a group of people as enthusiastic about interdisciplinary research as I was. Astrobiology supports this sense of excitement and community. Faculty and graduate students are learning together to be more interdisciplinary in their scientific knowledge and their research. For example, during the Astrobiology Topics class, at least four faculty members were usually in attendance, asking questions alongside the students. Similarly, at the Friday Harbor workshop, I watched a biologist tutor an astronomer on how to dissect a sea urchin, then watched the astronomer give the biologist a lesson on the pattern of "Newton's rings" visible on an oily slide. Experiences such as these help scientists grow by both teaching and learning. They also seed a future science whose direction is likely to be both unexpected and rewarding.

---

group, to discuss science, the program, education, politics, music, the details of graduate student life ... While the benefits of promoting such interactions may seem uncertain (but see Box 28.3), tangible results have accrued, including interdepartmental research collaborations, the writing of this book, and, in fact, the joint involvement of faculty and students in redesigning the astrobiology program itself.

28.4 The future of astrobiology

What will the future hold for astrobiology? While we dispute the importance of developing exclusive identities as astrobiologists, there can be little doubt that the scientific community will evaluate astrobiology in part according to those it identifies as astrobiologists. For this reason, students of astrobiology must maintain expected disciplinary standards. Training in astrobiology should be understood to represent an additional commitment, extra work, extra thought, and extra time dedicated to pursuing education in astrobiology alongside normal expectations of mastery in an established field. *Substitute* standards should not be invoked for astrobiologists; instead, *additional* standards should be expected. The student of astrobiology trained in oceanography should be as competent an oceanographer as any other; but in addition she or he should have considerable background and experience in other fields pertinent to astrobiological questions. The value of people trained in astrobiology to the scientific community and the public will reflect the degree to which such additional standards are attained.

Adherence to disciplinary standards, astrobiological work notwithstanding, does not imply satisfaction with the disciplinary barriers pervading modern science, however. In the long term, astrobiology has the potential to serve as an alternative to the segregation within science and within science education. Is it truly necessary for biology and physics to be taught as separate disciplines in high school, or to develop as separate departments in universities? How much specialization in science reflects historical inertia, sociological boundaries, and the apparatus by which science is learned, rather than true limits of disciplinary knowledge or ability? Why not anchor research and education to questions, rather than disciplines? Baruch Blumberg, the first director of NAI, has termed astrobiology "generational science" because of the future-looking and long-term experiments envisioned for it. But astrobiology is generational science not merely because of its research projects. As an educational program, astrobiology has the potential to raise a distinctive cohort of scientists and to help create a scientifically literate public. This too will require long-term commitments and generational adjustments. An initial group of scientists trained exclusively in a disciplinary mold will give way to a more interdisciplinary group; and that group in turn will be succeeded by a yet more broadly trained community. Such transitions will take time.

A meaningful *discipline* of astrobiology will be defined only through the development of its educational potential, its opportunity to break down the walls isolating disciplines. Integrative, interdisciplinary, and already familiar to the public, astrobiology is well positioned to advance this goal. At the same time, astrobiology faces many challenges, including the risks of dilettantism, sociological and economic whims, and reversion to the same disciplinary insularity to which it is currently a reaction. In this chapter we have outlined a general educational approach addressing some of those challenges. We emphasize in particular the need to cultivate basic interdisciplinary competence, to specialize in more than one field, and to interact as much as possible, in as many ways as possible, with individuals in many fields. While developed in the context of graduate school, this general framework may also be extended to earlier educational stages, when students are less constrained by specialized ignorance. This is a final important component of an astrobiological discipline: public outreach and expansion into undergraduate or high school programs. Already seven undergraduate textbooks and high school curricula based on astrobiology have appeared.[2]

In the end, astrobiology may bring us back to the integrative style of someone like Johannes Kepler four hundred years ago. He was not content merely to bequeath the world a detailed quantitative theory of orbital motion and a theory of vision. In an age of witchcraft and persecution, Kepler wrote a great synthesis of the Universe. He understood it as a musical masterpiece, an orchestra of resonating orbits and celestial movements. His intemperate vision of the Cosmos is now largely forgotten, his three Laws coldly distilled from it. Since the end of Kepler's reign as Imperial Mathematician, science has progressively distanced itself from such creative and all-encompassing syntheses. Grand unified theories remain future prospects, while details are toiled upon. Details can be inspirational – after all, it was Kepler's attention to detail in interpreting Tycho Brahe's decades of accurate observations that led to his famous Laws. Details,

however, are not ends unto themselves. Why are we scientists? – why are we teachers? – why are we curious? – if not to ask the grand questions that we cannot answer. Astrobiology is an opportunity for science to reincorporate some of the intemperance and ambition of earlier, more comprehensive approaches. No guarantee exists that these types of approach will necessarily yield "the truth" or even an enduring approximation of it. But three laws may yet be distilled from the next great synthesis – hardly a poor legacy.[3]

REFERENCES

Bennett, J., and Shostak, S. (2007). *Life in the Universe*, second edn. New York: Addison Wesley.

Fruton, J. S. (1999). *Proteins, Enzymes, Genes: the Interplay of Chemistry and Biology*. New Haven: Yale University Press.

Gilmour, I. and Sephton, M. A. (eds.) (2003). *An Introduction to Astrobiology*. Cambridge: Cambridge University Press.

Goldsmith, D. and Tobias Owen, T. (2002). *The Search for Life in the Universe*, third edn. Sausalito, CA: University Science.

Jakosky, B. (1998). *The Search for Life on Other Planets*. Cambridge: Cambridge University Press.

Lunine, J. (2005). *Astrobiology: a Multi-Disciplinary Approach*. San Francisco: Pearson/Addison Wesley.

Sagan, C. (1995). *The Demon-Haunted World: Science as a Candle in the Dark*. New York: Random House.

Staley, J. T. (2003). Astrobiology, the transcendent science: the promise of astrobiology as an integrative approach for science and engineering education and research. *Curr. Opin. Biotech.*, **14**, 347–354.

Summers, W. C. (1999). *Felix d'Herelle and the Origins of Molecular Biology*. New Haven: Yale University Press.

Ward, P. D. and Brownlee, D. (2000). *Rare Earth: Why Complex Life is Uncommon in the Universe*. New York: Copernicus.

[2] There are now five undergraduate textbooks for astrobiology. Four designed for non-science students are Jakosky (1998), Goldsmith & Owen (2002), Bennett and Shostak (2007), and Gilmour & Sephton (2003), while Lunine (2005) is for science majors. At the high school level two year-long astrobiology curricula have been developed: *Voyages Through Time: a High School Integrated Science Curriculum* (2003; www.voyagesthroughtime.org) and *Astrobiology: an Integrated Science Approach* (2005; www.terc.edu).

[3] We thank UW astrobiology graduate students Nicolas Pinel, Diane Carney, and Diane Nielsen for the boxes, and Jeremy Dodsworth and Brian Kristall for the figures. Others who contributed to the ideas in this chapter were David Allen, Karen Junge, Randy Perry, and Steve Vance.

Epilogue

Christopher F. Chyba
Princeton University

"Astrobiology" was originally defined as "the consideration of life in the Universe elsewhere than on Earth" (Lafleur, 1941). But as the field has advanced, we have learned to place no artificial barrier between the study of life on Earth and life that may exist elsewhere in the Cosmos. Astrobiology today is "the study of the living Universe" (NAI, 2004), be it here or elsewhere. It would be foolish to narrow the definition, for the approaches we take in searching for extraterrestrial life are strongly informed by our understanding of life on Earth, and our understanding of the origin and evolution of terrestrial life is informed both by the study of other planetary environments and by Earth's environment within the Solar System and Galaxy. As Carl Sagan (1974) remarked decades ago, we are able for the first time in human history to assess life on Earth "in a cosmic context." The assessment is still nascent and inchoate, but as the chapters in this book illustrate, the floodgates have opened and our knowledge is expanding quickly now. We will soon know much more.

Besides "astrobiology," the study of life in the Universe has also been called "cosmobiology" (Bernal, 1952), "exobiology" (Lederberg, 1960), and "bioastronomy" (IAU, 2004) (see Sections 2.3.1 and 2.4 for discussion). Under its exobiological label, the entire field was famously criticized by the biologist George Gaylord Simpson (1964), "in view of the fact that this 'science' has yet to demonstrate that its subject matter exists!" If astrobiology meant only the study of extraterrestrial life itself, Simpson's criticism would still have weight, four decades later. But his criticism would nevertheless seem peculiar to many physicists and astronomers. After all, much of the important research in physics and astronomy concerns exactly those objects or phenomena whose existence remains uncertain – consider the search for high-temperature superconductors, the Higgs boson, or proton decay. Each of these could turn out not to exist, and the last (Georgi and Glashow, 1974) has not been detected, despite dedicated attempts. Even black holes were studied for decades before compelling evidence of their existence accumulated (Melia and Falcke, 2001). In this sense, astrobiology simply extends into biology a circumstance that is familiar, even commonplace, to a number of its sister sciences.

Astrobiology is posed on the brink of remarkable and historic discoveries. Barring catastrophic failure, NASA's upcoming Kepler mission should determine later this decade the statistical frequency of Earth-size worlds around other stars (NASA, 2004). The question whether other planets like Earth exist has been asked for 2,400 years or more, at least since Aristotle posed the question in *De Caelo* (Aristotle, 1941 trans.) – and ruled out the possibility on theoretical grounds (Section 1.2.1). It is extraordinary that within a few years humanity will no longer have to grope for answers to this question. Instead we will *know* how common are other earths, around what kinds of stars they form, and in what heliocentric orbits. We should not let our civilization sleepwalk through this extraordinary transition in our knowledge of Earth's place in the Universe.

We are on the verge of extraordinary discoveries in other areas as well. We are coming to know Mars as a planet that had, and may still have, substantial liquid water (Chapter 18). Intriguingly, methane has been observed in the martian atmosphere at a concentration of 10 ppb (Section 18.8). Since the CH_4 photochemical lifetime in the martian atmosphere is only $\sim$300 yr, and the observed CH_4 is inhomogeneously distributed, there must be an active source. Krasnopolsky *et al.* (2004) favor methanogenesis by oases of living subterranean microorganisms as a hypothesis, though abiogenic sources are the conservative explanation until further evidence supports the biological claim. This is not a new caution. Decades ago, Pollack and Sagan (1967) warned against the temptation to attribute

Planets and Life: The Emerging Science of Astrobiology, eds. Woodruff T. Sullivan, III and John A. Baross. Published by Cambridge University Press. © Cambridge University Press 2007.

enigmatic martian features to biology: "The varieties of life on Earth and of possible organisms that we can imagine on Mars are so large that virtually any change in surface features can be 'explained' by a necessarily vague attribution to biological processes." But even if biological explanations should be preferred only when all non-biological ones fail, our growing knowledge of terrestrial extremophiles, together with discoveries about contemporary Mars, now make martian life seem to be a very open question. Planetary protection is therefore critical (Space Studies Board, 2006).

The exciting discoveries hardly stop at Mars. On Earth we are elucidating the deep hot biosphere (Chapter 14). At Jupiter, we have collected strong evidence that a second ocean of liquid water exists in the solar system, beneath the surface of Europa, and perhaps also for Ganymede and Callisto (Chapter 19). Whether Europa's ocean is sterile or not is a question we can hope to answer later this century.

Beyond possible microbial life, the continuum of questions posed by astrobiology extends to quantitative investigations of the evolution of intelligence on Earth (Marino *et al.*, 2004), its future technical evolution, and, relatedly, theoretical and empirical investigations of the prospects for intelligence and technology elsewhere (Chyba and Hand, 2005). Since early in the past century, far-seeing observers have warned that humanity may soon enter a realm where technical intelligence and its own directed evolution advances so quickly that our successor civilization can hardly be grasped (Bernal, 1929; see also Hart, 1985; Vinge, 1993). Yet well before then, exponentiating computation and biotechnology will put great power in the hands of small groups of the technically competent (Carlson, 2003), and it is not clear how we can manage this new world (Chyba and Greninger, 2004). The same technological prowess that radically advances our capacities in the search for extraterrestrial life drives new challenges to human civilization. If astrobiology is the study of the living Universe, the discipline must also learn to speak to the human future, a thing uniquely precious regardless of whether we are entirely alone or part of a grand biological Universe.

REFERENCES

Aristotle (1941). *De Caelo*. Trans. by J. L. Stocks in *The Basic Works of Aristotle*, ed. R. McKeon, 276^{a30}–276^{b24}. New York: Random House.

Bernal, J. D. (1929). *The World, the Flesh, and the Devil*. Second Edition 1969. Bloomington: University of Indiana Press.

Bernal, J. D. (1952). Lecture to the British Interplanetary Society, described in Slater A. E., *Journal of the British Interplanetary Society*, **12**, 114–118.

Carlson, R. (2003). The pace and proliferation of biological technologies. *Biosecurity and Bioterrorism*, **1**, 203–214.

Chyba, C. F. and Greninger, A. L. (2004). Biotechnology and bioterrorism: an unprecedented world. *Survival*, **46**, 143–162.

Chyba, C. F. and Hand, K. (2005). Astrobiology: The study of the living Universe. *Ann. Rev. Astron. Astrophys.*, **43**, 31–74.

Georgi, H. and Glashow, S. L. (1974). Unity of all elementary-particle forces. *Phys. Rev. Lett.*, **32**, 438–441.

Hart, M. H. (1985). Interstellar migration, the biological revolution, and the future of the Galaxy. In *Interstellar Migration and the Human Experience*, eds. B. R. Finney and E. M. Jones, pp. 278–291. Berkeley: University of California Press.

IAU (2004). *International Astronomical Union*. www.ifa. hawaii.edu/~meech/iau/

Krasnopolsky, V. A., Maillard, J. P., and Owen, T. C. (2004). Detection of methane in the martian atmosphere: evidence for life? *Icarus*, **172**, 537–547.

Lafleur, L. J. (1941). Astrobiology. *Leaflets of the Astronomical Soc. Pacific*. No. **143**, 333–340.

Lederberg, J. (1960). Exobiology: approaches to life beyond the Earth. *Science*, **132**, 393–398.

Marino, L., McShea, D., and Uhen, M. D. (2004). The origin and evolution of large brains in toothed whales. *Anatomical Record*, **281A**, 1247–1255.

Melia, F. and Falcke, H. (2001). The supermassive black hole at the galactic center. *Annu. Rev. Astron. Astrophys.*, **39**, 309–352.

NAI (2004). NASA Astrobiology Institute. www.nai.arc. nasa.gov/institute/about_nai.cfm#astrobiology

NASA (2004). http://www.kepler.arc.nasa.gov/summary.html.

Pollack, J. B. and Sagan, C. (1967). Secular changes and dark-area regeneration on Mars. *Icarus*, **6**, 434–439.

Sagan, C. (1974). The origin of life in a cosmic context. *Orig. Life Evol. Biosph.*, **5**, 497–505.

Simpson, G. G. (1964). The nonprevalence of humanoids. *Science*, **143**, 769–775.

Space Studies Board, National Research Council (2006). *Preventing the Forward Contamination of Mars*. Washington DC: National Academy Press, www.nas.edu.

Vinge, V. (1993). The coming singularity: How to survive in the post-human era. In *VISION-21 Symposium*, NASA Conf. Pub. 10129, pp. 11–22.

Part VIII
Appendices

Appendix A: Units & Usages

Communication across the many disciplines involved in astrobiology is fraught with difficulty on many levels, including even the seemingly simple matter of units and usage of terms and abbreviations. When first introduced in any chapter, unusual units often not known to those outside the field are defined. In this appendix we give conventions and conversions for various units used throughout the book.

A.1 Units

Time: $Gyr = 10^9 \, yr$; $Myr = 10^6 \, yr$

Past time: $Ga = $ "$10^9 \, yr$ ago" or "$10^9 \, yr$ old"

$\qquad Ma = $ "$10^6 \, yr$ ago" or "$10^6 \, yr$ old"

‰ = parts per thousand

ppm = parts per million; ppb = parts per billion (10^9)

Distance: parsec (pc) = 3.26 light years (lt-yr)
$\qquad = 2.06 \times 10^5 \, AU = 3.1 \times 10^{16} \, m$

Astronomical unit (AU) $= 1.50 \times 10^{11} \, m$

Absolute temperature scale (kelvins): $0 \, K = -273.15 \, °C$

Pressure units: $1 \, nt \, m^{-2} = 1$ pascal $= 10^{-5}$ bar
$\qquad = 10^{-5}$ atmosphere (Earth)

Wavelength: $1 \, nm = 10 \, Å$ (Ångstrom)

Molarity: moles (mol) per liter $= M$

A.2 Usages

Archean is the geological era before 2.5 Ga, while *Archaea* is one of the three domains of life (adjective: *Achaeal*).

Although metal is a well-defined chemical concept in all of science, in astronomy *metals* unfortunately refers to all elements other than H and He!

Terran refers to "of the planet Earth," as opposed to extraterrestrial (e.g., "terran life").

Terrestrial can mean "of the planet Earth"; "of the inner, rocky planets"; or "of the land" (as opposed to "marine").

Evolution: The astronomer's usage of the term *evolution* refers to the change with time of an entity (such as a star), analogous to the ageing of an individual organism. The biologist's meaning is in the context of a specific theory of how life as a whole changes over time.

Planets and Life: The Emerging Science of Astrobiology, eds. Woodruff T. Sullivan, III and John A. Baross. Published by Cambridge University Press. © Cambridge University Press 2007.

Appendix B: Planetary properties

TABLE B.1 Planetary Properties

Planet	a	P	e	ε	T
Earth	1.5×10^{11} m	1.00 yr			1.00 day
Mercury	0.39 AU	0.24	0.206	$0.1°$	176
Venus	0.72	0.62	0.007	$177°$	117
Earth	$\equiv 1.00$	$\equiv 1.00$	0.017	$23.4°$	$\equiv 1.0$
Mars	1.52	1.88	0.093	$25.2°$	1.03
Jupiter	5.20	11.9	0.048	$3.1°$	0.41
Saturn	9.54	29.4	0.056	$26.7°$	0.45
Uranus	19.19	84.0	0.046	$98°$	0.72
Neptune	30.07	165	0.009	$30°$	0.67
Pluto	39.48	248	0.249	$123°$	6.4
Moon	0.0026	0.075	0.055	$7°$	29.5
Sun				$7°$	25–36

Planet	r	M	ρ	g	v_{esc}	A
Earth	6370 km	6.0×10^{24} kg		$9.8\,\mathrm{m\,s^{-2}}$	$11.2\,\mathrm{km\,s^{-1}}$	
Mercury	0.38	0.06	$5.4\,\mathrm{g\,cm^{-3}}$	0.4	0.4	0.11
Venus	0.95	0.82	5.2	0.9	0.9	0.65
Earth	$\equiv 1.00$	$\equiv 1.00$	5.5	$\equiv 1.0$	$\equiv 1.0$	0.37
Mars	0.53	0.11	3.9	0.4	0.45	0.15
Jupiter	11.2	320	1.3	2.4	5.3	0.52
Saturn	9.5	95	0.7	0.9	3.2	0.47
Uranus	4.0	15	1.3	0.9	1.9	0.51
Neptune	3.9	17	1.6	1.1	2.1	0.41
Pluto	~ 0.19	0.002	~ 1.8	0.06	0.1	0.6
Moon	0.27	0.012	3.4	0.17	0.2	0.12
Sun	110	330,000	1.4	28	55	

For notes see the following page.

Planets and Life: The Emerging Science of Astrobiology, eds. Woodruff T. Sullivan, III and John A. Baross. Published by Cambridge University Press. © Cambridge University Press 2007.

Notes:

– Many values are given as ratios to the value for Earth; in such cases the value for Earth is given in the first row. Parameters are:

a = semimajor axis of orbit = mean distance from Sun (AU)

P = period of orbit

e = eccentricity of orbit (zero for a circular orbit, higher means more elliptical)

ε = obliquity to the ecliptic = tilt angle of rotation axis from the normal to the *ecliptic plane* (plane of the Earth's orbit); a value greater than 90° indicates *retrograde* rotation, i.e., rotation opposite in sense to the great majority of Solar System rotations (and orbits)

T = length of day (*synodic period* of rotation) as experienced by an observer on the planet's surface; this is the cycle to which an organism would have to adapt. This is not the true rotation period (*sidereal period*) as measured with respect to distant stars.

r = mean radius of planet

M = mass

ρ = mean density

g = acceleration due to gravity at planet's surface

v_{esc} = escape velocity from planet's surface

A = visual geometrical albedo (Sec. 4.1.1) = fraction of sunlight reflected by the planet

– Most data are from nssdc.gsfc.nasa.gov/planetary/planetfact.html. Listed values are designed to give the reader the general picture, not to equip him/her with values to use in accurate calculations.

– Many values are compromises, e.g., a planet's equatorial radius is generally greater than its polar radius.

– Controversy has arisen in recent years as to whether Pluto should be "demoted" from being a planet, and only become the largest of the Kuiper Belt Objects (KBOs). The waters were further muddied in 2004 with the discovery of 2003 UB313, a KBO that appears definitely to be larger than Pluto – is it or is it not a 10th planet? A nice discussion of the issues is given at the discoverer's website www.gps.caltech.edu/~mbrown/planetlila/index.html. Whatever its ontological status, its properties are $a = 68$ AU, $P = 560$ yr, $e = 0.44$ (therefore its solar distance varies from 38 to 98 AU), and $r \sim 1.5 \times$ the value for Pluto. The object also has an orbit highly inclined to the ecliptic plane (an inclination of 44°, versus Pluto at 17° and all other planets with much lower values). The controversy culminated in 2006 with heated discussions at a meeting of the International Astronomical Union, where a vote on the definition of the term *planet* was taken. Science, however, is not a democracy.

Appendix C: The geological timescale

S. M. Awramik, *University of California*
K. J. McNamara, *Western Australian Museum*

The geological timescale is one of science's great triumphs. It represents "deep time" (McPhee, 1982), millions and billions of years ago, time beyond human comprehension. Geologists use two different ways to discuss geological time: (a) absolute time, and (b) relative time.

Relative time was developed first, and is based on relative age relationships among rock units as determined by geometric relationships, fossils, and other distinctive attributes of the rock. Layered, sedimentary rocks (*strata*) are the most widely used, and during the nineteenth century allowed geologists to establish a relative geological timescale. The scale used strata with their contained fossils, and applied four fundamental principles of relative time: (1) original horizontality, (2) superposition, (3) original lateral continuity, and (4) fossil succession. *Fossil succession* refers to the particular vertical (stacked) order that fossils occur in strata. William Smith recognized this in 1799 and employed it to great effect in his geological map of England and Wales published in 1815. The other three principles were established much earlier by the Danish scholar Nicolaus Steno (1669). *Original horizontality* states that sediments are originally laid down in a horizontal or nearly horizontal manner. *Superposition* means that the oldest stratum is at the bottom and the youngest at the top. *Original lateral continuity* refers to the way strata extend laterally in all directions. The geometric relationship of bodies of rock to one another is also used; for example, an intrusion that cuts layered rock is younger than the layered rock it cuts. This was another of Steno's principles: cross-cutting relationships. It is from these principles that successions of sedimentary rocks with their distinguishing fossils were given names (e.g., Cambrian), arranged from oldest to youngest, and organized into a hierarchy (eon, era, period, and epoch), producing the first modern (relative) Geological Time Scale (Berry, 1987) (Figures C.1 and C.2).

This relative timescale was a major advancement, but confined to the Phanerozoic Eon because it was restricted to strata with animal fossils, and, to a lesser extent, plant fossils. The relative timescale had another major deficiency: it did not provide the age of a rock or fossil *in years*. Without an *absolute timescale*, it was impossible to measure *quantitatively* the rates of geological and evolutionary processes on the planet. This all changed, however, about a century ago with the discovery of radioactivity by Henri Becquerel in 1896. Eight years later Ernst Rutherford suggested that helium trapped in radioactive minerals could be used to determine the age of the Earth (Wilson, 1983). Using helium, Rutherford obtained an age of 40 Ma for one rock. By 1907, however, Bertram Boltwood had used lead, a stable end-product of uranium decay, to obtain ages ranging up to 2.2 Ga. These were the first absolute ages (Burchfield, 1990).

Radiometric dating makes use of radioactive elements trapped in minerals and rocks, and their spontaneous disintegration into stable, non-radioactive daughter products. The rate of decay is known from lab measurements and the ratio of the radioactive element (parent) to its daughter product (stable, non-radioactive) in the mineral can be measured. From that, the amount of time since the mineral crystallized can be determined. When a mineral crystallizes, a radioactive element can be 'locked' into the crystal lattice if it can chemically substitute for another element forming the mineral. There are several parent-daughter combinations that are used to date minerals and rocks. The most widely used are uranium-lead ($^{238}U/^{206}Pb$, $^{235}U/^{207}Pb$), potassium-argon ($^{40}K/^{40}Ar$), and rubidium-strontium ($^{87}Rb/^{87}Sr$). Methods using the ratios of different daughter products that have unique parent isotopes are also used, such as lead-lead ($^{206}Pb/^{207}Pb$), uranium-uranium ($^{235}U/^{238}U$), and argon-argon ($^{40}Ar/^{39}Ar$). Minerals that formed during the

Planets and Life: The Emerging Science of Astrobiology, eds. Woodruff T. Sullivan, III and John A. Baross. Published by Cambridge University Press. © Cambridge University Press 2007.

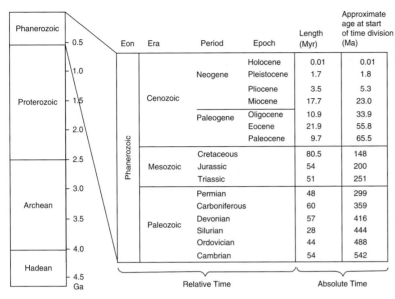

Eon	Era	Period	Epoch	Length (Myr)	Approximate age at start of time division (Ma)
Phanerozoic	Cenozoic	Neogene	Holocene	0.01	0.01
			Pleistocene	1.7	1.8
			Pliocene	3.5	5.3
			Miocene	17.7	23.0
		Paleogene	Oligocene	10.9	33.9
			Eocene	21.9	55.8
			Paleocene	9.7	65.5
	Mesozoic	Cretaceous		80.5	148
		Jurassic		54	200
		Triassic		51	251
	Paleozoic	Permian		48	299
		Carboniferous		60	359
		Devonian		57	416
		Silurian		28	444
		Ordovician		44	488
		Cambrian		54	542

Relative Time   Absolute Time

A   B

FIGURE C.1 The left-hand column (A) shows the eons comprising the entire geological timescale. The most recent eon, the Phanerozoic, is blown up on the right (B). An eon is divided into eras and periods, and the periods of the Cenozoic Era are further subdivided into epochs. Based on the International Commission on Stratigraphy's International Stratigraphic Chart (2004).

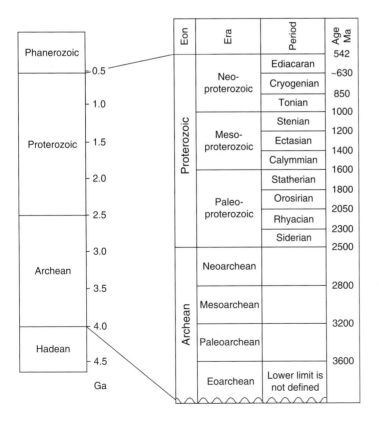

Eon	Era	Period	Age Ma
Proterozoic	Neoproterozoic	Ediacaran	542
		Cryogenian	~630
		Tonian	850
			1000
	Mesoproterozoic	Stenian	
			1200
		Ectasian	
			1400
		Calymmian	
			1600
	Paleoproterozoic	Statherian	
			1800
		Orosirian	
			2050
		Rhyacian	
			2300
		Siderian	
			2500
Archean	Neoarchean		
			2800
	Mesoarchean		
			3200
	Paleoarchean		
			3600
	Eoarchean	Lower limit is not defined	

FIGURE C.2 The Archean and Proterozoic eons have recently been subdivided into eras, and the eras of the Proterozoic divided into periods. All of the eras and periods are based on absolute age, except for the Ediacaran Period, which was recently established based on fossils and strata (Knoll et al., 2004), and is the first period since 1879 to be established without defined radiometric ages. Based on the International Commission on Stratigraphy's International Stratigraphic Chart (2004).

cooling of molten rock are usually used. Thus, the absolute age of an igneous intrusion that cuts sedimentary strata implies that the sediments can be no younger than that age (based on cross-cutting relationships).

The most important mineral for calibration of the geological timescale with absolute ages is zircon. Zircon ($ZrSiO_4$) can incorporate uranium when it crystallizes (uranium substitutes for zirconium). Uranium's daughter products, such as lead, normally remain trapped in the crystal structure, allowing for accurate dating. In addition to being found in granites and similar igneous rocks, zircon is also found in certain kinds of volcanic rocks and in volcanic ash. For the geological timescale, the radiometric dating of zircons from volcanic ash deposits interbedded with fossiliferous strata has proven most useful (Bowring et al., 1993). The ash represents molten material that was spewed out of the volcano and quickly cooled and crystallized. Zircon crystals are among the products. The ash is deposited and can become part of the sedimentary rock record. Radiometric dating of the zircons produces an age that is a good estimate of the age when the sediments were deposited. These techniques then provide an impressively accurate absolute calibration of the relative geological timescale (Gradstein et al., 2004), as shown in Figures C.1 and C.2. They also allow an extension into Precambrian times (before 542 Ma), where the eons of Proterozoic, Archean, and Hadean have been defined.

The oldest Earth rocks known to date are igneous rocks from northern Canada that have ages of 4.03 Ga (Bowring and Williams, 1999) and define the boundary between the Archean and Hadean. Although no rocks are known from the Hadean (by definition, the Hadean is that eon before the oldest known rock), zircons that eroded from a no-longer-existant rock have ages up to 4.4 Ga (Wilde et al.,

2001). Radiometric ages of meteorites provide a well-determined age of 4.55 Ga for the formation of the entire Solar System, including the Sun and Earth (Allègre et al., 1995).

References

Allègre, C. J., Manhès, G. and Gopel, C. (1995). The age of the Earth. *Geochim. Cosmochim. Acta*, **59**, 1445–1456.

Berry, W. B. N. (1987). *Growth of a Prehistoric Time Scale*. Palo Alto: Blackwell Scientific.

Bowring, S. A. and Williams, I. S. (1999). Priscoan (4.00–4.03 Ga) orthogneisses from northwestern Canada. *Contrib. Mineral. Petrol.*, **134**, 3–16.

Bowring, S. A. et al. (1993). Calibrating rates of early Cambrian evolution. *Science*, **261**, 1293–1298.

Burchfield, J. D. (1990). *Lord Kelvin and the age of the Earth*. Chicago: University of Chicago Press.

Gradstein, F., Ogg, J., and Smith, A. (2004). *A Geologic Time Scale 2004*. Cambridge: Cambridge University Press.

International Commission on Stratigraphy (2004). International Stratigraphic Chart. http://www.stratigraphy.org/chus.pdf.

Knoll, A. H., Walter, M. R., Narbonne, G. M., and Christie-Blick, N. (2004). A new period for the geologic time scale. *Science*, **305**, 621–622.

McPhee, J. A. (1981). *Basin and Range*. New York: Farrar, Straus, Giroux.

Steno, N. (1669). *De solido intra solidum naturaliter contento dissertationis prodomus [Preliminary discourse to a dissertation on a solid body naturally contained within a solid]*. Facsimilie Edition (1904). Florence: Berlin Junk.

Wilde, S. A., Valley, J. W., Peck, W. H., and Graham, C. M. (2001). Evidence from detrital zircons for the existence of continental crust and oceans on the Earth 4.4 Gyr ago. *Nature*, **409**, 175–178.

Wilson, D. (1983). *Rutherford*. London: Hodder and Stoughton.

Appendix D: Astrobiological destinations on planet Earth

Jelte P. Harnmeijer (Ed.)
University of Washington

Introduction

Astrobiology is a discipline that is best enjoyed in the field. What follows is a series of short descriptions by University of Washington students and faculty of selected astrobiological destinations that our planet offers. We cannot hope to provide a comprehensive list – with more space and time we might have included the Burgess shale of Canada; the Atacama desert of Chile; the Cretacious/Tertiary boundary at Gubbio, Italy; Louis Pasteur's home and lab in Paris; Witwatersrand mine in South Africa; the channelled scabland of eastern Washington state, to name but a few. Nevertheless, the ten selected locales have played primary roles in determining how we have come to view the phenomenon of life, and how we have placed constraints on its potential occurrence both on our own planet and elsewhere.

From boiling microbial ponds in Yellowstone to frozen wastes of Greenland harboring Earth's oldest sedimentary rocks, a lifetime of exploration awaits you. And for the astrobiologist, the Earth is only a small part of the laboratory...

D.1 Iceland

18° W 65° N
103,000 km^2
Steven Vance and Mark Claire

One of the most geologically fascinating regions on Earth, Iceland offers visitors an exceptional way to experience and visualize thermal processes at work within the Earth's crust. In addition, icy glaciers and near-boiling water meet on the surface to provide unique environments for microorganisms.

Iceland is the only place on Earth where a mid-ocean ridge spreading center lies above sea level. A plume of hot mantle material feeds a "hotspot" of intense volcanic activity, which lies directly underneath and pushes up the mid-Atlantic ridge. The island is continually (in geological terms) being reshaped by these volcano-tectonic processes.

Iceland's exceptional geothermal activity affords the inhabitants free hot water and renewable clean electricity. It also creates impressive geysers and boiling mud pots and hot springs, with abundant hyperthermophiles existing on the chemical and thermal gradients. Soaking in public hydrothermal baths is a daily ritual for many Icelanders.

Iceland's location close to the Arctic Circle causes intense seasonal variation. Much of the island is covered in glaciers. The immense temperature contrast arising from glaciers on volcanic mounds has led researchers to look for life in sub-glacial lakes fed by geothermal heat. Gaidos *et al.* (2004) have discovered bacterial communities under 300 m of ice in a volcanic caldera. Astrobiologists with interests in Europa, the Moon, and early Earth environments will find numerous analogues for study in Iceland.

Resources

Gaidos, E., Lanoil, B., *et al.* (2004). A viable microbial community in a subglacial volcanic crater lake, Iceland. *Astrobiology*, **4**, 327–344.

Thordarson, T. and Hoskuldsson, A. (2002). *Iceland.* (Vol. 4 of *Classic Geology in Europe*). New York: Terrabooks.

Planets and Life: The Emerging Science of Astrobiology, eds. Woodruff T. Sullivan, III and John A. Baross. Published by Cambridge University Press. © Cambridge University Press 2007.

FIGURE D.1 Steam rises from the Hveradular Valley, Kverkfjöll, Iceland, located at 16°72′ W, 64°65′ N. (Steven Vance.)

D.2 Yellowstone

Wyoming (96%), Montana (3%) and Idaho (1%); western United States
110°30′ E   44°36′ N
9,000 km^2
Nicolas Pinel

Yellowstone National Park, established as the world's first national park in 1872, hosts more than half of Earth's geothermal features. With more than 300 geysers, and over 10,000 fumaroles, hotsprings, and mudpots (most of them undisturbed by humans), the Park provides an opportunity for studying the interplay between extreme geology and its microbial extremophilic denizens.

The land sits atop a mantle hot spot that has led to the formation of three calderas within the Park, including one of the world's largest at 70 by 50 km. Calderas are formed by the collapse of magma chambers following catastrophic explosions (and consequent evacuation of the chamber) of overpressurized rhyolite lava. Caldera-forming events within the Park over the last 2

Myr have led to the extrusion of ash-flows with volumes of as much as 2500 km^3, visible as welded ash-flow tuffs up to 400 m thick.

At present, extensive hydrothermal activity is the dominant manifestation of the subsurface magmatic activity. The hydrothermal environments within Yellowstone allow one to explore diverse extremophilic microbial groups and their physiology. Various deeply branching and novel archaeal and bacterial groups have been found in Yellowstone's hotsprings, providing insights into the nature of life on the early Earth. Pools with pH values ranging from as acidic as 1 to as basic as 11 create niches for diverse communities of microbial extremophiles of the pH persuasion. Sometimes, even within the same hotspring, strong temperature gradients can lead on the outside to tightly integrated cyanobacterial mats at "lower" temperatures (<72 °C), then pink streamers attached to higher temperature portions of the outflow channels, and finally planktonic extremophiles living at near-boiling temperatures within the inner pool.

FIGURE D.2 A near-boiling pool densely inhabited by hyperthermophiles at Yellowstone National Park. (Jelte Harnmeijer.)

Resources

Brock, T. D. (1978). *Thermophilic Microorganisms and Life at High Temperatures*. New York: Springer Verlag. Detailed accounts of pioneering microbiological research on thermophilic organisms, much of it conducted within the Park.

volcanoes.usgs.gov/yvo/history.html. Description of the volcanism within the park and surrounding areas.

D.3 Antarctica

90° S
14,000,000 km²
Randall Perry

More than any other continent, Antarctica represents the closest that an astrobiologist can get to interplanetary travel. Almost twice as large as the continental United States, Antarctica harbors over 80% of the world's fresh water and is more than 98% covered with ice. Yet it is ironically amongst the driest places on Earth with very little precipitation and only a minimal amount of flowing water during a few weeks in the summer. The few areas not covered by ice, such as the Dry Valleys and the McMurdo Oasis in southern Victoria Land, provide stark and exciting research areas for microbiologists. A major allure for astrobiologists lies in the potential to study terrestrial analogues for cold planetary deserts such as those on Mars. Antarctica has also proven to be a treasure chest of exogenous material – fully 80% of all meteorites in history have been found in Antarctica, and almost all of those presently found. Meteorites, such as the controversial Martian meteorite ALH84001 (Section 18.5.3), are preserved and expeditiously found within the slow-moving glaciers and low-snowfall ice sheets.

Microorganisms play a more dominant role in Antarctica than on any other continent. Lichens, which represent symbiotic associations of algae and fungi, are important weathering agents of rocks. Cryptoendolithic lichens growing under the surface in sandstones thrive and biogenically weather the rocks. The ice and permafrost also contain microorganisms (Chapter 15). At these very low temperatures, microbes barely survive, change slowly, and presumably also evolve slowly: they may provide an excellent analogue for slow-growing organisms or bacterial spores that might be present on planets such as Mars.

FIGURE D.3 Searching for meteorites during the annual summer hunt in Antarctica. These blue ice fields are at Otway Massif, Grosvenor Range, Transantarctic Mountains. (Monika Kress and ANSMET/NSF.)

Comparison of Antarctica can also be made to other bodies such as Europa, with its probable deep, ice-covered ocean (Chapter 19). Antarctica has ~150 known sub-glacial lakes, with Lake Vostok, now under 4 km of ice, being possibly the most isolated body of water on Earth – it is estimated not to have had any exchange with the surface for millions of years. With a size of ~50 by 200 km and a depth of at least 0.9 km, Lake Vostok is one of the largest freshwater lakes on Earth, comparable to Lake Ontario. For years astrobiologists have been poised with their drills stopped just above the lake, afraid to sample the unknown microbial community for fear of contaminating it in the process.

Resources

www.ldeo.columbia.edu/~mstuding/vostok.html. Very informative site by a Lake Vostok researcher.

geology.cwru.edu/~ansmet. The US Antarctic meteorite hunting program.

D.4 Hawai'i

County of Hawai'i, USA ("the Big Island")
154°58′ W   19°22′ N (Kalapana, where a lava flow enters the ocean)
10,000 km²
Roger Buick

If you want to be transported back in time to when the land first rose above the sea and life was perhaps just starting, Hawai'i is the place for you. You can watch new land being formed, as basaltic lava from the Pu'u

O'o vent of the Kilauea volcano pours into the ocean and solidifies upon quenching. Where it cools on land, it forms wrinkled flows with ropy surfaces, called "pahoehoe." Where it hits the sea, the outer surface shatters to form glassy shards, which along with olivine crystals give rise to spectacular dark beach sands. The lava that continues flowing into the sea solidifies on the outside but stays molten inside. Cracks in the solid surface then allow large blobs of molten rock to be squeezed out and rapidly frozen, to form "pillows." Some of the oldest rocks on the planet show the same features, indicating that they formed in a similar environment (Section D.7).

Hawai'i has other features of astrobiological interest besides the formation of new crust. The 4205 m summit of Mauna Kea hosts an army of telescopes, including the twin 10 m diameter Keck Telescopes. These are the world's most powerful ground-based optical telescopes, and have been responsible for discoveries as diverse as extrasolar planets (Chapter 21) and distant young galaxies formed early in the history of the Universe. But don't drive a rental car all the way to the top – you'll be voiding your insurance if you go past the excellent visitor center at the 2700 m level.

Resources

www.ifa.hawaii.edu. The Institute for Astronomy of the University of Hawai'i manages all of the telescopes on the summit, in exchange for a share of the observing time.

www.ifa.hawaii.edu/UHNAI. One of the nodes of the NASA Astrobiology Institute is at the University of Hawai'i, and has the research theme "Water and Life in the Universe."

FIGURE D.4 Hot pahoehoe lava from the Kilauea volcano on the Big Island of Hawai'i.

FIGURE D.5 Charles Darwin's armchair (note the rollers) and rotatable desk in his study, at his former residence in Downe, England. Inset: Darwin's actual finches from the Galápagos Islands, also on display. (Woody Sullivan.)

hvo.wr.usgs.gov. The Hawaiian Volcano Observatory is a US Geological Survey institution that conducts research on active volcanoes. In addition, the Hawai'i Volcanoes National Park provides visitor facilities, hiking trails, etc.

D.5 Darwin's home

Downe, Kent, England
0°03′ E   51°19′ N
~10 hectares
John Edwards and Woody Sullivan

Down House was Charles Darwin's home and workplace for the forty years during which he revolutionized biological thought. Situated just outside the village of Downe, about 25 km south of London, it is accessible by train and bus, although by car is easiest. The site is now managed as an excellent museum and park by English Heritage.

Darwin purchased Down House in 1842 at the age of 33 after his return from the circumnavigating voyage of the *Beagle* (1831–36), on which his biological specimen collecting and eclectic observations led to his conception of organic evolution by natural selection (Chapter 10). He remained at Down House with his wife Emma and seven children for the rest of his life (he died in 1882), choosing to live in peaceful rural exile where he could do research and write.

The ground floor of Down House has been accurately restored to its appearance during Darwin's occupation. Especially inspiring is to stand in his restored study and imagine him laboring over *Origin of Species* (1859) surrounded by his unique rolling armchair and rotating desk, specimen and chemistry cabinets, bookshelves, and portraits of his close friends and supporters geologist Charles Lyell and botanist Joseph Hooker. Other rooms nicely evoke the family life of a Victorian gentleman, while upstairs the British Museum has installed an instructive set of displays on the birth of the concept of the origin of species by natural selection (including some of the actual famous Galápagos finches (Appendix D.6)), as well as the consequences of that pivotal event (Section 1.6.2).

The large grounds are beautifully maintained, with greenhouses and gardens filled with the same species of plants that Darwin studied. Do not leave without walking the famous "thinking path" Darwin religiously took every day to exercise and collect his thoughts – you may well have a great thought of your own!

Resources

williamcalvin.com/bookshelf/down_hse.htm. Details of a visit by a noted popular science writer.

www.english-heritage.org.uk. Basic visitor information.

Darwin, C. (1859). *On the Origin of Species*. London: John Murray.

D.6 Galápagos Islands

Galápagos Islands, Ecuador
91° W   1° S
8,000 km^2
Julie Huber

For natural history, there are few better places to visit than the Galápagos Islands. Located in the Pacific Ocean ~1,000 km west of Ecuador, the archipelago consists of 13 main islands and 6 minor ones. Best known for their unusual wildlife and volcanic activity,

FIGURE D.6 Giant tortoises cooling off in the Galápagos Archipelago. A small tortoise picked up by Charles Darwin in 1835 only recently died in an Australian zoo at an estimated age of 175 years! (Trina Litchendorf.)

the Galápagos played key roles in two major scientific discoveries of great relevance to astrobiology. First, Charles Darwin (Appendix D.5) visited for two months in 1835 and used observations of the Islands' unique animals to formulate his theory of evolution by natural selection (Chapter 10).

Secondly, 142 years later (in 1977) and ~400 km northeast of the islands and 2500 m below, in 1977 oceanographers discovered seafloor hydrothermal vents teeming with biological activity, forcing scientists to re-evaluate how and where life on Earth could exist (Corliss *et al.*, 1979). The hyperthermophilic microorganisms found at these vent sites give evidence of a subsurface biota about which we know very little (Chapter 14). While visiting the hydrothermal vents may prove a challenge for most visitors, the islands themselves are a popular destination for scientists and tourists alike. Wildlife attractions include giant tortoises (in Spanish *galápagos* means "saddle," for the saddleback tortoise), marine iguanas, sea lions, finches, and penguins. In order to protect the islands, there are strict limits on where visitors are allowed to go.

One of the best ways to view the diversity of life is underwater. Because no commercial fishing is allowed and the islands constitute a Marine Reserve, scuba diving is amongst the world's best.

Resources

www.darwinfoundation.org. The Charles Darwin Foundation is dedicated to the conservation of the Galápagos ecosystems, and conducts scientific research and environmental education.

Darwin, C. (1831–6). *The Voyage of the Beagle: Charles Darwin's Journal of Researches.* (1989, Browne, J. and Neve, M., eds.). New York: Penguin. A good entry, although an abridgement, to the full journals.

Corliss, J. B., Dymond, J., *et al.* (1979). Submarine thermal springs on the Galapagos rift. *Science*, **203**, 1073–6.

D.7 Isua

Isua, Greenland
50°00 W   65°10′ N
~35 km long, 1- to 3-km-thick belt of rock-outcrop
Jelte P. Harnmeijer

The remote 3.7–3.8 Ga rock package in the southwest of Greenland at Isua (Fig. 12.1) represents the oldest known and most complete fragment of early Archean crust, making it one of the prime astrobiological destinations on our planet. A large variety of rocks – including banded iron formations, conglomerates, and a variety of volcanics – lie exposed for studying conditions on the early Earth.

The Isua rocks form a 1–3 km thick, ~35-km-long curved belt opening towards the north. The western and eastern parts of the arc are separated by a large lake, fed by melt water from the ice-sheet that covers most of Greenland. A 1-hour helicopter trip through steep-sided fjords to the west brings you to the nearest airport, at the town of Nuuk (pop. 30,000). Stock up on food, or enjoy a beer with local fishermen and try to stay out of fights in one of Nuuk's pubs.

The majority of Isua rocks have unfortunately been affected by severe changes in pressure and temperature (*metamorphism*) following their deposition ~3.7–3.8 Ga. Another complication in their interpretation is widespread deformation through faulting and

FIGURE D.7 Geologists search for clues of early life in the snow-covered wastes of Isua, southwest Greenland. (Jelte Harnmeijer.)

folding. Nevertheless, most geologists agree that the Isua rocks were deposited on the surface of primordial oceanic crust, under a deep liquid-water ocean. One argument for this scenario is the presence of *pillow-basalt*, ancient lava that cooled rapidly when it first came into contact with colder ocean water (see Section D.4).

Of particular interest to astrobiologists is whether the Isua rocks contain evidence of life. Although no undisputed fossils have been found, some workers argue that $^{13}C/^{12}C$ isotope ratios of graphite in some of the Isua sedimentary rocks provide evidence for biological fractionation (Section 12.3). Active research is continuing to try to find other proxies for life in the most ancient surface rocks on our planet.

Resources

Mojzsis, S. J., Arrhenius, G., *et al.* (1996). Evidence for life on earth before 3,800 million years ago. *Nature*, **384**, 55–9.

van Zuilen, M. A., Lepland, A., *et al.* (2002). Reassessing the evidence for the earliest traces of life. *Nature*, **418**, 627–30.

www.ess.washington.edu/~jelte/Greenland/Greenland.html.

D.8 Pilbara

Western Australia
119° E   21° S
300 × 200 km
Jelte P. Harnmeijer

In the northwest corner of Australia (Fig. 12.1), inland from Port Hedland, one finds the Pilbara *craton* (an ancient stable region of continental crust), one of the planet's best windows into early Earth history. Fieldwork here is rugged and hot, despite one region deliriously named North Pole!

The Pilbara's remarkably well-preserved succession of sedimentary and volcanic rocks provides geologists with essential clues about surface processes 3.5 Ga. Certain evaporite minerals, for example, allow us to constrain the Archean ocean temperature, while turbidite sequences provide information about Archean sedimentary transport mechanisms.

Perhaps even more importantly, several units exposed in the Pilbara exhibit some of the most convincing evidence for primordial life yet, in the form of *stromatolites* (see Fig. 16.2 and Section 12.2.2). These structures, often cone-shaped in cross section, represent ancient fossilized remnants of algal mats that once dominated our biosphere. Many researchers believe that oxygenic photosynthetic prokaryotes were responsible for the formation of the Pilbara stromatolites. If true, these structures may represent the oldest evidence of oxygen production.

Finally, also from this area are the oldest (3.5 Ga) claimed fossils (by William Schopf), but their authenticity has been challenged by Martin Brazier (see discussion in Section 12.2.1).

Resources

Schopf, J. W. and Packer, B. M. (1987). Early Archean (3.3-billion to 3.5-billion-year-old) microfossils from Warrawoona Group, Australia. *Science*, **237**, 70–73.

Brasier, M. D., Green, O. R., *et al.* (2002). Questioning the evidence for Earth's oldest fossils. *Nature*, **416**, 76–81.

www.ess.washington.edu/~jelte/FirstLife.html.

FIGURE D.8 Rare rain clouds build over the rocks of the Pilbara, some of the oldest sediments on Earth. (Jelte Harnmeijer.)

D.9 Shark Bay

Western Australia
117° E   28° S
1,600 km²
Jelte P. Harnmeijer

Living stromatolites, particularly in marine environments, are exceedingly rare today. The turquoise waters of Shark Bay, on Australia's west coast, provide one of the few locales where astrobiologists can study living stromatolites. Stromatolite mats once dominated Earth's shallow waters, but today are found only in extremely saline or protected environments.

Stromatolites are not individual organisms; rather, a single dome represents a complex colony of many different species of microorganisms, all with different environmental and metabolic requirements (see Fig. 16.2 and Section 12.2.2). One species' waste-products are another species' primary source of carbon and/or energy, giving rise to a dynamic and intricate hierarchy of phototrophs and chemotrophs. Slice through a stromatolite, and you'll immediately see evidence of this rich hierarchy exhibited as different colored layers parallel to the outer surface of the dome.

Campsites provide accommodation at Hamelin Pool, where wooden piers allow visitors to view stromatolites from above the water. Shark Bay, which has a World Heritage listing and is a Marine Park, is also worth visiting for its abundant marine mammals and bird species.

Resources

www.sharkbaywa.com.au. Information on Shark Bay and visitor activities.

D.10 Arecibo radio telescope

Arecibo, Puerto Rico
66°45′ W   18°21′ N
305 m diameter
Woody Sullivan

Situated in the karst hills of central Puerto Rico, the Arecibo radio telescope (Fig. 26.3) is the world's largest dish (and has been since the 1960s) with a collecting area of 7 hectares, more than the rest of the world's radio telescopes combined! In order to achieve this huge size,

FIGURE D.9 Living stromatolites in Australia's Shark Bay. (Jelte Harnmeijer.)

its steerability is limited to regions within 20° of the zenith, which means that only ~35% of the sky can be observed at any one time. The surface (which is accurate to 2 mm) is a small sector of a sphere of radius 265 m, rather than the usual paraboloid for telescopes; this allows the dish to look off-zenith, but also creates various optical distortions that require correction.

Steering is effected by moving the focus (or "feed", where the receiving electronics are located) on a huge triangular superstructure that towers above the dish and is supported by three 130-m-tall towers. The feed structure, including a suspended secondary reflector of 25 m diameter, is hidden inside the odd-shaped radome at the top. A covering provides weather protection. The telescope can be used either in a passive mode, monitoring the sky for faint signals, or in an active mode (radar), for which a 1 MW beam of radio waves is transmitted and then received after reflecting off a solar system target. Ionospheric research is also conducted in the radar mode. The dish's significance to astrobiology is twofold: (1) its major role in many SETI projects (in particular for Phoenix, SERENDIP, and seti@home – Chapter 26), and (2) important radar observations of Venus and Mars, near-Earth asteroids, Saturn's rings, Titan, etc.

The observatory has an excellent new visitor's center, accessible from town via a scary, narrow, winding road.

Resources

www.naic.edu. Technical and popular-level information on both the technology and recent scientific results.

Appendix E: The micro*scope web tool

David J. Patterson and M. L. Sogin
Marine Biological Laboratory at Woods Hole

E.1 Exploring knowledge space for microbes

Astrobiology and biology more generally are integrating new visions of biodiversity with evolutionary and ecological processes. The body of knowledge about hundreds of thousands of microbial species is huge, and involves data on ontogenetic transitions and intraspecific variation; encompasses scales of biology ranging from molecular variation in a particular kind of cell to the role of individuals in complex ecosystems; and accommodates the biology of individuals whose identity and role change as a function of time and place, as well as in response to biotic and abiotic interactions. The knowledge may be either digital or in traditional media, such as often found in libraries, museums, and herbaria. Researchers need solutions that lead to a comprehensive and evolving "knowledge space" (which includes information *and* its interpretation). The solutions will include tools to empower experts to transfer knowledge from traditional to contemporary media, as well as allow them to integrate old with new. An approach that is universal, inclusive, scalable, and flexible can evolve into a comprehensive Encyclopedia of Life (Wilson, 2003).

The absence of websites offering comprehensive treatments of microbial diversity is a serious impediment for students and investigators who are in need of morphological, physiological, and lifestyle information about microbes. In fact, this is a problem not only for microbial life but for all life. There are no robust standards for indexing phenotypic and biodiversity data, it is difficult to parse and recompile existing relevant data resources. Informatics facilities that handle phenotypic information about organisms lag considerably behind tools to handle molecular, environmental, or bibliographic data (Edwards *et al.*, 2000). The situation is aggravated by a worldwide decline of taxonomic expertise – the living source of expert information about organisms (Godfray, 2002).

Various initiatives address the lack of biodiversity information, but deal with only selected aspects, e.g., the Australian Biodiversity Information Facility (ABIF, www.abif.org/), Tree of Life (TOL, tolweb.org/tree/), or Fishbase (www.fishbase.org). Centralized systems are not very flexible, may have duplicated content elsewhere, and face major organizational problems if they seek to ensure that the content remains up to date. An alternative approach is to link (or *federate*) independent distributed databases, possibly with agreed input and output interface protocols (such as the DiGiR or TAPIR protocols digir. sourceforge.net/, wiki.gbif.org/dadiwiki/wikka.php? wakka = TAPIR). Examples of federated systems include the Ocean Biogeographic Information System (OBIS, www.iobis.org/) or Specify (www. specifysoftware.org/Specify). The logical extreme of the distributed approach is a virtual meta-knowledge environment that uses internet services to bring knowledge together with evolving tools and standards (Kim *et al.*, 1998; Page, 2006; vsmith.info/node/17965). But such systems face problems with variable levels of quality control, and structures that would permit loosely defined communities to co-operate have not yet emerged.

We have been exploring some of these challenges with the web site **micro*scope** (microscope.mbl.edu) (Fig. E.1). micro*scope is a vehicle for traditional knowledge about microbial diversity (descriptions and images), and embodies the distributed approach in that it complements local knowledge with information distributed on the internet using outlinks to remote sites. The remote sites are carefully chosen as authoritative – examples include molecular databases (e.g., GenBank, www.ncbi.nih.gov/Genbank/index.html), culture collections (e.g., the American Type Culture Collection,

Planets and Life: The Emerging Science of Astrobiology, eds. Woodruff T. Sullivan, III and John A. Baross. Published by Cambridge University Press. © Cambridge University Press 2007.

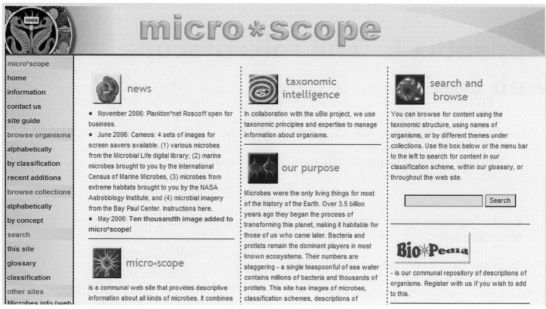

FIGURE E.1 Home page of micro*scope (microscope.mbl.edu).

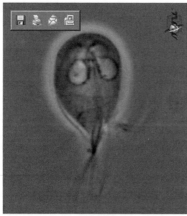

nuclei

Giardia (gee-arr-dee-a), a diplomonad flagellate that is a common inhabitant of the human intestine. These are diplomonads, they have two nuclei (anterior) and two sets with four flagella each arising from the front end of the cell but extending to the side or posteriorly. Normally they clamp onto cells of the

Giardia

From the collection *American Type Culture Collection*

Description of Giardia: Diplomonad flagellates, with flattened ventral face occupied by the adhesive disk which attaches the parasite to the intestinal mucosa of its host; cell tapers posteriorly and giving rise to 2 caudal flagella; all flagella directed to the rear; feeds by pinocytosis occuring on the dorsal face; adheres on the brush border of the intestinal epithelial cells and the sucking force is generated by the beating of the ventral enlarged flagella; includes common and widespread parasites of man.

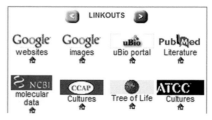

FIGURE E.2 Representative dynamic species sheet for diplomonas *Giardia*. Pages can be customized. This one includes an image, an image legend, a taxon description, an array of taxon-specific outlinks, both 'blind' and 'content certified', links to the classification, other resources within the website, access to large versions of the image, and a mechanism to comment on the page.

www.atcc.org/), the Provasoli-Guillard National Center for Culture of Marine Phytoplankton (ccmp.bigelow. org/), sources of published information (PubMed, www.ncbi.nlm.nih.gov/entrez), and the Tree of Life Project. Outlinks include broadly based discovery tools such as the Google search engine. Outlinks may either be blind and search remote sites in the hope of finding complementary data, or pre-index remote sites to offer links only when complementary data are available. The user can quickly assemble outlinks and customize them with keywords or by associating them with clades in the underlying classification.

Resources accessed through micro*scope are presented in dynamic fact sheets (Fig. E.2). Typically a sheet may include images, descriptions of the taxon, description of the assets, a classification scheme (or access to such a scheme), outlinks to associated information at remote sites, as well as an ability to add to or comment on the database.

The key organizing element within micro*scope is the classification structure and its content of names. From the time of Linnaeus, names have been the common denominator of all biodiversity information. A compilation of all synonyms, colloquial, or alternative names would provide a large metadata vocabulary that could find, collate, and index all taxon-related data through the internet (Patterson, 2003). However, such systems have not been used before for two reasons. The first is that organisms may have more than one name, and a detailed indexing system must be informed with all alternative names. Second, traditional systematics organizes names within hierarchies (classification) and users and providers have strong loyalties to particular classifications. Users who do not adopt a preferred classification will not succeed in finding what they want. The solution to this problem is to separate the objective elements of taxonomy (the names, which are the primary indexing metadata) from the subjective elements, such as the relationships between organisms (classifications). By separating the subjective from objective indexing elements, we can allow an unlimited number of alternative classifications to coexist, thereby

satisfying incompatible needs of different experts. This has been achieved for millions of species within the uBio project (Universal Biological Indexer and Organizer, www.ubio.org). Currently, micro*scope uses about 1,500,000 names as taxonomic metadata, including the 15,000 genera of protists and all genera of prokaryotes named in compliance with the International Code of Nomenclature of Bacteria. As discussed at the end of this section, this system is being extended to any kind of organism and will call upon the full names content of NameBank (with 6,000,000 publicly accessible names records at the time of writing).

There are two strategies to interrogate micro*scope. Search functions – whether of the entire site, of classifications, or of the internet – allow users to find particular assets linked to a biological name or to a term. Names reconciliation behind the scenes allows a search that begins with, for example, "people" to find content under "*Homo sapiens.*" Search functions are well suited to users who understand both the site's content and biology, but even nonexperts can be successful after being guided to resources through browsing functions. These are pre-constructed pathways providing visitors with options that progressively refine the process by which resources (referred to as assets) are selected. By selecting from various choices, the visitor eliminates resources until their needs are met. A website can have many browse pathways. For example, the classification structures of micro*scope offer one browse pathway to reach a specific microbe (Fig. E.3). In addition, we

Jump to: Eukaryotes & protists Eubacteria Archaea
           Fungi
           Animals Fish Birds Mammals Reptiles
           Higher plants Brown Algae Red Algae

Cyanidium

Classification by CU*STAR

Search

search options

Life
Cellular life
Eukaryota
Rhodophyta
Cyanidiophytes

Cyanidium, from Cyanidium at Yellowstone

Cyanidium, from Yellowstone National Park (Wyoming, USA)

Cyanidium, from Yellowstone National Park (Wyoming, USA)

Cyanidium, from Cyanidium at Yellowstone

Description of Cyanidium:
Unusual unicellular red alga (that is usually coloured blue green). Among eukaryotes, distinguished by its tolerance of extreme conditions such as acidity and high temperature.

FIGURE E.3 A classification page within micro*scope. It includes a navigable, phylogenetically informed classification that serves as a taxonomic browse function, if alternative names for the taxon are available, they too will be shown. In addition to showing thumbnails of some of the images, the outlink modules are also incorporated in this page.

Marine benthic dinoflagellates - NW Australia

The micro-algae and protists of north western Australia have not been studied to any great extent. The marine dinoflagellates included here were found in samples taken from sandy and muddy sediments at Cable Beach, Town Beach and Roebuck Bay (and adjacent sites) near Broome. This survey was carried out by Shauna Murray, Mona Hoppenrath and Jacob Larsen in 2003. This survey was supported by the Australian Biological Resources Study. Image of sunset by D. J. Patterson. Image copyright: D. J. Patterson, used under license to MBL (micro*scope).

2 3 4

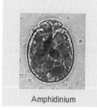

Amphidiniopsis

Amphidiniopsis hexagona

Amphidiniopsis hexagona

Amphidinium

Amphidinium boggayum

FIGURE E.4 Browsing by collections. This page has been reached by selecting microbes from marine habitats, then by selecting resources that relate to sediments, and then by selecting a particular collection.

assemble assets in collections, and the collections are placed within a concept hierarchy; this allows for another effective browse structure (Fig. E.4).

micro*scope is committed to contributing to education, and is integrated with the US National Science Digital Library and the Microbial Life Educational Resources (mler.mbl.edu/). It also provides "matrix identification keys" that students find simpler and superior to traditional "pathway keys." Pathway keys present a question, and the answer directs users to a subset of the key. This process continues until a terminus (or identity) is reached. For such keys to lead to the correct identification, however, users must correctly answer all questions in a single strict sequence. For microbes, this strategy is plagued by a high probability of failure because of unfamiliar taxa and the absence of "easy" identifying characteristics. LUCID (www.lucidcentral.com) and X:ID (www.ubio.org/index.php?pagename=XID/key) are matrix identification keys allowing the assembly of more tolerant and flexible identification guides (Fig. E.5). Characteristics are mapped against taxa in a matrix. To identify an organism, users select those characteristics which apply to the organism in question. Each entry eliminates taxa that do not comply with the characteristic. The characteristics can be entered in any sequence, each entry reducing the number of taxa that meet the criteria that have been applied. The software even can allow for incorrect answers. Matrix keys can be delivered through the internet, and taxa can be mapped against names within the classification to access relevant information in micro*scope.

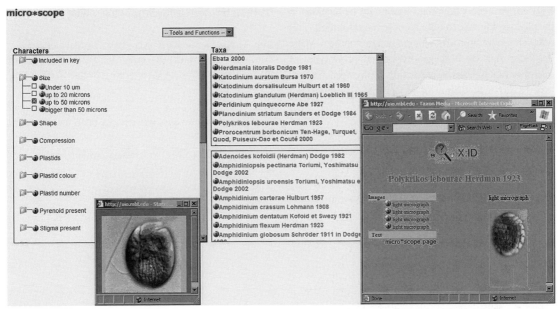

FIGURE E.5 The X:ID key. To the left are the characters that can be used to identify an organism. A size of "up to 50 microns" has been selected and the user is checking "plastid colour." In the lower center are organisms which have been eliminated because they do not have the selected characteristics so far, while the upper center has the remaining taxa. From each taxon there exist links to associated data, including the appropriate page of micro*scope and its resources.

The current release of micro*scope incorporates an index of 20,000 microbial names, descriptions of 3,500 microbial taxa, 10,000 images and associated descriptive text, and various educational resources. The next generation, referred to as "star," will underpin similar and expanded services for *any* category of organism. Star will be a downloadable template with access to an index of millions of names from the uBio project, will access assets assembled by other "star*sites," and will be part of large-scale cooperative environments called "star*nets." This is achieved by enabling each site to obtain index information from other sites. Consequently, each site can choose to point to data at any other site. Furthermore, by maintaining an archive that can be mirrored at any participating site, all of the knowledge acquired in these systems will remain accessible even after some of the participating sites become dysfunctional. The inaugural star*net was plankton* net – a multi-site initiative that seeks to develop a knowledge resource about marine phytoplankton (e.g., planktonnet. awi.de/).

E.2 Bridging phenotype and genotype databases

Biology knowledge space is distinguished in part by a gulf between traditional phenotypic knowledge and new and mostly molecular knowledge. We must bridge this gulf to avoid duplication and to use traditional knowledge to inform the exploratory process and to provide context to new insights. micro*scope makes accessible phenotypic data on microbial diversity accumulated over 200 years, as well as the fast-increasing volume of environmental genomics data. micro*scope, in combination with molecular inventories such as the Ribosomal Database Project (RDP, rdp.cme.msu.edu/) and the NASA Science Organizer (sciencedesk.arc.nasa.gov/organizer/), is exploring mechanisms of associating taxon-related data with environmental genomics. For example, NASA's Ames Research Center has developed a collaborative web-based tool, Science Organizer, to support field and laboratory studies of cyanobacterial mats. Collaborators will use Science Organizer to share data and analytical results. By linking Science Organizer and micro*scope, it will be possible to interpret environmental isolates in the context of phenotypic information from diverse microbes represented in micro*scope. Hyperlinks between close relatives of environmental isolates with specific taxa in micro*scope will allow the user to immediately gain information about the potential phenotype of the organism represented by the environmental isolate. Reverse links from micro*scope based upon its extensive taxonomic database will then query either GenBank or

RDP for sequences of species that might be included in the phylogenetic tree to obtain improved resolution. This research model produces a dynamic environment where the latest data on both genotypes and phenotypes of microbes can be interpreted within a phylogenetic context. It allows prediction of phylogenetic affinities and phenotype for environmental isolates. Many other types of data can be analyzed in a similar fashion.

References

Edwards, J. L. *et al.* (2000). Interoperability of biodiversity databases: biodiversity information on every desktop. *Science*, **289**, 2312–2314.

Godfray, H. C. (2002). Challenges for taxonomy. *Nature*, **417**, 17–19.

Kim, T., Ku, K.-M., and Kim, S. (1998). Virtual unification of distributed databases via web. In *Proceedings of 8th International CODATA Conference CODATA/DSAO 97*, pp. 96–101.

Page, R. D. M. (2006). A taxonomic search engine: federated taxonomic databases using web services. *BMC Bioinformatics*, **6**, 48.

Patterson, D. J. (2003). Progressing towards a biological names register. *Nature*, **422**, 661.

Wilson, E. O. (2003). The encyclopedia of life. *TREE*, **18**, 77–80.

Index

Page numbers in *italic* denote figures. Page numbers in **bold** denote tables.